W9-DIR-858

10. $\dfrac{d}{dx} e^x = e^x$, $\qquad \dfrac{d}{dx} e^{f(x)} = f'(x)e^{f(x)}$

11. $\dfrac{d}{dx} \ln x = \dfrac{1}{x}$, $\qquad x > 0$; $\qquad \dfrac{d}{dx} \ln f(x) = f'(x) \cdot \dfrac{1}{f(x)}$, $\qquad f(x) > 0$

12. $\dfrac{d}{dx} \ln |x| = \dfrac{1}{x}$, $\qquad x < 0$; $\qquad \dfrac{d}{dx} \ln |f(x)| = f'(x) \cdot \dfrac{1}{f(x)}$, $\qquad f(x) < 0$

13. If $\dfrac{dP}{dt} = kP$, then $P(t) = P_0 e^{kt}$.

14. If $\dfrac{dP}{dt} = -kP$, then $P(t) = P_0 e^{-kt}$.

15. $\dfrac{d}{dx} a^x = (\ln a)a^x$

16. $\dfrac{d}{dx} \log_a x = \dfrac{1}{\ln a} \cdot \dfrac{1}{x}$, $\qquad x > 0$; $\qquad \dfrac{d}{dx} \log_a |x| = \dfrac{1}{\ln a} \cdot \dfrac{1}{x}$, $\qquad x < 0$

17. $a^x = e^{x(\ln a)}$

APPLIED CALCULUS

MARVIN L. BITTINGER

BERNARD B. MORREL
Indiana University–Purdue University
at Indianapolis

ADDISON-WESLEY PUBLISHING COMPANY
Reading, Massachusetts ▪ Menlo Park, California
London ▪ Amsterdam ▪ Don Mills, Ontario ▪ Sydney

Sponsoring Editor:	Wayne Yuhasz
Production Manager:	Karen Guardino
Production Editor:	Herbert Merritt
Text and Cover Designer:	Vanessa Piñeiro
Illustrator:	VAP Group, Limited
Cover Photographer:	Benjamin Mendlowitz
Art Coordinator:	Joseph Vetere
Manufacturing Supervisor:	Ann Delacey

To my ultimate exponent, Elaine
M.L.B.

For Kate and Virginia
B.B.M.

Library of Congress Cataloging in Publication Data

Bittinger, Marvin L.
   Applied calculus.

   Includes index.
   1. Calculus.   I. Morrel, Bernard B.   II. Title.
QA303.B6447   1984   515   84-362
ISBN 0-201-12200-6

Copyright ©1984 by Addison-Wesley Publishing Company, Inc. All rights reserved. No part of this publication may be reproduced, stored in a retrieval system, or transmitted, in any form or by any means, electronic, mechanical, photocopying, recording, or otherwise, without the prior written permission of the publisher. Printed in the United States of America. Published simultaneously in Canada.

GHIJ-HA-89876

PREFACE

This book is a two-semester introduction to calculus as applied to business, economics, the behavioral sciences, the social sciences, biology, medicine. This book is an expansion of the one-term text, *Calculus: A Modeling Approach,* 3rd edition, by Marvin L. Bittinger. The additional material includes total differentials, more on multiple integration, Taylor polynomials and infinite series, numerical techniques and Newton's method, and an extended full chapter on differential equations. Prerequisite for the text is a basic course in algebra, although Chapter 1 provides sufficient review to unify the diverse backgrounds of most students.

The style, format, and approach of *Calculus: A Modeling Approach,* 3rd edition, have been retained in this book, along with the functional use of a second color where theorems and definitions are carefully labeled. Summaries of important formulas of certain chapters are provided inside the front and rear covers. Other significant features of the book are described below.

1. New Supplements

Several new supplements are now available for the book:

- *Computer Software Supplement.* Adapted from Cactusplot Software by John Losse of Scottsdale Community College, there is now a computer software supplement. This consists of an Apple II diskette, and a Computer Manual for the student. The software performs many tasks for the student. For example, it will create a graph of virtually any function considered in the course, together with secant lines and area under a curve. It also computes an input-output table and solves equations. To maximize the usefulness of this material, codes CSS appear throughout the text to inform the student that the computer material is appropriate and useful for the indicated subject matter.

■ *Video Tapes.* Video tape reviews have now been prepared which cover important topics for the book. A lecturer (John Jobe, of Oklahoma State University) speaks to students and works out examples with lucid explanations. While the videotapes do not provide an entire course on TV, they have many uses, among which are to supplement lectures, provide partial lectures, and to provide self-study opportunities for students.

■ *Student's Guide to Exercises.* Written by Judith A. Beecher, this supplement contains complete worked-out solutions, together with hints and suggestions, to selected exercises.

2. Content Changes

Major content changes from *Calculus: A Modeling Approach*, 3rd edition, are as follows:

■ The total differential and certain of its applications to approximation problems has been added to Chapter 7. The section on multiple integration has been greatly expanded and completely rewritten.

■ Chapter 9 deals with differential equations. It contains sections on first-order linear equations, higher-order linear equations with constant coefficients, and systems of linear equations (and their applications).

■ Chapter 10 of *Applied Calculus* is entirely new material. A gradual, heuristic introduction to Taylor's Series has been added. Emphasis here is on the computational, rather than the theoretical aspects of the subject. This chapter also contains a section on indeterminate forms and L'Hôpital's Rule.

■ Chapter 11 is also completely new. The first two sections show the student how to use the method of iteration and Newton's Method to solve equations. Section 11.3 applies these methods to the very practical problems of determining internal rates of return, or yield rates. Section 11.4 covers numerical integration, including the Trapezoidal Rule and Simpson's Rule. An optional section on extrapolation is followed by a rather detailed section on numerical solution of differential equations (including Euler's method for systems of differential equations).

3. Challenge Problems

An extensive number of challenge problems were used in the third edition of *Calculus: A Modeling Approach,* and even more have been added to *Applied Calculus.* Challenge questions have also been included in the chapter tests.

Among the salient features of *Calculus: A Modeling Approach* that have been retained or adapted in this book are the following.

1. Intuitive approach While the word "intuitive" has many meanings and interpretations, its use here, for the most part, means "experience based." That is, when a concept is taught, the learning is based on the student's prior experience or new experience given just before the concept is formalized. For example, in a maximum problem involving volume, a function is derived that is to be maximized. Instead of forging ahead with the standard calculus solution, the student is asked to stop, compute function values, graph the equation, and estimate the maximum value. This experience provides students with more insight into the problem—not only that different dimensions yield different volumes, but also that the dimensions yielding the maximum volume might even be conjectured or estimated as a result of the calculations.

2. Design and format Each page has an outer margin that is used in several ways. (1) In the margin, sample, developmental, and exploratory exercises are placed near the related text material so that the student can become actively involved in the development of the topic. These margin exercises have proved to be extremely beneficial. (2) As each new section begins, its behavioral objectives are stated in the margin. These can be easily spotted by the student; and when the typical question arises, "What material am I responsible for?" these objectives provide an answer. They may also help take the fear out of the word "Calculus."

3. Calculator exercises Exercises and examples geared to the use of a calculator are included throughout the text. While students who do not have a calculator can still accomplish the learning in the book, the book has been written with the point of view that the calculator is a common possession. These exercises are highlighted by the symbol ▦.

4. Applications Relevant and factual applications are included throughout the text. When the exponential model is studied, other applications, such as continuously compounded interest and the demand for natural resources, are also considered. The notions of total revenue, cost, and profit, together with their derivatives (marginal functions) are threads that run through the text, providing continued reinforcement and unification.

5. Tests Each chapter ends with a chapter test. Challenge questions are included in these tests. There is a cumulative review which can also serve as a final exam. All the answers to these tests are in the book. Four additional forms (of various types) of the tests appear, classroom ready, in the *Instructor's Manual*.

All margin exercises have answers in the text. It is recommended that students do all of these, stopping to do them when the text so indicates.

ACKNOWLEDGMENTS

The authors wish to express their appreciation to many people who helped with the development of the book; to their own students for providing suggestions and criticisms so willingly during the preceding editions; to Judy Beecher of Indiana University-Purdue University at Indianapolis for her helpful suggestions, proofreading, preparation of the *Instructor's Manual*, and writing of the *Students Guide To Exercises*; to John Losse of Scottsdale Community College for preparation of the Computer Software Supplement; to John Jobe of Oklahoma State University for preparation of the videotapes; to Kathy Hannon of IUPUI for a superb job of typing part of the manuscript; to Mike Penna of IUPUI for his help with the computer graphics; to Judy Penna of IUPUI for her precise proofreading of the manuscript, and to Rick Haston of *Latent Images* of Carmel, Indiana for taking many of the photographs.

In addition, we wish to thank: James W. Newsom, Tidewater Community College, Virginia Beach; Thomas J. Hill, University of Oklahoma, Norman; John Mathews, California State University, Fullerton; Richard Semmler, Northern Virginia Community College, Annandale; E. Robert Heal, Utah State University, Logan; and Shirley Goldman, University of California, Davis, for their thorough reviewing.

Indianapolis, Indiana   M. L. B.
January 1984   B. B. M.

CONTENTS

ALGEBRA REVIEW, FUNCTIONS, AND MODELING

1

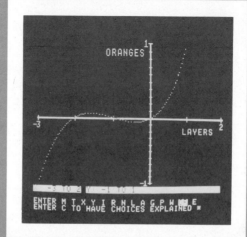

The number of oranges in a pile of this type is approximated by the function

$$y = \frac{1}{6}x^3 + \frac{1}{2}x^2 + \frac{1}{3}x,$$

where y is the number of oranges, and x is the number of layers. (Fred Bodin: Stock, Boston)

A graph of the function as found by the *Cactusplot Computer Software Supplement.*

OBJECTIVES

You should be able to

a) **Rename an exponential expression without exponents.**

b) **Multiply exponential expressions by adding exponents.**

c) **Divide exponential expressions by subtracting exponents.**

d) **Raise a power to a power by multiplying exponents.**

e) **Multiply algebraic expressions.**

f) **Factor algebraic expressions.**

g) **Solve applied problems involving the comparison of a power like $(3.1)^2$ with 3^2.**

h) **Solve applied problems involving compound interest.**

Rename without exponents.

1. 3^4 **2.** $(-3)^2$

3. $(1.02)^3$ **4.** $\left(\frac{1}{4}\right)^2$

Rename without exponents.

5. $(5t)^0$ **6.** $(5t)^1$

7. k^0 **8.** m^1

9. $\left(\frac{1}{4}\right)^1$ **10.** $\left(\frac{1}{4}\right)^0$

1.1 EXPONENTS, MULTIPLYING, AND FACTORING

Exponential Notation

Let us review the meaning of an expression

$$a^n,$$

where a is any real number and n is an integer; that is, n is a number in the set $\ldots, -3, -2, -1, 0, 1, 2, 3, \ldots$. The number a above is called the *base* and n is called the *exponent*. When n is larger than 1, then

$$a^n = \underbrace{a \cdot a \cdot a \cdots a}_{n \text{ factors}}.$$

In other words, a^n is the product of n factors, each of which is a.

In later sections of the book we will consider a^n when n is any real number.

Examples Rename without exponents.

a) $4^3 = 4 \cdot 4 \cdot 4$, or 64

b) $(-2)^5 = (-2)(-2)(-2)(-2)(-2)$, or -32

c) $(1.08)^2 = 1.08 \times 1.08$, or 1.1664

d) $\left(\frac{1}{2}\right)^3 = \frac{1}{2} \cdot \frac{1}{2} \cdot \frac{1}{2}$, or $\frac{1}{8}$

DO EXERCISES 1–4. (EXERCISES ARE IN THE MARGIN.)

We define an exponent of 1 as follows:

$$a^1 = a, \quad \text{for any real number } a.$$

That is, any real number to the first power is that number itself. We define an exponent of 0 as follows:

$$a^0 = 1, \quad \text{for any nonzero real number } a.$$

That is, any nonzero real number a to the 0 power is 1.

Examples Rename without exponents.

a) $(-2x)^0 = 1$ **b)** $(-2x)^1 = -2x$ **c)** $\left(\frac{1}{2}\right)^0 = 1$

d) $e^0 = 1$ **e)** $e^1 = e$ **f)** $\left(\frac{1}{2}\right)^1 = \frac{1}{2}$

DO EXERCISES 5–10.

Rename without negative exponents.

11. 2^{-4}

12. 10^{-2}

13. $\left(\dfrac{1}{4}\right)^{-3}$

14. t^{-7}

15. e^{-t}

16. M^{-1}

17. $(x + 1)^{-2}$

The meaning of a negative integer as an exponent is as follows:

$$a^{-n} = \frac{1}{a^n}, \quad \text{for any nonzero real number } a.$$

That is, any nonzero real number a to the $-n$ power is the reciprocal of a^n.

Examples Rename without negative exponents.

a) $2^{-5} = \dfrac{1}{2^5} = \dfrac{1}{2 \cdot 2 \cdot 2 \cdot 2 \cdot 2} = \dfrac{1}{32}$

b) $10^{-3} = \dfrac{1}{10^3} = \dfrac{1}{10 \cdot 10 \cdot 10} = \dfrac{1}{1000}$, or 0.001

c) $\left(\tfrac{1}{4}\right)^{-2} = \dfrac{1}{\left(\tfrac{1}{4}\right)^2} = \dfrac{1}{\tfrac{1}{4} \cdot \tfrac{1}{4}} = \dfrac{1}{\tfrac{1}{16}} = 1 \cdot \dfrac{16}{1} = 16$

d) $x^{-5} = \dfrac{1}{x^5}$

e) $e^{-k} = \dfrac{1}{e^k}$

f) $t^{-1} = \dfrac{1}{t^1} = \dfrac{1}{t}$

DO EXERCISES 11–17.

Properties of Exponents

Note the following:

$$b^5 \cdot b^{-3} = (b \cdot b \cdot b \cdot b \cdot b)\frac{1}{b \cdot b \cdot b} = \frac{b \cdot b \cdot b}{b \cdot b \cdot b} \cdot b \cdot b = 1 \cdot b \cdot b = b^2.$$

The result could have been obtained by adding the exponents. This is true in general.

THEOREM 1

For any real number a and any integers n and m,

$$a^n \cdot a^m = a^{n+m}.$$

(To multiply when the bases are the same, add the exponents.)

Multiply.

18. $t^4 \cdot t^5$

19. $t^{-4} \cdot t$

20. $10e^{-4} \cdot 5e^{-9}$

21. $t^{-3} \cdot t^{-4} \cdot t$

22. $4b^5 \cdot 6b^{-2}$

Divide.

23. $\dfrac{x^6}{x^2}$

24. $\dfrac{x^2}{x^6}$

25. $\dfrac{e^t}{e^t}$

26. $\dfrac{e^2}{e^k}$

27. $\dfrac{e^5}{e^{-7}}$

28. $\dfrac{e^{-5}}{e^{-7}}$

Examples Multiply.

a) $x^5 \cdot x^6 = x^{5+6} = x^{11}$

b) $x^{-5} \cdot x^6 = x^{-5+6} = x$

c) $2x^{-3} \cdot 5x^{-4} = 10x^{-3+(-4)} = 10x^{-7}$

d) $r^2 \cdot r = r^{2+1} = r^3$

DO EXERCISES 18–22.

Note the following:

$$b^5 \div b^2 = \frac{b^5}{b^2} = \frac{b \cdot b \cdot b \cdot b \cdot b}{b \cdot b} = \frac{b \cdot b}{b \cdot b} \cdot b \cdot b \cdot b = 1 \cdot b \cdot b \cdot b = b^3.$$

The result could have been obtained by subtracting the exponents. This is true in general.

THEOREM 2
_____

For any nonzero real number a and any integers n and m,

$$\frac{a^n}{a^m} = a^{n-m}.$$

(To divide when the bases are the same, subtract the exponent in the denominator from the exponent in the numerator.)

_____

Examples Divide.

a) $\dfrac{a^3}{a^2} = a^{3-2} = a^1 = a$

b) $\dfrac{x^7}{x^7} = x^{7-7} = x^0 = 1$

c) $\dfrac{e^3}{e^{-4}} = e^{3-(-4)} = e^{3+4} = e^7$

d) $\dfrac{e^{-4}}{e^{-1}} = e^{-4-(-1)} = e^{-4+1} = e^{-3}$, or $\dfrac{1}{e^3}$

DO EXERCISES 23–28.

Note the following:

$$(b^2)^3 = b^2 \cdot b^2 \cdot b^2 = b^{2+2+2} = b^6.$$

The result could have been obtained by multiplying the exponents. This is true in general.

Simplify.

29. $(x^{-4})^3$

30. $(e^2)^2$

31. $(e^x)^3$

32. $(5x^3y^5)^2$

33. $(4x^{-5}y^{-6}z^2)^{-4}$

Multiply.

34. $2(x + 7)$

35. $P(1 - i)$

36. $(x - 4)(x + 7)$

37. $(a - b)(a - b)$

38. $(a - b)(a + b)$

THEOREM 3

For any real number a and any integers n and m,

$$(a^n)^m = a^{nm}.$$

(To raise a power to a power, multiply the exponents.)

Examples Simplify.

a) $(x^{-2})^3 = x^{-2 \cdot 3} = x^{-6}$, or $\dfrac{1}{x^6}$

b) $(e^x)^2 = e^{2x}$

c) $(2x^4y^{-5}z^3)^{-3} = 2^{-3}(x^4)^{-3}(y^{-5})^{-3}(z^3)^{-3} = \dfrac{1}{2^3}x^{-12}y^{15}z^{-9}$, or $= \dfrac{y^{15}}{8x^{12}z^9}$

DO EXERCISES 29–33.

Multiplication

The distributive laws are important in multiplying. The laws are as follows:

For any numbers A, B, and C,

$$A(B + C) = AB + AC \quad \text{and} \quad A(B - C) = AB - AC.$$

Examples Multiply.

a) $3(x - 5) = 3 \cdot x - 3 \cdot 5 = 3x - 15$

b) $P(1 + i) = P \cdot 1 + P \cdot i = P + Pi$

c) $(x - 5)(x + 3) = (x - 5)x + (x - 5)3$
$$= x \cdot x - 5x + 3x - 5 \cdot 3$$
$$= x^2 - 2x - 15$$

d) $(a + b)(a + b) = (a + b)a + (a + b)b$
$$= a \cdot a + ba + ab + b \cdot b$$
$$= a^2 + 2ab + b^2$$

DO EXERCISES 34–38.

The following formulas, which are obtained using the distributive laws, are useful in multiplying.

Multiply.

39. $(x - h)^2$

$x^2 - 2xh + h^2$

40. $(3x + t)^2$

$9x^2 + 6xt + t^2$

41. $(5t - m)(5t + m)$

$25t^2 - m^2$

Factor.

42. $P - Pi$

$P(1 - i)$

43. $x^2 + 10xy + 25y^2$

$(x + 5y)(x + 5y)$

44. $4x^2 + 28x + 40$

$4(x^2 + 7x + 10)$
$4(x + 5)(x + 2)$

45. $25c^2 - d^2$

$(5c + d)(5c - d)$

46. $3x^2h + 3xh^2 + h^3$

$h(3x^2 + 3xh + h^2)$

$$(A + B)^2 = A^2 + 2AB + B^2 \tag{1}$$

$$(A - B)^2 = A^2 - 2AB + B^2 \tag{2}$$

$$(A - B)(A + B) = A^2 - B^2 \tag{3}$$

Examples Multiply.

a) $(x + h)^2 = x^2 + 2xh + h^2$

b) $(2x - t)^2 = (2x)^2 - 2(2x)t + t^2 = 4x^2 - 4xt + t^2$

c) $(3c + d)(3c - d) = (3c)^2 - d^2 = 9c^2 - d^2$

DO EXERCISES 39–41.

Factoring

Factoring is the reverse of multiplication. That is, to factor an expression, we find an equivalent expression that is a product. Always remember to look first for a common factor.

Examples Factor.

a) $P + Pi = P \cdot 1 + P \cdot i = P(1 + i)$ We used a distributive law.

b) $2xh + h^2 = h(2x + h)$

c) $x^2 - 6xy + 9y^2 = (x - 3y)^2$

d) $x^2 - 5x - 14 = (x - 7)(x + 2)$ Here we looked for factors of -14 whose sum is -5.

e) $x^2 - 9t^2 = (x - 3t)(x + 3t)$ We used $(A - B)(A + B) = A^2 - B^2$.

DO EXERCISES 42–46.

In later work we will consider expressions like

$$(x + h)^2 - x^2.$$

To simplify this, first note that

$$(x + h)^2 = x^2 + 2xh + h^2.$$

Subtracting x^2 on both sides of this equation, we get

$$(x + h)^2 - x^2 = 2xh + h^2.$$

47. How close is $(5.1)^2$ to 5^2?

Factoring out an h on the right side, we get

$$(x + h)^2 - x^2 = h(2x + h). \qquad (4)$$

Let us now use this result to compare two squares.

Example How close is $(3.1)^2$ to 3^2?

Solution Substituting $x = 3$ and $h = 0.1$ in Eq. (4), we get

$$(3.1)^2 - 3^2 = 0.1(2 \cdot 3 + 0.1) = 0.1(6.1) = 0.61.$$

So $(3.1)^2$ differs from 3^2 by 0.61.

DO EXERCISE 47.

Compound Interest

Suppose we invest P dollars at interest rate i, compounded annually. The amount A_1 in the account at the end of 1 year is given by

$$A_1 = P + Pi = P(1 + i) = Pr,$$

where, for convenience,

$$r = 1 + i.$$

Going into the second year we have Pr dollars, so by the end of the second year we would have the amount A_2 given by

$$A_2 = A_1 \cdot r = (Pr)r = Pr^2 = P(1 + i)^2.$$

Going into the third year we have Pr^2 dollars, so by the end of the third year we would have the amount A_3 given by

$$A_3 = A_2 \cdot r = (Pr^2)r = Pr^3 = P(1 + i)^3.$$

In general, we have the following.

THEOREM 4

> If an amount P is invested at interest rate i, compounded annually, in t years it will grow to the amount A given by
>
> $$A = P(1 + i)^t.$$

Example 1 Suppose $1000 is invested at 16% compounded annually. How much is in the account at the end of 2 years?

There is a formula for finding the amount in a savings account after a certain period of time. (*Marshall Henrichs*)

48. Suppose $1000 is invested at 14% compounded annually. How much is in the account at the end of 2 years?

[handwritten calculations]

49. Suppose $1000 is invested at 11% compounded semiannually ($n = 2$). How much is in the account at the end of 3 years?

[handwritten calculations]

Solution We substitute into the equation $A = P(1 + i)^t$ and get

$$A = 1000(1 + 0.16)^2 = 1000(1.16)^2 = 1000(1.3456) = \$1345.60.$$

DO EXERCISE 48.

If interest is compounded quarterly, we can find a formula like the one above as follows:

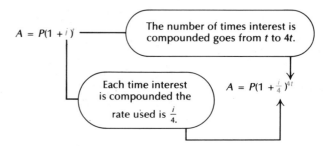

In general, the following theorem applies.

THEOREM 5

> If a principal P is invested at interest rate i, compounded n times a year, in t years it will grow to the amount A given by
>
> $$A = P\left(1 + \frac{i}{n}\right)^{nt}.$$

Example 2 Suppose $1000 is invested at 16% compounded quarterly. How much is in the account at the end of 2 years?

Solution We use the equation $A = P(1 + i/n)^{nt}$, substituting 1000 for P, 0.16 for i, 4 for n (compounding quarterly), and 2 for t; and get

$$A = 1000\left(1 + \frac{0.16}{4}\right)^{4\times2} = 1000(1 + 0.04)^8 = 1000(1.04)^8,$$

$$\approx 1000(1.368569050)$$

$$= 1368.569050$$

$$\approx 1368.57.^*$$

DO EXERCISE 49.

---

*A calculator note: A calculator with a y^x key and a ten-digit readout was used to find $(1.04)^8$ in Example 2. The number of places on the calculator may affect the accuracy of the answer. Thus you may occasionally find your answers do not agree with those in the key, which were found on a calculator with a ten-digit readout. In general, if you are using a calculator, do all your computations, and round only at the end, as in Example 2. Usually, your answer should agree to at least four digits. It might be wise to consult with your instructor on the accuracy required.

EXERCISE SET 1.1

Rename without exponents.

1. 5^3 **2.** 7^2 **3.** $(-7)^2$ **4.** $(-5)^3$ **5.** $(1.01)^2$

6. $(1.01)^3$ **7.** $\left(\dfrac{1}{2}\right)^4$ **8.** $\left(\dfrac{1}{4}\right)^3$ **9.** $(6x)^0$ **10.** $(6x)^1$

11. t^1 **12.** t^0 **13.** $\left(\dfrac{1}{3}\right)^0$ **14.** $\left(\dfrac{1}{3}\right)^1$

Rename without negative exponents.

15. 3^{-2} **16.** 4^{-2} **17.** $\left(\dfrac{1}{2}\right)^{-3}$ **18.** $\left(\dfrac{1}{2}\right)^{-2}$ **19.** 10^{-1}

20. 10^{-4} **21.** e^{-b} **22.** t^{-k} **23.** b^{-1} **24.** h^{-1}

Multiply.

25. $x^2 \cdot x^3$ **26.** $t^3 \cdot t^4$ **27.** $x^{-7} \cdot x$ **28.** $x^5 \cdot x$ **29.** $5x^2 \cdot 7x^3$

30. $4t^3 \cdot 2t^4$ **31.** $x^{-4} \cdot x^7 \cdot x$ **32.** $x^{-3} \cdot x \cdot x^3$ **33.** $e^{-t} \cdot e^t$ **34.** $e^k \cdot e^{-k}$

Divide.

35. $\dfrac{x^5}{x^2}$ **36.** $\dfrac{x^7}{x^3}$ **37.** $\dfrac{x^2}{x^5}$ **38.** $\dfrac{x^3}{x^7}$ **39.** $\dfrac{e^k}{e^k}$

40. $\dfrac{t^k}{t^k}$ **41.** $\dfrac{e^t}{e^4}$ **42.** $\dfrac{e^k}{e^3}$ **43.** $\dfrac{t^6}{t^{-8}}$ **44.** $\dfrac{t^5}{t^{-7}}$

45. $\dfrac{t^{-9}}{t^{-11}}$ **46.** $\dfrac{t^{-11}}{t^{-7}}$

Simplify.

47. $(t^{-2})^3$ **48.** $(t^{-3})^4$ **49.** $(e^x)^4$ **50.** $(e^x)^5$ **51.** $(2x^2y^4)^3$

52. $(2x^2y^4)^5$ **53.** $(3x^{-2}y^{-5}z^4)^{-4}$ **54.** $(5x^3y^{-7}z^{-5})^{-3}$ **55.** $(-3x^{-8}y^7z^2)^2$ **56.** $(-5x^4y^{-5}z^{-3})^4$

Multiply.

57. $5(x - 7)$ **58.** $4(x - 3)$

59. $x(1 - t)$ **60.** $x(1 + t)$

61. $(x - 5)(x - 2)$ **62.** $(x - 4)(x - 3)$

63. $(a - b)(a - b)(a^2 + ab + b^2)$ **64.** $(x^2 - xy - y^2)(x + y)$

65. $(2x + 5)(x - 1)$ **66.** $(3x + 4)(x - 1)$

67. $(a - 2)(a + 2)$ **68.** $(3x - 1)(3x + 1)$

69. $(5x + 2)(5x - 2)$ **70.** $(t - 1)(t + 1)$

71. $(a - h)^2$ **72.** $(a + h)^2$

73. $(5x + t)^2$ **74.** $(7a - c)^2$

75. $5x(x^2 + 3)^2$ **76.** $-3x^2(x^2 - 4)(x^2 + 4)$

Use the following equation (Eq. 1) for Exercises 77–80.

$$(x + h)^3 = (x + h)(x + h)^2 = (x + h)(x^2 + 2xh + h^2)$$
$$= (x + h)x^2 + (x + h)2xh + (x + h)h^2$$
$$= x^3 + x^2h + 2x^2h + 2xh^2 + xh^2 + h^3$$
$$= x^3 + 3x^2h + 3xh^2 + h^3 \tag{1}$$

77. $(a + b)^3$ **78.** $(a - b)^3$ **79.** $(x - 5)^3$ **80.** $(2x + 3)^3$

Factor.

81. $x - xt$ **82.** $x + xh$ **83.** $x^2 + 6xy + 9y^2$ **84.** $x^2 - 10xy + 25y^2$

85. $x^2 - 2x - 15$ **86.** $x^2 + 8x + 15$ **87.** $x^2 - x - 20$ **88.** $x^2 - 9x - 10$

89. $49x^2 - t^2$ **90.** $9x^2 - b^2$ **91.** $36t^2 - 16m^2$ **92.** $25y^2 - 9z^2$

93. $a^3b - 16ab^3$ **94.** $2x^4 - 32$ **95.** $a^8 - b^8$ **96.** $36y^2 + 12y - 35$

97. $10a^2x - 40b^2x$ **98.** $x^3y - 25xy^3$ **99.** $2 - 32x^4$ **100.** $2xy^2 - 50x$

101. $x^3 + 8$ (*Hint:* See Exercise 64.)   **102.** $a^3 - 27b^3$ (*Hint:* See Exercise 63.)

Use the following for Exercises 103 and 104: $(x + h)^2 - x^2 = h(2x + h)$.

103. a) How close is $(4.1)^2$ to 4^2?
     b) How close is $(4.01)^2$ to 4^2?
     c) How close is $(4.001)^2$ to 4^2?

104. a) How close is $(2.1)^2$ to 2^2?
     b) How close is $(2.01)^2$ to 2^2?
     c) How close is $(2.001)^2$ to 2^2?

From Eq. (1) it follows that $(x + h)^3 - x^3 = h(3x^2 + 3xh + h^2)$. Use this for Exercises 105 and 106.

105. a) How close is $(2.1)^3$ to 2^3?
     b) How close is $(2.01)^3$ to 2^3?
     c) How close is $(2.001)^3$ to 2^3?

106. a) How close is $(4.1)^3$ to 4^3?
     b) How close is $(4.01)^3$ to 4^3?
     c) How close is $(4.001)^3$ to 4^3?

The symbol ▦ indicates an exercise designed to be done using a calculator.

Business: Compound interest

107. Suppose $1000 is invested at 16%. How much is in the account at the end of 1 year, if interest is compounded:
     a) annually?
     b) semiannually?
     c) quarterly?
     d) daily? (▦ with y^x key)
     e) hourly?

108. Suppose $1000 is invested at 10%. How much is in the account at the end of 1 year, if interest is compounded:
     a) annually?
     b) semiannually?
     c) quarterly?
     d) daily? (▦ with y^x key)
     e) hourly?

OBJECTIVES

You should be able to

a) Solve equations like

$$-5x + 7 = 8x + 4,$$

   and

$$2t^2 = 9 + t.$$

b) Solve inequalities like

$$-5x + 7 < 8x + 4.$$

c) Solve applied problems.

d) Write interval notation for a given graph or inequality.

1. Solve $-\dfrac{7}{8}x + 5 = \dfrac{1}{4}x - 2.$

$$-\frac{7}{8}x + 7 = \frac{1}{4}x$$

$$7 = \frac{7}{8} + \frac{2}{8}x$$

$$7 = \frac{9}{8}x$$

$$\frac{7}{9} = \frac{9}{8}x$$

$$\frac{56}{9} = x$$

$$6\frac{2}{9} = x$$

1.2 EQUATIONS, INEQUALITIES, AND INTERVAL NOTATION

Equations

Basic to the solution of many equations are these two simple principles.

THE ADDITION PRINCIPLE

If an equation $a = b$ is true, then the equation $a + c = b + c$ is true for any number c.

THE MULTIPLICATION PRINCIPLE

If an equation $a = b$ is true, then the equation $ac = bc$ is true for any number c.

Example 1 Solve $-\frac{5}{6}x + 10 = \frac{1}{2}x + 2.$

Solution We first multiply on both sides by 6 to clear of fractions.

$$6\left(-\frac{5}{6}x + 10\right) = 6\left(\frac{1}{2}x + 2\right) \qquad \text{Multiplication Principle}$$

$$6\left(-\frac{5}{6}x\right) + 6 \cdot 10 = 6\left(\frac{1}{2}x\right) + 6 \cdot 2 \qquad \text{Distributive Law}$$

$$-5x + 60 = 3x + 12 \qquad \text{Simplifying}$$

$$60 = 8x + 12 \qquad \text{Addition Principle: We add } 5x \text{ on both sides to get the variable by itself.}$$

$$48 = 8x \qquad \text{We add } -12 \text{ on both sides.}$$

$$\frac{1}{8} \cdot 48 = \frac{1}{8} \cdot 8x \qquad \text{We multiply by } \frac{1}{8} \text{ on both sides.}$$

$$6 = x$$

The number 6 checks when it is substituted into the original equation; thus it is the solution.

DO EXERCISE 1.

To solve applied problems, we first translate to mathematical language, usually an equation. Then we solve the equation and check to see if the solution of the equation is a solution of the problem.

2. An investment is made at 14%, compounded annually. It grows to $826.50 at the end of 1 year. How much was invested originally?

[handwritten: A·14% = 826.50]

[handwritten: x + 14% (x) = 826.50]

[handwritten: 1.14x = 826.50]

[handwritten: x = $725]

Example 2 After a 5% gain in weight an animal weighs 693 lb. What was its original weight?

Solution We first translate to an equation:

$$\underbrace{\text{(Original weight)}}_{w} + \underbrace{5\%}_{+\ 5\%} \underbrace{\text{(Original weight)}}_{w} = 693$$
$$= 693.$$

Now we solve the equation:

$$w + 5\%w = 693$$
$$1 \cdot w + 0.05w = 693$$
$$(1 + 0.05)w = 693$$
$$1.05w = 693$$
$$w = \frac{693}{1.05} = 660.$$

Check: $660 + 5\% \times 660 = 660 + 0.05 \times 660 = 660 + 33 = 693$

DO EXERCISE 2.

The third principle for solving equations is the *Principle of Zero Products.*

THE PRINCIPLE OF ZERO PRODUCTS

> For any numbers *a* and *b*, if *ab* = 0, then a = 0 or b = 0; and if *a* = 0 or *b* = 0, then *ab* = 0.

An equation being solved by this principle *must* have a 0 on one side and a product on the other. The solutions are then obtained by setting each factor equal to 0 and solving the resulting equations.

Example 3 Solve $3x(x - 2)(5x + 4) = 0$.

Solution $3x(x - 2)(5x + 4) = 0$

$3x = 0$ or $x - 2 = 0$ or $5x + 4 = 0$		Principle of Zero Products
$\frac{1}{3} \cdot 3x = \frac{1}{3} \cdot 0$ or $x = 2$ or $5x = -4$		Solve each separately.
$x = 0$ or $x = 2$ or $x = -\dfrac{4}{5}$		

3. Solve $5x(x + 2)(2x - 3) = 0$.

$x = 0$

$x = -2$

$x = \frac{3}{2}$

4. Solve $x^2 + x = 12$.

$x^2 + x - 12 = 0$

$(x - 4)(x + 3) = 0$

$x = 4$

$x = -3$

5. Solve $x^3 = x$.

$x^3 - x = 0$

$x(x^2 - 1) = 0$

$x = 0$ $x^2 - 1 = 0$

$x^2 = 1$

$x = +1$

$x = -1$

The solutions are 0, 2, and $-\frac{4}{5}$.

DO EXERCISE 3.

Example 4 Solve $4x^3 = x$.

Solution

$$4x^3 = x$$

$$4x^3 - x = 0 \qquad \text{Adding } -x$$

$$x(4x^2 - 1) = 0$$

$$x(2x - 1)(2x + 1) = 0 \qquad \text{Factoring}$$

$$x = 0 \quad \text{or} \quad 2x - 1 = 0 \quad \text{or} \quad 2x + 1 = 0 \qquad \begin{array}{l}\text{Principle of}\\ \text{Zero Products}\end{array}$$

$$x = 0 \quad \text{or} \quad 2x = 1 \quad \text{or} \quad 2x = -1$$

$$x = 0 \quad \text{or} \quad x = \frac{1}{2} \quad \text{or} \quad x = -\frac{1}{2}$$

The solutions are 0, $\frac{1}{2}$, and $-\frac{1}{2}$.

DO EXERCISES 4 AND 5.

Inequalities

Principles for solving inequalities are similar to those for solving equations. We can add the same number on both sides of an inequality. We can also multiply on both sides by the same nonzero number; but if that number is negative, we must reverse the inequality sign. The following is a reformulation of the inequality-solving principles.

INEQUALITY-SOLVING PRINCIPLES

If the inequality $a < b$ is true, then:

i) $a + c < b + c$ is true, for any c;

ii) $a \cdot c < b \cdot c$, for any *positive* c;

iii) $a \cdot c > b \cdot c$, for any *negative* c.

Similar principles hold when $<$ is replaced by $\leq$ and $>$ is replaced by $\geq$.

6. Solve $3x < 11 - 2x$.

7. Solve $16 - 7x \leq 10x - 4$.

8. In Example 8, determine the number of suits the firm must sell so that its total revenue will be more than $40,000.

Intervals between houses are analogous to intervals on a number line. (*Daniel S. Brody: Stock, Boston*)

Example 5 Solve $17 - 8x \geq 5x - 4$.

Solution

$$17 - 8x \geq 5x - 4$$
$$-8x \geq 5x - 21 \qquad \text{Adding } -17$$
$$-13x \geq -21 \qquad \text{Adding } -5x$$
$$-\frac{1}{13}(-13x) \leq -\frac{1}{13}(-21) \quad \text{Multiplying by } -\frac{1}{13}, \text{ and}$$
$$\qquad\qquad\qquad\qquad\qquad \text{reversing the inequalty sign}$$
$$x \leq \frac{21}{13}$$

Any number less than or equal to $\frac{21}{13}$ is a solution.

DO EXERCISES 6 AND 7.

Example 6 Raggs, Ltd., a clothing firm, determines that its total revenue, in dollars, from the sale of x suits is given by

$$2x + 50.$$

Determine the number of suits the firm must sell so that its total revenue will be more than $70,000.

Solution We translate to an inequality and solve:

$$2x + 50 > 70{,}000$$
$$2x > 69{,}950 \qquad \text{Adding } -50$$
$$x > 34{,}975. \qquad \text{Multiplying by } \tfrac{1}{2}$$

Thus the company's total revenue will exceed $70,000 when it sells more than 34,975 suits.

DO EXERCISE 8.

Interval Notation

The set of real numbers corresponds to the set of points on a line.

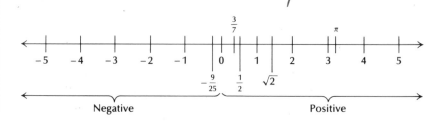

9. Write interval notation for each graph.

a)

b)

10. Write interval notation for:

a) the set of all numbers x such that $-1 < x < 4$;

b) the set of all numbers x such that $-\frac{1}{4} < x < \frac{1}{4}$.

11. Write interval notation for each graph.

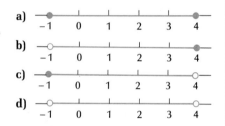

a)

b)

c)

d)

12. Write interval notation for:

a) the set of all numbers x such that $-\sqrt{2} < x < \sqrt{2}$;

b) the set of all numbers x such that $0 \le x < 1$;

c) the set of all numbers x such that $-6.7 < x \le -4.2$;

d) the set of all numbers x such that $3 \le x \le 7\frac{1}{2}$.

For real numbers a and b such that $a < b$ (a is to the left of b on a number line), we define the *open interval* (a, b) to be the set of numbers between, but not including, a and b. That is,

$$(a, b) = \text{the set of all numbers } x \text{ such that } a < x < b.$$

The graph of (a, b) is shown above. The open circles and the parentheses indicate that a and b are *not* included. The numbers a and b are called *endpoints*.

DO EXERCISES 9 AND 10.

The *closed interval* $[a, b]$ is the set of numbers between and including a and b. That is,

$$[a, b] = \text{the set of all numbers } x \text{ such that } a \le x \le b.$$

The graph of $[a, b]$ is shown above. The solid circles and the brackets indicate that a and b are included.

There are two kinds of *half-open intervals* defined as follows:

$$(a, b] = \text{the set of all numbers } x \text{ such that } a < x \le b.$$

The open circle and the parenthesis indicate that a is not included. The solid circle and the bracket indicate that b is included. Also,

$$[a, b) = \text{the set of all numbers } x \text{ such that } a \le x < b.$$

The solid circle and the bracket indicate that a is included. The open circle and the parenthesis indicate that b is not included.

DO EXERCISES 11 AND 12.

13. Write interval notation for each graph.

a)
5

b)
4

c)
4.8

d)
3

14. Write interval notation for:
  a) the set of all numbers x such that x̂ ≥ 8;
  b) the set of all numbers x such that x < −7;
  c) the set of all numbers x such that x > 10;
  d) the set of all numbers x such that x ≤ −0.78.

Some intervals are of unlimited extent in one or both directions. In such cases we use the infinity symbol ∞. For example,

$$[a, \infty) = \text{the set of all numbers } x \text{ such that } x \geq a.$$

a

Note that ∞ is not a number.

$$(a, \infty) = \text{the set of all numbers } x \text{ such that } x > a.$$

a

$$(-\infty, b] = \text{the set of all numbers } x \text{ such that } x \leq b.$$

b

$$(-\infty, b) = \text{the set of all numbers } x \text{ such that } x < b.$$

b

We can name the entire set of real numbers using $(-\infty, \infty)$.

$(-\infty, \infty)$

DO EXERCISES 13 AND 14.

Any point in an interval that is not an endpoint is an *interior* point.

Interior point
x
a
b
Endpoints

Note that all of the points in an open interval are interior points.

EXERCISE SET 1.2

Solve.

1. $-7x + 10 = 5x - 11$

2. $-8x + 9 = 4x - 70$

3. $5x - 17 - 2x = 6x - 1 - x$

4. $5x - 2 + 3x = 2x + 6 - 4x$

5. $x + 0.8x = 216$

6. $x + 0.5x = 210$

7. $x + 0.08x = 216$

8. $x + 0.05x = 210$

Applied problems

9. *Biology.* After a 6% gain in weight an animal weighs 508.8 lb. What was its original weight?

10. *Biology.* After a 7% gain in weight an animal weighs 363.8 lb. What was its original weight?

11. *Business.* An investment is made at 11%, compounded annually. It grows to $721.50 at the end of 1 year. How much was invested originally?

12. *Business.* An investment is made at 13%, compounded annually. It grows to $904 at the end of 1 year. How much was invested originally?

13. *Sociology.* After a 2% increase, the population of a city is 826,200. What was the former population?

14. *Sociology.* After a 3% increase, the population of a city is 741,600. What was the former population?

Solve

15. $2x(x + 3)(5x - 4) = 0$

16. $7x(x - 2)(2x + 3) = 0$

17. $x^2 + 1 = 2x + 1$

18. $2t^2 = 9 + t^2$

19. $t^2 - 2t = t$

20. $6x - x^2 = x$

21. $6x - x^2 = -x$

22. $2x - x^2 = -x$

23. $9x^3 = x$

24. $16x^3 = x$

25. $(x - 3)^2 = x^2 + 2x + 1$

26. $(x - 5)^2 = x^2 + x + 3$

Solve.

27. $3 - x \le 4x + 7$

28. $x + 6 \le 5x - 6$

29. $5x - 5 + x > 2 - 6x - 8$

30. $3x - 3 + 3x > 1 - 7x - 9$

31. $-7x < 4$

32. $-5x \ge 6$

33. $5x + 2x \le -21$

34. $9x + 3x \ge -24$

35. $2x - 7 < 5x - 9$

36. $10x - 3 \ge 13x - 8$

37. $8x - 9 < 3x - 11$

38. $11x - 2 \ge 15x - 7$

39. $8 < 3x + 2 < 14$

40. $2 < 5x - 8 \le 12$

41. $3 \le 4x - 3 \le 19$

42. $9 \le 5x + 3 < 19$

43. $-7 \le 5x - 2 \le 12$

44. $-11 \le 2x - 1 < -5$

Applied problems

45. *Business.* A firm determines that the total revenue, in dollars, from the sale of x units of a product is

$$3x + 1000.$$

Determine the number of units that must be sold so that its total revenue will be more than $22,000.

46. *Business.* A firm determines that the total revenue, in dollars, from the sale of x units of a product is

$$5x + 1000.$$

Determine the number of units that must be sold so that its total revenue will be more than $22,000.

47. To get a B in a course a student's average must be greater than or equal to 80% (at least 80%) and less than 90%. On the first three tests the student scores 78%, 90%, and 92%. Determine the scores on the 4th test that will yield a B.

Write interval notation for each graph in Exercises 49–56.

49.

51.

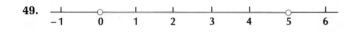

53.

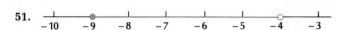

55.

48. To get a C in a course a student's average must be greater than or equal to 70% and less than 80%. On the first three tests the student scores 65%, 83%, and 82%. Determine the scores on the 4th test that will yield a C.

50.

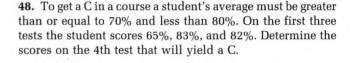

52.

54.

56.

Write interval notation for Exercises 57–62.

57. The set of all numbers x such that $-3 \leq x \leq 3$

59. The set of all numbers x such that $-14 \leq x < -11$

61. The set of all numbers x such that $x \leq -4$

58. The set of all numbers x such that $-4 < x < 4$

60. The set of all numbers x such that $6 < x \leq 20$

62. The set of all numbers x such that $x > -5$

OBJECTIVES

You should be able to

a) **Given a function and several inputs, find the outputs.**

b) **Graph a given function.**

c) **Decide if a graph is that of a function.**

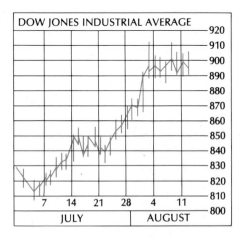

A graph of the Dow Jones Industrial Average.

1. Plot the ordered pairs (2, 0), (0, 2), (−1, 3), (4, 3), and (−2, −3).

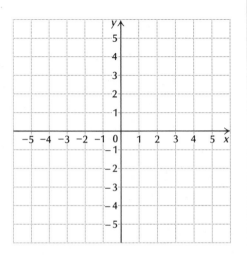

1.3 GRAPHS AND FUNCTIONS

Graphs

Each point in the plane corresponds to an ordered pair of numbers. Note that the pair (2, 5) is different from the pair (5, 2). This is why we call (2, 5) an *ordered pair*. The first member 2 is called the *first coordinate* of the point, and the second member 5 is called the *second coordinate*. Together these are called the *coordinates of the point*. The vertical line is called the *y-axis* and the horizontal line is called the *x-axis*.

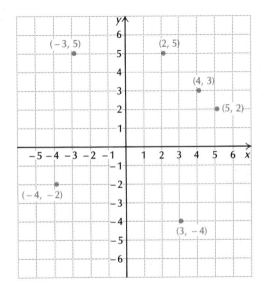

DO EXERCISE 1.

Graphs of Equations

A *solution* of an equation in two variables is an ordered pair of numbers that, when substituted alphabetically for the variables, gives a true sentence. For example, (−1, 2) is a solution of the equation $3x^2 + y = 5$, because when we substitute −1 for x and 2 for y we get a true sentence:

$$3x^2 + y = 5$$

$$\begin{array}{c|c} 3(-1)^2 + 2 & 5 \\ 3 + 2 & \\ 5 & \end{array}$$

2. Decide whether each pair is a solution of

$$x^2 - 2y = 6.$$

a) $(-2, -1)$ **b)** $(3, 0)$

3. Graph $y = -2x + 1$.

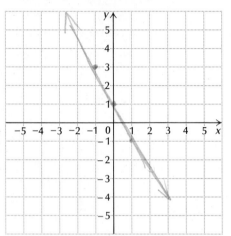

4. Graph $y = x^2 - 3$.

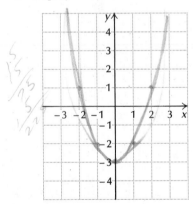

Remember, this symbol refers to the *Computer Software Supplement*. You do not need the Supplement to understand material where this symbol occurs, but if you have it, then you should go to it for material to study.

CSS

DO EXERCISE 2.

The *graph* of an equation is a geometric representation of all of its solutions. It is obtained by plotting enough ordered pairs (which are solutions) to see a pattern. The graph could be a line, curve (or curves), or some other configuration.

Example 1 Graph $y = 2x + 1$.

x	0	−1	−2	1	2
y	1	−1	−3	3	5

← We first choose these numbers at random (since y is expressed in terms of x).
← We find these numbers by substituting in the equation.

For example, when $x = -2$, $y = 2(-2) + 1 = -3$. This yields the pair $(-2, -3)$. We plot all the pairs from the table and, in this case, draw a line to complete the graph.

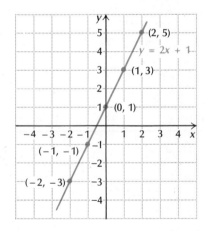

DO EXERCISE 3.

Example 2 Graph $y = x^2 - 1$.

x	0	1	2	−1	−2
y	−1	0	3	0	3

← We first choose these numbers at random (since y is expressed in terms of x).
← We find these numbers by substituting in the equation.

DO EXERCISE 4.

5. Graph $x = y^2 + 1$.

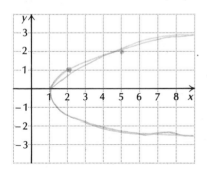

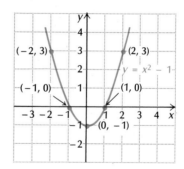

Example 3   Graph $x = y^2$.

x	0	1	4	1	4
y	0	1	2	-1	-2

⟵ We find these numbers by substituting in the equation.

⟵ This time we first choose these numbers at random since x is expressed in terms of y.

We plot these points, keeping in mind that x is still the first coordinate and y the second.

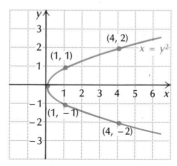

DO EXERCISE 5.

Functions

A *relation* is a set of ordered pairs. Thus the solutions of an equation in two variables form a relation.

A *function* is a special kind of relation. Such relations are of fundamental importance in calculus.

6. The operation of "taking the reciprocal" is a function. That is, the operation of going from x to $1/x$ is a function defined for all numbers except 0. Thus the domain is the set of all nonzero real numbers.

Complete this table.

Input	Output
5	$\frac{1}{5}$
$-\frac{2}{3}$	$-\frac{3}{2}$
$\frac{1}{4}$	4
$\frac{1}{a}$	a
k	$\frac{1}{k}$
$1 + t$	$\frac{1}{1+t}$

A Function as an Input–Output Relation

DEFINITION

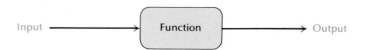

A *function* is a relation that assigns to each "input," or first coordinate, a unique "output," or second coordinate. The set of all input numbers is called the *domain*. The set of all output numbers is called the *range*.

Example 4 Squaring numbers is a function. We can take any number x as an input. We square that number to find the output, x^2.

Input	Output
-3	9
1.73	2.9929
k	k^2
$\sqrt{a}$	a
$1 + t$	$(1 + t)^2$, or $1 + 2t + t^2$

The domain of this function is the set of all real numbers, because any real number can be squared.

DO EXERCISE 6.

It is customary to use letters such as f and g to represent functions. Suppose f is a function and x is a number in its domain. For the input x, we can name the output as

$f(x)$, read "f of x," or "the value of f at x."

If f is the squaring function, then $f(3)$ is the output for the input 3. Thus $f(3) = 3^2 = 9$.

7. The reciprocal function is given by

$$f(x) = \frac{1}{x}.$$

Find $f(5)$, $f(-2)$, $f(1/4)$, $f(1/a)$, $f(k)$, $f(1 + t)$, and $f(x + h)$.

Example 5 The squaring function is given by

$$f(x) = x^2.$$

Find $f(-3)$, $f(1)$, $f(k)$, $f(\sqrt{k})$, $f(1 + t)$, and $f(x + h)$.

Solution

$$f(-3) = (-3)^2 = 9,$$
$$f(1) = 1^2 = 1,$$
$$f(k) = k^2,$$
$$f(\sqrt{k}) = (\sqrt{k})^2 = k,$$
$$f(1 + t) = (1 + t)^2 = 1 + 2t + t^2,$$
$$f(x + h) = (x + h)^2 = x^2 + 2xh + h^2$$

To find $f(x + h)$, remember what the function does—it squares the input. Thus $f(x + h) = (x + h)^2 = x^2 + 2xh + h^2$. This amounts to replacing x on both sides of $f(x) = x^2$, by $x + h$

DO EXERCISE 7.

Example 6 A function f subtracts the square of an input from the input. A description of f is given by

$$f(x) = x - x^2.$$

Find $f(4)$ and $f(x + h)$.

8. A function t is given by

$$t(x) = x + x^2.$$

Find $t(5)$, $t(-5)$, and $t(x + h)$.

Solution We replace the x's on both sides by the inputs. Thus

$$f(4) = 4 - 4^2 = 4 - 16 = -12;$$
$$f(x + h) = (x + h) - (x + h)^2 = x + h - (x^2 + 2xh + h^2)$$
$$= x + h - x^2 - 2xh - h^2.$$

DO EXERCISE 8.

Taking square roots is *not* a function, because an input can have more than one output. For example, the input 4 has two outputs 2 and -2.

When a function is given by a formula, its domain is understood to be the set of all numbers that can be substituted into the formula. For example, consider the reciprocal function

$$f(x) = \frac{1}{x}.$$

9. Subtracting 3 from a number and then taking the reciprocal is a function f given by

$$f(x) = \frac{1}{x - 3}.$$

a) What is the domain of this function? Explain.

b) Find $f(5)$, $f(4)$, $f(2.5)$, and $f(x + h)$.

The number of people at the beach at a given time is a function of the temperature, although a formula may not be readily available. (*Ellis Herwig: Stock, Boston*)

The only number that cannot be substituted into the formula is 0. We say that f is *not defined at* 0, or $f(0)$ *does not exist*. The domain consists of all nonzero real numbers.

Example 7 Taking principal square roots (nonnegative roots) is a function. Let g be this function. Then g can be described as

$$g(x) = \sqrt{x}.$$

(Recall from algebra that the symbol $\sqrt{a}$ represents the nonnegative square root of a.) a) Find the domain of this function. b) Find $g(0)$, $g(2)$, $g(a)$, $g(16)$, and $g(t + h)$.

Solution

a) The domain consists of numbers that can be substituted into the formula. We can take the principal square root of any nonnegative number, but we cannot take the principal square root of a negative number. Thus the domain consists of all nonnegative numbers.

b)

$$g(0) = \sqrt{0} = 0,$$
$$g(2) = \sqrt{2},$$
$$g(a) = \sqrt{a},$$
$$g(16) = \sqrt{16} = 4,$$
$$g(t + h) = \sqrt{t + h}$$

DO EXERCISE 9.

A Function as a Mapping

We can also think of a function as a "mapping" of one set to another.

DEFINITION

A *function* is a mapping that associates with each number x in one set (called the domain) a unique number y in another set.

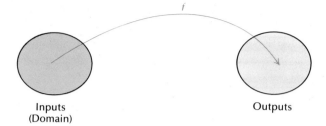

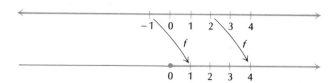

For example, the squaring function maps members of the set of real numbers to members of the set of nonnegative numbers.

The statement

$$y = f(x)$$

means that the number x is mapped to the number y by the function f. Functions are often implicit in certain equations. For example, consider

$$xy = 2.$$

For any nonzero x there is a unique number y satisfying the equation. This yields a function that is given explicitly by

$$y = f(x) = \frac{2}{x}.$$

On the other hand, consider the equation

$$x = y^2.$$

A number x would be related to two values of y, namely $\sqrt{x}$ and $-\sqrt{x}$. Thus this equation is not an implicit description of a function that maps inputs x to outputs y.

Graphs of Functions

Consider again the squaring function. The input 3 is associated with the output 9. The input–output pair (3, 9) is one point on the *graph* of this function.

10. Graph $f(x) = -2x + 1$.

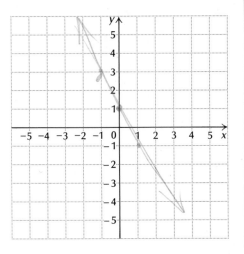

11. Graph $g(x) = x^2 - 3$.

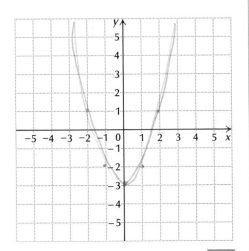

CSS

DEFINITION

The **graph** of a function f is a geometric representation of all of its input–output pairs $(x, f(x))$. In cases where the function is given by an equation, the graph of a function is the graph of the equation $y = f(x)$.

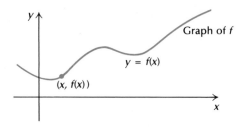

It is customary to locate input values (the domain) on the horizontal axis and output values on the vertical axis.

Example 8 Graph $f(x) = x^2 - 1$.

Solution

x	0	1	2	-1	-2
$f(x)$	-1	0	3	0	3

←— We first choose these inputs at random.

←— We compute these outputs.

Next we plot the input–output pairs from the table and, in this case, draw a curve to complete the graph.

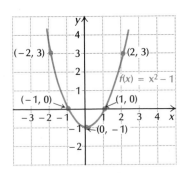

DO EXERCISES 10 AND 11.

The following figure illustrates how the idea of a mapping is connected with the graph of a function.

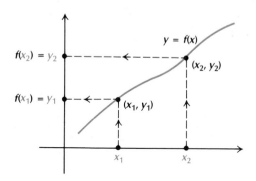

Let us now determine how we can look at a graph and decide whether it is a graph of a function. We already know that

$$x = y^2$$

does not yield a function that maps a number x to a unique number y. Look at its graph.

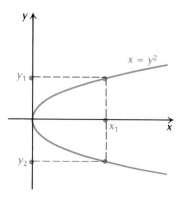

Note that there is a point x_1 that has two outputs. Equivalently, we have a vertical line that meets the graph more than once.

VERTICAL LINE TEST

A graph is that of a function provided no vertical line meets the graph more than once.

12. Which of the following are graphs of functions?

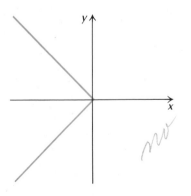

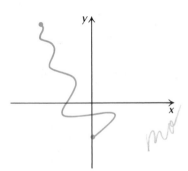

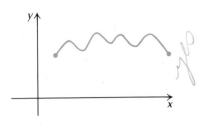

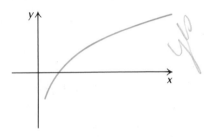

Examples Which of the following are graphs of functions?

a)

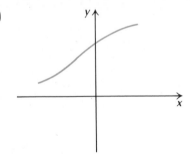

b)

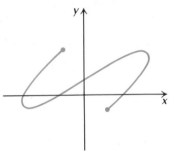

c)

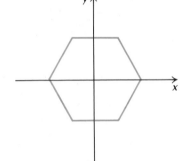

d)

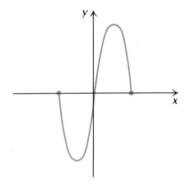

Solution

a) A function. No vertical line meets the graph more than once.

b) Not a function. A vertical line (in fact, many) meets the graph more than once.

c) Not a function.

d) A function.

DO EXERCISE 12.

Functions Defined Piecewise

Sometimes functions are defined piecewise. That is, we have different output formulas for different parts of the domain.

13. Graph the function defined as follows:

$$f(x) = \begin{cases} x + 3 & \text{for } x \leq -2, \\ 1 & \text{for } -2 < x \leq 3, \\ x^2 - 10 & \text{for } 3 < x. \end{cases}$$

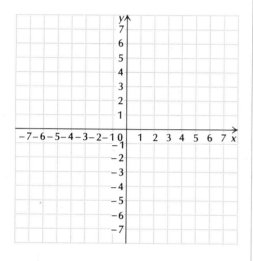

Example 9 Graph the function defined as follows.

$$f(x) = \begin{cases} 4 & \text{for } x \leq 0 \\ & \text{(This means that for any input } x \text{ less than or equal to } 0 \text{ the output is 4.)} \\ 4 - x^2 & \text{for } 0 < x \leq 2 \\ & \text{(This means that for any input } x \text{ greater than 0 and less than or equal to 2, the output is } 4 - x^2.) \\ 2x - 6 & \text{for } x > 2 \\ & \text{(This means that for any input } x \text{ greater than 2, the output is } 2x - 6.) \end{cases}$$

Solution See the graph below.

a) We graph $f(x) = 4$ for inputs less than or equal to 0 (that is, $x \leq 0$).

b) We graph $f(x) = 4 - x^2$ for inputs greater than 0 and less than or equal to 2 (that is, $0 < x \leq 2$).

c) We graph $f(x) = 2x - 6$ for inputs greater than 2 (that is, $x > 2$).

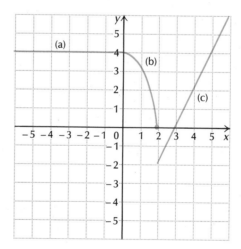

DO EXERCISE 13.

Some Final Remarks

Almost all the functions in this text can be described by equations. Some functions, however, cannot. We sometimes use the terminology *y is a function of* x. This means that x is an input and y is an output. We sometimes refer to x as, the *independent* variable when it represents inputs, and y as the *dependent* variable when it represents outputs. We may refer to "the function, $y = x^2$," without naming it with a letter f. We may simply refer to x^2 (alone) as a function.

EXERCISE SET 1.3

1. A function f is given by

$$f(x) = 2x + 3.$$

This function takes a number x, multiplies it by 2, and adds 3.

a) Complete this table. b) Find $f(5)$, $f(-1)$, $f(k)$, $f(1 + t)$, and $f(x + h)$.

Input	Output
4.1	
4.01	
4.001	
4	

CSS

2. A function f is given by

$$f(x) = 3x - 1.$$

This function takes a number x, multiplies it by 3, and subtracts 1.

a) Complete this table. b) Find $f(4)$, $f(-2)$, $f(k)$, $f(1 + t)$, and $f(x + h)$.

Input	Output
5.1	
5.01	
5.001	
5	

3. A function g is given by

$$g(x) = x^2 - 3.$$

This function takes a number x, squares it, and subtracts 3. Find $g(-1)$, $g(0)$, $g(1)$, $g(5)$, $g(u)$, $g(a + h)$, and $g(1 - h)$.

5. A function f is given by

$$f(x) = (x - 3)^2.$$

CSS

This function takes a number x, subtracts 3 from it, and squares the result.

a) Find $f(4)$, $f(-2)$, $f(0)$, $f(a)$, $f(t + 1)$, $f(t + 3)$, and $f(x + h)$.

b) Note that f could also be given by

$$f(x) = x^2 - 6x + 9.$$

Explain what this does to an input number x.

4. A function g is given by

$$g(x) = x^2 + 4.$$

This function takes a number x, squares it, and adds 4. Find $g(-3)$, $g(0)$, $g(-1)$, $g(7)$, $g(v)$, $g(a + h)$, and $g(1 - t)$.

6. A function f is given by

$$f(x) = (x + 4)^2.$$

This function takes a number x, adds 4 to it, and squares the result.

a) Find $f(3)$, $f(-6)$, $f(0)$, $f(k)$, $f(t - 1)$, $f(t - 4)$, and $f(x + h)$.

b) Note that f could also be given by

$$f(x) = x^2 + 8x + 16.$$

Explain what this does to an input number x.

Graph the following functions.

CSS

7. $f(x) = 2x + 3$

8. $f(x) = 3x - 1$

9. $g(x) = -4x$

10. $g(x) = -2x$

11. $f(x) = x^2 - 1$

12. $f(x) = x^2 + 4$

13. $g(x) = x^3$

14. $g(x) = \dfrac{1}{2}x^3$

Which of the following are graphs of functions?

15.

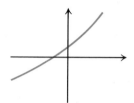

16.

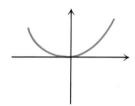

17.

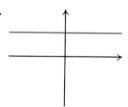

18.

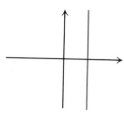

19.

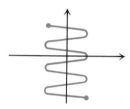

20.

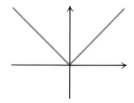

21.

22.

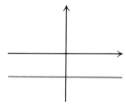

23. a) Graph $x = y^2 - 1$.
   b) Is this a function?

24. a) Graph $x = y^2 - 3$.
   b) Is this a function?

25. For $f(x) = x^2 - 3x$, find $f(x + h)$.

26. For $f(x) = x^2 + 4x$, find $f(x + h)$.

27. *Business: Revenue.* Raggs, Ltd., a clothing firm, determines that its total revenue (money coming in) from the sale of x suits is given by the function

$$R(x) = 2x + 50,$$

where $R(x)$ is the revenue, in dollars, from the sale of x suits. Find $R(10)$ and $R(100)$.

28. *Business: Compound interest.* The amount of money in a savings account at 14% compounded annually depends on the initial investment x and is given by the function

$$A(x) = x + 14\%x,$$

where $A(x) =$ the amount in the account at the end of one year. Find $A(100)$ and $A(1000)$.

Graph.

29. $f(x) = \begin{cases} 1 & \text{for } x < 0, \\ -1 & \text{for } x \geq 0 \end{cases}$ **CSS**

30. $f(x) = \begin{cases} 2 & \text{for } x \text{ an integer,} \\ -2 & \text{for } x \text{ not an integer} \end{cases}$

31. $f(x) = \begin{cases} -3 & \text{for } x = -2 \\ x^2 & \text{for } x \neq -2 \end{cases}$

32. $f(x) = \begin{cases} -2x - 6 & \text{for } x \leq -2, \\ 2 - x^2 & \text{for } -2 < x < 2, \\ 2x - 6 & \text{for } x \geq 2 \end{cases}$ **CSS**

Solve each of the following for y in terms of x. Decide whether each of the resulting equations represents a function.

33. $2x + y - 16 = 4 - 3y + 2x$

34. $2y^2 + 3x = 4x + 5$

35. $(4y^{2/3})^3 = 64x$

36. $(3y^{3/2})^2 = 72x$

1. **a)** Graph $y = 3$.
   b) Decide if it is a function.

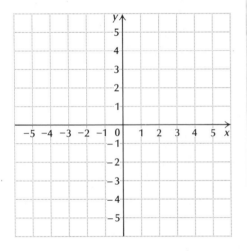

1.4 STRAIGHT LINES AND LINEAR FUNCTIONS

Horizontal and Vertical Lines

Let us consider graphs of equations $y = b$ and $x = a$.

Example 1

a) Graph $y = 4$.

b) Decide if the relation is a function.

Solution

a) The graph consists of all ordered pairs whose second coordinate is 4. To see how a pair such as $(-2, 4)$ could be a solution, we can consider the above equation in the form

$$0x + y = 4.$$

Then $(-2, 4)$ is a solution because

$$0(-2) + 4 = 4$$

is true.

b) The vertical line test holds. Thus this is a function.

DO EXERCISE 1.

Example 2

a) Graph $x = -3$.

b) Decide if it is a function.

2. a) Graph x = 1.
   b) Decide if it is a function.

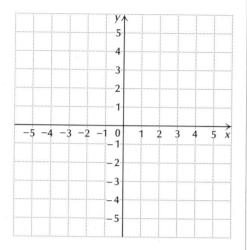

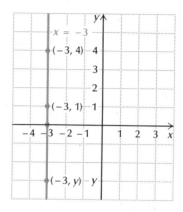

Solution

a) The graph consists of all ordered pairs whose first coordinate is −3.

b) This is *not* a function. It fails the vertical line test. The line itself meets the graph more than once, in fact, infinitely many times.

DO EXERCISE 2.

In general, we have the following.

THEOREM 6

The graph of **y = b**, a horizontal line, is the graph of a function.
The graph of **x = a**, a vertical line, is not the graph of a function.

The Equation y = mx

Consider the following table of numbers and look for a pattern.

x	1	−1	$-\frac{1}{2}$	2	−2	3	−7	5
y	3	−3	$-\frac{3}{2}$	6	−6	9	−21	15

3. a) Graph $y = -2x$.

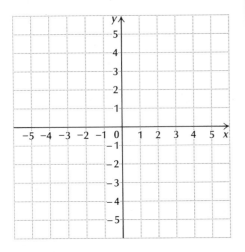

b) Is this a function?
c) What is the slope?

Lines of various slopes. (*Frank Siteman: Stock, Boston*)

Note that the ratio of the bottom number to the top one is 3. That is,

$$\frac{y}{x} = 3, \quad \text{or} \quad y = 3x.$$

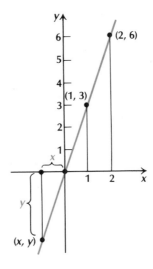

Ordered pairs from the table can be used to graph the equation $y = 3x$. Note that this is a function.

THEOREM 7

The graph of the function given by

$$y = mx \quad \text{or} \quad f(x) = mx$$

is the straight line through the origin (0, 0) and the point (1, m). The constant m is called the *slope* of the line.

DO EXERCISE 3.

Various graphs of $y = mx$ for positive m are shown as follows. Note that such graphs rise from left to right. A line with large positive slope rises faster than a line with smaller positive slope.

4. Consider these lines.

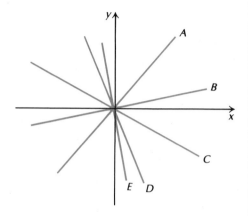

a) Which lines have positive slope?
b) Which lines have negative slope?
c) Which line has the largest slope?
d) Which line has the smallest slope?

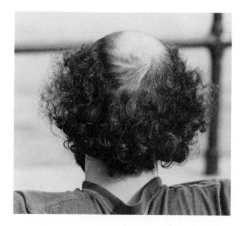

Hair will grow 6 inches in 12 months (*Ira Kirschenbaum: Stock, Boston*)

a)

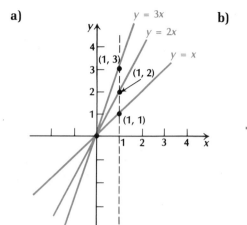

b)

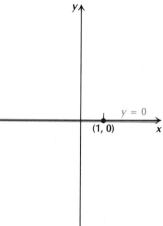

c)

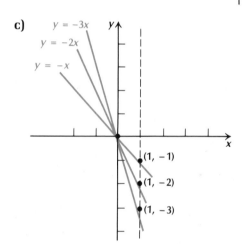

When $m = 0$, $y = 0x$, or $y = 0$. Graph (b) is a graph of $y = 0$. Note that this is the x-axis and is a horizontal line. Graphs of $y = mx$ for negative m are shown in (c). Note that such graphs fall from left to right.

DO EXERCISE 4.

Direct Variation

There are many applications involving equations like $y = mx$, where m is some positive number. In such situations we say that we have *direct variation*, and m (the slope) is called the *variation constant*,

5. *Ecology: Newspaper recycling.* The number T of trees saved by recycling is directly proportional to the height h of a stack of recycled newspaper.

a) It is known that a stack of newspaper 36 in. high will save 1 tree. Find an equation of variation expressing T as a function of h.

b) How many trees are saved by a stack of paper 162 in. (13.5 ft) high?

or *constant of proportionality.* Generally, only positive values of x and y are considered.

DEFINITION

> The variable y *varies directly* as x if there is some positive constant m such that $y = mx$. We also say that y is *directly proportional* to x.

Example 3 *Biomedical: Hair growth.* The number N of inches that human hair will grow is directly proportional to the time t in months. Hair will grow 6 inches in 12 months.

a) Find an equation of variation.

b) How many months does it take for hair to grow 10 inches?

Solution

a) $N = mt$, so $6 = m(12)$ and $\frac{1}{2} = m$. Thus $N = \frac{1}{2}t$.

b) To find how many months it takes for hair to grow 10 inches we solve

$$10 = \frac{1}{2}t$$

and get

$$20 = t.$$

Thus it takes 20 months for hair to grow 10 inches.

DO EXERCISE 5.

6. a) Using the same axes, graph

$$y = 3x$$

and

$$y = 3x + 1.$$

b) How can the graph of $y = 3x + 1$ be obtained from the graph of $y = 3x$?

The Equation $y = mx + b$

Compare the graphs (next page) of the equations

$$y = 3x \quad \text{and} \quad y = 3x - 2.$$

Note that the graph of $y = 3x - 2$ is a shift downward 2 units of the graph of $y = 3x$ and $y = 3x - 2$ has y-intercept $(0, -2)$. Note also that the graph of $y = 3x - 2$ is a graph of a function.

DO EXERCISE 6.

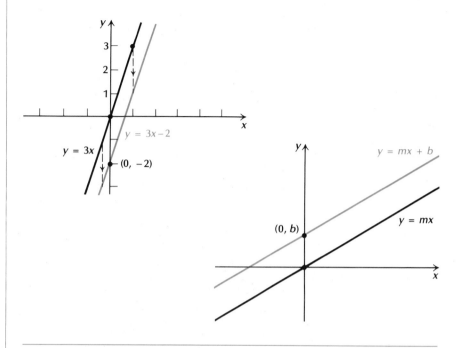

A *linear function* is given by

$$y = mx + b \quad \text{or} \quad f(x) = mx + b$$

and has a graph that is the straight line parallel to $y = mx$ with y-intercept $(0, b)$. The constant m is called the *slope*.

When $m = 0$, $y = 0x + b = b$, and we have what is known as a *constant function*. The graph of such a function is a horizontal line.

The Slope-Intercept Equation

Any nonvertical line l is uniquely determined by its slope m and its y-intercept $(0, b)$. In other words, the slope describes the "slant" of the line and the y-intercept is the point at which it crosses the y-axis. Thus we have the following definition.

DEFINITION

$y = mx + b$ is called the *slope-intercept* equation of a line.

Example 4 Find the slope and y-intercept of $2x - 4y - 7 = 0$.

7. Find the slope and y-intercept of $2x + 3y - 6 = 0$.

Solution We solve for y:

$$-4y = -2x + 7$$

$$y = \frac{1}{2}x - \frac{7}{4}$$

Slope: $\frac{1}{2}$ y-intercept: $\left(0, -\frac{7}{4}\right)$

DO EXERCISE 7.

8. Find an equation of the line with slope -4 containing the point $(2, -7)$.

The Point-Slope Equation

Suppose we know the slope of a line and some point of the line other than the y-intercept. We can still find an equation of the line.

Example 5 Find an equation of the line with slope 3 containing the point $(-1, -5)$.

Solution From the slope-intercept equation we have

$$y = 3x + b,$$

so we must determine b. Since $(-1, -5)$ is on the line, it follows that

$$-5 = 3(-1) + b,$$

so

$$-2 = b \quad \text{and} \quad y = 3x - 2.$$

DO EXERCISE 8.

Lines of the same slope. (*Anna Kaufman Moon: Stock Boston*)

If a point (x_1, y_1) is on the line

$$y = mx + b, \tag{1}$$

it must follow that

$$y_1 = mx_1 + b. \tag{2}$$

9. Find an equation of the line with slope -4 containing the point $(2, -7)$.

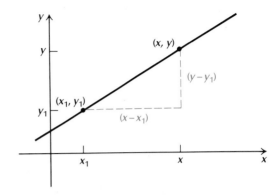

Subtracting Eq. (2) from (1) gets rid of the b's and we have

$$y - y_1 = (mx + b) - (mx_1 + b)$$
$$= mx + b - mx_1 - b$$
$$= mx - mx_1 = m(x - x_1).$$

DEFINITION

$y - y_1 = m(x - x_1)$ is called the *point-slope* equation of a line.

This definition allows us to write an equation of a line given its slope and the coordinates of *any* point on it.

Example 6 Find an equation of the line with slope 3 containing the point $(-1, -5)$.

Solution Substituting in

$$y - y_1 = m(x - x_1),$$

we get

$$y - (-5) = 3[x - (-1)].$$

Simplifying and solving for y, we get the slope-intercept equation as found in Example 5:

$$y + 5 = 3(x + 1)$$
$$y = 3x + 3 - 5$$
$$y = 3x - 2.$$

DO EXERCISE 9.

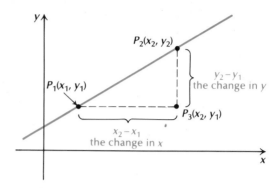

We now determine a way to compute the slope of a line when we know the coordinates of two of its points. Suppose (x_1, y_1) and (x_2, y_2) are the coordinates of two different points, P_1 and P_2, on a line that is not parallel to an axis. Consider a right triangle as shown, with legs parallel to the axes. The point P_3 with coordinates (x_2, y_1) is the third vertex of the triangle. As we move from P_1 to P_2, y changes from y_1 to y_2. The change in y is $y_2 - y_1$. Similarly, the change in x is $x_2 - x_1$. The ratio of these changes is the slope. To see this, consider the point-slope equation

$$y - y_1 = m(x - x_1).$$

Since (x_2, y_2) is on the line, it must follow that

$$y_2 - y_1 = m(x_2 - x_1).$$

Since the line is not vertical, the two x-coordinates must be different, so $x_2 - x_1$ is nonzero and we can divide by it to get the following theorem.

THEOREM 8

$$m = \frac{y_2 - y_1}{x_2 - x_1} = \frac{\text{change in } y}{\text{change in } x} = \text{slope of line containing points } (x_1, y_1) \text{ and } (x_2, y_2).$$

Example 7 Find the slope of the line containing the points $(-2, 6)$ and $(-4, 9)$.

Solution

$$m = \frac{y_2 - y_1}{x_2 - x_1} = \frac{6 - 9}{-2 - (-4)} = \frac{-3}{2} = -\frac{3}{2}$$

Note that it does not matter which point is taken first, so long as we subtract coordinates in the same order. In this example we can also find

Find the slope of the line containing each pair of points.

10. $(1, 3)(2, 5)$

11. $(-6, 4), (2, 5)$

12. $(4, 7), (6, -10)$

13. $(3, 5), (-1, 5)$

Find the slope, if it exists, of the line containing each pair of points.

14. $(4, -7), (-2, -7)$

15. $(4, -7), (4, -9)$

m as follows:

$$m = \frac{9 - 6}{-4 - (-2)} = \frac{3}{-2} = -\frac{3}{2}.$$

DO EXERCISES 10–13.

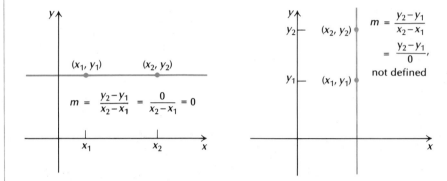

If a line is horizontal, the change in y for any two points is 0. Thus a horizontal line has slope 0. If a line is vertical, the change in x for any two points is 0. Thus the slope is not defined because we cannot divide by 0. A vertical line has no slope. Thus "0 slope" and "no slope" are two very distinct concepts.

DO EXERCISES 14 AND 15.

Applications of Linear Functions

Many applications are modeled by linear functions.

Example 8 *Business: Total cost.* Raggs, Ltd., a clothing firm, has *fixed costs* of $10,000 per year. These costs, such as rent, maintenance, and so on, must be paid no matter how much the company produces. To produce x units of a certain kind of suit it costs $20 per unit in addition to the fixed costs. That is, the *variable costs* are 20x dollars. These are costs that are directly related to production, such as material, wages, fuel, and so on. Then the *total cost*, $C(x)$, of producing x suits in a year is given by a function C:

$$C(x) = \text{(variable costs)} + \text{(fixed costs)} = 20x + 10{,}000.$$

a) Graph the variable cost, fixed cost, and total cost functions.

b) What is the total cost of producing 100 suits? 400 suits?

c) How much more does it cost to produce 400 suits than 100 suits?

16. Rework Example 9, given that variable costs = 30x, fixed costs = $15,000, and total costs = C(x) = 30x + 15,000.

Solution

a) The variable cost and fixed cost functions appear below left; the total cost function, below right. From a practical standpoint, the domains of these functions are nonnegative integers 0, 1, 2, 3, and so on. This is because it does not make sense to make a negative number of suits or a fractional number of suits. Nevertheless, it is common practice to draw the graphs as though the domains were the entire set of nonnegative real numbers.

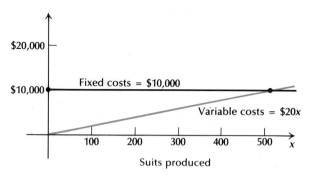

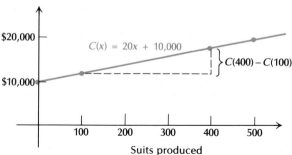

b) The total cost of producing 100 suits is

$$C(100) = 20 \cdot 100 + 10,000 = \$12,000.$$

The total cost of producing 400 suits is

$$C(400) = 20 \cdot 400 + 10,000 = \$18,000.$$

c) The extra cost of producing 400 rather than 100 suits is given by

$$C(400) - C(100) = \$18,000 - \$12,000 = \$6000.$$

DO EXERCISE 16.

Example 9 *Business: Profit and loss analysis.* In reference to Example 8, Raggs, Ltd. determines that its total revenue (money coming in) from the sale of x suits is $80 per suit. That is, total revenue $R(x)$ is given by the function

$$R(x) = 80x.$$

a) Graph $R(x)$ and $C(x)$ using the same axes.

b) The total profit $P(x)$ is given by a function P:

$$P(x) = (\text{total revenue}) - (\text{total costs}) = R(x) - C(x).$$

Determine $P(x)$ and draw its graph using the same axes.

c) The company will *break even* at that value of x for which $P(x) = 0$ (that is, no profit and no loss). This is where $R(x) = C(x)$. Find the break-even value of x.

Solution

a) The graphs of $R(x) = 80x$ and $C(x) = 20x + 10,000$ are shown here.

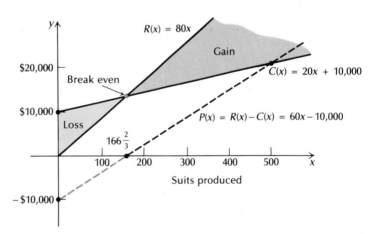

b) We see that $P(x) = R(x) - C(x) = 80x - (20x + 10,000) = 60x - 10,000$. The graph of $P(x)$ is shown.

c) To find the break-even value we solve $R(x) = C(x)$:

$$R(x) = C(x)$$

$$80x = 20x + 10,000$$

$$60x = 10,000$$

$$x = 166\frac{2}{3}.$$

17. Rework Example 10, given that

$$C(x) = 30x + 15{,}000$$

and

$$R(x) = 90x.$$

How do we interpret the fractional answer, since it is not possible to produce $\frac{2}{3}$ of a suit? We simply round to 167. Estimates of break-even points are usually sufficient since companies want to operate well away from break-even points where profit is maximized.

DO EXERCISE 17.

EXERCISE SET 1.4

Graph.

1. $y = -4$ **2.** $y = -3.5$ **3.** $x = 4.5$ **4.** $x = 10$

Graph. Find the slope and y-intercept.

5. $y = -3x$ **6.** $y = -0.5x$ **7.** $y = 0.5x$ **8.** $y = 3x$

9. $y = -2x + 3$ **10.** $y = -x + 4$ **11.** $y = -x - 2$ **12.** $y = -3x + 2$

Find the slope and y-intercept.

13. $2x + y - 2 = 0$ **14.** $2x - y + 3 = 0$ **15.** $2x + 2y + 5 = 0$ **16.** $3x - 3y + 6 = 0$

Find an equation of the line:

17. with $m = -5$, containing $(1, -5)$.   **18.** with $m = 7$, containing $(1, 7)$.

19. with $m = -2$, containing $(2, 3)$.   **20.** with $m = -3$ containing $(5, -2)$.

21. with y-intercept $(0, -6)$ and slope $\frac{1}{2}$.   **22.** with y-intercept $(0, 7)$ and slope $\frac{4}{3}$.

23. with slope 0, containing $(2, 3)$.   **24.** with slope 0, containing $(4, 8)$.

Find the slope of the line containing each pair of points.

25. $(-4, -2), (-2, 1)$ **26.** $(-2, 1) (6, 3)$ **27.** $(2, -4), (4, -3)$.

28. $(-5, 8), (5, -3)$ **29.** $(3, -7), (3, -9)$ **30.** $(-4, 2), (-4, 10)$

31. $(2, 3), (-1, 3)$ **32.** $\left(-6, \frac{1}{2}\right), \left(-7, \frac{1}{2}\right)$ **33.** $(x, 3x), (x + h, 3(x + h))$

34. $(x, 4x), (x + h, 4(x + h))$ **35.** $(x, 2x + 3), (x + h, 2(x + h) + 3)$ **36.** $(x, 3x - 1), (x + h, 3(x + h) - 1)$

DEFINITION

The *two-point* equation of the nonvertical line containing the points (x_1, y_1) and (x_2, y_2) is given by

$$y - y_1 = \frac{y_2 - y_1}{x_2 - x_1}(x - x_1). \qquad \textit{Two-point equation}$$

This can be proved by replacing m in the point-slope equation $y - y_1 = m(x - x_1)$ by $(y_2 - y_1)/(x_2 - x_1)$.

37.–48. Find an equation of the line containing each pair of points in Exercises 25–36.

Applied problems

49. *Ecology: Energy conservation.* The R-factor of home insulation is directly proportional to its thickness T.

a) Find an equation of variation where $R = 12.51$ when $T = 3$ in.

b) What is the R-factor for insulation that is 6 inches thick?

51. *Biomedical: Brain weight.* The weight B of a human's brain is directly proportional to its body weight W.

a) It is known that a person who weighs 200 lb has a brain that weighs 5 lb. Find an equation of variation expressing B as a function of W.

b) Express the variation constant as a percent, and interpret the resulting equation.

c) What is the weight of the brain of a person who weighs 120 lb?

53. *Business: Investment.* A person makes an investment of P dollars at 14%. After 1 year it grows to an amount A.

a) Show that A is directly proportional to P.

b) Find A when $P = \$100$.

c) Find P when $A = \$273.60$.

54. *Sociology: Urban population.* The population of a town is P. After a growth of 2%, its new population is N.

a) Assuming that N is directly proportional to P, find an equation of variation.

b) Find N when $P = 200{,}000$.

c) Find P when $N = 367{,}200$.

55. *Stopping distance on glare ice.* The stopping distance (at some fixed speed) of regular tires is given by a linear function of the air temperature F,

$$D(F) = 2F + 115,$$

where $D(F) =$ stopping distance, in feet, when the air temperature is F, in degrees Fahrenheit.

a) Find $D(0°)$, $D(-20°)$, $D(10°)$, and $D(32°)$.

b) Graph $D(F)$.

c) Explain why the domain should be restricted to the interval $[-57.5°, 32°]$.

50. *Biomedical: Nerve impulse speed.* Impulses in nerve fibers travel at a speed of 293 ft/sec. The distance D traveled in t sec is given by $D = 293t$. How long would it take an impulse to travel from the brain to the toes of a person who is 6 ft tall?

52. *Biomedical: Muscle weight.* The weight M of the muscles in a human is directly proportional to body weight W.

a) It is known that a person who weighs 200 lb has 80 lb of muscles. Find an equation of variation expressing M as a function of W.

b) Express the variation constant as a percent and interpret the resulting equation.

c) What is the muscle weight of a person who weighs 120 lb?

The muscle weight is directly proportional to body weight. (*Anna Kaufman Moon: Stock, Boston*)

56. *Sociology: Percentage of the population in college.* The percentage of the population in college is given by a linear function

$$P(t) = 1.25t + 15,$$

where $P(t) =$ the percentage in college the tth year after 1940. Plus $P(0)$ is the percentage in college in 1940, $P(30)$ is the percentage in college in 1970, and so on.

a) Find $P(0)$, $P(1)$, $P(30)$, and $P(40)$.

b) What percentage of the population will be in college in 1986?

c) Graph $P(t)$.

57. *Biology: Spread of an organism.* A certain kind of organism is released over an area of 2 sq mi. It grows and spreads over more area. The area covered by the organism after time *t* is given by a linear function

$$A(t) = 1.1t + 2,$$

where $A(t)$ = the area covered, in square miles, after time *t*, in years.

a) Find $A(0)$, $A(1)$, $A(4)$, and $A(10)$.

b) Graph $A(t)$.

c) Why should the domain be restricted to the interval $[0, \infty)$?

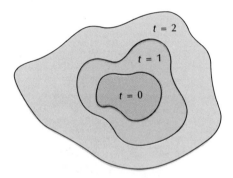

58. *Business: Profit and loss analysis.* A ski manufacturer is planning a new line of skis. For the first year, the fixed costs for setting up the new production line are $22,500. Variable costs for producing each pair of skis are estimated to be $40. The sales department projects that 3000 pairs can be sold during the first year at a price of $85 per pair.

a) Formulate a function $C(x)$ for the total cost of producing *x* pairs of skis.

b) Formulate a function $R(x)$ for the total revenue from the sale of *x* pairs of skis.

c) Formulate a function $P(x)$ for the total profit from the production and sale of *x* pairs of skis.

d) What profit or loss will the company realize if expected sales of 3000 pairs occur?

e) How many pairs must the company sell in order to break even?

59. *Business: Profit and loss analysis.* Boxowitz, Inc., a computer firm, is planning to sell a new minicalculator. For the first year, the fixed costs for setting up the new production line are $100,000. Variable costs for producing each calculator are estimated to be $20. The sales department projects that 150,000 calculators can be sold during the first year at a price of $45 each.

a) Formulate a function $C(x)$ for the total cost of producing *x* calculators.

b) Formulate a function $R(x)$ for the total revenue from the sale of *x* calculators.

c) Formulate a function $P(x)$ for the total profit from the production and sale of *x* calculators.

d) What profit or loss will the firm realize if the expected sales of 150,000 calculators occur?

e) How many calculators must the firm sell in order to break even?

60. *Business: Straight-line depreciation.* A company buys an office machine for $5200 on January 1 of a given year. The machine is expected to last for 8 years, at the end of which time its *trade-in*, or *salvage*, *value* will be $1100. If the company figures the decline in value to be the same each year, then the *book value*, or *salvage value*, after *t* years, $0 \le t \le 8$, is given by the linear function

$$V(t) = C - t\left(\frac{C - S}{N}\right),$$

where C = the original cost of the item ($5200), N = the years of expected life (8), and S = the salvage value ($1100).

a) Find the linear function for the straight-line depreciation of the office machine.

b) Find the salvage value after 0 years, 1 year, 2 years, 3 years, 4 years, 7 years, 8 years.

61. *Anthropology: Estimating heights.* An anthropologist can use certain linear functions to estimate the height of a male or female, given the length of certain bones. A *humerus* is the bone from the elbow to the shoulder. Let x = the length of the humerus in centimeters. Then the height, in centimeters, of a male with a humerus of length x is given by
$$M(x) = 2.89x + 70.64.$$
The height, in centimeters, of a female with a humerus of length x is given by $F(x) = 2.75x + 71.48.$
A 45-cm humerus was uncovered in a ruins.

a) If we assume it was from a male, how tall was he?

b) If we assume it was from a female, how tall was she?

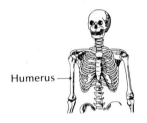

Humerus →

62. *Business: Sales commissions.* A person applying for a sales position is offered alternative salary plans:
Plan A: A base salary of $600 per month plus a commission of 4% of the gross sales for the month.
Plan B: A base salary of $700 per month plus a commission of 6% of the gross sales for the month in excess of $10,000.

a) For each plan, formulate a function that expresses monthly earnings as a function of gross sales x.

b) For what gross sales values is plan B preferable?

OBJECTIVES

You should be able to

a) **Graph a given function.**

b) **Convert from radical notation to fractional exponents, and from fractional exponents to radical notation.**

c) **Determine the domain of a rational function.**

d) **Find the equilibrium point, given a demand and supply function.**

1.5 OTHER TYPES OF FUNCTIONS

Quadratic Functions

DEFINITION

A *quadratic function f* is given by
$$f(x) = ax^2 + bx + c, \quad \text{where } a \neq 0.$$

We have already considered some such functions, for example, $f(x) = x^2$ and $g(x) = x^2 - 1$. Graphs of quadratic functions are always cup-shaped, like those in Example 1. Each has a dashed line of symmetry.

Example 1 Graph $y = x^2 - 2x - 3$ and $y = -2x^2 + 4x + 1$.

Solutions $y = x^2 - 2x - 3$ $\qquad\qquad\qquad y = -2x^2 + 4x + 1$

x	0	1	2	3	4	−1	−2
y	−3	−4	−3	0	5	0	5

x	0	1	2	3	−1
y	1	3	1	−5	−5

1. Using the same axes, graph $y = x^2$ and $y = -x^2$. (Note: $-x^2$ means $-1 \cdot x^2$.) **CSS**

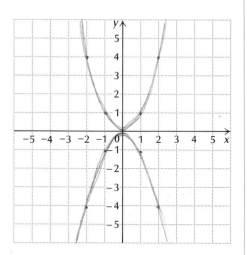

2. a) Using the same axes, graph $y = x^2$ and $y = (x - 3)^2$.

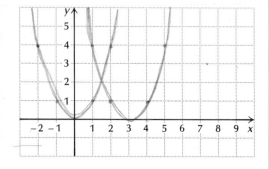

b) How could the graph of $y = (x - 3)^2$ be obtained from the graph of $y = x^2$? **CSS**

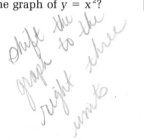

Ohift the graph to the right three units.

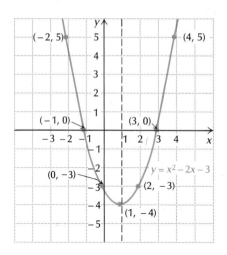

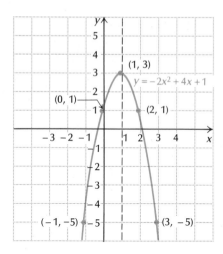

If the coefficient a is positive, the graph opens upward. If a is negative, the graph opens downward.

DO EXERCISES 1 AND 2.

First coordinates of points where a quadratic function intersects the x-axis (x-intercepts), if they exist, can be found by solving the quadratic equation $ax^2 + bx + c = 0$. If real-number solutions exist, they can be found using the quadratic formula.

THEOREM 9   The Quadratic Formula

> The solutions of any quadratic equation $ax^2 + bx + c = 0$, $a \neq 0$, are given by
>
> $$x = \frac{-b \pm \sqrt{b^2 - 4ac}}{2a}.$$

Example 2 Solve $3x^2 - 4x = 2$.

Solution First find standard form $ax^2 + bx + c = 0$, and determine a, b, and c.

$$3x^2 - 4x - 2 = 0,$$

$$a = 3, \quad b = -4, \quad c = -2$$

3. Solve $3x^2 = 7 - 2x$.

$$3x^2 + 2x - 7 = 0$$

$$\frac{-2 \pm \sqrt{4 - 4(3)(-7)}}{2(3)}$$

$$\frac{-2 \pm \sqrt{4 + 84}}{6}$$

$$\frac{-2 \pm \sqrt{88}}{6}$$

$$\frac{-2 \pm 2\sqrt{22}}{6}$$

$$\frac{-1 \pm \sqrt{22}}{3}$$

$$x =$$

Then use the quadratic formula:

$$x = \frac{-b \pm \sqrt{b^2 - 4ac}}{2a} = \frac{-(-4) \pm \sqrt{(-4)^2 - 4(3)(-2)}}{2 \cdot 3}$$

$$= \frac{4 \pm \sqrt{16 + 24}}{6} = \frac{4 \pm \sqrt{40}}{6}$$

$$= \frac{4 \pm \sqrt{4 \cdot 10}}{6} = \frac{4 \pm 2\sqrt{10}}{6}$$

$$= \frac{2(2 \pm \sqrt{10})}{2 \cdot 3} = \frac{2 \pm \sqrt{10}}{3}.$$

The solutions are $(2 + \sqrt{10})/3$ and $(2 - \sqrt{10})/3$.

DO EXERCISE 3.

Polynomial Functions

Linear and quadratic functions are part of a general class of polynomial functions.

DEFINITION

A *polynomial function f* is given by

$$f(x) = a_n x^n + a_{n-1} x^{n-1} + \cdots + a_2 x^2 + a_1 x^1 + a_0,$$

where *n* is a nonnegative integer, and a_n, a_{n-1}, . . . , a_1, a_0 are real numbers.

The following are some examples:

$f(x) = -5,$ (A constant function)

$f(x) = 4x + 3,$ (A linear function)

$f(x) = -x^2 + 2x + 3,$ (A quadratic function)

$f(x) = 2x^3 - 4x^2 + x + 1.$ (A cubic function)

In general, graphing polynomial functions other than linear and quadratic is difficult. We use calculus to sketch such graphs in Section 3.3. Some *power* functions, such as

$$y = ax^n,$$

are relatively easy to graph.

4. Using the same set of axes, graph $y = x^2$ and $y = x^4$.

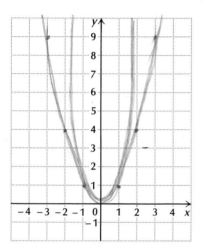

Example 3 Using the same set of axes, graph $y = x^2$ and $y = x^3$.

Solution

x	-2	-1	$-\dfrac{1}{2}$	0	$\dfrac{1}{2}$	1	2
x^2	4	1	$\dfrac{1}{4}$	0	$\dfrac{1}{4}$	1	4
x^3	-8	-1	$-\dfrac{1}{8}$	0	$\dfrac{1}{8}$	1	8

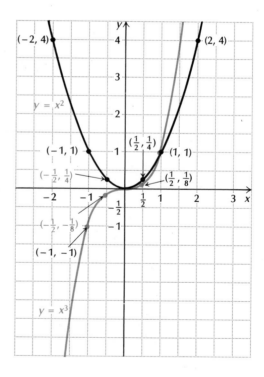

DO EXERCISE 4.

5. Determine the domain of each function.

a) $g(x) = \dfrac{x^2 - 16}{x + 4}$

$\dfrac{(x - 4)(x + 4)}{(x + 4)} = 0$

$x - 4 = 0$

$x = 4$

all real except 4

b) $t(x) = \dfrac{x + 7}{x^2 + 4x - 5}$

$\dfrac{x + 7}{(x + 5)(x - 1)} = 0$

$\dfrac{x}{(x+5)(x-1)} = \dfrac{-7}{(x+5)(x-1)}$

$x = -7$

c) $k(x) = \dfrac{1}{x - 5}$

$(k(x))(x - 5) = 1$

$k(x)(x) = 6$

$x = 6$

Rational Functions

DEFINITION

Functions given by the ratio of two polynomials are called *rational*.

The following are examples of rational functions:

$$f(x) = \frac{x^2 - 9}{x - 3},$$

$$g(x) = \frac{x^2 - 16}{x + 4},$$

$$h(x) = \frac{x - 3}{x^2 - x - 2}.$$

The domain of a rational function is restricted to those input values that do not result in division by 0. Thus for f the domain consists of all real numbers except 3. To determine the domain of h, we set the denominator equal to 0 and solve:

$$x^2 - x - 2 = 0,$$

$$(x + 1)(x - 2) = 0,$$

$$x = -1 \quad \text{or} \quad x = 2.$$

Thus -1 and 2 are not in the domain. The domain consists of all real numbers except -1 and 2.

DO EXERCISE 5.

One important class of rational functions is given by $y = k/x$.

Example 4 Graph $y = 1/x$.

Solution

x	-3	-2	-1	$-\frac{1}{2}$	$-\frac{1}{4}$	$\frac{1}{4}$	$\frac{1}{2}$	1	2	3
y	$-\frac{1}{3}$	$-\frac{1}{2}$	-1	-2	-4	4	2	1	$\frac{1}{2}$	$\frac{1}{3}$

6. Graph $y = \dfrac{-1}{x}$.

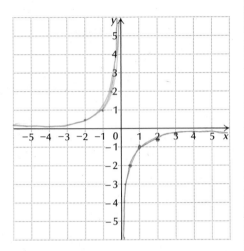

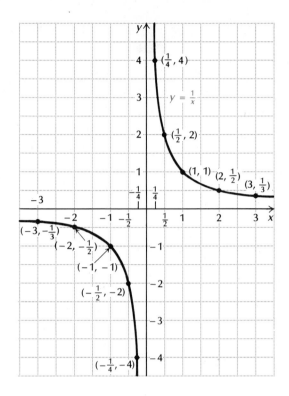

Note that 0 is not in the domain of this function since it would yield a denominator of 0. This function is decreasing over the intervals $(-\infty, 0)$ and $(0, \infty)$. It is an example of inverse variation.

DEFINITION

> **y varies inversely** as **x** if there is some positive number **k** such that **y = k/x**. We also say that **y** is **inversely proportional** to **x**.

DO EXERCISE 6.

Example 5 *Stocks and gold.* Certain economists theorize that stock prices are inversely proportional to the price of gold. That is, when the price of gold goes up, the prices of stock go down; and when the price of gold goes down, the prices of stock go up. Let us assume that the Dow-Jones Industrial Average, D, an index of the overall price of stock, is inversely proportional to the price of gold, G, in dollars per ounce. One day the Dow-Jones was 818 and the price of gold was \$520 per ounce. What will the Dow-Jones be if the price of gold drops to \$490?

7. *Demand.* The price p of a certain kind of radio is found to be inversely proportional to the number sold, x. It was found that 240,000 radios will be sold when the price per radio is $12.50. How many will be sold if the price is $18.75?

$P = \dfrac{k}{x}$

$12.50 = \dfrac{k}{240,000}$

$k = 3,000,000$

$18.75 = \dfrac{3,000,000}{x}$

$x = 160,000$

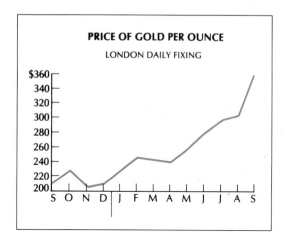

PRICE OF GOLD PER OUNCE

LONDON DAILY FIXING

Solution

a) We know that $D = k/G$, so $818 = k/520$ and $k = 425{,}360$. Thus $D = 425{,}360/G$.

b) We substitute 490 for G and compute D:

$$D = \frac{425{,}360}{490} \approx 868.1.$$

WARNING! Do not put too much "stock" in the equation of Example 5. It is meant to give us an idea of economic relationships. An equation to accurately predict the stock market has not been found.

DO EXERCISE 7.

Absolute Value Functions

The following is an example of an absolute value function and its graph. The absolute value of a number is its distance from 0. We denote the absolute value of a number x as $|x|$.

Example 6 Graph $f(x) = |x|$.

Solution

x	-3	-2	-1	0	1	2	3
$f(x)$	3	2	1	0	1	2	3

8. Graph $f(x) = |x - 1|$. To find an output, take an input, subtract 1 from it, and then take the absolute value.

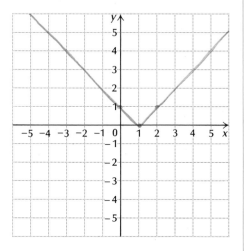

9. Graph $f(x) = \sqrt{x}$.

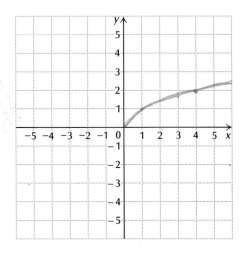

10. Find the domain of $y = \sqrt{2x + 3}$. (*Hint:* Solve $2x + 3 \geq 0$.)

$$2x + 3 \geq 0$$
$$2x \geq -3$$
$$x \geq -\tfrac{3}{2}$$

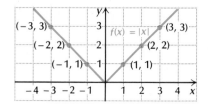

DO EXERCISE 8.

Square Root Functions

The following is an example of a square root function and its graph.

Example 7 Graph $f(x) = -\sqrt{x}$.

Solution The domain of this function is just the nonnegative numbers—the interval $[0, \infty)$. Table 1 at the back of the book contains approximate values of square roots of certain numbers.

x	0	1	2	3	4	5	10
$f(x)$, or $-\sqrt{x}$	0	−1	−1.4	−1.7	−2	−2.2	−3.2

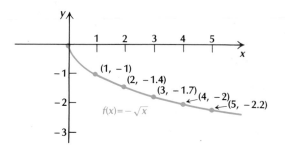

DO EXERCISES 9 AND 10.

Power Functions with Fractional Exponents

We are motivated to define fractional exponents so that the laws we discussed in Section 1.1 still hold. For example, if the laws of exponents are to hold, we would have

$$a^{1/2} \cdot a^{1/2} = a^{1/2+1/2} = a^1 = a.$$

Convert to fractional exponents.

11. $\sqrt[4]{t^3}$ $t^{3/4}$

12. $\sqrt[5]{y}$ $y^{1/5}$

13. $\dfrac{1}{\sqrt[5]{x^2}}$ $x^{-2/5}$

14. $\dfrac{1}{\sqrt[3]{t}}$ $t^{-1/3}$

15. $\sqrt{x^6}$ x^3

16. $\sqrt{x^7}$ $x^{7/2}$

Convert to radical notation.

17. $y^{1/7}$ $\sqrt[7]{y}$

18. $x^{3/2}$ $\sqrt[2]{x^3}$

19. $t^{-3/2}$ $\dfrac{1}{\sqrt{t^3}}$

20. $b^{-1/2}$ $\dfrac{1}{\sqrt{b}}$

Simplify.

21. $4^{5/2}$ $\sqrt{4^5} = 32$

22. $27^{2/3}$ $= 9$

Thus we are led to define $a^{1/2}$ as $\sqrt{a}$. Similarly, we are led to define $a^{1/3}$ as the cube root of a, $\sqrt[3]{a}$. In general,

$$a^{1/n} = \sqrt[n]{a}.$$

Again, if the laws of exponents are to hold, we would have

$$\sqrt[n]{a^m} = (a^m)^{1/n} = (a^{1/n})^m = a^{m/n}.$$

An expression $a^{-m/n}$ is defined by

$$a^{-m/n} = \frac{1}{a^{m/n}} = \frac{1}{\sqrt[n]{a^m}}.$$

Examples Convert to fractional exponents.

a) $\sqrt[3]{x^2} = x^{2/3}$

b) $\sqrt[4]{y} = y^{1/4}$

c) $\dfrac{1}{\sqrt[3]{b^5}} = \dfrac{1}{b^{5/3}} = b^{-5/3}$

d) $\dfrac{1}{\sqrt{x}} = \dfrac{1}{x^{1/2}} = x^{-1/2}$

e) $\sqrt{x^8} = x^{8/2}$, or x^4

DO EXERCISES 11–16.

Examples Convert to radical notation.

a) $x^{1/3} = \sqrt[3]{x}$

b) $t^{6/7} = \sqrt[7]{t^6}$

c) $x^{-2/3} = \dfrac{1}{x^{2/3}} = \dfrac{1}{\sqrt[3]{x^2}}$

d) $e^{-1/4} = \dfrac{1}{e^{1/4}} = \dfrac{1}{\sqrt[4]{e}}$

DO EXERCISES 17–20.

Examples Simplify.

a) $8^{5/3} = (8^{1/3})^5 = (\sqrt[3]{8})^5 = 2^5 = 32$

b) $81^{3/4} = (81^{1/4})^3 = (\sqrt[4]{81})^3 = 3^3 = 27$

DO EXERCISES 21 AND 22.

Earlier when we graphed $f(x) = \sqrt{x}$, we were also graphing $f(x) = x^{1/2}$, or $f(x) = x^{0.5}$. The power functions

$$f(x) = ax^k, \quad k \text{ fractional},$$

do arise in application. For example, the *home range* of an animal is defined as the region to which it confines its movements. It has been

Caribou in their territory area. (© *Charlie Ott from National Audubon Society*)

hypothesized in statistical studies* that the area H of that region can be approximated using the body weight W of an animal by the function

$$H = W^{1.41}$$

W	0	10	20	30	40	50
H	0	26	68	121	182	249

Note that

$$H = W^{1.41} = W^{141/100} = \sqrt[100]{W^{141}}.$$

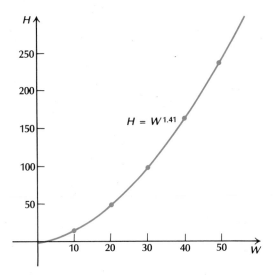

In the graph shown here, note that this is an increasing function. As body weight increases, the area over which the animal moves increases. It would be hard, however, to find accurate function values unless we used a calculator; we are simply illustrating that such functions do have application.

Supply and Demand Functions

Supply and demand in economics are modeled by increasing and decreasing functions. Although specific scientific formulas for these

*See J. M. Emlen, *Ecology: An Evolutionary Approach*, p. 200 (Reading, Mass.: Addison-Wesley, 1973).

concepts are not usually known, the notions of increasing and decreasing yield understanding of the ideas.

Demand functions   Look at the following table.

DEMAND SCHEDULE

Quantity (x) of 5-lb Bags in Millions	Price (p) per Bag
4	$5
5	4
7	3
10	2
15	1

The table shows the relationship between the price p per bag of sugar and the quantity x of 5-lb bags that the consumer will buy at that price. Note that as the price per bag increases, the quantity demanded by the consumer decreases; and as the price per bag decreases, the quantity demanded by the consumer increases. Thus it is natural to think of x as a function of p. In our later work it will be more convenient to think of p as a function of x. Thus, for a *demand* function D, $D(x)$ is the price per unit of an item when x units are demanded by the consumer. The following figure is the graph of a demand function for sugar (using the preceding table.)

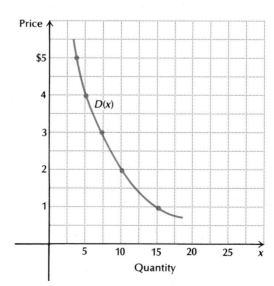

Supply functions   Look at the following table.

SUPPLY SCHEDULE	
Quantity (x) of 5-lb Bags in Millions	Price (p) per Bag
0	$1
10	2
15	3
20	4
24	5

The table shows the relationship between the price p per bag of sugar and the quantity x of 5-lb bags that the seller is willing to supply at that price. Note that as the price per bag increases, the more the seller is willing to supply; and as the price per bag decreases, the less the seller is willing to supply. Again, it is natural to think of x as a function of p, but for our later work it is more convenient to think of p as a function of x. Thus, for a *supply* function S, S(x) is the price per unit of an item at which the seller is willing to supply x units of a product to the consumer. The following figure is the graph of a supply function for sugar (using the preceding table).

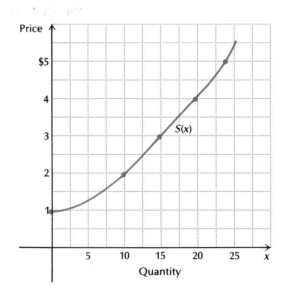

Let us now look at these curves together. Note that as supply increases, demand decreases; and as supply decreases, demand increases. The point of intersection of the two curves (x_E, p_E) is called the *equilibrium point*. The equilibrium price p_E (in this case $2 per bag) is the point at which the amount x_E (in this case 10 million bags) that the seller willingly supplies is the same as the amount that the consumer willingly demands. The situation is analogous to a buyer and seller haggling over the sale of an item. The equilibrium point or selling price is what they finally agree on.

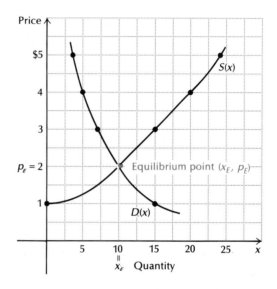

Example 8 Find the equilibrium point for the demand and supply functions

$$D(x) = (x - 6)^2 \quad \text{and} \quad S(x) = x^2 + x + 10.$$

Solution To find the equilibrium point, we set $D(x) = S(x)$ and solve:

$$(x - 6)^2 = x^2 + x + 10$$

$$x^2 - 12x + 36 = x^2 + x + 10$$

$$-12x + 36 = x + 10$$

$$-13x = -26$$

$$x = \frac{-26}{-13}$$

$$x = 2.$$

23. Given $D(x) = (x - 5)^2$ and $S(x) = x^2 + x + 3$, find the equilibrium point. **CSS**

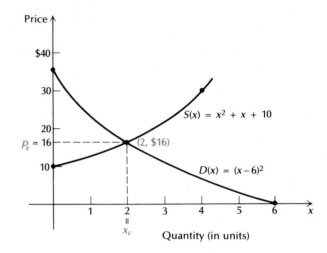

Thus $x_E = 2$ (units). To find p_E we substitute x_E into either $D(x)$ or $S(x)$. We use $D(x)$. Then

$$p_E = D(x_E) = D(2) = (2 - 6)^2 = (-4)^2 = \$16.$$

Thus the equilibrium price is $16 per unit and the equilibrium point is $(2, \$16)$.

DO EXERCISE 23.

EXERCISE SET 1.5

Using the same set of axes, graph each pair of equations.   **CSS**

1. $y = \dfrac{1}{2}x^2$, $y = -\dfrac{1}{2}x^2$ **2.** $y = \dfrac{1}{4}x^2$, $y = -\dfrac{1}{4}x^2$ **3.** $y = x^2$, $y = (x - 1)^2$ **4.** $y = x^2$, $y = (x - 3)^2$

5. $y = x^2$, $y = (x + 1)^2$ **6.** $y = x^2$, $y = (x + 3)^2$ **7.** $y = |x|$, $y = |x + 3|$ **8.** $y = |x|$, $y = |x + 1|$

9. $y = x^3$, $y = x^3 + 1$ **10.** $y = x^3$, $y = x^3 - 1$ **11.** $y = \sqrt{x}$, $y = \sqrt{x + 1}$ **12.** $y = \sqrt{x}$, $y = \sqrt{x - 2}$

Graph.

13. $y = x^2 - 4x + 3$ **14.** $y = x^2 - 6x + 5$ **15.** $y = -x^2 + 2x - 1$ **16.** $y = -x^2 - x + 6$

17. $y = \dfrac{2}{x}$ **18.** $y = \dfrac{3}{x}$ **19.** $y = \dfrac{-2}{x}$ **20.** $y = \dfrac{-3}{x}$

21. $y = \dfrac{1}{x^2}$ **22.** $y = \dfrac{1}{x - 1}$ **23.** $y = \sqrt[3]{x}$ **24.** $y = \dfrac{1}{|x|}$

(*Hint:* ▦ or use Table 1 at the end of the book)

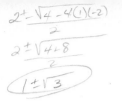

$\dfrac{2 \pm \sqrt{4 - 4(1)(-2)}}{2}$

$\dfrac{2 \pm \sqrt{4+8}}{2}$

$\boxed{1 \pm \sqrt{3}}$

$\dfrac{2 \pm 2\sqrt{5}}{2}$ $\boxed{1 \pm \sqrt{5}}$

$\dfrac{2 \pm \sqrt{4 - 4(1)(-4)}}{2}$

Solve. $x^2 - 2x - 2$ $x^2 - 2x - 4 = 0$

25. $x^2 - 2x = 2$ **26.** $x^2 - 2x + 1 = 5$ **27.** $x^2 + 6x = 1$ **28.** $x^2 + 4x = 3$

29. $4x^2 = 4x + 1$ **30.** $-4x^2 = 4x - 1$ **31.** $3y^2 + 8y + 2 = 0$ **32.** $2p^2 - 5p = 1$

Convert to fractional exponents.

33. $\sqrt{x^3}$ $x^{3/2}$ **34.** $\sqrt{x^5}$ $x^{5/2}$ **35.** $\sqrt[5]{a^3}$ $a^{3/5}$ **36.** $\sqrt[4]{b^2}$ $b^{1/2}$ $\sqrt{b^p} = b^{p/q}$ **37.** $\sqrt[7]{t}$ $t^{1/7}$ **38.** $\sqrt[8]{c}$ $c^{1/8}$

39. $\dfrac{1}{\sqrt[3]{t^4}}$ $t^{-4/3}$ **40.** $\dfrac{1}{\sqrt[5]{b^6}}$ $b^{-6/5}$ **41.** $\dfrac{1}{\sqrt{t}}$ $t^{-1/2}$ **42.** $\dfrac{1}{\sqrt{m}}$ $m^{-1/2}$ **43.** $\dfrac{1}{\sqrt{x^2 + 7}}$ **44.** $\sqrt{x^3 + 4}$

Convert to radical notation.

45. $x^{1/5}$ **46.** $t^{1/7}$ **47.** $y^{2/3}$ **48.** $t^{2/5}$ **49.** $t^{-2/5}$ **50.** $y^{-2/3}$

51. $b^{-1/3}$ **52.** $b^{-1/5}$ **53.** $e^{-17/6}$ **54.** $m^{-19/6}$ **55.** $(x^2 - 3)^{-1/2}$ $\dfrac{1}{\sqrt{x^2 - 3}}$

Simplify.

57. $9^{3/2}$ $= 27$ **58.** $16^{5/2}$ $= 1024$ **59.** $64^{2/3}$ $= 16$ **60.** $8^{2/3}$ $= 4$ **61.** $16^{3/4}$ $= 8$ **62.** $25^{5/2}$ $= 3125$

Determine the domain of each function.

63. $f(x) = \dfrac{x^2 - 25}{x - 5}$ **64.** $f(x) = \dfrac{x^2 - 4}{x + 2}$ **65.** $f(x) = \dfrac{x^3}{x^2 - 5x + 6}$

66. $f(x) = \dfrac{x^4 + 7}{x^2 + 6x + 5}$ **67.** $f(x) = \sqrt{5x + 4}$ $\boxed{CSS}$ $x \geq -\frac{4}{5}$ **68.** $f(x) = \sqrt{2x - 6}$

Find the equilibrium point for each of the following demand and supply functions. $\boxed{CSS}$

69. $D(x) = -2x + 8$, $S(x) = x + 2$ $-2x + 8 = x + 2$ $-3x = -6$ $x = 2$

70. $D(x) = -\dfrac{5}{6}x + 10$, $S(x) = \dfrac{1}{2}x + 2$

71. $D(x) = (x - 3)^2$, $S(x) = x^2 + 2x + 1$

72. $D(x) = (x - 4)^2$, $S(x) = x^2 + 2x + 6$

73. $D(x) = (x - 4)^2$, $S(x) = x^2$

74. $D(x) = (x - 6)^2$, $S(x) = x^2$

75. ▦ It is theorized that the dividends paid on utilities stocks are inversely proportional to the prime (interest) rate. Recently, the dividends D on the stock of Indianapolis Power and Light were \$2.09 per share and the prime rate was 19%. The prime rate R dropped to 17.5%. What dividends would be paid if the assumption of inverse proportionality is correct?

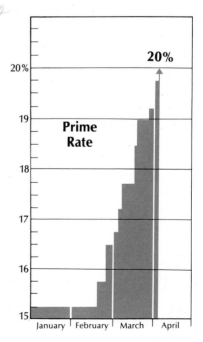

76. (⌨ with y^x key) *Biology.* The *territory area* of an animal is defined to be its defended region, or exclusive region. For example, a lion has a certain region over which it is ruler. It has been hypothesized in statistical studies* that the area T of that region is approximated using body weight W by the power function

$$T = W^{1.31}.$$ `CSS`

Complete the table of approximate function values and graph the function.

W	0	10	20	30	40	50	100	150
T	0	20						

*See footnote on p. 56.

*See footnote on p. 56.

OBJECTIVES

You should be able to use curve fitting to find a model for a set of data, then use the model to make predictions.

1.6 MATHEMATICAL MODELING

What is a Mathematical Model?

When the essential parts of a problem situation are described in mathematical language, we say that we have a *mathematical model*. For example, the arithmetic of the natural numbers constitutes a mathematical model for situations in which counting is the essential ingredient. Situations in which calculus can be brought to bear often require the use of equations and functions, and typically there is concern with the way a change in one variable effects a change in another.

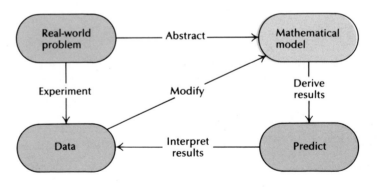

Mathematical models are abstracted from real-world situations. Procedures within the mathematical model then give results that allow one to predict what will happen in that real-world situation. To the extent that these predictions are inaccurate or the results of experimentation

Use the model for the world record in the mile run.

1. What will the world record be in 1988?

2. When will the world record be 3:48.0? (*Hint:* 3:48.0 = 3.80)

do not conform to the model, the model is in need of modification. This is shown in the diagram.

The diagram seems to indicate that mathematical modeling is an ongoing, possibly everchanging, process. This is often the case. For example, finding a mathematical model that will enable accurate prediction of population growth is not a simple problem. Surely any population model one might devise will have to be altered as further relevant information is acquired.

While models can reveal worthwhile information, one must always be cautious in their use. An interesting case in point is a study* showing that world records in *any* running race can be modeled by a linear function. In particular, for the mile run,

$$R = -0.00582x + 15.3476,$$

where R is the world record in minutes, and x is the year. Roger Bannister shocked the world in 1954 by breaking the 4-minute mile. Had people been aware of this model they would not have been shocked, for when we substitute 1954 for x we get

$$R = -0.00582(1954) + 15.3476 = 3.97532 \approx 3:58.5.$$

The actual record was 3:59.4. Although this model will continue for 40 to 50 years to be worthwhile in predicting the world record in the mile, upon using some common sense, we see that we can't get meaningful answers to some questions. For example, we could use the model to find when the 1-minute mile will be broken. We set $R = 1$ and solve for x:

$$1 = -0.00582x + 15.3476$$

$$2465 = x.$$

Most track people would assure us that the 1-minute mile is beyond human capability. In fact, at the time of this writing, experienced runners think it will never reach 3:40.0, the current world record being 3:48.8. Going to an even further extreme, we see that the model predicts that the 0-minute mile will be run in 2637. In conclusion, one must be careful in the use of any model. (You will develop this model in Exercise Set 7.4.)

DO EXERCISES 1 AND 2.

---

*H. W. Ryder, H. J. Carr, and P. Herget, "Future Performance in Footracing," *Scientific American*, 234 (June 1976): 109–119.

In Section 1.5 we saw an example of a mathematical model utilizing supply and demand functions. In general, the idea is to find a function that fits observations and theoretical reasoning (including common sense) as well as possible. In Section 7.4, we will see how calculus can be used to develop and analyze models. For now we will consider one type of modeling procedure using a somewhat oversimplified procedure that we call *curve fitting*.

Curve Fitting

The following four functions fit many situations.

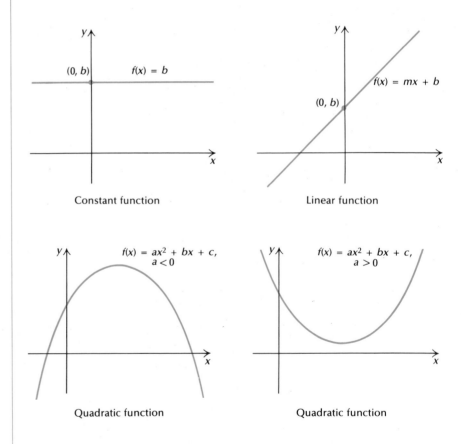

Constant function

Linear function

Quadratic function

Quadratic function

The following is a procedure that sometimes works for finding mathematical models.

> **Curve Fitting**
>
> Given a set of data:
>
> 1. **Graph the data.**
> 2. **Look at the data and determine whether a known function seems to fit.**
> 3. **Find a function that fits the data by using data points to derive the constants.**

The following problem is based on factual data.

Example 1 *Business: Taxes from each dollar earned.* From the given set of data, (a) find a model and (b) use the model to find that part of each dollar earned in 1985 that will go for taxes.

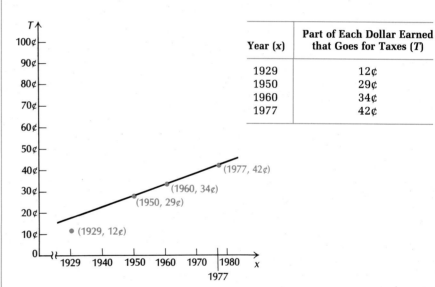

Year (x)	Part of Each Dollar Earned that Goes for Taxes (T)
1929	12¢
1950	29¢
1960	34¢
1977	42¢

Solution.

a) The graph is shown above. It looks as though a linear function fits this data fairly well:

$$T = mx + b. \qquad (1)$$

To derive the constants (or parameters) m and b, we pick two data points. This is a subjective matter, but, of course, you should not pick a point that deviates greatly from the general pattern. We pick the points (1950, 29¢) and (1977, 42¢). Since the points are to be

3. Rework Example 1, but this time use the data points (1950, 29¢) and (1960, 34¢). How does the answer to (b) compare to (b) in Example 1?

solutions of Eq. (1) it follows that

$$42 = m \cdot 1977 + b \quad \text{or} \quad 42 = 1977m + b, \qquad (2)$$

$$29 = m \cdot 1950 + b \quad \text{or} \quad 29 = 1950m + b. \qquad (3)$$

This is a system of equations. We subtract Eq. (3) from Eq. (2) to get rid of b:

$$13 = 27m.$$

Then we have

$$\frac{13}{27} = m.$$

Substituting $\frac{13}{27}$ for m in Eq. (3), we get

$$29 = 1950 \cdot \frac{13}{27} + b$$

$$29 = 938.9 + b \qquad \text{We estimate } 1950 \cdot \frac{13}{27}.$$

$$-909.9 = b.$$

Substituting these values of m and b into Eq. (1), we get the function (model) given by

$$T = \frac{13}{27}x - 909.9.$$

Since we are only interested in estimates, we use an approximation for $\frac{13}{27}$ and get

$$T = 0.481x - 909.9. \qquad (4)$$

b) That part of the earned dollar that will go for taxes in 1985 is found by letting $x = 1985$ in Eq. (4):

$$T = 0.481(1985) - 909.9 \approx 45¢.$$

DO EXERCISE 3.

To repeat, the curve fitting technique illustrated here is an over-simplified method of finding models. Other techniques such as the Least Squares Method (Section 7.4) and computer simulations are used for more thorough research.

Example 2 For the given set of factual data, (a) find a model and (b) use the model to find the death rate for those who sleep 2 hr; 8 hr; 10 hr.

4.

Age (x) of Driver in Years	Number (A) of Daytime Accidents Committed by Driver of Age x
20	420
40	150
60	210
70	400

a) For the given set of data, graph the data and find a quadratic function that fits the data. Use the data points (20, 420), (40, 150), and (70, 400). First give exact fractional values for a, b, and c. Then give decimal values rounded to the nearest thousandth.

b) How many daytime accidents are committed by a driver of age 16?

Average Number of Hours of Sleep (x)	Death Rate per 100,000 Males (y)
5	1121
7	626
9	967

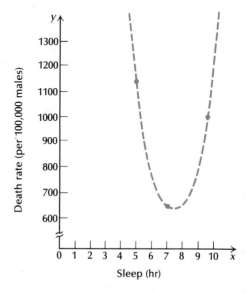

Solution

a) The graph is shown above. It looks as if a quadratic function might fit:

$$y = ax^2 + bx + c. \tag{1}$$

To derive the constants (or parameters) a, b, and c, we use the three data points (5, 1121), (7, 626), and (9, 967). Since these points are to be solutions of Eq. (1), it follows that

$$1121 = a \cdot 5^2 + b \cdot 5 + c \quad \text{or} \quad 1121 = 25a + 5b + c,$$

$$626 = a \cdot 7^2 + b \cdot 7 + c \quad \text{or} \quad 626 = 49a + 7b + c,$$

$$967 = a \cdot 9^2 + b \cdot 9 + c \quad \text{or} \quad 967 = 81a + 9b + c.$$

We solve this system using procedures studied in this chapter to get

$$a = 104.5, \quad b = -1501.5, \quad \text{and} \quad c = 6016.$$

Substituting these values of a, b, and c into Eq. (1), we get the function (model) given by

$$y = 104.5x^2 - 1501.5x + 6016.$$

b) The death rate for 2 hr is given by

$$y = 104.5(2^2) - 1501.5(2) + 6016 = 3431.$$

The death rate for 8 hr is given by

$$y = 104.5(8^2) - 1501.5(8) + 6016 = 692.$$

The death rate for 10 hr is given by

$$y = 104.5(10^2) - 1501.5(10) + 6016 = 1451.$$

DO EXERCISE 4 (on preceding page).

EXERCISE SET 1.6

For each exercise, (a) and (b), are as follows:

a) Graph the data given.

b) Look at the data and determine whether one of the six models discussed in this section seems to fit.

1. *Business.* Raggs, Ltd., keeps track of its total costs of producing x items of a certain suit. These data are shown below.

Number of Suits (x)	Total Cost (C) of Producing x Suits
0	$10,000
1	10,030
2	10,059
3	10,094

c) Use the data points (1, $10,030) and (3, $10,094) to find a linear function that fits the data.

d) Predict the total cost of producing 4 suits; 10 suits.

e) Use the data points (0, $10,000) and (2, $10,059) to find a linear function that fits the data.

f) Use the model of (e) to predict the total cost of producing 4 suits; 10 suits.

2. *Business.* Pizza, Unltd., keeps track of its total costs of producing x pizzas. These data are shown below.

Number of Pizzas (x)	Total Cost (C) of Producing x Pizzas
0	$1000
1	1001
2	1001.80
3	1002.50

c) Use the data points (1, $1001) and (3, $1002.50) to find a linear function that fits the data.

d) Predict the total cost of producing 4 pizzas; 100 pizzas.

e) Use the data points (0, $1000) and (2, $1001.80) to find a linear function that fits the data.

f) Use the model of (e) to predict the total cost of producing 4 pizzas; 100 pizzas.

The problems in Exercises 3 and 4 are based on factual data.

3.

Travel Speed (x) in mph	Number (D) of Vehicles Involved in an Accident in Daytime (for Every 100 Million Miles of Travel)
20	10,000
30	1,000
40	200
50	150
60	95
70	90
80	190

c) Use the data points (30, 1000), (50, 150), and (70, 90) to find a quadratic function that fits the data.

d) Use the model to find the number of vehicles involved in an accident at 60 mph. Check this with the data.

5. *Business*

Year (t)	Total Sales (S) in Dollars
1	$100,310
2	100,290
3	100,305
4	100,280

c) Use the data points (1, $100,310) and (2, $100,290) to find a linear function that fits the data.

d) This data set approximates a constant function. What procedure, apart from that of (c), could you use to find the constant?

4.

Travel Speed (x) in mph	Number (N) of Vehicles Involved in an Accident in Nighttime (for Every 100 Million Miles of Travel)
20	10,000
30	2,000
40	400
50	250
60	250
70	350
80	1,500

c) Use the data points (20, 10,000), (50, 250), and (80, 1500) to find a quadratic function fitting the data.

d) Use the model to find the number of vehicles involved in an accident at 30 mph. Check this with the data.

6. *Business*

Year (t)	Sales (S) in Dollars
1	$10,000
2	21,000
3	27,000
4	37,000

c) Use the data points (1, $10,000) and (4, $37,000) to find a linear function that fits the data.

d) Predict the sales of the company in the 5th year.

CHAPTER 1 TEST

1. Rename without a negative exponent: e^{-k}.

2. Divide: $\dfrac{e^{-5}}{e^{8}}$.

3. Multiply: $(x + h)^2$.

4. Factor: $25x^2 - t^2$.

5. A person makes an investment at 13% compounded annually. It grows to $1039.60 at the end of 1 year. How much was originally invested?

6. Solve: $-3x < 12$.

7. A function is given by $f(x) = x^2 - 4$. Find (a) $f(-3)$; (b) $f(x + h)$.

8. What is the slope and y-intercept of $y = -3x + 2$?

9. Find an equation of the line with slope $\frac{1}{4}$, containing the point $(8, -5)$.

10. Find the slope of the line containing the points $(-2, 3)$ and $(-4, -9)$.

11. The weight F of fluids in a human is directly proportional to body weight W. It is known that a person who weighs 180 lb has 120 lb of fluids. Find an equation of variation expressing F as a function of W.

12. A record company has fixed costs of $10,000 for producing a record master. Thereafter, the variable costs are $0.50 per record for duplicating from the record master. Revenue from each record is expected to be $1.30.

a) Formulate a function $C(x)$ for the total cost of producing x records.

b) Formulate a function $R(x)$ for the total revenue from the sale of x records.

c) Formulate a function $P(x)$ for the total profit from the production and sale of x records.

d) How many records must the company sell to break even?

13. Find the equilibrium point for the demand and supply functions

$$D(x) = (x - 7)^2 \quad \text{and} \quad S(x) = x^2 + x + 4.$$

14. Graph: $y = 4/x$.

15. Convert to fractional exponents: $1/\sqrt{t}$.

16. Convert to radical notation: $t^{-3/5}$.

Determine the domain of each function.

17. $f(x) = \dfrac{x^2 + 20}{(x - 2)(x + 7)}$

18. $f(x) = \sqrt{5x + 10}$

19. Find a linear function that fits the data points $(1, 3)$ and $(2, 7)$,

20. Find a quadratic function that fits the data points $(1, 5)$, $(2, 9)$, and $(3, 4)$.

21. Write interval notation for this graph.

$$c \qquad\qquad\qquad d$$

22. Graph.

$$f(x) = \begin{cases} x^2 + 2 & \text{for } x \geq 0, \\ x^2 - 2 & \text{for } x < 0 \end{cases}$$

23.   The demand function for a product is given by

$$p = D(x) = \sqrt[3]{800 - x}.$$

a) Find the price per unit when 9 units are sold.

b) Find the number of units sold when the price per unit is $6.50.

DIFFERENTIATION

The path of the loose ski follows a tangent line to the ski chute at the point where it breaks loose. (Jonathan Rawle: Stock, Boston)

2

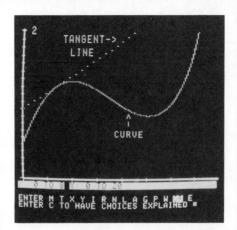

The *Computer Software Supplement* can be used to show tangent lines to curves.

You should be able to

a) **Determine whether a graph is of a continuous function.**

b) **Determine whether a function is continuous at a given point a.**

c) **Find**
$$\lim_{x \to a} f(x),$$
if it exists.

d) **Find a limit like**
$$\lim_{h \to 0} (3x^2 + 3xh + h^2).$$

A river forms a continuous curve. (*Pro Pix from Monkmeyer*)

2.1 LIMITS AND CONTINUITY

In this section we give an intuitive (meaning "based on prior and present experience") treatment of two important concepts: continuity and limits.

Continuity

The following are graphs of functions that are *continuous* over the whole real line $(-\infty, \infty)$.

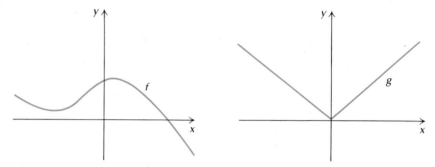

Note that there are no "jumps" or holes in the graphs. For now we will use a somewhat intuitive definition of continuity, which we will refine later. We say that a function is *continuous* over, or on, some interval of the real line if its graph can be traced without lifting a pencil from the paper. The following are graphs of functions that are *not* continuous over the whole real line.

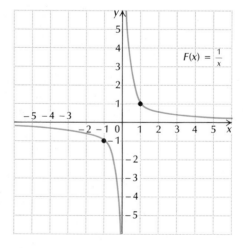

$$F(x) = \frac{1}{x}$$

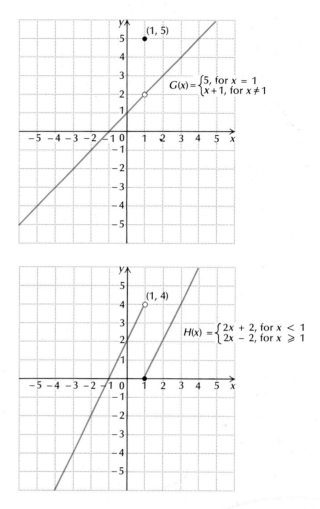

For functions *G* and *H*, the open circle indicates that the circled point is not part of the graph.

In each case the graph *cannot* be traced without lifting the pencil from the paper. However, each case represents a different situation. Let us discuss why each case fails to be continuous over the whole real line.

The function F fails to be continuous over the *whole* real line $(-\infty, \infty)$. Since *F* is not defined at $x = 0$, the point $x = 0$ is not part of the domain, so $f(0)$ does not exist and there is no point $(0, f(0))$ on the graph. Thus there is no point to trace at $x = 0$. However, *F* is continuous on the intervals $(-\infty, 0)$ and $(0, \infty)$.

The function G is not continuous over the whole real line since it is not continuous at $x = 1$. Let us trace the graph of *G* to the left of $x = 1$. As *x* approaches 1, *G*(x) seems to approach 2; but at $x = 1$, *G*(x) *jumps*

1. Which functions are continuous?

a)

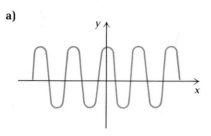

b)

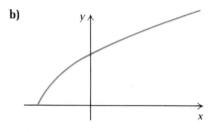

c)

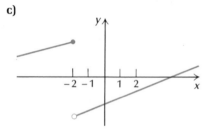

d)

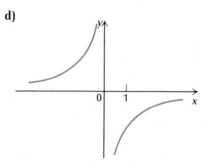

2. a) Decide whether the function in Margin Exercise 1(c) is continuous at −2; at 1.

   b) Decide whether the function in Margin Exercise 1(d) is continuous at 1; at 0.

up to 5, while to the right of x = 1, *G*(x) *jumps* back to some value close to 2. Thus *G* is discontinuous at x = 1.

　　The function H is not continuous over the whole real line since it is not continuous at x = 1. Let us trace the graph of *H* starting to the left of x = 1. As x approaches 1, *H*(x) seems to approach 4; but at x = 1, *H*(x) *jumps* down to 0, while just to the right of x = 1, *H*(x) is close to 0. Thus *H*(x) is discontinuous at x = 1.

DO EXERCISES 1 AND 2.

Limits

　　We can formalize these notions about *continuity* by introducing the concepts of *limits*. The study of limits has application not just to continuity but throughout calculus.

　　Consider the function *f* given by

$$f(x) = 2x + 3. \qquad \boxed{\textbf{CSS}}$$

Suppose we select input numbers x closer and closer to the number 4, and look at the output numbers 2x + 3. Study the following input–output table and graph.

These inputs approach 4 from the left.	4	These inputs approach 4 from the right.

x	2	3.6	3.9	3.99	3.999	4.001	4.01	4.1	4.8	5
f(x)	7	10.2	10.8	10.98	10.998	11.002	11.02	11.2	12.6	13

These outputs approach 11.	11	These outputs approach 11.

In the table and the graph, as input numbers approach 4 from the left, output numbers approach 11. As input numbers approach 4 from the right, output numbers approach 11. Thus we say:

　　　　As x *approaches* 4,　2x + 3 *approaches* 11.

An arrow, →, is often used for the word "approaches." Thus the above can be written

　　　　As x → 4,　2x + 3 → 11.

The number 11 is said to be the *limit* of 2x + 3 as x approaches 4. We can abbreviate this statement as follows:

$$\lim_{x \to 4} (2x + 3) = 11.$$

3. Consider

$$f(x) = 3x - 1.$$

a) Complete this table. (▦ helpful, though not necessary)

x	f(x)
5	14
5.8	16.4
5.9	16.7
5.99	16.97
5.999	16.997
6 ←	17 →?
6.001	17.003
6.01	17.03
6.1	17.3
6.4	18.2
7	20

b) Find $\lim\limits_{x \to 6} f(x).$ = 17

c) Graph $f(x) = 3x - 1.$

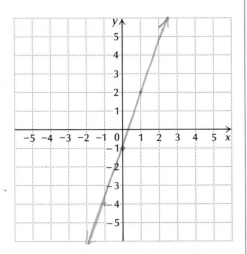

Use the graph to find the following limits.

d) $\lim\limits_{x \to -1} f(x)$ = −4

e) $\lim\limits_{x \to 2} f(x)$ = 5

f) $\lim\limits_{x \to 0} f(x)$ = −1

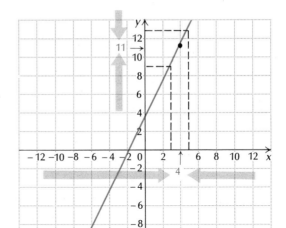

$f(x) = 2x + 3$

This is read, "The limit, as x approaches 4, of 2x + 3 is 11."

DEFINITION

A function *f* has the *limit L* as *x* approaches *a*, written

$$\lim_{x \to a} f(x) = L,$$

if we can get *f(x)* as close to *L* as we wish by restricting *x* to a sufficiently small interval about *a* but excluding *a*.

DO EXERCISE 3.

Example 1 Consider the function

$$F(x) = \frac{1}{x}.$$

Find the following limits, if they exist.

a) $\lim\limits_{x \to 3} F(x)$ = 1/3

b) $\lim\limits_{x \to 0} F(x)$ = undefined

Solution

a) From the graph on the next page we see that as inputs x approach 3 from either the left or right, outputs 1/x approach 1/3. We can also check this on an input–output table. Thus we have

$$\lim_{x \to 3} F(x) = \frac{1}{3}.$$

4. Consider the function

$$g(x) = \frac{1}{x - 3}.$$

a) Complete this table. (⊞ helpful)

Inputs, x	Outputs, g(x)
1	-1/2
2	-1
2.9	-10
2.99	-100
2.999	-1000
3 ⟵ ⟶ ?	
3.001	1000
3.01	100
3.1	10
3.8	80
4	1

b) Find $\lim_{x \to 3} g(x)$. =does not exist

c) Graph $g(x) = \frac{1}{x - 3}$.

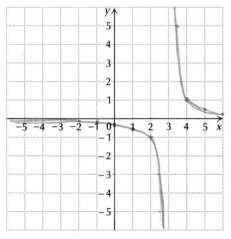

Use the graph to find these limits, if they exist.

d) $\lim_{x \to 3} g(x)$ do not exist

e) $\lim_{x \to 1} g(x)$ = 1/2

f) $\lim_{x \to 4} g(x)$ = 1

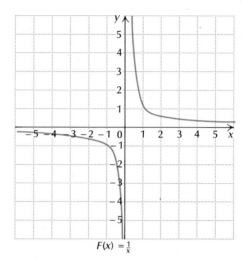

$F(x) = \frac{1}{x}$

b) From the graph, or from an input–output table, we see that as inputs x approach 0 from the left, outputs get smaller and smaller without bound. Similarly, as inputs x approach 0 from the right, outputs get larger and larger without bound. We say that

$$\lim_{x \to 0} \frac{1}{x} \text{ does not exist.}$$

DO EXERCISE 4.

The following is very important to keep in mind in determining whether a limit exists.

In order for a limit to exist, the limits from the left and right must both exist and be the same.

Example 2 Consider the function H defined as follows (we considered it earlier):

$$H(x) = \begin{cases} 2x + 2 & \text{for } x < 1, \\ 2x - 2 & \text{for } x \geq 1. \end{cases}$$

Graph the function and find the following limits, if they exist.

a) $\lim_{x \to 1} H(x)$ **b)** $\lim_{x \to -3} H(x)$

Solution The graph is shown on the next page.

a) As inputs x approach 1 from the left, outputs H(x) approach 4. Thus the limit from the left is 4. But as inputs x approach 1 from the right,

5. Consider the following graph.

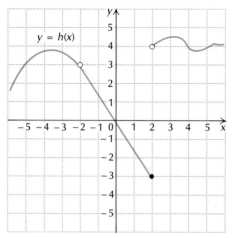

Find the following limits, if they exist.

a) $\lim\limits_{x \to 2} h(x)$ **b)** $\lim\limits_{x \to 0} h(x) = 0$

do not exist

c) $\lim\limits_{x \to -2} h(x)$

do not exist

6. Consider the following function.

$$f(x) = \begin{cases} 2 - x^2 & \text{for } x \geq 0, \\ x^2 - 2 & \text{for } x < 0 \end{cases}$$

a) Graph $y = f(x)$.

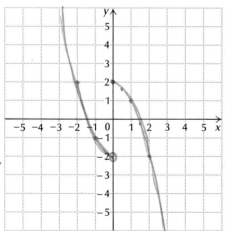

Find the following limits, if they exist.

b) $\lim\limits_{x \to 0} f(x)$ **c)** $\lim\limits_{x \to -2} f(x) = 2$

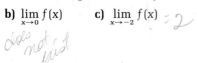

does not exist

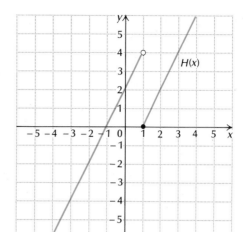

outputs $H(x)$ approach 0. Thus the limit from the right is 0. Since the limit from the left, 4, is not the same as the limit from the right, 0, we say

$$\lim\limits_{x \to 1} H(x) \text{ does not exist.}$$

b) As inputs x approach -3 from the left, outputs $H(x)$ approach -4, so the limit from the left is -4. As inputs x approach -3 from the right, outputs $H(x)$ approach -4, so the limit from the right is -4. Since the limits from the left and from the right exist and are the same, we have

$$\lim\limits_{x \to -3} H(x) = -4.$$

DO EXERCISES 5 AND 6.

The following is also important in finding limits.

> **The limit at a point a *does not depend* on the function value at a, $f(a)$, should that exist.** That is, whether or not a limit exists at a has nothing to do with its function value at a, $f(a)$.

Example 3 Consider the function G defined as follows:

$$G(x) = \begin{cases} 5 & \text{for } x = 1, \\ x + 1 & \text{for } x \neq 1. \end{cases}$$

Graph the function and find the following limits, if they exist.

a) $\lim\limits_{x \to 1} G(x)$ **b)** $\lim\limits_{x \to 3} G(x)$

7. Consider the following function:

$$f(x) = \begin{cases} -3 & \text{for } x = -2, \\ x^2 & \text{for } x \neq -2. \end{cases}$$

a) Graph $y = f(x)$.

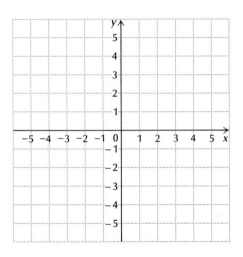

Find the following limits, if they exist.

b) $\lim\limits_{x \to -2} f(x)$

c) $\lim\limits_{x \to 2} f(x)$

d) Does $\lim\limits_{x \to -2} f(x) = f(-2)$?

e) Does $\lim\limits_{x \to 2} f(x) = f(2)$?

Solution The graph is shown below.

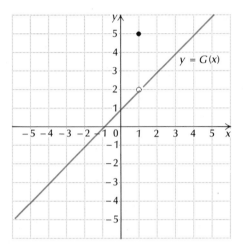

a) As inputs x approach 1 from the left, outputs $G(x)$ approach 2, so the limit from the left is 2. As inputs x approach 1 from the right, outputs $G(x)$ approach 2, so the limit from the right is 2. Since the limit from the left, 2, is the same as the limit from the right, 2, we have

$$\lim_{x \to 1} G(x) = 2.$$

Note that the limit, 2, is not the same as the function value at 1, $G(1)$, which is 5.

b) We have $\lim_{x \to 3} G(x) = 4 = G(3)$. In this case the function value and the limit are the same.

DO EXERCISES 7 AND 8.

Limits and Continuity

After working through Example 3, we might ask, "When can we substitute to find a limit?" The answer lies in the following definition, which also happens to provide a more formal definition of continuity.

8. Consider the following graph.

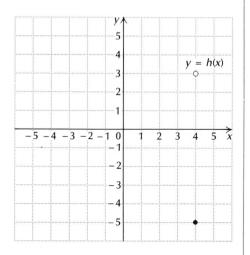

Find the following limits, if they exist.

a) $\lim_{x \to 1} h(x)$

b) $\lim_{x \to 4} h(x)$

c) Does $\lim_{x \to 1} h(x) = h(1)$?

d) Does $\lim_{x \to 4} h(x) = h(4)$?

9. Consider

$$f(x) = 3x - 1.$$

(See Margin Exercise 3.)

a) Does $f(6)$ exist? If so, what is it?

b) Does $\lim_{x \to 6} f(x)$ exist? If so, what is it?

c) Does

$$\lim_{x \to 6} f(x) = f(6)?$$

d) Is f continuous at 6?

DEFINITION

A function f is *continuous* at $x = a$ if:

1. $f(a)$ exists,
2. $\lim_{x \to a} f(x)$ exists, and
3. $\lim_{x \to a} f(x) = f(a)$.

A function is *continuous over an interval I* if it is continuous at each point in *I*.

Example 4 Determine whether the function given by

$$f(x) = 2x + 3$$

is continuous at $x = 4$.

Solution This function is continuous at 4 because

1. $f(4)$ exists, $f(4) = 11$,
2. $\lim_{x \to 4} f(x)$ exists, $\lim_{x \to 4} f(x) = 11$ (as shown earlier),
3. $\lim_{x \to 4} f(x) = 11 = f(4)$.

In fact, $f(x) = 2x + 3$ is continuous at any point on the real line.

DO EXERCISE 9.

Example 5 Determine whether the function in Example 1 is continuous at $x = 0$.

Solution The function is *not* continuous at $x = 0$ because $f(0)$ does not exist.

Example 6 Determine whether the function in Example 2 is continuous at $x = 1$.

Solution The function is not continuous at $x = 1$ because $\lim_{x \to 1} H(x)$ does not exist.

Example 7 Determine whether the function in Example 3 is continuous at $x = 1$.

Solution The function is not continuous at $x = 1$ because $G(1) = 5$, but $\lim_{x \to 1} G(x) = 2$.

10. Determine whether the function in Margin Exercise 4 is continuous at

   a) $x = 3$;

   b) $x = -2$.

11. Determine whether the function in Margin Exercise 6 is continuous at $x = 0$.

12. Determine whether the function in Margin Exercise 7 is continuous at $x = -2$.

13. Provide an argument to show that the function given by

$$f(x) = \frac{\sqrt[3]{x} - 7x^2}{x - 2}$$

is continuous so long as $x \neq 2$.

DO EXERCISES 10–12.

CONTINUITY PRINCIPLES

> The following continuity principles, which we will not prove, allow us to build up continuous functions.
>
> i) Any constant function is continuous (such a function never varies).
>
> ii) For any positive integer n, x^n and $\sqrt[n]{x}$ are continuous. When n is even, the inputs of $\sqrt[n]{x}$ are restricted to $[0, \infty)$.
>
> iii) If $f(x)$ and $g(x)$ are continuous, then so are $f(x) + g(x)$, $f(x) - g(x)$, and $f(x) \cdot g(x)$.
>
> iv) If $f(x)$ is continuous, so is $1/f(x)$, so long as the inputs x are not such that the outputs $f(x) = 0$.

Example 8 Provide an argument to show that

$$f(x) = x^2 - 3x + 2$$

is continuous.

Solution Now x^2 is continuous, by (ii). The constant function 3 is continuous by (i), and the function x is continuous by (ii), so the product $3x$ is continuous by (iii). Thus $x^2 - 3x$ is continuous by (iii), and since the constant 2 is continuous, we can apply (iii) again to show that $x^2 - 3x + 2$ is continuous.

In similar fashion, we can show that any polynomial such as

$$f(x) = x^4 - 5x^3 + x^2 - 7$$

is continuous. A rational function is a quotient of two polynomials

$$r(x) = \frac{f(x)}{q(x)}.$$

Thus by (iv), a rational function is continuous so long as the inputs x are not such that $q(x) = 0$.

DO EXERCISE 13.

More on Limits

If a function is continuous at a, we can substitute to find the limit.

Example 9 Find $\lim_{x \to 2} (x^4 - 5x^3 + x^2 - 7)$.

Find each limit, if it exists.

14. $\lim\limits_{x \to -2} (x^4 - 5x^3 + x^2 - 7)$

Solution It follows from the continuity principles that $x^4 - 5x^3 + x^2 - 7$ is continuous. Thus the limit can be found by substitution:

$$\lim_{x \to 2} (x^4 - 5x^3 + x^2 - 7) = 2^4 - 5 \cdot 2^3 + 2^2 - 7$$

$$= 16 - 40 + 4 - 7 = -27.$$

Example 10 Find $\lim_{x \to 0} \sqrt{x^2 - 3x + 2}$.

Solution By using the continuity principles, we have shown that $x^2 - 3x + 2$ is continuous; and so long as x is restricted to values for which $x^2 - 3x + 2$ is nonnegative, it follows from principle (ii) that $\sqrt{x^2 - 3x + 2}$ is continuous. Thus we can substitute to find the limit:

$$\lim_{x \to 0} \sqrt{x^2 - 3x + 2} = \sqrt{0^2 - 3 \cdot 0 + 2} = \sqrt{2}.$$

DO EXERCISES 14 AND 15.

There are limit principles that correspond to the continuity principles. We can use them to find limits when we are uncertain of the continuity of a function at a given point.

15. $\lim\limits_{x \to 1} \sqrt{x^2 + 3x + 4}$

LIMIT PRINCIPLES

If $\lim_{x \to a} f(x) = L$ and $\lim_{x \to a} g(x) = M$, then we have the following.

L1. $\lim\limits_{x \to a} c = c$.
(The limit of a constant is the constant.)

L2. $\lim\limits_{x \to a} x^n = a^n$, $\lim\limits_{x \to a} \sqrt[n]{x} = \sqrt[n]{a}$, for any positive integer n.
(When n is even, the inputs of $\sqrt[n]{x}$ must be restricted to $[0, \infty)$.)

L3. $\lim\limits_{x \to a} [f(x) \pm g(x)] = \lim\limits_{x \to a} f(x) \pm \lim\limits_{x \to a} g(x) = L \pm M$.
(The limit of a sum or difference is the sum or difference of the limits.)

$$\lim_{x \to a} [f(x) \cdot g(x)] = [\lim_{x \to a} f(x)] \cdot [\lim_{x \to a} g(x)] = L \cdot M.$$

(The limit of a product is the product of the limits.)

L4. $\lim\limits_{x \to a} \dfrac{1}{f(x)} = \dfrac{1}{\lim\limits_{x \to a} f(x)} = \dfrac{1}{L}$, provided $L \neq 0$.

(The limit of a reciprocal is the reciprocal of the limit.)

Example 11 Find $\lim_{x \to -3} (x^2 - 9)/(x + 3)$.

Solution The function $(x^2 - 9)/(x + 3)$ is not continuous at $x = -3$.

Find each limit, if it exists.   CSS

16. $\lim\limits_{x \to -4} \dfrac{x^2 - 16}{x + 4}$

17. $\lim\limits_{x \to 3} \dfrac{x - 3}{x^2 - 9}$

18. a) Complete the table.

h	2x + h
1	
0.7	
0.4	
0.1	
0.01	
0.001	

b) Find

$$\lim\limits_{h \to 0} (2x + h).$$

We use some algebraic simplification and then some limit principles.

$$\lim_{x \to -3} \frac{x^2 - 9}{x + 3} = \lim_{x \to -3} \frac{(x + 3)(x - 3)}{x + 3}$$

$$= \lim_{x \to -3} (x - 3), \qquad \text{assuming } x \neq -3$$

$$= \lim_{x \to -3} x - \lim_{x \to -3} 3, \qquad \text{(by L3)}$$

$$= -3 - 3 = -6$$

DO EXERCISES 16 AND 17.

In Section 2.2 we encounter expressions with two variables, x and h; our interest is in limits where x is fixed as a constant and $h \to 0$.

Example 12 Find $\lim_{h \to 0} (3x^2 + 3xh + h^2)$.

Solution If we treat x as a constant, using the limit principles, it follows that

$$\lim_{h \to 0} (3x^2 + 3xh + h^2) = 3x^2 + 3x(0) + 0^2 = 3x^2.$$

The reader can check any limit about which there is uncertainty by using an input–output table. Below is a table for this limit.

h	$3x^2 + 3xh + h^2$	
1	$3x^2 + 3x \cdot 1 + 1^2$, or	$3x^2 + 3x + 1$
0.8	$3x^2 + 3x(0.8) + (0.8)^2$, or	$3x^2 + 2.4x + 0.64$
0.5	$3x^2 + 3x(0.5) + (0.5)^2$, or	$3x^2 + 1.5x + 0.25$
0.1	$3x^2 + 3x(0.1) + (0.1)^2$, or	$3x^2 + 0.3x + 0.01$
0.01	$3x^2 + 3x(0.01) + (0.01)^2$, or	$3x^2 + 0.03x + 0.0001$
0.001	$3x^2 + 3x(0.001) + (0.001)^2$, or $3x^2 + 0.003x + 0.000001$	

From the pattern in the table, it appears that

$$\lim_{h \to 0} (3x^2 + 3xh + h^2) = 3x^2.$$

DO EXERCISE 18.

19. *Earned-run average.* A pitcher's earned-run average (the average number of runs given up every 9 innings or 1 game) is given by

$$A = 9 \cdot \frac{n}{i},$$

where n = the number of earned runs allowed and i = the number of innings pitched. Suppose we fix the number of earned runs allowed at 4 and let i vary. We get a function given by

$$A(i) = 9 \cdot \frac{4}{i}.$$

a) Complete the following table, rounding to two decimal places.

Innings Pitched (i)	Earned-run Average (A)
9	
8	
7	
6	
5	
4	
3	
2	
1	
$\frac{2}{3}$ (2 outs)	
$\frac{1}{3}$ (1 out)	

b) Find

$$\lim_{i \to 0} A(i).$$

c) On the basis of (a) and (b), what might a pitcher's earned run average be if 4 runs were allowed and there were 0 outs?

Limits and Infinity

We discussed in Example 1 that the limit

$$\lim_{x \to 0} \frac{1}{x} \text{ does not exist.}$$

Go back and look at the graph. As x approaches 0 from the right, the outputs get larger and larger. These numbers do not approach any real number, though it might be said that the limit from the right is ∞ (infinity). As x approaches 0 from the left, the outputs get smaller and smaller. These numbers do not approach any real number, though it might be said that the limit from the left is $-\infty$ (negative infinity).

DO EXERCISE 19.

Limits at infinity We sometimes need to determine limits when the inputs get larger and larger, that is, when they approach infinity. In such cases we are finding *limits at infinity*. Such a limit would be expressed as

$$\lim_{x \to \infty} f(x).$$

Example 13 Find

$$\lim_{x \to \infty} \left(\frac{3x - 1}{x} \right).$$

Solution One way to find such a limit is to use an input–output table as follows.

Inputs, x	1	10	50	100	2000
Outputs, $\frac{3x - 1}{x}$	2.0	2.9	2.98	2.99	2.9995

As the inputs get larger and larger, the outputs get closer and closer to 3. Thus,

$$\lim_{x \to \infty} \left(\frac{3x - 1}{x} \right) = 3.$$

Another way to do this is to use some algebra, and the fact that as $x \to \infty$, $b/ax^n \to 0$, for any positive integer n.

20. Consider

$$f(x) = \frac{2x + 5}{x}.$$ `CSS`

a) Complete this table (▦helpful).

Inputs, x	Outputs, $\frac{2x + 5}{x}$
4	
20	
80	
200	
1,000	
10,000	

b) Find $\lim\limits_{x \to \infty} \dfrac{2x + 5}{x}$.

21. Find $\lim\limits_{x \to \infty} \dfrac{2x^2 + x - 7}{3x^2 - 4x + 1}$.

We multiply by 1, using $(1/x) \div (1/x)$. This amounts to dividing both numerator and denominator by x:

$$\lim_{x \to \infty} \frac{3x - 1}{x} = \lim_{x \to \infty} \frac{3x - 1}{x} \cdot \frac{(1/x)}{(1/x)}$$

$$= \lim_{x \to \infty} \frac{(3x - 1)\dfrac{1}{x}}{x \cdot \dfrac{1}{x}}$$

$$= \lim_{x \to \infty} \frac{3x \cdot \dfrac{1}{x} - 1 \cdot \dfrac{1}{x}}{1}$$

$$= \lim_{x \to \infty} \left(3 - \frac{1}{x}\right) = 3 + 0 = 3.$$

DO EXERCISE 20.

Example 14 Find $\lim_{x \to \infty} (3x^2 - 7x + 2)/(7x^2 + 5x + 1)$.

Solution The highest power of x is x^2. We divide both numerator and denominator by x^2:

$$\lim_{x \to \infty} \frac{3x^2 - 7x + 2}{7x^2 + 5x + 1} = \lim_{x \to \infty} \frac{3 - \dfrac{7}{x} + \dfrac{2}{x^2}}{7 + \dfrac{5}{x} + \dfrac{1}{x^2}}$$

$$= \frac{3 - 0 + 0}{7 + 0 + 0} = \frac{3}{7}.$$

DO EXERCISE 21.

EXERCISE SET 2.1

Which of the following are continuous?

1.

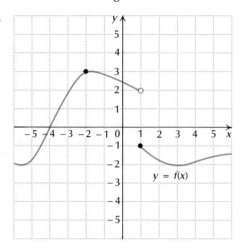

2.

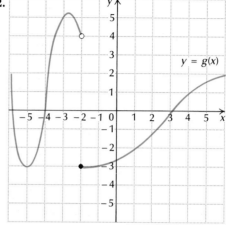

3.

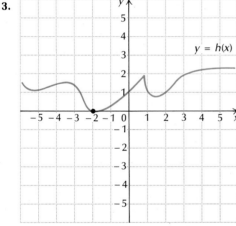

4.

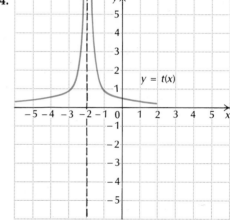

Use the graphs and functions in Exercises 1–4 to answer the following.

5. a) Find $\lim\limits_{x \to 1} f(x)$.

   b) Find $f(1)$.

   c) Is f continuous at $x = 1$?

   d) Find $\lim\limits_{x \to -2} f(x)$.

   e) Find $f(-2)$.

   f) Is f continuous at $x = -2$?

6. a) Find $\lim\limits_{x \to 1} g(x)$.

   b) Find $g(1)$.

   c) Is g continuous at $x = 1$?

   d) Find $\lim\limits_{x \to -2} g(x)$.

   e) Find $g(-2)$.

   f) Is g continuous at $x = -2$?

7. a) Find $\lim\limits_{x\to 1} h(x)$.

    b) Find $h(1)$.

    c) Is h continuous at $x = 1$?

    d) Find $\lim\limits_{x\to -2} h(x)$.

    e) Find $h(-2)$.

    f) Is h continuous at $x = -2$?

8. a) Find $\lim\limits_{x\to 1} t(x)$.

    b) Find $t(1)$.

    c) Is t continuous at $x = 1$?

    d) Find $\lim\limits_{x\to -2} t(x)$.

    e) Find $t(-2)$.

    f) Is t continuous at $x = -2$?

The postage function. Postal rates are as follows: 20¢ for the first ounce and 17¢ for each additional ounce or fraction thereof. Formally, if x is the weight of a letter in ounces, then $p(x)$ is the cost of mailing the letter, where

$$p(x) = 20¢, \quad \text{if } 0 < x \le 1,$$

$$p(x) = 37¢, \quad \text{if } 1 < x \le 2,$$

$$p(x) = 54¢, \quad \text{if } 2 < x \le 3,$$

and so on, up to 12 ounces (at which point postal cost also depends on distance). The graph of p is shown at the right.

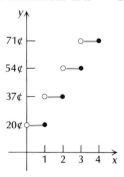

9. Is p continuous at 1? at $1\frac{1}{2}$? at 2? at 2.53?

10. Is p continuous at 3? at $3\frac{1}{4}$? at 4? at 3.98?

Using the graph, find each limit, if it exists.

11. $\lim\limits_{x\to 1} p(x)$

12. $\lim\limits_{x\to 1/2} p(x)$

13. $\lim\limits_{x\to 2.3} p(x)$

14. $\lim\limits_{x\to 2} p(x)$

Find each of the following limits. Use any method: algebra, graphs, or input–output tables.   `CSS`

15. $\lim\limits_{x\to 1} x^2 - 3$

16. $\lim\limits_{x\to 1} x^2 + 4$

17. $\lim\limits_{x\to 0} \dfrac{3}{x}$

18. $\lim\limits_{x\to 0} \dfrac{-4}{x}$

19. $\lim\limits_{x\to 3} 2x + 5$

20. $\lim\limits_{x\to 4} 5 - 3x$

21. $\lim\limits_{x\to -5} \dfrac{x^2 - 25}{x + 5}$

22. $\lim\limits_{x\to -4} \dfrac{x^2 - 16}{x + 4}$

23. $\lim\limits_{x\to -2} \dfrac{5}{x}$

24. $\lim\limits_{x\to -5} \dfrac{-2}{x}$

25. $\lim\limits_{x\to 2} \dfrac{x^2 + x - 6}{x - 2}$

26. $\lim\limits_{x\to -4} \dfrac{x^2 - x - 20}{x + 4}$

27. $\lim\limits_{x\to 5} \sqrt[3]{x^2 - 17}$

28. $\lim\limits_{x\to 2} \sqrt{x^2 + 5}$

29. $\lim\limits_{x\to 1} x^4 - x^3 + x^2 + x + 1$

30. $\lim\limits_{x\to 2} 2x^5 - 3x^4 + x^3 - 2x^2 + x + 1$

31. $\lim\limits_{x\to 2} \dfrac{1}{x - 2}$

32. $\lim\limits_{x\to 1} \dfrac{1}{(x - 1)^2}$

33. $\lim\limits_{x\to 2} \dfrac{3x^2 - 4x + 2}{7x^2 - 5x + 3}$

34. $\lim\limits_{x\to -1} \dfrac{4x^2 + 5x - 7}{3x^2 - 2x + 1}$

35. $\lim\limits_{x\to 2} \dfrac{x^2 + x - 6}{x^2 - 4}$

36. $\lim\limits_{x\to 4} \dfrac{x^2 - 16}{x^2 - x - 12}$

37. ▦ $\lim\limits_{x\to 1} \dfrac{1 - \sqrt{x}}{1 - x}$

38. ▦ $\lim\limits_{x\to 4} \dfrac{\sqrt{x} - 2}{x - 4}$

39. $\lim\limits_{h\to 0} 6x^2 + 6xh + 2h^2$

40. $\lim\limits_{h\to 0} 10x + 5h$

41. $\lim\limits_{h\to 0} \dfrac{-2x - h}{x^2(x + h)^2}$

42. $\lim\limits_{h\to 0} \dfrac{-5}{x(x + h)}$

43. $\lim\limits_{x\to \infty} \dfrac{2x - 4}{5x}$

44. $\lim\limits_{x\to \infty} \dfrac{3x + 1}{4x}$

45. $\lim\limits_{x\to \infty} 5 - \dfrac{2}{x}$

46. $\lim\limits_{x\to \infty} 7 + \dfrac{3}{x}$

47. $\lim\limits_{x \to \infty} \dfrac{2x - 5}{4x + 3}$

48. $\lim\limits_{x \to \infty} \dfrac{6x + 1}{5x - 2}$

49. $\lim\limits_{x \to \infty} \dfrac{2x^2 - 5}{3x^2 - x + 7}$

50. $\lim\limits_{x \to \infty} \dfrac{4 - 3x - 12x^2}{1 + 5x + 3x^2}$

51. Consider

$$f(x) = \begin{cases} 1 & \text{for } x \neq 2, \\ -1 & \text{for } x = 2. \end{cases}$$

Find:

a) $\lim\limits_{x \to 0} f(x)$;   b) $\lim\limits_{x \to 2} f(x)$.

c) Is f continuous at 0? at 2?

52. Consider

$$g(x) = \begin{cases} -4 & \text{for } x = 3, \\ 2x + 5 & \text{for } x \neq 3. \end{cases}$$

Find:

a) $\lim\limits_{x \to 3} g(x)$;   b) $\lim\limits_{x \to 2} g(x)$.

c) Is g continuous at 3? at 2?

53. *Business: Depreciation.* A new conveyor system costs $10,000. In any year it depreciates 8% of its value at the beginning of that year.

a) What is the annual depreciation in each of the first five years?

b) What is the total depreciation at the end of ten years?

c) What is the limit of the sum of the annual depreciation costs?

54. *Business: Depreciation.* A new car costs $6000. In any year it depreciates 30% of its value at the beginning of that year.

a) What is the annual depreciation in each of the first five years?

b) What is the total depreciation at the end of ten years?

c) What is the limit of the sum of the annual depreciation costs?

55. Inside its own 5-yd line, a defensive football team is penalized half the distance to the goal. Suppose a defensive team keeps getting penalized. What is the limit of the distance of the offensive team from the goal? Can the offensive team ever score a touchdown in this manner?

Find each limit, if it exists.

56. $\lim\limits_{x \to 0} \dfrac{|x|}{x}$

57. $\blacksquare\lim\limits_{x \to 1} \dfrac{2 - \sqrt{x + 3}}{x - 1}$

58. $\lim\limits_{x \to 1} \dfrac{x^3 - 1}{x^2 - 1}$

59. $\lim\limits_{x \to \infty} \dfrac{4 - 3x}{5 - 2x^2}$

60. $\lim\limits_{x \to \infty} \dfrac{6x^5 - x^4}{4x^2 - 3x^3}$

OBJECTIVES

You should be able to

a) **Compute an average rate of change of one variable with respect to another.**

b) **Find a simplified difference quotient.**

1. Refer to the graph of suits produced by Raggs, Ltd.

   a) Find the number of suits produced per hour from

   8 A.M. to 9 A.M., *20*
   9 A.M. to 10 A.M., *35*
   10 A.M. to 11 A.M., *9*
   11 A.M. to 12 P.M. *36*

   b) Which interval in (a) had the highest number? *11Am-12pm*

   c) Why do you think this happened?

   d) Which interval in (a) had the lowest number? *10Am-11Am*

   e) Why do you think this happened?

   f) What was the average number of suits produced per hour from 8 A.M. to 12 P.M.? *25*

2.2 AVERAGE RATES OF CHANGE

The graph below shows the total production of suits by Raggs, Ltd., during one morning of work. Industrial psychologists have found curves like this typical of the production of factory workers.

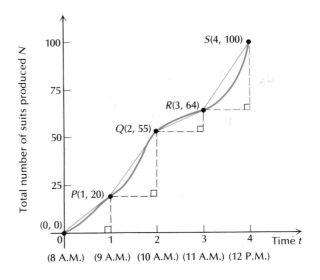

Example 1 What is the number of suits produced from 9 A.M. to 10 A.M.?

Solution At 10 A.M., 55 suits had been produced. At 9 A.M., 20 suits had been produced. In the hour from 9 A.M. to 10 A.M., the number of suits produced was

$$55 \text{ suits} - 20 \text{ suits}, \quad \text{or} \quad 35 \text{ suits}.$$

Note that this is the slope of the line from P to Q.

Example 2 What was the average number of suits produced per hour from 9 A.M. to 11 A.M.?

Solution

$$\frac{64 \text{ suits} - 20 \text{ suits}}{11 \text{ A.M.} - 9 \text{ A.M.}} = \frac{44 \text{ suits}}{2 \text{ hr}} = 22 \frac{\text{suits}}{\text{hr}} \text{ (suits per hour)}$$

This is the slope of the line from P to R. It is not shown in the graph.

DO EXERCISE 1.

Let us consider a function $y = f(x)$ and two inputs x_1 and x_2. The *change in input*, or the *change in x*, is

$$x_2 - x_1.$$

The *change in output*, or the *change in y*, is

$$y_2 - y_1.$$

DEFINITION

>The *average rate of change of y with respect to x,* as *x* changes from x_1 to x_2, is the ratio of the change in output to the change in input:
>
>$$\frac{y_2 - y_1}{x_2 - x_1}.$$

If we look at a graph of the function, we see that

$$\frac{y_2 - y_1}{x_2 - x_1} = \frac{f(x_2) - f(x_1)}{x_2 - x_1}$$

and that this is the slope of the line from $P(x_1, y_1)$ to $Q(x_2, y_2)$. The line $\overleftrightarrow{PQ}$ is called a *secant* line.

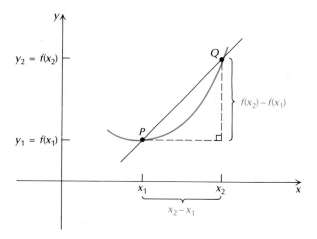

Example 3 For $y = f(x) = x^2$, find the average rates of change as:

a) x changes from 1 to 3; 4

b) x changes from 1 to 2; 3

c) x changes from 2 to 3. 5

2. For

$$f(x) = x^3,$$

find the average rates of change
and sketch the secant lines as:

a) x changes from 1 to 4; $(1,1)(4,64) = 21$
b) x changes from 1 to 2; $(1,1)(2,8) = 7$
c) x changes from 2 to 4; $(2,8)(4,64) = 24$
d) x changes from −1 to −4. $(-1,-1)(-4,-64) = 21$

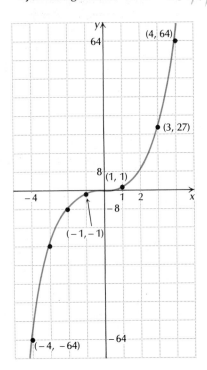

Solution The following graph is not necessary to the computations, but
gives us a look at two of the secant lines whose slopes are being computed.

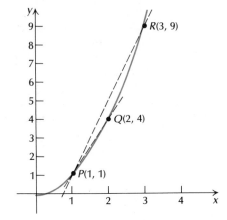

a) When $x_1 = 1$,

$$y_1 = f(x_1) = f(1) = 1^2 = 1,$$

and when $x_2 = 3$,

$$y_2 = f(x_2) = f(3) = 3^2 = 9.$$

The average rate of change is

$$\frac{y_2 - y_1}{x_2 - x_1} = \frac{f(x_2) - f(x_1)}{x_2 - x_1} = \frac{9 - 1}{3 - 1} = \frac{8}{2} = 4.$$

b) When $x_1 = 1$,

$$y_1 = f(x_1) = f(1) = 1^2 = 1,$$

and when $x_2 = 2$,

$$y_2 = f(x_2) = f(2) = 2^2 = 4.$$

The average rate of change is

$$\frac{4 - 1}{2 - 1} = \frac{3}{1} = 3.$$

c) When $x_1 = 2$,

$$y_1 = f(x_1) = f(2) = 2^2 = 4,$$

and when $x_2 = 3$,

$$y_2 = f(x_2) = f(3) = 3^2 = 9.$$

3. For

$$f(x) = \frac{1}{2}x + 1,$$

find the average rates of change and sketch the secant lines as:

a) x changes from 2 to 4;

b) x changes from 2 to 3;

c) x changes from −1 to 4.

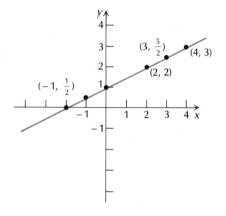

The average rate of change is

$$\frac{9 - 4}{3 - 2} = \frac{5}{1} = 5.$$

DO EXERCISES 2 AND 3.

For a linear function average rates of change are the same for any choice of x_1 and x_2, being equal to the slope m of the line. As we saw in Example 3 and in Margin Exercise 2, a function that is not linear has average rates of change that vary with the choice of x_1 and x_2.

Difference Quotients

Let us now simplify our notation a bit, by doing away with subscripts. Instead of x_1, we will simply write x.

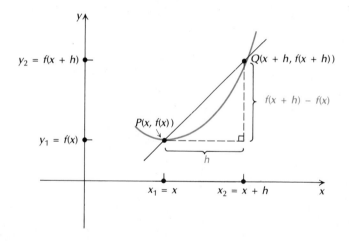

To get from x_1, or x, to x_2 we move a distance h. Thus $x_2 = x + h$. Then the average rate of change, also called a *difference quotient*, is given by

$$\frac{y_2 - y_1}{x_2 - x_1} = \frac{f(x_2) - f(x_1)}{x_2 - x_1}$$

$$= \frac{f(x + h) - f(x)}{(x + h) - x}$$

$$= \frac{f(x + h) - f(x)}{h}.$$

4. For

$$f(x) = 4x^2,$$

complete the following table to find the difference quotients.

x	h	$x + h$	$f(x)$	$f(x + h)$	$f(x + h) - f(x)$	$\dfrac{f(x + h) - f(x)}{h}$
3	2					
3	1					
3	0.1					
3	0.01					
3	0.001					

DEFINITION

The average rate of change of f with respect to x is also called the *difference quotient*. It is given by

$$\frac{f(x + h) - f(x)}{h}.$$

The difference quotient is equal to the slope of the line from $P(x, f(x))$ to $Q(x + h, f(x + h))$.

Example 4 For $f(x) = x^2$, find the difference quotient when:

a) $x = 5$ and $h = 3$;

b) $x = 5$ and $h = 0.1$.   CSS

Solution

a) We substitute $x = 5$ and $h = 3$ into the formula:

$$\frac{f(x + h) - f(x)}{h} = \frac{f(5 + 3) - f(5)}{3} = \frac{f(8) - f(5)}{3}.$$

Now $f(8) = 8^2 = 64$ and $f(5) = 5^2 = 25$, and we have

$$\frac{f(8) - f(5)}{3} = \frac{64 - 25}{3} = \frac{39}{3} = 13.$$

b) We substitute $x = 5$ and $h = 0.1$ into the formula:

$$\frac{f(x + h) - f(x)}{h} = \frac{f(5 + 0.1) - f(5)}{0.1} = \frac{f(5.1) - f(5)}{0.1}.$$

Now $f(5.1) = (5.1)^2 = 26.01$ and $f(5) = 25$, and we have

$$\frac{f(5.1) - f(5)}{0.1} = \frac{26.01 - 25}{0.1} = \frac{1.01}{0.1} = 10.1.$$

DO EXERCISE 4.

For the function in Example 4, let us find a general form of the difference quotient. This will allow more efficient computations.

Example 5 For $f(x) = x^2$, find a simplified form of the difference quotient. Then find the value of the difference quotient when $x = 5$ and $h = 0.1$.

5. For

$$f(x) = 4x^2,$$

find a simplified form of the difference quotient by completing steps (a) through (c). Then complete the table in (d) using the simplified form.

a) Find $f(x + h)$.

b) Find $f(x + h) - f(x)$.

c) Find $[f(x + h) - f(x)]/h$ and simplify.

d) Complete the following table.

CSS

x	h	$\dfrac{f(x + h) - f(x)}{h}$
6	−3	
6		
6	−1	
6	−0.1	
6	−0.01	
6	−0.001	

6. a) For

$$f(x) = 4x^3,$$

find a simplified difference quotient.

b) Complete the following table.

x	h	$\dfrac{f(x + h) - f(x)}{h}$
−2	1	
−2	0.1	
−2	0.01	
−2	0.001	

Solution We have

$$f(x) = x^2,$$

so

$$f(x + h) = (x + h)^2 = x^2 + 2xh + h^2.$$

Then

$$f(x + h) - f(x) = (x^2 + 2xh + h^2) - x^2 = 2xh + h^2.$$

So

$$\frac{f(x + h) - f(x)}{h} = \frac{2xh + h^2}{h} = \frac{h(2x + h)}{h} = 2x + h.$$

It is important to note that a difference quotient is defined *only* when $h \neq 0$. The simplification above is valid only for nonzero values of h. When $x = 5$ and $h = 0.1$,

$$\frac{f(x + h) - f(x)}{h} = 2x + h = 2 \cdot 5 + 0.1 = 10 + 0.1 = 10.1.$$

DO EXERCISE 5.

Example 6 For $f(x) = x^3$, find a simplified form of the difference quotient.

Solution Now $f(x) = x^3$, so

$$f(x + h) = (x + h)^3 = x^3 + 3x^2h + 3xh^2 + h^3.$$

This is shown in Exercise Set 1.1. Then

$$f(x + h) - f(x) = (x^3 + 3x^2h + 3xh^2 + h^3) - x^3$$
$$= 3x^2h + 3xh^2 + h^3.$$

So

$$\frac{f(x + h) - f(x)}{h} = \frac{3x^2h + 3xh^2 + h^3}{h}$$

$$= \frac{h(3x^2 + 3xh + h^2)}{h} = 3x^2 + 3xh + h^2.$$

Again, this is true *only* for $h \neq 0$.

DO EXERCISE 6.

7. a) For

$$f(x) = \frac{1}{x}$$

find a simplified difference quotient.

b) Complete the following table.

x	h	$\dfrac{f(x+h) - f(x)}{h}$
2	3	
2	1	
2	0.1	
2	0.01	
2	0.001	

Example 7 For $f(x) = 3/x$, find a simplifed form of the difference quotient.

Solution Now

$$f(x) = \frac{3}{x},$$

so

$$f(x + h) = \frac{3}{x + h}.$$

Then

$$f(x + h) - f(x) = \frac{3}{x + h} - \frac{3}{x}$$

$$= \frac{3}{x + h} \cdot \frac{x}{x} - \frac{3}{x} \cdot \frac{x + h}{x + h}$$

Here we are multiplying by 1 to get a common denominator.

$$= \frac{3x - 3(x + h)}{x(x + h)}$$

$$= \frac{3x - 3x - 3h}{x(x + h)} = \frac{-3h}{x(x + h)}.$$

So

$$\frac{f(x + h) - f(x)}{h} = \frac{\dfrac{-3h}{x(x + h)}}{h} = \frac{-3h}{x(x + h)} \cdot \frac{1}{h} = \frac{-3}{x(x + h)}.$$

This is true only for $h \neq 0$.

DO EXERCISE 7.

EXERCISE SET 2.2

1. *Economics: Utility.* Utility is a type of function that arises in economics. When a consumer receives x units of a certain product, a certain amount of pleasure, or utility, *U*, is derived from them. Below is a typical graph of a utility function.

a) Find the average rate of change of *U* as x changes from 0 to 1; 1 to 2; 2 to 3; 3 to 4.

b) Why do you think the average rates of change are decreasing?

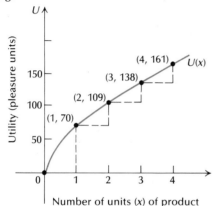

Number of units (x) of product

3. *Psychology: Memory.* The total number of words *M(t)* that a person can memorize in time *t*, in minutes, is shown in the graph below.

a) Find the average rate of change of *M* as *t* changes from 0 to 8; 8 to 16; 16 to 24; 24 to 32; 32 to 36.

b) Why do the average rates of change become 0 after 24 minutes?

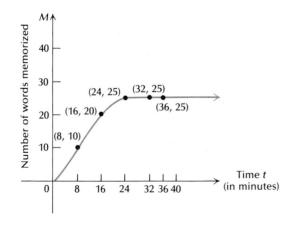

2. *Business: Advertising results.* The graph below shows a typical response to advertising. After an amount *a* is spent on advertising, the company sells *N(a)* units of a product.

a) Find the average rate of change of *N*, as *a* changes from 0 to 1; 1 to 2; 2 to 3; 3 to 4.

b) Why do you think the average rates of change are decreasing?

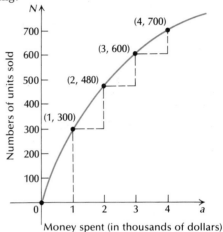

Money spent (in thousands of dollars)

4. *Biomedical: Temperature during an illness.* The °F temperature *T* of a patient during an illness is given by the graph below, where *t* = time in days.

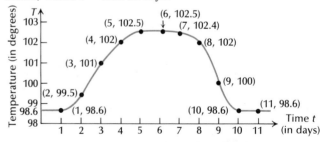

a) Find the average rate of change of *T* as *t* changes from 1 to 10. Using this rate of change, would you know that the person was sick?

b) Find the average rate of change of *T* with respect to *t*, as *t* changes from 1 to 2; 2 to 3; 3 to 4; 4 to 5; 5 to 6; 6 to 7; 7 to 8; 8 to 9; 9 to 10; 10 to 11.

c) When do you think the temperature began to rise?

d) When do you think the temperature reached its peak?

e) When do you think the temperature began to subside?

f) When was the temperature back to normal?

5. *Sociology: Population growth.* The two curves at the right describe the number of people in each of two countries, and, at time *t*, in years.

a) Find the average rate of change of each population (number of people in the population) with respect to time *t*, as *t* changes from 0 to 4. This is often called an *average growth rate*.

b) If the calculation in (a) were the only one made, would we detect the fact that the populations were growing differently?

c) Find the average rates of change of each population as *t* changes from 0 to 1; 1 to 2; 2 to 3; 3 to 4.

d) For which population does the statement "the population grew 125 million each year" convey the least information about what really took place?

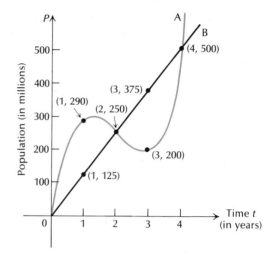

6. *Business: Cost.* A firm determines that the total cost *C* of producing *x* units of a certain product is given by

$$C(x) = -0.05x^2 + 50x,$$

where $C(x)$ is in dollars.

a) Find $C(301)$.

b) Find $C(300)$.

c) Find $C(301) - C(300)$.

d) Find $\dfrac{C(301) - C(300)}{301 - 300}$.

7. *Business: Revenue.* A firm determines that the total revenue (money coming in) from the sale of *x* units of a certain product is given by

$$R(x) = -0.01x^2 + 1000x,$$

where $R(x)$ is in dollars.

a) Find $R(301)$.

b) Find $R(300)$.

c) Find $R(301) - R(300)$.

d) Find $\dfrac{R(301) - R(300)}{301 - 300}$.

8. *Average velocity.* A car is at a distance *s* (in miles) from its starting point in *t* hours, given by

$$s(t) = 10t^2.$$

a) Find $s(2)$ and $s(5)$.

b) Find $s(5) - s(2)$. What does this represent?

c) Find the average rate of change of distance with respect to time as *t* changes from $t_1 = 2$ to $t_2 = 5$. This is known as *average velocity* or *speed*.

9. *Average velocity.* An object is dropped from a certain height. It is known that it will fall a distance *s* (in feet) in *t* seconds, given by

$$s(t) = 16t^2.$$

a) How far will the object fall in 3 seconds?

b) How far will the object fall in 5 seconds?

c) What is the average rate of change of distance with respect to time during the time from 3 to 5 seconds? This is also *average velocity* or *speed*.

10. *Sociology: Divorce rate.* It is known that in 1960 there were 400,000 divorces. In 1980 there were 990,000 divorces. Find the average rate of change in the number of divorces with respect to time. This is called an *average divorce rate*.

11. *Sociology: Marriage rate.* It is known that in 1960 there were 1,450,000 marriages. In 1980 there were 2,900,000 marriages. Find the average rate of change of the number of marriages with respect to time. This is called an *average marriage rate*.

12. At the beginning of a trip, the odometer on a car reads 30,680 and the car has a full tank of gas. At the end of the trip the odometer reads 30,970. It takes 20 gallons of gas to fill the tank again.

a) What is the average rate of consumption (rate of change of the number of miles with respect to the number of gallons)?

b) What is the average rate of change of the number of gallons with respect to the number of miles?

13. *Business: Total revenue.* In 1979 the total revenue of *Chi-Chi's, Inc.,* a national Mexican food franchise, was $1,047,000. In 1982 it was $40,500,000. Find the average rate of change of total revenue.

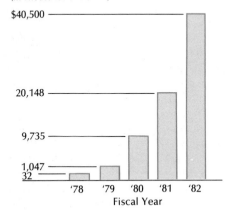

TOTAL REVENUES
(In Thousands of Dollars)

For the functions in each of Exercises 14–25, (a) find a simplified difference quotient; (b) complete the table to the right.

14. $f(x) = 5x^2$

15. $f(x) = 7x^2$

16. $f(x) = -5x^2$

17. $f(x) = -7x^2$

18. $f(x) = 5x^3$

19. $f(x) = 7x^3$

20. $f(x) = \dfrac{4}{x}$

21. $f(x) = \dfrac{5}{x}$

22. $f(x) = 2x + 3$

23. $f(x) = -2x + 5$

24. $f(x) = x^2 + x$

25. $f(x) = x^2 - x$

x	h	$\dfrac{f(x+h) - f(x)}{h}$
4	2	
4	1	
4	0.1	
4	0.01	

CSS

Find the simplified difference quotient.

26. $f(x) = mx + b$

27. $f(x) = ax^2 + bx + c$

28. $f(x) = ax^3 + bx^2$

29. $f(x) = \sqrt{x}$

30. $f(x) = x^4$

31. $f(x) = \dfrac{1}{x^2}$

32. $f(x) = \dfrac{1}{1 - x}$

33. $f(x) = \dfrac{x}{1 + x}$

OBJECTIVES

Given a formula of a function, you should be able to find a formula for its derivative, and then find various values of the derivative.

2.3 DIFFERENTIATION USING LIMITS

Tangent Lines

A line tangent to a circle is a line that touches the circle exactly once.

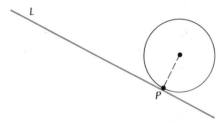

This definition becomes unworkable with other curves. For example, consider the following curve. Line L touches the curve at point P but meets the curve at other places. It will be considered a tangent line, but "touching at one point" cannot be its definition.

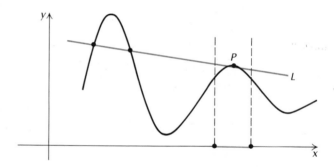

Note in the preceding figure that, over a suitably small interval containing P, line L does touch the curve exactly once. This is still not a suitable definition of a *tangent line* because it allows a line like M in the following figure to be a tangent, which we will not accept.

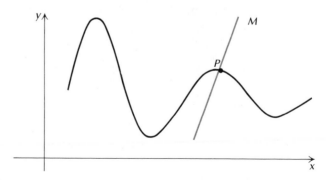

1. a) Which appear to be tangent lines?

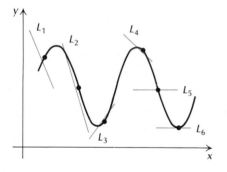

b) Below is a graph of $y = x^2$. Tangent lines are drawn at various points on the graph. Let $m(x)$ = the slope at the point $(x, f(x))$.

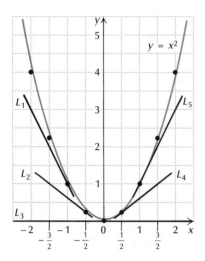

Estimate the slope of each line and complete this table.

Lines	x	$m(x)$
L_1	-1	-2
L_2	$-\frac{1}{2}$	-1
L_3	0	0
L_4	$\frac{1}{2}$	1
L_5	1	2

c) Derive a formula for $m(x)$.

Later we will give a definition of a tangent line, but for now we will rely on intuition (experience). In the following figure, L_1 and L_2 are not tangents. All the others are tangent lines.

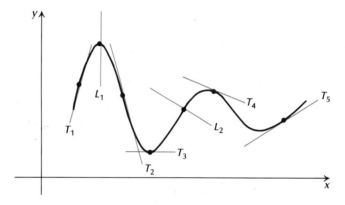

DO EXERCISE 1.

Why Do We Study Tangent Lines?

The reason for this will become evident in Chapter 3. For now, look at the following graph of a total profit function.

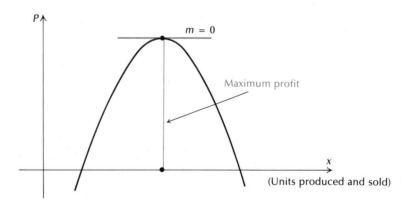

Note that the largest (or maximum) value of the function occurs where the graph has a horizontal tangent; that is, where the tangent line has slope 0.

Differentiation Using Limits

We shall define *tangent* line in such a way that it makes sense for *any* curve. To do this we use the notion of limit.

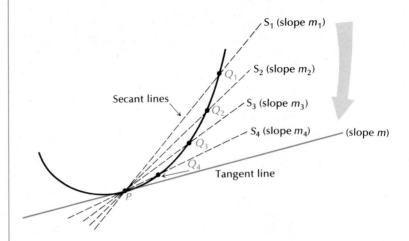

In the figure above, we obtain the line tangent to the curve at point P by considering secant lines through P and neighboring points Q_1, Q_2, and so on. As the points Q approach P, the secant lines approach the tangent line. Each secant has a slope. The slopes of the secant lines approach the slope of the tangent line. In fact, we *define* the *tangent line* to be the line that contains the point P and has slope m, where m is the limit of the slopes of the secant lines as the points Q approach P.

How might we calculate the limit m?

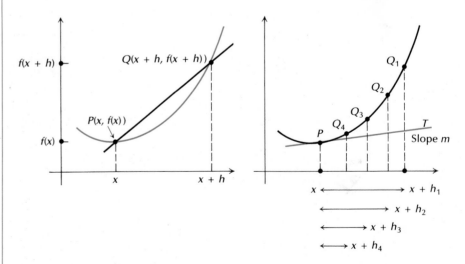

Suppose P has coordinates $(x, f(x))$. Then the first coordinate of Q is x plus some number h, or $x + h$. The coordinates of Q are $(x + h, f(x + h))$. Now from Section 2.2, we know that the slope of the secant line $\overleftrightarrow{PQ}$ is given by the difference quotient

$$\frac{f(x + h) - f(x)}{h}.$$

Now, as we see in the figure on the right above, as the points Q approach P, $x + h$ approaches x. That is, h approaches 0. Thus we have the following.

$$\text{The slope of the tangent line} = m = \lim_{h \to 0} \frac{f(x + h) - f(x)}{h}.$$

The formal definition of the *derivative of a function f* can now be given. We will designate the derivative at x, $f'(x)$, rather than $m(x)$.

DEFINITION

For a function $y = f(x)$, its *derivative* at x is defined as follows:

$$f'(x) = \lim_{h \to 0} \frac{f(x + h) - f(x)}{h},$$

provided the limit exists.

This is the basic definition of *differential calculus*.

Let us now calculate some formulas for derivatives. That is, given a formula for a function f, we will be trying to find a formula for f'.

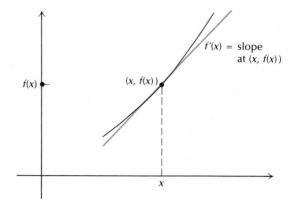

> "Nothing in this world is so powerful as an idea whose time has come."
>
> *Victor Hugo*

There are three steps in calculating a derivative.

1. **Write down the difference quotient $[f(x + h) - f(x)]/h$.**
2. **Simplify the difference quotient.**
3. **Find the limit as $h \to 0$.**

A formula for the derivative of a linear function

$$f(x) = mx + b$$

is

$$f'(x) = m.$$

Let us verify it using the definition.

Example 1 For $f(x) = mx + b$, find $f'(x)$.

Solution We have

1. $\dfrac{f(x + h) - f(x)}{h} = \dfrac{[m(x + h) + b] - (mx + b)}{h}$

2. $\qquad = \dfrac{mx + mh + b - mx - b}{h} = \dfrac{mh}{h} = m$

3. $\displaystyle\lim_{h \to 0} \dfrac{f(x + h) - f(x)}{h} = \lim_{h \to 0} m = m,$

since m does not involve h. Thus

$$f'(x) = m.$$

In Margin Exercise 1 you may have conjectured that the function

$$f(x) = x^2$$

has derivative

$$f'(x) = 2x.$$

This would mean that the tangent line at $x = 4$ has slope $f'(4) = 8$. Let us verify this particular case and then the general formula.

Example 2 For $f(x) = x^2$, find $f'(4)$. **CSS**

Solution We have

1. $\dfrac{f(4 + h) - f(4)}{h} = \dfrac{(4 + h)^2 - 4^2}{h}$

2. $\dfrac{f(4 + h) - f(4)}{h} = \dfrac{16 + 8h + h^2 - 16}{h} = \dfrac{8h + h^2}{h}$

$\qquad\qquad = \dfrac{h(8 + h)}{h} = 8 + h$

2. For $f(x) = x^2$, find $f'(5)$ using the definition of a derivative.

$\dfrac{f(5+h) - f(5)}{h} = \dfrac{(5+h)^2 - 5^2}{h}$

$\dfrac{f(5+h) - f(5)}{h} = \dfrac{25 + h^2 - 25 + 10h}{h}$

$\dfrac{f(5+h) - f(5)}{h} = \dfrac{h(h+10)}{h}$

$\lim_{x \to 0} \dfrac{f(5+h) - f(5)}{h} = h + 10 = 10$

3. For $f(x) = 4x^2$, find $f'(x)$. Then find $f'(5)$ and interpret the meaning.

$\dfrac{f(5+h) - f(5)}{h} = \dfrac{4(5+h)^2 - 4(5)^2}{h}$

$\dfrac{f(5+h) - f(5)}{h} = \dfrac{4(25 + 10h + h^2) - 100}{h}$

$\dfrac{f(5+h) - f(5)}{h} = \dfrac{100 + 40h + 4h^2 - 100}{h}$

$\dfrac{f(5+h) - f(5)}{h} = \dfrac{4k(h+10)}{k}$

$\lim_{x \to 0} \dfrac{f(5+h) - f(5)}{h} = 4h + 40 = 40$

$\dfrac{f(x+h) - f(x)}{h} = \dfrac{4(x+h)^2 - 4x^2}{h}$

$= \dfrac{4(x^2 + 2xh + h^2) - 4x^2}{h}$

$= \dfrac{4x^2 + 8xh + h^2 - 4x^2}{h}$

$= \dfrac{h(8x+h)}{h} = 8x$

3. $\displaystyle\lim_{h \to 0} \frac{f(4+h) - f(4)}{h} = \lim_{h \to 0}(8 + h) = 8.$

Thus $f'(4) = 8$.

DO EXERCISE 2.

Example 3 For $f(x) = x^2$, find (the general formula) $f'(x)$.

Solution

1. We have

$$\frac{f(x+h) - f(x)}{h} = \frac{(x+h)^2 - x^2}{h}.$$

2. In Example 5 of Section 2.2, we showed how this difference quotient can be simplified as follows:

$$\frac{f(x+h) - f(x)}{h} = 2x + h.$$

3. We want to find

$$\lim_{h \to 0} \frac{f(x+h) - f(x)}{h} = \lim_{h \to 0}(2x + h).$$

As $h \to 0$, we see that $2x + h \to 2x$. Thus

$$\lim_{h \to 0}(2x + h) = 2x,$$

and we have

$$f'(x) = 2x,$$

which tells us, for example, that at $x = -3$, the curve has a tangent line whose slope is

$$f'(-3) = 2(-3), \quad \text{or} \quad -6.$$

We may say, simply, "The curve has slope -6."

DO EXERCISE 3.

Example 4 For $f(x) = x^3$, find $f'(x)$. Then find $f'(-1)$ and $f'(10)$.

4. For $f(x) = 4x^3$, find $f'(x)$. Then find $f'(-5)$ and $f'(0)$.

$$\frac{f(x+h)-f(x)}{h} = \frac{4(x+h)^3 - 4x^3}{h}$$

$$= \frac{4x^3 + 12x^2h + 12xh^2 + 4h^3 - 4x^3}{h}$$

$$= \frac{4h(3x^2 + 3xh + h^2)}{h}$$

$$= 12x^2 + 12xh + 4h^2$$

$$f'(x) = 12x^2$$
$$\lim_{h \to 0} = 12x^2$$

$$f'(-5) = 300$$
$$f'(0) = 0$$

Solution

1. We have

$$\frac{f(x+h) - f(x)}{h} = \frac{(x+h)^3 - x^3}{h}.$$

2. In Example 6 of Section 2.2, we showed how this difference quotient can be simplified as follows:

$$\frac{f(x+h) - f(x)}{h} = 3x^2 + 3xh + h^2.$$

3. We then have

$$\lim_{h \to 0} \frac{f(x+h) - f(x)}{h} = \lim_{h \to 0} (3x^2 + 3xh + h^2) = 3x^2.$$

An input–output table for this is shown in Section 2.1. Thus for $f(x) = x^3$, we have $f'(x) = 3x^2$. Then

$$f'(-1) = 3(-1)^2 = 3 \quad \text{and} \quad f'(10) = 3(10)^2 = 300.$$

DO EXERCISE 4.

Example 5 For $f(x) = 3/x$, find $f'(x)$. Then find $f'(1)$ and $f'(2)$.

Solution

1. We have

$$\frac{f(x+h) - f(x)}{h} = \frac{[3/(x+h)] - (3/x)}{h}.$$

2. In Example 7 of Section 2.2, we showed that this difference quotient can be simplified as follows:

$$\frac{f(x+h) - f(x)}{h} = \frac{-3}{x(x+h)}.$$

3. We want to find

$$\lim_{h \to 0} \frac{f(x+h) - f(x)}{h} = \lim_{h \to 0} \frac{-3}{x(x+h)}.$$

As $h \to 0$, $x + h \to x$, so we have

$$f'(x) = \lim_{h \to 0} \frac{-3}{x(x+h)} = \frac{-3}{x^2}.$$

5. For

$$f(x) = \frac{1}{x},$$

find $f'(x)$. Then find $f'(-10)$ and $f'(-2)$.

$$f'(x) = \frac{-1}{x^2}$$

$$f'(-10) = \frac{-1}{100} = -.01$$

$$f'(-2) = \frac{-1}{4} = .25$$

Then

$$f'(1) = \frac{-3}{1^2} = -3 \quad \text{and} \quad f'(2) = \frac{-3}{2^2} = -\frac{3}{4}.$$

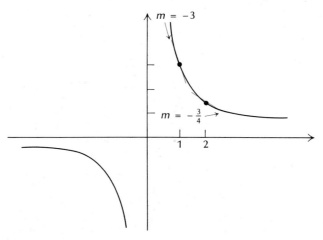

Note that $f'(0)$ does not exist because $f(0)$ does not exist. We say, "f is not differentiable at 0." When a function is not defined at a point, it is not differentiable at that point. In fact, if a function is discontinuous at a point, it is not differentiable at that point.

DO EXERCISE 5.

It can happen that a function f is defined and continuous at a point but that its derivative f' is not. The function f given by

$$f(x) = |x|$$

is an example. Note that

$$f(0) = |0| = 0,$$

so the function is defined at 0.

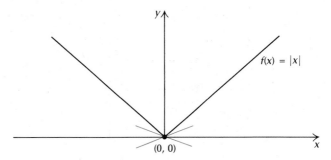

Suppose we tried to draw a tangent line at (0, 0). A function like this with a corner (not smooth) would seem to have many tangents at (0, 0), and hence many slopes. The derivative at such a point would not be unique. Let us try to calculate the derivative at 0.

Now

$$f'(x) = \lim_{h \to 0} \frac{|x + h| - |x|}{h}$$

Thus at x = 0, we have

$$f'(0) = \lim_{h \to 0} \frac{|0 + h| - |0|}{h} = \lim_{h \to 0} \frac{|h|}{h}.$$

h	$\dfrac{\lvert h \rvert}{h}$	h	$\dfrac{\lvert h \rvert}{h}$
2	$\dfrac{\lvert 2 \rvert}{2}$, or $\dfrac{2}{2}$, or 1	−2	$\dfrac{\lvert -2 \rvert}{-2}$, or $\dfrac{2}{-2}$, or −1
1	1	−1	−1
0.1	1	−0.1	−1
0.01	1	−0.01	−1
0.001	1	−0.001	−1

Look at the input–output tables. Note that as h approaches 0 from the right, $|h|/h$ approaches 1, but as h approaches 0 from the left, $|h|/h$ approaches −1. Thus

$$\lim_{h \to 0} \frac{|h|}{h} \text{ does not exist,}$$

so

$$f'(0) \text{ does not exist.}$$

If a function has a "sharp point" or "corner", it will not have a derivative at that point.

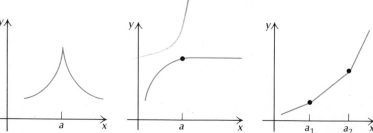

6. List the points at which the function is not differentiable.

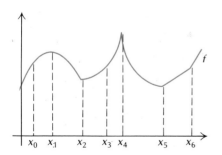

DO EXERCISE 6.

A function may also fail to be differentiable at a point by having a vertical tangent at that point. The following function has a vertical tangent at point a.

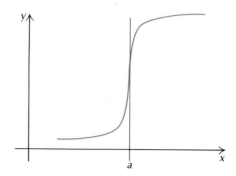

Recall that vertical lines have no slope, hence there is no derivative at such a point.

Each of the preceding examples, including $f(x) = |x|$, is continuous at each point in an interval I but not differentiable at each point in I. That is, continuity does not imply differentiability. On the other hand, if we know that a function is differentiable at each point in an interval I, then it is continuous over I. The function $f(x) = x^2$ is an example. Also, if a function is discontinuous at some point a, then it is not differentiable at a. Thus when we know a function is differentiable over an interval, it is *smooth* in the sense that there are no "sharp points," "corners," or "breaks" in the graph.

EXERCISE SET 2.3

For each function, find $f'(x)$. Then find $f'(-2)$, $f'(-1)$, $f'(0)$, $f'(1)$, and $f'(2)$, if they exist.   CSS

1. $f(x) = 5x^2$ **2.** $f(x) = 7x^2$ **3.** $f(x) = -5x^2$ **4.** $f(x) = -7x^2$

5. $f(x) = 5x^3$ **6.** $f(x) = 7x^3$ **7.** $f(x) = 2x + 3$ **8.** $f(x) = -2x + 5$

9. $f(x) = -4x$ **10.** $f(x) = \dfrac{1}{2}x$ **11.** $f(x) = x^2 + x$ **12.** $f(x) = x^2 - x$

13. $f(x) = \dfrac{4}{x}$ **14.** $f(x) = \dfrac{5}{x}$ **15.** $f(x) = mx$ **16.** $f(x) = ax^2 + bx + c$

17. List the points in the graph below at which the function is not differentiable.

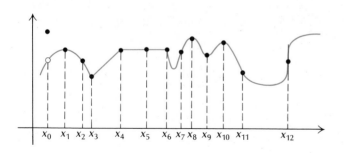

18. *The postage function.* Consider the postage function defined in Exercise Set 2.1. At what values is the function not differentiable?

19. Consider the function f given by

$$f(x) = \frac{x^2 - 9}{x + 3}.$$

For what values is this function not differentiable?

Find $f'(x)$.

20. $f(x) = x^4$

21. $f(x) = \dfrac{1}{x^2}$

22. $f(x) = \dfrac{1}{1 - x}$

23. $f(x) = \dfrac{x}{1 + x}$

24. $f(x) = \sqrt{x}$ $\left(\text{Multiply by 1, using } \dfrac{\sqrt{x + h} + \sqrt{x}}{\sqrt{x + h} + \sqrt{x}}.\right)$

OBJECTIVES

You should be able to

a) **Differentiate using the Power Rule, the Sum–Difference Rule, or the rule for differentiating a constant or a constant times a function.**

b) **Find the points on the graph of a function where the tangent line has a given slope.**

2.4 DIFFERENTIATION TECHNIQUES: POWER AND SUM–DIFFERENCE RULES

Leibniz's Notation

When y is a function of x, we will also designate the derivative, $f'(x)$, as*

$$\frac{dy}{dx},$$

which is read "the derivative of y with respect to x." This notation was invented by the German mathematician Leibniz. It does *not* mean dy divided by dx! (That is, we cannot interpret dy/dx as a quotient until meanings are given to dy and dx, which we will not do here.) For example, if $y = x^2$, then

$$\frac{dy}{dx} = 2x.$$

---

*The notation $D_x y$ is also used.

1. For

$$y = x^3,$$

use the results of previous work to find:

a) $\dfrac{dy}{dx}$; $3x^2$

b) $\dfrac{d}{dx}x^3$;

c) $\dfrac{dy}{dx}\bigg|_{x=4}$.

The German mathematician and philosopher Gottfried Wilhelm von Leibniz (1646–1716) and the English mathematician, philosopher, and physicist Sir Isaac Newton (1642–1727) are both credited with the invention of the calculus, though each made the invention independent of the other. Newton used the dot notation $\dot{y}$ for dy/dt, where y is a function of time, and this notation is still used, though it is not as prevalent as Leibniz's notation.

Find $\dfrac{dy}{dx}$ (differentiate).

We may also write

$$\frac{d}{dx}f(x)$$

to denote the derivative of f with respect to x. For example,

$$\frac{d}{dx}x^2 = 2x.$$

The value of dy/dx when $x = 5$ can be denoted by

$$\frac{dy}{dx}\bigg|_{x=5}.$$

Thus for $dy/dx = 2x$,

$$\frac{dy}{dx}\bigg|_{x=5} = 2 \cdot 5, \quad \text{or} \quad 10.$$

In general, for $y = f(x)$,

$$\frac{dy}{dx}\bigg|_{x=a} = f'(a).$$

DO EXERCISE 1.

The Power Rule

In the remainder of this section we will develop rules and techniques for efficient differentiation.

This table contains functions and derivatives that we have found in previous work. Look for a pattern.

Function	Derivative
x^2	$2x^1$
x^3	$3x^2$
x^{-1}, or $\dfrac{1}{x}$	$-1 \cdot x^{-2}$, or $-\dfrac{1}{x^2}$

Perhaps you have discovered the following.

2. $y = x^6$

$6x^5$

3. $y = x^{-7}$

$-7x^{-6}$ or

$-\dfrac{7}{x^8}$

4. $y = \sqrt[3]{x}$

$x^{\frac{1}{3}} = \dfrac{1}{3}x^{-\frac{2}{3}}$

$\dfrac{1}{3x^{2/3}} =$

$\dfrac{1}{3\sqrt[3]{x^2}}$

5. $y = x^{-1/4}$

$-\dfrac{1}{4}x^{-5/4}$

$= \dfrac{-1}{4x^{5/4}}$

$= \dfrac{-1}{4\sqrt[4]{x^5}}$

$= \dfrac{-1}{4x\sqrt[4]{x}}$

THEOREM 1 Power Rule

For any real number a,

$$\frac{d}{dx}x^a = a \cdot x^{a-1}.$$

Note that this rule holds no matter what the exponent. That is, to differentiate x^a, write down the exponent a, followed by x with an exponent 1 less than a.

① ① Bring down the exponent as a factor.

$$x^a$$
$$a \cdot x^{a-1}$$
② ② Subtract 1 from the exponent.

Example 1 $\dfrac{d}{dx}x^5 = 5x^4$

Example 2 $\dfrac{d}{dx}x = 1 \cdot x^{1-1}$

$$= 1 \cdot x^0 = 1$$

Example 3 $\dfrac{d}{dx}x^{-4} = -4 \cdot x^{-4-1}$

$$= -4x^{-5}, \quad \text{or} \quad -4 \cdot \frac{1}{x^5}, \quad \text{or} \quad -\frac{4}{x^5}$$

The Power Rule allows us to differentiate $\sqrt{x}$.

Example 4 $\dfrac{d}{dx}\sqrt{x} = \dfrac{d}{dx}x^{1/2} = \dfrac{1}{2} \cdot x^{1/2-1}$

$$= \frac{1}{2}x^{-1/2}, \quad \text{or} \quad \frac{1}{2} \cdot \frac{1}{x^{1/2}}, \quad \text{or} \quad \frac{1}{2} \cdot \frac{1}{\sqrt{x}}, \quad \text{or} \quad \frac{1}{2\sqrt{x}}$$

Example 5 $\dfrac{d}{dx}x^{-2/3} = -\dfrac{2}{3}x^{(-2/3)-1}$

$$= -\frac{2}{3}x^{-5/3}, \quad \text{or} \quad -\frac{2}{3}\frac{1}{x^{5/3}}, \quad \text{or} \quad -\frac{2}{3\sqrt[3]{x^5}}$$

DO EXERCISES 2–5.

6. Find $g'(x)$, if

$$g(x) = -14.$$

g'(x) = 0

The Derivative of a Constant Function

Look at the graph of the constant function $F(x) = c$. What is the slope of each point on the graph?

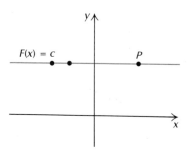

We now have the following.

THEOREM 2   *CSS*

The derivative of a constant function is 0.

Proof. Let F be the function given by $F(x) = c$. Then

$$\frac{F(x + h) - F(x)}{h} = \frac{c - c}{h}$$

$$= \frac{0}{h}$$

$$= 0.$$

The difference quotient is always 0. Thus, as $h \to 0$, the limit of the difference quotient approaches 0, so $F'(x) = 0$.

DO EXERCISE 6.

The Derivative of a Constant Times a Function

Now let us consider differentiating functions like

$$f(x) = 5x^2 \quad \text{and} \quad g(x) = -7x^4.$$

Note that we already know how to differentiate x^2 and x^4. Let us again look for a pattern in the results of Exercise Set 2.3.

Function	Derivative
$5x^2$	$10x$
$-4x$	-4
$-7x^2$	$-14x$
$5x^3$	$15x^2$

Perhaps you have discovered the following.

THEOREM 3

The derivative of a constant times a function is the constant times the derivative of the function. Using derivative notation, we can write this as

$$\frac{d}{dx}[c \cdot f(x)] = c \cdot f'(x).$$

Proof. Let F be the function given by $F(x) = cf(x)$. Then

$$\frac{F(x + h) - F(x)}{h} = \frac{cf(x + h) - cf(x)}{h} = c\left[\frac{f(x + h) - f(x)}{h}\right].$$

As $h \to 0$, the limit on the right is the same as c times $f'(x)$. Thus $F'(x) = cf'(x)$.

Combining this rule with the Power Rule allows us to find many derivatives.

Example 6 $\dfrac{d}{dx}5x^4 = 5\dfrac{d}{dx}x^4 = 5 \cdot 4 \cdot x^{4-1} = 20x^3$

Example 7 $\dfrac{d}{dx} - 9x = -9\dfrac{d}{dx}x = -9 \cdot 1 = -9$

With practice you will be able to differentiate many such functions in one step.

Example 8 $\dfrac{d}{dx}\dfrac{-4}{x^2} = \dfrac{d}{dx} - 4x^{-2} = -4 \cdot \dfrac{d}{dx}x^{-2} = -4(-2)x^{-2-1}$

$$= 8x^{-3}, \quad \text{or} \quad \frac{8}{x^3}$$

Example 9 $\dfrac{d}{dx} - x^{0.7} = -1 \cdot \dfrac{d}{dx}x^{0.7} = -1 \cdot 0.7 \cdot x^{0.7-1} = -0.7x^{-0.3}$

[handwritten margin note: The Constant can be placed in front of the derivative function]

Find $\dfrac{dy}{dx}$.

7. $y = 5x^{20}$

$100x^{19}$

8. $y = -\dfrac{3}{x}$

$\dfrac{-3}{1} = -3$

$2x^{-\frac{1}{2}}$

$-2x^{-3/2}$

$6x^{-3/2}$

$\dfrac{6}{\sqrt[3]{x^2}}$

9. $y = -8\sqrt{x}$

$-8 \cdot x^{\frac{1}{2}}$

$-8\left(\frac{1}{2}\right)x^{-\frac{1}{2}}$

$-16x^{-\frac{1}{2}}$

$\dfrac{-16}{\sqrt{x}}$

10. $y = 0.16x^{6.25}$

DO EXERCISES 7–10.

The Derivative of a Sum or Difference

In Exercise 11 of Exercise Set 2.3, you found that for

$$f(x) = x^2 + x$$

the derivative is

$$f'(x) = 2x + 1.$$

Note that the derivative of x^2 is $2x$, and the derivative of x is 1; and the sum of these derivatives is $f'(x)$. This illustrates the following.

THEOREM 4 **Sum–Difference Rule**

a) The derivative of a sum is the sum of the derivatives:*

If $F(x) = f(x) + g(x)$, then $F'(x) = f'(x) + g'(x)$.

b) The derivative of a difference is the difference of the derivatives:

If $F(x) = f(x) - g(x)$, then $F'(x) = f'(x) - g'(x)$.

Any function that is a sum or difference of several terms can be differentiated term by term.

Example 10 $\dfrac{d}{dx}(3x + 7) = \dfrac{d}{dx}(3x) + \dfrac{d}{dx}(7)$

$$= 3\dfrac{d}{dx}x + 0 = 3 \cdot 1 = 3$$

Example 11 $\dfrac{d}{dx}(5x^3 - 3x^2) = \dfrac{d}{dx}(5x^3) - \dfrac{d}{dx}(3x^2)$

$$= 5\dfrac{d}{dx}x^3 - 3\dfrac{d}{dx}x^2$$

$$= 5 \cdot 3x^2 - 3 \cdot 2x = 15x^2 - 6x$$

**Proof.* We have

$$\dfrac{F(x + h) - F(x)}{h} = \dfrac{[f(x + h) + g(x + h)] - [f(x) + g(x)]}{h}$$

$$= \dfrac{f(x + h) - f(x)}{h} + \dfrac{g(x + h) - g(x)}{h}.$$

As $h \to 0$, the two terms on the right approach $f'(x)$ and $g'(x)$, respectively, so their sum approaches $f'(x) + g'(x)$. Thus $F'(x) = f'(x) + g'(x)$.

Find $\frac{dy}{dx}$ (differentiate).

Example 12 $\dfrac{d}{dx}\left(24x - \sqrt{x} + \dfrac{2}{x}\right) = \dfrac{d}{dx}(24x) - \dfrac{d}{dx}(\sqrt{x}) + \dfrac{d}{dx}\left(\dfrac{2}{x}\right)$

$$= 24 \cdot \frac{d}{dx}x - \frac{d}{dx}x^{1/2} + 2 \cdot \frac{d}{dx}x^{-1}$$

$$= 24 \cdot 1 - \frac{1}{2}x^{(1/2)-1} + 2(-1)x^{-1-1}$$

$$= 24 - \frac{1}{2}x^{-1/2} - 2x^{-2}$$

$$= 24 - \frac{1}{2\sqrt{x}} - \frac{2}{x^2}$$

11. $y = -\dfrac{1}{4}x - 9$

DO EXERCISES 11–13.

A word of caution! The derivative of

$$f(x) + c,$$

a function plus a constant, is just the derivative of the function

$$f'(x).$$

The derivative of

$$c \cdot f(x),$$

a function times a constant, is the constant times the derivative

$$c \cdot f'(x).$$

That is, for a product the constant is retained, but for a sum it is not.

It is important to be able to determine points at which the tangent line to a curve has a certain slope—that is, points at which the derivative attains a certain value.

12. $y = 7x^4 + 6x^2$

13. $y = 15x^2 + \dfrac{4}{x} + \sqrt{x}$

Example 13 Find the points on the graph of $y = -x^3 + 6x^2$ at which the tangent line is horizontal.

Solution A horizontal tangent has slope 0. Thus we seek the values of x for which $dy/dx = 0$. That is, we want to find x such that

$$-3x^2 + 12x = 0.$$

14. Find the points on the graph of

$$y = \frac{1}{3}x^3 - 2x^2 + 4x$$

at which the tangent line is horizontal.   *CSS*

We factor and solve:

$$x(-3x + 12) = 0$$
$$x = 0 \quad \text{or} \quad -3x + 12 = 0$$
$$x = 0 \quad \text{or} \quad -3x = -12$$
$$x = 0 \quad \text{or} \quad x = 4.$$

We are to find the points *on the graph*, so we have to determine the second coordinates from the original equation $y = -x^3 + 6x^2$.

For $x = 0$, $\quad y = -0^3 + 6 \cdot 0^2 = 0.$

For $x = 4$, $\quad y = -(4)^3 + 6 \cdot 4^2 = -64 + 96 = 32.$

Thus the points we are seeking are $(0, 0)$ and $(4, 32)$.

DO EXERCISE 14.

Example 14 Find the points on the graph of $y = -x^3 + 6x^2$ at which the tangent has slope 6.

Solution We want to find values of x for which $dy/dx = 6$. That is, we want to find x such that

$$-3x^2 + 12x = 6.$$

To solve, we add -6 and get

$$-3x^2 + 12x - 6 = 0.$$

We can simplify this equation by multiplying by $-\frac{1}{3}$, since each term has a common factor of -3. We get

$$x^2 - 4x + 2 = 0.$$

This is a quadratic equation, not readily factorable, so we use the quadratic formula, where $a = 1$, $b = -4$, and $c = 2$:

$$x = \frac{-b \pm \sqrt{b^2 - 4ac}}{2a} = \frac{-(-4) \pm \sqrt{(-4)^2 - 4 \cdot 1 \cdot 2}}{2 \cdot 1} = \frac{4 \pm \sqrt{8}}{2}$$

$$= \frac{2 \cdot 2 \pm 2\sqrt{2}}{2 \cdot 1}$$

$$= \frac{2}{2} \cdot \frac{2 \pm \sqrt{2}}{1}$$

$$= 2 \pm \sqrt{2}.$$

15. Find the points on the graph of

$$y = \frac{1}{3}x^3 - 2x^2 + 4x$$

at which the tangent line has slope 3.

The solutions are $2 + \sqrt{2}$ and $2 - \sqrt{2}$. We determine the second coordinates from the original equation. For $x = 2 + \sqrt{2}$,

$$
\begin{aligned}
y &= -(2 + \sqrt{2})^3 + 6(2 + \sqrt{2})^2 \\
&= -[(2 + \sqrt{2})^2(2 + \sqrt{2})] + 6(4 + 4\sqrt{2} + 2) \\
&= -[(6 + 4\sqrt{2})(2 + \sqrt{2})] + 6(6 + 4\sqrt{2}) \\
&= -[12 + 6\sqrt{2} + 8\sqrt{2} + 8] + 36 + 24\sqrt{2} \\
&= -[20 + 14\sqrt{2}] + 36 + 24\sqrt{2} \\
&= -20 - 14\sqrt{2} + 36 + 24\sqrt{2} = 16 + 10\sqrt{2}.
\end{aligned}
$$

Similarly, for $x = 2 - \sqrt{2}$,

$$y = 16 - 10\sqrt{2}.$$

Thus the points we are seeking are $(2 + \sqrt{2}, 16 + 10\sqrt{2})$ and $(2 - \sqrt{2}, 16 - 10\sqrt{2})$.

DO EXERCISE 15.

We illustrate the results of Examples 13 and 14 in the following graph. You will not be asked to sketch such graphs at this time.

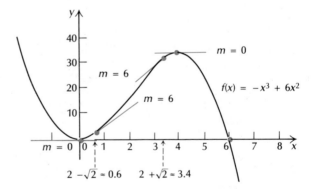

EXERCISE SET 2.4

Find $\dfrac{dy}{dx}$.

1. $y = x^7$

2. $y = x^8$

3. $y = 15$

4. $y = 78$

5. $y = 4x^{150}$

6. $y = 7x^{200}$

7. $y = x^3 + 3x^2$

8. $y = x^4 - 7x$

9. $y = 8\sqrt{x}$

10. $y = 4\sqrt{x}$

11. $y = x^{0.07}$

12. $y = x^{0.78}$

13. $y = \dfrac{1}{2}x^{4/5}$

14. $y = -4.8x^{1/3}$

15. $y = x^{-3}$

16. $y = x^{-4}$

17. $y = 3x^2 - 8x + 7$

18. $y = 4x^2 - 7x + 5$

19. $y = \sqrt[4]{x} - \dfrac{1}{x}$

20. $y = \sqrt[5]{x} - \dfrac{2}{x}$

Find $f'(x)$.

21. $f(x) = 0.64x^{2.5}$

22. $f(x) = 0.32x^{12.5}$

23. $f(x) = \dfrac{5}{x} - x$

24. $f(x) = \dfrac{4}{x} - x$

25. $f(x) = 4x - 7$

26. $f(x) = 7x + 11$

27. $f(x) = 4x + 9$

28. $f(x) = 7x - 14$

29. $f(x) = \dfrac{x^4}{4}$

30. $f(x) = \dfrac{x^3}{3}$

31. $f(x) = -0.01x^2 - 0.5x + 70$

32. $f(x) = -0.01x^2 + 0.4x + 50$

33. $f(x) = 3x^{-2/3} + x^{3/4} + x^{6/5} + \dfrac{8}{x^3}$

34. $f(x) = x^{-3/4} - 3x^{2/3} + x^{5/4} + \dfrac{2}{x^4}$

For each function, find the points on the graph at which the tangent line is horizontal.   `CSS`

35. $y = x^2$

36. $y = -x^2$

37. $y = -x^3$

38. $y = x^3$

39. $y = 3x^2 - 5x + 4$

40. $y = 5x^2 - 3x + 8$

41. $y = -0.01x^2 - 0.5x + 70$

42. $y = -0.01x^2 + 0.4x + 50$

43. $y = 2x + 4$

44. $y = -2x + 5$

45. $y = 4$

46. $y = -3$

47. $y = -x^3 + x^2 + 5x - 1$

48. $y = -\dfrac{1}{3}x^3 + 6x^2 - 11x - 50$

49. $y = \dfrac{1}{3}x^3 - 3x + 2$

50. $y = x^3 - 6x + 1$

For each function, find the points on the graph at which the tangent line has slope 1.

51. $y = 20x - x^2$

52. $y = 6x - x^2$

53. $y = -0.025x^2 + 4x$

54. $y = -0.01x^2 + 2x$

55. $y = \dfrac{1}{3}x^3 + 2x^2 + 2x$

56. $y = \dfrac{1}{3}x^3 - x^2 - 4x + 1$

57. Find the points on the graph of

$$y = x^4 - \frac{4}{3}x^2 - 4$$

at which the tangent line is horizontal.

58. Find the points on the graph of

$$y = 2x^6 - x^4 - 2$$

at which the tangent line is horizontal.

Find dy/dx. Each of the following can be differentiated using the rules developed in this section, but some algebra may be required beforehand.

59. $y = x(x - 1)$

60. $y = (x - 1)(x + 1)$

61. $y = (x - 2)(x + 3)$

62. $y = \dfrac{5x^2 - 8x + 3}{8}$

63. $y = \dfrac{x^5 + x}{x^2}$

64. $y = (5x)^2$

65. $y = (-4x)^3$

66. $y = \sqrt{7x}$

67. $y = \sqrt[3]{8x}$

68. $y = (x - 3)^2$

69. $y = (x + 1)^3$

70. $y = (x - 2)^3(x + 1)$

71. Prove Theorem 4(b).

OBJECTIVES

You should be able to

a) **Given a distance function $s(t)$, find a formula for the velocity $v(t)$ and the acceleration $a(t)$, and evaluate $s(t)$, $v(t)$, and $a(t)$ for given values of t.**

b) **Given y as a function of x, find the rate of change of y with respect to x, and evaluate this rate of change for values of x.**

An instantaneous velocity. (*Robert V. Fuschetto: Photo Researchers, Inc.*)

2.5 APPLICATIONS AND RATES OF CHANGE

Instantaneous Rate of Change

A car travels 100 miles in 2 hours. Its *average* speed (or velocity) is 100 mi/2 hr, or 50 mi/hr. This is the *average rate of change* of distance with respect to time. At various times during the trip the speedometer did not read 50, however. Thus we say that 50 is the *average*. A snapshot of the speedometer taken at any instant would indicate *instantaneous* speed, or rate of change.

Average rates of change are given by difference quotients. If distance s is a function of time t, then average velocity is given by

$$\text{Average velocity} = \frac{s(t + h) - s(t)}{h}.$$

Instantaneous rates of change are found by letting $h \to 0$. Thus

$$\text{Instantaneous velocity} = \lim_{h \to 0} \frac{s(t + h) - s(t)}{h} = s'(t).$$

Example 1 An object travels in such a way that distance s (in miles) from the starting point is a function of time t (in hours) as follows:

$$s(t) = 10t^2.$$

a) Find the average velocity between the times $t = 2$ and $t = 5$.

b) Find the (instantaneous) velocity when $t = 4$.

Rates of Change in Economics

In the study of economics we are frequently interested in how such quantities as cost, revenue, and profit change with an increase in product quantity. In particular, we are interested in what is called *marginal** cost or profit (or whatever). This term is used to signify *rate of change with respect to quantity*. Thus, if

$$C(x) = \text{the } total \ cost \text{ of producing x units of a product}$$
$$\text{(usually considered in some time period),}$$

then

$$C'(x) = \text{the } marginal \ cost$$

$$= \text{the rate of change of the total cost with respect to}$$
$$\text{the number of units, x, produced.}$$

Let us think about these interpretations. The total cost of producing 5 units of a product is $C(5)$. The rate of change $C'(5)$ is the cost per unit at that stage in the production process. That this cost per unit does not include fixed costs is seen in this example.

$$C(x) = \underbrace{(x^2 + 4x)}_{\text{Variable costs}} + \underbrace{\$10,000}_{\text{Fixed costs (constant)}}$$

Then

$$C'(x) = 2x + 4.$$

This is because the derivative of a constant is 0. This verifies an economic principle that says the fixed costs of a company have no effect on marginal cost.

Following are some other marginal functions. Recall that

$$R(x) = \text{the } total \ revenue \text{ from the sale of x units.}$$

Then

$$R'(x) = \text{the } marginal \ revenue$$

$$= \text{the rate of change of the total revenue with respect}$$
$$\text{to the number x of units sold.}$$

---

*The term "marginal" comes from the Marginalist School of Economic Thought, which originated in Austria for the purpose of applying mathematics and statistics to the study of economics.

Also

$$P(x) = \text{the } \textit{total profit} \text{ from the production and sale}$$
$$\text{of x units of a product}$$
$$= R(x) - C(x).$$

Then

$$P'(x) = \text{the } \textit{marginal profit}$$
$$= \text{the rate of change of the total profit with respect}$$
$$\text{to the number of units x produced and sold}$$
$$= R'(x) - C'(x).$$

Example 5 Given

$$R(x) = 50x,$$
$$C(x) = 2x^3 - 12x^2 + 40x + 10,$$

find:

a) $P(x)$;

b) $R(2), C(2), P(2)$;

c) $R'(x), C'(x), P'(x)$;

d) $R'(2), C'(2), P'(2)$.

Solution

a) $P(x) = R(x) - C(x) = 50x - (2x^3 - 12x^2 + 40x + 10)$

$$= -2x^3 + 12x^2 + 10x - 10$$

b) $R(2) = 50 \cdot 2 = \$100$ (the total revenue from the sale of the first 2 units)

$C(2) = 2 \cdot 2^3 - 12 \cdot 2^2 + 40 \cdot 2 + 10 = \$58$ (the total cost of producing the first 2 units)

$P(2) = R(2) - C(2) = \$100 - \$58 = \$42$ (the total profit from the production and sale of the first 2 units)

c) $R'(x) = 50$,

$C'(x) = 6x^2 - 24x + 40$,

$P'(x) = R'(x) - C'(x) = 50 - (6x^2 - 24x + 40)$

$$= -6x^2 + 24x + 10$$

6. Given

$$R(x) = 50x - 0.5x^2,$$

$$C(x) = 10x + 3,$$

find:

a) $P(x)$;

b) $R(40)$, $C(40)$, $P(40)$;

c) $R'(x)$, $C'(x)$, $P'(x)$;

d) $R'(40)$, $C'(40)$, $P'(40)$.

e) Is the marginal revenue constant?

d) $R'(2) = \$50$ per unit,

$\quad C'(2) = 6 \cdot 2^2 - 24 \cdot 2 + 40 = \$16$ per unit,

$\quad P'(2) = \$50 - \$16 = \$34$ per unit

Note that marginal revenue is constant. No matter how much is produced and sold, the revenue per unit stays the same. This may not always be the case. Also note that $C'(2)$, or $16 per unit, is not the average cost per unit, which is given by

$$\frac{\text{Total cost of producing 2 units}}{2 \text{ units}} = \frac{\$58}{2} = \$29 \text{ per unit.}$$

In general,

$$A(x) = \text{the } \textit{average cost} \text{ of producing } x \text{ units} = \frac{C(x)}{x}.$$

DO EXERCISE 6.

Let us look at a typical marginal cost function C' and its associated total cost function C.

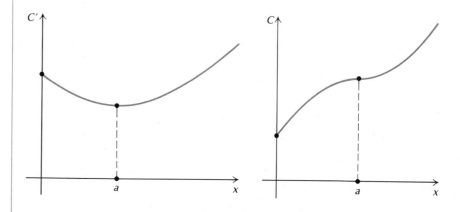

Marginal cost normally decreases as more units are produced until it reaches some minimum value at a, and then it increases. This is probably due to something like having to pay overtime or buying more machinery. Since $C'(x)$ represents the slope of $C(x)$ and is positive and decreasing up to a, the graph turns downward as x goes from 0 to a. Then past a it turns upward.

EXERCISE SET 2.5

1. Given

$$s(t) = t^3 + t,$$

where s is measured in feet and t in seconds, find:

a) $v(t)$,

b) $a(t)$,

c) the velocity and acceleration when $t = 4$ sec.

2. Given

$$s(t) = 3t + 10,$$

where s is measured in miles and t in hours, find:

a) $v(t)$,

b) $a(t)$,

c) the velocity and acceleration when $t = 2$ hr. When the distance function is given by a linear function, we have what is called *uniform motion*.

3. *View to the horizon.* The view, or distance, in miles, which one can see to the horizon from a height h, in feet, is given by

$$V = 1.22\sqrt{h}.$$

a) Find the rate of change of V with respect to h.

b) How far can one see to the horizon out an airplane window from a height of 40,000 ft?

c) Find the rate of change at $h = 40,000$.

4. *Stopping distance on glare ice.* The stopping distance (at some fixed speed) of regular tires is given by a linear function of the air temperature F,

$$D(F) = 2F + 115,$$

where $D(F)$ = the stopping distance, in feet, when the air temperature is F, in degrees Fahrenheit. Find the rate of change of the stopping distance D with respect to the air temperature F.

5. *Sociology: Percentage of the population in college.* The percentage of the population in college is given by a linear function

$$P(t) = 1.25t + 15,$$

where $P(t)$ = the percentage in college the t-th year after 1940. Find the rate of change of the percentage P with respect to time t.

6. *Biomedical: Healing wound.* The circular area A, in square centimeters, of a healing wound is given by

$$A = \pi r^2,$$

where r is the radius, in centimeters. Find the rate of change of the area with respect to the radius.

7. *Biomedical: Healing wound.* The circumference C, in centimeters, of a healing wound is given by

$$C = 2\pi r,$$

where r is the radius, in centimeters. Find the rate of change of the circumference with respect to the radius.

8. *Sociology: Population growth rate.* The population of a city grows from an initial size of 100,000 to an amount P given by

$$P = 100{,}000 + 2000t^2,$$

where t is measured in years.

a) Find the growth rate.

b) Find the number of people in the city after 10 years (at $t = 10$ yr).

c) Find the growth rate at $t = 10$ yr.

9. *Biomedical: Fever.* The temperature T of a person during an illness is given by

$$T(t) = -0.1t^2 + 1.2t + 98.6,$$

where T is the temperature (°F) at time t, measured in days.

a) Find the rate of change of the temperature with respect to time.

b) Find the temperature at $t = 1.5$ days.

c) Find the rate of change at $t = 1.5$ days.

11. *Biomedical: Blood pressure.* For a certain dosage of x cc (cubic centimeters) of a drug, the resultant blood pressure B is given by

$$B(x) = 0.05x^2 - 0.3x^3.$$

Find the rate of change of the blood pressure with respect to the dosage.

12. *Ecology: Home range.* The home range H of an animal is defined as the region to which it confines its movements. The area of that region is related to its body weight by

$$H = W^{1.41}$$

(see Section 1.5). Find dH/dW.

14. Given

$$R(x) = 50x - 0.5x^2,$$

$$C(x) = 4x + 10,$$

find:

a) $P(x)$;

b) $R(20)$, $C(20)$, $P(20)$;

c) $R'(x)$, $C'(x)$, $P'(x)$;

d) $R'(20)$, $C'(20)$, $P'(20)$.

10. *Business: Advertising.* A firm estimates that it will sell N units of a product after spending a dollars on advertising, where

$$N(a) = -a^2 + 300a + 6,$$

and a is measured in thousands of dollars.

a) What is the rate of change of the number of units sold with respect to the amount spent on advertising?

b) How many units will be sold after spending $10 thousand on advertising?

c) What is the rate of change at $a = 10$?

These lions may be determining territory area. (*Ian Cleghorn: Photo Researchers, Inc.*)

13. *Ecology: Territory area.* The territory area T of an animal is defined to be its defended, or exclusive, region. The area T of that region is related to its body weight by

$$T = W^{1.31}$$

(see Section 1.5). Find dT/dW.

15. Given

$$R(x) = 5x,$$

$$C(x) = 0.001x^2 + 1.2x + 60,$$

find:

a) $P(x)$;

b) $R(100)$, $C(100)$, $P(100)$;

c) $R'(x)$, $C'(x)$, $P'(x)$;

d) $R'(100)$, $C'(100)$, $P'(100)$.

You should be able to differentiate using the Product and Quotient Rules.

Use the Product Rule to find $f'(x)$.

1. $f(x) = 3x^8 \cdot x^{10}$

$3x^8 \cdot 10x^9 + x^{10} \cdot 24x^7$

$30x^{17} + 24x^{17}$

$54x^{17}$

3^{15x^4}

2. $f(x) = (9x^3 + 4x^2 + 10)(-7x^2 + x^4)$

$(9x^3 + 4x^2 + 10)(-14x + 3x^3) +$

$(-7x^2 + x^4)(27x^2 + 8x) =$

$-126x^4 - 56x^3 - 140x + 27x^6 + 12x^5 + 30x^3$

$-189x^4 - 56x^3 + 27x^6 + 8x^5 =$

$54x^6 + 20x^5 - 315x^4 - 82x^3 - 140x$

$x(54x^5 + 20x^4 - 315x^3 - 82x^2 - 140)$

2.6 DIFFERENTIATION TECHNIQUES: PRODUCT AND QUOTIENT RULES

The derivative of a sum is the sum of the derivatives, but the derivative of a product is *not* the product of the derivatives. To see this, consider x^2 and x^5. The product is x^7, and the derivative of this product is $7x^6$. The individual derivatives are $2x$ and $5x^4$, and the product of these derivatives is $10x^5$, which is not $7x^6$.

The following is the rule for finding the derivative of a product.

THEOREM 5 Product Rule

If $F(x) = f(x) \cdot g(x)$, then

$$F'(x) = f(x) \cdot g'(x) + f'(x) \cdot g(x).$$

The derivative of a product is the first factor times the derivative of the second factor, plus the derivative of the first factor times the second factor.

Let us check this for $x^2 \cdot x^5$. There are five steps.

1. Write down the first factor.

2. Multiply it by the derivative of the second factor.

3. Write the derivative of the first factor.

4. Multiply it by the second factor.

5. Add the result of steps 1 and 2 to the result of steps 3 and 4.

$\overset{x^2}{\underset{①\ ②}{\vert}} \overset{x^5}{\underset{③\ ④}{\ }}$

$x^2 \cdot 5x^4 + 2x \cdot x^5$
$= 5x^6 + 2x^6 \qquad ⑤$
$= 7x^2$

Example 1

$$\frac{d}{dx}(x^4 - 2x^3 - 7)(3x^2 - 5x) = (x^4 - 2x^3 - 7)(6x - 5)$$
$$+ (4x^3 - 6x^2)(3x^2 - 5x)$$

Note that we could have multiplied the polynomials and then differentiated, avoiding the use of the Product Rule, but this would have been more work.

DO EXERCISES 1 AND 2.

The derivative of a quotient is *not* the quotient of the derivatives. To see why, consider x^5 and x^2. The quotient x^5/x^2 is x^3, and the derivative of this quotient is $3x^2$. The individual derivatives are $5x^4$ and $2x$, and the quotient of these derivatives $5x^4/2x$ is $(5/2)x^3$, which is not $3x^2$. The rule for differentiating quotients is as follows.

3. For

$$f(x) = \frac{x^9}{x^5},$$

find $f'(x)$ using the Quotient Rule.

$$\frac{x^5 \cdot 9x^8 - x^9 \cdot 5x^4}{(x^5)^2}$$

$$\frac{9x^{13} - 5x^{13}}{x^{10}}$$

$$\frac{4x^{13}}{x^{10}}$$

$$4x^3$$

THEOREM 6   **Quotient Rule**

If

$$q(x) = \frac{f(x)}{g(x)},$$

then

$$q'(x) = \frac{g(x) \cdot f'(x) - g'(x) \cdot f(x)}{[g(x)]^2}$$

The derivative of a quotient is the denominator times the derivative of the numerator, minus the derivative of the denominator times the numerator, all divided by the square of the denominator.

Another way to remember this is shown below. It starts with squaring the denominator. The denominator is also used as the first factor of the first term above.

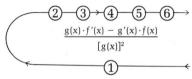

1. Square the denominator.
2. Write down the denominator.
3. Multiply the denominator by the derivative of the numerator.
4. Write a minus sign.
5. Find the derivative of the denominator.
6. Multiply it by the numerator.

Example 2 For $q(x) = x^5/x^3$, find $q'(x)$.

Solution

$$q'(x) = \frac{x^3 \cdot 5x^4 - 3x^2 \cdot x^5}{[x^3]^2}$$

$$= \frac{5x^7 - 3x^7}{x^6}$$

$$= \frac{2x^7}{x^6} = 2x$$

DO EXERCISE 3.

Differentiate.

4. $f(x) = \dfrac{1 - x^2}{x^5}$

Example 3 Differentiate $f(x) = (1 + x^2)/x^3$.

Solution

$$f'(x) = \frac{x^3 \cdot 2x - 3x^2(1 + x^2)}{(x^3)^2} = \frac{2x^4 - 3x^2 - 3x^4}{x^6} = \frac{-x^4 - 3x^2}{x^6}$$

$$= \frac{-1 \cdot x^2 \cdot x^2 - 3x^2}{x^6} = \frac{x^2(-x^2 - 3)}{x^6} = \frac{-x^2 - 3}{x^4}.$$

Example 4 Differentiate $f(x) = (x^2 - 3x)/(x - 1)$.

Solution

$$f'(x) = \frac{(x - 1)(2x - 3) - 1(x^2 - 3x)}{(x - 1)^2}$$

$$= \frac{2x^2 - 5x + 3 - x^2 + 3x}{(x - 1)^2}$$

$$= \frac{x^2 - 2x + 3}{(x - 1)^2}$$

It is not necessary to multiply out $(x - 1)^2$.

DO EXERCISES 4 AND 5.

5. $f(x) = \dfrac{x^2 - 1}{x^3 + 1}$

An Application

We discussed earlier that it is more typical for a total revenue function to vary depending on the number x of units sold. Let us see what can determine this. Recall the consumer's demand function $p = D(x)$, discussed in Section 1.5. It is the price p a seller must charge in order to sell exactly x units of a product. This is typically a decreasing function.

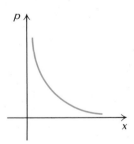

The total revenue from the sale of x units is then given by

$$R(x) = (\text{number of units sold}) \cdot (\text{price charged to sell the units}),$$

6. A company determines that the demand function for a certain product is given by

$$p = D(x) = 200 - x.$$

a) Find an expression for total revenue $R(x)$.

b) Find the marginal revenue $R'(x)$.

or

$$R(x) = x \cdot p = xD(x).$$

A typical graph of a revenue function is shown below.

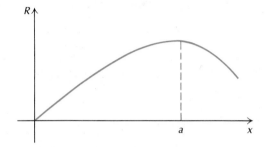

To sell more units, $D(x)$ decreases. Because we have a product $x \cdot D(x)$, the revenue typically rises for a while as x increases, but tapers off as $D(x)$ gets smaller and smaller.

Using the Product Rule, we can obtain an expression for the marginal revenue $R'(x)$ in terms of x and $D'(x)$ as

$$R(x) = xD(x),$$

so

$$R'(x) = 1 \cdot D(x) + x \cdot D'(x) = D(x) + xD'(x).$$

You need not memorize this. You can merely repeat the Product Rule where necessary.

DO EXERCISE 6.

EXERCISE SET 2.6

Differentiate.

1. $y = x^3 \cdot x^8$; two ways

2. $y = x^4 \cdot x^9$; two ways.

3. $y = \dfrac{-1}{x}$; two ways

4. $y = \dfrac{1}{x}$; two ways

5. $y = \dfrac{x^8}{x^5}$; two ways

6. $y = \dfrac{x^9}{x^5}$; two ways

7. $y = (8x^5 - 3x^2 + 20)(8x^4 - 3\sqrt{x})$

8. $f(x) = (7x^6 + 4x^3 - 50)(9x^{10} - 7\sqrt{x})$

9. $f(x) = x(300 - x)$

10. $f(x) = x(400 - x)$

11. $f(x) = \dfrac{x}{300 - x}$

12. $f(x) = \dfrac{x}{400 - x}$

13. $f(x) = \dfrac{3x - 1}{2x + 5}$

14. $f(x) = \dfrac{2x + 3}{x - 5}$

15. $y = \dfrac{x^2 + 1}{x^3 - 1}$

16. $y = \dfrac{x^3 - 1}{x^2 + 1}$

17. $y = \dfrac{x}{1 - x}$

18. $y = \dfrac{x}{3 - x}$ **19.** $y = \dfrac{x - 1}{x + 1}$ **20.** $y = \dfrac{x + 2}{x - 2}$ **21.** $f(x) = \dfrac{1}{x - 3}$ **22.** $f(x) = \dfrac{1}{x + 2}$

23. $f(x) = \dfrac{3x^2 + 2x}{x^2 + 1}$ **24.** $f(x) = \dfrac{3x^2 - 5x}{x^2 - 1}$ **25.** $f(x) = \dfrac{3x^2 - 5x}{x^8}$ **26.** $f(x) = \dfrac{3x^2 + 2x}{x^5}$

In each of Exercises 27–30, a demand function $p = D(x)$ is given. Find (a) the total revenue $R(x)$, and (b) the marginal revenue $R'(x)$.

27. $D(x) = 400 - x$ **28.** $D(x) = 500 - x$ **29.** $D(x) = \dfrac{4000}{x} + 3$ **30.** $D(x) = \dfrac{3000}{x} + 5$

31. In Section 2.5, we defined the average cost of producing x units of a product in terms of the total cost $C(x)$ by

$$A(x) = \frac{C(x)}{x}.$$

Use the Quotient Rule to find a general expression for *marginal average cost* $A'(x)$.

32. In this section we determined that

$$R(x) = xD(x).$$

Then

$$D(x) = \frac{R(x)}{x} = \text{the average revenue from the sale of } x \text{ units.}$$

Use the Quotient Rule to find a general expression for *marginal average revenue* $D'(x)$.

▶

Differentiate each function.

33. $f(x) = \dfrac{x^3}{\sqrt{x} - 5}$ **34.** $g(t) = \dfrac{1 + \sqrt{t}}{t^5 + 3}$ **35.** $f(v) = \dfrac{3}{1 + v + v^2}$ **36.** $g(z) = \dfrac{1 + z + z^2}{1 - z + z^2}$

37. $p(t) = \dfrac{t}{1 - t + t^2 - t^3}$ **38.** $f(x) = \dfrac{\dfrac{2}{3x} - 1}{\dfrac{3}{x^2} + 5}$ **39.** $h(x) = \dfrac{x^3 + 5x^2 - 2}{\sqrt{x}}$ **40.** $y(t) = 5t(t - 1)(2t + 3)$

41. $f(x) = x(3x^3 + 6x - 2)(3x^4 + 7)$ **42.** $g(x) = (x^3 - 8) \cdot \dfrac{x^2 + 1}{x^2 - 1}$ **43.** $f(t) = (t^5 + 3) \cdot \dfrac{t^3 - 1}{t^3 + 1}$

44. $f(x) = \dfrac{(x^2 + 3x)(x^5 - 7x^2 - 3)}{x^4 - 3x^3 - 5}$ **45.** $f(x) = \dfrac{(2x^2 + 3)(4x^3 - 7x + 2)}{x^7 - 2x^6 + 9}$ **46.** $s(t) = \dfrac{5t^8 - 2t^3}{(t^5 - 3)(t^4 + 7)}$

OBJECTIVE

You should be able to differentiate using the Extended Power Rule.

2.7 THE EXTENDED POWER RULE AND THE CHAIN RULE

The Extended Power Rule

How do we differentiate more complicated functions such as

$$y = (1 + x^2)^3, \qquad y = (1 + x^2)^{89}, \quad \text{or} \quad y = (1 + x^2)^{1/3}?$$

For $(1 + x^2)^3$ we can expand and then differentiate, but while this could be done for $(1 + x^2)^{89}$, it would certainly be time-consuming, and such an expansion of the Power Rule would not work for $(1 + x^2)^{1/3}$. Not knowing this, we might conjecture that the derivative of the function $y = (1 + x^2)^3$ is

$$3(1 + x^2)^2. \tag{1}$$

Differentiate.

1. $f(x) = (1 + x^2)^{10}$

[handwritten:] $10(1+x^2)^9 \cdot 2x$

$20x(1+x^2)^9$

2. $y = (1 - x^2)^{1/2}$

[handwritten:] $\frac{1}{2}(1-x^2)^{-1/2} \cdot -2x$

$\frac{-x}{\sqrt{(1+x^2)^{1/2}}}$

To check this, we expand $(1 + x^2)^3$ and then differentiate. From Section 1.1, we know that $(a + h)^3 = a^3 + 3a^2h + 3ah^2 + h^3$, so

$$(1 + x^2)^3 = 1^3 + 3 \cdot 1^2 \cdot (x^2)^1 + 3 \cdot 1 \cdot (x^2)^2 + (x^2)^3$$

$$= 1 + 3x^2 + 3x^4 + x^6.$$

(We could also have done this by finding $(1 + x^2)^2$ and then multiplying again by $1 + x^2$.) It follows that

$$\frac{dy}{dx} = 6x + 12x^3 + 6x^5 = (1 + 2x^2 + x^4)6x$$

$$= 3(1 + x^2)^2 \cdot 2x. \qquad (2)$$

Comparing this with Eq. (1), we see that the Power Rule is not sufficient for such a differentiation. Note that the factor $2x$ in the actual derivative Eq. (2) is the derivative of the "inside" function, $1 + x^2$. This is consistent with the following new rule.

THEOREM 7   The Extended Power Rule

Suppose $g(x)$ is a function of x. Then for any real number a,

$$\frac{d}{dx}[g(x)]^a = a[g(x)]^{a-1} \cdot \frac{d}{dx}g(x).$$

[handwritten:] $5(1+x^3)^4 \cdot \frac{d}{dx}x^2$

Let us differentiate $(1 + x^3)^5$. There are three steps to carry out.

$\boxed{1 + x^3}\ ^5$ **1.** Mentally block out the "inside" function $1 + x^3$.

$5\boxed{1 + x^3}\ ^4$ **2.** Differentiate the "outside" function $\boxed{1 + x^3}\ ^5$.

$5(1 + x^3)^4 \cdot 3x^2$ **3.** Multiply by the derivative of the "inside"
$= 15x^2(1 + x^3)^4$ function.

Step 3 is most commonly overlooked. Try not to forget it!

Example 1

$$\frac{d}{dx}(1 + x^3)^{1/2} = \frac{1}{2}(\,\boxed{1 + x^3}\,)^{1/2-1} \cdot 3x^2 = \frac{1}{2}(1 + x^3)^{-1/2} \cdot 3x^2$$

$$= \frac{3x^2}{2\sqrt{1 + x^3}}$$

DO EXERCISES 1 AND 2.

Example 2 Differentiate $y = (1 - x^2)^3 - (1 - x^2)^2$.

3. Differentiate:

$$f(x) = (1 + x^2)^2 - (1 + x^2)^3.$$

Solution Here we combine the Difference Rule and the Extended Power Rule:

$$\frac{dy}{dx} = 3(1 - x^2)^2(-2x) - 2(1 - x^2)(-2x)$$

We differentiate each term using the Extended Power Rule.

$$= -6x(\,1 - x^2\,)^2 + 4x(\,1 - x^2\,)$$

$$= x(\,1 - x^2\,)[-6(1 - x^2) + 4]$$

Here we factor out $x(1 - x^2)$.

$$= x(1 - x^2)[-6 + 6x^2 + 4]$$

$$= x(1 - x^2)(6x^2 - 2) = 2x(1 - x^2)(3x^2 - 1).$$

DO EXERCISE 3.

Example 3 Differentiate $f(x) = (x - 5)^4(7 - x)^{10}$.

Solution Here we combine the Product Rule and the Extended Power Rule:

$$f'(x) = (x - 5)^4 10(7 - x)^9(-1) + 4(x - 5)^3(7 - x)^{10}$$

$$= -10(x - 5)^4(7 - x)^9 + 4(x - 5)^3(7 - x)^{10}$$

$$= (x - 5)^3(7 - x)^9[-10(x - 5) + 4(7 - x)]$$

We factored out $(x - 5)^3(7 - x)^9$.

$$= (x - 5)^3(7 - x)^9[-10x + 50 + 28 - 4x]$$

$$= (x - 5)^3(7 - x)^9(78 - 14x)$$

$$= 2(x - 5)^3(7 - x)^9(39 - 7x).$$

DO EXERCISE 4.

4. Differentiate:

$$y = (x - 4)^5(6 - x)^3.$$

Example 4 Differentiate $f(x) = \sqrt[4]{\dfrac{x + 3}{x - 1}}$.

Solution We must use the Quotient Rule to differentiate the inside function $(x + 3)/(x - 1)$:

$$\frac{d}{dx}\sqrt[4]{\frac{x + 3}{x - 1}} = \frac{d}{dx}\left(\frac{x + 3}{x - 1}\right)^{1/4} = \frac{1}{4}\left(\frac{x + 3}{x - 1}\right)^{1/4 - 1}\left[\frac{(x - 1)1 - 1(x + 3)}{(x - 1)^2}\right]$$

$$= \frac{1}{4}\left(\frac{x + 3}{x - 1}\right)^{-3/4}\left[\frac{x - 1 - x - 3}{(x - 1)^2}\right]$$

$$= \frac{1}{4}\left(\frac{x + 3}{x - 1}\right)^{-3/4} \cdot \frac{-4}{(x - 1)^2}$$

$$= \left(\frac{x + 3}{x - 1}\right)^{-3/4} \cdot \frac{-1}{(x - 1)^2}.$$

5. Differentiate:

$$y = \sqrt[3]{\frac{x + 5}{x - 4}}.$$

DO EXERCISE 5.

The Chain Rule

The Extended Power Rule is a special case of a more general rule called the *Chain Rule*. Before discussing it, we shall define the *composition* of functions. Consider the following, for example:

$$f(x) = x^3 \qquad \text{This function cubes each input.}$$

and

$$g(x) = 1 + x^2 \qquad \text{This function adds 1 to the square of each input.}$$

We define a new function that first does what g does (adds 1 to the square) and then does what f does (cubes). The new function is called the *composition* of f and g and is symbolized $f(g(x))$. We can visualize the composition of functions as follows.

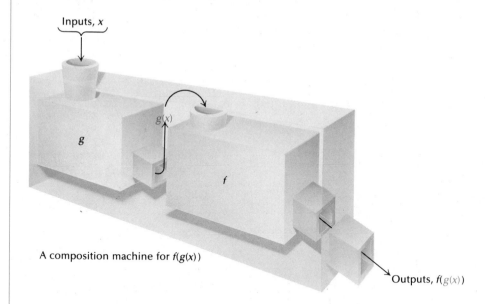

A composition machine for $f(g(x))$

Example 5 Given $f(x) = x^3$ and $g(x) = 1 + x^2$, find $f(g(x))$ and $g(f(x))$.

Solution We find $f(g(x))$ by substituting g(x) for x:

$$f(g(x)) = f(1 + x^2) \qquad \text{Substituting } 1 + x^2 \text{ for } x$$

$$= (1 + x^2)^3 = 1 + 3x^2 + 3x^4 + x^6.$$

6. Given $f(x) = 3x$ and $g(x) = x^2 - 1$, find $f(g(x))$ and $g(f(x))$.

We find $g(f(x))$ by substituting $f(x)$ for x:

$$g(f(x)) = g(x^3) \qquad \text{Substituting } x^3 \text{ for } x$$
$$= 1 + (x^3)^2$$
$$= 1 + x^6.$$

DO EXERCISE 6.

Example 6 Given $f(x) = \sqrt{x}$ and $g(x) = x - 1$, find $f(g(x))$ and $g(f(x))$.

Solution

$$f(g(x)) = f(x - 1) = \sqrt{x - 1}, \qquad g(f(x)) = g(\sqrt{x}) = \sqrt{x} - 1$$

DO EXERCISE 7.

THEOREM 8   The Chain Rule

The derivative of the composition $f(g(x))$ is given by

$$\frac{d}{dx} f(g(x)) = f'(g(x)) \cdot \frac{d}{dx} g(x).$$

7. Given $f(x) = 4x + 5$ and $g(x) = \sqrt[3]{x}$, find $f(g(x))$ and $g(f(x))$.

Note how the Extended Power Rule is a special case.

$$\frac{d}{dx} [g(x)]^a = a[g(x)]^{a-1} \cdot \frac{d}{dx} g(x)$$

The Chain Rule often appears in another form. Suppose $y = f(u)$ and $u = g(x)$. Then

$$\frac{dy}{dx} = \frac{dy}{du} \cdot \frac{du}{dx}.$$

For example, if $y = 2 + \sqrt{u}$ and $u = x^3 + 1$, then

$$\frac{dy}{du} = \frac{1}{2} u^{-1/2} \quad \text{and} \quad \frac{du}{dx} = 3x^2,$$

so

$$\frac{dy}{dx} = \frac{dy}{du} \cdot \frac{du}{dx}$$

$$= \frac{1}{2\sqrt{u}} \cdot 3x^2$$

$$= \frac{3x^2}{2\sqrt{x^3 + 1}}. \qquad \text{Substituting } x^3 + 1 \text{ for } u$$

EXERCISE SET 2.7

Differentiate.

1. $y = (1 - x)^{55}$ **2.** $y = (1 - x)^{100}$ **3.** $y = \sqrt{1 + 8x}$ **4.** $y = \sqrt{1 - x}$ **5.** $y = \sqrt{3x^2 - 4}$

6. $y = \sqrt{4x^2 + 1}$ **7.** $y = (3x^2 - 6)^{-40}$ **8.** $y = (4x^2 + 1)^{-50}$ **9.** $y = x\sqrt{2x + 3}$ **10.** $y = x\sqrt{4x - 7}$

11. $y = x^2\sqrt{x - 1}$ **12.** $y = x^3\sqrt{x + 1}$ **13.** $y = \dfrac{1}{(3x + 8)^2}$ **14.** $y = \dfrac{1}{(4x + 5)^2}$

15. $f(x) = (1 + x^3)^3 - (1 + x^3)^4$ **16.** $f(x) = (1 + x^3)^5 - (1 + x^3)^4$ **17.** $f(x) = x^2 + (200 - x)^2$

18. $f(x) = x^2 + (100 - x)^2$ **19.** $f(x) = (x + 6)^{10}(x - 5)^4$ **20.** $f(x) = (x - 4)^8(x + 3)^9$

21. $f(x) = (x - 4)^8(3 - x)^4$ **22.** $f(x) = (x + 6)^{10}(5 - x)^9$ **23.** $f(x) = -4x(2x - 3)^3$

24. $f(x) = -5x(3x + 5)^6$ **25.** $f(x) = \sqrt{\dfrac{1 - x}{1 + x}}$ **26.** $f(x) = \sqrt{\dfrac{3 + x}{2 - x}}$

27. Consider

$$f(x) = \frac{x^2}{(1 + x)^5}.$$

a) Find $f'(x)$ using the Quotient Rule and the Extended Power Rule.

b) Note that $f(x) = x^2(1 + x)^{-5}$. Find $f'(x)$ using the Product Rule and the Extended Power Rule.

c) Compare answers to (a) and (b).

28. Consider

$$g(x) = (x^3 + 5x)^2.$$

a) Find $g'(x)$ using the Extended Power Rule.

b) Note that $g(x) = x^6 + 10x^4 + 25x^2$. Find $g'(x)$.

c) Compare answers to (a) and (b).

29. A total cost function is given by

$$C(x) = 1000\sqrt{x^3 + 2}.$$

Find the marginal cost $C'(x)$.

30. A total revenue function is given by

$$R(x) = 2000\sqrt{x^2 + 3}.$$

Find the marginal revenue $R'(x)$.

31. *Business: Compound interest.* If $1000 is invested at interest rate i, compounded annually, in 3 years it will grow to amount A given by

$$A = \$1000(1 + i)^3.$$

(See Section 1.1.) Find the rate of change dA/di.

32. *Business: Compound interest.* If $1000 is invested at interest rate i, compounded quarterly, in 5 years it will grow to amount A given by

$$A = \$1000\left(1 + \frac{i}{4}\right)^{20}.$$

Find the rate of change dA/di.

▶

Differentiate the following functions.

33. $y = \sqrt[3]{x^3 - 6x + 1}$

34. $s = \sqrt[4]{t^4 + 3t^2 + 8}$

35. $y = \dfrac{x}{\sqrt{x - 1}}$

36. $y = \dfrac{(x + 1)^2}{(x^2 + 1)^3}$

37. $u = \dfrac{(1 + 2v)^4}{v^4}$

38. $y = x\sqrt{1 + x^2}$

39. $y = \dfrac{\sqrt{1 - x^2}}{1 - x}$

40. $w = \dfrac{u}{\sqrt{1 + u^2}}$

41. $y = \left(\dfrac{x^2 - x - 1}{x^2 + 1}\right)^3$

42. $y = \sqrt{1 + \sqrt{x}}$

43. $s = \dfrac{\sqrt{t} - 1}{\sqrt{t} + 1}$

44. $y = x^{2/3} \cdot \sqrt[3]{1 + x^2}$

OBJECTIVE

You should be able to find a higher-order derivative.

1. Find the first six derivatives of

$$f(x) = 2x^6 - x^5 + 10.$$

$'\quad 12x^5 - 5x^4$

$''\quad 60x^4 - 20x^3$

$'''\quad 240x^3 - 60x^2$

$''''\quad 720x^2 - 120x$

$'''''\quad 1440x - 120$

$''''''\quad 1440$

2.8 HIGHER DERIVATIVES

Consider the function given by

$$y = f(x)$$
$$= x^5 - 3x^4 + x.$$

Its derivative f' is given by

$$y' = f'(x)$$
$$= 5x^4 - 12x^3 + 1.$$

This function f' can be differentiated. We use the notation f'' for the derivative $(f')'$. We call f'' the *second derivative* of f. It is given by

$$y'' = f''(x) = 20x^3 - 36x^2.$$

Continuing in this manner, we have

$$f'''(x) = 60x^2 - 72x, \qquad \text{The third derivative of } f$$
$$f''''(x) = 120x - 72, \qquad \text{The fourth derivative of } f$$
$$f'''''(x) = 120. \qquad \text{The fifth derivative of } f$$

When notation, like $f'''''(x)$, gets lengthy we can abbreviate it using a numeral in parentheses. Thus $f^{(n)}(x)$ is the nth derivative. For the above function,

$$f^{(4)}(x) = 120x - 72, \qquad f^{(5)}(x) = 120, \qquad f^{(6)}(x) = 0,$$

and $f^{(n)}(x) = 0$, for any $n \geq 6$.

DO EXERCISE 1.

Leibniz's notation for the second derivative of a function given by $y = f(x)$ is

$$\frac{d^2y}{dx^2} \quad \text{or} \quad \frac{d}{dx}\left(\frac{dy}{dx}\right),$$

read "the second derivative of y with respect to x." The 2's in this notation are *not* exponents. If $y = x^5 - 3x^4 + x$, then

$$\frac{d^2y}{dx^2} = 20x^3 - 36x^2.$$

Leibniz's notation for the third derivative is d^3y/dx^3, for the fourth

2. For

$$y = x^7 - x^3,$$

find:

a) $\dfrac{dy}{dx}$;

b) $\dfrac{d^2y}{dx^2}$;

c) $\dfrac{d^3y}{dx^3}$;

d) $\dfrac{d^4y}{dx^4}$.

3. For $y = \dfrac{2}{x}$, find $\dfrac{d^2y}{dx^2}$.

4. For $y = (x^2 - 12x)^{30}$, find y' and y''.

derivative d^4y/dx^4, and so on:

$$\frac{d^3y}{dx^3} = 60x^2 - 72x, \qquad \frac{d^4y}{dx^4} = 120x - 72, \qquad \frac{d^5y}{dx^5} = 120.$$

DO EXERCISE 2.

Example 1 For $y = 1/x$, find d^2y/dx^2.

Solution We have $y = x^{-1}$, so

$$\frac{dy}{dx} = -1 \cdot x^{-1-1} = -x^{-2}, \quad \text{or} \quad -\frac{1}{x^2}.$$

Then

$$\frac{d^2y}{dx^2} = (-2)(-1)x^{-2-1} = 2x^{-3}, \quad \text{or} \quad \frac{2}{x^3}.$$

DO EXERCISE 3.

Example 2 For $y = (x^2 + 10x)^{20}$, find y' and y''.

Solution To find y', we use the Extended Power Rule:

$$y' = 20(x^2 + 10x)^{19}(2x + 10) = 20(x^2 + 10x)^{19} \cdot 2(x + 5)$$

$$= 40(x^2 + 10x)^{19}(x + 5)$$

To find y'', we use the Extended Power Rule and the Product Rule:

$$y'' = 19 \cdot 40(x^2 + 10x)^{18}(2x + 10)(x + 5) + 40(x^2 + 10x)^{19}(1)$$

$$= 760(x^2 + 10x)^{18} \cdot 2(x + 5)(x + 5) + 40(x^2 + 10x)^{19}$$

$$= 1520(x^2 + 10x)^{18}(x + 5)^2 + 40(x^2 + 10x)^{19}$$

$$= 40(x^2 + 10x)^{18}[38(x + 5)^2 + (x^2 + 10x)]$$

$$= 40(x^2 + 10x)^{18}[38(x^2 + 10x + 25) + x^2 + 10x]$$

$$= 40(x^2 + 10x)^{18}[39x^2 + 390x + 950].$$

DO EXERCISE 4.

Acceleration can be thought of as a second derivative. As an object moves, its distance from a fixed point after time t is some function of the

5. For $s(t) = 3t + t^4$, find the acceleration $a(t)$.

time, say $s(t)$. Then

$$v(t) = s'(t) = \text{the velocity at time } t,$$

and

$$a(t) = v'(t) = s''(t) = \text{the acceleration at time } t.$$

Whenever a quantity is a function of time, the first derivative gives the rate of change with respect to time and the second derivative gives the acceleration. For example, if $y = P(t)$ gives the number of people in a population at time t, then $P'(t)$ gives how fast the size of the population is changing and $P''(t)$ gives the acceleration in the size of the population.

DO EXERCISE 5.

EXERCISE SET 2.8

Find d^2y/dx^2 for each of the following.

1. $y = 3x + 5$

2. $y = -4x + 7$

3. $y = -\dfrac{1}{x}$

4. $y = -\dfrac{3}{x}$

5. $y = x^{1/4}$

6. $y = \sqrt{x}$

7. $y = x^4 + \dfrac{4}{x}$

8. $y = x^3 - \dfrac{3}{x}$

9. $y = x^{-3}$

10. $y = x^{-4}$

11. $y = x^n$

12. $y = x^{-n}$

13. $y = x^4 - x^2$

14. $y = x^4 + x^3$

15. $y = \sqrt{x - 1}$

16. $y = \sqrt{x + 1}$

17. $y = ax^2 + bx + c$

18. $y = (x^3 + 15x)^{20}$

19. For $y = x^4$, find d^4y/dx^4.

20. For $y = x^5$, find d^4y/dx^4.

21. For $y = x^6 - x^3 + 2x$, find d^5y/dx^5.

22. For $y = x^7 - 8x^2 + 2$, find d^6y/dx^6.

23. For $y = (x^2 - 5)^{10}$, find d^2y/dx^2.

24. For $y = x^k$, find d^5y/dx^5.

25. If s is a distance given by $s(t) = t^3 + t^2 + 2t$, find the acceleration.

26. If s is a distance given by $s(t) = t^4 + t^2 + 3t$, find the acceleration.

27. A population grows from an initial size of 100,000 to an amount $P(t)$ given by

$$P(t) = 100{,}000(1 + 0.6t + t^2).$$

What is the acceleration in the size of the population?

28. A population grows from an initial size of 100,000 to an amount $P(t)$ given by

$$P(t) = 100{,}000(1 + 0.4t + t^2).$$

What is the acceleration in the size of the population?

Find y', y'', and y'''.

29. $y = x^{-1} + x^{-2}$

30. $y = \dfrac{1}{1 - x}$

31. $y = x\sqrt{1 + x^2}$

32. $y = 3x^5 + 8\sqrt{x}$

33. $y = \dfrac{3x - 1}{2x + 3}$ **34.** $y = \dfrac{1}{\sqrt{x - 1}}$ **35.** $y = \dfrac{x}{\sqrt{x - 1}}$ **36.** $y = \dfrac{\sqrt{x} - 1}{\sqrt{x} + 1}$

Find $f''(x)$.

37. $f(x) = \dfrac{x}{x - 1}.$ **38.** $f(x) = \dfrac{1}{1 + x^2}$

A summary of the important formulas for this chapter is given on the inside front cover.

CHAPTER 2 TEST

Which functions are continuous?

1.

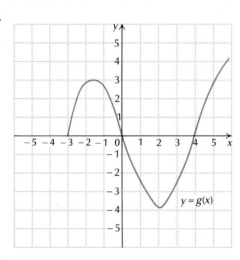

2.

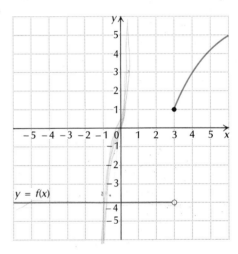

For the function in Question 2, answer the following.

3. Find $\lim\limits_{x \to 3} f(x)$. *does not exist*

4. Find $f(3)$. *1*

5. Is f continuous at 3? *N*

6. Find $\lim\limits_{x \to 4} f(x)$. *3*

7. Find $f(4)$. *3*

8. Is f continuous at 4? *Y*

Find each limit, if it exists.

9. $\lim\limits_{x \to 1} 3x^4 - 2x^2 + 5$ *6*   *3 - 2 + 5 = 6*

10. $\lim\limits_{x \to 1} \dfrac{x - 1}{x^2 - 1}$ *= 0*

11. $\lim\limits_{x \to 0} \dfrac{7}{x}$ *∅*

12. $\lim\limits_{x \to \infty} \dfrac{4x - 3}{x}$

13. Find a simplified difference quotient for

$$f(x) = 3x^2 + 1.$$

14. Find the points on the graph of

$$y = x^3 - 3x^2$$

at which the tangent line is horizontal.

Find dy/dx for each of the following

15. $y = x^{84}$ **16.** $y = 10\sqrt{x}$ **17.** $y = \dfrac{-10}{x}$ **18.** $y = x^{5/4}$ **19.** $y = -0.5x^2 + 0.61x + 90$

Differentiate.

20. $y = \dfrac{1}{3}x^3 - x^2 + 2x + 4$ **21.** $y = \dfrac{2x - 5}{x^4}$ **22.** $f(x) = \dfrac{x}{5 - x}$

23. $f(x) = (x + 3)^4(7 - x)^5$ **24.** $y = (x^5 - 4x^3 + x)^{-5}$ **25.** $f(x) = x\sqrt{x^2 + 5}$

26. For $y = x^4 - 3x^2$, find d^3y/dx^3.

27. Given $R(x) = 50x$ and $C(x) = 0.001x^2 + 1.2x + 60$, find:

a) $P(x)$;

b) $R(10)$, $C(10)$, $P(10)$;

c) $R'(x)$, $C'(x)$, $P'(x)$;

d) $R'(10)$, $C'(10)$, $P'(10)$.

28. In a certain memory experiment a person is able to memorize M words after t minutes, where

$$M = -0.001t^3 + 0.1t^2.$$

a) Find the rate of change of the number of words memorized with respect to time.

b) How many words are memorized the first 10 minutes (at $t = 10$)?

c) What is the memory rate at $t = 10$ minutes?

29. Find $\lim\limits_{x \to 2} \dfrac{x^3 - 8}{x - 2}$.

30. Differentiate $y = (1 - 3x)^{2/3}(1 + 3x)^{1/3}$.

APPLICATIONS OF
DIFFERENTIATION

3

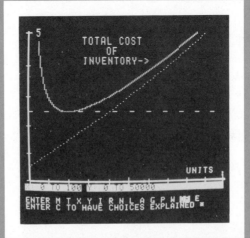

There are costs involved in keeping an inventory of products to sell.
(Cary Wolinsky: Stock, Boston)

This is the graph of a typical inventory cost function. Its minimum occurs where there is a horizontal tangent.

OBJECTIVES

You should be able to find maximum and minimum values of functions.

3.1 THE SHAPE OF A GRAPH: FINDING MAXIMUM AND MINIMUM VALUES

First and second derivatives give us information about the shape of a graph that may be relevant in finding maximum and minimum values of functions. Throughout this section we will assume that the functions are continuous.

Increasing and Decreasing Functions

If the graph of a function rises from left to right, it is said to be *increasing*. If the graph drops from left to right, it is said to be *decreasing*.

Examples

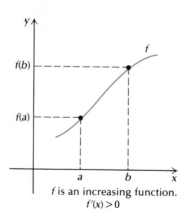

f is an increasing function.
$f'(x) > 0$

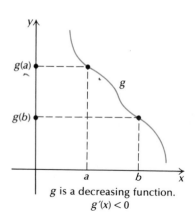

g is a decreasing function.
$g'(x) < 0$

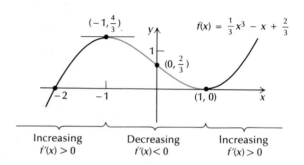

1. Consider the following graph.

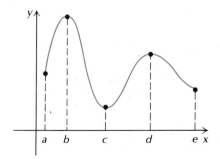

a) Over what intervals is the function increasing?

b) Over what intervals is the derivative positive? To determine this, locate a straightedge at points on the graph and decide whether slopes of tangent lines are positive.

c) Over what intervals is the function decreasing?

d) Over what intervals is the derivative negative?

Look for a pattern.

2. Consider the following graph.

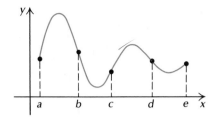

a) Over what intervals is the derivative increasing?

b) Over what intervals is the derivative decreasing?

We have seen in Chapter 1 how the slope of a linear function determines whether it is increasing or decreasing (or neither). For a general function, the derivative yields similar information. Let us investigate in the margin exercise how this happens.

DO EXERCISE 1.

The following is how we can use derivatives to determine whether a function is increasing or decreasing.

THEOREM 1
_____

If $f'(x) > 0$, for all x in an interval I, then f is increasing over I.
If $f'(x) < 0$, for all x in an interval I, then f is decreasing over I.

_____

Concavity: Increasing and Decreasing Derivatives

The following are two functions. The graph on the left is turning upward and the other is turning downward. Let's see if we can relate this to their derivatives.

Consider the graph of f. Take a ruler, or straightedge, and move along the curve from left to right. What happens to the slopes of the tangent lines? Do the same for the graph of g. Look for a pattern.

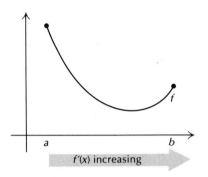

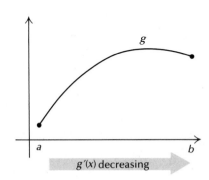

DO EXERCISE 2.

3. Consider the following graph.

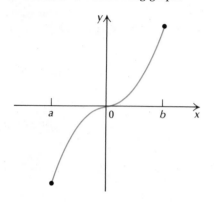

a) Over what intervals is the graph concave up?

b) Over what intervals is the graph concave down?

4. Which are points of inflection?

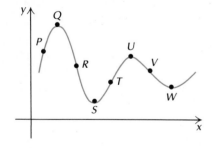

We have the following.

THEOREM 2

1. If $f''(x) > 0$ on an interval I, then f is turning upward on I (since f' is increasing on I). Such a graph is said to be *concave up over I*.

2. If $f''(x) < 0$ on an interval I, then f is turning downward on I (since f' is decreasing on I). Such a graph is said to be *concave down over I*.

The following is a helpful memory device.

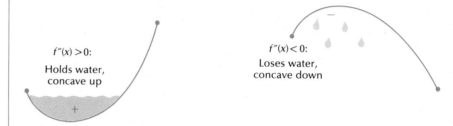

DO EXERCISE 3.

A *point of inflection*, or an *inflection point*, is a point across which the direction of concavity changes. For example, in the figure below, point P is an inflection point of the graph on the left. Points P, Q, R, and S are inflection points of the graph on the right. In Margin Exercise 3, the point $(0, 0)$ is an inflection point.

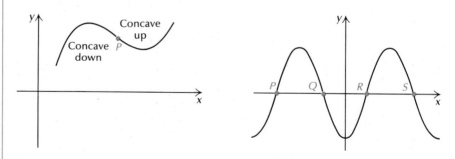

DO EXERCISE 4.

Just knowing the values of f' and f'' at some specific point x_0 can yield a lot of information about the shape of the graph over some (possibly small) interval containing x_0 as an interior point (assuming f'' exists and is continuous over the interval).

	Derivatives	Effect on the Graph at $x = x_0$	Shape of the Graph at $x = x_0$
1.	$f'(x_0) > 0,$ $f''(x_0) > 0$	f is increasing at x_0. The graph of f is concave up (over some interval containing x_0).	
2.	$f'(x_0) > 0,$ $f''(x_0) < 0$	f is increasing at $x = x_0$. The graph of f is concave down.	
3.	$f'(x_0) < 0,$ $f''(x_0) > 0$	f is decreasing at $x = x_0$. The graph of f is concave up.	
4.	$f'(x_0) < 0,$ $f''(x_0) < 0$	f is decreasing at $x = x_0$. The graph of f is concave down.	

	Derivatives	Effect on the Graph at $x = x_0$	Shape of the Graph at $x = x_0$
5.	$f'(x_0) = 0,$ $f''(x_0) > 0$	f' is negative to the left of x_0 and positive to the right of x_0. The graph of f is concave up.	
6.	$f'(x_0) = 0,$ $f''(x_0) < 0$	f' is positive to the left of x_0 and negative to the right of x_0. The graph of f is concave down.	

Critical Points

DEFINITION

A *critical point* of a function is an interior point c of its domain at which the function has a horizontal tangent, or at which the derivative does not exist. That is, c is a critical point if

$$f'(c) = 0 \quad \text{or} \quad f'(c) \text{ does not exist.}$$

Consider the following graph.

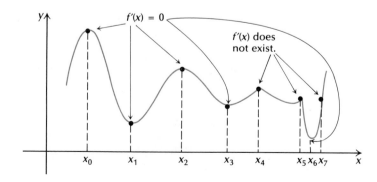

5. Consider this graph.

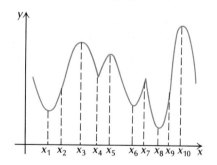

a) At which points are there horizontal tangents?

b) At which points does the derivative not exist?

c) Which are critical points?

6. Try to draw a graph of a continuous function from P to Q that increases on part or parts of $[a, b]$ and decreases on part or parts of $[a, b]$.

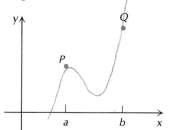

Does the function you drew have any critical points between a and b?

7. Now try to draw a graph of a continuous function from P to Q that increases on part or parts of $[a, b]$, and decreases on part or parts of $[a, b]$, but in such a way that no critical points occur between a and b.

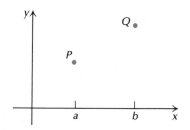

The points x_0, x_1, x_2, x_3 and x_6 are all critical points because the derivative is 0 at each of these points. The points x_4, x_5, and x_7 are all critical points because the derivative does not exist at these points.

DO EXERCISE 5.

The Shape of a Graph Between Critical Points and Endpoints

Suppose we have a continuous function defined over an interval $[a, b]$.

DO EXERCISES 6 AND 7.

Consider the following graph.

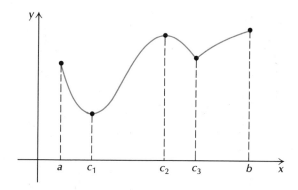

We have three critical points: c_1, c_2, and c_3. We will refer to these, together with the endpoints, as *key* points. That is, the key points are

$$a, \quad b, \quad c_1, \quad c_2, \quad c_3.$$

Note, in the foregoing graph, that between any two consecutive key points the function is either increasing or decreasing.

This graph and the experience with Margin Exercises 6 and 7 lead us to the following principle.

SHAPE PRINCIPLE

Suppose f is a continuous function over an interval $[a, b]$. Then between any two consecutive key points (a, b, plus critical points c_1, c_2, c_3, . . . c_n) the function is either increasing or decreasing.

---

8. In each graph find the points at which maximum and minimum values occur.

a)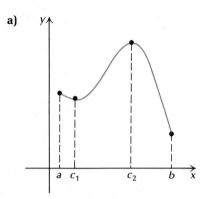

b)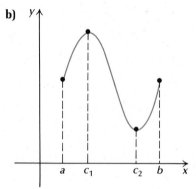

Finding Maximum and Minimum Values

Consider the function f whose graph over the interval $[a, b]$ is shown in Fig. 1. The function value $f(c_1)$ is called a *minimum* value of the function, and $f(b)$ is called a *maximum value* of the function.

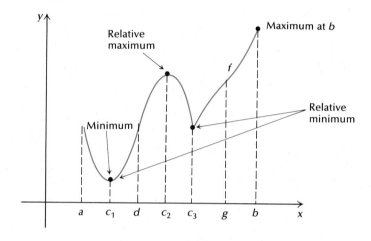

Figure 1

DEFINITION

A function f on an interval $[a, b]$ has a *maximum* at x_0 if

$$f(x_0) \geq f(x) \quad \text{for all } x \text{ in } [a, b].$$

A function f on an interval $[a, b]$ has a *minimum* at x_0 if

$$f(x_0) \leq f(x) \quad \text{for all } x \text{ in } [a, b].$$

DO EXERCISE 8.

In Fig. 1, note that at $x = c_2$ the graph changes from increasing to decreasing. We say that the function has a *relative maximum* at $x = c_2$. At $x = c_3$ the graph changes from decreasing to increasing. We say that the function has a *relative minimum* at $x = c_3$. Note that the function also has a relative minimum at $x = c_1$, the same point at which it has a minimum.

9. Consider the graphs in Margin Exercise 8.

a) At which points in the first graph does the function have a relative maximum?

b) At which points in the first graph does the function have a relative minimum?

c) At which points in the second graph does the function have a relative maximum?

d) At which points in the second graph does the function have a relative minimum?

DEFINITION

A function f has a *relative maximum* at $x = x_0$ if the graph changes from

increasing to decreasing at $x = x_0$ and x_0 is a critical point.

A function has a *relative minimum* at $x = x_0$ if the graph changes from

decreasing to increasing at $x = x_0$ and x_0 is a critical point.

DO EXERCISE 9.

You may have discovered two theorems. We will not consider their proofs. The first is as follows.

THEOREM 3

A continuous function f defined on a closed interval $[a, b]$ must have a maximum and minimum value at points in $[a, b]$.

The second is a modification of the Shape Principle.

THEOREM 4   Maximum–Minimum Principle 1*

Suppose f is a continuous function over an interval $[a, b]$. To find the maximum and minimum values of the function,

a) First find $f'(x)$.

b) Then find the critical points. That is, find all points c for which

$$f'(c) = 0 \quad \text{or} \quad f'(c) \text{ does not exist.}$$

c) Determine the key points. If $c_1, c_2, \ldots, c_n$ are the critical points and the interval is $[a, b]$, then the key points are

$$a, b, c_1, c_2, \ldots, c_n.$$

d) Find the function values at the key points:

$$f(a), f(b), f(c_1), f(c_2), \ldots, f(c_n).$$

The largest of these is the *maximum* of f on the interval $[a, b]$.
The smallest of these is the *minimum* of f on the interval $[a, b]$.

This follows from the Shape Principle because between two key points the function is either increasing or decreasing. Thus whatever the maximum and minimum values are, they occur among function values of the key points.

---

*Also called the *First Derivative Test*.

Example 1 Find the maximum and minimum values of

$$f(x) = x^3 - 3x + 2$$

on the interval $[-2, \frac{3}{2}]$.

Solution

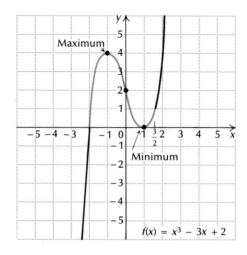

$f(x) = x^3 - 3x + 2$

a) Find $f'(x)$:

$$f'(x) = 3x^2 - 3.$$

b) Find the critical points. The derivative exists for all real numbers. Thus we merely solve $f'(x) = 0$:

$$3x^2 - 3 = 0$$
$$3x^2 = 3$$
$$x^2 = 1$$
$$x = \pm 1.$$

c) Determine the key points. The key points are $-2, \frac{3}{2}, -1, 1$.

10. Find the maximum and minimum values of

$$f(x) = x^3 - x^2 - x + 2$$

on the interval $[-1, 2]$.   `CSS`

d) Find the function values at the key points:

$$f(-2) = (-2)^3 - 3(-2) + 2 = -8 + 6 + 2 = 0; \qquad \text{Minimum}$$

$$f\left(\frac{3}{2}\right) = \left(\frac{3}{2}\right)^3 - 3\left(\frac{3}{2}\right) + 2 = \frac{27}{8} - \frac{9}{2} + 2 = \frac{27}{8} - \frac{36}{8} + \frac{16}{8} = \frac{7}{8};$$

$$f(-1) = (-1)^3 - 3(-1) + 2 = -1 + 3 + 2 = 4; \qquad \text{Maximum}$$

$$f(1) = \qquad (1)^3 - 3(1) + 2 = 1 - 3 + 2 = 0. \qquad \text{Minimum}$$

The largest of these values, 4, is the maximum. It occurs at $x = -1$. The smallest of these values is 0. It occurs twice at $x = -2$ and $x = 1$. Thus the

$$\text{Maximum} = 4 \text{ at } x = -1$$

and the

$$\text{Minimum} = 0 \text{ at } x = -2 \text{ and } x = 1.$$

DO EXERCISE 10.

Example 2 Find the maximum and minimum values of

$$f(x) = x^3 - 3x + 2$$

on the interval $\left[-3, -\frac{3}{2}\right]$.

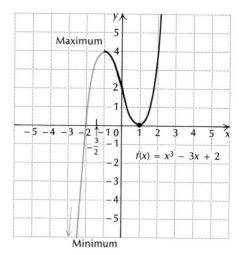

Solution As in Example 1, the derivative is 0 at -1 and 1. But neither -1 nor 1 is in the interval $\left[-3, -\frac{3}{2}\right]$, so there are no critical points in this

11. Find the maximum and minimum values of

$$f(x) = x^3 - x^2 - x + 2$$

on the interval $[5, 6]$.

interval. Thus the maximum and minimum values occur at the endpoints:

$$f(-3) = (-3)^3 - 3(-3) + 2 = -27 + 9 + 2 = -16; \qquad \text{Min}$$

$$f\left(-\frac{3}{2}\right) = \left(-\frac{3}{2}\right)^3 - 3\left(-\frac{3}{2}\right) + 2 = -\frac{27}{8} + \frac{9}{2} + 2 = \frac{25}{8} = 3\frac{1}{8}. \qquad \text{Max}$$

Thus the

$$\text{Maximum} = 3\frac{1}{8} \text{ at } x = -\frac{3}{2}$$

and the

$$\text{Minimum} = -16 \text{ at } x = -3.$$

DO EXERCISE 11.

When there is only one critical point c_0 in I, we may not need to check endpoint values to determine whether the function has a maximum or a minimum at that point. Consider these cases.

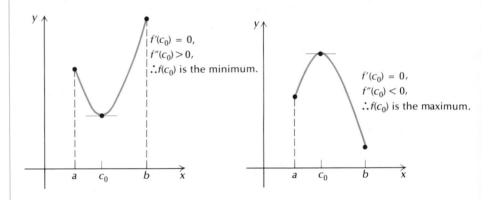

When $f'(c_0) = 0$ and $f''(c_0) > 0$, $f'(x)$ changes from negative to positive as x goes from the left of c_0 to the right. That is, the function f is decreasing to the left of c_0 and increasing to the right of c_0. It follows that $f(c_0)$ is the minimum value of f on I. Similarly, if $f'(c_0) = 0$ and $f''(c_0) < 0$, $f'(x)$ changes from positive to negative as x goes from the left of c_0 to the right. That is, the function f is increasing to the left of c_0 and decreasing to the right of c_0. It follows that $f(c_0)$ is the maximum value of f on I. The above turns out to hold no matter what the interval I, whether it is open, closed, or extends to infinity.

THEOREM 5   **Maximum–Minimum Principle 2***

Suppose f is a function such that $f'(x)$ exists for every x in an interval I, and that there is *exactly one* (critical) point c_0, interior to I, for which $f'(c_0) = 0$. Then

$$f(c_0) \text{ is the maximum value on } I \text{ if } f''(c_0) < 0$$

or

$$f(c_0) \text{ is the minimum value on } I \text{ if } f''(c_0) > 0.$$

If $f''(c_0) = 0$, we would have to use Maximum–Minimum Principle 1, or we would have to know more about the behavior of the function on the given interval.

Example 3 Find the maximum and minimum values of

$$f(x) = 4x - x^2.$$

Solution When no interval is specified, we consider the entire domain of the function. In this case the domain is the set of all real numbers.

a) Find $f'(x)$:

$$f'(x) = 4 - 2x.$$

b) Find the critical points. The derivative exists for all real numbers. Thus we merely solve $f'(x) = 0$:

$$4 - 2x = 0$$
$$-2x = -4$$
$$x = 2.$$

Since there is only one critical point, we can use the second derivative:

$$f''(x) = -2.$$

Now the second derivative is constant, so $f''(2) = -2$, and since this is negative, we have the

$$\text{Maximum} = f(2) = 4 \cdot 2 - 2^2 = 8 - 4 = 4 \text{ at } x = 2.$$

The function has no minimum, as the graph indicates.

---

*Also called the *Second Derivative Test*.

12. Find the maximum and minimum values of

$$f(x) = x^2 - 4x.$$

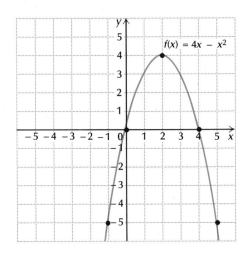

DO EXERCISE 12.

Example 4 Find the maximum and minimum values of $f(x) = 4x - x^2$ on the interval $[0, 4]$.

Solution By the reasoning in Example 3 we know that the maximum value is $f(2)$, or 4. We know this here also, without checking the endpoints. This time we have to check for the minimum:

$$f(0) = 4 \cdot 0 - 0^2 = 0 \quad \text{and} \quad f(4) = 4 \cdot 4 - 4^2 = 0.$$

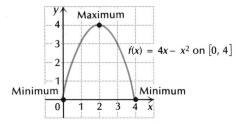

Thus the minimum is 0. It occurs twice at $x = 0$ and $x = 4$. Thus the

$$\text{Maximum} = 4 \text{ at } x = 2$$

and the

$$\text{Minimum} = 0 \text{ at } x = 0 \text{ and } x = 4.$$

13. Find the maximum and minimum values of

$$f(x) = x^2 - 4x$$

on the interval $[0, 4]$.

DO EXERCISE 13.

Keep the following in mind.

> **To find maximum and minimum values:**
>
> a) On a closed interval, you can use Maximum–Minimum Principle 1.
>
> b) On an interval with no endpoints, you can use Maximum–Minimum Principle 2 when there is only one critical point.*
>
> c) On an interval with endpoints, you can use Maximum–Minimum Principle 2 when there is only one critical point and you want a faster method of determining whether a maximum or minimum occurs at the critical point.

Example 5 Find the maximum and minimum values of

$$f(x) = (x - 1)^3 + 2.$$

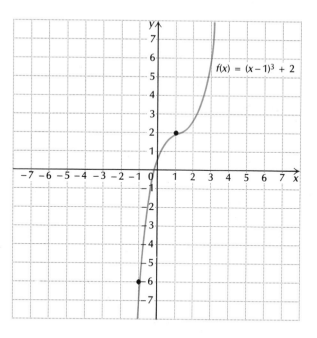

$$f(x) = (x - 1)^3 + 2$$

---

*The case of finding absolute maximum–minimums when more than one critical point occurs on such an interval is beyond the scope of this book.

APPLICATIONS OF DIFFERENTIATION

14. Find the maximum and minimum values of

$$f(x) = x^3.$$

15. Find the maximum and minimum values of

$$f(x) = x^3$$

on the interval $[-2, 2]$. (*Hint:* This function must have maximum and minimum values because it is restricted to a closed interval.) What are the only numbers at which these can occur?

Solution

a) Find $f'(x)$:

$$f'(x) = 3(x - 1)^2.$$

b) Find the critical points. The derivative exists for all real numbers. Thus we solve $f'(x) = 0$:

$$3(x - 1)^2 = 0$$

$$(x - 1)^2 = 0$$

$$x - 1 = 0$$

$$x = 1.$$

Since there is only one critical point and there are no endpoints, we can use the second derivative:

$$f''(x) = 6(x - 1).$$

Now

$$f''(1) = 6(1 - 1) = 0,$$

so Maximum–Minimum Principle 2 fails. We cannot use Maximum–Minimum Principle 1 because there are no endpoints. But note that $f'(x) = 3(x - 1)^2$ is never negative. Thus it is increasing everywhere except at $x = 1$, so there is no maximum or minimum. At $x = 1$, the function has a *point of inflection*.

DO EXERCISES 14 AND 15.

Example 6 Find the maximum and minimum values of $f(x) = 5x + (35/x)$ on the interval $(0, \infty)$.

Solution

a) Find $f'(x)$. We first express $f(x)$ as

$$f(x) = 5x + 35x^{-1}.$$

Then

$$f'(x) = 5 - 35x^{-2} = 5 - \frac{35}{x^2}.$$

b) Now $f'(x)$ exists for all values of x in $(0, \infty)$. Thus the only critical

16. Find the maximum and minimum values of

$$f(x) = 10x + \frac{1}{x}$$

on $(0, \infty)$.   `CSS`

points are those for which $f'(x) = 0$:

$$5 - \frac{35}{x^2} = 0$$

$$5 = \frac{35}{x^2}$$

$$5x^2 = 35 \qquad \text{Multiplying by } x^2, \text{ since } x \neq 0$$

$$x^2 = 7$$

$$x = \pm\sqrt{7}.$$

The only critical point in $(0, \infty)$ is $\sqrt{7}$. Thus we can use the second derivative,

$$f''(x) = 70x^{-3}$$

$$= \frac{70}{x^3},$$

to determine whether we have a maximum or minimum. Now $f''(x)$ is positive for all values of x in $(0, \infty)$, so $f''(\sqrt{7}) > 0$, and the

$$\text{Minimum} = f(\sqrt{7})$$

$$= 5 \cdot \sqrt{7} + \frac{35}{\sqrt{7}} \text{ at } x$$

$$= \sqrt{7}.$$

The function has no maximum value.

THEOREM 6

> **Suppose a function has only one critical point c in an interval that does not have endpoints or does not contain its endpoints, such as $(-\infty, \infty)$, $(0, \infty)$, or (a, b). Then, if the function has a maximum, it will have no minimum; and if it has a minimum, it will have no maximum.**

See Examples 3 and 6.

DO EXERCISE 16.

EXERCISE SET 3.1

The curves on the graph at the right show the gasoline mileage obtained when traveling at a constant speed, for an average-size car and a compact car.

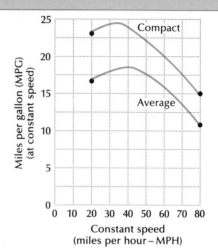

1. Consider the graph for the average-size car over the interval [20, 80].

a) Estimate the speed at which the maximum gasoline mileage is obtained.

b) Estimate the speed at which the minimum gasoline mileage is obtained.

c) What is the mileage obtained at 70 mph?

d) What is the mileage obtained at 55 mph?

e) What percent increase in mileage is there by traveling at 55 mph rather than at 70 mph?

2. Answer the same questions as in Exercise 1 for the compact car.

For the following functions, find the maximum and minimum values, if they exist, over the indicated interval. When no interval is specified, use the real line $(-\infty, \infty)$.   **CSS**

3. $f(x) = 5 + x - x^2$; [0, 2]

4. $f(x) = 4 + x - x^2$; [0, 2]

5. $f(x) = x^3 - x^2 - x + 2$; [0, 2]

6. $f(x) = x^3 + \frac{1}{2}x^2 - 2x + 5$; [0, 1]

7. $f(x) = x^3 - x^2 - x + 2$; [-1, 0]

8. $f(x) = x^3 + \frac{1}{2}x^2 - 2x + 5$; [-2, 0]

9. $f(x) = 3x - 2$; [-1, 1]

10. $f(x) = 2x + 4$; [-1, 1]

11. $f(x) = 3x - 2$

12. $f(x) = 2x + 4$

13. $f(x) = x(70 - x)$

14. $f(x) = x(50 - x)$

15. $f(x) = 2x^2 - 40x + 400$

16. $f(x) = 2x^2 - 20x + 100$

17. $f(x) = x - \frac{4}{3}x^3$; $(0, \infty)$

18. $f(x) = 16x - \frac{4}{3}x^3$; $(0, \infty)$

19. $f(x) = 17x - x^2$

20. $f(x) = 27x - x^2$

21. $f(x) = \frac{1}{3}x^3 - 3x$; [-2, 2]

22. $f(x) = \frac{1}{3}x^3 - 5x$; [-3, 3]

23. $f(x) = -0.001x^2 + 4.8x - 60$

24. $f(x) = -0.01x^2 + 1.4x - 30$

25. $f(x) = -\frac{1}{3}x^3 + 6x^2 - 11x - 50$; (0, 3)

26. $f(x) = -x^3 + x^2 + 5x - 1$; $(0, \infty)$

27. $f(x) = 15x^2 - \frac{1}{2}x^3$; [0, 30]

28. $f(x) = 4x^2 - \frac{1}{2}x^3$; [0, 8]

29. $f(x) = 2x + \frac{72}{x}$; $(0, \infty)$

30. $f(x) = x + \frac{3600}{x}$; $(0, \infty)$

31. $f(x) = x^2 + \frac{432}{x}$; $(0, \infty)$

32. $f(x) = x^2 + \frac{250}{x}$; $(0, \infty)$

33. $f(x) = 2x^4 - x$; $[-1, 1]$ **34.** $f(x) = 2x^4 + x$; $[-1, 1]$ **35.** $f(x) = \sqrt[3]{x}$; $[0, 8]$

36. $f(x) = \sqrt{x}$; $[0, 4]$ **37.** $f(x) = (x + 1)^3$ **38.** $f(x) = (x - 1)^3$

39. See Exercise 10 in Exercise Set 2.5. What is the maximum number of units sold? What must be spent on advertising in order to sell that number of units?

40. See Exercise 9 in Exercise Set 2.5. What is the maximum temperature during the illness and on what day does it occur?

41. See Exercise 61 in Exercise Set 1.4.

a) What is the maximum distance it takes to stop on glare ice? At what air temperature does this occur?

b) What is the minimum distance it takes to stop on glare ice? At what air temperature does this occur?

42. See Exercise 5 in Exercise Set 2.5. Consider the function over the interval $[0, 40]$, that is, the years 1940 to 1980.

a) What is the maximum percentage in college and in what year does it occur?

b) What is the minimum percentage in college and in what year does it occur?

43. In Exercise 3 of Exercise Set 1.6, we determined that at travel speed (constant velocity) x there are y accidents in daytime for every 100 million miles of travel, where y is given by

$$y = x^2 - 122.5x + 3775.$$

At what travel speed do the fewest accidents occur?

44. At travel speed (constant velocity) x, the cost y, in cents per mile, of operating a car is given by

$$y = 0.02x^2 - 1.3x + 30.$$

At what travel speed is the cost of operating a car a minimum?

▶

Find the maximum and minimum values, if they exist, over the indicated interval. When no interval is specified, use the real line $(-\infty, \infty)$.

45. $g(x) = x\sqrt{x + 3}$; $[-3, 3]$

46. $h(x) = x\sqrt{1 - x}$; $[0, 1]$

47. $f(x) = x^{2/3}$; $[-1, 1]$

48. $g(x) = x^{2/3}$

49. $f(x) = \frac{1}{3}x^3 - x + \frac{2}{3}$

50. $f(x) = \frac{1}{3}x^3 - \frac{1}{2}x^2 - 2x + 1$

51. $f(x) = \frac{1}{3}x^3 - 2x^2 + x$; $[0, 4]$

52. $g(x) = \frac{1}{3}x^3 + 2x^2 + x$; $[-4, 0]$

53. $t(x) = x^4 - 2x^2$

54. $f(x) = 2x^4 - 4x^2 + 2$

55. *Business.* Several costs in a business environment can be separated into two components: those that increase with volume and those that decrease with volume. Quality of customer service, although more expensive as it is increased, has part of its increased cost offset by customer goodwill. A firm has determined that its cost of service is the following function of "quality units,"

$$C(x) = (2x + 4) + \left(\frac{2}{x - 6}\right), \quad x > 6.$$

Find the number of "quality units" the firm should use to minimize its total cost of service.

56. Let

$$y = (x - a)^2 + (x - b)^2.$$

For what value of x is y a minimum?

*3.2 DERIVATIVES AND GRAPH SKETCHING

OBJECTIVE

You should be able to use derivatives to sketch the graphs of functions. Use the information obtained to determine relative maxima and minima.

Let us see how we can use derivatives to sketch graphs. We will use the following procedure.   CSS

PROCEDURE FOR GRAPH SKETCHING

a) Find $f'(x)$ and $f''(x)$.

b) Find the critical points of f by finding where $f'(x)$ does not exist and by solving $f'(x) = 0$. Find the function values at these points.

c) Use the critical points. Find the intervals where f is increasing, by solving $f'(x) > 0$. Find the intervals where f is decreasing, by solving $f'(x) < 0$.

d) Find the critical points of f' by finding where $f''(x)$ does not exist and by solving $f''(x) = 0$. Find the function values at these points.

e) Find the intervals where f is concave up by solving $f''(x) > 0$. Find the intervals where f is concave down by solving $f''(x) < 0$.

f) Sketch the graph using the information (a) through (e), plotting extra points (computing them with your calculator) if the need arises.

Example 1   Sketch the graph $f(x) = x^3 - 3x + 2$. Use the information obtained to determine relative maxima and minima.

Solution

a) Find $f'(x)$ and $f''(x)$:

$$f'(x) = 3x^2 - 3, \qquad f''(x) = 6x.$$

b) Find the critical points of f by finding where $f'(x)$ does not exist and by solving $f'(x) = 0$. Now $f'(x) = 3x^2 - 3$ exists for all values of x, so the only critical points are where

$$3x^2 - 3 = 0$$

$$3x^2 = 3$$

$$x^2 = 1$$

$$x = \pm 1.$$

Now $f(-1) = 4$ and $f(1) = 0$. These give the points $(-1, 4)$ and $(1, 0)$ on the graph.

---

*This section can be omitted without loss of continuity.

c) Use the critical points. Find the intervals where f is increasing by solving $f'(x) > 0$. Find the intervals where f is decreasing by solving $f'(x) < 0$. Whether $f'(x) = 3x^2 - 3 = 3(x + 1)(x - 1)$ is positive or negative depends on the positiveness or negativeness (signs) of the factors $x + 1$ and $x - 1$. We can determine this efficiently with the following diagram.

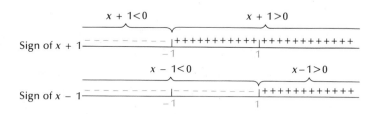

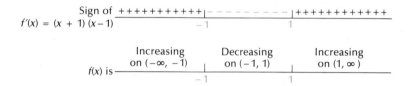

d) Find the critical points of f' by finding where $f''(x)$ does not exist and by solving $f''(x) = 0$. Now $f''(x) = 6x$ exists for all values of x, so the only critical points of f' are where

$$6x = 0$$

$$x = 0.$$

Now $f(0) = 2$. This gives another point $(0, 2)$ on the graph.

e) Find the intervals where f is concave up by solving $f''(x) > 0$. Find the intervals where f is concave down by solving $f''(x) < 0$. We see that $6x < 0$ when $x < 0$ and that $6x > 0$ when $x > 0$, so f is concave down on the interval $(-\infty, 0)$ and concave up on the interval $(0, \infty)$. Since $(0, 2)$ is a point across which the concavity changes, it is a point of inflection.

f) Sketch the graph using the preceding information.

1. Sketch the graph of

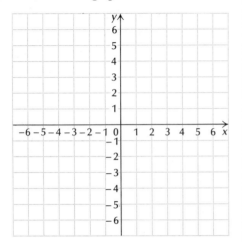

$$f(x) = \frac{1}{3}x^3 - \frac{1}{2}x^2 - 2x + 1.$$

Use the information obtained to determine relative maxima and minima.   CSS

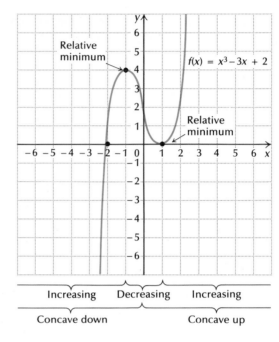

Since $f'(x)$ is positive in an interval to the left of -1 (f increasing) and negative in an interval to the right of -1 (f decreasing), the function has a relative maximum at $(-1, 4)$. Since $f'(x)$ is negative in an interval to the left of 1 (f decreasing) and positive in an interval to the right (f increasing), the function has a relative minimum at $(1, 0)$.

DO EXERCISE 1.

Example 2 Sketch the graph of $f(x) = x^4 - 2x^2$. Use the information obtained to determine relative maxima and minima.

Solution

a) Find $f'(x)$ and $f''(x)$:

$$f'(x) = 4x^3 - 4x, \qquad f''(x) = 12x^2 - 4.$$

b) Now $f'(x) = 4x^3 - 4x$ exists for all values of x, so the only critical points are where

$$4x^3 - 4x = 0$$

$$4x(x^2 - 1) = 0$$

$$4x = 0 \quad \text{or} \quad x^2 - 1 = 0$$

$$x = 0 \quad \text{or} \qquad x^2 = 1$$

$$x = \pm 1.$$

Now $f(0) = 0$, $f(-1) = -1$, and $f(1) = -1$. These give the points $(0, 0)$, $(-1, -1)$, and $(1, -1)$ on the graph.

c) Whether $f'(x) = 4x^3 - 4x = 4x(x + 1)(x - 1)$ is positive or negative depends on the signs of the factors $4x$, $x + 1$, and $x - 1$. We can determine this efficiently with a diagram as follows.

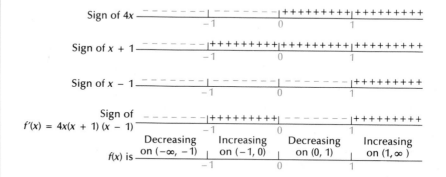

d) Now $f''(x) = 12x^2 - 4$ exists for all values of x, so the only critical points of f' are where

$$12x^2 - 4 = 0$$

$$4(3x^2 - 1) = 0$$

$$3x^2 - 1 = 0$$

$$3x^2 = 1$$

$$x^2 = \frac{1}{3}$$

$$x = \pm\sqrt{\frac{1}{3}} = \pm\frac{1}{\sqrt{3}}.$$

Now

$$f\left(\frac{1}{\sqrt{3}}\right) = \left(\frac{1}{\sqrt{3}}\right)^4 - 2\left(\frac{1}{\sqrt{3}}\right)^2 = \frac{1}{9} - \frac{2}{3} = -\frac{5}{9} \text{ and } f\left(-\frac{1}{\sqrt{3}}\right) = -\frac{5}{9}.$$

These give the points

$$\left(-\frac{1}{\sqrt{3}}, -\frac{5}{9}\right) \text{ and } \left(\frac{1}{\sqrt{3}}, -\frac{5}{9}\right)$$

on the graph. These are approximately, $(-0.6, -0.6)$ and $(0.6, -0.6)$.

e) Whether $f''(x) = 12x^2 - 4 = 4(3x^2 - 1) = 4(\sqrt{3}x + 1)(\sqrt{3}x - 1)$ is positive or negative depends on the signs of the factors $\sqrt{3}x + 1$ and $\sqrt{3}x - 1$. We can determine this with a diagram as follows.

2. Sketch the graph of

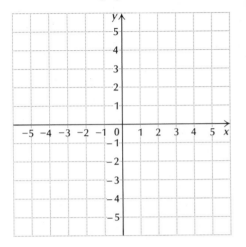

$$f(x) = 2x^4 - 4x^2 + 2.$$

Use the information obtained to determine relative maxima and minima.

Sign of $\sqrt{3}x + 1$ $\underline{|++++++++++|+++++++++}$
$ -1/\sqrt{3} \approx -0.6 1/\sqrt{3} \approx 0.6$

Sign of $\sqrt{3}x - 1$ $\underline{|+++++++++}$
$ -1/\sqrt{3} 1/\sqrt{3}$

Sign of
$f''(x) = (\sqrt{3}x + 1)(\sqrt{3}x - 1)$ $\underline{+++++++++|-----------|+++++++++}$
$ -1/\sqrt{3} 1/\sqrt{3}$

We see that f is concave up on the intervals

$$\left(-\infty, -\frac{1}{\sqrt{3}}\right) \quad \text{and} \quad \left(\frac{1}{\sqrt{3}}, \infty\right)$$

and concave down on the interval

$$\left(-\frac{1}{\sqrt{3}}, \frac{1}{\sqrt{3}}\right).$$

Since each of the points

$$\left(-\frac{1}{\sqrt{3}}, -\frac{5}{9}\right) \quad \text{and} \quad \left(\frac{1}{\sqrt{3}}, -\frac{5}{9}\right)$$

is a point on the graph across which the concavity changes, each is a point of inflection.

f) Sketch the graph using the preceding information. By solving $x^4 - 2x^2 = 0$ we can find the x-intercepts easily. They are $(-\sqrt{2}, 0)$, $(0, 0)$, and $(\sqrt{2}, 0)$. This also aids the graphing.

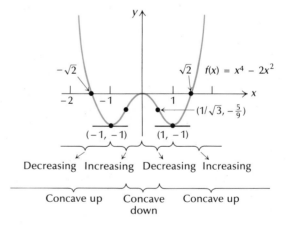

There is a relative maximum at $(0, 0)$ and two relative minima at $(-1, -1)$ and $(1, -1)$.

DO EXERCISE 2.

3. Sketch the graph of

$$f(x) = 3x^{2/3}.$$

Determine the relative maxima and minima.   `CSS`

Example 3 Sketch the graph of $f(x) = x^{2/3}$. Determine the relative maxima and minima.

Solution First note that $f(x) = x^{2/3} = \sqrt[3]{x^2} = (\sqrt[3]{x})^2$ and that $f(x)$ exists for all x in $(-\infty, \infty)$.

a) Find $f'(x)$ and $f''(x)$:

$$f'(x) = \frac{2}{3}x^{-1/3} = \frac{2}{3\sqrt[3]{x}}, \qquad f''(x) = -\frac{2}{9}x^{-4/3} = -\frac{2}{9\sqrt[3]{x^4}}.$$

b) Since $f'(0)$ does not exist, 0 is a critical point. The equation $f'(x) = 0$ has no solution, so the only critical point is 0. Now $f(0) = 0^{2/3} = 0$. This gives the point $(0, 0)$ on the graph.

c) Now when $x < 0$, $\sqrt[3]{x} < 0$, so $f'(x) = (2/3)\sqrt[3]{x} < 0$. When $x > 0$, $\sqrt[3]{x} > 0$, so

$$f'(x) = \frac{2}{3\sqrt[3]{x}} > 0.$$

Thus f is decreasing on the interval $(-\infty, 0)$ and increasing on the interval $(0, \infty)$.

d) Since $f''(0)$ does not exist, 0 is a critical point of f'. The equation $f''(x) = 0$ has no solution, so the only critical point of f' is 0. We have already found $f(0)$ in step (b).

e) For any $x \neq 0$, $x^4 > 0$, so $\sqrt[3]{x^4} > 0$. Thus

$$f''(x) = -\frac{2}{9\sqrt[3]{x^4}} < 0$$

for any $x \neq 0$. Thus f is concave down on each of the intervals $(-\infty, 0)$ and $(0, \infty)$, and there is no point of inflection.

f) Sketch the graph using the preceding information. It helps to use the values $f(1) = 1$ and $f(-1) = 1$ and the resulting points $(1, 1)$ and $(-1, 1)$ to draw the graph. There is a relative minimum at $(0, 0)$.

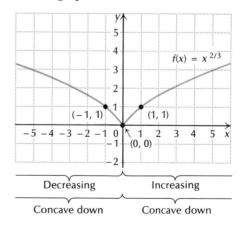

DO EXERCISE 3.

Example 4 Sketch a graph of $f(x) = x + 4/x$. Determine the relative maxima and minima.

Solution First note that 0 is not in the domain of the function.

a) Find $f'(x)$ and $f''(x)$:

$$f'(x) = 1 - 4x^{-2} = 1 - \frac{4}{x^2}, \qquad f''(x) = 8x^{-3} = \frac{8}{x^3}.$$

b) Since 0 is not in the domain of f, it is not a critical point. It is also not in the domain of f' or f''. Thus the only critical points are where

$$1 - \frac{4}{x^2} = 0$$

$$1 = \frac{4}{x^2}$$

$$x^2 = 4$$

$$x = \pm 2.$$

Now $f(-2) = -4$ and $f(2) = 4$. These give the points $(-2, -4)$ and $(2, 4)$ on the graph.

c) Note that

$$f'(x) = 1 - \frac{4}{x^2} = \frac{x^2 - 4}{x^2}.$$

For any $x \neq 0$, $x^2 > 0$, so the positiveness or negativeness of $f'(x)$ depends on $x^2 - 4$ and its factors $x + 2$ and $x - 2$.

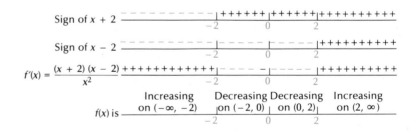

d) Since 0 is not in the domain of f' and f'', the only critical points of f' are where

$$\frac{8}{x^3} = 0.$$

But this equation has no solution. Thus f' has no critical points.

e) For any $x < 0$, $x^3 < 0$, so

$$f''(x) = \frac{8}{x^3} < 0.$$

For any $x > 0$, $x^3 > 0$, so

$$f''(x) = \frac{8}{x^3} > 0.$$

Thus f is concave down on the interval $(-\infty, 0)$ and concave up on the interval $(0, \infty)$. There is no point of inflection since 0 is not in the domain of f.

f) Sketch the graph using the preceding information. There is a relative minimum at $(2, 4)$ and a relative maximum at $(-2, -4)$. Note

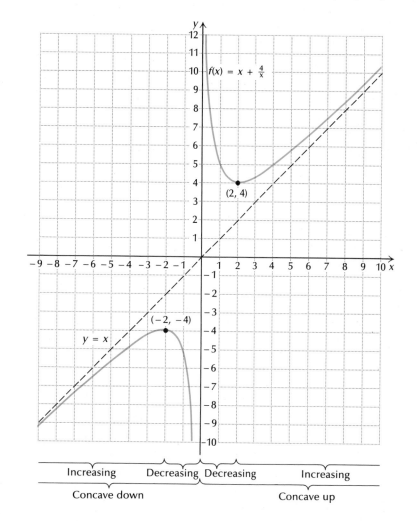

4. Sketch a graph of

$$f(x) = x + \frac{1}{x}.$$

that as $x \to \infty$, $4/x \to 0$, and $x + (4/x)$ gets closer and closer to the line $y = x$. A similar thing happens as $x \to -\infty$. Note also that as $x \to 0$ from the left, $f(x)$ approaches $-\infty$, and that as $x \to 0$ from the right, $f(x)$ approaches ∞. Because of these limiting characteristics, lines such as $y = x$ and the y-axis, in this example, are called *asymptotes*.

DO EXERCISE 4.

EXERCISE SET 3.2

Sketch the graph of each function. Determine the relative maxima and minima.

CSS

1. $f(x) = 2 - x^2$ **2.** $f(x) = 3 - x^2$ **3.** $f(x) = x^2 + x - 1$ **4.** $f(x) = x^2 - x$

5. $f(x) = \frac{8}{3}x^3 - 2x + \frac{1}{3}$ **6.** $f(x) = x^3 - 3x + 2$ **7.** $f(x) = (x - 1)^3$ **8.** $f(x) = (x + 2)^3$

9. $f(x) = (x + 1)^{2/3}$ **10.** $f(x) = (x - 1)^{2/3}$ **11.** $f(x) = x^4 - 6x^2$ **12.** $f(x) = 2x^2 - x^4$

13. $f(x) = x + (9/x)$ **14.** $f(x) = x + (2/x)$ **15.** $f(x) = x^3 - 2x^2 - 4x + 3$ **16.** $f(x) = x^3 - 6x^2 + 9x + 1$

17. $f(x) = 3x^4 + 4x^3$ **18.** $f(x) = x^4 - 2x^3$

Using the same set of axes, sketch the graph of the total revenue, total cost, and total profit functions.

19. $R(x) = 50x - 0.5x^2$, $C(x) = 4x + 10$ **20.** $R(x) = 50x - 0.5x^2$, $C(x) = 10x + 3$

Sketch the graph of each function.

21. $f(x) = x - \sqrt{x}$ **22.** $f(x) = x^2 + \frac{1}{x^2}$ **23.** $f(x) = \frac{1}{x^2 - 1}$ **24.** $f(x) = \frac{1}{x^2 + 1}$

25. $y = \frac{2x^2}{x^2 - 16}$ **26.** $y = (x - 1)^{2/3} - (x + 1)^{2/3}$

OBJECTIVE

You should be able to solve maximum–minimum problems.

3.3 MAXIMUM–MINIMUM PROBLEMS

One very important application of the differential calculus is the solving of maximum–minimum problems, that is, finding the maximum or minimum value of some varying quantity Q and the point at which that maximum or minimum occurs.

Example 1 A hobby store has 20 ft of fencing to fence off a rectangular electric train area in one corner of its display room. The two sides up against the wall require no fence. What dimensions of the rectangle will maximize the area? What is the maximum area?

1. *Exploratory exercises*

a) Complete this table.

x	y 20 − x	A x(20 − x)
0		
4		
6.5		
8		
10		
12		
13.2		
20		

b) Make a graph of x versus A, that is, of points (x, A) from the table; and connect them with a smooth curve.

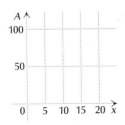

c) Does it matter what dimensions we use?

d) Make a conjecture about what the maximum might be and where it would occur.

How can the electric train area be maximized? (*Owen Franken: Stock, Boston*).

Exploratory Solution Intuitively, one might think that it does not matter what dimensions one uses; they will all yield the same area. To show that this is not true, as well as to conjecture a possible solution, consider the exploratory exercises in Margin Exercise 1. But, before doing those exercises let us express the area in terms of one variable. If we let x = the length of one side and y = the length of the other, then since the sum of the lengths must be 20 ft, we have

$$x + y = 20 \quad \text{and} \quad y = 20 - x.$$

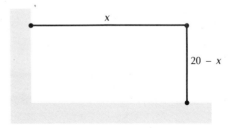

Then the area is given by

$$A = xy$$
$$A = x(20 - x) = 20x - x^2.$$

DO EXERCISE 1.

Calculus Solution We are trying to find the maximum value of

$$A = 20x - x^2 \quad \text{on the interval } (0, 20).$$

2. A rancher has 50 ft of fencing to fence off a rectangular animal pen in the corner of a barn. What dimensions of the rectangle will yield the maximum area? What is the maximum area?

We consider the interval $(0, 20)$ because x is the length of one side and cannot be negative. Since there is only 20 ft of fencing, x cannot be greater than 20. Also, x cannot be 20 because the length of y would be 0.

a) We first find $A'(x)$, where $A(x) = 20x - x^2$:

$$A'(x) = 20 - 2x.$$

b) This derivative exists for all values of x in $(0, 20)$. Thus the only critical points are where

$$A'(x) = 20 - 2x = 0$$

$$-2x = -20$$

$$x = 10.$$

Since there is only one critical point in the interval, we can use the second derivative to determine whether we have a maximum. Note that

$$A''(x) = -2,$$

which is a constant. Thus $A''(10)$ is negative, so $A(10)$ is a maximum. Now

$$A(10) = 10(20 - 10) = 10 \cdot 10 = 100.$$

Thus the maximum area of 100 sq ft is obtained using 10 ft for the length of one side, and $20 - 10$, or 10 ft, for the other. Note that while you may have conjectured this in Margin Exercise 1, the tools of calculus allowed us to prove it.

DO EXERCISE 2.

Example 2 A stereo manufacturer determines that in order to sell x units of a new stereo its price per unit must be

$$p = D(x) = 1000 - x.$$

It also determines that the total cost of producing x units is given by

$$C(x) = 3000 + 20x.$$

a) Find the total revenue $R(x)$.

b) Find the total profit $P(x)$.

c) How many units must the company produce and sell to maximize profit?

d) What is the maximum profit?

e) What price per unit must be charged to make this maximum profit?

3. A company determines that in order to sell x units of a certain product its price per unit must be

$$p = D(x) = 200 - x.$$

It also determines that its total cost of producing x units is given by

$$C(x) = 5000 + 8x.$$

a) Find the total revenue $R(x)$.

b) Find the total profit $P(x)$.

c) How many units must the company produce and sell in order to maximize profit?

d) What is the maximum profit?

e) What price per unit must be charged to make this maximum profit?

Solution

a) $R(x)$ = Total revenue = (number of units) · (price per unit)

$$= \quad x \quad \cdot \quad p$$

$$= x(1000 - x) = 1000x - x^2$$

b) $P(x) = R(x) - C(x) = (1000x - x^2) - (3000 + 20x)$

$$= -x^2 + 980x - 3000$$

c) To find the maximum value of $P(x)$ we first find $P'(x)$:

$$P'(x) = -2x + 980.$$

This is defined for all real numbers (actually we are interested in numbers x in $[0, \infty)$ only, since we cannot produce a negative number of stereos). Thus we solve

$$P'(x) = -2x + 980 = 0$$

$$-2x = -980$$

$$x = 490.$$

Since there is only one critical point, we can try to use the second derivative to determine whether we have a maximum. Note that

$$P''(x) = -2, \quad \text{a constant.}$$

Thus $P''(490)$ is negative, so $P(490)$ is a maximum.

d) The maximum profit is given by

$$P(490) = -(490)^2 + 980 \cdot 490 - 3000 = \$237,100.$$

Thus the stereo manufacturer makes a maximum profit of $237,100 by producing and selling 490 stereos.

e) The price per unit to make the maximum profit is

$$p = 1000 - 490 = \$510.$$

CSS

DO EXERCISE 3.

Marginal Analysis

Let us take a general look at the total profit function and its related functions.

In the first graph we have the total cost and total revenue functions. We can estimate what the maximum profit might be by looking for the widest gap between $R(x)$ and $C(x)$. Points B_0 and B_2 are "break-even" points.

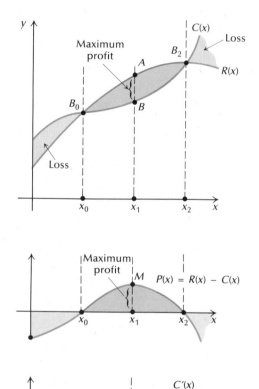

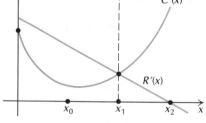

In the second graph we have the total profit function. Note that when production is too low ($<x_0$) there is a loss because of high fixed or initial costs and low revenue. When production is too high ($>x_2$), there is also a loss due to high marginal costs and low marginal revenues, as seen in the third graph.

   The business operates at a profit everywhere between x_0 and x_2. Note that maximum profit occurs at a critical point x_1 of $P(x)$. If we assume that $P'(x)$ exists for all x in some interval, usually $[0, \infty)$, this critical point occurs at some number x such that

$$P'(x) = 0.$$

Since $P(x) = R(x) - C(x)$, it follows that

$$P'(x) = R'(x) - C'(x).$$

Thus the maximum profit occurs at some number x such that

$$R'(x) - C'(x) = 0$$

or

$$R'(x) = C'(x).$$

In summary, we have the following theorem.

THEOREM 7

Maximum profit is achieved when marginal revenue equals marginal cost:

$$R'(x) = C'(x).$$

Here is a general strategy for solving maximum–minimum problems. While it may not guarantee success, it should certainly enhance one's chances.

1. **Read the problem carefully. If relevant, draw a picture.**
2. **Label the picture with appropriate variables and constants, noting what varies and what stays fixed.**
3. **Translate the problem to an equation, involving a quantity Q to be maximized or minimized.**
4. **Try to express Q as a function of *one* variable. Use the procedures developed in Section 3.2 to determine the maximum or minimum values and the points at which they occur.**

Example 3 From a thin piece of cardboard 8 in. by 8 in., square corners are cut out so that the sides can be folded up to make a box. What dimensions will yield a box of maximum volume? What is the maximum volume?

Exploratory Solution One might again think that it does not matter what the dimensions are, but our experience with Example 1 should lead us to think otherwise. We make a drawing as shown below.

4. *Exploratory exercises*

a) Complete this table.

x	h $\frac{1}{2}(8-x)$	V $x \cdot x \cdot \frac{1}{2}(8-x)$
0		
1		
2		
3		
4		
4.6		
5		
6		
6.8		
7		
8		

b) Make a graph of x versus V.

c) Make a conjecture about what the maximum might be and where it would occur.

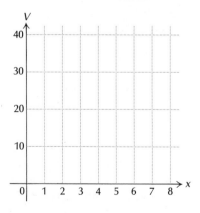

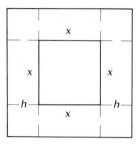

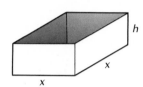

When squares of length h on a side are cut out of the corners, we are left with a square base of length x. The volume of the resulting box is

$$V = lwh = x \cdot x \cdot h.$$

We want to express V in terms of one variable. Note that the overall length of a side of the cardboard is 8 in. We see from the drawing that

$$h + x + h = 8,$$

or

$$x + 2h = 8.$$

Solving for h we get

$$2h = 8 - x$$

$$h = \frac{1}{2}(8 - x) = \frac{1}{2} \cdot 8 - \frac{1}{2}x = 4 - \frac{1}{2}x.$$

Thus

$$V = x \cdot x \cdot \left(4 - \frac{1}{2}x\right) = x^2\left(4 - \frac{1}{2}x\right) = 4x^2 - \frac{1}{2}x^3.$$

In Margin Exercise 4 you will compute some values of V.

DO EXERCISE 4.

Calculus Solution You probably noted in Margin Exercise 4 that it was a bit more difficult than in Example 1 to conjecture where the maximum occurs. At the least it seems reasonable that it occurs for some x between 5 and 6. Let us find out for certain, using calculus. We are trying to find the maximum value of

$$V(x) = 4x^2 - \frac{1}{2}x^3 \quad \text{on the interval } (0, 8).$$

We first find $V'(x)$:

$$V'(x) = 8x - \frac{3}{2}x^2.$$

5. Repeat Example 3, but for a piece of cardboard that is 10 in. by 10 in.

Now $V'(x)$ exists for all x in the interval $(0, 8)$, so we set it equal to 0 to find the critical values:

$$V'(x) = 8x - \frac{3}{2}x^2 = 0$$

$$x\left(8 - \frac{3}{2}x\right) = 0$$

$$x = 0 \quad \text{or} \quad 8 - \frac{3}{2}x = 0$$

$$x = 0 \quad \text{or} \quad -\frac{3}{2}x = -8$$

$$x = 0 \quad \text{or} \quad x = -\frac{2}{3}(-8) = \frac{16}{3}.$$

The only critical point in $(0, 8)$ is $\frac{16}{3}$. Thus we can use the second derivative,

$$V''(x) = 8 - 3x,$$

to determine whether we have a maximum. Since

$$V''\left(\frac{16}{3}\right) = 8 - 3 \cdot \frac{16}{3} = -8,$$

$V''(\frac{16}{3})$ is negative, so $V(\frac{16}{3})$ is a maximum, and

$$V\left(\frac{16}{3}\right) = 4 \cdot \left(\frac{16}{3}\right)^2 - \frac{1}{2}\left(\frac{16}{3}\right)^3 = \frac{1024}{27} = 37\frac{25}{27}.$$

The maximum volume is $37\frac{25}{27}$ cu. in. The dimensions that yield this maximum volume are

$$x = \frac{16}{3} = 5\frac{1}{3} \text{ in.,} \quad \text{by } x = 5\frac{1}{3} \text{ in.,} \quad \text{by } h = 4 - \frac{1}{2}\left(\frac{16}{3}\right) = 1\frac{1}{3} \text{ in.}$$

It would surely have been difficult to guess this from Margin Exercise 4.

DO EXERCISE 5.

In the following problem, an open-top container of fixed volume is to be constructed. We want to determine the dimensions that will allow it to be built with the least amount of material. Such a problem could be important from an ecological standpoint.

Example 4 A container firm is designing an open-top rectangular box, with a square base, that will hold 108 cubic centimeters (cc). What dimensions yield the minimum surface area? What is the minimum surface area?

Solution The surface area of the box is

$$S = x^2 + 4xy.$$

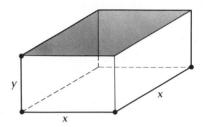

The volume must be 108 cc, and is given by

$$V = x^2y = 108.$$

To express S in terms of one variable, we solve $x^2y = 108$ for y:

$$y = \frac{108}{x^2}.$$

Then

$$S = x^2 + 4x\left(\frac{108}{x^2}\right) = x^2 + \frac{432}{x}.$$

Now S is defined only for positive numbers, and the problem dictates that the length x be positive, so we are minimizing S on the interval $(0, \infty)$. We first find dS/dx:

$$\frac{dS}{dx} = 2x - \frac{432}{x^2}.$$

Since dS/dx exists for all x in $(0, \infty)$, the only critical points are where $dS/dx = 0$. Thus we solve the following equation:

$$2x - \frac{432}{x^2} = 0$$

$$x^2\left(2x - \frac{432}{x^2}\right) = x^2 \cdot 0 \qquad \text{We multiply by } x^2 \text{ to clear of fractions.}$$

$$2x^3 - 432 = 0$$

$$2x^3 = 432$$

$$x^3 = 216$$

$$x = 6.$$

6. Repeat Example 4, but for a fixed volume of 500 cc.

This is the only critical point, so we can use the second derivative to determine whether we have a minimum:

$$\frac{d^2S}{dx^2} = 2 + \frac{864}{x^3}.$$

Note that this is positive for all positive values of x. Thus we have a minimum at $x = 6$. When $x = 6$, it follows that $y = 3$:

$$y = \frac{108}{6^2} = \frac{108}{36} = 3.$$

Thus the surface area is minimized when $x = 6$ cm (centimeters) and $y = 3$ cm. The minimum surface area is

$$S = 6^2 + 4 \cdot 6 \cdot 3 = 108 \text{ cm}^2.$$

This, by coincidence, is the same number as the fixed volume.

DO EXERCISE 6.

Example 5 *Determining a ticket price.* Fight promoters ride a thin line between profit and loss, especially in determining the price to charge for admission to closed-circuit television showings in local theaters. By keeping records, a theater determines that if the admission price is $20, it averages 1000 people in attendance. But, for every increase of $1, it loses 100 customers from the average. Every customer spends an average of $0.80 on concessions. What admission price should the theater charge to maximize total revenue?

Solution Let $x =$ the amount by which the price of $20 should be increased (if x is negative, the price would be decreased). We first express total revenue R as a function of x. Note that

$$R(x) = (\text{Revenue from tickets}) + (\text{Revenue from concessions})$$

$$= (\text{Number of people}) \cdot (\text{Ticket price}) + \$0.80(\text{Number of people})$$

$$= (1000 - 100x)(20 + x) + 0.80(1000 - 100x)$$

$$= 20{,}000 - 2000x + 1000x - 100x^2 + 800 - 80x$$

$$R(x) = -100x^2 - 1080x + 20{,}800.$$

We are trying to find the maximum value of R over the set of all real numbers. To find x such that $R(x)$ is a maximum, we first find $R'(x)$:

$$R'(x) = -200x - 1080.$$

This derivative exists for all real numbers x; thus the only critical points are where $R'(x) = 0$, so we solve that equation:

7. Transit companies also ride a thin line between profit and loss. A company determines that at a fare of 30¢, it will average 10,000 fares a day. For every increase of 10¢, it loses 2000 customers. What fare should be charged to maximize revenue? (*Hint:* Let x = the number of 10¢ fare increases (if x is negative, the fare would be decreased). Then the new fare would be 30 + 10x.)

$$-200x - 1080 = 0$$

$$-200x = 1080$$

$$x = -5.4 = -\$5.40.$$

Since this is the only critical point, we can use the second derivative,

$$R''(x) = -200,$$

to determine whether we have a maximum. Since $R''(-5.4)$ is negative, $R(-5.4)$ is a maximum. Thus to maximize revenue the theater should charge

$$\$20 + (-\$5.40) \quad \text{or} \quad \$14.60 \text{ per ticket.}$$

That is, this reduced ticket price will get more people into the theater,

$$1000 - 100(-5.4), \quad \text{or} \quad 1540,$$

and will result in maximum revenue.

DO EXERCISE 7.

EXERCISE SET 3.3

1. Of all the numbers whose sum is 50, find the two that have the maximum product. That is, maximize $Q = xy$, where $x + y = 50$.

2. Of all the numbers whose sum is 70, find the two that have the maximum product. That is, maximize $Q = xy$, where $x + y = 70$.

3. In Exercise 1, can there be a minimum product? Explain.

4. In Exercise 2, can there be a minimum product? Explain.

5. Of all numbers whose difference is 4, find the two that have the minimum product.

6. Of all numbers whose difference is 6, find the two that have the minimum product.

7. Maximize $Q = xy^2$, where x and y are positive numbers, such that $x + y^2 = 1$.

8. Maximize $Q = xy^2$, where x and y are positive numbers, such that $x + y^2 = 4$.

9. Minimize $Q = x^2 + y^2$, where $x + y = 20$.

10. Minimize $Q = x^2 + y^2$, where $x + y = 10$.

11. Maximize $Q = xy$, where x and y are positive numbers, such that $(4/3)x^2 + y = 16$.

12. Maximize $Q = xy$, where x and y are positive numbers, such that $x + (4/3)y^2 = 1$.

13. A rancher wants to build a rectangular fence next to a river, using 120 yd of fencing. What dimensions of the rectangle will maximize the area? What is the maximum area? Note that the rancher does not have to fence in the side next to the river.

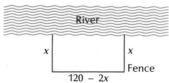

14. A rancher wants to enclose two rectangular areas near a river, one for sheep and one for cattle. There are 240 yd of fencing available. What is the largest total area that can be enclosed?

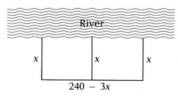

15. A carpenter is building a room with a fixed perimeter of 54 ft. What are the dimensions of the largest room that can be built? What is its area?

16. Of all rectangles that have a perimeter of 34 ft, find the dimensions of the one with the largest area. What is its area?

Business: Maximizing profit. Find the maximum profit and the number of units that must be produced and sold to yield the maximum profit.

17. $R(x) = 50x - 0.5x^2$, $C(x) = 4x + 10$

18. $R(x) = 50x - 0.5x^2$, $C(x) = 10x + 3$

19. $R(x) = 2x$, $C(x) = 0.01x^2 + 0.6x + 30$

20. $R(x) = 5x$, $C(x) = 0.001x^2 + 1.2x + 60$

21. $R(x) = 9x - 2x^2$, $C(x) = x^3 - 3x^2 + 4x + 1$; $R(x)$ and $C(x)$ are in thousands of dollars, and x is in thousands of units.

22. $R(x) = 100x - x^2$, $C(x) = \frac{1}{3}x^3 - 6x^2 + 89x + 100$; $R(x)$ and $C(x)$ are in thousands of dollars, and x is in thousands of units.

23. Raggs, Ltd., a clothing firm, determines that to sell x suits its price per suit must be

$$p = D(x) = 150 - 0.5x.$$

It also determines that its total cost of producing x suits is given by

$$C(x) = 4000 + 0.25x^2.$$

a) Find the total revenue $R(x)$.

b) Find the total profit $P(x)$.

c) How many suits must the company produce and sell to maximize profit?

d) What is the maximum profit?

e) What price per suit must be charged to make this maximum profit?

24. An appliance firm is marketing a new refrigerator. It determines that to sell x refrigerators its price per refrigerator must be

$$p = D(x) = 280 - 0.4x.$$

It also determines that its total cost of producing x refrigerators is given by

$$C(x) = 5000 + 0.6x^2.$$

a) Find the total revenue $R(x)$.

b) Find the total profit $P(x)$.

c) How many refrigerators must the company produce and sell to maximize profit?

d) What is the maximum profit?

e) What price per refrigerator must be charged to make this maximum profit?

25. From a thin piece of cardboard 30 in. by 30 in., square corners are cut out so the sides can be folded up to make a box. What dimensions will yield a box of maximum volume? What is the maximum volume?

26. From a thin piece of cardboard 20 in. by 20 in., square corners are cut out so the sides can be folded up to make a box. What dimensions will yield a box of maximum volume? What is the maximum volume?

27. A container company is designing an open-top, square-based, rectangular box that will have a volume of 62.5 cu. in. What dimensions yield the minimum surface area? What is the minimum surface area?

28. A soup company is constructing an open-top, rectangular, metal tank with a square base, that will have a volume of 32 cu. ft. What dimensions yield the minimum surface area? What is the minimum surface area?

29. A university is trying to determine what price to charge for football tickets. At a price of $6 per ticket it averages 70,000 per game. For every increase of $1 it loses 10,000 people from the average. Every person at the game spends an average of $1.50 on concessions. What price per ticket should be charged to maximize revenue? How many people will attend at that price?

30. Suppose you are the owner of a 30-unit motel. All units are occupied when you charge $20 a day per unit. For every increase of x dollars in the daily rate, there are x units vacant. Each occupied room costs $2 per day to service and maintain. What should you charge per unit to maximize profit?

31. An apple farm yields an average of 30 bushels of apples per tree when 20 trees are planted on an acre of ground. Each time 1 more tree is planted per acre, the yield decreases 1 bu per tree due to the extra congestion. How many trees should be planted to get the highest yield?

32. When a theater owner charges $3 for admission, there is an average attendance of 100 people. For every $0.10 increase in admission, there is a loss of 1 customer from the average. What admission should be charged to maximize revenue?

33. The postal service places a limit of 84 in. on the combined length and girth (distance around) of a package to be sent parcel post. What dimensions of a rectangular box with square cross section will contain the largest volume that can be mailed? (*Hint:* There are two different girths.)

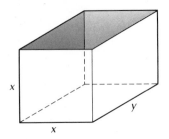

34. A rectangular play area is to be laid out in a person's back lot, and is to contain 48 square yards. The neighbor agrees to pay half the cost of the side of the play area that lines the lot. What dimensions will minimize the cost of the fence?

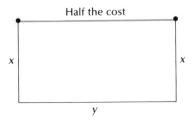

35. For what positive number is the sum of its reciprocal and five times its square a minimum?

36. For what positive number is the sum of its reciprocal and four times its square a minimum?

37. A rectangular box with a volume of 320 cubic feet is to be constructed with a square base and top. The cost per square foot for the bottom is 15¢, for the top is 10¢, and for the sides is 2.5¢. What dimensions will minimize the cost?

38. A merchant who was purchasing a display sign from a salesclerk said, "I want a sign 10 ft by 10 ft." The salesman responded, "That's just what we'll give you; only to make to more aesthetic, why don't we change it to 7 ft by 13 ft?" Comment.

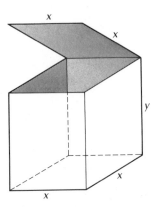

39. A Norman window is a rectangle with a semicircle on top. Suppose the perimeter of a particular Norman window is to be 24 ft. What should its dimensions be in order to allow the maximum amount of light to enter through the window?

40. Solve Exercise 39, but this time the semicircle is to be stained glass, which transmits only half as much light as the semicircle in Exercise 39.

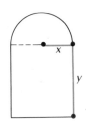

A Norman window.
(*Owen Franken: Stock, Boston*)

41. The amount of money deposited in a financial institution in savings accounts is directly proportional to the interest rate the financial institution pays on the money. Suppose a financial institution can loan *all* the money it takes in on its savings accounts at an interest rate of 18%. What interest rate should it pay on its savings accounts to maximize profits?

43. A 24-in. piece of string is cut in two pieces. One piece is used to form a circle and the other to form a square. How should the string be cut so the sum of the areas is a minimum? a maximum?

42. ▦ A page in this book is 73.125 sq in. On the average there is a 0.75-in. margin at the top and at the bottom of each page, and a 0.5-in. margin on each of the sides. What should the outside dimensions of each page be so the printed area is a maximum? Measure the outside dimensions to see whether the actual dimensions maximize the printed area.

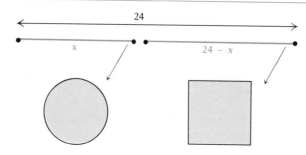

44. A power line is to be constructed from a power station at point *A* (see the figure) to an island at point *C*, which is directly 1 mile out in the water from a point *B* on the shore. Point *B* is 4 miles downshore from the power station at *A*. It costs $5000 per mile to lay the power line under water and $3000 per mile to lay the line under ground. At what point *S* downshore from *A* should the line come to the shore to minimize cost? Note that *S* could very well be *B* or *A*. (*Hint:* The length of *CS* is $\sqrt{1 + x^2}$.)

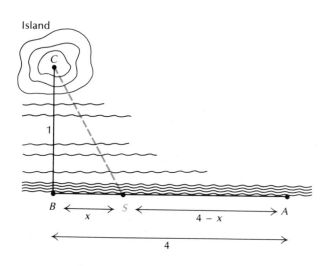

45. ▦ *Biology: Flights of homing pigeons.* It is known that homing pigeons tend to avoid flying over water in the daytime, perhaps because the downdrafts of air over water make flying difficult. Suppose a homing pigeon is released on an island at point *C*, which is directly 3 miles out in the water from a point *B* on shore. Point *B* is 8 miles downshore from the pigeon's home loft at point *A*. Assume a pigeon requires 1.28 times the rate of energy over land to fly over water. Toward what point *S* downshore from *A* should the pigeon fly to minimize the total energy required to get to home loft *A*? Assume (Total energy) = (Energy rate over water) · (Distance over water) + (Energy rate over land) · (Distance over land).   **CSS**

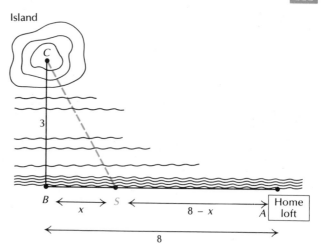

46. A road is to be built between two cities, C_1 and C_2, on opposite sides of a river of uniform width r. Because of the river, a bridge must be built. C_1 is a units from the river, and C_2 is b units from the river; $a \leq b$. Where should the bridge be located to minimize the total distance between the cities? Give a general solution using the constants a, b, p, and r in the drawing to the right.

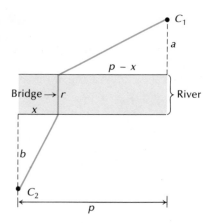

47. The total cost function for producing x units of a certain product is given by

$$C(x) = 8x + 20 + \frac{x^3}{100}.$$

a) Find the marginal cost $C'(x)$.

b) Find the average cost $A(x) = C(x)/x$.

c) Find the *marginal average cost* $A'(x)$.

d) Find the minimum of $A(x)$ and the value x_0 at which it occurs. Find the marginal cost at x_0.

e) Compare $A(x_0)$ and $C'(x_0)$.

49. Minimize $Q = x^3 + 2y^3$, where x and y are positive numbers, such that $x + y = 1$.

48. Consider $A(x) = C(x)/x$.

a) Find $A'(x)$ in terms of $C'(x)$ and $C(x)$.

b) Show that $A(x)$ has a minimum at that value of x_0 such that

$$C'(x_0) = A(x_0) = \frac{C(x_0)}{x_0}.$$

This shows that when marginal cost and average cost are the same, a product is being produced at the least average cost.

50. Minimize $Q = 3x + y^3$, where $x^2 + y^2 = 2$.

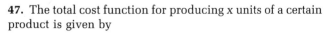

OBJECTIVES

You should be able, given certain inventory costs, to find how many times a year a store should reorder a product, and in what lot size, to minimize total inventory costs.

3.4 BUSINESS APPLICATIONS: MINIMIZING INVENTORY COSTS

A retail outlet of a business is concerned about inventory costs. Suppose, for example, an appliance store sells 2500 television sets per year. One way it could operate is to order all the sets at once. But then the owners would face the carrying costs (insurance, building space, and so on) of storing them all. Thus they might make several smaller orders, say 5, so that the largest number they would ever have to store is 500. On the other hand, each time they reorder there are certain costs such as paperwork, delivery charges, manpower, and so on. It would, therefore, seem that there is some balance between carrying costs and reorder costs. We will see how calculus can help to determine what that

How can inventory costs be minimized? (*Jan Lukas: Photo Researchers, Inc.*)

balance might be. We will be trying to minimize the following function:

$$\text{Total inventory costs} = \left(\begin{array}{c}\text{Yearly carrying} \\ \text{costs}\end{array}\right) + \left(\begin{array}{c}\text{Yearly reorder} \\ \text{costs}\end{array}\right)$$

The *lot size* x refers to the largest amount ordered each reordering period. Note the following graphs. Thus if the lot size is x, then x/2 represents the average amount held in stock over the course of the year.

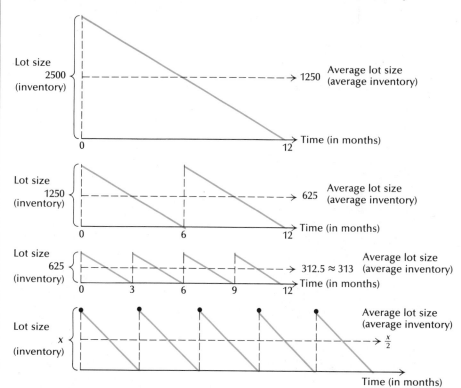

1. ▦ Without a knowledge of calculus one might make a rough estimate of the lot size that will minimize total inventory costs by completing a table like the following. Complete the table and make such an estimate.

Lot Size, x	Number of Reorders, $\dfrac{2500}{x}$	Average Inventory, $\dfrac{x}{2}$	Carrying Costs, $10 \cdot \dfrac{x}{2}$	Cost of each Order, $20 + 9x$	Reorder Costs, $(20 + 9x)\dfrac{2500}{x}$	Total Inventory Costs $C(x)$, $10 \cdot \dfrac{x}{2} + (20 + 9x)\dfrac{2500}{x}$
2500	1	1250	$12,500	$22,520	$22,520	$35,020
1250	2	625	$6,250	$11,270	$22,540	
500	5	250	$2,500	$4,520		
250	10	125				
167	15	84				
125	20					
100	25					
90	28					
50	50					

Example 1 A retail appliance store sells 2500 television sets per year. It costs $10 to store one set for a year. To reorder, there is a fixed cost of $20 plus $9 for each set. How many times per year should the store reorder, and in what lot size, in order to minimize inventory costs?

Solution Let x = the lot size. Now inventory costs are given by

$$C(x) = \text{(Yearly carrying costs)} + \text{(Yearly reorder costs)}.$$

We consider each separately.

a) *Yearly carrying costs.* The average amount held in stock is x/2, and it costs $10 per set for storage. Thus

$$\text{Yearly carrying costs} = \begin{pmatrix} \text{Yearly cost} \\ \text{per item} \end{pmatrix}\begin{pmatrix} \text{Average number} \\ \text{of items} \end{pmatrix}$$

$$= 10 \cdot \frac{x}{2}.$$

b) *Yearly reorder costs.* Now x = the lot size, and suppose there are N reorders each year. Then Nx = 2500, and N = 2500/x. Thus

$$\text{Yearly reorder costs} = \begin{pmatrix} \text{Cost of each} \\ \text{order} \end{pmatrix}\begin{pmatrix} \text{Number of} \\ \text{reorders} \end{pmatrix}$$

$$= (20 + 9x)\frac{2500}{x}.$$

c) Hence

$$C(x) = 10 \cdot \frac{x}{2} + (20 + 9x)\frac{2500}{x}$$

$$C(x) = 5x + \frac{50,000}{x} + 22,500.$$

DO EXERCISE 1.

d) We want to find a minimum value of C on the interval [1, 2500]. We first find $C'(x)$:

$$C'(x) = 5 - \frac{50,000}{x^2}.$$

e) Now $C'(x)$ exists for all x in [1, 2500], so the only critical points are

2. An appliance store sells 600 refrigerators per year. It costs $30 to store one refrigerator for one year. To reorder refrigerators there is a fixed cost of $40, plus $11 for each refrigerator. How many times per year should the store order refrigerators, and in what lot size, in order to minimize inventory costs?

CSS

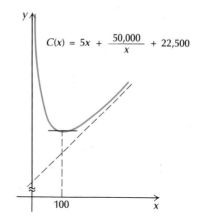

$$C(x) = 5x + \frac{50{,}000}{x} + 22{,}500$$

those x such that $C'(x) = 0$. We solve $C'(x) = 0$:

$$5 - \frac{50{,}000}{x^2} = 0$$

$$5 = \frac{50{,}000}{x^2}$$

$$5x^2 = 50{,}000$$

$$x^2 = 10{,}000$$

$$x = \pm 100.$$

Now there is only one critical point in the interval $[1, \ 2500]$, $x = 100$, so we can use the second derivative to see if we have a maximum or minimum:

$$C''(x) = \frac{100{,}000}{x^3}.$$

Now $C''(x)$ is positive for all x in $[1, 2500]$, so we do have a minimum at $x = 100$. Thus to minimize inventory costs, the store should order sets $(2500/100)$, or 25, times per year. The lot size is 100.

DO EXERCISE 2.

What happens in such problems when the answer is not a whole number? For functions of this type, we consider the two whole numbers closest to the answer, and substitute them into $C(x)$. The value that yields the smallest $C(x)$ is the lot size.

Example 2 Repeat Example 1, using all the data given, but change the $10 storage cost to $20. How many times per year should the store reorder television sets, and in what lot size, in order to minimize inventory costs?

3. Repeat Margin Exercise 2, using all the data given, but change the $30 storage cost to $50.

Solution Comparing this with Example 1, we find that the inventory cost function becomes

$$C(x) = 20 \cdot \frac{x}{2} + (20 + 9x)\frac{2500}{x} = 10x + \frac{50,000}{x} + 22,500.$$

Then we find $C'(x)$, set it equal to 0, and solve for x:

$$C'(x) = 10 - \frac{50,000}{x^2} = 0$$

$$10 = \frac{50,000}{x^2}$$

$$10x^2 = 50,000$$

$$x^2 = 5000$$

$$x = \sqrt{5000} \approx 70.7. \quad \text{⊞ or Table 1}$$

Since it does not make sense to reorder 70.7 sets each time, we consider the two numbers closest to 70.7, which are 70 and 71. Now

$$C(70) \approx \$23,914.29 \quad \text{and} \quad C(71) \approx \$23,914.23.$$

It follows that the lot size that will minimize cost is 71, although the difference, $0.06, is not significant. (*Note:* Such a procedure will not work for all types of functions, but will work for the type we are considering here. The number of times an order should be placed is $2500/71 \approx 35$, so there is still some estimating involved.)

DO EXERCISE 3.

The value of the lot size that minimizes total inventory costs is often referred to as the *economic ordering quantity*. There are three assumptions made in using the foregoing method to determine the economic ordering quantity. The first is that the demand for the product is the same throughout the year. For television sets this may be reasonable, but for seasonal items such as clothing or skis, this assumption may not be reasonable. The second assumption is that the time between the placing of an order and its receipt should be consistent throughout the year. The third assumption is that the various costs involved, such as storage, shipping charges, and so on, do not vary. This may not be reasonable in a time of inflation, although one may account for them by anticipating what they might be and using average costs. Nevertheless, the model described above can be useful, and it allows us to analyze a seemingly difficult problem using the calculus.

EXERCISE SET 3.4

1. A sporting goods store sells 100 pool tables per year. It costs $20 to store one pool table for one year. To reorder pool tables there is a fixed cost of $40, plus $16 for each pool table. How many times per year should the store order pool tables, and in what lot size, in order to minimize inventory costs?

2. A pro shop in a bowling alley sells 200 bowling balls per year. It costs $4 to store one bowling ball for one year. To reorder bowling balls there is a fixed cost of $1, plus $0.50 for each bowling ball. How many times per year should the shop order bowling balls, and in what lot size, in order to minimize inventory costs?

3. A retail outlet for Boxowitz Calculators sells 360 calculators per year. It costs $8 to store one calculator for one year. To reorder calculators, there is a fixed cost of $10, plus $8 for each calculator. How many times per year should the store order calculators, and in what lot size, in order to minimize inventory costs?

4. A sporting goods store in southern California sells 720 surfboards per year. It costs $2 to store one surfboard for one year. To reorder surfboards there is a fixed cost of $5, plus $2.50 for each surfboard. How many times per year should the store order surfboards, and in what lot size, in order to minimize inventory costs?

5. Repeat Exercise 3, using all the data given, but change the $8 storage charge to $9.

6. Repeat Exercise 4, using all the data given, but change the $5 fixed cost to $4.

7. *Minimizing inventory costs: A general solution.* A store sells Q units of a product per year. It costs a dollars to store one unit for one year. To reorder units, there is a fixed cost of b dollars, plus c dollars for each unit. In what lot size should the store reorder in order to minimize inventory costs?

8. Use the general solution found in Exercise 7 to find how many times per year a store should reorder, and in what lot size, when $Q = 2500$, $a = \$10$, $b = \$20$, and $c = \$9$.

OBJECTIVES

You should be able to

a) Given a reproduction curve and an initial population P_0, locate population values for subsequent years.

b) Given a reproduction curve described by a formula, find the population at which the maximum sustainable harvest occurs and the maximum sustainable harvest.

3.5 BIOLOGICAL APPLICATION: MAXIMUM SUSTAINABLE HARVEST

Reproduction Curves

In certain situations biologists are able to determine what is called a *reproduction curve*. This is a function

$$y = f(P)$$

such that if P is the population at a certain time t, then the population one year later, at time $t + 1$, is $f(P)$. Such a curve is shown in Fig. 1. The line $y = P$ is significant for two reasons. First, if $y = P$ is a description of f, then we know the population stays the same from year to year. But the graph of f in Fig. 1 lies mostly above the line. Thus the population is increasing. Now if we start with a population P, we know that $f(P)$ is the population one year later. Given $f(P)$ we move horizontally from $(P, f(P))$ until we hit the line $y = P$; then we move down to the horizontal axis. This locates $f(P)$ on the horizontal axis, so we can find the population two years later. In this way we can generate a sequence of population values over a period of years.

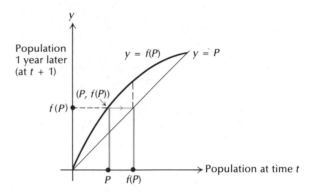

Figure 1

To see how this happens, look at Fig 2. Suppose P_0 is some initial population. The population one year later is given by

$$P_1 = f(P_0).$$

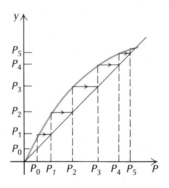

Figure 2

The population two years later is given by

$$P_2 = f(P_1),$$

and so on. We thus obtain a sequence of population values

$$P_0, P_1, P_2, P_3, \ldots.$$

If we transfer values from the horizontal axis of Fig. 2 to the vertical axis of Fig. 3 and plot points (t, P_t), where t represents time, we obtain a graph of population versus time.

1. Below is a reproduction curve.

   a) Using P_0 as an initial population, locate P_1, P_2, P_3, P_4, and P_5 on both the vertical and horizontal axes.

   b) Use a ruler to transfer these values to the vertical axis of the second graph. Plot the points (t, P_t) and connect them with a smooth curve. Estimate the equilibrium level and draw a line to represent it.

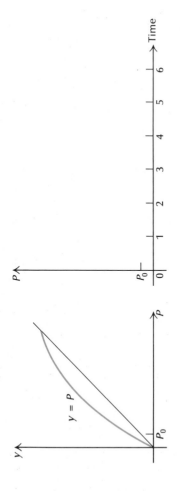

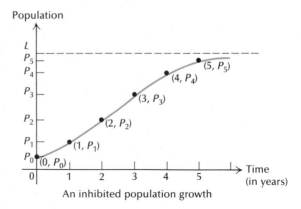

An inhibited population growth

Figure 3

Note in Fig. 3 that the population tends toward a *limiting value L*, or *equilibrium level*. This curve is called an *S-shaped curve*, or *logistic curve*, and is encountered in situations where the growth is inhibited by certain environmental factors, such as resources or the size of an ecosystem. For example, a colony of bacteria in a petri dish will grow to a certain size and then stop growing due to waste contamination. We will consider this type of growth in more detail in Chapter 7.

DO EXERCISE 1.

In Fig. 4 the reproduction curve falls below the line $y = P$. Note that as time passes, the population curve oscillates about the equilibrium level.

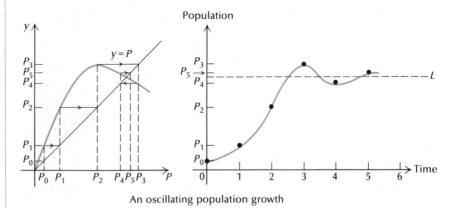

An oscillating population growth

Figure 4   An oscillating population growth.

2. a) Repeat the steps of Margin Exercise 1 for the reproduction curve below.
  b) Is this an oscillating or cyclic population?

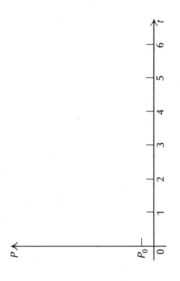

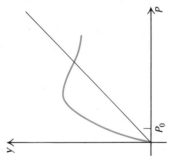

In Fig. 5 the reproduction curve falls even further below line $y = P$. The population-versus-time line then repeats itself in cycles, in this case every four years, never approaching an equilibrium level. A population of blow flies confined in a laboratory will exhibit such growth, as will certain predator–prey interactions.

We have drawn a line-segment graph of population versus time, since this is often done in practice.

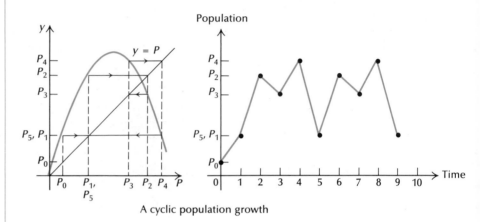

A cyclic population growth

Figure 5 A cyclic population growth.

DO EXERCISE 2.

Suppose a certain population is growing in such a way that if P is the population at a particular time, then one year later there will be a 2% increase. The reproduction curve is given by

$$\begin{aligned} f(P) &= P + 2\%P \\ &= 1 \cdot P + 0.02P \\ &= (1 + 0.02)P \\ &= 1.02P. \end{aligned}$$

DO EXERCISE 3.

The reproduction curve $f(P) = 1.02P$ is shown in Fig. 6 along with the population-versus-time curve.

3. ▦ The population of the world in 1976 was 4.0 billion. Suppose the reproduction curve is given by

$$f(P) = 1.019P.$$

Complete the table.

Year	Population
1976	4.0 billion
1977	
1978	
1979	
1980	
1981	
1982	
1983	
1984	
1985	
1986	

4. An amount P is invested at $8\frac{1}{4}\%$ compounded annually. Find the reproduction curve.

5. A certain population is growing in such a way that if P is the population, then one year later it will have a 1% increase. Find the reproduction curve.

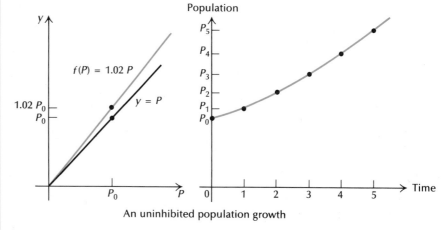

Figure 6   An uninhibited population growth.

Note that the population increases indefinitely. We will study growth similar to this in Chapter 4. It is interesting that the above also models the growth of an amount, or "population," of money invested at 2% (simple interest) compounded annually.

Example 1   An amount P is invested at $7\frac{1}{2}\%$ compounded annually. Find the reproduction curve.

Solution

$$f(P) = P + 7\tfrac{1}{2}\%P = 1 \cdot P + 0.075P = (1 + 0.075)P = 1.075P$$

DO EXERCISES 4 AND 5.

Maximum Sustainable Harvest

We know that a population P will grow to $f(P)$ in one year. If this were a population of fur-bearing animals, then one could "harvest" the amount

$$f(P) - P$$

each year without depleting the initial population P. Now suppose we wanted the value of P_0 that would allow the harvest to be the largest. If we could determine that P_0, then we would let the population grow until it reached that level, and then would begin harvesting year after year the amount $f(P_0) - P_0$.

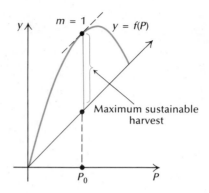

Let the harvest function H be given by

$$H(P) = f(P) - P.$$

Then

$$H'(P) = f'(P) - 1.$$

Now if we assume that $H'(P)$ exists for all values of P and that there is only one critical point, it follows that the maximum sustainable harvest occurs at that value P_0 such that

$$H'(P_0) = f'(P_0) - 1 = 0$$

and

$$H''(P_0) = f''(P_0) < 0.$$

Or, equivalently, we have the following.

THEOREM 8

The maximum sustainable harvest occurs at P_0 such that

$$f'(P_0) = 1 \quad \text{and} \quad f''(P_0) < 0,$$

and is given by

$$H(P_0) = f(P_0) - P_0.$$

Example 2 A certain population of fur-bearing animals has the reproduction curve

$$f(P) = P(10 - P),$$

where P is measured in thousands. Find the population at which the maximum sustainable harvest occurs. Find the maximum sustainable harvest.

CSS

6. A certain population of fur-bearing animals has the reproduction curve

$$f(P) = P(8 - P),$$

where P is measured in thousands. Find the population at which the maximum sustainable harvest occurs. Find the maximum sustainable harvest.

Solution Now

$$f(P) = 10P - P^2,$$

so

$$f'(P) = 10 - 2P$$

and

$$f''(P) = -2.$$

We set $f'(P) = 1$ and solve:

$$10 - 2P = 1$$
$$-2P = -9$$
$$P = 4.5.$$

There is a maximum since the second derivative is negative for all values of P. We find the maximum sustainable harvest by substituting 4.5 into the equation $H(P) = f(P) - P = (10P - P^2) - P$:

$$H(4.5) = [10 \cdot 4.5 - (4.5)^2] - 4.5 = 24.75 - 4.5 = 20.25.$$

Thus the maximum sustainable harvest is 20,250 at $P = 4500$.

DO EXERCISE 6.

EXERCISE SET 3.5

For each reproduction curve, (a) find the population at which the maximum sustainable harvest occurs, and (b) find the maximum sustainable harvest.

1. $f(P) = P(20 - P)$, where P is measured in thousands.

2. $f(P) = P(6 - P)$, where P is measured in thousands.

3. $f(P) = -0.025P^2 + 4P$, where P is measured in thousands. This is the reproduction curve in the Hudson Bay area for the *snowshoe hare*, a fur-bearing animal.

4. $f(P) = -0.01P^2 + 2P$, where P is measured in thousands. This is the reproduction curve in the Hudson Bay area for the *lynx*, a fur-bearing animal.

A snowshoe hare. (© *Ed Cesar from National Audubon Society*)

A lynx. (© *Russ Kinne: Photo Researchers, Inc.*)

5. $f(P) = 40\sqrt{P}$, where P is measured in thousands. Assume this is the production curve for the brown trout population in a large lake.

7. $f(P) = 1.08P$

6. $f(P) = 30\sqrt{P}$, where P is measured in thousands. Assume this is the production curve for the blue gill population in a large lake.

8. $f(P) = 1.075P$

OBJECTIVES

You should be able to

a) Given a function $y = f(x)$ and a value for Δx, find Δy.

b) Given a function $y = f(x)$ and a value for Δx or dx, find dy.

c) Use differentials to make approximations of numbers like

$$\sqrt{27} \quad \text{or} \quad \sqrt[3]{10}.$$

3.6 APPROXIMATION

Delta Notation

Recall the difference quotient,

$$\frac{f(x + h) - f(x)}{h},$$

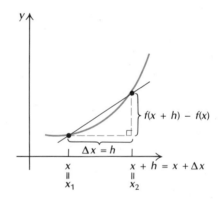

which is used to define the derivative of a function at x. The number h was considered to be a *change* in x. Another notation for such a change is Δx, read "delta x." The expression Δx is *not* the product of Δ and x, but is an entity unto itself; that is, it is a new type of variable that represents the *change* in the value of x from a *first* value to a *second*. Thus

$$\Delta x = (x + h) - x = h.$$

If subscripts are used for the first and second values of x, we would have

$$\Delta x = x_2 - x_1 \quad \text{or} \quad x_2 = x_1 + \Delta x.$$

Now Δx can be positive or negative.

Examples

a) If $x_1 = 4$ and $\Delta x = 0.7$, then $x_2 = 4.7$.

b) If $x_1 = 4$ and $\Delta x = -0.7$, then $x_2 = 3.3$.

1. For $y = x^2$, $x = 3$, and $\Delta x = -0.1$, find Δy.

We usually omit the subscripts and use x and $x + \Delta x$.

Now suppose we have a function given by $y = f(x)$. A change in x from x to $x + \Delta x$ yields a change in y from $f(x)$ to $f(x + \Delta x)$. The change in y is given by

$$\Delta y = f(x + \Delta x) - f(x).$$

Example 1 For $y = x^2$, $x = 4$, and $\Delta x = 0.1$, find Δy.

Solution

$$\Delta y = (4 + 0.1)^2 - 4^2 = (4.1)^2 - 4^2 = 16.81 - 16 = 0.81$$

Example 2 For $y = x^3$, $x = 2$, and $\Delta x = -0.1$, find Δy.

Solution

$$\Delta y = [2 + (-0.1)]^3 - 2^3 = (1.9)^3 - 2^3 = 6.859 - 8 = -1.141$$

DO EXERCISES 1 AND 2.

2. For $y = x^3$, $x = 2$, and $\Delta x = 1$, find Δy.

Using delta notation, the difference quotient

$$\frac{f(x + h) - f(x)}{h}$$

becomes

$$\frac{f(x + \Delta x) - f(x)}{\Delta x} = \frac{\Delta y}{\Delta x}.$$

We can then express the derivative as

$$\frac{dy}{dx} = \lim_{\Delta x \to 0} \frac{\Delta y}{\Delta x}.$$

Note how the delta notation resembles the Leibniz notation.

For values of Δx close to 0 we have the approximation

$$\frac{dy}{dx} \approx \frac{\Delta y}{\Delta x}, \quad \text{or} \quad f'(x) \approx \frac{\Delta y}{\Delta x}.$$

Multiplying both sides of the second expression by Δx we get

$$\Delta y \approx f'(x)\, \Delta x.$$

We can see this in the graph below.

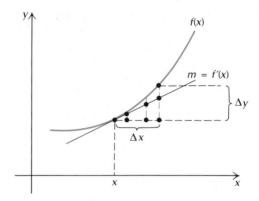

Recall that the derivative is a limit of slopes $\Delta y/\Delta x$ of secant lines. Thus as Δx gets smaller, the ratio $\Delta y/\Delta x$ gets closer to dy/dx. Note too that over small intervals the tangent line is a good approximation, linearly, to the function. Thus it is reasonable to assume that average rates of change $\Delta y/\Delta x$ of the function are approximately the same as the slope of the tangent line.

Let us use the fact that

$$\Delta y \approx f'(x)\, \Delta x$$

to make certain approximations, such as square roots.

Example 3 Approximate $\sqrt{27}$ using $\Delta y \approx f'(x)\, \Delta x$.   `CSS`

Solution We first think of the number closest to 27 that is a perfect square. This is 25. What we will do is approximate how y, or $\sqrt{x}$, changes when 25 changes by $\Delta x = 2$. Let

$$y = f(x) = \sqrt{x}.$$

Then

$$\Delta y = \sqrt{x + \Delta x} - \sqrt{x} = \sqrt{x + \Delta x} - y,$$

so

$$y + \Delta y = \sqrt{x + \Delta x}.$$

Now

$$\Delta y \approx f'(x)\, \Delta x = \frac{1}{2}x^{-1/2}\, \Delta x = \frac{1}{2\sqrt{x}}\, \Delta x.$$

Let $x = 25$ and $\Delta x = 2$. Then

$$\Delta y \approx f'(x)\, \Delta x = \frac{1}{2\sqrt{25}} \cdot 2 = \frac{1}{\sqrt{25}} = \frac{1}{5} = 0.2.$$

3. Approximate $\sqrt{67}$ using $\Delta y \approx f'(x) \, \Delta x$. See how close the approximation is by finding $\sqrt{67}$ in Table 1 at the back of the book.

So

$$\sqrt{27} = \sqrt{x + \Delta x} = y + \Delta y \approx \sqrt{25} + 0.2 = 5 + 0.2 = 5.2.$$

To five decimal places, $\sqrt{27} = 5.19615$. Thus our approximation is fairly good.

DO EXERCISE 3.

Suppose we have a total cost function $C(x)$. When $\Delta x = 1$, we have

$$\Delta C \approx C'(x).$$

Whether this is a good approximation depends on the function and on the values of x. Let us consider an example.

Example 4 For the total cost function

$$C(x) = 2x^3 - 12x^2 + 30x + 200:$$

a) Find ΔC and $C'(x)$ when $x = 2$ and $\Delta x = 1$;

b) Find ΔC and $C'(x)$ when $x = 100$ and $\Delta x = 1$.

Solution

a) $\Delta C = C(2 + 1) - C(2) = C(3) - C(2) = \$236 - \$228 = \$8$.   Recall that $C(2)$ is the total cost of producing 2 units, and $C(3)$ is the total cost of producing 3 units, so $C(3) - C(2)$, or \$8, is the cost of the third unit. Now

$$C'(x) = 6x^2 - 24x + 30, \quad \text{so} \quad C'(2) = \$6.$$

b) $\Delta C = C(100 + 1) - C(100) = C(101) - C(100) = \$58,220$.   Note that this is the cost of the 101st unit. Now

$$C'(100) = \$57,630.$$

Note that in (a) we might not consider the approximation between ΔC and $C'(x)$ to be too good, while in (b) the approximation might be considered quite good since the numbers are so large. We have purposely used $\Delta x = 1$ to illustrate the following.

$$C'(x) \approx C(x + 1) - C(x)$$

Marginal cost is (approximately) the cost of the $(x + 1)$st, or next, unit.

This is the historical definition that economists have given to marginal cost.   CSS

4. Consider the total cost function

$$C(x) = 0.01x^2 + 4x + 500.$$

a) Find ΔC and $C'(x)$ when $x = 5$ and $\Delta x = 1$.

b) Find ΔC and $C'(x)$ when $x = 100$ and $\Delta x = 1$.

Similarly, the following is true.

$$R'(x) \approx R(x + 1) - R(x)$$

Marginal revenue is (approximately) the revenue from the sale of the $(x + 1)$st, or next, unit.

And

$$P'(x) \approx P(x + 1) - P(x)$$

Marginal profit is (approximately) the profit from the production and sale of the $(x + 1)$st, or next, unit.

DO EXERCISE 4.

EXERCISE SET 3.6

In Exercises 1–8, find Δy and $f'(x)\,\Delta x$.

1. For $y = f(x) = x^2$, $x = 2$, and $\Delta x = 0.01$

2. For $y = x^3$, $x = 2$, and $\Delta x = 0.01$

3. For $y = f(x) = x + x^2$, $x = 3$, and $\Delta x = 0.04$

4. For $y = f(x) = x - x^2$, $x = 3$, and $\Delta x = 0.02$

5. For $y = f(x) = 1/x^2$, $x = 1$, and $\Delta x = 0.5$

6. For $y = f(x) = 1/x$, $x = 1$, and $\Delta x = 0.2$

7. For $y = f(x) = 3x - 1$, $x = 4$, and $\Delta x = 2$

8. For $y = f(x) = 2x - 3$, $x = 8$, and $\Delta x = 0.5$

9. For the total cost function

$$C(x) = 0.01x^2 + 0.6x + 30,$$

find ΔC and $C'(x)$ when $x = 70$ and $\Delta x = 1$.

10. For the total cost function

$$C(x) = 0.01x^2 + 1.6x + 100,$$

find ΔC and $C'(x)$ when $x = 80$ and $\Delta x = 1$.

11. For the total revenue function

$$R(x) = 2x,$$

find ΔR and $R'(x)$ when $x = 70$ and $\Delta x = 1$.

12. For the total revenue function

$$R(x) = 3x,$$

find ΔR and $R'(x)$ when $x = 80$ and $\Delta x = 1$.

13. a) Using $C(x)$ of Exercise 9 and $R(x)$ of Exercise 11, find the total profit $P(x)$.

b) Find ΔP and $P'(x)$ when $x = 70$ and $\Delta x = 1$.

14. a) Using $C(x)$ of Exercise 10 and $R(x)$ of Exercise 12, find the total profit $P(x)$.

b) Find ΔP and $P'(x)$ when $x = 80$ and $\Delta x = 1$.

Approximate, using $\Delta y \approx f'(x)\,\Delta x$.

15. $\sqrt{19}$ **16.** $\sqrt{10}$ **17.** $\sqrt{102}$ **18.** $\sqrt{103}$ **19.** $\sqrt[3]{10}$ **20.** $\sqrt[3]{28}$

21. The spherical volume of a cancer tumor is given by

$$V = \frac{4}{3}\pi r^3,$$

where r is the radius in centimeters. By approximately how much does the volume increase when the radius is increased from 1 cm to 1.2 cm? Use 3.14 for π.

22. The circular area of a healing wound is given by

$$A = \pi r^2,$$

where r is the radius in centimeters. By approximately how much does the area decrease when the radius is decreased from 2 cm to 1.9 cm? Use 3.14 for π.

OBJECTIVES

You should be able to

a) **Differentiate implicitly and find the slope of a curve at a given point.**

b) **Solve related rate problems.**

*3.7 IMPLICIT DIFFERENTIATION AND RELATED RATES

Implicit Differentiation

Consider the equation

$$y^3 = x.$$

This equation *implies* that y is a function of x, for if we solve for y, we get

$$y = \sqrt[3]{x}$$
$$= x^{1/3}.$$

We know from our work in this chapter that

$$\frac{dy}{dx} = \frac{1}{3}x^{-2/3}. \tag{1}$$

A method known as *implicit differentiation* allows us to find dy/dx *without* solving for y. We use the Chain Rule, treating y as a function of x. We use the Extended Power Rule, and differentiate both sides of

$$y^3 = x$$

with respect to x:

$$\frac{d}{dx}y^3 = \frac{d}{dx}x.$$

The derivative on the left side is found using the Extended Power Rule:

$$3y^2\frac{dy}{dx} = 1.$$

---

*This section can be omitted without loss of continuity.

1. Use implicit differentiation. Find

$$\frac{dy}{dx}.$$

Leave the answer expressed in terms of y:

$$y^5 = 2x.$$

Then

$$\frac{dy}{dx} = \frac{1}{3y^2}, \quad \text{or} \quad \frac{1}{3}y^{-2}.$$

We can show that this indeed gives us the same answer as Eq. (1) by replacing y by $x^{1/3}$:

$$\frac{dy}{dx} = \frac{1}{3}y^{-2} = \frac{1}{3}(x^{1/3})^{-2} = \frac{1}{3}x^{-2/3}.$$

DO EXERCISE 1.

Often, it is difficult or impossible to solve for y, obtaining an explicit expression in terms of x. For example, the equation

$$y^3 + x^2y^5 - x^4 = 27$$

determines y as a function of x, but it would be difficult to solve for y. We can nevertheless find a formula for the derivative of y *without* solving for y. This involves computing dy^n/dx for various integers n, and hence involves the Extended Power Rule in the form

$$\frac{dy^n}{dx} = ny^{n-1} \cdot \frac{dy}{dx}.$$

Example 1 For

$$y^3 + x^2y^5 - x^4 = 27,$$

a) Find dy/dx using implicit differentiation;

b) Find the slope of the tangent line to the curve at the point $(0, 3)$.

Solution

a) We differentiate the term x^2y^5 using the Product Rule. Note that any time an expression involving y is differentiated, dy/dx must be a factor of the answer. When an expression involving just x is differentiated, there is no factor dy/dx.

$$\frac{d}{dx}(y^3 + x^2y^5 - x^4) = \frac{d}{dx}27$$

2. For

$$y^3 + x^2y^4 + x^3 = 8,$$

a) Find

$$\frac{dy}{dx}$$

using implicit differentiation.

b) Find the slope of the tangent line at (0, 2).

$$\frac{d}{dx}y^3 + \frac{d}{dx}x^2y^5 - \frac{d}{dx}x^4 = 0$$

$$3y^2 \cdot \frac{dy}{dx} + \overbrace{x^2 \cdot 5y^4 \cdot \frac{dy}{dx} + 2x \cdot y^5} - 4x^3 = 0$$

Get all terms
involving dy/dx

$$3y^2 \cdot \frac{dy}{dx} + 5x^2y^4 \cdot \frac{dy}{dx} = 4x^3 - 2xy^5$$

alone on one
side.

$$(3y^2 + 5x^2y^4)\frac{dy}{dx} = 4x^3 - 2xy^5$$

Solve for

$$\frac{dy}{dx} = \frac{4x^3 - 2xy^5}{3y^2 + 5x^2y^4}$$

dy/dx. Leave
answer in terms
of x and y.

b) To find the slope of the tangent line to the curve at (0, 3), we replace x by 0 and y by 3:

$$\frac{dy}{dx} = \frac{4 \cdot 0^3 - 2 \cdot 0 \cdot 3^5}{3 \cdot 3^2 + 5 \cdot 0^2 \cdot 3^4} = 0.$$

DO EXERCISE 2.

The demand function for a product (see Sections 1.5 and 2.6) is often given implicitly.

3. For the following equation, differentiate implicitly to find dp/dx:

$$100\sqrt{p} = 800 - x.$$

Example 2 For the following demand equation, differentiate implicitly to find dp/dx:

$$x = \sqrt{200 - p^3}.$$

Solution

$$\frac{d}{dx}x = \frac{d}{dx}(\sqrt{200 - p^3})$$

$$1 = \frac{1}{2}(200 - p^3)^{-1/2} \cdot (-3p^2) \cdot \frac{dp}{dx}$$

$$1 = \frac{-3p^2}{2\sqrt{200 - p^3}} \cdot \frac{dp}{dx}$$

$$\frac{2\sqrt{200 - p^3}}{-3p^2} = \frac{dp}{dx}.$$

DO EXERCISE 3.

4. A stone is thrown in a pond. A circular ripple is spreading over the pond in such a way that its radius r is increasing at the rate of 3 feet per second at the moment when r goes through the value $r = 4$ ft. At that moment how fast is the disturbed area increasing?

The radius of each ripple is a function of time. (Marshall Henrichs)

Related Rates

Suppose y is a function of x, say

$$y = f(x),$$

and x varies with time t (as a function of time t). Since y depends on x and x depends on t, y also depends on t. That is, y is also a function of time t. The Chain Rule gives the following:

$$\frac{dy}{dt} = \frac{dy}{dx} \cdot \frac{dx}{dt}.$$

Thus the rate of change of y is *related* to the rate of change of x. Let us see how this comes up in problems. It helps to keep in mind that any variable can be thought of as a function of time t, even though a specific expression in terms of t may not be given.

Example 3 A restaurant supplier services the restaurants in a circular area in such a way that its radius r is increasing at the rate of 2 miles per year at the moment when r goes through the value $r = 5$ mi. At that moment how fast is the area increasing?

Solution The area A and the radius r are always related by the equation for the area of a circle

$$A = \pi r^2.$$

We take the derivative of both sides with respect to t:

$$\frac{dA}{dt} = 2\pi r \cdot \frac{dr}{dt}.$$

At the moment in question, $dr/dt = 2$ mi/yr (miles per year) and $r = 5$ mi, so

$$\frac{dA}{dt} = 2\pi(5 \text{ mi})\left(2\frac{\text{mi}}{\text{yr}}\right)$$

$$= 20\pi\frac{\text{mi}^2}{\text{yr}} \approx 63 \text{ square miles per year.}$$

DO EXERCISE 4.

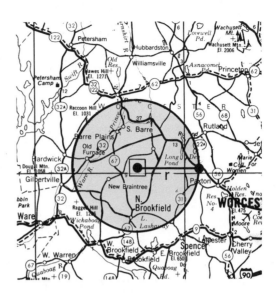

Example 4 *Business: Rate of change of revenue, cost, and profit.* For a company making stereos, total revenue from the sale of x stereos is given by

$$R(x) = 1000x - x^2,$$

and total cost is given by

$$C(x) = 3000 + 20x.$$

Suppose the company is producing and selling stereos at the rate of 10 stereos per day at the moment when the 400th stereo is produced. At that same moment, what is the rate of change of total (a) revenue? (b) cost? (c) profit?

Solution

a) $\dfrac{dR}{dt} = 1000 \cdot \dfrac{dx}{dt} - 2x \cdot \dfrac{dx}{dt}$ Differentiating with respect to time

$\phantom{\dfrac{dR}{dt}} = 1000 \cdot 10 - 2(400)10$ Substituting 10 for dx/dt and 400 for x

$\phantom{\dfrac{dR}{dt}} = \$2000$ per day

5. For a certain product a company determines that total revenue from the sale of x units is given by

$$R(x) = 200x - x^2,$$

and total cost is given by

$$C(x) = 5000 + 8x.$$

Suppose the company is producing and selling x units at a rate of 8 per day at the moment the 100th unit is produced. At the same moment, what is the rate of change of total (a) revenue? (b) cost? (c) profit?

b) $\dfrac{dC}{dt} = 20 \cdot \dfrac{dx}{dt}$ Differentiating with respect to time

$$= 20(10)$$

$$= \$200 \text{ per day}$$

c) Since $P = R - C$,

$$\frac{dP}{dt} = \frac{dR}{dt} - \frac{dC}{dt} = \$2000 \text{ per day} - \$200 \text{ per day}$$

$$= \$1800 \text{ per day.}$$

DO EXERCISE 5.

EXERCISE SET 3.7

Differentiate implicitly to find dy/dx. Then find the slope of the curve at the given point.

1. $xy - x + 2y = 3; \left(-5, \dfrac{2}{3}\right)$ **2.** $xy + y^2 - 2x = 0; (1, -2)$ **3.** $x^2 + y^2 = 1; \left(\dfrac{1}{2}, \dfrac{\sqrt{3}}{2}\right)$

4. $x^2 - y^2 = 1; (\sqrt{3}, \sqrt{2})$ **5.** $x^2y - 2x^3 - y^3 + 1 = 0; (2, -3)$ **6.** $4x^3 - y^4 - 3y + 5x + 1 = 0; (1, -2)$

Differentiate implicitly to find dy/dx.

7. $2xy + 3 = 0$ **8.** $x^2 + 2xy = 3y^2$ **9.** $x^2 - y^2 = 16$ **10.** $x^2 + y^2 = 25$

11. $y^5 = x^3$ **12.** $y^3 = x^5$ **13.** $x^2y^3 + x^3y^4 = 11$ **14.** $x^3y^2 - x^5y^3 = -19$

For the following demand equations, differentiate implicitly to find dp/dx.

15. $p^2 + p + 2x = 40$ **16.** $xp^3 = 24$ **17.** $(p + 4)(x + 3) = 48$ **18.** $1000 - 300p + 25p^2 = x$

19. *Biomedical: Rate of change of a cancer tumor.* The volume of a cancer tumor is given by

$$V = \frac{4}{3}\pi r^3.$$

The radius is increasing at the rate of 0.03 centimeter per day (cm/day) at the moment when $r = 1.2$ cm. How fast is the volume changing at that moment?

20. *Biomedical: Rate of change of a healing would.* The area of a healing wound is given by

$$A = \pi r^2.$$

The radius is decreasing at the rate of 1 millimeter per day (-1 mm/day) at the moment when $r = 25$ mm. How fast is the area decreasing at that moment?

Business: Rate of change of total revenue, cost, and profit. Find the rates of change of total revenue, cost, and profit for each of the following.

21. $R(x) = 50x - 0.5x^2,$

$C(x) = 4x + 10,$

when $x = 30$ and $dx/dt = 20$ units per day

22. $R(x) = 50x - 0.5x^2,$

$C(x) = 10x + 3,$

when $x = 10$ and $dx/dt = 5$ units per day

23. $R(x) = 2x,$

$C(x) = 0.01x^2 + 0.6x + 30,$

when $x = 20$ and $dx/dt = 8$ units per day

24. $R(x) = 280x - 0.4x^2,$

$C(x) = 5000 + 0.6x^2,$

when $x = 200$ and $dx/dt = 300$ units per day

25. Two cars start from the same point at the same time. One travels north at 25 miles per hour (mph), and the other travels east at 60 mph. How fast is the distance between them increasing at the end of 1 hr? (*Hint:* $D^2 = x^2 + y^2$. To find D after 1 hr, solve $D^2 = 25^2 + 60^2$.)

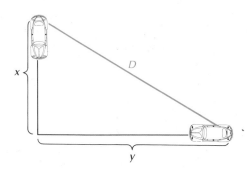

Biomedical: Poiseuille's Law. The flow of blood in a blood vessel is faster toward the center of the vessel and slower toward the outside. The speed of the blood V is given by

$$V = \frac{p}{4Lv}(R^2 - r^2),$$

where

R = the radius of the blood vessel,

r = the distance of the blood from center of the vessel,

and p, L, v are physical constants related to pressure, length, and viscosity of the blood vessel.

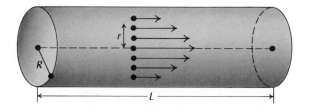

26. A ladder 26 ft long leans against a vertical wall. If the lower end is being moved away from the wall at the rate of 5 ft/sec, how fast is the height of the top decreasing (this will be a negative rate) when the lower end is 10 ft from the wall?

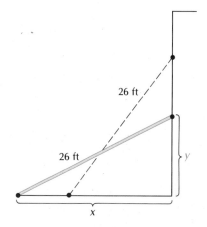

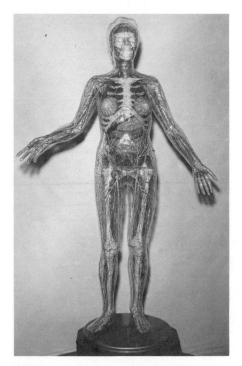

The flow of blood in a blood vessel can be modeled by Poiseuille's Law. (*Museum of Science, Boston*)

27. Assume r is a constant as well as p, L, and v.

a) Find the rate of change dV/dt in terms of R and dR/dt when $L = 1$ mm, $p = 100$, and $v = 0.05$.

b) A person goes out into the cold to shovel snow. Cold air has the effect of contracting blood vessels far from the heart. Suppose a blood vessel contracted at a rate of

$$\frac{dR}{dt} = -0.0015 \text{ mm/min}$$

at a place in the blood vessel where the radius is $R = 0.0075$ mm. Find the rate of change dV/dt at that location.

28. Assume r is a constant as well as p, L, and v.

a) Find the rate of change dV/dt in terms of R and dR/dt when $L = 1$ mm, $p = 100$, and $v = 0.05$.

b) When shoveling snow in cold air, a person with a history of heat trouble can develop angina (chest pains) due to contracting blood vessels. To counteract this, a person may take a nitroglycerin tablet, which dilates the blood vessels. Suppose, after a nitroglycerin tablet is taken, a blood vessel dilates at a rate of

$$\frac{dR}{dt} = 0.0025 \text{ mm/min}$$

at a place in the blood vessel where the radius is $R = 0.02$ mm. Find the rate of change dV/dt.

29. Two variable quantities A and B are found to be related by the equation

$$A^3 + B^3 = 9.$$

What is the rate of change dA/dt at the moment when $A = 2$ and $dB/dt = 3$?

30. Two variable quantities G and H, nonnegative, are found to be related by the equation

$$G^2 + H^2 = 25.$$

What is the rate of change dH/dt when $dG/dt = 3$ and $G = 0$? $G = 1$? $G = 3$?

▶ Differentiate implicitly to find dy/dx.

31. $\sqrt{x} + \sqrt{y} = 1$

32. $\dfrac{1}{x^2} + \dfrac{1}{y^2} = 5$

33. $y^3 = \dfrac{x - 1}{x + 1}$

34. $y^2 = \dfrac{x^2 - 1}{x^2 + 1}$

35. $x^{3/2} + y^{2/3} = 1$

36. $(x - y)^3 + (x + y)^3 = x^5 + y^5$

Differentiate implicitly to find dy/dx and d^2y/dx^2.

37. $xy + x - 2y = 4$

38. $y^2 - xy + x^2 = 5$

39. $x^2 - y^2 = 5$

40. $x^3 - y^3 = 8$

CHAPTER 3 TEST

1. For $y = x^4 - 3x^2$, find d^3y/dx^3.

Find the maximum and minimum values, if they exist, over the indicated interval. Where no interval is specified, use the real line.

2. $f(x) = x(6 - x)$

3. $f(x) = x^3 + x^2 - x + 1; \left[-2, \dfrac{1}{2}\right]$

4. $f(x) = -x^2 + 8.6x + 10$

5. $f(x) = -2x + 5; [-1, 1]$

6. $f(x) = -2x + 5$

7. $f(x) = 3x^2 - x - 1$

8. $f(x) = x^2 + \dfrac{128}{x}; (0, \infty)$

9. Of all numbers whose difference is 8, find the two that have the minimum product.

10. Minimum $Q = x^2 + y^2$, where $x - y = 10$.

11. Find the maximum profit and the number of units that must be produced and sold to yield the maximum profit.

$$R(x) = x^2 + 110x + 60,$$

$$C(x) = 1.1x^2 + 10x + 80$$

12. From a piece of cardboard 60 in. by 60 in., square corners are cut out so the sides can be folded up to make a box. What dimensions will yield a box of maximum volume? What is the maximum volume?

13. A sporting goods store sells 1225 tennis rackets per year. It costs $2 to store one tennis racket for one year. To reorder tennis rackets, there is a fixed cost of $1, plus $0.50 for each tennis racket. How many times per year should the sporting goods store order tennis rackets, and in what lot size, in order to minimize inventory costs?

14. Consider the reproduction curve $f(P) = P(100 - P)$, where P is measured in thousands. Find the population at which the maximum sustainable harvest occurs. Find the maximum sustainable harvest.

15. For $y = f(x) = x^2 - 3$, $x = 5$, and $\Delta x = 0.1$, find Δy and $f'(x)\,\Delta x$.

16. Approximate $\sqrt{104}$, using $\Delta y \approx f'(x)\,\Delta x$.

17. Sketch the graph of

$$f(x) = 3x^4 + 4x^3 - 6x^2 - 12x + 1.$$

18. Differentiate implicitly to find dy/dx. Then find the slope of the curve at the given point.

$$x^3 + y^3 = 9; \ (1, 2)$$

19. A board 13 ft long leans against a vertical wall. If the lower end is being moved away from the wall at the rate of 0.4 ft/sec, how fast is the upper end coming down when the lower end is 12 ft from the wall?

▶

20. Find the maximum and minimum values, if they exist, over the indicated interval.

$$f(x) = \frac{x^2}{1 + x^3}; \ [0, \infty)$$

21. The total cost of producing x units of a product is given by

$$C(x) = 100x + 100\sqrt{x} + \frac{\sqrt{x^3}}{100}.$$

a) Find the average cost $A(x)$.

b) Find the minimum value of $A(x)$.

EXPONENTIAL
AND
LOGARITHMIC
FUNCTIONS

4

The purchase of Manhattan Island from the Indians. (*The Bettmann Archive, Inc.*)

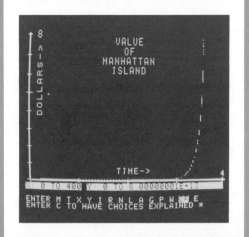

This graph shows the inflated value of Manhattan Island. Time is in hundreds of years and value is in ten trillion dollars.

OBJECTIVES

You should be able to

a) **Graph certain exponential functions.**

b) **Differentiate functions involving e.**

c) **Solve applied problems involving exponential functions.**

1. (▦ with a $\boxed{y^x}$ key.) Complete this table. Round to six decimal places.

r	2^r
3	
3.1	
3.14	
3.141	
3.1415	
3.14159	

What seems to be the value of 2^π to two decimal places?

4.1 EXPONENTIAL FUNCTIONS

Graphs of Exponential Functions

In Chapter 1 we reviewed definitions of expressions of the type a^x, where x was a rational number. For example,

$$a^{2.34}, \quad \text{or} \quad a^{234/100},$$

means "raise a to the 234th power and take the 100th root."

What about expressions with irrational exponents, such as $2^{\sqrt{2}}$, 2^π, or $2^{-\sqrt{3}}$? An irrational number is a number named by an infinite, non-repeating decimal. Let us consider 2^π. We know that π is irrational with infinite, nonrepeating decimal expansion:

$$3.141592654\ldots.$$

This means that π is approached as a limit by the rational numbers

$$3, \quad 3.1, \quad 3.14, \quad 3.141, \quad 3.1415, \ldots,$$

so it seems reasonable that 2^π should be approached as a limit by the rational powers

$$2^3, \quad 2^{3.1}, \quad 2^{3.14}, \quad 2^{3.141}, \quad 2^{3.1415}, \ldots,$$

DO EXERCISE 1.

In general, a^x is approximated by the values of a^r for rational numbers r near x; a^x is the limit of a^r as r approaches x through rational values. In summary, for $a > 0$, the definition of a^x for rational numbers x can be extended to arbitrary real numbers x in such a way that the usual laws of exponents, such as

$$a^x \cdot a^y = a^{x+y}, \quad a^x \div a^y = a^{x-y}, \quad (a^x)^y = a^{xy}, \quad \text{and} \quad a^{-x} = \frac{1}{a^x},$$

still hold. Moreover, the function so obtained,

$$f(x) = a^x,$$

is continuous.

DEFINITION

An **exponential function** f is given by

$$f(x) = a^x, \quad a > 0.$$

2. Consider $y = f(x) = 3^x$.   `CSS`

a) Complete this table of function values.

x	0	$\frac{1}{2}$	1	2	−1	−2
3^x	1	1.7	3	9	.333	.1111

b) Graph $f(x) = 3^x$.

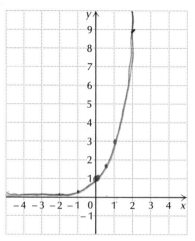

The following are examples of exponential functions:

$$f(x) = 2^x, \qquad f(x) = \left(\frac{1}{2}\right)^x, \qquad f(x) = (0.4)^x.$$

Note, in contrast to power functions like $y = x^2$ or $y = x^3$, that the variable in an exponential function is in the exponent. Exponential functions have extensive application. Let us consider their graphs.

Example 1   Graph $y = f(x) = 2^x$.

Solution

a)   First we find some function values.

x	0	$\frac{1}{2}$	1	2	3	−1	−2
$y = f(x)$ (or 2^x)	1	1.4	2	4	8	$\frac{1}{2}$	$\frac{1}{4}$

Note: For

$x = 0$, $y = 2^0 = 1$;

$x = \frac{1}{2}$, $y = 2^{1/2} = \sqrt{2} \approx 1.4$;

$x = 1$, $y = 2^1 = 2$;

$x = 2$, $y = 2^2 = 4$;

$x = 3$, $y = 2^3 = 8$;

$x = -1$, $y = 2^{-1} = \frac{1}{2}$;

$x = -2$, $y = 2^{-2} = \frac{1}{2^2} = \frac{1}{4}$.

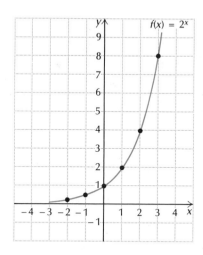

b)   Next, we plot the points and connect them with a smooth curve as shown in the figure.

DO EXERCISE 2.

3. Consider $y = f(x) = \left(\dfrac{1}{3}\right)^x$. CSS

a) Complete this table of function values.

x	0	$\frac{1}{2}$	1	2	−1	−2
y	1	.6	$\frac{1}{3}$	.111	3	9

b) Graph $f(x) = \left(\dfrac{1}{3}\right)^x$.

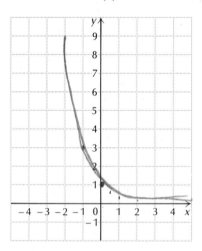

Example 2 Graph $y = f(x) = (\frac{1}{2})^x$.

Solution

a) We first find some function values. Before we do this, note that

$$y = f(x) = \left(\frac{1}{2}\right)^x = (2^{-1})^x = 2^{-x}.$$

This will ease our work.

x	0	$\frac{1}{2}$	1	2	−1	−2	−3
y	1	0.7	$\frac{1}{2}$	$\frac{1}{4}$	2	4	8

Note: For

$x = 0, y = 2^{-0} = 1;$

$x = \dfrac{1}{2}, y = 2^{-1/2}$

$= \dfrac{1}{\sqrt{2}} \approx \dfrac{1}{1.4} \approx 0.7;$

$x = 1, y = 2^{-1} = \dfrac{1}{2};$

$x = 2, y = 2^{-2} = \dfrac{1}{4};$

$x = -1, y = 2^{-(-1)} = 2;$

$x = -2, y = 2^{-(-2)} = 4;$

$x = -3, y = 2^{-(-3)} = 8.$

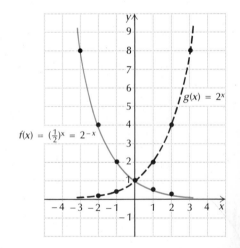

b) We plot these points and connect them with a smooth curve as shown by the solid curve in the figure. The dashed curve shows $g(x) = 2^x$ for comparison.

DO EXERCISE 3.

The following are some properties of the exponential function for various bases.

1. The function $f(x) = a^x$, where $a > 1$, is a positive, increasing, continuous function; and as x gets smaller, a^x approaches 0.

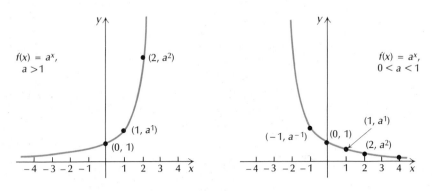

2. The function $f(x) = a^x$, where $0 < a < 1$, is a positive, decreasing, continuous function; and as x gets larger, a^x approaches 0. When $a = 1$, $f(x) = a^x = 1^x = 1$, and is a constant function.

Let us consider finding derivatives of exponential functions.

The Derivative of a^x, the Number e

Let us consider finding the derivative of the exponential function

$$f(x) = a^x.$$

The derivative is given by

$$f'(x) = \lim_{h \to 0} \frac{f(x + h) - f(x)}{h}$$ Definition of the derivative

$$= \lim_{h \to 0} \frac{a^{x+h} - a^x}{h}$$ Substituting a^{x+h} for $f(x + h)$ and a^x for $f(x)$

$$= \lim_{h \to 0} \frac{a^x \cdot a^h - a^x \cdot 1}{h}$$

$$= \lim_{h \to 0} a^x \cdot \left(\frac{a^h - 1}{h} \right)$$

$$= a^x \cdot \lim_{h \to 0} \frac{a^h - 1}{h}$$ Since the variable is h and $h \to 0$, we treat a^x as a constant, and the limit of a constant times a function is the constant times the limit.

4. In order to investigate ` CSS `

$$\lim_{h \to 0} \frac{2^h - 1}{h},$$

we choose a sequence of numbers h, approaching 0, and compute

$$\frac{2^h - 1}{h}.$$

a) Complete this table. Values of 2^h can be found using a calculator with a square-root key. Just take successive square roots.

h	$\dfrac{2^h - 1}{h}$
$\frac{1}{2}$	.828
$\frac{1}{4}$	.756
$\frac{1}{8}$	.724
$\frac{1}{16}$	.708
$\frac{1}{32}$	.700

b) To the nearest tenth, what is the value of

$$\lim_{h \to 0} \frac{2^h - 1}{h}? = .7$$

5. a) Complete this table. ` CSS `

h	$\dfrac{3^h - 1}{h}$
$\frac{1}{2}$	1.464
$\frac{1}{4}$	1.264
$\frac{1}{8}$	1.177
$\frac{1}{16}$	1.137
$\frac{1}{32}$	1.117

We get

$$f'(x) = a^x \cdot \lim_{h \to 0} \frac{a^h - 1}{h}. \tag{1}$$

In particular, for $g(x) = 2^x$,

$$g'(x) = 2^x \cdot \lim_{h \to 0} \frac{2^h - 1}{h}.$$

Note that the limit does not depend on the value of x at which we are evaluating the derivative. For $g'(x)$ to exist, we must determine whether

$$\lim_{h \to 0} \frac{2^h - 1}{h} \text{ exists.}$$

Let us investigate this question.

DO EXERCISE 4.

Margin Exercise 4 suggests that $(2^h - 1)/h$ has a limit as h approaches 0, and that its approximate value is 0.7, so that

$$g'(x) \approx (0.7)2^x.$$

In other words, the derivative is a constant times the function value 2^x. Similarly, for $t(x) = 3^x$,

$$t'(x) = 3^x \cdot \lim_{h \to 0} \frac{3^h - 1}{h}.$$

Again we can find an approximation for the limit that does not depend on the value of x at which we are evaluating the derivative.

DO EXERCISE 5.

Margin Exercise 5 suggests that $(3^h - 1)/h$ has a limit as h approaches 0, and that its approximate value is 1.1, so that

$$t'(x) \approx (1.1)3^x.$$

b) To the nearest tenth, what is the value of

$$\lim_{h \to 0} \frac{3^h - 1}{h}?$$ = 1.1

6. a) Complete this table.

x	−3	−2	−1	0	1	2	3
2^x	.125	.25	0.5	1	2	4	8
$(0.7)2^x$	.0875	.175	0.35	.7	1.4	2.8	5.6

b) Using the same set of axes, graph $g(x) = 2^x$ with a solid curve, and $g'(x) \approx (0.7)2^x$ with a dashed curve.

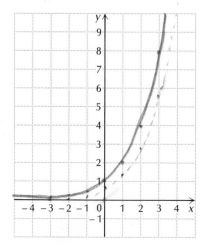

In other words, the derivative is a constant times the function value 3^x.

DO EXERCISE 6 (THIS PAGE) AND EXERCISE 7 (PAGE 216).

In Fig. 1 we have graphed $g(x) = 2^x$ and $g'(x) \approx (0.7)2^x$. Note that the graph of g' lies *below* the graph of g.

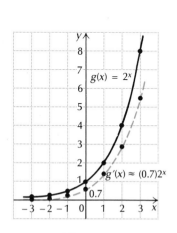

Figure 1

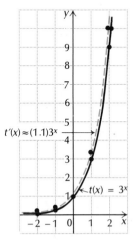

Figure 2

In Fig. 2 we have graphed $t(x) = 3^x$ and $t'(x) \approx (1.1)3^x$. Note that the graph of t' lies *above* the graph of t.

We might expect that there is exactly one base a between 2 and 3 for which a^x and its derivative have the same graph. This conjecture can be proved (though we will not do it here).

DEFINITION

We define the number e to be the unique positive real number for which

$$\lim_{h \to 0} \frac{e^h - 1}{h} = 1.$$

It follows that for the exponential function $f(x) = e^x$,

$$f'(x) = e^x \cdot \lim_{h \to 0} \frac{e^h - 1}{h} = e^x \cdot 1 = e^x.$$

7. a) Complete this table.

x	−2	−1	0	1	2
3^x	$\frac{1}{9}$	$\frac{1}{3}$	1	3	9
$(1.1)3^x$	$\frac{1.1}{9}$	$\frac{1.1}{3}$	1.1	3.3	9.9

.122   .366

b) Using the same set of axes, graph $h(x) = 3^x$ with a solid line, and $h'(x) \approx (1.1)3^x$ with a dashed line.

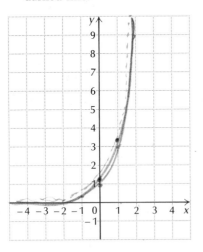

In Margin Exercise 8 you will not only consider an application of e, but also find decimal approximations.

DO EXERCISE 8 (PAGE 217).

Suppose we were to have the compounding periods n increase indefinitely. The amount in the investment of Margin Exercise 8 would be growing at interest compounded continuously, and would approach about $2.718. It can be shown that the number e can be described by a limit.

THEOREM 1

$$e = \lim_{n \to \infty} \left(1 + \frac{1}{n}\right)^n$$

That is, e is that number which

$$\left(1 + \frac{1}{n}\right)^n$$

CSS

approaches as n gets larger without bound. To ten decimal places, e is given by

$$e = 2.7182818284 \ldots .$$

We have established that for the function $f(x) = e^x$, we also have $f'(x) = e^x$, or, simply, the following.

THEOREM 2

$$\frac{d}{dx}e^x = e^x$$

Note that this says that the derivative (the slope of the tangent line) at any x is the same as the function value. Let us find some other derivatives.

Example 3

$$\frac{d}{dx}3e^x = 3e^x$$

8. ($\blacksquare$ with a $\boxed{y^x}$ key). The compound interest formula, which we developed in Chapter 1, is

$$A = P\left(1 + \frac{i}{n}\right)^{nt},$$

where A is the amount an initial investment P will be worth after t years at interest rate i, compounded n times per year.

Suppose $1 is an initial investment at 100% interest for 1 year (no bank would pay this). The above formula becomes

$$A = \left(1 + \frac{1}{n}\right)^{n}.$$

Complete this table. Round to six decimal places.

n	$\left(1 + \frac{1}{n}\right)^{n}$
1 (compounding annually)	
2 (compounding semiannually)	
3	
4 (compounding quarterly)	
5	
100	
365 (compounding daily)	
8760 (compounding hourly)	

Differentiate.

9. $y = 6e^x$

10. $y = x^3e^x$

11. $f(x) = \dfrac{e^x}{x^2}$

Example 4

$$\frac{d}{dx}x^2e^x = x^2 \cdot e^x + 2x \cdot e^x \qquad \text{Product Rule}$$

$$= e^x(x^2 + 2x), \text{ or } xe^x(x + 2) \qquad \text{Factoring}$$

Example 5

$$\frac{d}{dx}\left(\frac{e^x}{x^3}\right) = \frac{x^3 \cdot e^x - e^x(3x^2)}{x^6} \qquad \text{Quotient Rule}$$

$$= \frac{x^2e^x(x - 3)}{x^6} \qquad \text{Factoring}$$

$$= \frac{e^x(x - 3)}{x^4} \qquad \text{Simplifying}$$

DO EXERCISES 9–11.

The following rule (a form of the Chain Rule) allows us to find many other derivatives.

THEOREM 3

$$\frac{d}{dx}e^{f(x)} = f'(x)e^{f(x)}$$

The following gives us a way to remember this rule.

$$g(x) = e^{x^2 - 5x}$$

① Take the derivative of the exponent.

$$g'(x) = (2x - 5)e^{x^2 - 5x}$$

② Multiply the derivative of the exponent by the original function.

Example 6

$$\frac{d}{dx}e^{3x} = 3e^{3x}$$

Example 7

$$\frac{d}{dx}e^{-x^2 + 4x - 7} = (-2x + 4)e^{-x^2 + 4x - 7}$$

Differentiate.

12. $f(x) = e^{-4x}$

$-4e^{-4x}$

13. $y = e^{x^3+8x}$

$3x^2 + 8e^{x^3+8x}$

14. $f(x) = e^{\sqrt{x^2+5}}$

$\frac{1}{2}(x^2+5)^{-\frac{1}{2}} e^{\sqrt{x^2+5}}$

15. Graph $f(x) = 2e^{-x}$. Use Table 4. For example, for $x = 3$, $f(3) = 2e^{-3} = 2(0.0498) \approx 0.1$.

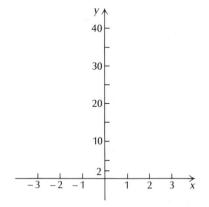

Example 8

$$\frac{d}{dx}e^{\sqrt{x^2-3}} = \frac{1}{2}(x^2-3)^{-1/2}\cdot 2x\cdot e^{\sqrt{x^2-3}}$$

$$= x(x^2-3)^{-1/2}\cdot e^{\sqrt{x^2-3}}$$

$$= \frac{xe^{\sqrt{x^2-3}}}{\sqrt{x^2-3}}$$

DO EXERCISES 12–14.

Graphs of e^x, e^{-x}, and $1 - e^{-kx}$

We use a calculator with an $\boxed{e^x}$ key or Table 4 (in the back of the book) to find approximate values of e^x and e^{-x}. With these we can draw graphs (Figs. 3 and 4) of the functions.

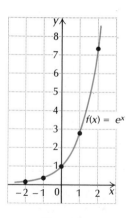

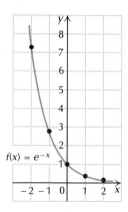

Figure 3   Figure 4

Note that the graph of e^{-x} is a reflection or mirror image of the graph of e^x across the y-axis.

DO EXERCISE 15.

Functions of the type $f(x) = 1 - e^{-kx}$ are also important.

Example 9 Graph $f(x) = 1 - e^{-2x}$, for nonnegative values of x.

Solution We obtain these values using a calculator with an $\boxed{e^x}$ key or Table 4 at the back of the book.

16. a) Complete this table for

$$f(x) = 1 - e^{-x}.$$

x	0	$\frac{1}{2}$	1	2	3	4
$f(x)$						

b) Graph $f(x) = 1 - e^{-x}$.   `CSS`

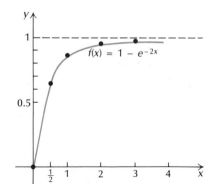

x	0	$\frac{1}{2}$	1	2	3
$f(x)$	0	0.63	0.86	0.98	0.998

For example,

$$f(1) = 1 - e^{-2(1)}$$
$$= 1 - e^{-2(1)}$$
$$= 1 - 0.135335 \approx 0.86.$$

DO EXERCISE 16.

In general, the graph of $f(x) = 1 - e^{-kx}$, for $k > 0$, increases from 0, since $f'(x) = ke^{-kx} > 0$, and approaches 1 as x gets larger; that is, $\lim_{x \to \infty} (1 - e^{-kx}) = 1$.

A word of caution! Functions of the type a^x (for example, 2^x, 3^x, and e^x) are different from functions of the type x^a (for example, x^2, x^3, $x^{1/2}$). For a^x the variable is in the exponent. For x^a the variable is in the base. The derivative of a^x is not xa^{x-1}. In particular, we have the following.

$$\frac{d}{dx}e^x \neq xe^{x-1}, \quad \text{but} \quad \frac{d}{dx}e^x = e^x$$

EXPONENTIAL AND LOGARITHMIC FUNCTIONS

EXERCISE SET 4.1

Graph. *CSS*

1. $y = 4^x$ **2.** $y = 5^x$ **3.** $y = (0.4)^x$ **4.** $y = (0.2)^x$ **5.** $x = 4^y$ **6.** $x = 5^y$

Differentiate.

7. $f(x) = e^{3x}$ **8.** $f(x) = e^{2x}$ **9.** $f(x) = 5e^{-2x}$ **10.** $f(x) = 4e^{-3x}$ **11.** $f(x) = 3 - e^{-x}$

12. $f(x) = 2 - e^{-x}$ **13.** $f(x) = -7e^x$ **14.** $f(x) = -4e^x$ **15.** $f(x) = \frac{1}{2}e^{2x}$ **16.** $f(x) = \frac{1}{4}e^{4x}$

17. $f(x) = x^4 e^x$ **18.** $f(x) = x^5 e^x$ **19.** $f(x) = \frac{e^x}{x^4}$ **20.** $f(x) = \frac{e^x}{x^5}$ **21.** $f(x) = e^{-x^2+7x}$

22. $f(x) = e^{-x^2+8x}$ **23.** $f(x) = e^{-x^2/2}$ **24.** $f(x) = e^{x^2/2}$ **25.** $y = e^{\sqrt{x-7}}$ **26.** $y = e^{\sqrt{x-4}}$

27. $y = \sqrt{e^x - 1}$ **28.** $y = \sqrt{e^x + 1}$ **29.** $y = xe^{-2x} + e^{-x} + x^3$ **30.** $y = e^x + x^3 - xe^x$

31. $y = 1 - e^{-x}$ **32.** $y = 1 - e^{-3x}$ **33.** $y = 1 - e^{-kx}$ **34.** $y = 1 - e^{-mx}$

Graph, using Table 4. *CSS*

35. $f(x) = e^{2x}$ **36.** $f(x) = e^{(1/2)x}$ **37.** $f(x) = e^{-2x}$ **38.** $f(x) = e^{-(1/2)x}$

39. $f(x) = 1 - e^{-x}$, for nonnegative values of x.   **40.** $f(x) = 2(1 - e^{-x})$, for nonnegative values of x.

41. *Business.* A company's total cost, in millions of dollars, is given by

$$C(t) = 100 - 50e^{-t},$$

where t = time. Find:

a) the marginal cost $C'(t)$;

b) $C'(0)$;

c) $C'(4)$.

42. *Business.* A company's total cost, in millions of dollars, is given by

$$C(t) = 200 - 40e^{-t},$$

where t = time. Find:

a) the marginal cost $C'(t)$;

b) $C'(0)$;

c) $C'(5)$.

▶ Differentiate. *CSS*

43. $y = (e^{3x} + 1)^5$ **44.** $y = (e^{x^2} - 2)^4$ **45.** $y = \dfrac{e^{3t} - e^{7t}}{e^{4t}}$ **46.** $y = \sqrt[3]{e^{3t} + t}$

47. $y = \dfrac{e^x}{x^2 + 1}$ **48.** $y = \dfrac{e^x}{1 - e^x}$ **49.** $f(x) = e^{\sqrt{x}} + \sqrt{e^x}$ **50.** $f(x) = \dfrac{1}{e^x} + e^{1/x}$

51. $f(x) = e^{x/2} \cdot \sqrt{x - 1}$ **52.** $f(x) = \dfrac{xe^{-x}}{1 + x^2}$ **53.** $f(x) = \dfrac{e^x - e^{-x}}{e^x + e^{-x}}$ **54.** $f(x) = e^{e^x}$

▦ with $\boxed{y^x}$ key. Each of the following is an expression for e. Find the function values that are approximations for e. Round to five decimal places.

55. $e = \lim_{t \to 0} f(t); f(t) = (1 + t)^{1/t}$. Find $f(1)$, $f(0.5)$, $f(0.2)$, $f(0.1)$, and $f(0.001)$.

56. $e = \lim_{t \to 1} g(t); g(t) = t^{1/(t-1)}$. Find $g(0.5)$, $g(0.9)$, $g(0.99)$, $g(0.999)$, and $g(0.9998)$.

57. Find the maximum value of $f(x) = x^2 e^{-x}$ on $[0, 4]$.   **58.** Find the minimum value of $f(x) = xe^x$ on $[-2, 0]$.

59. Sketch the graph of $y = x^2 e^{-x}$.   **60.** Sketch the graph of $y = e^{-x^2}$.

$$\frac{d\left(\ln\left(f(x)\right)\right)}{dx} = f'(x) \cdot \frac{1}{f(x)}$$

4.2 LOGARITHMIC FUNCTIONS

Graphs of Logarithmic Functions

Suppose we want to solve the equation

$$10^x = 1000.$$

We are trying to find that power of 10 which will give 1000. We can see that the answer is 3. The number 3 is called the "logarithm, base 10, of 1000."

DEFINITION

The definition of a *logarithm* is as follows:

$$y = \log_a x \quad \text{means} \quad x = a^y.$$

The number a is called the *logarithmic base*.

Thus, for logarithms base 10, $\log_{10} x$ is that number y such that $x = 10^y$. A logarithm can thus be thought of as an exponent. We can convert from a logarithmic equation to an exponential equation, and conversely, as follows.

LOGARITHMIC EQUATION	EXPONENTIAL EQUATION
$\log_a M = N$	$a^N = M$
$\log_{10} 100 = 2$	$10^2 = 100$
$\log_{10} 0.01 = -2$	$10^{-2} = 0.01$
$\log_{49} 7 = \dfrac{1}{2}$	$49^{1/2} = 7$

DO EXERCISES 1 AND 2.

  $a^c = b$

In order to graph a logarithmic equation, we can graph its equivalent exponential equation.

Example 1 Graph $y = \log_2 x$.

Solution We first write the equivalent exponential equation

$$x = 2^y.$$

We select values for y and find the corresponding values of 2^y.

x (or 2^y)	1	2	4	8	$\dfrac{1}{2}$	$\dfrac{1}{4}$
y	0	1	2	3	-1	-2

OBJECTIVES

You should be able to

a) Differentiate functions involving natural logarithms.

b) Given an exponential equation, write an equivalent logarithmic equation.

c) Given a logarithmic equation, write an equivalent exponential equation.

d) Given $\log_a 3 = 3 = 1.099$ and $\log_a 5 = 1.609$, find logarithms like $\log_a 15$ and $\log_a 5a$.

e) Solve an equation like $e^t = 40$, for t.

f) Solve problems involving exponential and natural logarithm functions.

1. Write equivalent exponential equations.

   a) $\log_b P = T$ $b^T = P$

   b) $\log_9 3 = \dfrac{1}{2}$ $9^{1/2} = 3$

   c) $\log_{10} 1000 = 3$ $10^3 = 1000$

   d) $\log_{10} 0.1 = -1$ $10^{-1} = .1$

2. Write equivalent logarithmic equations.

   a) $e^k = T$ $\log_e T = k$

   b) $16^{1/4} = 2$ $\log_{16} 2 = 1/4$

   c) $10^4 = 10,000$ $\log_{10} 10000 = 4$

   d) $10^{-3} = 0.001$ $\log_{10} .001 = -3$

$3^y = x$

3. Graph $y = \log_3 x$.

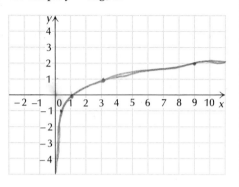

4. Consider

$$f(x) = 10^x$$

and

$$g(x) = \log_{10} x.$$

a) Find $f(3)$. _1000_

b) Find $g(1000)$. _= 3_

c) Use ▦. Find $g(5)$. _.6981_

d) Use ▦. Find $f(0.699)$. _5_

Next, we plot points, remembering that x is still the first coordinate.

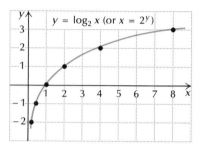

DO EXERCISE 3.

The graphs of $f(x) = 2^x$ and $g(x) = \log_2 x$ are shown on the same axes. Note that we can obtain the graph of g by reflecting the graph of f across the line $y = x$. Graphs obtained in this manner are known as *inverses* of each other.

While we cannot develop inverses in detail, it is of interest to note that they "undo" each other. For example,

$$f(3) = 2^3 = 8 \qquad \text{The input 3 gives the output 8.}$$

and

$$g(8) = \log_2 8 = 3. \qquad \text{The input 8 gets us back to 3.}$$

DO EXERCISE 4.

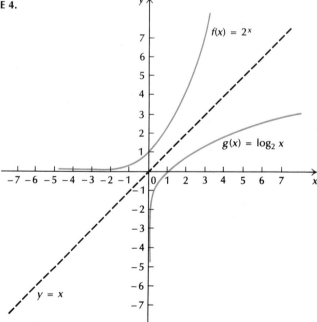

Basic Properties of Logarithms

The following are some basic properties of logarithms. The proofs follow from properties of exponents.

THEOREM 4

PROPERTY 1. $\log_a MN = \log_a M + \log_a N$

PROPERTY 2. $\log_a \dfrac{M}{N} = \log_a M - \log_a N$

PROPERTY 3. $\log_a M^k = k \cdot \log_a M$

PROPERTY 4. $\log_a a = 1$

PROPERTY 5. $\log_a a^k = k$

PROPERTY 6. $\log_a 1 = 0$

Proof of Properties 1 and 2. Let $X = \log_a M$ and $Y = \log_a N$. Then, writing the equivalent exponential equations, we have

$$M = a^X \quad \text{and} \quad N = a^Y.$$

Then by properties of exponents (see Section 1.1), we have

$$MN = a^X \cdot a^Y = a^{X+Y},$$

so

$$\log_a MN = X + Y$$
$$= \log_a M + \log_a N,$$

and

$$\frac{M}{N} = a^X \div a^Y = a^{X-Y},$$

so

$$\log_a \frac{M}{N} = X - Y$$
$$= \log_a M - \log_a N.$$

Proof of Property 3. Let $X = \log_a M$. Then

$$a^x = M,$$

so

$$(a^X)^k = M^k, \quad \text{or} \quad a^{Xk} = M^k.$$

Thus

$$\log_a M^k = Xk = k \cdot \log_a M.$$

Given

$$\log_a 2 = 0.301,$$

$$\log_a 5 = 0.699,$$

find each of the following.

5. $\log_a 4$

.602

6. $\log_a 10$

1.000

7. $\log_a \frac{2}{5}$

-.398

8. $\log_a \frac{5}{2}$

.398

9. $\log_a \frac{1}{5}$

-.699

10. $\log_a \sqrt{a^3}$

$\frac{3}{2}$

11. $\log_a 5a$

1.699

12. $\log_a 16$

1.204

Proof of Property 4. $\log_a a = 1$ because $a^1 = a$.
Proof of Property 5. $\log_a a^k = k$ because $(a^k) = a^k$.
Proof of Property 6. $\log_a 1 = 0$ because $a^0 = 1$.

Let us illustrate these properties.

Examples Given

$$\log_a 2 = 0.301,$$

$$\log_a 3 = 0.477,$$

find each of the following.

a) $\log_a 6$ $\log_a 6 = \log_a (2 \cdot 3) = \log_a 2 + \log_a 3$ Property 1

$$= 0.301 + 0.477$$

$$= 0.778$$

b) $\log_a \frac{2}{3}$ $\log_a \frac{2}{3} = \log_a 2 - \log_a 3$ Property 2

$$= 0.301 - 0.477$$

$$= -0.176$$

c) $\log_a 81$ $\log_a 81 = \log_a 3^4 = 4 \log_a 3$ Property 3

$$= 4(0.477)$$

$$= 1.908$$

d) $\log_a \frac{1}{3}$ $\log_a \frac{1}{3} = \log_a 1 - \log_a 3$ Property 2

$$= 0 - 0.477$$ Property 6

$$= -0.477$$

e) $\log_a \sqrt{a}$ $\log_a \sqrt{a} = \log_a a^{1/2} = \frac{1}{2}$ Property 5

f) $\log_a 2a$ $\log_a 2a = \log_a 2 + \log_a a$ Property 1

$$= 0.301 + 1$$ Property 4

$$= 1.301$$

g) $\log_a 5$ *No way to find using these properties.*
         $(\log_a 5 \neq \log_a 2 + \log_a 3)$

h) $\dfrac{\log_a 3}{\log_a 2}$ $\dfrac{\log_a 3}{\log_a 2} = \dfrac{0.477}{0.301} = 1.58$.

We simply divided, not using any of the properties.

DO EXERCISES 5–12.

13. ▦ Find each logarithm. Round to four decimal places.

a) log 31,456 4.4977

b) log 0.9080701 -4.1881

c) log 78.6 1.8954

d) log 7.86 $.8954$

e) log 0.786 $-.1046$

f) log 0.0786 -1.1046

g) log 0.00786 -2.1046

Common Logarithms (Optional)

Logarithms to the base 10 are called *common logarithms*. When we write

$$\log M,$$

with no base indicated, base 10 is to be understood. Note the following comparison of common logarithms and powers of 10.

$1000 = 10^3$	The common	$\log 1000 = 3$
$100 = 10^2$	logarithms	$\log 100 = 2$
$10 = 10^1$	at the right	$\log 10 = 1$
$1 = 10^0$	follow from	$\log 1 = 0$
$0.1 = 10^{-1}$	the powers at	$\log 0.1 = -1$
$0.01 = 10^{-2}$	the left.	$\log 0.01 = -2$
$0.001 = 10^{-3}$		$\log 0.001 = -3$

Before calculators and computers became so readily available, common logarithms were used extensively to do certain kinds of computations. In fact, this is why logarithms were developed. Since the standard notation we use for numbers is based on 10, it is logical that base 10, or common, logarithms were used for computations. Common logarithms can be found using a calculator or Table 3. Today, computations with common logarithms are mainly of historical interest; the logarithm functions, base e, are of modern importance.

DO EXERCISE 13

Natural Logarithms

The number e, which is approximately 2.718282, was developed in Section 4.1. It has extensive application in many fields. The number $\log_e x$ is called the *natural logarithm* of x and is abbreviated ln x; that is,

DEFINITION

$$\ln x = \log_e x.$$

The following is a restatement of the basic properties of logarithms in terms of natural logarithms.

THEOREM 5

PROPERTY 1. $\ln MN = \ln M + \ln N$

PROPERTY 2. $\ln \dfrac{M}{N} = \ln M - \ln N$

PROPERTY 3. $\ln a^k = k \cdot \ln a$

PROPERTY 4. $\ln e = 1$

PROPERTY 5. $\ln e^k = k$

PROPERTY 6. $\ln 1 = 0$

Let us illustrate these properties.

Examples Given

$$\ln 2 = 0.6931,$$

$$\ln 3 = 1.0986,$$

find each of the following.

a) $\ln 6$ $\ln 6 = \ln(2 \cdot 3) = \ln 2 + \ln 3$ Property 1

$$= 0.6931 + 1.0986$$

$$= 1.7917$$

b) $\ln 81$ $\ln 81 = \ln (3^4)$

$$= 4 \ln 3 \qquad \text{Property 3}$$

$$= 4(1.0986)$$

$$= 4.3944$$

c) $\ln \dfrac{2}{3}$ $\ln \dfrac{2}{3} = \ln 2 - \ln 3$ Property 2

$$= 0.6931 - 1.0986$$

$$= -0.4055$$

d) $\ln \dfrac{1}{3}$ $\ln \dfrac{1}{3} = \ln 1 - \ln 3$ Property 2

$$= 0 - 1.0986 \qquad \text{Property 6}$$

$$= -1.0986$$

e) $\ln 2e$ $\ln 2e = \ln 2 + \ln e$ Property 1

$$= 0.6931 + 1 \qquad \text{Property 4}$$

$$= 1.6931$$

Given

$$\ln 2 = 0.6931,$$
$$\ln 5 = 1.6094,$$

find each of the following.

14. $\ln 10$ 2.3025

15. $\ln \frac{2}{5}$ $-.9163$

16. $\ln \frac{5}{2}$ $.9163$

17. $\ln 16$ 2.7724

18. $\ln 5e$ 2.6094

19. $\ln \sqrt{e}$ $\frac{1}{2}$

20. $\ln \frac{1}{5}$ -1.6094

▦ Find each logarithm. Round to six decimal places.

21. $\ln 2$ $.693147$

22. $\ln 20$ 2.995732

23. $\ln 100$ 4.605170

24. $\ln 0.07432$ -2.599375

25. $\ln 1.08$ 7.696104

26. $\ln 0.9999$ -1.000050

f) $\ln \sqrt{e^3}$ $\ln \sqrt{e^3} = \ln e^{3/2}$

$$= \frac{3}{2} \qquad \text{Property 5}$$

DO EXERCISES 14–20.

Finding Natural Logarithms Using a Calculator

If you have a $\boxed{\ln}$ key on your calculator, you can find natural logarithms directly.

Examples Find each logarithm on your calculator. Round to six decimal places.

a) $\ln 5.24 = 1.656321$

b) $\ln 0.00001277 = -11.268412$

DO EXERCISES 21–26.

Finding Natural Logarithms Using a Table (Optional)

If you do not have a calculator with a natural logarithm key, you can use Table 3 at the back of the book. Part of that table is shown below. It shows some values of ln x.

Example 2 Find ln 5.24. Use Table 3.

Solution To find ln 5.24, locate the row headed 5.24; then move across to the column headed 0.04. Note the shaded number in the table. Thus

$$\ln 5.24 = 1.6563.$$

We find natural logarithms of numbers not in Table 3 as follows. We first express in scientific notation the number whose natural logarithm we are finding; that is, as a product $M \times 10^k$, where $1 \leq M < 10$.

TABLE 3

x	0.00	0.01	0.02	0.03	0.04	0.05	0.06	0.07	0.08	0.09
5.0	1.6094	1.6114	1.6134	1.6154	1.6174	1.6194	1.6214	1.6233	1.6253	1.6273
5.1	1.6292	1.6312	1.6332	1.6351	1.6371	1.6390	1.6409	1.6429	1.6448	1.6467
5.2	1.6487	1.6506	1.6525	1.6544	1.6563	1.6582	1.6601	1.6620	1.6639	1.6658
5.3	1.6677	1.6696	1.6715	1.6734	1.6752	1.6771	1.6790	1.6808	1.6827	1.6845
5.4	1.6864	1.6882	1.6901	1.6919	1.6938	1.6956	1.6974	1.6993	1.7011	1.7029

Using Table 3, find each logarithm.

27. ln 8.13

28. ln 81,300

29. ln 0.0813

30. ln 2000

31. ln 0.0001

Example 3 Find ln 5240. Use Table 3.

Solution

$$
\begin{aligned}
\ln 5240 &= \ln (5.24 \times 1000) & \\
&= \ln (5.24 \times 10^3) & 1000 = 10^3 \\
&= \ln 5.24 + \ln 10^3 & \text{Property 1} \\
&= \ln 5.24 + 3 \ln 10 & \text{Property 3} \\
&= 1.6563 + 6.9078 & \text{Find ln 5.24 in the body of} \\
& & \text{Table 3 and 3 ln 10 at the bottom.} \\
&= 8.5641 &
\end{aligned}
$$

Example 4 Find ln 0.000524. Use Table 3.

Solution

$$
\begin{aligned}
\ln 0.000524 &= \ln (5.24 \times 0.0001) & \\
&= \ln (5.24 \times 10^{-4}) & 0.0001 = 10^{-4} \\
&= \ln 5.24 + \ln 10^{-4} & \text{Property 1} \\
&= \ln 5.24 - 4 \ln 10 & \text{Property 3} \\
&= 1.6563 - 9.2103 & \text{Find ln 5.24 in the body of Table} \\
& & \text{3 and 4 ln 10 at the bottom.} \\
&= -7.554 &
\end{aligned}
$$

A number like ln 5.243 cannot be found in Table 3. In such cases we round and then use Table 3:

$$\ln 5.243 \approx \ln 5.24 \approx 1.6563.$$

DO EXERCISES 27–31.

Exponential Equations

In an equation where a variable occurs in an exponent, we call the equation *exponential*. We can use logarithms to manipulate or solve exponential equations.

Example 5 Solve for t: $e^t = 40$.

Solution

$$
\begin{aligned}
\ln e^t &= \ln 40 & \text{Taking the natural logarithm on both sides} \\
t &= \ln 40 & \text{Property 5} \\
t &= 3.688879 & \text{▦ or Table 3} \\
t &\approx 3.7 &
\end{aligned}
$$

Solve for t.

32. $e^t = 80$

$$\ln e^t = \ln 80$$
$$t = \ln 80$$
$$t = 4.3820$$

33. $e^{-0.06t} = 0.07$

$$-.06t = \ln .07$$
$$t = \frac{\ln .07}{-.06}$$
$$t = 44.3210$$

It should be noted that this is an approximation for t even though an equal sign is often used.

Example 6 Solve for t: $e^{-0.04t} = 0.05$.

Solution

$$\ln e^{-0.04t} = \ln 0.05 \qquad \text{Taking the natural logarithm on both sides}$$
$$-0.04t = \ln 0.05 \qquad \text{Property 5}$$
$$t = \frac{\ln 0.05}{-0.04}$$
$$t = \frac{-2.995732}{-0.04} \qquad \blacksquare \text{ or Table 3}$$
$$t \approx 75$$

Calculator note. For purposes of space and explanation, we have rounded the value of $\ln 0.05$ to -2.995732 in an intermediate step. On a calculator you should find

$$\frac{\ln 0.05}{-0.04},$$

obtaining

$$\frac{-2.995732274}{-0.04}.$$

Divide, and round at the end. Answers in the key are found in this manner. Remember, the number of places in a table or on a calculator may affect the accuracy of the answer. Usually, your answer should agree to at least three digits.

DO EXERCISES 32 AND 33.

Graphs of Natural Logarithm Functions

There are two ways we might obtain the graph of $y = f(x) = \ln x$. One is by writing its equivalent equation $x = e^y$.

Then we select values for y, and use a calculator or Table 4 to find the corresponding values of e^y. We then plot points, remembering that x still is the first coordinate.

x (or e^y)	0.1	0.4	1	2.7	7.4	20
y	-2	-1	0	1	2	3

34. a) Complete, using ▦ or Table 3.

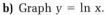

x	0.5	1	2	3	4
ln x			0.7		

b) Graph y = ln x.　　　　*CSS*

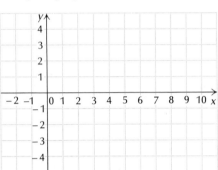

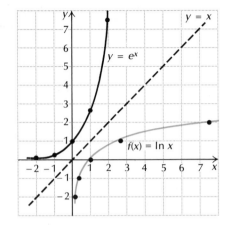

Figure 5

(How does this procedure in Fig. 5 compare with that used in plotting Fig. 3?) Note again that *f* and *g* are inverses of each other. That is, the graph of y = ln x is a reflection, or mirror image, across the line y = x, of the graph of y = e^x.

The second way of graphing y = ln x is by using a calculator or Table 3 at the back of the book. For example, ln 2 = 0.6931 ≈ 0.7.

DO EXERCISE 34.

These properties follow.

THEOREM 6

ln x exists only for positive numbers x.
ln x < 0 for 0 < x < 1.
ln x > 0 for x > 1.

The Derivative of ln x

Consider $f(x) = \ln x$.

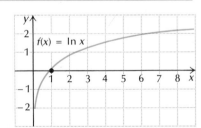

We can show that $f'(x) = 1/x$ (the slope of the tangent line at x is just the reciprocal of x). We are trying to find the derivative of

$$f(x) = \ln x. \tag{1}$$

We first write its equivalent exponential equation:

$$e^{f(x)} = x.$$

$\ln x = \log_e x = f(x)$, so $e^{f(x)} = x$, by the definition of logarithms

(2)

Now we differentiate both sides of this equation:

$$\frac{d}{dx} e^{f(x)} = \frac{d}{dx} x \qquad \text{This is implicit differentiation.}$$

$$f'(x) \cdot e^{f(x)} = 1$$

$$f'(x) \cdot x = 1 \qquad \text{Substituting } x \text{ for } e^{f(x)} \text{ from Eq. (2)}$$

$$f'(x) = \frac{1}{x}.$$

Thus we have the following.

CSS

THEOREM 7

$$\frac{d}{dx} \ln x = \frac{1}{x}$$

This is true only for positive values of x, since ln x is defined only for positive numbers. (For negative numbers x, this derivative formula becomes

$$\frac{d}{dx} \ln |x| = \frac{1}{x},$$

but we will consider such a case very little in this chapter.) Let us find some derivatives.

Example 7

$$\frac{d}{dx} 3 \ln x = \frac{3}{x}$$

Example 8

$$\frac{d}{dx}(x^2 \ln x + 5x) = x^2 \cdot \frac{1}{x} + 2x \cdot \ln x + 5 \qquad \text{Product Rule on } x^2 \ln x$$

$$= x + 2x \cdot \ln x + 5$$

Simplifying

$$= x(1 + 2 \ln x) + 5$$

Differentiate.

35. $y = 5 \ln x$

36. $f(x) = x^3 \ln x + 4x$

37. $f(x) = \dfrac{\ln x}{x^2}$

Example 9

$$\frac{d}{dx}\left(\frac{\ln x}{x^3}\right) = \frac{x^3 \cdot (1/x) - (\ln x)(3x^2)}{x^6} \qquad \text{Quotient Rule}$$

$$= \frac{x^2 - 3x^2 \ln x}{x^6}$$

$$= \frac{x^2(1 - 3 \ln x)}{x^6} \qquad \text{Factoring}$$

$$= \frac{1 - 3 \ln x}{x^4} \qquad \text{Simplifying}$$

DO EXERCISES 35–37.

The following rule (a form of the Chain Rule) allows us to find many other derivatives.

THEOREM 8

$$\frac{d}{dx} \ln f(x) = f'(x) \cdot \frac{1}{f(x)}$$

The following gives us a way of remembering this rule.

$$g(x) = \ln (x^2 - 8x)$$

$$g'(x) = (2x - 8) \cdot \frac{1}{x^2 - 8x}$$

① Differentiate the "inside" function.

② Multiply by the reciprocal of the "inside" function.

Example 10

$$\frac{d}{dx} \ln 3x = 3 \cdot \frac{1}{3x} = \frac{1}{x}$$

Note that we could have done this another way using Property 1:

$$\ln 3x = \ln 3 + \ln x;$$

then

$$\frac{d}{dx} \ln 3x = \frac{d}{dx} \ln 3 + \frac{d}{dx} \ln x = 0 + \frac{1}{x} = \frac{1}{x}.$$

Example 11

$$\frac{d}{dx} \ln (x^2 - 5) = 2x \cdot \frac{1}{x^2 - 5} = \frac{2x}{x^2 - 5}$$

Differentiate.

38. $f(x) = \ln 5x$

39. $f(x) = \ln (3x^2 + 4)$

40. $y = \ln (\ln 5x)$

41. $y = \ln \left(\dfrac{x^5 - 2}{x}\right)$

Example 12

$$\frac{d}{dx} \ln (\ln x) = \frac{1}{x} \cdot \frac{1}{\ln x} = \frac{1}{x \ln x}$$

Example 13

$$\frac{d}{dx} \ln \left(\frac{x^3 + 4}{x}\right) = \frac{d}{dx} [\ln (x^3 + 4) - \ln x]$$

Property 2. This avoids using the Quotient Rule

$$= 3x^2 \cdot \frac{1}{x^3 + 4} - \frac{1}{x} = \frac{3x^2}{x^3 + 4} - \frac{1}{x}$$

$$= \frac{3x^2}{x^3 + 4} \cdot \frac{x}{x} - \frac{1}{x} \cdot \frac{x^3 + 4}{x^3 + 4}$$

$$= \frac{(3x^2)x - (x^3 + 4)}{x(x^3 + 4)} = \frac{3x^3 - x^3 - 4}{x(x^3 + 4)} = \frac{2x^3 - 4}{x(x^3 + 4)}$$

DO EXERCISES 38–41.

Applications

Example 14 *Forgetting.* In a psychological experiment students were shown a set of nonsense syllables, such as POK, and asked to recall them every second thereafter. The percentage $R(t)$ who retained the syllables after t seconds was found to be given by

$$R(t) = 80 - 27 \ln t, \quad \text{for } t \geq 1.$$

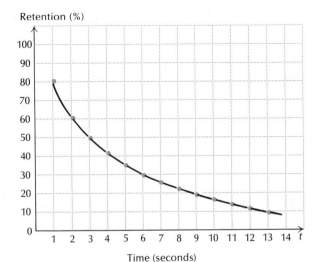

Time (seconds)

42. *Advertising.* A model for advertising response is given by

$$N(a) = 500 + 200 \ln a, \qquad a \geq 1,$$

where

$N(a)$ = the number of units sold and

a = the amount spent on advertising, in thousands of dollars.

a) How many units were sold after spending $1000? (Substitute 1 for a, not 1000.)

b) Find $N'(a)$.

c) Find maximum and minimum values, if they exist. `CSS`

Conduct your own memory experiment. Study this photograph carefully. Then put it aside and write down as many items as you can. Wait a half-hour and again write down as many as you can. Do this five more times. Make a graph of the number of items you remember versus time. Does the graph appear to be logarithmic? (*Cliff Garboden: Stock, Boston*)

Strictly speaking, the function is not continuous, but in order to use calculus, we "fill in" the graph with a smooth curve, considering $R(t)$ to be defined for any number $t \geq 1$. This is not unreasonable, since we are now able to find the percentage who retained the syllables after $t = 3.417$ seconds, instead of just after integer values such as 1, 2, 3, 4, and so on.

a) What percentage retained the syllables after 1 second?

b) Find $R'(t)$, the rate of change of R with respect to t.

c) Find maximum and minimum values, if they exist.

Solution

a) $R(1) = 80 - 27 \cdot \ln 1 = 80 - 27 \cdot 0 = 80\%$

b) $R'(t) = -27 \cdot \dfrac{1}{t} = -\dfrac{27}{t}$

c) Now $R'(t)$ exists for all values of t in the interval $[1, \infty)$. Note that for $t \geq 1$, $-27/t < 0$. Thus there are no critical points and R is decreasing. Then R has a maximum value at the endpoint 1. This maximum value is $R(1)$, or 80%.

DO EXERCISE 42.

Example 15 *Business.* A company begins a radio advertising campaign in New York City to market a new product. The percentage of the "target market" that buys a product is normally a function of the length of the advertising campaign. The radio station estimates this percentage as $(1 - e^{-0.04t})$ for this type of product, where t = the number of days of the campaign. The target market is estimated to be 1,000,000 people and the price per unit is $0.50. The costs of advertising are $1000 per day. Find the length of the advertising campaign that will result in maximum profit.

Solution That the percentage of the target market that buys the product can be modeled by $f(t) = 1 - e^{-0.04t}$ is justified by looking at its graph.

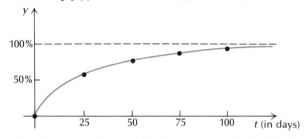

The function increases from 0 (0%) toward 1 (100%). The longer the advertising campaign, the larger the percentage of the market that has bought the product. (See also the discussion in Chapter 6.)

Recall the profit function (here expressed in terms of time, t):

$$\text{Profit} = \text{Revenue} - \text{Cost}$$

$$P(t) = R(t) - C(t).$$

a) Find $R(t)$.

$$R(t) = (\text{Price per unit}) \cdot (\text{Target market}) \cdot (\text{Percentage buying})$$

$$R(t) = 0.5(1,000,000)(1 - e^{-0.04t}) = 500,000 - 500,000e^{-0.04t}$$

b) Find $C(t)$.

$$C(t) = (\text{Advertising costs per day}) \cdot (\text{Number of days})$$

$$C(t) = 1000t$$

c) Find $P(t)$, and take its derivative.

$$P(t) = R(t) - C(t)$$

$$P(t) = 500,000 - 500,000e^{-0.04t} - 1000t$$

$$P'(t) = (-0.04)(-500,000e^{-0.04t}) - 1000$$

$$P'(t) = 20,000e^{-0.04t} - 1000$$

43. Solve the problem in Example 15 when the price per unit is $0.80.

d) Set the first derivative equal to 0 and solve.

$$20{,}000e^{-0.04t} - 1000 = 0$$

$$20{,}000e^{-0.04t} = 1{,}000$$

$$e^{-0.04t} = \frac{1{,}000}{20{,}000} = 0.05 \qquad (1)$$

$$\ln e^{-0.04t} = \ln 0.05$$

$$-0.04t = \ln 0.05$$

$$t = \frac{\ln 0.05}{-0.04}$$

$$t = \frac{-2.995732}{-0.04} \qquad \blacksquare \ \text{or Table 3}$$

$$t \approx 75$$

e) We have only one critical point. So we can use the second derivative to determine whether we have a maximum.

$$P''(t) = -0.04(20{,}000e^{-0.04t}) = -800e^{-0.04t}$$

Now since exponential functions are positive, $e^{-0.04t} > 0$ for all numbers t. Thus, since $-800e^{-0.04t} < 0$ for all numbers t, $P''(t)$ is less than 0 for $t = 75$ and we have a maximum.

The length of the advertising campaign must be 75 days to result in maximum profit.

DO EXERCISE 43.

EXERCISE SET 4.2

Write equivalent exponential equations.

1. $\log_2 8 = 3$ **2.** $\log_3 81 = 4$ **3.** $\log_8 2 = \frac{1}{3}$ **4.** $\log_{27} 3 = \frac{1}{3}$

5. $\log_a K = J$ **6.** $\log_a J = K$ **7.** $\log_b T = v$ **8.** $\log_c Y = t$

Write equivalent logarithmic equations.

9. $e^M = b$ **10.** $e^t = p$ **11.** $10^2 = 100$ **12.** $10^3 = 1000$

13. $10^{-1} = 0.1$ **14.** $10^{-2} = 0.01$ **15.** $M^p = V$ **16.** $Q^n = T$

Given $\log_b 3 = 1.099$ and $\log_b 5 = 1.609$, find each of the following. Do not use tables.

17. $\log_b 15$ **18.** $\log_b \frac{3}{5}$ **19.** $\log_b \frac{5}{3}$ **20.** $\log_b \frac{1}{3}$ **21.** $\log_b \frac{1}{5}$ **22.** $\log_b \sqrt{b}$

23. $\log_b \sqrt{b^3}$ **24.** $\log_b 3b$ **25.** $\log_b 5b$ **26.** $\log_b 9$ **27.** $\log_b 25$ **28.** $\log_b 75$

Given $\ln 4 = 1.3863$ and $\ln 5 = 1.6094$, find each of the following. Do not use tables.

29. $\ln 20$ **30.** $\ln \dfrac{4}{5}$ **31.** $\ln \dfrac{5}{4}$ **32.** $\ln \dfrac{1}{5}$ **33.** $\ln \dfrac{1}{4}$ **34.** $\ln 5e$

35. $\ln 4e$ **36.** $\ln \sqrt{e^6}$ **37.** $\ln \sqrt{e^8}$ **38.** $\ln 25$ **39.** $\ln 16$ **40.** $\ln 100$

⊞ Find each logarithm. Round to six decimal places.

41. $\ln 5894$ **42.** $\ln 99{,}999$ **43.** $\ln 0.0182$ **44.** $\ln 0.00087$ **45.** $\ln 1.88$ **46.** $\ln 18.8$

47. $\ln 0.0188$ **48.** $\ln 0.188$ **49.** $\ln 906$ **50.** $\ln 8100$ **51.** $\ln 0.011$ **52.** $\ln 0.00056$

Solve for t.

53. $e^t = 100$ **54.** $e^t = 1000$ **55.** $e^t = 60$ **56.** $e^t = 90$

57. $e^{-t} = 0.1$ **58.** $e^{-t} = 0.01$ **59.** $e^{-0.02t} = 0.06$ **60.** $e^{0.07t} = 2$

Differentiate.

61. $y = -6 \ln x$ **62.** $y = -4 \ln x$ **63.** $y = x^4 \ln x - \dfrac{1}{2}x^2$ **64.** $y = x^5 \ln x - \dfrac{1}{4}x^4$

65. $y = \dfrac{\ln x}{x^4}$ **66.** $y = \dfrac{\ln x}{x^5}$ **67.** $y = \ln \dfrac{x}{4}$ **68.** $y = \ln \dfrac{x}{2}$

$$\left[\text{Hint: } \ln \frac{x}{4} = \ln x - \ln 4 \right]$$

69. $f(x) = \ln (5x^2 - 7)$ **70.** $f(x) = \ln (7x^3 + 4)$ **71.** $f(x) = \ln (\ln 4x)$ **72.** $f(x) = \ln (\ln 3x)$

73. $f(x) = \ln \left(\dfrac{x^2 - 7}{x} \right)$ **74.** $f(x) = \ln \left(\dfrac{x^2 + 5}{x} \right)$ **75.** $f(x) = e^x \ln x$ **76.** $f(x) = e^{2x} \ln x$

77. $f(x) = \ln (e^x + 1)$ **78.** $f(x) = \ln (e^x - 2)$ **79.** $f(x) = (\ln x)^2$ **80.** $f(x) = (\ln x)^3$
(Hint: The Extended Power Rule)

Applications

CSS

81. *Psychology: Forgetting.* Students in college botany took a final exam. They took equivalent forms of the exam in monthly intervals thereafter. The average score, $S(t)$ in percent, after t months was found to be given by

$$S(t) = 68 - 20 \ln (t + 1), \qquad t \geq 0.$$

a) What was the average score when they initially took the test, $t = 0$?

b) What was the average score after 4 months?

c) What was the average score after 24 months?

d) What percentage of the initial score did they retain after 2 years (24 months)?

e) Find $S'(t)$.

f) Find maximum and minimum values, if they exist.

82. *Psychology: Forgetting.* Students in college zoology took a final exam. They took equivalent forms of the exam in monthly intervals thereafter. The average score, $S(t)$ in percent, after t months was found to be given by

$$S(t) = 78 - 15 \ln (t + 1), \qquad t \geq 0.$$

a) What was the average score when they initially took the test, $t = 0$?

b) What was the average score after 4 months?

c) What was the average score after 24 months?

d) What percentage of the initial score did they retain after 2 years (24 months)?

e) Find $S'(t)$.

f) Find maximum and minimum values, if they exist.

83. *Business: Advertising.* A model for advertising response is given by

$$N(a) = 1000 + 200 \ln a, \qquad a \geq 1,$$

where

$N(a)$ = the number of units sold, and

a = the amount spent on advertising, in thousands of dollars.

a) How many units were sold after spending $1000 ($a = 1$) on advertising?

b) Find $N'(a)$ and $N'(10)$.

c) Find maximum and minimum values, if they exist.

84. *Business: Advertising.* A model for advertising response is given by

$$N(a) = 2000 + 500 \ln a, \qquad a \geq 1,$$

where

$N(a)$ = the number of units sold, and

a = the amount spent on advertising, in thousands of dollars.

a) How many units were sold after spending $1000 dollars ($a = 1$) on advertising?

b) Find $N'(a)$ and $N'(10)$.

c) Find maximum and minimum values, if they exist.

85. *Psychology: Walking speed.* Bornstein and Bornstein found in a study that the average walking speed v of a person living in a city of population p, in thousands, is given by

$$v(p) = 0.37 \ln p + 0.05,$$

where v is in feet per second.

a) The population of Seattle is 531,000. What is the average walking speed of a person living in Seattle? Find $v(531)$.

b) The population of New York is 7,900,000. What is the average walking speed of a person living in New York?

c) Find $v'(p)$. Interpret $v'(p)$.

86. *Biomedical: The Reynolds number.* For many kinds of animals the Reynolds number R is given by

$$R = A \ln r - Br,$$

where A and B are positive constants and r is the radius of the aorta. Find the maximum value of R.

87. *Business.* Solve Example 15 given that the costs of advertising are $2000 per day.

89. *Biomedical: Acceptances of a new medicine.* The percentage P of doctors who accept a new medicine is given by

$$P(t) = 1 - e^{-0.2t}$$

where t = time, in months.

a) Find $P(1)$ and $P(6)$.

b) Find $P'(t)$.

c) How many months will it take for 90% of the doctors to become aware of the new medicine?

88. *Business.* Solve Example 15 given that the costs of advertising are $4000 per day.

90. *Psychology: Hullian learning model.* A typist learns to type W words per minute after t weeks of practice, where W is given by

$$W(t) = 100(1 - e^{-0.3t}).$$

a) Find $W(1)$ and $W(8)$.

b) Find $W'(t)$.

c) After how many weeks will the typist's speed be 95 words per minute?

read 276, 277, 278
try 1–5
page 283

91. *Business: Growth of a stock.* The value of a stock is modeled by

$$V(t) = \$58(1 - e^{-1.1t}) + \$20,$$

where V is the value of the stock after time t, in months.

a) Find $V(1)$ and $V(12)$.

b) Find $V'(t)$.

c) After how many months will the value of the stock be $75?

92. *Business: Marginal revenue.* The demand function for a certain product is given by

$$p = D(x) = 800e^{-0.125x}.$$

Recall that total revenue is given by $R(x) = xD(x)$.

a) Find $R(x)$.

b) Find the marginal revenue, $R'(x)$.

c) At what value of x will the revenue be maximum?

Differentiate.

93. $y = (\ln x)^{-4}$

94. $y = (\ln x)^n$

95. $f(t) = \ln (t^3 + 1)^5$

96. $f(t) = \ln (t^2 + t)^3$

97. $f(x) = [\ln (x + 5)]^4$

98. $f(x) = \ln [\ln (\ln 3x)]$

99. $f(t) = \ln [(t^3 + 3)(t^2 - 1)]$

100. $f(t) = \ln \dfrac{1 - t}{1 + t}$

101. $y = \ln \dfrac{x^5}{(8x + 5)^2}$

102. $y = \ln \sqrt{5 + x^2}$

103. $f(t) = \dfrac{\ln t^2}{t^2}$

104. $f(x) = \dfrac{1}{5}x^5 \left(\ln x - \dfrac{1}{5}\right)$

105. $y = \dfrac{x^{n+1}}{n + 1}\left(\ln x - \dfrac{1}{n + 1}\right)$

106. $y = \dfrac{x \ln x - x}{x^2 + 1}$

107. $y = \ln (t + \sqrt{1 + t^2})$

108. $f(x) = \ln \dfrac{1 + \sqrt{x}}{1 - \sqrt{x}}$

109. $f(x) = \ln [\ln x]^3$

110. $f(x) = \dfrac{\ln x}{1 + (\ln x)^2}$

111. Find $\displaystyle\lim_{h \to 0} \dfrac{\ln (1 + h)}{h}$.

112. ▦ Which is larger, e^π or π^e?

113. ▦ Find $\sqrt[e]{e}$. Compare it to other expressions of the type $\sqrt[x]{x}$, $x > 0$. What can you conclude?

114. Find the minimum value of $f(x) = x \ln x$.

CSS

115. Find the minimum value of $f(x) = x^2 \ln x$.

116. Sketch the graph of $y = \dfrac{\ln x}{x^2}$.

Solve for t.

117. $P = P_0 e^{kt}$

118. $P = P_0 e^{-kt}$

▦ Use input–output tables. Find each limit.

119. $\displaystyle\lim_{x \to \infty} \ln x$

120. $\displaystyle\lim_{x \to 1} \ln x$

Verify each of the following.

121. $\ln x = \dfrac{\log x}{\log e} \approx 2.3026 \log x$

122. $\log x = \dfrac{\ln x}{\ln 10} \approx 0.4343 \ln x$

OBJECTIVES

You should be able to

a) **State the solution of an equation**

$$\frac{dP}{dt} = kP,$$

as

$$P(t) = P_0 e^{kt}.$$

b) **Given a growth rate, find the doubling time.**

c) **Given the doubling time, find the growth rate.**

d) **Solve applied problems involving exponential growth.**

4.3 APPLICATIONS: THE UNINHIBITED GROWTH MODEL, $dP/dt = kP$

What will the world population be in 1986? (*Peter Vandermark: Stock, Boston*)

Consider the function

$$f(x) = 2e^{3x}.$$

Differentiating, we get

$$f'(x) = 3 \cdot 2e^{3x} = 3 \cdot f(x).$$

This, graphically, says that the derivative, or slope of the tangent line, is simply the constant 3 times the function value.

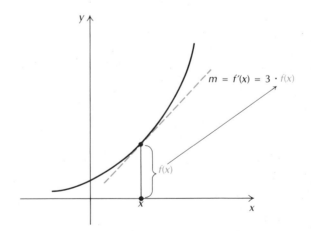

1. **a)** Differentiate $y = 5e^{4x}$.

   b) Express dy/dx in terms of y.

2. **a)** State the solution of
   $$\frac{dN}{dt} = kN.$$

   b) State the solution of
   $$f'(t) = k \cdot f(t).$$

DO EXERCISE 1.

In general, we have the following.

THEOREM 9

A function $y = f(x)$ satisfies the equation

$$\frac{dy}{dx} = ky \qquad [f'(x) = k \cdot f(x)]$$

if and only if

$$y = ce^{kx} \qquad [f(x) = ce^{kx}]$$

for some constant c.

No matter what the variables, you should be able to write the solution.

Example 1 The solution of $dA/dt = kA$ is $A = ce^{kt}$, or $A(t) = ce^{kt}$.

Example 2 The solution of $dP/dt = kP$ is $P = ce^{kt}$, or $P(t) = ce^{kt}$.

Example 3 The solution of $f'(Q) = k \cdot f(Q)$ is $f(Q) = ce^{kQ}$.

DO EXERCISE 2.

The equation

$$\frac{dP}{dt} = kP, \quad k > 0 \qquad [P'(t) = k \cdot P(t), \quad k > 0]$$

is the basic model of uninhibited population growth, whether it be a population of humans, a bacteria culture, or money invested at interest compounded continuously. Neglecting special inhibiting and stimulating factors, we know that a population normally reproduces itself at a rate proportional to its size, and this is exactly what the equation $dP/dt = kP$ says. The solution of the equation is

$$P(t) = ce^{kt}, \tag{1}$$

where $t =$ the time. At $t = 0$, we have some "initial" population $P(0)$ that we will represent by P_0. We can rewrite Eq. (1) in terms of P_0 as follows:

$$P_0 = P(0) = ce^{k \cdot 0} = ce^0 = c \cdot 1 = c.$$

3. *Exploratory exercises: Growth.* Use a sheet of $8\frac{1}{2} \times 11$ paper. Cut it into two equal pieces. Then cut these into four equal pieces. Then cut these into eight equal pieces, and so on, performing five cutting steps.

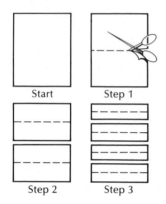

Start   Step 1

Step 2   Step 3

a) Place all the pieces in a stack and measure the thickness.

b) A piece of paper is typically 0.004 in. thick. Check the calculation in (a) by completing this table.

	t	$0.004 \cdot 2^t$
Start	0	$0.004 \cdot 2^0$, or 0.004
Step 1	1	$0.004 \cdot 2^1$, or 0.008
Step 2	2	$0.004 \cdot 2^2$, or 0.016
Step 3	3	
Step 4	4	
Step 5	5	

c) Compute the thickness of the paper (in miles) after 25 steps.

Thus $P_0 = c$, so we can express $P(t)$ as

$$P(t) = P_0 e^{kt}.$$

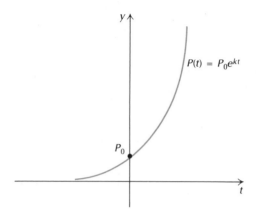

Its graph is the curve shown here, which shows how uninhibited growth results in a "population explosion."

DO EXERCISE 3.

The constant k is called the *rate of exponential growth*, or simply the *growth rate*. This is not the rate of change of the population size, which is

$$\frac{dP}{dt} = kP,$$

but the constant that P must be multiplied by to get its rate of change. It is thus a different use of the word *rate*. It is like the *interest rate* paid by a bank. If the interest rate is 12%, or 0.12, we do not mean that your bank balance P is growing at the rate of 0.12 dollars per year, but at the rate of $0.12P$ dollars per year. We therefore express the rate as 12% per year, rather than 0.12 dollars per year. We could say that the rate is 0.12 dollars *per dollar* per year. When interest is compounded continuously, the interest rate *is* a true exponential growth rate.

Example 4 *Business: Interest compounded continuously.* Suppose an amount P_0 is invested in a savings account where interest is compounded continuously at 12% per year. That is, the balance P grows at the rate given by

$$\frac{dP}{dt} = 0.12P.$$

4. *Business.* Suppose an amount P_0 is invested in a savings account where interest is compounded continuously at 13% per year. That is, the balance P grows at the rate given by

$$\frac{dP}{dt} = 0.13P.$$

a) Find the solution of the equation in terms of P_0 and 0.13.

b) Suppose $1000 is invested. What is the balance after 1 year?

c) After what period of time will an investment of $1000 double itself? CSS

Under ideal conditions, the growth rate of this rabbit population might be 11.7% per day. When will this population of rabbits double? (*Julie O'Neil: Stock, Boston*)

a) Find the solution of the equation in terms of P_0 and 0.12.

b) Suppose $100 is invested. What is the balance after 1 year?

c) After what period of time will an investment of $100 double itself?

Solution

a) $P(t) = P_0 e^{0.12t}$

b) $P(1) = 100e^{0.12(1)} = 100e^{0.12} = 100(1.127497)$

$$\approx \$112.75 \qquad \text{📱 or Table 4}$$

c) We are asking at what time T does $P(T) = \$200$. The number T is called the *doubling time*. To find T, we solve the equation:

$$200 = 100e^{0.12 \cdot T}$$

$$2 = e^{0.12T}. \qquad \text{Multiplying by } \tfrac{1}{100}$$

We use natural logarithms to solve this equation:

$$\ln 2 = \ln e^{0.12T}$$

$$\ln 2 = 0.12T \qquad \text{Property 5: } \ln e^k = k$$

$$\frac{\ln 2}{0.12} = T$$

$$\frac{0.693147}{0.12} = T \qquad \text{📱 or Table 3}$$

$$5.8 \approx T.$$

Thus $100 will double itself in 5.8 years.

DO EXERCISE 4.

Let us consider another development of the formula

$$P(t) = P_0 e^{kt}$$

for interest compounded continuously. First, consider the compound interest formula

$$A = P_0 \left(1 + \frac{k}{n}\right)^{nt},$$

letting $P = P_0$ and $k = i$. We are interested in what happens as n gets very large, that is, as $n \to \infty$. To determine this limit we first make a substitution:

$$h = \frac{n}{k}.$$

Then

$$hk = n \quad \text{and} \quad \frac{1}{h} = \frac{k}{n}.$$

Note that k is a positive constant. Thus, as n gets large so does h, and as h gets large so does n. To find a formula for continuously compounded interest, we evaluate the following limit:

$$P(t) = \lim_{h \to \infty} \left[P_0 \left(1 + \frac{k}{n} \right)^{nt} \right] \qquad \text{Letting the number of compounding periods become infinite}$$

$$= P_0 \lim_{h \to \infty} \left[\left(1 + \frac{1}{h} \right)^{hkt} \right] \qquad \text{The limit of a constant times a function is the constant times the limit. We also substitute } 1/h \text{ for } k/n \text{ and } hk \text{ for } n.$$

$$= P_0 \left[\lim_{h \to \infty} \left(1 + \frac{1}{h} \right)^{h} \right]^{kt} \qquad \text{The limit of a power is the power of the limit— a form of L2 and L3 in Section 2.1.}$$

$$= P_0 [e]^{kt}. \qquad \text{Definition of } e$$

We can find a general expression relating the growth rate k and the doubling time T by solving the following equation:

$$2P_0 = P_0 e^{kT}$$

$$2 = e^{kT} \qquad \text{Multiplying by } 1/P_0$$

$$\ln 2 = \ln e^{kT}$$

$$\ln 2 = kT.$$

THEOREM 10

The growth rate k and the doubling time T are related by

$$kT = \ln 2 = 0.693147,$$

or

$$k = \frac{\ln 2}{T} = \frac{0.693147}{T}$$

and

$$T = \frac{\ln 2}{k} = \frac{0.693147}{k}.$$

Note that this relationship between k and T does not depend on P_0.

Example 5 At one time in Canada, a bank advertised that it would

5. Complete this table relating growth rate k and doubling time T.

Growth Rate k (% per year)	Doubling Time T (in years)
2%	
	10
14%	
	15
1%	

6. Upon completion of the 1980 census, it was determined that the population of the United States was 225 million. It was estimated that the population P was growing exponentially at the rate of 0.8% per year. That is,

$$\frac{dP}{dt} = 0.008P$$

where t = the time in years.

a) Find the solution of the equation assuming $P_0 = 225$ and $k = 0.008$.

b) Estimate the U.S. population in 1989 ($t = 9$).

c) After what period of time will the population be double that in 1980? CSS

double your money in 6.6 years. What is the interest rate on such an account, assuming interest to be compounded continuously?

Solution

$$k = \frac{\ln 2}{T} = \frac{0.693147}{6.6} = 0.105 = 10.5\%$$

DO EXERCISE 5.

Example 6 *Ecology: World population growth.* The population of the world passed 4 billion on March 28, 1976. On the basis of data available at that time it was estimated that the population P was growing exponentially at the rate of 1.9% per year. That is, $dP/dt = 0.019P$, where t = the time, in years, from 1976. (To facilitate computations we assume the population was 4 billion at the start of 1976.)

a) Find the solution of the equation assuming $P_0 = 4$ and $k = 0.019$.

b) Estimate the world population in 1986 ($t = 10$).

c) After what period of time will the population be double that in 1976?

Solution

a) $P(t) = 4e^{0.019t}$

b) $P(10) = 4e^{0.019(10)} = 4e^{0.19} = 4(1.209250)$ ▦ or Table 4

$$\approx 4.8 \text{ billion}$$

c) $T = \dfrac{\ln 2}{k} = \dfrac{0.693147}{0.019} \approx 36.5 \text{ yr}$

Thus, according to this model, the 1976 population will double by the year 2012. No wonder ecologists are alarmed!

DO EXERCISE 6.

The Rule of 70

The relationship between doubling time T and interest rate k is the basis of a rule often used in the investment world, called the *Rule of 70:* To estimate how long it will take to double your money at varying rates of return, divide 70 by the rate of return. To see how this works, let the interest rate $k = r\%$. Then,

$$T = \frac{\ln 2}{k} = \frac{0.693147}{r\%} = \frac{0.693147}{r \times 0.01} = \frac{0.693147}{r \times 0.01} \cdot \frac{100}{100} = \frac{69.3147}{r} \approx \frac{70}{r}.$$

(Donald C. Dietz: Stock, Boston)

Some myths about alcohol (Indianapolis Alcohol Safety Action Project).
It's a fact—the blood alcohol concentration (BAC) in the human body is
measurable. And there's no cure for its
effect on the central nervous system except time. It takes time for the body's
metabolism to recover.

That means a cup of coffee, a cold
shower, and fresh air can't erase the effect of several drinks.

There are variables, of course: a
person's body weight, how many
drinks have been consumed in a given
time, how much has been eaten, etc.
These account for different BAC levels.
But the myth that some people can
"handle their liquor" better than others
is a gross rationalization—especially
when it comes to driving. Some people
can act more sober than others. But an
automobile doesn't act; it reacts.

Modeling Other Phenomena

Example 7 *Biomedical: Alcohol absorption and the risk of having
an accident.* Extensive research has provided data relating the risk R
(%) of having an automobile accident to the blood alcohol level b (%).
Note that these data are not a perfect fit (see the part between $b = 0$ and
$b = 0.05$), but we can approximate the data with an exponential function. The modeling assumption is that the rate of change of the risk R
with respect to the blood alcohol level b is given by

$$\frac{dR}{db} = kR.$$

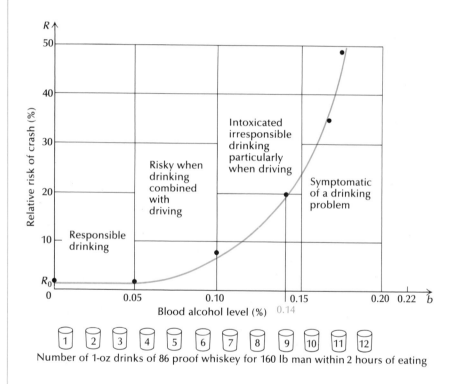

Number of 1-oz drinks of 86 proof whiskey for 160 lb man within 2 hours of eating

a) Find the solution of the equation, assuming $R_0 = 1\%$.

b) Find k, using the data point $R(0.14) = 20$. (This is how one might
fit the data to an exponential equation.)

c) Rewrite $R(b)$ in terms of k.

d) At what blood alcohol level will the risk of having an accident be
100%? Round to the nearest hundredth.

7. *Ecology: Electrical energy demand.* Past data on electrical energy demand in the United States are shown in the graph.

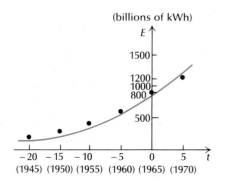

(billions of kWh)

It appears that we can fit an exponential function to the data. We accept the modeling assumption that the rate of change of electrical energy need E (in billion kilowatthours, kWh) with respect to time is given by

$$\frac{dE}{dt} = kE.$$

a) Find the solution of the equation, assuming $E_0 = 800$ billion kWh. That is, at $t = 0(1965)$, $E = 800$.

b) Find k using the data point $E(5) = 1200$ billion kWh. That is, in 1970, 1200 billion kWh were used. Round to the nearest hundredth.

c) Rewrite $E(t)$ in terms of k.

d) How much electrical energy will be needed in 1995?

Solution

a) Since both R and b are percents, we omit the % symbol for ease of computation. The solution is

$$R(b) = e^{kb}, \qquad \text{since } R_0 = 1.$$

b) We solve this equation for k:

$$20 = e^{k(0.14)} = e^{0.14k}.$$

We use natural logarithms to solve this equation:

$$\ln 20 = \ln e^{0.14k}$$

$$\ln 20 = 0.14k$$

$$\frac{\ln 20}{0.14} = k$$

$$\frac{2.995732}{0.14} = k \qquad \text{▦ or Table 3}$$

$$21.4 = k. \qquad \text{Rounding to the nearest tenth}$$

c) $R(b) = e^{21.4b}$

d) We solve this equation for b:*

$$100 = e^{21.4b}$$

$$\ln 100 = \ln e^{21.4b}$$

$$\ln 100 = 21.4b$$

$$\frac{\ln 100}{21.4} = b$$

$$\frac{4.605170}{21.4} = b \qquad \text{▦ or Table 3}$$

$$0.22 = b. \qquad \text{Rounded to the nearest hundredth}$$

Thus when the blood alcohol level is 0.22%, according to this model, the risk of an accident is 100%. From the graph, this would occur after 12 1-oz drinks of 86 proof whiskey. "Theoretically" the model tells us that after 12 drinks of whiskey one is "sure" to have an accident. This might be questioned in actuality, since a person who has had 12 drinks might not be able to drive at all.

DO EXERCISE 7.

---

*The answer could be found by leaving the value k in your calculator, unrounded, and then making this computation. Answers in the exercises have been found that way.

EXERCISE SET 4.3

1. State the solution of $dQ/dt = kQ$ in terms of Q_0.

3. *Business: Compound interest.* Suppose P_0 is invested in a savings account where interest is compounded continuously at 9% per year. That is, the balance P grows at the rate given by $\boxed{CSS}$

$$\frac{dP}{dt} = 0.09P.$$

a) Find the solution of the equation in terms of P_0 and 0.09.

b) Suppose $1000 is invested. What is the balance after 1 year? after 2 years?

c) When will an investment of $1000 double itself?

5. *Ecology: Population growth.* The growth rate of the population of Central America is 3.5% per year (one of the highest in the world). What is the doubling time?

7. *Business: Annual interest rate.* A bank advertises that it compounds interest continuously and that it will double your money in 10 years. What is its annual interest rate?
$\boxed{CSS}$

9. *Ecology: Population growth.* The population of the USSR was 209 million in 1959. It was estimated that the population P was growing exponentially at the rate of 1% per year, that is,

$$\frac{dP}{dt} = 0.01P.$$

a) Find the solution of the equation, assuming $P_0 = 209$ and $k = 0.01$.

b) Estimate the population of the USSR in 1999.

c) After what period of time will the population be double that of 1959?

11. *Biomedical: Blood alcohol level.* In Example 7 (on alcohol absorption), at what blood alcohol level will the risk of an accident by 80%?

2. State the solution of $dR/dt = kR$ in terms of R_0.

4. *Business: Compound interest.* Suppose P_0 is invested in a savings account where interest is compounded continuously at 10% per year. That is, the balance P grows at the rate given by

$$\frac{dP}{dt} = 0.10P.$$

a) Find the solution of the equation in terms of P_0 and 0.10.

b) Suppose $20,000 is invested? What is the balance after 1 year? after 2 years?

c) When will an investment of $20,000 double itself?

6. *Ecology: Population growth.* The growth rate of the population of Alaska is 2.8% per year (one of the largest of the fifty states). What is the doubling time?

8. *Business: Annual interest rate.* A bank advertises that it compounds interest continuously and that it will double your money in 12 years. What is its annual interest rate?

10. *Ecology: Population growth.* The population of Europe west of the USSR was 430 million in 1961. It was estimated that the population was growing exponentially at the rate of 1% per year, that is,

$$\frac{dP}{dt} = 0.01P.$$

a) Find the solution of the equation, assuming $P_0 = 430$ and $k = 0.01$.

b) Estimate the population of Europe in 1991.

c) After what period of time will the population be double that of 1961?

12. *Biomedical: Blood alcohol level.* In Example 7 (on alcohol absorption), at what blood alcohol level will the risk of an accident by 90%?

13. *Business: Franchise expansion.* A national hamburger firm is selling franchises throughout the country. The president estimates that the number of franchises N will increase at a rate of 10% per year, that is,

$$\frac{dN}{dt} = 0.10N.$$

a) Find the solution of the equation, assuming the number of franchises at $t = 0$ is 50.

b) How many franchises will there be in 20 years?

c) After what period of time will the initial number of 50 franchises double?

15. *Ecology: Oil demand.* The growth rate of the demand for oil in the United States is 10% per year. When will the demand be double that of 1980?

In Exercises 17–22, find k to six decimal places.

17. *Ecology: Population growth.* The population of Tempe, Arizona, was 25 thousand in 1960. In 1969 it was 52 thousand. Assuming the exponential model:

a) Find the value k ($P_0 = 25$). Use natural logarithms. Write the equation.

b) Estimate the population of Tempe in 1986.

19. *Business: Wine sales in the U.S.* The total number of dollars spent on wine in the U.S. in 1934 was $90 million. In 1974 the amount spent was $1480 million. Assuming the exponential model:

a) Find the value k ($P_0 = 90$). Use natural logarithms. Write the equation.

b) Estimate the amount spent on wine in 1984.

c) After what period of time will the amount spent on wine be double that spent in 1974?

21. *Business: Consumer price index.* The *consumer price index* compares the costs of goods and services over various years, where 1967 is used as a base (P_0). The same goods and services that cost $100 in 1967 cost $184.50 in 1977. Assuming the exponential model:

a) Find the value k ($P_0 = \$100$), and write the equation.

b) Estimate what the same goods and services will cost in 1987.

c) After what period of time will the same goods and services cost double that of 1967?

14. *Business: Franchise expansion.* Pizza, Unltd., a national pizza firm, is selling franchises throughout the country. The president estimates that the number of franchises N will increase at a rate of 15% per year, that is,

$$\frac{dN}{dt} = 0.15N.$$

a) Find the solution of the equation, assuming the number of franchises at $t = 0$ is 40.

b) How many franchises will there be in 20 years?

c) After what period of time will the initial number of 40 franchises double?

16. *Ecology: Coal demand.* The growth rate of the demand for coal in the world is 4% per year. When will the demand be double that of 1980?

18. *Ecology: Population growth.* The population of Kansas City was 475 thousand in 1960. In 1970 it was 507 thousand. Assuming the exponential model:

a) Find the value of k ($P_0 = 475$). Use natural logarithms. Write the equation.

b) Estimate the population of Kansas City in 2000.

20. *Business: Cost of a double-dip ice cream cone.* In 1970 the cost of a double-dip ice cream cone was 52¢. In 1978 it was 66¢. Assuming the exponential model:

a) Find the value k ($P_0 = 52$). Use natural logarithms. Write the equation.

b) Estimate the cost of a cone in 1986.

c) After what period of time will the cost of a cone be twice that of 1978?

22. *Business: Job opportunities.* It is estimated that there were 714,000 accountants employed in 1972 and it is projected that there will be 935,000 accountants needed in 1985. Assuming the exponential model:

a) Find the value k ($P_0 = 714,000$), and write the equation.

b) Estimate the number of accountants needed in 1990.

c) After what period of time will the need for accountants be double that of 1972?

▶

Business: Effective annual yield. Suppose $100 is invested at 12% compounded continuously for 1 year. We know from Example 4 that the balance will be $112.75. This is the same as if $100 were invested at 12.75% and compounded once a year (simple interest). The 12.75% is called the *effective annual yield*. In general, if P_0 is invested at $k(\%)$ compounded continuously, then the effective annual yield is that number i satisfying $P_0(1 + i) = P_0 e^k$. Then $1 + i = e^k$, or

$$\text{Effective annual yield} = i = e^k - 1.$$

23. An amount is invested at 14% per year compounded continuously. What is the effective annual yield?

24. An amount is invested at 8% per year compounded continuously. What is the effective annual yield?

25. The effective annual yield on an investment compounded continuously is 9.42%. At what rate was it invested?

26. The effective annual yield on an investment compounded continuously is 10.52%. At what rate was it invested?

27. Find an expression relating the growth rate k and the tripling time T_3.

28. Find an expression relating the growth rate k and the quadrupling time T_4.

29. Gather data concerning population growth in your city. Estimate its population in 1984; in 2000.

30. A quantity Q_1, grows exponentially with a doubling time of 1 year. A quantity Q_2 grows exponentially with a doubling time of 2 years. If the initial amounts of Q_1 and Q_2 are the same, when will Q_1 be twice the size of Q_2?

31. ▦ *Business: Value of Manhattan Island.* Peter Minuit, of the Dutch West India Company, purchased Manhattan Island from the Indians in 1626 for $24 worth of merchandise. Assuming an exponential rate of inflation of 8%, how much would Manhattan be worth in 1984?

32. ▦ *Ecology: Population growth in the Virgin Islands.* The U.S. Virgin Islands have one of the highest growth rates in the world, 9.6%. In 1970 the population was 75,150. The land area of the Virgin Islands is 3,097,600 square yards. Assuming this growth rate continues and is exponential, after what period of time will the population of the Virgin Islands be such that there is one person for every square yard of land?

33. ▦ *Ecology: Bicentennial growth of the United States.* The population of the United States in 1776 was about 2,508,000. In its bicentennial year the population was about 216,000,000. Assuming the exponential model, what was the growth rate of the United States through its bicentennial years?

34. ▦ *Business: Cost of a first-class postage stamp.* The cost of a first-class postage stamp in 1962 was 4¢. In 1982 it was 20¢. This was exponential growth. What was the growth rate? Round to six decimal places. What will be the cost of a first-class postage stamp in 1987? in 1997?

35. ▦ *Business: Cost of a prime-rib dinner.* The average cost of a prime rib dinner in 1962 was $4.65, and was increasing at an exponential growth rate of 5.1%. What will the cost of such a dinner be in 1987? in 1997?

36. ▦ *Business: Cost of a Hershey bar.* The cost of a Hershey bar in 1962 was $0.05, and was increasing at an exponential growth rate of 9.7%. What will the cost of a Hershey bar be in 1987? in 1997?

37. ▦ A growth rate of 100% per day corresponds to what exponential growth rate per hour?

38. Show that any two measurements of an exponentially growing population will determine k. That is, show that if y has the values y_1 at t_1 and y_2 at t_2, then

$$k = \frac{\ln (y_2/y_1)}{t_2 - t_1}$$

OBJECTIVES

You should be able to

a) State the solution of an equation

$$\frac{dP}{dt} = -kP$$

as

$$P(t) = P_0 e^{-kt}.$$

b) Given a decay rate, find the half-life.

c) Given a half-life, find the decay rate.

d) Solve applied problems involving decay.

e) Solve applied problems involving Newton's law of cooling.

1. Using the same set of axes, graph $y = e^{2x}$ and $y = e^{-2x}$.

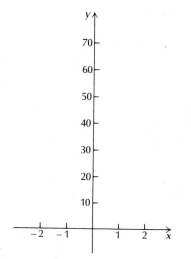

4.4 APPLICATIONS: DECAY

DO EXERCISE 1.

In the equation of population growth $dP/dt = kP$, the constant k is actually given by

$$k = \text{(Birth rate)} - \text{(Death rate)}.$$

Thus a population "grows" only when the *birth rate* is greater than the *death rate*. When the birth rate is less than the death rate, k will be negative so the population will be decreasing, or "decaying," at a rate proportional to its size. The equation

$$\frac{dP}{dt} = -kP \qquad \text{(where } k > 0)$$

shows P to be *decreasing* as a function of time, and the solution

$$P(t) = P_0 e^{-kt}$$

shows it to be decreasing exponentially. This is exponential *decay*. The amount present initially at $t = 0$ is again P_0.

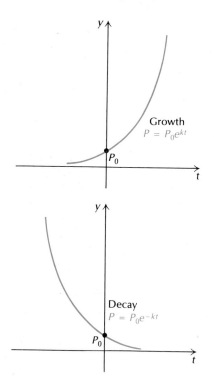

Radioactive Decay

Radioactive elements decay exponentially; that is, they disintegrate at a rate that is proportional to the amount present.

Example 1 Strontium-90 has a decay rate of 2.8% per year. The rate of change of an amount N is given by

$$\frac{dN}{dt} = -0.028N.$$

a) Find the solution of the equation in terms of N_0 (the amount present at $t = 0$).

b) Suppose 1000 grams of strontium-90 is present at $t = 0$. How much will remain after 100 years?

c) After what amount of time will half of the 1000 grams remain?

Solution   CSS

a) $N(t) = N_0 e^{-0.028t}$

b) $N(100) = 1000 e^{-0.028(100)} = 1000 e^{-2.8}$

$$= 1000(0.060810) \qquad \text{⊞ or Table 4}$$

$$\approx 60.8 \text{ grams}$$

c) We are asking at what time T will $N(T) = 500$. The number T is called the *half-life*. To find T we solve the equation:

$$500 = 1000 e^{-0.028T}$$

$$\frac{1}{2} = e^{-0.028T}$$

$$\ln \frac{1}{2} = \ln e^{-0.028T}$$

$$\ln 1 - \ln 2 = -0.028T$$

$$0 - \ln 2 = -0.028T$$

$$\frac{-\ln 2}{-0.028} = T$$

$$\frac{\ln 2}{0.028} = T$$

$$\frac{0.693147}{0.028} = T$$

$$25 \approx T.$$

Thus the half-life of strontium-90 is 25 years.

2. Xenon-133 has a decay rate of 14% per day. The rate of change of an amount N is given by

$$\frac{dN}{dt} = -0.14N.$$

a) Find the solution of the equation in terms of N_0.

b) Suppose 1000 grams of xenon-133 is present at $t = 0$. How much will remain after 10 days?

c) After what time will half of the 1000 grams remain?

DO EXERCISE 2.

We can find a general expression relating the decay rate k and the half-life T by solving the equation:

$$\frac{1}{2}P_0 = P_0 e^{-kT}$$

$$\frac{1}{2} = e^{-kT}$$

$$\ln \frac{1}{2} = \ln e^{-kT}$$

$$\ln 1 - \ln 2 = -kT$$

$$0 - \ln 2 = -kT$$

$$-\ln 2 = -kT$$

$$\ln 2 = kT.$$

Again, we have the following.

THEOREM 11

The *decay rate k* and the *half-life T* are related by

$$kT = \ln 2 = 0.693147,$$

or

$$k = \frac{\ln 2}{T} \quad \text{and} \quad T = \frac{\ln 2}{k}.$$

Thus the half-life T depends only on the decay rate k. In particular, it is independent of the initial population size.

The effect of half-life is shown in this radioactive decay curve.

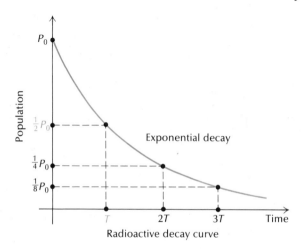

Radioactive decay curve

3. The decay rate of cesium-137 is 2.3% per year. What is its half-life?

The exponential function gets closer to 0 as t gets larger, but never reaches 0. Thus, in theory, a radioactive substance never completely decays.

Example 2 Plutonium, a common product and ingredient of nuclear reactors, is of great concern to those who are against the building of nuclear reactors. Its decay rate is 0.003% per year. What is its half-life?

Solution

$$T = \frac{\ln 2}{k} = \frac{0.693147}{0.00003} = 23,105 \text{ years}$$

DO EXERCISES 3 AND 4.

Example 3 *Carbon dating.* The radioactive element carbon-14 has a half-life of 5750 years. The percentage of carbon-14 present in the remains of plants and animals can be used to determine age. How old is an animal bone that has lost 30% of its carbon-14?

4. The half-life of barium-140 is 13 days. What is its decay rate?

Solution

a) Find the decay rate k.

$$k = \frac{\ln 2}{T} = \frac{0.693147}{5750} = 0.00012, \quad \text{or } 0.012\% \text{ per year}$$

b) Find the exponential equation for the amount $N(t)$ that remains from an initial amount N_0 after t years.

$$N(t) = N_0 e^{-0.00012t}$$

(*Note:* This equation can be used for any subsequent carbon dating problem.)

c) If an animal bone has lost 30% of its carbon-14 from an initial amount P_0, then 70% (P_0) is the amount present. To find the age t of the bone, we solve the following equation for t:

$$70\% \ P_0 = P_0 e^{-0.00012t}$$

$$0.7 = e^{-0.00012t}$$

$$\ln 0.7 = \ln e^{-0.00012t}$$

$$\ln 0.7 = -0.00012t$$

$$-0.356675 = -0.00012t \quad \blacksquare \text{ or Table 3}$$

$$\frac{0.356675}{0.00012} = t$$

$$2973 \approx t.$$

How can scientists determine that an animal bone has lost 30% of its carbon-14? The assumption is that the percentage of carbon-14 in the atmosphere and in living plants and animals is the same. When a plant or animal dies, the amount of carbon-14 decays exponentially. The scientist burns the animal bone and uses a geiger counter to determine the percentage of the smoke that is carbon-14. It is the amount this varies from the percentage in the atmosphere that tells how much carbon-14 has been lost.

5. How old is a skeleton that has lost 80% of its carbon-14?

Exploratory Exercises: Cooling. Draw a glass of hot tap water. Place a thermometer in the glass and check the temperature. Check the temperature every 30 minutes thereafter. Plot your data on this graph and connect the points with a smooth curve.

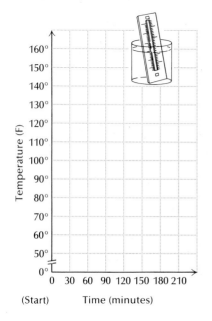

a) What was the temperature at the start?

b) At what temperature does there seem to be a leveling off of the graph?

c) What is the difference between your answers to (a) and (b)?

d) How does the temperature in (b) compare with the room temperature?

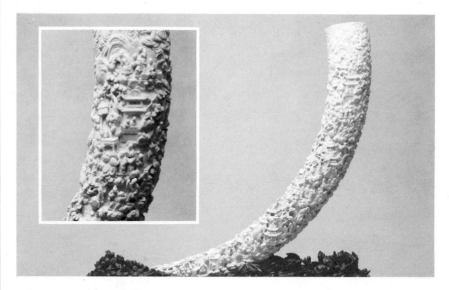

This tusk is 136 years old. It has lost 2% of its carbon-14. (*M. W. Sexton: Peabody Museum of Salem*)

Thus an animal bone that has lost 30% of its carbon-14 is about 2973 years old.

DO EXERCISE 5.

Newton's Law of Cooling

Before you study the following, do the exploratory exercises in the margin.

NEWTON'S LAW OF COOLING

> The temperature T of a cooling object drops at a rate which is proportional to the difference $T - C$, where C is the constant temperature of the surrounding medium. Thus
>
> $$\frac{dT}{dt} = -k(T - C). \tag{1}$$
>
> The solution of Eq. (1) is
>
> $$T = T(t) = ae^{-kt} + C. \tag{2}$$

We can check this by differentiating:

$$\frac{dT}{dt} = -kae^{-kt} = -k(ae^{-kt}) = -k(T - C).$$

Example 4 The temperature of a cup of freshly brewed coffee is 200° and the room temperature is 70°. The temperature cools to 190° in 5 minutes.

a) What is the temperature after 10 minutes?

b) How long does it take for the temperature to cool to 90°?

Solution

a) We first find the value of a in Eq. (2). At $t = 0$, $T = 200°$. We solve the following equation for a:

$$200 = ae^{-k \cdot 0} + 70$$

$$200 = a \cdot 1 + 70$$

$$130 = a.$$

Now we find k using the fact that at $t = 5$, $T = 190°$. We solve the following equation for k:

$$190 = 130e^{-k \cdot 5} + 70$$

$$120 = 130e^{-5k}$$

$$\frac{12}{13} = e^{-5k}$$

$$\ln \frac{12}{13} = \ln e^{-5k}$$

$$= -5k$$

$$-5k = \ln 12 - \ln 13$$

$$k = -\frac{1}{5}(\ln 12 - \ln 13)$$

$$= -\frac{1}{5}(2.484907 - 2.564949) \qquad \blacksquare \text{ or Table 3}$$

$$\approx 0.016.$$

Now we find the temperature at $t = 10$:

$$T(10) = 130e^{-0.016(10)} + 70 = 130e^{-0.16} + 70$$

$$= 130(0.852144) + 70$$

$$\approx 181°.$$

b) To find how long it will take for the temperature to be 90°, we solve

6. The temperature of a hot cup of soup is 200°. The room temperature is 70°. The temperature cools to 190° in 8 minutes.

   a) Find the value of the constant a in Newton's law of cooling.

   b) Find the value of the constant k.

   c) What is the temperature after 10 minutes?

   d) How long does it take for the soup to cool to 80°?

7. Return to the data you found in the exploratory exercises. Find an equation that "fits" the data. Use this equation to check values of other data points. How do they compare? Is it ever "theoretically" possible for the temperature of the water to be the same as the room temperature?

for t: **CSS**

$$90 = 130e^{-0.016t} + 70$$

$$20 = 130e^{-0.016t}$$

$$\frac{2}{13} = e^{-0.016t}$$

$$\ln \frac{2}{13} = \ln e^{-0.016t} = -0.016t$$

$$-0.016t = \ln 2 - \ln 13$$

$$t = -\frac{1}{0.016}(\ln 2 - \ln 13)$$

$$t = -\frac{1}{0.016}(0.693147 - 2.564949) \quad \text{▦ or Table 3}$$

$$t \approx 117 \text{ minutes.}$$

DO EXERCISES 6 AND 7.

The graph of $T(t) = ae^{-kt} + C$ is as follows. Note that $\lim_{t \to \infty} T(t) = C$. The temperature of the object decreases toward the temperature of the surrounding medium.

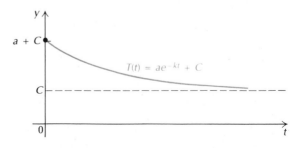

Mathematically, this model tells us that the temperature never reaches C, but in practice this happens eventually. At least, the temperature of the cooling object gets so close to that of the surrounding medium that no device could detect a difference. Let us now see how Newton's law of cooling could be used in solving a crime.

Example 5 *When was the murder committed?* The police discover the body of a calculus professor. Critical to solving the crime is determining when the murder was committed. The police call the coroner, who arrives at 12:00 P.M. He immediately takes the temperature of the body and finds it to be 94.6°. He waits 1 hour, takes the temperature again, and finds it to be 93.4°. He also notes that the temperature of the room is 70°. When was the murder committed?

Solution We first find a in the equation $T(t) = ae^{-kt} + C$. Assuming the temperature of the body was normal when the murder occurred, we have $T = 98.6°$ at $t = 0$. Thus

$$98.6° = ae^{-k \cdot 0} + 70°,$$

so

$$a = 28.6°.$$

Thus T is given by $T(t) = 28.6e^{-kt} + 70$.

We want to find the number of hours N since the murder was committed. To find N we must first determine k. From the two temperature readings the coroner made, we have

$$94.6 = 28.6e^{-kN} + 70, \quad \text{or} \quad 24.6 = 28.6e^{-kN}; \qquad (1)$$

$$93.4 = 28.6e^{-k(N+1)} + 70, \quad \text{or} \quad 23.4 = 28.6e^{-k(N+1)}. \qquad (2)$$

Dividing Eq. (1) by Eq. (2), we get

$$\frac{24.6}{23.4} = \frac{28.6e^{-kN}}{28.6e^{-k(N+1)}} = e^{-kN + k(N+1)} = e^{-kN + kN + k} = e^k.$$

We solve this equation for k:

$$\ln \frac{24.6}{23.4} = \ln e^k \qquad \text{Taking the natural logarithm on both sides}$$

$$\ln 24.6 - \ln 23.4 = k$$

$$3.202746 - 3.152736 \approx k$$

$$0.05 \approx k.$$

Now we substitute back into Eq. (1) and solve for N:

$$24.6 = 28.6e^{-0.05N}$$

$$\ln 24.6 = \ln 28.6 \, e^{-0.05N}$$

$$\ln 24.6 = \ln 28.6 + \ln e^{-0.05N}$$

$$\ln 24.6 - \ln 28.6 = -0.05N$$

$$\frac{-0.150660}{-0.05} = N$$

$$3 \text{ hr} \approx N.$$

8. A butcher is murdered and the body thrown in a cooler where the temperature is 40°. The coroner, arriving at 3:00 P.M., takes the temperature of the body, finds it to be 73.8°, waits 1 hour, takes the temperature again, and finds it to be 70.3°. When was the murder committed?

The coroner arrived at 12:00 P.M., so the murder was committed at about 9:00 A.M.

DO EXERCISE 8.

Models have many applications. This article, which is actually an advertisement, describes the effects of melting ice on the alcoholic content of Scotch whisky. The graphs are of the equation describing Newton's law of cooling. (*Nadler & Larimer, Inc.*)

EXERCISE SET 4.4

1. The decay rate of iodine-131 is 9.6% per day. What is its half-life?

2. The decay rate of krypton-85 is 6.3% per year. What is its half-life?

3. The half-life of polonium is 3 minutes. What is its decay rate?

4. The half-life of lead is 22 yr. What is its decay rate?

5. Of an initial amount of 1000 grams of polonium, how much will remain after 20 minutes? See Exercise 3 for the value of k.

6. Of an initial amount of 1000 grams of lead, how much will remain after 100 yr? See Exercise 4 for the value of k.

7. *Carbon dating.* How old is a piece of wood that has lost 90% of its carbon-14?

8. *Carbon dating.* How old is an ivory tusk that has lost 40% of its carbon-14?

9. *Carbon dating.* How old is a Chinese artifact that has lost 60% of its carbon-14?

10. *Carbon dating.* How old is a skeleton that has lost 50% of its carbon-14?

11. In a *chemical reaction* substance A decomposes at a rate proportional to the amount of A present.

a) Write an equation relating A to the amount left of an initial amount A_0 after time t.

b) It is found that 8 grams of A will reduce to 4 grams in 3 hours. After what time will there be only 1 gram left?

12. In a *chemical reaction* substance A decomposes at a rate proportional to the amount of A present.

a) Write an equation relating A to the amount left of an initial amount A_0 after time t.

b) It is found that 10 lb of A will reduce to 5 lb in 3.3 hr. After what time will there be only 1 lb left?

13. *Weight loss.* The initial weight of a starving animal is W_0. Its weight W after t days is given by

$$W = W_0 e^{-0.008t}.$$

a) What percentage of its weight does it lose each day?

b) What percentage of its initial weight remains after 30 days?

14. *Weight loss.* The initial weight of a starving animal is W_0. Its weight W after t days is given by

$$W = W_0 e^{-0.009t}.$$

a) What percentage of its weight does it lose each day?

b) What percentage of its initial weight remains after 30 days?

15. *Satellite power.* The power supply of a satellite is a radioisotope. The power output P, in watts, decreases at a rate proportional to the amount present. P is given by

$$P = 50e^{-0.004t},$$

where $t =$ the time in days.

a) How much power will be available after 375 days?

b) What is the half-life of the power supply?

c) The satellite's equipment cannot operate on fewer than 10 watts of power. How long can the satellite stay in operation?

d) How much power did the satellite have to begin with?

16. *Atmospheric pressure.* Atmospheric pressure P at altitude a is given by

$$P = P_0 e^{-0.00005a},$$

where $P_0 =$ the pressure at sea level. Assume $P_0 = 14.7$ lb/in^2 (pounds per square inch).

a) Find the pressure at an altitude of 1000 ft.

b) Find the pressure at 20,000 ft.

c) At what altitude is the pressure 1.47 lb/in^2?

17. *Salvage value.* A business estimates that the salvage value V of a piece of machinery after t years is given by

$$V(t) = \$40{,}000e^{-t}.$$

a) What did the machinery cost initially?

b) What is the salvage value after 2 years?

18. *Supply and demand.* The supply and demand for the sale of stereos by a sound company are given by

$$S(x) = e^x, \qquad D(x) = 163{,}000e^{-x},$$

where $S(x) =$ the price at which the company is willing to supply x stereos and $D(x) =$ the demand price for a quantity of x stereos. Find the equilibrium point. For reference, see Section 1.5.

BEER–LAMBERT LAW

A beam of light enters a medium, such as water or smoky air, with initial intensity I_0. Its intensity is decreased depending on the thickness (or concentration) of the medium. The intensity I at a depth (or concentration) of x units is given by

$$I = I_0 e^{-\mu x}.$$

The constant μ ("mu"), called the *coefficient of absorption*, varies with the medium.

---

19. *Light through sea water* has $\mu = 1.4$ when x is measured in meters (m).

a) What percentage of I_0 remains at a depth of sea water that is 1 m? 2 m? 3 m?

b) Plant life cannot exist below 10 meters. What percentage of I_0 remains at 10 meters?

21. The temperature of a hot liquid is 100° and the room temperature is 75°. The liquid cools to 90° in 10 minutes.

a) Find the value of the constant a in Newton's law of cooling.

b) Find the value of the constant k. Round to the nearest hundredth.

c) What is the temperature after 20 minutes?

d) How long does it take for the liquid to cool to 80°?

23. The coroner arrives at the scene of a murder at 11 P.M. She takes the temperature of the body and finds it to be 85.9°. She waits 1 hour, takes the temperature again, and finds it to be 83.4°. She notes that the room temperature is 60°. When was the murder committed?

25. *Ecology: Population decrease of Cincinnati.* The population of Cincinnati was 503,000 in 1960 and 453,000 in 1970. Assuming the population is decreasing according to the exponential decay model:

a) Find the value k and write the equation.

b) Estimate the population of Cincinnati in 1990.

c) After what period of time will Cincinnati have just 1 person?

27. *Business: Present value.* A parent, following the birth of a child, wants to make an initial investment of P_0 that will grow to $10,000 by the child's 20th birthday. Interest is compounded continuously at 8%. What should the initial investment be? Such an amount is called the *present value* of $10,000 due 20 years from now.

20. *Light through smog.* Particulate concentrations of pollution reduce sunlight. In a smoggy area $\mu = 0.01$ and x = the concentration of particulates measured in micrograms per cubic meter. What percentage of an initial amount I_0 of sunlight passes through smog that has a concentration of 100 micrograms per cubic meter?

22. The temperature of a hot liquid is 100° and is placed in a refrigerator where the temperature is 40°. The liquid cools to 90° in 5 minutes.

a) Find the value of the constant a in Newton's law of cooling.

b) Find the value of the constant k. Round to the nearest hundredth.

c) What is the temperature after 10 minutes?

d) How long does it take for the liquid to cool to 41°?

24. The coroner arrives at the scene of a murder at 2 A.M. He takes the temperature of the body and finds it to be 61.6°. He waits 1 hour, takes the temperature again, and finds it to be 57.2°. The body is in a meat freezer, where the temperature is 10°. When was the murder committed?

26. *Ecology: Population decrease of Panama.* The population of Panama was 1,464,000 in 1970 and 1,260,000 in 1980. Assuming the population is decreasing according to the exponential decay model:

a) Find the value k and write the equation.

b) Estimate the population of Panama in 2000.

c) After what period of time will the population of Panama be 100,000?

28. *Ecology: Population of the USSR in a preceding year.* The population of the USSR was 258 million in 1980 and was growing at the rate of 1% per year. What was the population in 1970? in 1940?

29. ▦ *Business: Consumer price index.* The consumer price index compares the costs of goods and services over various years, where 1967 is used as a base. The same goods and services that cost \$100 in 1967 cost \$42 in 1940. Assuming the exponential decay model:

a) Find the value k and write the equation. Round to the nearest hundredth.

b) Estimate what the same goods and services cost in 1900.

You should be able to

a) **Express a power like 2^3 as a power of e.**

b) **Differentiate functions involving a^x.**

c) **Differentiate functions involving $\log_a x$.**

*4.5 THE DERIVATIVES OF a^x AND $\text{LOG}_a x$

The Derivative of a^x

To find the derivative of a^x, for any base a, we first express it as a power of e. In order to do this we first prove the following.

PROPERTY 7.　$b^{\log_b x} = x$

To prove this, let

$$y = \log_b x.$$

Then, by the definition of a logarithm,

$$b^y = x.$$

Substituting $\log_b x$ for y, we have

$$b^{\log_b x} = x.$$

We can now express a^x as a power of e. Using Property 7 where $b = e$ and $x = a$, we have

$$a = e^{\ln a}.　　\text{Remember: } \ln a = \log_e a$$

Raising both sides to the power x, we get

$$a^x = (e^{\ln a})^x$$

$$= e^{x \cdot \ln a}.　　\text{Multiplying exponents}$$

Thus we have the following.

---

*This section can be omitted without loss of continuity.

Express as a power of e.

1. 4^5

2. 2^x

3. Differentiate.

$$y = 5^x$$

THEOREM 12

$$a^x = e^{x \cdot \ln a}$$

Examples Express as a power of e.

a) 3^2 　　　　　　$3^2 = e^{2 \cdot \ln 3}$

　　　　　　　　　　$\approx e^{2(1.098612)}$　▣ or Table 3

　　　　　　　　　　$= e^{2.1972}$

b) 10^x 　　　　　$10^x = e^{x \cdot \ln 10}$

　　　　　　　　　　$\approx e^{x(2.3026)}$

　　　　　　　　　　$= e^{2.3026x}$

DO EXERCISES 1 AND 2.

Now we can differentiate.

Example 1

$$\frac{d}{dx} 2^x = \frac{d}{dx} e^{x \cdot \ln 2}$$

$$= \left[\frac{d}{dx} (x \cdot \ln 2) \right] \cdot e^{x \cdot \ln 2}$$

$$= (\ln 2)(e^{\ln 2})^x$$

$$= \ln 2 \cdot 2^x$$

We completed this by taking the derivative of $x \ln 2$, and replacing $e^{x \cdot \ln 2}$ by 2^x. Note that $\ln 2 \approx 0.7$, so the above verifies our earlier approximation of the derivative of 2^x as $(0.7)2^x$ (see Section 4.1).

DO EXERCISE 3.

In general,

$$\frac{d}{dx} a^x = \frac{d}{dx} e^{x \cdot \ln a}$$

$$= \left[\frac{d}{dx} (x \cdot \ln a) \right] \cdot e^{x \ln a}$$

$$= \ln a \cdot a^x.$$

Thus we have the following.

THEOREM 13

$$\frac{d}{dx} a^x = (\ln a)\, a^x$$

Example 2

$$\frac{d}{dx} 3^x = (\ln 3)\, 3^x$$

Example 3

$$\frac{d}{dx} (1.4)^x = \ln 1.4 \cdot (1.4)^x$$

Compare these formulas:

$$\frac{d}{dx} a^x = \ln a \cdot a^x;$$

$$\frac{d}{dx} e^x = e^x.$$

It is the simplicity of the last formula that is a reason for the use of the base e in calculus. The many applications of e in natural phenomena provide other reasons.

One other result also follows from what we have done. If

$$f(x) = a^x,$$

we know that

$$f'(x) = a^x (\ln a).$$

We also showed in Section 4.1 that

$$f'(x) = a^x \left(\lim_{h \to 0} \frac{a^h - 1}{h} \right).$$

Since $a^x > 0$, we have the following.

THEOREM 14

$$\ln a = \lim_{h \to 0} \frac{a^h - 1}{h}$$

Differentiate.

4. $f(x) = 4^x$

DO EXERCISES 4 AND 5.

The Derivative of $\log_a x$

Just as the derivative of a^x is expressed in terms of $\ln a$, so is the derivative of $\log_a x$. To find this derivative we first express $\log_a x$ in terms of $\ln a$ using Property 7:

$$a^{\log_a x} = x.$$

Then

$$\ln a^{\log_a x} = \ln x$$

$$(\log_a x) \cdot \ln a = \ln x \qquad \text{Property 3, treating } \log_a x \text{ as an exponent}$$

and

$$\log_a x = \boxed{\frac{1}{\ln a}} \cdot \ln x.$$

$$\underline{\qquad} \text{constant}$$

The derivative of $\log_a x$ follows.

5. $f(x) = (4.3)^x$

THEOREM 15

$$\frac{d}{dx} \log_a x = \frac{1}{\ln a} \cdot \frac{1}{x}$$

Comparing this with

$$\frac{d}{dx} \ln x = \frac{1}{x},$$

we again see a reason for the use of the base e in calculus.

Example 4

$$\frac{d}{dx} \log_3 x = \frac{1}{\ln 3} \cdot \frac{1}{x}$$

Example 5

$$\frac{d}{dx} \log x = \frac{1}{\ln 10} \cdot \frac{1}{x} \qquad \log x = \log_{10} x$$

Differentiate.

6. $y = \log_2 x$

7. $y = -7 \log x$

8. $f(x) = x^6 \log x$

Example 6

$$\frac{d}{dx} x^2 \log x = x^2 \frac{1}{\ln 10} \cdot \frac{1}{x} + 2x \log x \qquad \text{Product Rule}$$

$$= \frac{x}{\ln 10} + 2x \log x = x\left(\frac{1}{\ln 10} + 2 \log x\right)$$

DO EXERCISES 6–8.

EXERCISE SET 4.5

Express as a power of e.

1. 5^4 **2.** 2^3 **3.** $(3.4)^{10}$ **4.** $(5.3)^{20}$

5. 4^k **6.** 5^R **7.** 8^{kT} **8.** 10^{kR}

Differentiate.

9. $y = 6^x$ **10.** $y = 7^x$ **11.** $f(x) = 10^x$ **12.** $f(x) = 100^x$

13. $f(x) = x(6.2)^x$ **14.** $f(x) = x(5.4)^x$ **15.** $y = x^3 10^x$ **16.** $y = x^4 5^x$

Applied problems

Earthquake magnitude. The magnitude R (measured on the Richter scale) of an earthquake of intensity I is defined as

$$R = \log \frac{I}{I_0},$$

where I_0 is a minimum intensity used for comparison. When one earthquake is 10 times as intense as another, its magnitude on the Richter scale is 1 higher. If one earthquake is 100 times as intense as another, its magnitude on the Richter scale is 2 higher, and so on. Thus an earthquake whose magnitude is 7 on the Richter scale is 10 times as intense as an earthquake whose magnitude is 6. Earthquakes can be interpreted as multiples of the minimum intensity I_0.

17. The Mexico City earthquake of 1978 had an intensity of $10^{7.85} \cdot I_0$. What was its magnitude on the Richter scale?

18. The Anchorage, Alaska, earthquake on March 27, 1964, had an intensity of $10^{8.4} \cdot I_0$. What was its magnitude on the Richter scale?

This photograph shows part of the damage of the earthquake in Anchorage, Alaska, in 1964. (*Pro Pix, from Monkmeyer*)

19. *Earthquake intensity.* The intensity I of an earthquake is given by

$$I = I_0 10^R,$$

where R = the magnitude on the Richter scale; and I_0 = the minimum intensity, where $R = 0$, used for comparison.

a) Find I, in terms of I_0, for an earthquake of magnitude 7 on the Richter scale.

b) Find I, in terms of I_0, for an earthquake of magnitude 8 on the Richter scale.

c) Compare your answers to (a) and (b).

d) Find the rate of change dI/dR.

Differentiate.

21. $y = \log_4 x$

22. $y = \log_5 x$

25. $f(x) = \log \dfrac{x}{3}$

26. $f(x) = \log \dfrac{x}{5}$

29. *Earthquake magnitude.* The magnitude R (measured on the Richter scale) of an earthquake of intensity I is defined to be

$$R = \log \frac{I}{I_0},$$

where I_0 = the minimum intensity (used for comparison). (The exponential form of this definition is given in Exercise 19.) Find the rate of change dR/dI.

31. *Response to drug dosage.* The response y to a dosage x of a drug is given by

$$y = m \log x + b.$$

The response may be hard to measure with a number. The patient might sweat more, have an increase in temperature, or pass out. Find the rate of change dy/dx.

33. *Ecology: Recycling aluminum cans.* It is known that $\frac{1}{4}$ of all aluminum cans distributed will be recycled each year. A beverage company distributes 250,000 cans. The number still in use after time t, in years, is given by

$$N(t) = 250{,}000\left(\frac{1}{4}\right)^t.$$

Find $N'(t)$.

20. *Intensity of sound.* The intensity of a sound is given by

$$I = I_0 10^{0.1L},$$

where L = the loudness of the sound as measured in decibels; and I_0 = the minimum intensity detectable by the human ear.

a) Find I, in terms of I_0, for the loudness of a power mower, which is 100 decibels.

b) Find I, in terms of I_0, for the loudness of just audible sound, which is 10 decibels.

c) Compare your answers to (a) and (b).

(d) Find the rate of change dI/dL.

23. $f(x) = 2 \log x$

24. $f(x) = 5 \log x$

27. $y = x^3 \log_8 x$

28. $y = x \log_6 x$

30. *Loudness of sound.* The *loudness* L of a sound of intensity I is defined to be

$$L = 10 \log \frac{I}{I_0},$$

where I_0 = the minimum intensity detectable by the human ear; and L = the loudness measured in decibels. (The exponential form of this definition is given in Exercise 20.) Find the rate of change dL/dI.

32. *Business: Double declining-balance depreciation.* An office machine is purchased for \$5200. Under certain assumptions its salvage value V depreciates according to a method called double declining balance, basically 80% each year, and is given by

$$V(t) = \$5200(0.80)^t,$$

where t is time, in years. Find $V'(t)$.

34. Find $\displaystyle\lim_{h \to 0} \frac{3^h - 1}{h}$.

▶

Use the Chain Rule and other formulas given in this section to differentiate each of the following.

35. $f(x) = 3^{2x}$

36. $y = 2^{x^4}$

37. $y = x^x,\ x > 0$

38. $y = \log_3(x^2 + 1)$

39. $f(x) = x^{e^x},\ x > 0$

40. $y = a^{f(x)}$

41. $y = \log_a f(x),\ f(x)$ positive

42. $y = [f(x)]^{g(x)},\ f(x)$ positive

OBJECTIVE

Given a demand function, you should be able to find the elasticity and the value(s) of x for which total revenue is maximized.

*4.6 ECONOMIC APPLICATION: ELASTICITY OF DEMAND

Suppose x represents the quantity of goods sold and p is the price per unit of the goods. Recall that x and p are related by the demand function

$$p = D(x).$$

Suppose there is a change Δx in the quantity sold. The percent change in quantity is

$$\frac{\Delta x}{x}.$$

A change in the quantity sold produces a change Δp in the price. The percent change in price is

$$\frac{\Delta p}{p}.$$

The ratio of these percents is given by

$$\frac{(\Delta x/x)}{(\Delta p/p)},$$

which can be expressed as

$$\frac{p}{x} \cdot \frac{1}{\Delta p/\Delta x}. \qquad (1)$$

For continuous functions

$$\lim_{\Delta x \to 0} \frac{\Delta p}{\Delta x} = \frac{dp}{dx},$$

_____

*This section can be omitted without loss of continuity.

and the limit as $\Delta x \to 0$ of the expression in Eq. (1) becomes

$$\frac{p}{x} \cdot \frac{1}{dp/dx} = \frac{D(x)}{x} \cdot \frac{1}{D'(x)} = \frac{1}{x} \cdot \frac{D(x)}{D'(x)}.$$

DEFINITION

The *elasticity of demand E* is given by

$$E = -\frac{p}{x} \cdot \frac{1}{dp/dx} = -\frac{1}{x} \cdot \frac{D(x)}{D'(x)}.$$

The numbers x and p are always nonnegative. The slope of the demand curve dp/dx is always negative since the demand curve is decreasing. The minus sign makes E nonnegative and easier, for our purposes, to work with. We will find the second expression for elasticity the most useful for computations.

Example A company determines that the demand function for a certain product is given by

$$p = D(x) = 200 - x.$$

Find:

a) the elasticity as a function of x;

b) the elasticity at $x = 70$ and $x = 150$;

c) the value of x for which $E = 1$;

d) the total revenue function;

e) the value of x for which the revenue is a maximum.

Solution

a) To find the elasticity, we first find

$$D'(x) = -1.$$

Then we substitute -1 for $D'(x)$ and $200 - x$ for $D(x)$ in the second expression for elasticity:

$$E(x) = -\frac{1}{x} \cdot \frac{D(x)}{D'(x)} = -\frac{1}{x} \cdot \frac{200 - x}{-1} = \frac{200 - x}{x}.$$

b) $E(70) = \dfrac{200 - 70}{70} = \dfrac{130}{70} = \dfrac{13}{7}, \qquad E(150) = \dfrac{200 - 150}{150} = \dfrac{50}{150} = \dfrac{1}{3}$

1. A company determines that the demand function for a product is

$$p = D(x) = 300 - x.$$

Find:

a) the elasticity as a function of x;

b) the elasticity at $x = 100$ and $x = 200$;

c) the value of x for which $E = 1$;

d) the total revenue function;

e) the value of x for which the revenue is a maximum.

c) We set $E(x) = 1$ and solve for x:

$$\frac{200 - x}{x} = 1$$

$$200 - x = x \qquad \text{We multiply by } x \text{ assuming } x \neq 0.$$

$$200 = 2x$$

$$100 = x.$$

d) Recall that the total revenue $R(x)$ is given by $xD(x)$. Then

$$R(x) = xD(x)$$

$$= x(200 - x)$$

$$= 200x - x^2.$$

e) To find the value of x that maximizes total revenue, we find $R'(x)$:

$$R'(x) = 200 - 2x.$$

Now $R'(x)$ exists for all values of x in the interval $[0, \infty)$. Thus we solve:

$$R'(x) = 200 - 2x = 0$$

$$-2x = -200$$

$$x = 100.$$

Since there is only one critical point, we can try to use the second derivative to see if we have a maximum:

$$R''(x) = -2, \quad \text{a constant.}$$

Thus $R''(100)$ is negative, so $R(100)$ is a maximum.

DO EXERCISE 1.

Note in parts (c) and (e) of both Example 1 and Margin Exercise 1 that the value of x for which $E = 1$ is the same as the value of x for which total revenue is a maximum. This is always true.

THEOREM 16

Total revenue is a maximum at the value(s) of x for which $E = 1$.

We can prove this as follows. We know that

$$R(x) = xD(x),$$

so

$$R'(x) = 1 \cdot D(x) + xD'(x)$$

$$= D(x)\left[1 + \frac{xD'(x)}{D(x)}\right] \qquad \text{Check this by multiplying.}$$

$$= D(x)\left[1 - \frac{1}{E}\right]$$

$$R'(x) = 0 \quad \text{when } 1 - \frac{1}{E} = 0, \quad \text{or } E = 1.$$

It is of benefit to look at a typical demand curve in relation to elasticity and total revenue.

The demand curve is decreasing overall. For values of x for which $E > 1$, the total revenue is increasing. For values of x for which $E < 1$, the total revenue is decreasing. For the value of x for which $E = 1$, the total revenue is a maximum.

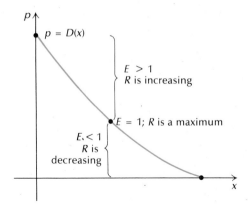

DEFINITION

_____

For a particular value of x:

1. **The demand is *inelastic* if $E < 1$;**
2. **The demand is *elastic* if $E > 1$;**
3. **The demand has *unit elasticity* if $E = 1$.**

_____

In summary, suppose a company puts more units of a product, say a new calculator, on the market and the total revenue increases. Then we say the demand is *elastic*. If the total revenue decreases, we say the demand is *inelastic*.

EXERCISE SET 4.6

For each demand function, find (a) the elasticity and (b) the value(s) of x for which total revenue is a maximum.

1. $p = D(x) = 400 - x$

2. $p = D(x) = 500 - x$

3. $p = D(x) = 200 - 4x$

4. $p = D(x) = 500 - 2x$

5. $p = D(x) = \dfrac{400}{x}$

6. $p = D(x) = \dfrac{3000}{x}$

7. $p = D(x) = \sqrt{500 - x}$

8. $p = D(x) = \sqrt{300 - x}$

9. $p = D(x) = 100e^{-0.25x}$

10. $p = D(x) = 200e^{-0.05x}$

11. $p = D(x) = \dfrac{100}{(x + 3)^2}$

12. $p = D(x) = \dfrac{300}{(x + 8)^2}$

13. *Constant elasticity curve*

a) Find the elasticity of the demand function

$$p = D(x) = \frac{k}{x^n},$$

where k is a positive constant and n is an integer greater than 0.

b) Is the value of the elasticity dependent on the quantity sold?

c) Does total revenue have a maximum? When?

14. *Exponential demand curve*

a) Find the elasticity of the demand function

$$p = D(x) = Ae^{-kx},$$

where A and k are positive constants.

b) Is the value of the elasticity dependent on the quantity sold?

c) Does total revenue have a maximum? At what value of x?

15. Let

$$L(x) = \ln D(x).$$

Describe the elasticity in terms of $L'(x)$.

A summary of important formulas for this chapter is given on the inside front cover.

CHAPTER 4 TEST

Differentiate.

1. $y = e^x$

2. $y = \ln x$

3. $f(x) = e^{-x^2}$

4. $f(x) = \ln \dfrac{x}{7}$

5. $f(x) = e^x - 5x^3$

6. $f(x) = 3e^x \ln x$

7. $y = \ln(e^x - x^3)$

8. $y = \dfrac{\ln x}{e^x}$

Given $\ln 2 = 0.6931$ and $\ln 7 = 1.9459$, find each of the following.

9. $\ln 14$

10. $\ln \dfrac{2}{7}$

11. $\ln 7e$

12. State the solution of $dM/dt = kM$, in terms of M_0.

13. The doubling time of a certain bacteria culture is 4 hours. What is the growth rate? Round to the nearest tenth of a percent.

15. The demand by airlines for fuel is increasing at the rate of 12% per year. That is,

$$\frac{dF}{dt} = 0.12F,$$

where F = the amount of fuel used and t = the time in years.

a) The airlines used 3 billion gallons of fuel in 1960. Find the solution of the equation, assuming $F_0 = 3$ and $k = 0.12$.

b) How much fuel will be needed in 1988?

c) After what period of time will the demand be double that in 1960?

17. The decay rate of zirconium is 1.1% per day. What is its half-life?

Differentiate.

19. $f(x) = 20^x$

20. $y = \log_{20} x$

14. An investment is made at 6.931% per year compounded continuously. What is the doubling time? Round to the nearest year.

16. A dose of a drug is injected into the body of a patient. The drug amount in the body decreases at the rate of 10% per hour. That is,

$$\frac{dA}{dt} = -0.1A,$$

where A = the amount in the body and t = the time in hours.

a) A dose of 3 cubic centimeters (cc) is administered. Assuming $A_0 = 3$ and $k = 0.1$, find the solution to the equation.

b) How much of the initial dose of 3 cc will remain after 10 hours?

c) After what period of time does half the original dose remain?

18. The half-life of tellurium is 1,000,000 years. What is its decay rate? As a percent, round to six decimal places.

21. Find:

a) the elasticity of the demand function

$$p = D(x) = 400e^{-0.2x};$$

b) the value of x for which total revenue is a maximum.

22. Differentiate $y = x (\ln x)^2 - 2x \ln x + 2x$.

23. Find the maximum and minimum values of $f(x) = x^4 e^{-x}$ on $[0, 10]$.

INTEGRATION

The area under a curve can be approximated by a sum of rectangular areas. (*Clif Garboden, Stock, Boston*)

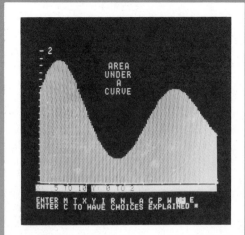

This is the same technique used by the *Computer Software Supplement*.

5

OBJECTIVES

You should be able to

a) **Find the indefinite integral (anti-derivative) of a given function.**

b) **Find a function f with a given derivative and function value.**

c) **Solve applied problems involving antiderivatives.**

5.1 THE ANTIDERIVATIVE

In Chapters 2, 3, and 4 we have considered several interpretations of the derivative. Some are listed below.

Function	Derivative
Distance	Velocity
Revenue	Marginal revenue
Cost	Marginal cost
Population	Rate of growth of population

For population we actually considered the derivative first and then the function. Many problems can be solved by doing the reverse of differentiation, called *antidifferentiation*.

The Antiderivative

Suppose that y is a function of x and that the derivative is the constant 8. Can we find y? It is easy to see that one such function is $8x$. That is, $8x$ is a function whose derivative is 8. Are there other functions whose derivative is 8? Yes. Here are some examples:

$$8x + 3, \qquad 8x - 10, \qquad 8x + \sqrt{2}.$$

All these functions are $8x$ plus some constant. There are no other functions having a derivative of 8 other than those of the form $8x + C$. Another way of saying this is that any two functions having a derivative of 8 must differ by a constant. This is true in general.

THEOREM 1

If two functions F and G have the same derivative on an interval, then

$$F(x) = G(x) + C, \quad \text{where } C \text{ is a constant.}$$

The reverse of differentiating is called *antidifferentiating*. The result of antidifferentiating is called an *antiderivative*. Above we found antiderivatives of the function 8. There are several of them, but they are all $8x$ plus some constant.

Example 1 Antidifferentiate (find the antiderivatives of) x^2. CSS

Find three antiderivatives.

1. $\dfrac{dy}{dx} = 7$ *[handwritten: $7x + C$]*

2. $\dfrac{dy}{dx} = -2$ *[handwritten: $-2x + C$]*

Find the general form of each antiderivative.

3. x *[handwritten: $\dfrac{x^2}{2} + C$]*

4. x^3 *[handwritten: $\dfrac{x^4}{4} + C$]*

5. e^x *[handwritten: $\dfrac{e^{x+1}}{x+1} + C$ $e^x + C$]*

6. $\dfrac{1}{x}$ *[handwritten: $= x^{-1}$ $\dfrac{x^0}{0} = 0$ $\ln x + C$]*

Integrate. Don't forget the constant of
integration!

7. $\displaystyle\int x^3 \, dx$

8. $\displaystyle\int x \, dx$

9. $\displaystyle\int \dfrac{1}{x} \, dx,\; x > 0$

Solution One antiderivative is $x^3/3$. All other antiderivatives differ
from this by a constant, so we can denote them as follows:

$$\frac{x^3}{3} + C$$

This is the *general form* of the antiderivative.

DO EXERCISES 1–6.

Integrals and Integration

The process of antidifferentiation is, in some contexts, called *integration*, and the general form of the antiderivative is referred to as an
indefinite integral. A common notation for the indefinite integral, from
Leibniz, is

$$\int f(x) \, dx.$$

The symbol $\int$ is called an *integral sign*. The symbol dx plays no appar-
ent role at this point in our development, but will be useful later. In this
context, $f(x)$ is called the *integrand*. We illustrate this notation using
the preceding example.

Example 2 Integrate $\int x^2 \, dx$.

Solution $\int x^2 \, dx = x^3/3 + C$

The symbol on the left is read "the integral of x^2, dx." (The "dx" is often
omitted in the reading.) In this case the integrand is x^2. The constant C
is called the *constant of integration*.

Example 3 Integrate $\int e^x \, dx$.

Solution $\int e^x \, dx = e^x + C$

DO EXERCISES 7–9.

To integrate (or antidifferentiate) we make use of differentiation formulas, in effect reading them in reverse. Below are some of these, stated in
reverse, as integration formulas. These can be checked by differentiating the right-hand side and noting that the result is, in each case,
the integrand.

THEOREM 2   Basic Integration Formulas

1. $\int k\ dx$, (k a constant) $= kx + C$

2. $\int x^r\ dx = \dfrac{x^{r+1}}{r+1} + C$ (provided $r \neq -1$), or

   $\int (r+1)x^r\ dx = x^{r+1} + C$

   (To integrate a power of x other than -1, increase the power by 1 and divide by the increased power.)

3. $\int x^{-1}\ dx = \int \dfrac{1}{x}\ dx = \ln x + C, \quad x > 0; \quad \int x^{-1}\ dx = \ln |x| + C,$
   $x < 0$ (We will generally consider $x > 0$.)

4. $\int e^x\ dx = e^x + C$

The following rules allow us to find many other integrals. They are obtainable by reversing two familiar differentiation rules.

THEOREM 3

Rule A. $\int kf(x)\ dx = k \int f(x)\ dx$

(The integral of a constant times a function is the constant times the integral.)

Rule B. $\int [f(x) + g(x)]\ dx = \int f(x)\ dx + \int g(x)\ dx$

(The integral of a sum is the sum of the integrals.)

Example 4

$$\int (5x + 4x^3)\ dx = \int 5x\ dx + \int 4x^3\ dx \qquad \text{Rule B}$$

$$= 5 \int x\ dx + \int 4x^3\ dx \qquad \text{Rule A}$$

(Note that we did not factor the 4 out of the second integral. This is because we can find the antiderivative of $4x^3$ directly as x^4, as shown in the second part of formula 2.) Then,

$$\int (5x + 4x^3)\ dx = 5 \cdot \frac{x^2}{2} + x^4 + C = \frac{5}{2}x^2 + x^4 + C.$$

(Don't forget the constant of integration!)

10. Integrate. Don't forget the constant of integration!

$$\int (7x^4 + 2x)\, dx$$

$$\frac{7x^5}{5} + \frac{2x^2}{2} + C$$

$$\frac{7x^5}{5} + x^2 + C = 0$$

Integrate. Don't forget the constant of integration.

11. $\int (e^x - x^{2/5})\, dx$

$$\int e^x\, dx - \int x^{2/5}\, dx$$

$$e^x - \frac{x^{7/5}}{7/5} + C$$

$$e^x - \frac{5}{7} x^{7/5} + C$$

12. $\int \left(\frac{5}{x} - 7 + \frac{1}{x^6} \right)\, dx$

$$\int \frac{5}{x}\, dx - \int 7\, dx + \int \frac{1}{x^6}\, dx$$

$$5\ln x - 7x + \int x^{-6}$$

$$5\ln x - 7x + \frac{x^{-5}}{-5}$$

$$5\ln x + 7x - \frac{x^{-5}}{5}$$

Note:

We can always check by differentiating.

Thus, in Example 4,

$$\frac{d}{dx}\left(\frac{5}{2}x^2 + x^4 + C \right) = 2 \cdot \frac{5}{2} \cdot x + 4x^3 = 5x + 4x^3.$$

DO EXERCISE 10.

Example 5 $\displaystyle\int (e^x - \sqrt{x})\, dx = \int e^x\, dx - \int \sqrt{x}\, dx$

$$= \int e^x\, dx - \int x^{1/2}\, dx$$

$$= e^x - \frac{x^{(1/2)+1}}{\frac{1}{2} + 1} + C$$

$$= e^x - \frac{x^{3/2}}{\frac{3}{2}} + C$$

$$= e^x - \frac{2}{3}x^{3/2} + C$$

Example 6 $\displaystyle\int \left(1 - \frac{3}{x} + \frac{1}{x^4} \right)\, dx = \int 1\, dx - 3\int \frac{dx}{x} + \int x^{-4}\, dx$

$$= x - 3\ln x + \frac{x^{-4+1}}{-4 + 1} + C$$

$$= x - 3\ln x - \frac{x^{-3}}{3} + C$$

DO EXERCISES 11 AND 12.

Another Look at Antiderivatives

The graphs of the antiderivatives of x^2 are the graphs of the functions

$$y = \int x^2\, dx = \frac{x^3}{3} + C$$

for the various values of the constant C.

13. Using the same set of axes, graph

$$y = \frac{x^3}{3}, \quad y = \frac{x^3}{3} + 1, \quad \text{and}$$

$$y = \frac{x^3}{3} - 1.$$

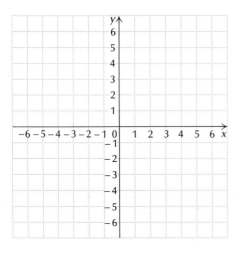

DO EXERCISE 13.

As shown in the following graphs, x^2 is the derivative of each function. That is, the tangent line at the point

$$\left(a, \frac{a^3}{3} + C\right)$$

has slope a^2. The curves $(x^3/3) + C$ fill up the plane, exactly one curve going through any given point (x_0, y_0).

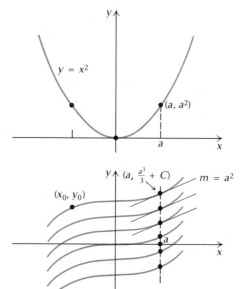

Suppose we look for an antiderivative of x^2 having a specified value at a certain point, say $f(-1) = 2$. We find that there is only one such function.

Example 7 Find the function f such that

$$f'(x) = x^2 \quad \text{and} \quad f(-1) = 2.$$

Solution

a) We find $f(x)$ by integrating:

$$f(x) = \int x^2 \, dx = \frac{x^3}{3} + C.$$

14. Find f such that

$$f'(x) = x^2 \quad \text{and} \quad f(-2) = 5.$$

$\int x^2\, dx = \dfrac{x^3}{3} + C$

$f(-2) = \dfrac{-8}{3} + C$

$\dfrac{-8}{3} + C = 5$

$C = 23/3$

$f(x) = \dfrac{x^3}{3} + 23/3$

15. Find g such that

$$g'(x) = 2x - 4 \quad \text{and} \quad g(2) = 9.$$

$\int 2x - 4\, dx = \dfrac{2x^2}{2} - 4x + C$

$= x^2 - 4x + C$

$g(2) = 4 - 8 + C$

$= -4 + C$

$9 = -4 + C = 9$

$C = 13$

$g(x) = x^2 - 4x + 13$

16. A company determines that the marginal cost C' of producing the xth unit of a certain product is given by

$$C'(x) = x^2 + 5x.$$

Find the total cost function C, assuming fixed costs to be \$35.

$\int (x^2 + 5x)\, dx = \dfrac{x^3}{3} + \dfrac{5x^2}{2} + K$

$C(0) = K$

$K = 35$

$\dfrac{x^3}{3} + \dfrac{5x^2}{2} + 35$

b) The condition $f(-1) = 2$ allows us to find C:

$$f(-1) = \frac{(-1)^3}{3} + C = 2,$$

and solving for C we get

$$-\frac{1}{3} + C = 2$$

$$C = 2 + \frac{1}{3}, \quad \text{or} \quad \frac{7}{3}.$$

Thus $f(x) = (x^3/3) + (7/3)$.

DO EXERCISES 14 AND 15.

Applied Problems

Example 8 A company determines that the marginal cost C' of producing the xth unit of a certain product is given by

$$C'(x) = x^3 + 2x.$$

Find the total cost function C, assuming fixed costs (costs when 0 units are produced) are \$45.

Solution

a) We integrate to find $C(x)$, using K for the integration constant to avoid confusion with the cost function C:

$$C(x) = \int C'(x)\, dx = \int (x^3 + 2x)\, dx = \frac{x^4}{4} + x^2 + K.$$

b) Fixed costs are \$45. This means that $C(0) = 45$. This allows us to determine the value of K:

$$C(0) = \frac{0^4}{4} + 0^2 + K = 45$$

$$K = 45.$$

Thus $C(x) = x^4/4 + x^2 + 45$.

DO EXERCISE 16.

17. Suppose $v(t) = 4t^3$ and $s(0) = 13$. Find $s(t)$.

Recall that the position coordinate, at time t, of an object moving along a number line is $s(t)$. Then

$$s'(t) = v(t) = \text{the } velocity \text{ at time } t,$$

$$v'(t) = a(t) = \text{the } acceleration \text{ at time } t.$$

Example 9 Suppose $v(t) = 5t^4$ and $s(0) = 9$. Find $s(t)$.

Solution

a) We find $s(t)$ by integrating:

$$s(t) = \int v(t)\, dt = \int 5t^4\, dt = t^5 + C.$$

b) The condition $s(0) = 9$ allows us to determine C:

$$s(0) = 0^5 + C = 9, \qquad C = 9.$$

Thus $s(t) = t^5 + 9$.

DO EXERCISE 17.

Example 10 Suppose $a(t) = 12t^2 - 6$, $v(0) = $ initial velocity $= 5$, and $s(0) = 10$. Find $s(t)$.

Solution

a) We find $v(t)$ by integrating $a(t)$:

$$v(t) = \int a(t)\, dt = \int (12t^2 - 6)\, dt = 4t^3 - 6t + C_1.$$

b) The condition $v(0) = 5$ allows us to find C_1:

$$v(0) = 4 \cdot 0^3 - 6 \cdot 0 + C_1 = 5, \qquad C_1 = 5.$$

Thus $v(t) = 4t^3 - 6t + 5$.

c) We find $s(t)$ by integrating $v(t)$:

$$s(t) = \int v(t)\, dt = \int (4t^3 - 6t + 5)\, dt = t^4 - 3t^2 + 5t + C_2.$$

d) The condition $s(0) = 10$ allows us to find C_2:

$$s(0) = 0^4 - 3 \cdot 0^2 + 5 \cdot 0 + C_2 = 10, \qquad C_2 = 10.$$

Thus $s(t) = t^4 - 3t^2 + 5t + 10$.

18. Suppose $a(t) = 24t^2 - 12$, $v(0) = 7$, and $s(0) = 8$. Find $s(t)$.

DO EXERCISE 18.

EXERCISE SET 5.1

Integrate.

1. $\int x^6 \, dx$

2. $\int x^7 \, dx$

3. $\int 2 \, dx$

4. $\int 4 \, dx$

5. $\int x^{1/4} \, dx$

6. $\int x^{1/3} \, dx$

7. $\int (x^2 + x - 1) \, dx$

8. $\int (x^2 - x + 2) \, dx$

9. $\int (t^2 - 2t + 3) \, dt$

10. $\int (3t^2 - 4t + 7) \, dt$

11. $\int 5e^x \, dx$

12. $\int 3e^x \, dx$

13. $\int (x^3 - x^{8/7}) \, dx$

14. $\int (x^4 - x^{6/5}) \, dx$

15. $\int \frac{1000}{x} \, dx$

16. $\int \frac{500}{x} \, dx$

17. $\int \frac{dx}{x^2} \left(\text{or} \int \frac{1}{x^2} \, dx \right)$

18. $\int \frac{dx}{x^3}$

Applied problems

23. A company determines that the marginal cost C' of producing the xth unit of a certain product is given by

$$C'(x) = x^3 - 2x.$$

Find the total cost function C, assuming fixed costs are $100.

24. A company determines that the marginal cost C' of producing the xth unit of a certain product is given by

$$C'(x) = x^3 - x.$$

Find the total cost function C, assuming fixed costs are $200.

25. A company determines that the marginal revenue R' from selling the xth unit of a certain product is given by

$$R'(x) = x^2 - 3.$$

a) Find the total revenue function R, assuming $R(0) = 0$.
b) Why is $R(0) = 0$ a reasonable assumption?

Find $s(t)$.

26. A company determines that the marginal revenue R' from selling the xth unit of a certain product is given by

$$R'(x) = x^2 - 1.$$

a) Find the total revenue function R, assuming $R(0) = 0$.
b) Why is $R(0) = 0$ a reasonable assumption?

Find $v(t)$.

27. $v(t) = 3t^2$, $s(0) = 4$ **28.** $v(t) = 2t$, $s(0) = 10$

29. $a(t) = 4t$, $v(0) = 20$ **30.** $a(t) = 6t$, $v(0) = 30$

Find $s(t)$.

31. $a(t) = -2t + 6$, $v(0) = 6$, and $s(0) = 10$.

32. $a(t) = -6t + 7$, $v(0) = 10$, and $s(0) = 20$.

33. For a freely falling object, $a(t) = -32$ ft/sec^2, $v(0) =$ initial velocity $= v_0$, and $s(0) =$ initial height $= s_0$. Find a general expression for $s(t)$ in terms of v_0 and s_0.

34. A ball is thrown from a height of 10 ft, $s(0) = 10$, at an initial velocity of 80 ft/sec, $v(0) = 80$. How long will it take to hit the ground? (See Exercise 33.)

35. A car with constant acceleration goes from 0 to 60 mph in $\frac{1}{2}$ minute. How far does the car travel during that time?

36. *Efficiency of a machine operator.* The rate at which a machine operator's efficiency E (expressed as a percentage) changes with respect to time is given by

$$\frac{dE}{dt} = 40 - 10t,$$

where $t =$ the number of hours the operator has been at work.

a) Find $E(t)$, given that the operator's efficiency after working 2 hr is 72%. That is, $E(2) = 72$.

b) Use the answer to (a) to find the operator's efficiency after 4 hr; after 8 hr.

37. *Efficiency of a machine operator.* The rate at which a machine operator's efficiency E (expressed as a percentage) changes with respect to time is given by

$$\frac{dE}{dt} = 30 - 10t,$$

where $t =$ the number of hours the operator has been at work.

a) Find $E(t)$, given that the operator's efficiency after working 2 hr is 72%. That is, $E(2) = 72$.

b) Use the answer to (a) to find the operator's efficiency after 3 hr; after 5 hr.

38. *Psychology: Memory.* In a certain memory experiment the rate of memorizing is given by

$$M'(t) = 0.2t - 0.003t^2,$$

where $M(t)$ is the number of Spanish words memorized in t minutes.

a) Find $M(t)$ if it is known that $M(0) = 0$.

b) How many words are memorized in 8 minutes?

A machine operator's efficiency changes with respect to time. (*Donald C. Dietz: Stock, Boston*)

39. *Biomedical.* The area A of a healing wound is decreasing at the rate given by

$$A'(t) = -43.4t^{-2}, \qquad 1 \le t \le 7,$$

where t is the time in days and A is in square centimeters.

a) Find $A(t)$ if $A(1) = 39.7$.

b) Find the area of the wound after 7 days.

Find f.

40. $f'(t) = \sqrt{t} + \dfrac{1}{\sqrt{t}}, \qquad f(4) = 0$

41. $f'(t) = t^{\sqrt{3}}, \qquad f(0) = 8$

Integrate.

42. $\displaystyle\int (5t + 4)^2 \, dt$

43. $\displaystyle\int (x - 1)^2 x^3 \, dx$

44. $\displaystyle\int (1 - t)\sqrt{t} \, dt$

45. $\displaystyle\int \frac{(t + 3)^2}{\sqrt{t}} \, dt$

46. $\displaystyle\int \frac{x^4 - 6x^2 - 7}{x^3} \, dx$

47. $\displaystyle\int (t + 1)^3 \, dt$

48. $\displaystyle\int \frac{1}{\ln 10} \frac{dx}{x}$

49. $\displaystyle\int b e^{ax} \, dx$

50. $\displaystyle\int (3x - 5)(2x + 1) \, dx$

51. $\displaystyle\int \sqrt[3]{64x^4} \, dx$

52. $\displaystyle\int \frac{x^2 - 1}{x + 1} \, dx$

53. $\displaystyle\int \frac{t^3 + 8}{t + 2} \, dt$

OBJECTIVES

You should be able to

a) **Find the area under a curve on a given closed interval.**

b) **Interpret the area under a curve in two other ways.**

1. Consider the constant function $f(x) = 3$.

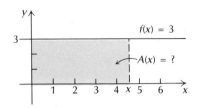

a) Find $A(x)$.

b) Find $A(1)$, $A(2)$, and $A(5)$.

c) Graph $A(x)$.

d) How do $f(x)$ and $A(x)$ compare?

2. A clothing firm, Raggs, Ltd., determines that the marginal cost of each suit it produces is $50.

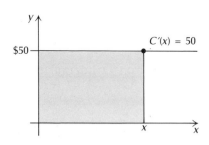

a) Find the total cost $C(x)$ of producing x suits, assuming fixed costs are $0 (ignore fixed costs).

b) Find the area of the shaded rectangle. Compare your answer to (a).

c) Graph $C(x)$. Why is this an increasing function?

5.2 AREA

We now consider the application of integration to finding areas of certain regions. Consider a function whose outputs are positive in an interval (the function might be 0 at one of the endpoints.) We wish to find the area of the region between the graph of the function and the x-axis on that interval.

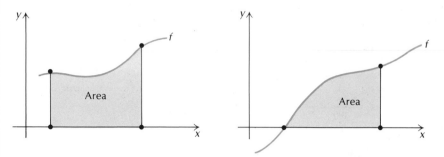

Let us first consider a constant function $f(x) = m$ on the interval from 0 to x, $[0, x]$.

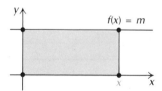

The figure formed is a rectangle, and its area is mx. Suppose we allow x to vary, giving us rectangles of different areas. The area of each rectangle is still mx. We have an area *function*:

$$A(x) = mx.$$

Its graph is shown below.

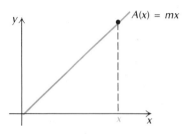

DO EXERCISES 1–3.

3. With better management Raggs, Ltd., of Margin Exercise 2, is able to decrease its production costs by $10 per suit for every hundred suits it produces. This is shown below.

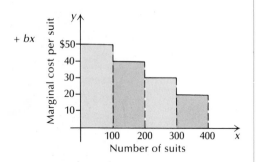

a) Find the total cost of producing 400 suits.

b) Find the total area of the rectangles. Compare your answer with (a).

4. Consider the function $f(x) = 3x$.

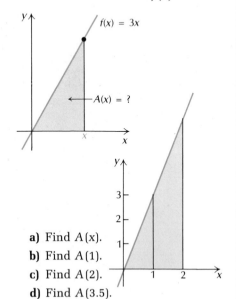

a) Find $A(x)$.
b) Find $A(1)$.
c) Find $A(2)$.
d) Find $A(3.5)$.
e) Graph $A(x)$.
f) How do $f(x)$ and $A(x)$ compare?

Let us next consider the linear function $f(x) = mx$ on the interval from 0 to x, $[0, x]$.

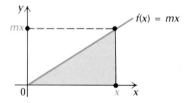

The figure formed this time is a triangle, and its area is $\frac{1}{2}$ the base times the height, $\frac{1}{2} \cdot x \cdot (mx)$, or $\frac{1}{2}mx^2$. If we allow x to vary, we again get an area function:

$$A(x) = \frac{1}{2}mx^2.$$

Its graph is as shown below.

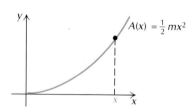

DO EXERCISE 4.

Now consider the linear function $f(x) = mx + b$ on the interval from 0 to x, $[0, x]$.

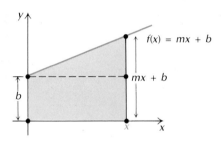

The figure formed this time is a trapezoid, and its area is $\frac{1}{2}$ the height

5. Raggs, Ltd., of Margin Exercise 3, installs new sewing machines. This allows the marginal cost per suit to decrease continually in such a way that

$$C'(x) = -0.1x + 50.$$

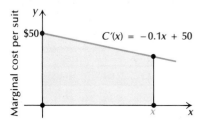

a) Find the total cost of producing x suits, ignoring fixed costs.

b) Find the area of the shaded trapezoid.

c) Find the total cost of producing 400 suits. Compare this answer with that of Margin Exercise 3.

times the sum of the lengths of its parallel sides (or, noting the dashed line, the area of the triangle plus the rectangle):

$$\frac{1}{2} \cdot x \cdot [b + (mx + b)],$$

or

$$\frac{1}{2} \cdot x \cdot (mx + 2b),$$

or

$$\frac{1}{2} mx^2 + bx.$$

If we allow x to vary, we again get an area function:

$$A(x) = \frac{1}{2} mx^2 + bx.$$

Its graph is as shown below.

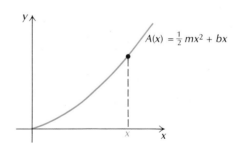

DO EXERCISE 5.

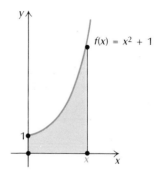

Now we consider the function $f(x) = x^2 + 1$ on the interval from 0 to x, $[0, x]$. The graph of the region in question is as shown, but it is not so easy this time to find the area function because the graph of $f(x)$ is not a straight line. Let us tabulate our previous results and look for a pattern.

$f(x)$	$A(x)$
$f(x) = 3$	$A(x) = 3x$
$f(x) = m$	$A(x) = mx$
$f(x) = 3x$	$A(x) = \dfrac{3}{2}x^2$
$f(x) = mx$	$A(x) = \dfrac{1}{2}mx^2$
$f(x) = mx + b$	$A(x) = \dfrac{1}{2}mx^2 + bx$

You may have conjectured that the area function $A(x)$ is an anti-derivative of $f(x)$. In the following exploratory exercises you will investigate further.

EXPLORATORY EXERCISES: FINDING AREAS

1. The region under the graph of $f(x) = x^2 + 1$, on the interval $[0, 2]$, is shown to the right.

   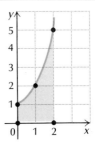

   a) Make a copy of the shaded region on thin paper.

   b) Cut up the shaded region in any way you wish in order to fill up squares in the grid below. Make an estimate of the total area.

   c) Using the antiderivative

   $$F(x) = \frac{x^3}{3} + x,$$

   find $F(2)$.

2. Repeat Exercise 1(a) and (b) for the shaded region of the graph at the right.

  c) Using the antiderivative

$$F(x) = \frac{x^3}{3} + x,$$

  find $F(3)$.

  d) Compare your answers to (b) and (c).

The conjecture concerning areas and antiderivatives (or integrals) is true. It is expressed as follows.

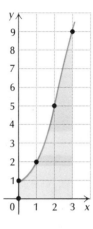

THEOREM 4

> Let f be a positive, continuous function on an interval $[a, b]$ and let $A(x)$ be the area of the region between the graph of f and the x-axis on the interval $[a, x]$. Then $A(x)$ is a differentiable function of x and
>
> $$A'(x) = f(x).$$

Proof. The situation described in the theorem is shown here.

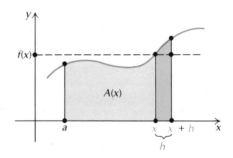

The derivative of $A(x)$ is, by definition of a derivative,

$$A'(x) = \lim_{h \to 0} \frac{A(x + h) - A(x)}{h}.$$

Note from the drawing that $A(x + h) - A(x)$ is the area of the small, shaded, vertical strip. The area of this small strip is approximately that of a rectangle of base h and height $f(x)$, especially for small values of h. Thus we have

$$A(x + h) - A(x) \approx f(x) \cdot h.$$

Now

$$A'(x) = \lim_{h \to 0} \frac{A(x+h) - A(x)}{h} = \lim_{h \to 0} \frac{f(x) \cdot h}{h} = \lim_{h \to 0} f(x) = f(x),$$

since $f(x)$ does not involve h.

The theorem above also holds if $f(x) = 0$ at one or both endpoints of the interval $[a, b]$.

Since the area function A is an antiderivative of f, and since any two antiderivatives differ by a constant, we easily conclude that the area function and any antiderivative differ by a constant.

We can think of the function A as given by

$$A(x) = \text{the area on the interval } [a, x],$$

where a is some fixed point and x varies.

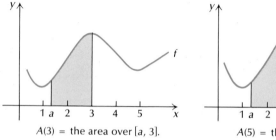

$A(3) =$ the area over $[a, 3]$. $A(5) =$ the area over $[a, 5]$.

Now let us find some areas.

Example 1 Find the area under the graph of $y = x^2 + 1$ on the interval $[-1, 2]$.

Solution

a) We first make a drawing. This includes a graph of the function and the region in question.

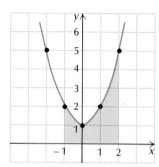
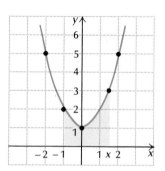

6. Find the area under the graph of $y = x^2 + 3$ on the interval $[1, 2]$.

b) Second, we make a drawing showing a portion of the region from -1 to x. Now $A(x)$ is the area of this portion, that is, in the interval $[-1, x]$.

c) Now

$$A(x) = \int (x^2 + 1)\, dx = \frac{x^3}{3} + x + C,$$

where C has to be determined. Since we know that $A(-1) = 0$ (there is no area above the number -1), we can substitute for x in $A(x)$, as follows:

$$A(-1) = \frac{(-1)^3}{3} + (-1) + C = 0$$

$$-\frac{1}{3} - 1 + C = 0$$

$$C = \frac{4}{3}.$$

This determines that $C = \frac{4}{3}$, so we have

$$A(x) = \frac{x^3}{3} + x + \frac{4}{3}.$$

Then the area in the interval $[-1, 2]$ is $A(2)$. We compute $A(2)$ as follows:

$$A(2) = \frac{2^3}{3} + 2 + \frac{4}{3}$$

$$= \frac{8}{3} + 2 + \frac{4}{3}$$

$$= \frac{12}{3} + 2$$

$$= 6.$$

DO EXERCISE 6.

Example 2 Find the area under the graph of $y = x^3$ on the interval $[0, 5]$.

Solution

a) We first make a drawing that includes a graph of the function and the region in question.

7. Find the area under the graph of $y = x^2 + x$ on the interval $[0, 3]$.

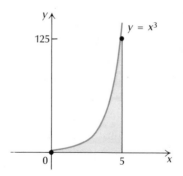
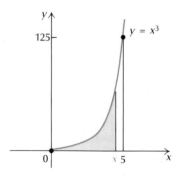

b) Second, we make a drawing showing a portion of that region from 0 to x. Now $A(x)$ is the area of this portion, that is, on the interval $[0, x]$.

c) Now

$$A(x) = \int x^3 \, dx = \frac{x^4}{4} + C.$$

where C has to be determined. Since we know that $A(0) = 0$, we can substitute 0 for x in $A(x)$, as follows:

$$A(0) = \frac{0^4}{4} + C = 0$$

$$C = 0.$$

This determines C. So

$$A(x) = \frac{x^4}{4}.$$

Then the area in the interval $[0, 5]$ is $A(5)$. We can compute $A(5)$ as follows:

$$A(5) = \frac{5^4}{4} = \frac{625}{4} = 156\frac{1}{4}.$$

DO EXERCISE 7.

Since the area under a curve, as in the preceding examples, is an anti-derivative, area can also be associated with various kinds of functions. For example, if we have a velocity function over an interval $[0, b]$, then the area under the curve in that interval is the total distance. Suppose the velocity function is

$$v(t) = t^3.$$

In 5 hours the total distance covered is $156\frac{1}{4}$. We can see this in Example 2, simply by changing the variable from x to t. For a marginal cost function over the interval $[0, x]$, the area under the curve is the total cost of producing x units, or the accumulated cost.

Example 3   Raggs, Ltd., goes even further to reduce production costs. In addition to purchasing new sewing machines, the president has air conditioning installed and takes a calculus course. These cause the marginal cost per suit to decrease rapidly in such a way that

$$C'(x) = 0.0003x^2 - 0.2x + 50.$$

Find the total cost of producing 400 suits. (Ignore fixed costs.)

Solution

a)   First, we make a drawing. This includes a graph of the function and the region in question.

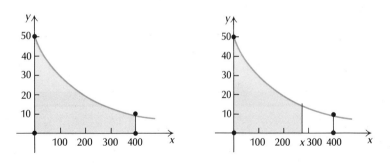

b)   Second, we make a drawing showing a portion of that region from 0 to x. Now $A(x)$ is the area of that portion, that is, on the interval $[0, x]$.

c)   Now,

$$C(x) = A(x) = \int (0.0003x^2 - 0.2x + 50) \, dx$$

$$= 0.0001x^3 - 0.1x^2 + 50x + K,$$

where K has to be determined. We are ignoring fixed costs, $K = 0$, and we have

$$C(x) = 0.0001x^3 - 0.1x^2 + 50x.$$

8. Referring to Example 3:

   a) Compare \$10,400 to your answer for Margin Exercise 5. Has the company reduced total costs?

   b) Find the total cost of producing 100 suits.

Then the area in the interval $[0, 400]$ is $A(400)$ or $C(400)$. We can compute $C(400)$ as folows:

$$C(400) = 0.0001 \cdot 400^3 - 0.1 \cdot 400^2 + 50 \cdot 400, \quad \text{or} \quad \$10,400.$$

DO EXERCISE 8.

EXERCISE SET 5.2

Find the area under the given curve on the interval indicated.

1. $y = 4$; $[1, 3]$ **2.** $y = 5$; $[1, 3]$ **3.** $y = 2x$; $[1, 3]$ **4.** $y = x^2$; $[0, 3]$

5. $y = x^2$; $[0, 5]$ **6.** $y = x^3$; $[0, 2]$ **7.** $y = x^3$; $[0, 1]$ **8.** $y = 1 - x^2$; $[-1, 1]$

9. $y = 4 - x^2$; $[-2, 2]$ **10.** $y = e^x$; $[0, 2]$ **11.** $y = e^x$; $[0, 3]$ **12.** $y = \dfrac{1}{x}$; $[1, 2]$

13. $y = \dfrac{1}{x}$; $[1, 3]$ **14.** $y = x^2 - 4x$; $[-4, -2]$ **15.** $y = x^2 - 4x$; $[-4, -1]$

In each case, give two interpretations of the shaded region.

16.

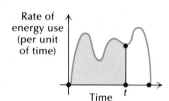

17.

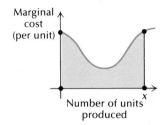

18.

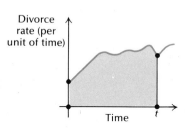

19.

20.

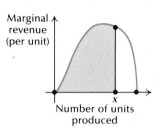

21.

22.

23.

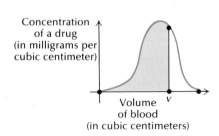

24.

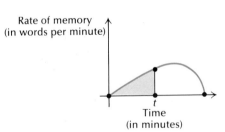

25. A particle starts out from the origin. Its velocity at time t is given by

$$v(t) = 3t^2 + 2t.$$

a) Find the distance the particle has traveled after t hours.

b) Find the distance the particle has traveled after 5 hours.

27. A sound company determines that the marginal cost of producing the xth stereo is given by

$$C'(x) = 100 - 0.2x, \qquad C(0) = 0.$$

It also determines that its marginal revenue from the sale of the xth stereo is given by

$$R'(x) = 100 + 0.2x, \qquad R(0) = 0.$$

a) Find the total cost of producing x stereos.

b) Find the total revenue from selling x stereos.

c) Find the total profit from the production and sale of x stereos.

d) Find the total profit from the production and sale of 1000 stereos.

26. A particle starts out from the origin. Its velocity at time t is given by

$$v(t) = 4t^3 + 2t.$$

a) Find the distance the particle has traveled after t hours.

b) Find the distance the particle has traveled after 2 hours.

28. A refrigeration company determines that the marginal cost of producing the xth refrigerator is given by

$$C'(x) = 50 - 0.4x, \qquad C(0) = 0.$$

It also determines that its marginal revenue from the sale of the xth refrigerator is given by

$$R'(x) = 50 + 0.4x, \qquad R(0) = 0.$$

a) Find the total cost of producing x refrigerators.

b) Find the total revenue from selling x refrigerators.

c) Find the total profit from the production and sale of x refrigerators.

d) Find the total profit from the production and sale of 1000 refrigerators.

▶

Find the area under the curve on the interval indicated.

29. $y = \dfrac{x^2 - 1}{x - 1}$; $[2, 3]$

30. $y = \dfrac{x^5 - x^{-1}}{x^2}$; $[1, 5]$

31. $y = (x - 1)\sqrt{x}$; $[4, 16]$

32. $y = (x + 2)^3$; $[0, 1]$

33. $y = \dfrac{\sqrt[3]{x^2} - 1}{\sqrt[3]{x}}$; $[1, 8]$

34. $y = \dfrac{x^3 + 8}{x + 2}$; $[0, 1]$

OBJECTIVES

You should be able to

a) **Evaluate a definite integral.**

b) **Find the area under a graph on [a, b].**

c) **Solve applied problems involving definite integrals.**

Integrate.

1. $\int_a^b 2x \, dx$

2. $\int_a^b e^x \, dx$

5.3 INTEGRATION ON AN INTERVAL: THE DEFINITE INTEGRAL

Let f be a positive continuous function on an interval $[a, b]$. We know that f has an antiderivative, namely $A(x)$. Let F and G be any two antiderivatives of f. Then

$$F(b) - F(a) = G(b) - G(a).$$

To understand this, recall that F and G differ by a constant. That is, $F(x) = G(x) + C$. Then

$$F(b) - F(a) = [G(b) + C] - [G(a) + C] = G(b) - G(a).$$

Thus the difference $F(b) - F(a)$ has the same value for all antiderivatives of f. It is called the *definite integral* of f from a to b.

Definite integrals are usually symbolized as follows:

$$\int_a^b f(x) \, dx.$$

This is read "the integral from a to b of $f(x) \, dx$" (the dx is sometimes omitted from the reading). From the preceding development we see that to find a definite integral $\int_a^b f(x) \, dx$, we first find an antiderivative $F(x)$. The simplest one is the one for which the constant of integration is 0. We evaluate F at b and at a and subtract.

DEFINITION

$\int_a^b f(x) \, dx$ **is defined to be** $F(b) - F(a)$**, where** F **is any antiderivative of** f**.**

Evaluating definite integrals is called *integrating*. The numbers a and b are known as the *limits of integration*.

Example 1 Integrate $\int_a^b x^2 \, dx$.

Solution Using the antiderivative $F(x) = x^3/3$, we have

$$\int_a^b x^2 \, dx = \frac{b^3}{3} - \frac{a^3}{3}.$$

DO EXERCISES 1 AND 2.

Integrate.

3. $\int_1^3 2x \, dx$

$\dfrac{2x^2}{2}$ $9-1=8$

4. $\int_{-2}^0 e^x \, dx$

$1-e^{-2}$

5. $\int_0^1 (2x - x^2) \, dx$

$x^2 - \dfrac{x^3}{3}$

$1 - \dfrac{1}{3} = \dfrac{2}{3} - 0 = \dfrac{2}{3}$

6. $\int_1^e \left(1 + 3x^2 - \dfrac{1}{x}\right) dx$

$x + x^3 - \ln x$

$e + e^3 - \ln e - (1+1)$

$e + e^3 - 1 - 2$

$e + e^3 - 3$

It is convenient to use an intermediate notation

$$\int_a^b f(x) \, dx = [F(x)]_a^b$$

$$= F(b) - F(a).$$

We now evaluate several definite integrals.

Example 2 $\int_{-1}^2 x^2 \, dx = \left[\dfrac{x^3}{3}\right]_{-1}^2 = \dfrac{2^3}{3} - \dfrac{(-1)^3}{3}$

$$= \dfrac{8}{3} - \left(-\dfrac{1}{3}\right) = \dfrac{8}{3} + \dfrac{1}{3} = 3$$

Example 3 $\int_0^3 e^x \, dx = [e^x]_0^3 = e^3 - e^0 = e^3 - 1$

Example 4 $\int_1^4 (x^2 - x) \, dx = \left[\dfrac{x^3}{3} - \dfrac{x^2}{2}\right]_1^4 = \left(\dfrac{4^3}{3} - \dfrac{4^2}{2}\right) - \left(\dfrac{1^3}{3} - \dfrac{1^2}{2}\right)$

$$= \left(\dfrac{64}{3} - \dfrac{16}{2}\right) - \left(\dfrac{1}{3} - \dfrac{1}{2}\right)$$

$$= \dfrac{64}{3} - 8 - \dfrac{1}{3} + \dfrac{1}{2} = 13\dfrac{1}{2}$$

Example 5 $\int_1^e \left(1 + 2x - \dfrac{1}{x}\right) dx = [x + x^2 - \ln x]_1^e$

$$= (e + e^2 - \ln e)$$
$$- (1 + 1^2 - \ln 1)$$
$$= (e + e^2 - 1) - (1 + 1 - 0)$$
$$= e + e^2 - 1 - 1 - 1$$
$$= e + e^2 - 3$$

It is important to note that in $\int_a^b f(x) \, dx$, $a < b$. That is, the larger number is on the top!

DO EXERCISES 3–6.

The area under a curve can be expressed by a definite integral.

THEOREM 5

Let f be a positive continuous function over the closed interval $[a, b]$. The area under the graph of f on the interval $[a, b]$ is

$$\int_a^b f(x)\ dx.$$

Proof

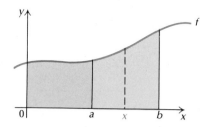

Let

$$A(x) = \text{the area of the region over } [0, x].$$

Then

$$A'(x) = f(x),$$

so $A(x)$ is an antiderivative of $f(x)$. Then

$$\int_a^b f(x)\ dx = A(b) - A(a).$$

But $A(b) - A(a)$ is the area over $[0, b]$ minus the area over $[0, a]$, which is the area over $[a, b]$.

Let us now find some areas.

Example 6   Find the area under $y = x^2 + 1$ on $[-1, 2]$.

Solution $\displaystyle\int_{-1}^{2} (x^2 + 1)\ dx = \left[\frac{x^3}{3} + x\right]_{-1}^{2}$

$$= \left(\frac{2^3}{3} + 2\right) - \left(\frac{(-1)^3}{3} + (-1)\right)$$

$$= \left(\frac{8}{3} + 2\right) - \left(-\frac{1}{3} - 1\right)$$

$$= \frac{8}{3} + 2 + \frac{1}{3} + 1 = 6$$

7. Find the area under $y = x^2 + 3$ on [1, 2]. Don't forget that it helps to draw the graph. (Compare the result with Margin Exercise 6 of Section 5.2.

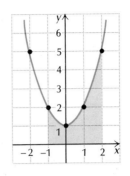

Compare this with Example 1 on p. 290.

DO EXERCISE 7.

Example 7 Find the area under $y = x^3$ on [0, 5].

Solution $\int_0^5 x^3\, dx = \left[\dfrac{x^4}{4}\right]_0^5 = \dfrac{5^4}{4} - \dfrac{0^4}{4} = \dfrac{625}{4}$

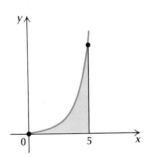

Compare this with Example 2 on p. 291.

DO EXERCISE 8.

8. Find the area under $y = x^2 + x$ on [0, 3]. (Compare the result with Margin Exercise 7 of Section 5.2.

Example 8 Find the area under $y = 1/x$ on [1, 4].

Solution $\int_1^4 \dfrac{dx}{x} = [\ln x]_1^4 = \ln 4 - \ln 1$

$= \ln 4$

≈ 1.3863 ▦ or Table 3

9. Find the area under $y = \dfrac{1}{x}$ on $[1, 7]$.

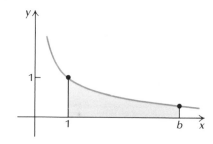

Example 9 Find the area under $y = 1/x^2$ on $[1, b]$.

Solution
$$\int_1^b \frac{dx}{x^2} = \int_1^b x^{-2}\,dx = \left[\frac{x^{-2+1}}{-2+1}\right]_1^b$$

$$= \left[\frac{x^{-1}}{-1}\right]_1^b = \left[-\frac{1}{x}\right]_1^b = \left(-\frac{1}{b}\right) - \left(-\frac{1}{1}\right)$$

$$= 1 - \frac{1}{b}$$

10. Find the area under $y = \dfrac{1}{x^4}$ on $[1, b]$.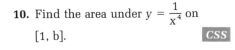

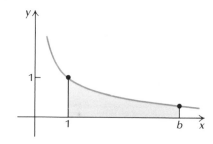

DO EXERCISES 9 AND 10.

The following properties of definite integrals can be derived rather easily from the definition of definite integral and from the properties of the indefinite integral.

PROPERTY 1

$$\int_a^b k \cdot f(x)\,dx = k \cdot \int_a^b f(x)\,dx$$

The integral of a constant times a function is the constant times the integral of the function. That is, we can "factor out" a constant from the integrand.

11. Integrate $\displaystyle\int_1^2 20x^3 \, dx$.

Example 10

$$\int_0^5 100e^x \, dx = 100 \int_0^5 e^x \, dx$$

$$= 100[e^x]_0^5$$

$$= 100(e^5 - e^0)$$

$$= 100(e^5 - 1)$$

DO EXERCISE 11.

$$\int_a^b [f(x) + g(x)] \, dx = \int_a^b f(x) \, dx + \int_a^b g(x) \, dx$$

The integral of a sum is the sum of the integrals.

PROPERTY 3

For $a < c < b$,

$$\int_a^b f(x) \, dx = \int_a^c f(x) \, dx + \int_c^b f(x) \, dx.$$

For any number c between a and b, the integral from a to b is the integral from a to c plus the integral from c to b.

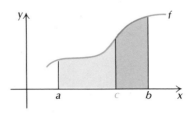

Property 3 has particular application when a function is defined in different ways over subintervals.

Example 11 Find the area under the graph of $y = f(x)$ from -4 to 5, where

$$f(x) = \begin{cases} 9 & \text{if } x < 3, \\ x^2 & \text{if } x \geq 3. \end{cases}$$

12. Find the area under the graph of $y = f(x)$ from -3 to 2, where

$$f(x) = \begin{cases} 4 - x^2 & \text{if } x < 0, \\ 4 - x^2 & \text{if } x \geq 0. \end{cases}$$

CSS

Solution

$$\int_{-4}^{5} f(x)\ dx = \int_{-4}^{3} f(x)\ dx + \int_{3}^{5} f(x)\ dx$$

$$= \int_{-4}^{3} 9\ dx + \int_{3}^{5} x^2\ dx$$

$$= 9 \int_{-4}^{3} dx + \int_{3}^{5} x^2\ dx$$

$$= 9[x]_{-4}^{3} + \left[\frac{x^3}{3}\right]_{3}^{5}$$

$$= 9[3 - (-4)] + \left(\frac{5^3}{3}\right) - \left(\frac{3^3}{3}\right)$$

$$= 95\frac{2}{3}$$

DO EXERCISE 12.

Applied Problem

Example 12 *Business: Accumulated sales.* The sales of a company are expected to grow continuously at a rate given by the function

$$S'(t) = 100e^t,$$

where $S'(t) =$ the sales rate, in dollars per day, at time t.

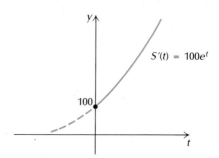

13. The sales of a company are expected to grow continuously at a rate given by the function

$$S'(t) = 200e^t,$$

where $S'(t) =$ the sales rate, in dollars per day, at time t.

a) Find the accumulated sales for the first 8 days.

b) On what day will accumulated sales exceed $300,000?

c) The accumulated sales from the 8th through the 10th day is given by the integral from 7 to 10,

$$\int_7^{10} S'(t)\, dt.$$

Find this.

a) Find the accumulated sales for the first 7 days.

b) On what day will accumulated sales exceed $810,000?

Solution

a) Accumulated sales through day 7 are

$$\int_0^7 S'(t)\, dt = \int_0^7 100e^t\, dt = 100 \int_0^7 e^t\, dt$$

$$= 100[e^t]_0^7 = 100(e^7 - e^0)$$

$$= 100(1096.633158 - 1)$$

$$= 100(1095.633158) \approx \$109{,}563.32.$$

b) Accumulated sales through day k are

$$\int_0^k S'(t)\, dt = \int_0^k 100e^t\, dt = 100 \int_0^k e^t\, dt 100(e^k - 1).$$

$$= 100[e^t]_0^k = 100(e^k - e^0) = 100(e^k - 1).$$

We set this equal to $810,000 and solve for k:

$$100(e^k - 1) = 810{,}000$$

$$e^k - 1 = 8100$$

$$e^k = 8101.$$

We solve this equation for k using natural logarithms:

$$e^k = 8101$$

$$\ln e^k = \ln 8101$$

$$k = 8.999743 \qquad \blacksquare \text{ or Table 3}$$

$$k \approx 9.$$

DO EXERCISE 13.

EXERCISE SET 5.3

Integrate.   **CSS**

1. $\int_0^1 (x - x^2)\, dx$ **2.** $\int_1^2 (x^2 - x)\, dx$ **3.** $\int_{-1}^1 (x^2 - x^4)\, dx$ **4.** $\int_0^b e^x\, dx$ **5.** $\int_a^b e^t\, dt$

6. $\int_0^a (ax - x^2)\, dx$ **7.** $\int_a^b 3t^2\, dt$ **8.** $\int_a^b 4t^3\, dt$ **9.** $\int_1^e \left(x + \frac{1}{x}\right) dx$ **10.** $\int_1^e \left(x - \frac{1}{x}\right) dx$

11. $\int_0^1 \sqrt{x}\, dx$ **12.** $\int_0^1 3\sqrt{x}\, dx$ **13.** $\int_0^1 \frac{10}{17} t^3\, dt$ **14.** $\int_0^1 \frac{12}{13} t^2\, dt$

Find the area under the graph on the interval indicated.

15. $y = x^3$; $[0, 2]$ **16.** $y = x^4$; $[0, 1]$ **17.** $y = x^2 + x + 1$; $[2, 3]$

18. $y = 2 - x - x^2$; $[-2, 1]$ **19.** $y = 5 - x^2$; $[-1, 2]$ **20.** $y = e^x$; $[-2, 3]$

21. $y = e^x$; $[-1, 5]$ **22.** $y = 2x + \frac{1}{x^2}$; $[1, 4]$ **23.** $y = 2x - \frac{1}{x^2}$; $[1, 3]$

Find the area under the graph on $[-2, 3]$ where

24. $f(x) = \begin{cases} x^2 & \text{if } x < 1, \\ 1 & \text{if } x \geq 1 \end{cases}$ **25.** $f(x) = \begin{cases} 4 - x^2 & \text{if } x < 0, \\ 4 & \text{if } x \geq 0 \end{cases}$

26. *Business: Accumulated sales.* Raggs, Ltd., estimates that its sales will grow continuously at a rate given by the function

$$S'(t) = 10e^t,$$

where $S'(t)$ = the sales rate, in dollars per day, at time t.

a) Find the accumulated sales for the first 5 days.

b) Find the sales from the 2nd through the 5th day. This is the integral from 1 to 5.

c) On what day will accumulated sales exceed $40,000?

28. A particle starts out from the origin. Its velocity at time t is given by

$$v(t) = 3t^2 + 2t.$$

How far does it travel from the 2nd through the 5th hour (from $t = 1$ to $t = 5$)?

30. *Business.* Raggs, Ltd., determines that the marginal cost per suit is given by

$$C'(x) = 0.0003x^2 - 0.2x + 50.$$

Ignoring fixed costs, find the total cost of producing the 101st through the 400th suit (integrate from $x = 100$ to $x = 400$).

27. *Business: Accumulated sales.* A company estimates that its sales will grow continuously at a rate given by the function

$$S'(t) = 20e^t,$$

where $S'(t)$ = the sales rate, in dollars per day, at time t.

a) Find the accumulated sales for the first 5 days.

b) Find the sales from the 2nd through the 5th day. This is the integral from 1 to 5.

c) On what day will accumulated sales exceed $20,000?

29. A particle starts out from the origin. Its velocity at time t is given by

$$v(t) = 4t^3 + 2t.$$

How far does it travel from the 1st through the 3rd hour (from $t = 0$ to $t = 3$)?

31. *Business.* In Exercise 30, find the cost of producing the 201st through the 400th suit (integrate from $x = 200$ to $x = 400$).

Psychology: Memorizing. In the psychological process of memorizing, the rate of memorizing (say, in words per minute) increases with respect to time, but eventually a maximum rate of memorizing is reached from which the memory rate decreases.

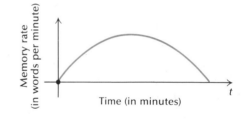

32. Suppose in a certain memory experiment the rate of memorizing is given by   `CSS`

$$M'(t) = -0.009t^2 + 0.2t,$$

where $M'(t)$ = the memory rate in words per minute. How many words are memorized in the first 10 minutes (from $t = 0$ to $t = 10$)?

34. *Psychology: Business.* A company is producing a new product. However, due to the nature of the product, it is felt that the time required to produce each unit will decrease as the workers become more familiar with the production procedure. It is determined that the function for the learning process is

$$T(x) = ax^b,$$

where

$T(x)$ = the cumulative average time to produce x units,

  x = the number of units produced,

  a = the hours required to produce the 1st unit, and

  b = the slope of the learning curve.

a) Find an expression for the total time required to produce 100 units.

b) Suppose $a = 100$ hr and $b = -0.322$. Find the total time required to produce 100 units. (*Hint:* $100^{0.678} \approx 22.7$.)

33. Suppose in a certain memory experiment the rate of memorizing is given by

$$M'(t) = -0.003t^2 + 0.2t,$$

where $M'(t)$ = the memory rate in words per minute. How many words are memorized in the first 10 minutes ($t = 0$ to $t = 10$)?

Integrate.   `CSS`

35. $\int_1^2 (4x + 3)(5x - 2)\, dx$

36. $\int_2^5 (t + \sqrt{3})(t - \sqrt{3})\, dt$

37. $\int_0^1 (t + 1)^3\, dt$

38. $\int_1^3 \left(x - \dfrac{1}{x}\right)^2 dx$

39. $\int_1^3 \dfrac{t^5 - t}{t^3}\, dt$

40. $\int_4^9 \dfrac{t + 1}{\sqrt{t}}\, dt$

41. $\int_3^5 \dfrac{x^2 - 4}{x - 2}\, dx$

42. $\int_0^1 \dfrac{t^3 + 1}{t + 1}\, dt$

OBJECTIVES

You should be able to

a) **Evaluate the definite integral of a continuous function.**

b) **Find the area of a region bounded by two graphs.**

c) **Solve applied problems involving area between two graphs.**

5.4 THE DEFINITE INTEGRAL: AREA BETWEEN CURVES

We have considered the definite integral for functions that are positive on an interval $[a, b]$ (the function might be 0 at one or both endpoints). Now we will consider functions that have negative values. First, let us evaluate the integral of a funtion without negative values:

$$\int_0^2 x^2 \, dx = \left[\frac{x^3}{3}\right]_0^2 = \frac{2^3}{3} - \frac{0^3}{3} = \frac{8}{3}.$$

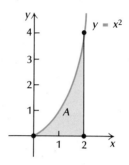

Thus the area of the shaded region is $\frac{8}{3}$.

Now let us consider the function $y = -x^2$ on the interval $[0, 2]$. Even though we have not defined the definite integral for functions with negative values, let us apply the evaluation procedures and see what we get:

$$\int_0^2 -x^2 \, dx = \left[-\frac{x^3}{3}\right]_0^2$$

$$= \left[-\frac{2^3}{3} - \left(-\frac{0^3}{3}\right)\right] = -\frac{8}{3}.$$

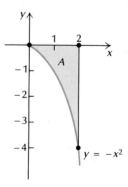

1. To find the area of the shaded region, integrate

$$\int_0^2 x^3 \, dx.$$

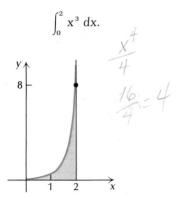

2. To find the area of the shaded region, integrate

$$\int_0^2 -x^3 \, dx.$$

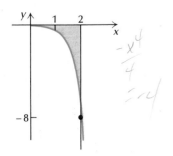

The graphs of these two functions are reflections of each other across the x-axis. Thus the areas of the shaded regions are the same—that is, $\frac{8}{3}$. The evaluation procedure in the second case gave us $-\frac{8}{3}$. This illustrates that for negative-valued functions, the definite integral gives us the additive inverse of the area between the curve and the x-axis.

DO EXERCISES 1 AND 2.

Now let us consider the function $x^2 - 1$ on $[-1, 2]$. It has both positive and negative values. We will apply the preceding evaluation procedures, even though function values are not all nonnegative. We will do this in two ways.

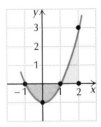

First,

$$\int_{-1}^2 (x^2 - 1) \, dx = \int_{-1}^1 (x^2 - 1) \, dx + \int_1^2 (x^2 - 1) \, dx$$

$$= \left[\frac{x^3}{3} - x \right]_{-1}^1 + \left[\frac{x^3}{3} - x \right]_1^2$$

$$= \left[\left(\frac{1^3}{3} - 1 \right) - \left(\frac{(-1)^3}{3} - (-1) \right) \right]$$

$$+ \left[\left(\frac{2^3}{3} - 2 \right) - \left(\frac{1^3}{3} - 1 \right) \right]$$

$$= \left[-\frac{4}{3} \right] + \left[\frac{4}{3} \right] = 0.$$

This shows that the area of the region under the x-axis is the same as the area of the region over the x-axis.

3. Integrate.

a) $\int_0^1 (x^2 - x)\, dx$

b) $\int_1^2 (x^2 - x)\, dx$

c) $\int_0^2 (x^2 - x)\, dx$

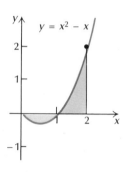

$y = x^2 - x$

4. Integrate.

a) $\int_{-2}^0 x^3\, dx$

b) $\int_0^2 x^3\, dx$

c) $\int_{-2}^2 x^3\, dx$

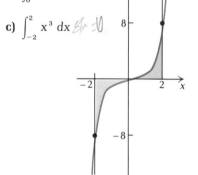

$y = x^3$

Now let us evaluate in another way:

$$\int_{-1}^2 (x^2 - 1)\, dx = \left[\frac{x^3}{3} - x\right]_{-1}^2 = \left(\frac{2^3}{3} - 2\right) - \left[\frac{(-1)^3}{3} - (-1)\right]$$

$$= \left(\frac{8}{3} - 2\right) - \left(-\frac{1}{3} + 1\right)$$

$$= \frac{8}{3} - 2 + \frac{1}{3} - 1$$

$$= 0.$$

This result is consistent with the first. Thus we are motivated to extend our definition of definite integral to include any continuous function, having positive, negative, or zero values.

DEFINITION

For any function, continuous on an interval $[a, b]$,

$$\int_a^b f(x)\, dx = F(b) - F(a),$$

where F is any antiderivative of f.

DO EXERCISES 3 AND 4.

The definite integral turns out to be the area above the x-axis minus the area below.

Examples Decide whether $\int_a^b f(x)\, dx$ is positive, negative, or zero.

a)

b)

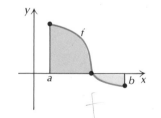

c)

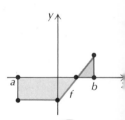

In each exercise:

a) decide whether $\int_a^b f(x)\,dx$ is positive, negative, or zero;

b) express $\int_a^b f(x)\,dx$ in terms of A.

Solution

a) $\int_a^b f(x)\,dx = 0.$

There is the same area above as below.

b) $\int_a^b f(x)\,dx > 0.$

There is more area above.

c) $\int_a^b f(x)\,dx < 0.$

There is more area below.

DO EXERCISES 5–8.

Area of a Region Bounded by Two Graphs

Suppose we want to find the area of a region bounded by the graphs of two functions $y = f(x)$ and $y = g(x)$. $\boxed{CSS}$

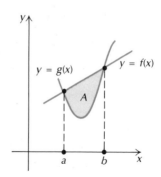

Note that the area of the region in question, I, is that of II minus that of III.

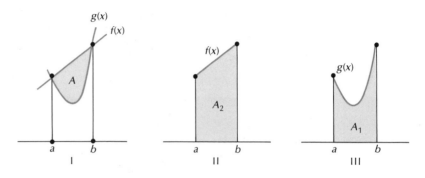

5.

6.

7.

8.

Thus,

$$A = \int_a^b f(x)\,dx - \int_a^b g(x)\,dx \quad \text{or} \quad A = \int_a^b [f(x) - g(x)]\,dx.$$

In general, we have the following.

THEOREM 6

Let f and g be continuous functions and suppose $f(x) \geq g(x)$ over the interval $[a, b]$. Then the area of the region between the two curves, from $x = a$ to $x = b$, is

$$\int_a^b [f(x) - g(x)]\,dx.$$

Example 1 Find the area of the region bounded by the graphs of $y = 2x + 1$ and $y = x^2 + 1$.

Solution

a) First, make a reasonably accurate sketch to ensure that you have the right configuration. Note which is the *upper* graph. Here it is $2x + 1 \geq x^2 + 1$ over the interval $[0, 2]$.

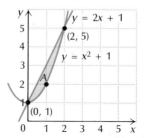

b) Second, if boundaries are not stated, determine the first coordinates of possible points of intersection. Usually you can do this just by looking at the graph. If not, you can solve the system of equations as follows. At the points of intersection, $y = x^2 + 1$ and $y = 2x + 1$, so

$$x^2 + 1 = 2x + 1$$

$$x^2 - 2x = 0$$

$$x(x - 2) = 0$$

$$x = 0 \quad \text{or} \quad x = 2.$$

9. Find the area of the region bounded by the graphs of

$$y = x \quad \text{and} \quad y = x^2. \quad \boxed{\textit{CSS}}$$

$[0, 1]$

$a = \int_0^1 [x - x^2]\, dx$

$= \int_0^1 (x - x^2)\, dx$

$= \left[\dfrac{x^2}{2} - \dfrac{x^3}{3} \right]_0^1$

$= \left[\left(\dfrac{1}{2} - \dfrac{1}{3} \right) - 0 \right] -$

$\dfrac{3}{6} \; \mathcal{A} = \dfrac{1}{2}$

Thus the interval with which we are concerned is $[0, 2]$.

c) Compute the area as follows:

$$\int_0^2 [(2x + 1) - (x^2 + 1)]\, dx = \int_0^2 (2x - x^2)\, dx$$

$$= \left[x^2 - \frac{x^3}{3} \right]_0^2$$

$$= \left(2^2 - \frac{2^3}{3} \right) - \left(0^2 - \frac{0^3}{3} \right)$$

$$= 4 - \frac{8}{3}$$

$$= \frac{4}{3}.$$

DO EXERCISE 9.

Application

Example 2 *Emission control.* A clever college student develops an engine that is believed to meet federal standards for emission control. The engine's rate of emission is given by

$$E(t) = 2t^2,$$

where $E(t)$ = emissions in billions of pollution particulates per year at time t, in years. The emission rate of a conventional engine is given by

$$C(t) = 9 + t^2.$$

The curves are seen here.

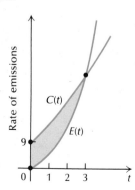

a) At what point in time will the emission rates be the same?

10. A company determines that its marginal revenue per unit is given by

$$R'(x) = 14x, \qquad R(0) = 0.$$

Its marginal cost per unit is given by

$$C'(x) = 4x - 10, \qquad C(0) = 0.$$

Total profit from the production and sale of k units is given by

$$P(k) = R(k) - C(k)$$

$$= \int_0^k [R'(x) - C'(x)] \, dx.$$

a) Find $P(k)$.

b) Find $P(5)$.

b) Before the time in (a), what is the reduction in emissions resulting from using the student's engine?

Solution

a) The rates of emission will be the same when $E(t) = C(t)$, or

$$2t^2 = 9 + t^2$$

$$t^2 - 9 = 0$$

$$(t - 3)(t + 3) = 0$$

$$t = 3 \quad \text{or} \quad t = -3.$$

Since negative time has no meaning in this problem, the emission rates will be the same when $t = 3$ years.

b) The reduction in emissions is represented by the area of the shaded region. It is the area between $y = 9 + t^2$ and $y = 2t^2$, from $t = 0$ to $t = 3$, and is computed as follows:

$$\int_0^3 [(9 + t^2) - 2t^2] \, dt = \int_0^3 (9 - t^2) \, dt$$

$$= \left[9t - \frac{t^3}{3} \right]_0^3$$

$$= \left(9 \cdot 3 - \frac{3^3}{3} \right) - \left(9 \cdot 0 - \frac{0^3}{3} \right)$$

$$= 27 - 9$$

$$= 18 \text{ billion pollution particulates.}$$

DO EXERCISE 10.

EXERCISE SET 5.4

Find the area of the region bounded by the given graphs.   `CSS`

1. $y = x$, $y = x^3$, $x = 0$, $x = 1$

2. $y = x$, $y = x^4$

3. $y = x + 2$, $y = x^2$

4. $y = x^2 - 2x$, $y = x$

5. $y = 6x - x^2$, $y = x$

6. $y = x^2 - 6x$, $y = -x$

7. $y = 2x - x^2$, $y = -x$

8. $y = x^2$, $y = \sqrt{x}$

9. $y = x$, $y = \sqrt{x}$

10. $y = 3$, $y = x$, $x = 0$

11. $y = 5$, $y = \sqrt{x}$, $x = 0$

12. $y = x^2$, $y = x^3$

13. $y = 4 - x^2$, $y = 4 - 4x$

14. $y = x^2 + 1$, $y = x^2$, $x = 1$, $x = 3$

15. $y = x^2 + 3$, $y = x^2$, $x = 1$, $x = 2$

16. *Business: Profit.* A company determines that its marginal revenue per day is given by

$$R'(t) = 100e^t, \qquad R(0) = 0,$$

where $R(t)$ = the revenue, in dollars, on the tth day. Also, its marginal cost per day is given by

$$C'(t) = 100 - 0.2t, \qquad C(0) = 0,$$

where $C(t)$ = the cost, in dollars, on the tth day. Find the total profit from $t = 0$ to $t = 10$ (the first 10 days). *Note:*

$$P(T) = R(T) - C(T) = \int_0^T [R'(t) - C'(t)]\, dt.$$

17. *Psychology: Memorizing.* In a certain memory experiment, subject A is able to memorize words at the rate

$$m'(t) = -0.009t^2 + 0.2t \quad \text{(words per minute)}.$$

In the same memory experiment subject B is able to memorize at the rate

$$M'(t) = -0.003t^2 + 0.2t \quad \text{(words per minute)}.$$

a) Which subject has the higher rate of memorization?

b) How many more words does that subject memorize from $t = 0$ to $t = 10$ (during the first 10 minutes)?

▶ Find the area of the region bounded by the given graphs.

18. $y = x^2$, $y = x^{-2}$, $x = 1$, $x = 5$

20. $y = x^2$, $y = \sqrt[3]{x^2}$, $x = 1$, $x = 8$

22. $x + 2y = 2$, $y - x = 1$, $2x + y = 7$

19. $y = e^x$, $y = e^{-x}$, $x = 0$, $x = 1$

21. $y = x^2$, $y = x^3$, $x = -1$, $x = 1$

23. $y = x + 6$, $y = -2x$, $y = x^3$

24. *Biomedical: Poiseuille's Law.* The flow of blood in a blood vessel is faster toward the center of the vessel and slower toward the outside. The speed of the blood is given by

$$V = \frac{p}{4Lv}(R^2 - r^2),$$

where R = the radius of the blood vessel, r = the distance of the blood from the center of the vessel, and p, v, L are physical constants related to pressure and viscosity of the blood and the length of the blood vessel. If R is constant, we can think of V as a function of r:

$$V(r) = \frac{p}{4Lv}(R^2 - r^2).$$

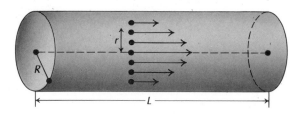

Total blood flow Q is given by

$$Q = \int_0^R 2\pi \cdot V(r) \cdot r \cdot dr.$$

Find Q.

OBJECTIVES

You should be able to

a) **Integrate using substitution.**

b) **Solve applied problems involving integration by substitution.**

5.5 INTEGRATION TECHNIQUES: SUBSTITUTION

The following formulas provide a basis for an integration technique called *substitution*.

A. $\int (r + 1) \cdot u^r \, du = u^{r+1} + C$

B. $\int e^u \, du = e^u + C$

C. $\int \dfrac{1}{u} \, du = \ln u + C, \ u > 0; \quad$ or $\quad \int \dfrac{1}{u} \, du = \ln |u| + C, \ u < 0.$

(We will generally consider $u > 0$.)

In the Leibniz notation dy/dx, we did not give specific definitions of dy and dx. Nevertheless it will be convenient to treat dy/dx as a quotient. Thus, from

$$\frac{dy}{dx} = f'(x)$$

we can derive

$$dy = f'(x) \, dx$$

It is possible to define dy and dx, but it is not necessary for our purposes.

Example 1 For $y = f(x) = x^3$, find dy.

Solution We have

$$\frac{dy}{dx} = f'(x) = 3x^2,$$

so

$$dy = f'(x) \, dx = 3x^2 \, dx.$$

Example 2 For $u = g(x) = \ln x$, find du.

Solution We have

$$\frac{du}{dx} = g'(x) = \frac{1}{x},$$

so

$$du = g'(x) \, dx = \frac{1}{x} \, dx, \quad \text{or} \quad \frac{dx}{x}.$$

1. For $y = f(x) = 6x^2 + x$, find dy.

2. For $u = g(x) = x + 3$, find du.

3. Integrate.

$$\int 3x^2 e^{x^3}\, dx$$

DO EXERCISES 1 AND 2.

So far the dx in

$$\int f(x)\, dx$$

has played no role in integrating other than to indicate the variable of integration. Now it will be convenient to make use of dx. Consider the integral

$$\int 2x \cdot e^{x^2}\, dx.$$

If we set

$$u = x^2,$$

then

$$du = 2x\, dx.$$

If we substitute u for x^2 and du for $2x\, dx$, the integral takes on the form

$$\int e^u\, du.$$

Since

$$\int e^u\, du = e^u + C,$$

it follows that

$$\int 2x \cdot e^{x^2}\, dx = \int e^u\, du$$
$$= e^u + C$$
$$= e^{x^2} + C.$$

We can check the result by differentiation. This procedure is referred to as *substitution*, or *change of variable*. It is a *trial-and-error* procedure; that is, if we try a substitution that doesn't result in an integrand that can be easily integrated, we try another. It will not always work! It *will* work if the integrand fits one of the rules A, B, or C.

DO EXERCISE 3.

Let us consider some further examples.

4. Integrate.

$$\int \frac{2x\ dx}{5 + x^2}$$

Example 3 $\int \frac{2x\ dx}{1 + x^2} = \int \frac{du}{u}$ ___Substitution___ | Let $u = 1 + x^2$; then $du = 2x\ dx$.

$$= \ln u + C$$

$$= \ln (1 + x^2) + C$$

Remember that this is a trial-and-error process. Suppose we had made the substitution

$$u = x^2.$$

Then we would have

$$du = 2x\ dx,$$

and the integral would become

$$\int \frac{du}{1 + u}.$$

This is still not easily integrated, so we would try another substitution.

DO EXERCISE 4.

5. Integrate.

$$\int \frac{2x\ dx}{(3 + x^2)^2}$$

Example 4 $\int \frac{2x\ dx}{(1 + x^2)^2} = \int \frac{du}{u^2}$ ___Substitution___ | $u = 1 + x^2$, $du = 2x\ dx$

$$= \int u^{-2}\ du = -u^{-1} + C$$

$$= -\frac{1}{u} + C = -\frac{1}{1 + x^2} + C$$

DO EXERCISE 5.

6. Integrate.

$$\int \frac{\ln x\ dx}{x}$$

Example 5 $\int \frac{\ln 3x\ dx}{x} = \int u\ du$ ___Substitution___ | $u = \ln 3x$, $du = \frac{1}{x}\ dx$

$$= \frac{u^2}{2} + C$$

$$= \frac{(\ln 3x)^2}{2} + C$$

DO EXERCISE 6.

7. Integrate.

$$\int x^2 \cdot e^{x^3} \, dx$$

Integrate.

8. $\int e^{5x} \, dx$

9. $\int e^{0.02x} \, dx$

10. $\int e^{-x} \, dx$

Example 6 Integrate $\int x e^{x^2} \, dx$.

Solution Suppose we try

$$u = x^2;$$

then we have

$$du = 2x \, dx.$$

We don't quite have $2x \, dx$. We have only $x \, dx$ and will need to supply a 2. We do this by multiplying by 1, using $\frac{1}{2} \cdot 2$:

$$\frac{1}{2} \cdot 2 \cdot \int x e^{x^2} \, dx = \frac{1}{2} \int 2x e^{x^2} \, dx$$

$$= \frac{1}{2} \int e^{x^2} (2x \, dx)$$

$$= \frac{1}{2} \int e^u \, du$$

$$= \frac{1}{2} e^u + C$$

$$= \frac{1}{2} e^{x^2} + C.$$

DO EXERCISE 7.

Example 7 $\quad \int b e^{ax} \, dx = \dfrac{b}{a} \int a e^{ax} \, dx$ ⎫
$$\left. = \frac{b}{a} \int e^u \, du \quad \right\} \quad \underline{\text{Substitution}} \quad \boxed{\begin{array}{l} u = ax, \\ du = a \, dx \end{array}}$$
$$= \frac{b}{a} e^u + C = \frac{b}{a} e^{ax} + C$$

Note that this gives us a formula for integrating $b e^{ax}$.

DO EXERCISES 8–10.

Example 8 $\quad \displaystyle\int \frac{dx}{x+3} = \int \frac{du}{u} \quad \underline{\text{Substitution}} \quad \boxed{\begin{array}{l} u = x+3, \\ du = dx \end{array}}$

$$= \ln u + C = \ln (x+3) + C$$

11. Integrate.

$$\int x^3(x^4 + 5)^{19} \, dx$$

With practice, you will make certain substitutions mentally and just write down the answer. Examples 7 and 8 are good illustrations of this.

Example 9

$$\int x^2(x^3 + 1)^{10} \, dx$$

$$= \frac{1}{3} \int 3x^2(x^3 + 1)^{10} \, dx \left. \begin{array}{l} \\ \\ \end{array} \right\} \quad \text{Substitution} \quad \boxed{\begin{array}{l} u = x^3 + 1, \\ du = 3x^2 \, dx \end{array}}$$

$$= \frac{1}{3} \int u^{10} \, du$$

$$= \frac{1}{3} \cdot \frac{u^{11}}{11} + C = \frac{1}{33}(x^3 + 1)^{11} + C.$$

DO EXERCISE 11.

Example 10 Evaluate $\int_0^1 x^2(x^3 + 1)^{10} \, dx$.

Solution

a) First we find the indefinite integral (shown in Example 9).

12. Evaluate.

$$\int_1^e \frac{\ln x \, dx}{x}$$

(See Margin Exercise 6.)

b) Then we evaluate the definite integral on [0, 1]:

$$\int_0^1 x^2(x^3 + 1)^{10} \, dx = \left[\frac{1}{33}(x^3 + 1)^{11} \right]_0^1$$

$$= \frac{1}{33}[(1^3 + 1)^{11} - (0^3 + 1)^{11}]$$

$$= \frac{1}{33}(2^{11} - 1^{11})$$

$$= \frac{2^{11} - 1}{33}.$$

DO EXERCISE 12.

EXERCISE SET 5.5

Integrate. (Be sure to check by differentiating!)

1. $\int \dfrac{3x^2\,dx}{7 + x^3}$ *$u = 7 + x^3$*
    $du = 3x^2$ $\int \dfrac{du}{u}$

2. $\int \dfrac{3x^2\,dx}{1 + x^3}$

3. $\int e^{4x}\,dx$

4. $\int e^{3x}\,dx$

5. $\int e^{x/2}\,dx$

6. $\int e^{x/3}\,dx$

7. $\int x^3 e^{x^4}\,dx$

8. $\int x^4 e^{x^5}\,dx$

9. $\int t^2 e^{-t^3}\,dx$

10. $\int t e^{-t^2}\,dt$

11. $\int \dfrac{\ln 4x\,dx}{x}$

12. $\int \dfrac{\ln 5x\,dx}{x}$

13. $\int \dfrac{dx}{1 + x}$

14. $\int \dfrac{dx}{5 + x}$

15. $\int \dfrac{dx}{4 - x}$

16. $\int \dfrac{dx}{1 - x}$

17. $\int t^2 (t^3 - 1)^7\,dt$

18. $\int t(t^2 - 1)^5\,dt$

19. $\int (x^4 + x^3 + x^2)^7 (4x^3 + 3x^2 + 2x)\,dx$

20. $\int (x^3 - x^2 - x)^9 (3x^2 - 2x - 1)\,dx$

21. $\int \dfrac{e^x\,dx}{4 + e^x}$

22. $\int \dfrac{e^t\,dt}{3 + e^t}$

23. $\int \dfrac{\ln x^2}{x}\,dx$

24. $\int \dfrac{(\ln x)^2}{x}\,dx$

25. $\int \dfrac{dx}{x \ln x}$

26. $\int \dfrac{dx}{x \ln x^2}$

27. $\int \sqrt{ax + b}\,dx$

28. $\int x\sqrt{ax^2 + b}\,dx$

29. $\int b e^{ax}\,dx$

30. $\int P_0 e^{kt}\,dt$

Integrate.

31. $\displaystyle\int_0^1 2x e^{x^2}\,dx$

32. $\displaystyle\int_0^1 3x^2 e^{x^3}\,dx$

33. $\displaystyle\int_0^1 x(x^2 + 1)^5\,dx$

34. $\displaystyle\int_1^2 x(x^2 - 1)^7\,dx$

35. $\displaystyle\int_1^3 \dfrac{dt}{1 + t}$

36. $\displaystyle\int_1^3 e^{2x}\,dx$

37. $\displaystyle\int_1^4 \dfrac{2x + 1}{x^2 + x - 1}\,dx$

38. $\displaystyle\int_1^3 \dfrac{2x + 3}{x^2 + 3x}\,dx$

39. $\displaystyle\int_0^b e^{-x}\,dx$

40. $\displaystyle\int_0^b 2e^{-2x}\,dx$

41. $\displaystyle\int_0^b m e^{-mx}\,dx$

42. $\displaystyle\int_0^b k e^{-kx}\,dx$

43. $\displaystyle\int_0^4 (x - 6)^2\,dx$

44. $\displaystyle\int_0^3 (x - 5)^2\,dx$

45. ▦ *Sociology: Divorce.* The U.S. divorce rate is approximated by

$$D(t) = 100,000e^{0.025t},$$

where $D(t)$ = the number of divorces occurring at time t and t = the number of years measured from 1900. That is, $t = 0$ corresponds to 1900, $t = 88\frac{9}{365}$ corresponds to January 9, 1988, and so on.

a) Find the total number of divorces from 1900 to 1988. Note that this is

$$\int_0^{88} D(t)\, dt.$$

b) Find the total number of divorces from 1980 to 1988. Note that this is

$$\int_{80}^{88} D(t)\, dt.$$

46. ▦ *Business: Value of an investment.* A company buys a new machine for $250,000. The marginal revenue from the sale of products produced by the machine is projected to be

$$R'(t) = 4000t.$$

The salvage value of the machine decreases at the rate of

$$V(t) = 25,000e^{-0.1t}.$$

The total profit from the machine after T yrs is given by

$$P(T) = \begin{pmatrix} \text{Revenue from} \\ \text{sale of prod.} \end{pmatrix} + \begin{pmatrix} \text{Revenue from} \\ \text{sale of mach.} \end{pmatrix} - \begin{pmatrix} \text{Cost of} \\ \text{machine} \end{pmatrix}$$

$$= \int_0^T R'(t)\, dt \quad + \quad \int_0^T V(t)\, dt \quad - \quad \$250,000$$

a) Find $P(T)$.

b) Find $P(10)$.

Integrate.

47. $\displaystyle\int 5x\sqrt{1 - 4x^2}\, dx$

48. $\displaystyle\int \frac{dx}{ax + b}$

49. $\displaystyle\int \frac{x^2}{e^{x^3}}\, dx$

50. $\displaystyle\int \frac{e^{\sqrt{t}}}{\sqrt{t}}\, dt$

51. $\displaystyle\int \frac{e^{1/t}}{t^2}$

52. $\displaystyle\int \frac{(\ln x)^{99}}{x}\, dx$

53. $\displaystyle\int \frac{dx}{x(\ln x)^4}$

54. $\displaystyle\int (e^t + 2)e^t\, dt$

55. $\displaystyle\int x^2\sqrt{x^3 + 1}\, dx$

56. $\displaystyle\int \frac{t^2}{\sqrt[4]{2 + t^3}}\, dt$

57. $\displaystyle\int \frac{x - 3}{(x^2 - 6x)^{1/3}}\, dx$

58. $\displaystyle\int \frac{[(\ln x)^2 + 3(\ln x) + 4]}{x}\, dx$

59. $\displaystyle\int \frac{t^3 \ln (t^4 + 8)}{t^4 + 8}\, dt$

60. $\displaystyle\int \frac{t^2 + 2t}{(t + 1)^2}\, dt$ Hint: $\dfrac{t^2 + 2t}{(t + 1)^2} = \dfrac{t^2 + 2t + 1 - 1}{t^2 + 2t + 1} = 1 - \dfrac{1}{(t + 1)^2}$

61. $\displaystyle\int \frac{x^2 + 6x}{(x + 3)^2}\, dx$

62. $\displaystyle\int \frac{x + 3}{x + 1}\, dx$ Hint: Divide $\dfrac{x + 3}{x + 1} = 1 + \dfrac{2}{x + 1}$.

63. $\displaystyle\int \frac{t - 5}{t - 4}\, dt$

64. $\displaystyle\int \frac{dx}{x(\ln x)^n}$

65. $\displaystyle\int \frac{dx}{e^x + 1}$ Hint: $\dfrac{1}{e^x + 1} = \dfrac{e^{-x}}{1 + e^{-x}}$

66. $\displaystyle\int \frac{e^x - e^{-x}}{e^x + e^{-x}}\, dx$

67. $\displaystyle\int \frac{(\ln x)^n}{x}\, dx$

OBJECTIVES

You should be able to

a) **Integrate using integration by parts.**

b) **Solve applied problems involving integration by parts.**

c) **Integrate using a table of integration formulas.**

5.6 INTEGRATION TECHNIQUES: INTEGRATION BY PARTS AND BY TABLES

Integration by Parts

Recall the product rule for derivatives:

$$\frac{d}{dx}\, uv = \frac{du}{dx}v + \frac{dv}{dx}u = u\frac{dv}{dx} + v\frac{du}{dx}.$$

Integrating both sides, we get

$$uv = \int u\frac{dv}{dx}\,dx + \int v\frac{du}{dx}\,dx$$

$$= \int u\,dv + \int v\,du.$$

Solving for $\int u\,dv$, we get the following.

THEOREM 7   **Integration by Parts Formula**

$$\boldsymbol{\int u\,dv = uv - \int v\,du}$$

This equation can be used as a formula for integrating in certain situations. These are situations in which an integrand is a product of two functions, and one of the functions can be integrated by using the techniques we have already developed. For example,

$$\int xe^x\,dx$$

can be considered as follows:

$$\int x(e^x\,dx) = \int u\,dv,$$

where $u = x$ and $dv = e^x\,dx$.

We already know how to integrate $e^x\,dx$, or dv. The simplest antiderivative is e^x. This is v. Now since $du = dx$, the formula gives us

$$\int \overset{u}{(x)}\overset{dv}{(e^x\,dx)} = \overset{u}{(x)}\overset{v}{(e^x)} - \int \overset{v}{(e^x)}\overset{du}{(dx)}$$

$$= xe^x - e^x + C.$$

This method of integrating is called *integration by parts*.

1. Integrate.

$$\int 3x \cdot e^{3x}\, dx$$

2. Integrate.

$$\int x \cdot \ln x\, dx$$

Note that integration by parts is a trial-and-error process, as is substitution. In the preceding example, suppose we had reversed the roles of x and e^x. We would have obtained

$$u = e^x, \qquad dv = x\, dx,$$

$$du = e^x\, dx, \qquad v = \frac{x^2}{2},$$

and

$$\int (e^x)(x\, dx) = (e^x)\left(\frac{x^2}{2}\right) - \int \left(\frac{x^2}{2}\right)(e^x\, dx).$$

Now the integrand on the right is more difficult to integrate than the one we started with. When we can integrate *both* factors of an integrand, and thus have a choice as to how to apply the integration-by-parts formula, it can happen that only one (and maybe none) of the possibilities will work.

DO EXERCISE 1.

Let us consider some further examples.

Example 1 Integrate $\int \ln x\, dx$.

Solution Note that $\int (dx/x) = \ln x + C$, but we do not yet know how to find $\int \ln x\, dx$. Let

$$u = \ln x \quad \text{and} \quad dv = dx.$$

Then

$$du = \frac{1}{x}\, dx \quad \text{and} \quad v = x.$$

Using the integration-by-parts formula gives

$$\overset{u\quad dv}{\int (\ln x)(dx)} = \overset{u\quad v}{(\ln x)x} - \int \overset{v\quad du}{x\left(\frac{1}{x}\, dx\right)}$$

$$= x \ln x - \int dx = x \ln x - x + C.$$

DO EXERCISE 2.

Example 2 Integrate $\int x\sqrt{x + 1}\, dx$.

3. Integrate.

$$\int x\sqrt{x + 3} \, dx$$

Solution We let

$$u = x \quad \text{and} \quad dv = (x + 1)^{1/2} \, dx.$$

Then

$$du = dx \quad \text{and} \quad v = \frac{2}{3}(x + 1)^{3/2}.$$

Note that we had to use substitution to integrate dv. Using the integration-by-parts formula gives us

$$\int x\sqrt{x + 1} \, dx = x \cdot \frac{2}{3}(x + 1)^{3/2} - \frac{2}{3}\int (x + 1)^{3/2} \, dx$$

$$= \frac{2}{3}x(x + 1)^{3/2} - \frac{2}{3} \cdot \frac{2}{5}(x + 1)^{5/2} + C$$

$$= \frac{2}{3}x(x + 1)^{3/2} - \frac{4}{15}(x + 1)^{5/2} + C.$$

DO EXERCISE 3.

Example 3 Integrate $\int_1^2 \ln x \, dx$.

Solution

a) First find the indefinite integral (Example 1).

b) Then evaluate the definite integral:

$$\int_1^2 \ln x \, dx = [x \ln x - x]_1^2$$

$$= (2 \ln 2 - 2) - (1 \cdot \ln 1 - 1)$$

$$= 2 \ln 2 - 2 + 1$$

$$= 2 \ln 2 - 1.$$

4. Integrate.

$$\int_1^2 x \ln x \, dx$$

(See Margin Exercise 2.)

DO EXERCISE 4.

Tables of Integration Formulas

You have probably noticed that, generally speaking, integration is more difficult and "tricky" than differentiation. Because of this, integral formulas that are reasonable and/or important have been gathered into tables. Table 5 at the back of the book, though quite brief, is such an example. Entire books of integration formulas are available in libraries, and lengthy tables are also available in mathematics handbooks. Such tables are usually classified by the form of the integrand. The idea is to

5. Using Table 5, integrate

$$\int \frac{1}{x^2 - 25} \, dx.$$

6. Using Table 5, integrate

$$\int x^3 e^x \, dx.$$

properly match the integral in question with a formula in the table. Sometimes a technique such as integration by parts may need to be applied before a table can be used.

Example 4 Integrate

$$\int \frac{dx}{x(3 - x)}.$$

Solution This integral fits *Formula 20* in Table 5:

$$\int \frac{1}{x(ax + b)} \, dx = \frac{1}{b} \ln \left(\frac{x}{ax + b} \right) + C.$$

In our integral, $a = -1$ and $b = 3$, so we have, by the formula,

$$\int \frac{1}{x(3 - x)} \, dx = \int \frac{dx}{x(-1 \cdot x + 3)}$$

$$= \frac{1}{3} \ln \left(\frac{x}{-1 \cdot x + 3} \right) + C$$

$$= \frac{1}{3} \ln \left(\frac{x}{3 - x} \right) + C.$$

DO EXERCISE 5.

Example 5 Integrate $\int (\ln x)^3 \, dx$.

Solution This integral fits *Formula 9* in Table 5:

$$\int (\ln x)^n \, dx = x(\ln x)^n - n \int (\ln x)^{n-1} \, dx + C, \quad n \neq -1.$$

We must apply the formula three times:

$$\int (\ln x)^3 \, dx = x(\ln x)^3 - 3 \int (\ln x)^2 \, dx + C \qquad \text{Formula 9}$$

$$= x(\ln x)^3 - 3[x(\ln x)^2 - 2 \int \ln x \, dx] + C$$

$$\text{Applying Formula 9 again}$$

$$= x(\ln x)^3 - 3[x(\ln x)^2 - 2(x \ln x - \int dx)] + C$$

$$\text{Applying Formula 9 for the third time}$$

$$= x(\ln x)^3 - 3x(\ln x)^2 + 6x \ln x - 6x + C.$$

DO EXERCISE 6.

EXERCISE SET 5.6

Integrate. Use integration by parts. Do not use Table 5. Check by differentiating.

1. $\int 5xe^{5x}\,dx$ **2.** $\int 2xe^{2x}\,dx$ **3.** $\int x^3(3x^2\,dx)$ **4.** $\int x^2(2x\,dx)$ **5.** $\int xe^{2x}\,dx$

6. $\int xe^{3x}\,dx$ **7.** $\int xe^{-2x}\,dx$ **8.** $\int xe^{-x}\,dx$ **9.** $\int x^2\ln x\,dx$ **10.** $\int x^3\ln x\,dx$

11. $\int x\ln x^2\,dx$ **12.** $\int x^2\ln x^3\,dx$ **13.** $\int \ln(x+3)\,dx$ **14.** $\int \ln(x+1)\,dx$ **15.** $\int (x+2)\ln x\,dx$

16. $\int (x+1)\ln x\,dx$ **17.** $\int (x-1)\ln x\,dx$ **18.** $\int (x-2)\ln x\,dx$ **19.** $\int x\sqrt{x+2}\,dx$ **20.** $\int x\sqrt{x+4}\,dx$

21. $\int x^3\ln 2x\,dx$ **22.** $\int x^2\ln 5x\,dx$ **23.** $\int x^2e^x\,dx$ **24.** $\int (\ln x)^2\,dx$ **25.** $\int x^2e^{2x}\,dx$

26. $\int x^{-5}\ln x\,dx$

Integrate. Use integration by parts. Do not use Table 5.

27. $\int_1^2 x^2\ln x\,dx$ **28.** $\int_1^2 x^3\ln x\,dx$ **29.** $\int_2^6 \ln(x+3)\,dx$

30. $\int_0^5 \ln(x+1)\,dx$ **31.** $\int_0^1 xe^x\,dx$ **32.** $\int_0^1 xe^{-x}\,dx$

Integrate. Use Table 5.

33. $\int xe^{-3x}\,dx$ **34.** $\int xe^{4x}\,dx$ **35.** $\int 5^x\,dx$ **36.** $\int \dfrac{1}{\sqrt{x^2-9}}\,dx$ **37.** $\int \dfrac{1}{16-x^2}\,dx$

38. $\int \dfrac{1}{x\sqrt{4+x^2}}\,dx$ **39.** $\int \dfrac{x}{5-x}\,dx$ **40.** $\int \dfrac{x}{(1-x)^2}\,dx$ **41.** $\int \dfrac{1}{x(5-x)^2}\,dx$ **42.** $\int \sqrt{x^2+9}\,dx$

43. *Ecology: Electrical energy use.* The rate of electrical energy used by a family in kilowatt hours per day is given by

$$K(t) = 10te^{-t},$$

where t is the time, in hours. That is, t is in the interval $[0, 24]$.

a) How many kilowatt hours does the family use in the first T hours of a day ($t = 0$ to $t = T$)?

b) How many kilowatt hours does the family use in the first 4 hours of the day?

44. *Biomedical: Drug dosage.* Suppose an oral dose of a drug is taken. From that time, the drug is assimilated in the body and excreted through the urine. The total amount of the drug that has passed through the body in time T is given by

$$\int_0^T E(t)\,dt,$$

where E is the rate of excretion of the drug through the urine. A typical rate of excretion function is

$$E(t) = te^{-kt},$$

where $k > 0$ and t is time, in hours.

a) Use integration by parts to find a formula for

$$\int_0^T E(t)\,dt.$$

b) ▦ Find

$$\int_0^{10} E(t)\,dt, \qquad \text{when } k = 0.2 \text{ mg/hr.}$$

Integrate by parts. Do not use Table 5.

45. $\int \sqrt{x} \ln x \, dx$ **46.** $\int x^n \ln x \, dx$ **47.** $\int \frac{te^t}{(t+1)^2} \, dt$ **48.** $\int x^2 (\ln x)^2 \, dx$ **49.** $\int \frac{\ln x}{\sqrt{x}} \, dx$ **50.** $\int x^n (\ln x)^2 \, dx$

51. Verify that, for any positive integer n,

$$\int x^n e^x \, dx = x^n e^x - n \int x^{n-1} e^x \, dx.$$

52. Verify that, for any positive integer n,

$$\int (\ln x)^n \, dx = x(\ln x)^n - n \int (\ln x)^{n-1} \, dx.$$

OBJECTIVES

You should be able to

a) **Approximate**

$$\int_a^b f(x) \, dx$$

by adding areas of rectangles.

b) **Find the average value of a function over a given interval.**

*5.7 THE DEFINITE INTEGRAL AS A LIMIT OF SUMS

We now consider approximating the area of a region by dividing it into subregions that are almost rectangles. In the figure below, $[a, b]$ has been divided into 4 subintervals, each having width Δx, or $(b - a)/4$.

The heights of the rectangles shown are

$$f(x_1), \quad f(x_2), \quad f(x_3), \quad \text{and} \quad f(x_4).$$

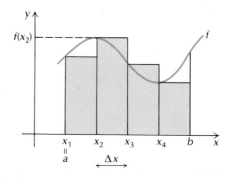

The area of the region under the curve is approximately the sum of the areas of the four rectangles:

$$f(x_1)\Delta x + f(x_2)\Delta x + f(x_3)\Delta x + f(x_4)\Delta x.$$

We can name this sum using *summation notation*, which utilizes the Greek capital letter sigma, Σ:

$$\sum_{i=1}^{4} f(x_i)\Delta x.$$

This is read "the sum of the numbers $f(x_i)\Delta x$ from $i = 1$ to $i = 4$." To recover the original expression, substitute the numbers 1 through 4 successively into $f(x_i)\Delta x$ and write plus signs between the results.

Example 1 Write summation notation for $2 + 4 + 6 + 8 + 10$.

*This section can be omitted without loss of continuity.

Write the summation notation.

1. $1 + 4 + 9 + 16 + 25 + 36$

2. $e + e^2 + e^3 + e^4$

3. $P(x_1)\Delta x + P(x_2)\Delta x + \cdots + P(x_{38})\Delta x$

Express without using summation notation.

4. $\sum_{i=1}^{3} 4^i$

5. $\sum_{i=1}^{5} ie^i$

6. $\sum_{i=1}^{20} t(x_i)\Delta x$

Solution

$$2 + 4 + 6 + 8 + 10 = \sum_{i=1}^{5} 2i.$$

Example 2 Write summation notation for

$$g(x_1)\Delta x + g(x_2)\Delta x + \cdots + g(x_{19})\Delta x.$$

Solution

$$g(x_1)\Delta x + g(x_2)\Delta x + \cdots + g(x_{19})\Delta x = \sum_{i=1}^{19} g(x_i)\Delta x$$

DO EXERCISES 1–3.

Example 3 Express $\sum_{i=1}^{4} 3^i$ without using summation notation.

Solution

$$\sum_{i=1}^{4} 3^i = 3^1 + 3^2 + 3^3 + 3^4, \quad \text{or} \quad 120$$

Example 4 Express $\sum_{i=1}^{30} h(x_i)\Delta x$ without using summation notation.

Solution

$$\sum_{i=1}^{30} h(x_i)\Delta x = h(x_1)\Delta x + h(x_2)\Delta x + \cdots + h(x_{30})\Delta x$$

DO EXERCISES 4–6.

Approximation of area by rectangles becomes more accurate as we use more rectangles and smaller subintervals, as we show in the figures below.

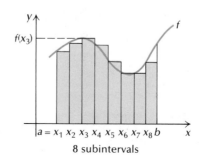

8 subintervals

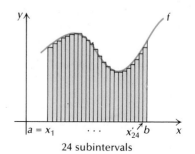

24 subintervals

In general, the interval $[a, b]$ is divided into n equal subintervals, each of width $\Delta x = (b - a)/n$.

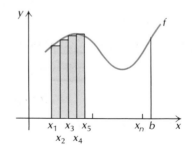

The heights of the rectangles are

$$f(x_1), f(x_2), \cdots, f(x_n).$$

The area of the region under the curve is approximated by the sum of the areas of the rectangles:

$$\sum_{i=1}^{n} f(x_i)\Delta x.$$

We now obtain the actual area by letting the number of intervals increase indefinitely and by taking the limit. The area is thus given by

$$A = \lim_{n \to \infty} \sum_{i=1}^{n} f(x_i)\Delta x.$$

The area is also given by a definite integral:

$$\int_a^b f(x)\, dx = \lim_{n \to \infty} \sum_{i=1}^{n} f(x_i)\Delta x.$$

The fact that we can so express the integral of a function (positive or otherwise) as a limit of a sum or in terms of an antiderivative is so important that it has a name: *The Fundamental Theorem of Integral Calculus.*

CSS

THE FUNDAMENTAL THEOREM OF INTEGRAL CALCULUS

If a function f has an antiderivative F on $[a, b]$, then

$$\int_a^b f(x)\, dx = F(b) - F(a) = \lim_{n \to \infty} \sum_{i=1}^{n} f(x_i)\,\Delta x.$$

It is interesting to envision that, as we take the limit on the right, the summation sign stretches into something reminiscent of an S (the inte-

7. Referring to Example 5, find

$$\sum_{i=1}^{8} C'(x_i)\Delta x,$$

where the interval [0, 400] is divided into 8 equal subintervals of length

$$\Delta x = \frac{400 - 0}{8} = 50.$$

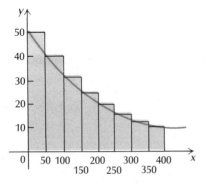

gral sign) and the Δx becomes dx. This is also a motivation for the use of dx in the integral notation.

This result allows us to approximate the value of a definite integral by a sum, making it as good as we please by taking n sufficiently large.

Example 5 Raggs, Ltd., determines that the marginal cost per suit is

$$C'(x) = 0.0003x^2 - 0.2x + 50.$$

Approximate the total cost of producing 400 suits by computing the sum $\sum_{i=1}^{4} C'(x_i)\,\Delta x$.

Solution The interval [0, 400] is divided into 4 subintervals, each of length $\Delta x = (400 - 0)/4 = 100$. Now x_i is varying from $x_1 = 0$ to $x_5 = 400$.

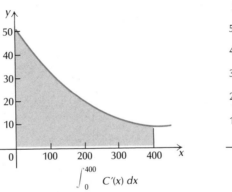

$$\int_{0}^{400} C'(x)\, dx$$

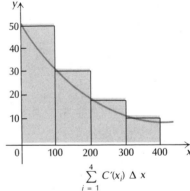

$$\sum_{i=1}^{4} C'(x_i)\, \Delta x$$

Thus we have

$$\sum_{i=1}^{4} C'(x_i)\Delta x$$

$$= C'(0) \cdot 100 + C'(100) \cdot 100 + C'(200) \cdot 100 + C'(300) \cdot 100$$

$$= 50 \cdot 100 + 33 \cdot 100 + 22 \cdot 100 + 17 \cdot 100$$

$$= \$12{,}200.$$

Now

$$\int_{0}^{400} C'(x)\, dx = \$10{,}400. \qquad \text{See Example 3 of Section 5.2.}$$

Thus the approximation is not too far off, even though the number of subintervals is small. In Margin Exercise 7 you will obtain a better approximation using 8 subintervals.

DO EXERCISE 7.

8. In graphs (a) and (b), compute the areas of each rectangle to four decimal places. Then add them to approximate the area under the curve $y = 1/x$ over $[1, 7]$.

a)

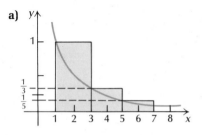

b)

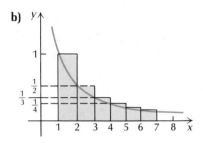

c) Evaluate

$$\int_1^7 \frac{1}{x}\, dx.$$

Find this answer in Table 3 and compare it to (a) and (b). (*Note:* Table 3 contains approximations of natural logarithms accurate to four decimal places. We could construct Table 3 using procedures like those in (a) and (b).)

DO EXERCISE 8.

The fact that an integral can be approximated by a sum is useful when the antiderivative of a function does not have an elementary formula. For example, for the function $e^{-x^2/2}$, important in probability, there is no formula for the antiderivative. So, tables of approximate values of its integral have been computed using summation methods.

The Average Value of a Function

Suppose that

$$T = f(t)$$

is the temperature at time t recorded at a weather station on a certain day. The station uses a 24-hour clock, so the domain of the temperature function is the interval $[0, 24]$. The function is continuous.

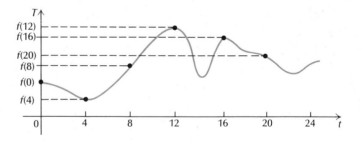

To find the average temperature for the given day, we might take six temperature readings at 4-hour intervals, starting at midnight:

$$T_0 = f(0), \quad T_1 = f(4), \quad T_2 = f(8), \quad T_3 = f(12), \quad T_4 = f(16),$$

$$T_5 = f(20).$$

The average reading would then be the sum of these six readings divided by 6:

$$T_{av} = \frac{T_0 + T_1 + T_2 + T_3 + T_4 + T_5}{6}.$$

This computation of the average temperature may not give the most useful answer. For example, suppose it is a hot summer day, and that at 2:00 in the afternoon (hour 14 on the 24-hour clock) there is a short thunderstorm that cools the air for an hour between our readings. This temporary dip would not show up in the average computed above.

What can we do? We could take 48 readings at $\frac{1}{2}$-hour intervals. This should give a better result. In fact, the shorter the time between readings, the better the result should be. It seems reasonable that we might define *the average value of T* over the interval $[0, 24]$ to be the limit, as $n \to \infty$, of the average of n values:

$$\text{Average value of } T = \lim_{n \to \infty} \frac{1}{n} \sum_{i=1}^{n} T_i = \lim_{n \to \infty} \frac{1}{n} \sum_{i=1}^{n} f(t_i).$$

Note that this is not too far from our definition of an integral. All we would need is to get $24/n$, which is Δt, into the summation. We do this by expressing $1/n$ as $(1/24) \cdot (24/n)$. Then

$$\text{Average value of } T = \lim_{n \to \infty} \frac{1}{n} \sum_{i=1}^{n} f(t_i)$$

$$= \lim_{n \to \infty} \frac{1}{24} \cdot \frac{24}{n} \sum_{i=1}^{n} f(t_i)$$

$$= \frac{1}{24} \lim_{n \to \infty} \sum_{i=1}^{n} f(t_i) \cdot \frac{24}{n}$$

$$= \frac{1}{24} \lim_{n \to \infty} \sum_{i=1}^{n} f(t_i) \Delta t \qquad \Delta t = \frac{24}{n}$$

$$= \frac{1}{24} \int_{0}^{24} f(t) \, dt.$$

DEFINITION

Let f be a continuous function over a closed interval $[a, b]$. Its *average value*, y_{av}, is given by

$$y_{av} = \frac{1}{b - a} \int_{a}^{b} f(x) \, dx.$$

Let us consider average value in another way. If we multiply on both sides of

$$y_{av} = \frac{1}{b - a} \int_{a}^{b} f(x) \, dx$$

by $b - a$, we get

$$(b - a)y_{av} = \int_{a}^{b} f(x) \, dx.$$

9. Find the average value of $f(x) = x^3$ over the interval $[0, 2]$.

10. The temperature over a 10-hr period is given by

$$f(t) = -t^2 + 5t + 40, \quad 0 \le t \le 10.$$

a) Find the average temperature.

b) Find the minimum temperature.

c) Find the maximum temperature.

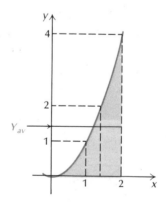

Now the expression on the left will give the area of a rectangle of length $b - a$ and height y_{av}. The area of such a rectangle is the same as the area bounded by $y = f(x)$ on the interval $[a, b]$.

Example 6 Find the average value of $f(x) = x^2$ over the interval $[0, 2]$.

Solution The average value is

$$\frac{1}{2} \int_0^2 x^2 \, dx = \frac{1}{2} \left[\frac{x^3}{3} \right]_0^2$$

$$= \frac{1}{2} \left(\frac{2^3}{3} - \frac{0^3}{3} \right)$$

$$= \frac{1}{2} \cdot \frac{8}{3} = \frac{4}{3}, \quad \text{or} \quad 1\frac{1}{3}.$$

Note that although the values of $f(x)$ increase from 0 to 4, we would not expect the average value to be 2, because we see from the graph that $f(x)$ is less than 2 over more than half the interval.

DO EXERCISES 9 AND 10.

11. The sales of a company are expected to grow according to the function

$$S(t) = 100t + t^2,$$

where $S(t)$ = the sales, in dollars, on the tth day. Find the average sales from $t = 1$ to $t = 4$ (from the 1st to the 4th day).

Example 7 The emissions of an engine are given by

$$E(t) = 2t^2,$$

where $E(t)$ = emissions in billions of pollution particulates at time t, in years. Find the average emissions from $t = 1$ to $t = 5$.

Solution The average emissions are

$$\frac{1}{5-1} \int_1^5 2t^2\, dt = \frac{1}{4}\left[\frac{2}{3}t^3\right]_1^5$$

$$= \frac{1}{4}\cdot\frac{2}{3}(5^3 - 1^3)$$

$$= \frac{1}{6}(125 - 1)$$

$$= 20\frac{2}{3} \text{ billion pollution particulates.}$$

DO EXERCISE 11.

EXERCISE SET 5.7

1. a) Approximate

$$\int_1^7 (dx/x^2)$$

by computing the area of each rectangle to four decimal places and adding.

b) Evaluate

$$\int_1^7 (dx/x^2).$$

Compare the answer to (a).

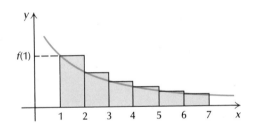

2. a) Approximate

$$\int_0^5 (x^2 + 1)\, dx$$

by computing the area of each rectangle and adding.

b) Evaluate

$$\int_0^5 (x^2 + 1)\, dx.$$

Compare the answer to (a).

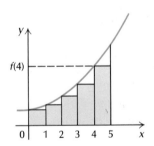

Find the average value over the given interval.

3. $y = 2x^3$; $[-1, 1]$ **4.** $y = 4 - x^2$; $[-2, 2]$ **5.** $y = e^x$; $[0, 1]$ **6.** $y = e^{-x}$; $[0, 1]$

7. $y = x^2 - x + 1$; $[0, 2]$ **8.** $y = x^2 + x - 2$; $[0, 4]$ **9.** $y = 3x + 1$; $[2, 6]$ **10.** $y = 4x + 1$; $[3, 7]$

11. $y = x^n$; $[0, 1]$ **12.** $y = x^n$; $[1, 2]$

13. *Psychology: Results of studying.* A student's score on a test is a function

$$S(t) = t^2, \quad t \text{ in } [0, 10],$$

where $S(t) = $ the score after t hours of study.

a) Find the maximum score the student can achieve and how many hours of study are required to attain it.

b) Find the average score over the 10-hour interval.

14. *Psychology: Results of practice.* A typist's speed over a 4-minute interval is given by

$$W(t) = -6t^2 + 12t + 90, \quad t \text{ in } [0, 4],$$

where $W(t) = $ the speed in words per minute at time t.

a) Find the speed at the beginning of the interval.

b) Find the maximum speed and when it occurs.

c) Find the average speed over the 4-minute interval.

16. *Sociology: Average population of a city.* The population of a city increased and then decreased over an 8-year period according to the function

$$P(t) = -0.1t^2 + t + 3, \quad 0 \le t \le 8,$$

where P is in millions, and t is time.

a) Find the average population.

b) Find the minimum population.

c) Find the maximum population.

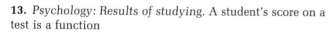

A student's test score is a function of the time spent studying. (*Hugh Rogers, from Monkmeyer*)

15. *Sociology: Average population.* The population of the United States is given by

$$P(t) = 216e^{0.008t},$$

where P is in millions, and t is the number of years since 1976. Find the average value of the population from 1976 to 1986.

17. *Biomedical: Average drug dosage.* The amount of a drug in the body at time t is given by

$$A(t) = 3e^{-0.1t},$$

where A is in cubic centimeters and t is time in hours.

a) What is the initial dosage of the drug?

b) What is the average amount in the body over a 2-hour period?

The Trapezoidal Rule. Another way to approximate an integral is to replace each rectangle in the sum (see the figure at the right) by a trapezoid, as shown in the second figure. The area of a trapezoid is $h(c_1 + c_2)/2$, where c_1 and c_2 are the lengths of the parallel sides. Thus, in the second figure,

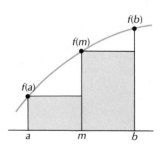

$$\int_a^b f(x)\,dx = \int_a^m f(x)\,dx + \int_m^b f(x)\,dx$$

$$\approx \Delta x \frac{f(a) + f(m)}{2} + \Delta x \frac{f(m) + f(b)}{2}$$

$$\approx \Delta x \left[\frac{f(a)}{2} + f(m) + \frac{f(b)}{2} \right].$$

For an interval $[a, b]$ subdivided into n equal subintervals of length $\Delta x = (b - a)/n$, we get the approximation

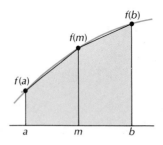

$$\int_a^b f(x)\,dx$$

$$\approx \Delta x \left[\frac{f(a)}{2} + f(x_2) + f(x_3) + \cdots + f(x_n) + \frac{f(b)}{2} \right],$$

where $x_1 = a$. This is called the *Trapezoidal Rule*.

18. Use the Trapezoidal Rule and the interval subdivision of Exercise 1 to approximate

$$\int_1^7 \frac{dx}{x^2}.$$

19. Use the Trapezoidal Rule and the interval subdivision of Exercise 2 to approximate

$$\int_0^5 (x^2 + 1)\,dx.$$

A summary of the important formulas for this chapter is given on the inside back cover.

CHAPTER 5 TEST

Integrate.

1. $\int dx$

2. $\int 1000x^4 \, dx$

3. $\int \left(e^x + \frac{1}{x} + x^{3/8} \right) dx$

Find the area under the curve on the interval indicated.

4. $y = x - x^2$; [0, 1]

5. $y = \frac{4}{x}$; [1, 3]

6. Give two interpretations of the shaded area.

Integrate.

7. $\int_{-1}^{2} (2x + 3x^2) \, dx$ **8.** $\int_{0}^{1} e^{-2x} \, dx$ **9.** $\int_{a}^{b} \frac{dx}{x}$

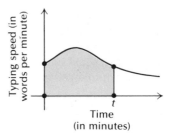

Time
(in minutes)

Decide if $\int_{a}^{b} f(x) \, dx$ is positive, negative, or zero.

10.

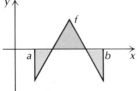

11.

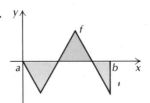

12.

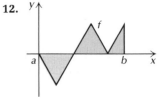

Integrate. Use substitution. Do not use Table 5.

13. $\int \frac{dx}{x + 8}$

14. $\int e^{-0.5x} \, dx$

15. $\int t^3 (t^4 + 1)^9 \, dt$

Integrate. Use integration by parts. Do not use Table 5.

16. $\int xe^{5x} \, dx$

17. $\int x^3 \ln x^4 \, dx$

Integrate. Use Table 5.

18. $\int 2^x \, dx$

19. $\int \frac{dx}{x(7 - x)}$

20. Find the average value of $y = 4t^3 + 2t$ over [−1, 2].

21. Find the area of the region bounded by $y = x$, $y = x^5$, $x = 0$, $x = 1$.

22. Approximate

$$\int_0^5 (25 - x^2)\, dx$$

by computing the area of each rectangle and adding.

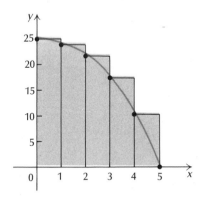

23. An air conditioning company determines that the marginal cost of the xth air conditioner is given by

$$C'(x) = -0.2x + 500, \qquad C(0) = 0.$$

Find the total cost of producing 100 air conditioners.

24. A typist's speed over a 4-minute interval is given by

$$W(t) = -6t^2 + 12t + 90, \qquad t \text{ in } [0, 4],$$

where $W(t)$ = the speed in words per minute at time t. How many words are typed during the second minute (from $t = 1$ to $t = 2$)?

▶ ──

Integrate. Use any method.

25. $\displaystyle \int \frac{[(\ln x)^3 - 4(\ln x)^2 + 5]}{x}\, dx$

26. $\displaystyle \int \ln\left(\frac{x + 3}{x + 5}\right) dx$

APPLICATIONS OF
INTEGRATION

6

This is Randy Stoughton, a professional bowler. In a recent tournament he averaged 204, with a standard deviation of 32. The area under a normal curve can be used to represent the probability of his bowling certain scores. (*Rich Haston, Latent Images, Carmel, Indiana*)

The probability is 14% that a score from 235 to 268 will be bowled.

OBJECTIVES

Given a demand function $D(x)$ and a supply function $S(x)$, you should be able to find the equilibrium point, the consumer's surplus, and the producer's surplus.

6.1 ECONOMIC APPLICATION: CONSUMER'S AND PRODUCER'S SURPLUS

Recall that the consumer's demand curve $D(x)$ gives the demand price per unit that the consumer is willing to pay for x units. The producer's supply curve $S(x)$ gives the price per unit at which the producer is willing to supply x units. The equilibrium point (x_E, p_E) is the intersection of the two curves.

Suppose the following figure represents the supply and demand of college students for movies.

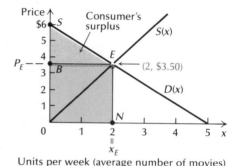

Units per week (average number of movies)

Here we might think of $6 as a price the consumer is willing to pay rather than see no movie at all, and 5 (on the x-axis) is the number of movies per week that the students would go to see if they were free. The area of rectangle $OBEN$ is $2 \cdot 3.50$, or $7.00. This represents the amount the consumer pays to see 2 movies a week at $3.50 each. Suppose a consumer is able to buy 2 movie tickets at the equilibrium price of $3.50 each. Money is saved over any higher-priced tickets which could be bought. The total amount saved is the area of triangle SEB. That amount is defined to be *consumer's surplus* and is

$$\frac{1}{2} \cdot 2 \cdot \$2.50 \quad \text{or} \quad \$2.50$$

and represents the bonus the consumer receives from living in a competitive society. For a producer, or seller, selling at the equilibrium price also is a gain, or surplus, over selling at a lower price. The total amount gained is the area of triangle OBE. That amount is defined to be *producer's surplus* and is

$$\frac{1}{2} \cdot 2 \cdot \$3.50 \quad \text{or} \quad \$3.50.$$

DEFINITION

Consumer's surplus is defined as

$$\int_0^{x_E} D(x)\,dx - x_E p_E,$$

where $D(x)$ is the demand curve.

Producer's surplus is defined as

$$x_E p_E - \int_0^{x_E} S(x)\,dx,$$

where $S(x)$ is the supply curve.

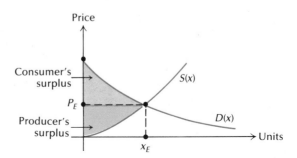

Example Given

$$D(x) = (x - 5)^2,$$
$$S(x) = x^2 + x + 3,$$

find:

a) the equilibrium point;

b) the consumer's surplus;

c) the producer's surplus.

Solution

a) To find the equilibrium point, we set $D(x) = S(x)$ and solve:

$$(x - 5)^2 = x^2 + x + 3$$

$$x^2 - 10x + 25 = x^2 + x + 3$$

$$-10x + 25 = x + 3$$

$$22 = 11x$$

$$\frac{22}{11} = x$$

$$2 = x.$$

Thus $x_E = 2$ units. To find p_E we substitute x_E into $D(x)$ or $S(x)$. We use $D(x)$. Then

$$p_E = D(x_E) = D(2) = (2 - 5)^2 = (-3)^2 = \$9 \text{ per unit.}$$

Thus the equilibrium point is (2, $9).

b) The consumer's surplus is

$$\int_0^{x_E} D(x) \, dx - x_E p_E,$$

or

$$\int_0^2 (x - 5)^2 \, dx - 2 \cdot 9 = \left[\frac{(x - 5)^3}{3} \right]_0^2 - 18$$

$$= \frac{(2 - 5)^3}{3} - \frac{(0 - 5)^3}{3} - 18$$

$$= \frac{(-3)^3}{3} - \frac{(-5)^3}{3} - 18$$

$$= -\frac{27}{3} + \frac{125}{3} - \frac{54}{3}$$

$$= \frac{44}{3} = \$14.67.$$

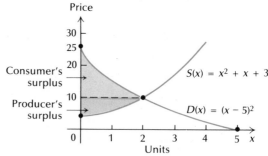

Given

$$D(x) = (x - 6)^2,$$

$$S(x) = x^2 + x + 10,$$

find each of the following.

1. the equilibrium point;

2. the consumer's surplus;

3. the producer's surplus.

c) The producer's surplus is

$$x_E p_E - \int_0^{x_E} S(x)\, dx,$$

or

$$2 \cdot 9 - \int_0^2 (x^2 + x + 3)\, dx$$

$$= 2 \cdot 9 - \left[\frac{1}{3}x^3 + \frac{1}{2}x^2 + 3x\right]_0^2$$

$$= 18 - \left[\left(\frac{1}{3} \cdot 2^3 + \frac{1}{2} \cdot 2^2 + 3 \cdot 2\right) - \left(\frac{1}{3} \cdot 0^3 + \frac{1}{2} \cdot 0^2 + 3 \cdot 0\right)\right]$$

$$= 18 - \left(\frac{8}{3} + 2 + 6\right)$$

$$= \frac{22}{3} = \$7.33.$$

DO EXERCISES 1–3.

EXERCISE SET 6.1

In each exercise, find (a) the equilibrium point, (b) the consumer's surplus, and (c) the producer's surplus.   `CSS`

1. $D(x) = -\frac{5}{6}x + 10$, $S(x) = \frac{1}{2}x + 2$ **2.** $D(x) = -2x + 8$, $S(x) = x + 2$ **3.** $D(x) = (x - 4)^2$, $S(x) = x^2 + 2x + 6$

4. $D(x) = (x - 3)^2$, $S(x) = x^2 + 2x + 1$ **5.** $D(x) = (x - 6)^2$, $S(x) = x^2$ **6.** $D(x) = (x - 8)^2$, $S(x) = x^2$

7. ▦ $D(x) = e^{-x + 4.5}$, $S(x) = e^{x - 5.5}$ **8.** ▦ $D(x) = \sqrt{56 - x}$, $S(x) = x$

OBJECTIVES

You should be able to

a) Find the balance in a savings account from an initial investment at a given interest rate, compounded continuously, for a given period of time t.

b) Find the total money flow over the period of time in (a).

c) Find the total use of a natural resource over a given period of time.

6.2 APPLICATIONS OF THE MODEL $\int_0^T P_0 e^{kt}\, dt$

In this chapter we will make frequent use of the integration formula

$$\int b e^{ax}\, dx = \frac{b}{a} e^{ax} + C.$$

You should memorize it. It was derived by substitution in Example 7 of Section 5.5.

Recall the basic model of exponential growth (Section 4.3):

$$P'(t) = k \cdot P(t) \quad \text{or} \quad \frac{dP}{dt} = kP.$$

1. Find the balance in a savings account after 2 years from an initial investment of $1000 at an interest rate of 7.5% compounded continuously.

The solution of the equation is

$$P(t) = P_0 e^{kt}. \tag{2}$$

Thus $P(t)$ is an antiderivative of $kP_0 e^{kt}$, as we can see by using Eq. (1):

$$\int kP_0 e^{kt}\, dt = \frac{kP_0}{k} e^{kt} = P_0 e^{kt}.$$

One application of Eq. (2) is to compute the balance of a savings account after t years, from an initial investment of P_0 at continuous interest rate k.

Example 1 Find the balance in a savings account after 3 years from an initial investment of $1000 at interest rate 8% compounded continuously.

Solution Using Eq. (2) with $k = 0.08$, $t = 3$, and $P_0 = \$1000$, we get

$$P(3) = 1000e^{0.08(3)} = 1000e^{0.24} = 1000(1.271249) \quad \text{▦ or Table 4}$$

$$\approx \$1271.25$$

DO EXERCISE 1.

The Integral $\int_0^T P_0 e^{kt}\, dt$

Consider the integral of $P_0 e^{kt}$ over the interval $[0, T]$:

$$\int_0^T P_0 e^{kt}\, dt = \left[\frac{P_0}{k} \cdot e^{kt} \right]_0^T = \frac{P_0}{k}(e^{kT} - e^{k \cdot 0}) = \frac{P_0}{k}(e^{kT} - 1).$$

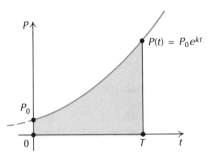

A series of equal payments made at equal time intervals is an annuity. (*Mahon from Monkmeyer*)

Thus

$$\int_0^T P_0 e^{kt}\, dt = \frac{P_0}{k}(e^{kT} - 1) \tag{3}$$

In the remainder of this section we consider two applications of this definite integral.

Business: The Amount of an Annuity

An *annuity* is a series of equal payments made at equal time intervals. Rent payments are an example of an annuity. Deposits in a savings account can also be an annuity. For example, suppose a person makes a deposit of $1000 annually in a savings account on which interest is compounded continuously at 8%. The amount in the account at the end of a certain time is called the *amount of the annuity*. Let us find the amount of the given annuity for a period of 5 years. The following time diagram can help. Note that the deposits are made at the end of each year.

Each $1000 grows over a different time period. The total amount in the account after 5 years, the amount of the annuity, is given by

$$1000e^{0.08(4)} + 1000e^{0.08(3)} + 1000e^{0.08(2)} + 1000e^{0.08(1)} + 1000e^{0.08(0)}$$

$$= \$1377.13 + \$1271.25 + \$1173.51 + \$1083.29 + \$1000$$

$$= \$5905.18.$$

The amount of an annuity is also the sum

$$\sum_{t=0}^{4} 1000e^{0.08t}\, \Delta t,$$

where $\Delta t = 1$. We can *approximate* this sum by integrating the function $1000e^{0.08t}$ over the interval $[0.5]$:

$$\int_0^5 1000e^{0.08t}\, dt = \left[\frac{1000}{0.08} e^{0.08t} \right]_0^5 = 12{,}500(e^{0.08 \cdot 5} - e^{0.08 \cdot 0})$$

$$= 12{,}500(e^{0.4} - 1)$$

$$= 12{,}500(1.491825 - 1)$$

$$\approx \$6147.81.$$

This is very close to the actual amount of the annuity. Note here that the integral provides an approximation to a sum, in contrast to a sum approximating an integral.

2. Find the amount of an annuity where $2000 per year is being invested at 7.5% compounded continuously for 10 years. [CSS]

In general, we have the following.

THEOREM 1

The *amount of an annuity* A_T, where P_0 dollars per year is being invested at interest rate k compounded continuously over T years, is *approximated* by the expression

$$A_T = \frac{P_0}{k}(e^{kT} - 1).$$

Example 2 Find the amount of an annuity where $1000 per year is being invested at 8% compounded continuously for 15 years.

Solution

$$A_{15} = \frac{1000}{0.08}(e^{0.08(15)} - 1) = 12{,}500(e^{1.2} - 1)$$

$$= 12{,}500(3.320116 - 1)$$

$$\approx \$29{,}001.46$$

DO EXERCISE 2.

3. What annual payment P_0 should be made so that the amount of an annuity over 20 years, at interest rate 7.5% compounded continuously, will be $10,000?

Example 3 What annual payment P_0 should be made so that the amount of an annuity over 20 years, at interest rate 8% compounded continuously, will be $10,000?

Solution We find P_0 such that

$$10{,}000 = \frac{P_0}{0.08}(e^{0.08(20)} - 1).$$

Solving, we get

$$800 = P_0(e^{1.6} - 1)$$

$$800 = P_0(4.953032 - 1)$$

$$800 = P_0(3.953032)$$

$$\$202.38 \approx P_0.$$

DO EXERCISE 3.

4. Money is invested in a savings account at the rate of $1000 each year for 6 years. Interest is compounded continuously at 7.5%. Find the amount of the continuous money flow.

Continuous Money Flow

We have said that the integral

$$\int_0^5 1000 e^{0.08t}\, dt$$

approximates the amount of an annuity that is the sum

$$\sum_{t=0}^4 1000 e^{0.08t} \Delta t.$$

If that is the case, just what does

$$\int_0^T P_0 e^{kt}\, dt$$

represent? Consider a continuous stream or flow of money into an investment at the rate of P_0 dollars per year. If an infinitesimal (very small) amount of time dt passes, we have accumulated

$$(P_0 \cdot dt) \text{ dollars.}$$

But during this small time interval, the money has been growing at interest rate k compounded continuously, so that $(P_0 \cdot dt)$ dollars has grown to

$$(P_0 \cdot dt) \cdot e^{kt} \text{ dollars.}$$

We find the accumulation of all these amounts in time T by the integral

$$\int_0^T P_0 e^{kt}\, dt,$$

which we might call the *amount of a continuous annuity*.

If the rate of flow of money into the investment is given by some variable function of time $R(t)$, then the *amount of the continuous money flow* is given by

$$\int_0^T R(t) e^{kt}\, dt.$$

DO EXERCISE 4.

Ecology: Depletion of Natural Resources

Another application of the integral of exponential growth concerns

$$P(t) = P_0 e^{kt}$$

5. The *demand for oil.* In 1980 ($t = 0$) the world use of oil was 66,164 million barrels, and the demand for it was growing exponentially at the rate of 10% per year. If the demand continues at this rate, how many barrels of oil will the world use from 1980 to 1990?

as a model of the demand for natural resources. Suppose P_0 represents the amount of a natural resource (such as coal, oil, and so forth) used at time $t = 0$, and suppose the growth rate for the use of this resource is k. Then, assuming exponential growth (which is the case for the use of many resources), the amount to be used at time t is $P(t)$, given by

$$P(t) = P_0 e^{kt}.$$

The total amount used during an interval $[0, T]$ is given by

$$\int_0^T P_0 e^{kt}\, dt = \frac{P_0}{k}(e^{kT} - 1). \tag{4}$$

Example 4 *The demand for copper.* In 1980 ($t = 0$) the world use of copper was

$$21{,}350{,}000 \text{ tons,}$$

and the demand for it was growing exponentially at the rate of 15% per year. If the growth continues at this rate, how many tons of copper will the world use from 1980 to 1990?

Solution Using Eq. (4), we have

$$\int_0^{10} 21{,}350{,}000 e^{0.15t}\, dt = \frac{21{,}350{,}000}{0.15}(e^{0.15 \cdot 10} - 1)$$

$$= 142{,}333{,}333(e^{1.5} - 1)$$

$$= 142{,}333{,}333(4.481689 - 1) \qquad \blacksquare \text{ or Table 4}$$

$$= 142{,}333{,}333(3.481689)$$

$$\approx 495{,}559{,}688.$$

Thus from 1980 to 1990 the world will use 495,559,688 tons of copper.

DO EXERCISE 5.

Example 5 *The depletion of copper.* The world reserves of copper are

$$689{,}000{,}000 \text{ tons.}$$

Assuming the growth rate in Example 4 continues and that no new reserves are discovered, when will the world reserves of copper be exhausted?

Bingham Canyon mine in Utah has produced more copper than any other mine in history. The grade of ore, however, has dropped from 1.93 percent copper in 1906 to 0.6 today. At present it is planned that mining will cease when the percentage of copper reaches 0.4. (*Steve Kahn for Sohio*)

Solution Using Eq. (4), we want to find T such that

$$689,000,000 = \frac{21,350,000}{0.15}(e^{0.15T} - 1).$$

We solve for T as follows:

$$689,000,000 = 142,333,333(e^{0.15T} - 1)$$

$$\frac{689,000,000}{142,333,333} = e^{0.15T} - 1$$

$$4.8 = e^{0.15T} - 1 \qquad \text{Rounding to the nearest tenth. You do not need to round if you are using a calculator.}$$

$$5.8 = e^{0.15T}$$

$$\ln 5.8 = \ln e^{0.15T} \qquad \text{Taking the natural logarithm on both sides.}$$

$$\ln 5.8 = 0.15T \qquad \text{Recall: } \ln e^k = k.$$

$$\frac{\ln 5.8}{0.15} = T$$

$$\frac{1.757858}{0.15} = T \qquad \text{▦ or Table 3}$$

$$12 \approx T \qquad \text{Rounding to the nearest one.}$$

6. *The depletion of oil.* The world reserves of oil are 670,700 million barrels. In 1980 ($t = 0$) the world use of oil was 66,164 million barrels, and the growth rate for the use of oil was 10%. Assuming this growth rate continues and that no new reserves are discovered, when will the world reserves of oil be exhausted?

Thus 12 years from 1980 (or by 1992), the world reserves of copper will be exhausted.

DO EXERCISE 6.

EXERCISE SET 6.2

CSS

1. Find the amount in a savings acount after 3 years from an initial investment of $100 at 9% compounded continuously.

2. Find the amount in a savings account after 4 years from an initial investment of $100 at 10% compounded continuously.

3. Find the amount of an annuity where $100 per year is being invested at 9% compounded continuously for 20 years.

4. Find the amount of an annuity where $100 per year is being invested at 10% compounded continuously for 20 years.

5. Find the amount of an annuity where $1000 per year is being invested at 8.5% compounded continuously for 40 years.

6. Find the amount of an annuity where $1000 per year is being invested at 7.5% compounded continuously for 40 years.

7. What annual payment should be made so that the amount of an annuity over 20 years, at interest rate 8.5% compounded continuously, will be $50,000?

8. What annual payment should be made so that the amount of an annuity over 20 years, at interest rate 7.5% compounded continuously, will be $50,000?

9. What annual payment should be made so that the amount of an annuity over 30 years, at interest rate 9% compounded continuously, will be $40,000?

10. What annual payment should be made so that the amount of an annuity over 30 years, at interest rate 10% compounded continuously, will be $40,000?

11. *The demand for aluminum ore.* In 1980 ($t = 0$) the world use of aluminum ore was

64,674,000 tons,

and the demand for it was growing exponentially at the rate of 12% per year. If the demand continues to grow at this rate, how many tons of aluminum ore will the world use from 1980 to 1990?

12. *The demand for natural gas.* In 1980 ($t = 0$) the world use of natural gas was

52,360 billion cubic feet,

and the demand for it was growing exponentially at the rate of 4% per year. If the demand continues to grow at this rate, how many cubic feet of natural gas will the world use from 1980 to 1990?

13. *The depletion of aluminum ore.* The world reserves of aluminum ore are

22,670,000,000 tons.

Assuming the growth rate of Exercise 11 continues and that no new reserves are discovered, when will the world reserves of aluminum ore be exhausted?

14. *The depletion of natural gas.* The world reserves of natural gas are

2,911,000 billion cubic feet.

Assuming the growth rate of Exercise 12 continues and that no new reserves are discovered, when will the world reserves of natural gas be exhausted?

▶

15. Suppose that P dollars are deposited at the end of *each day* in a savings account paying 8% interest compounded continuously. Estimate, as an integral, the amount of money that will be in the account at the end of 1 year. Evaluate the integral, in general, and when $P = \$1000$. Use 365 days for 1 year.

16. Repeat Exercise 15, but estimating the amount of money that will be in the account at the end of 2 years.

Stock dividends. The total dividends on stock $D(t)$ a company pays in time T is given by

$$D(T) = \int_0^T d_0 e^{pkt}\,dt$$

where $d_0 = $ the instantaneous dividend payment at time 0, $p = $ the percentage of the company's earnings that it retains, and $k = $ the rate of return that a company can earn on its assets if it were to invest them.

17. Find the total dividends when $d_0 = \$10$, $p = 80\%$, $k = 15\%$, and $T = 50$ years.

18. Find a general formula for $D(T)$.

Amount of a continuous money flow. The amount of a continuous money flow, as described in this section, is given by

$$\int_0^T R(t)e^{kt}\,dt.$$

Find the amount of a continuous money flow when:

19. $R(t) = \$1000$, $k = 8\%$, $T = 30$ years.

20. ▦ $R(t) = t^2$, $k = 7\%$, $T = 40$ years.

OBJECTIVES

You should be able to find
a) **The present value of an investment due t years later at a certain interest rate compounded continuously.**
b) **The capital value of a rental property.**

6.3 APPLICATIONS OF THE MODEL $\int_0^T Pe^{-kt}\,dt$

A representative of a financial institution is often asked to solve a problem like the following.

Example 1 A parent, following the birth of a child, wants to make an initial investment of P_0 that will grow to \$10,000 by the child's 20th birthday. Interest is compounded continuously at 8%. What should the initial investment be?

Solution Using the equation $P = P_0 e^{kt}$, we find P_0 such that

$$10,000 = P_0 e^{0.08 \cdot 20}, \quad \text{or} \quad 10,000 = P_0 e^{1.6}.$$

Now

$$\frac{10,000}{e^{1.6}} = P_0, \quad \text{or} \quad 10,000 e^{-1.6} = P_0,$$

and, using a calculator or Table 4, we have

$$P_0 = 10,000 e^{-1.6} = 10,000(0.201897) = \$2018.97$$

1. A parent, following the birth of a child, wants to make an initial investment P_0 that will grow to $10,000 by the child's 20th birthday. Interest is compounded continuously at 7.5%. What should this initial investment be?

2. Find the present value of $40,000 due 5 years later at 10%, compounded continuously.   `CSS`

Thus the parent must deposit $2018.97, which will grow to $10,000 by the child's 20th birthday.

Economists call $2018.97 the *present value* of $10,000 due 20 years from now at 8% compounded continuously.*

DO EXERCISE 1.

In general, the present value P_0 of an amount P due t years later is found by solving the following equation for P_0:

$$P_0 e^{kt} = P$$
$$P_0 = \frac{P}{e^{kt}} = Pe^{-kt}.$$

THEOREM 2

> The *present value P_0* of an amount P due t years later at interest rate k, compounded continuously, is given by
>
> $$P_0 = Pe^{-kt}.$$

Note that this can be interpreted as exponential decay from the future back to the present.

DO EXERCISE 2.

Suppose a person owns a rental property that earns $100 a month. The current interest rate (amount being charged for loans or being paid for investments) is 9% compounded continuously. The *capital value* of the property over some time period is the sum of all the present values of the rental payments.Therefore for 6 months the capital value is found as follows:

Payment	Present Value
1	$100e^{-0.09(1/12)} = 99.25
2	$100e^{-0.09(2/12)} = 98.51$
3	$100e^{-0.09(3/12)} = 97.78$
4	$100e^{-0.09(4/12)} = 97.04$
5	$100e^{-0.09(5/12)} = 96.32$
6	$100e^{-0.09(6/12)} = 95.60$
	Capital value = $584.50

*The process of computing the present value is called *discounting*.

Thus the capital value is the sum

$$\sum_{i=1}^{6} 100e^{-0.09t_i},$$

where $t_i = i/12$ and i runs from 1 to 6. Now $100 = 100 \cdot (12/12) = (100 \cdot 12) \cdot (1/12) = 1200 \cdot (1/12)$. So the preceding sum can be expressed as

$$\sum_{i=1}^{6} 1200e^{-0.09t_i}\, \Delta t,$$

where $\Delta t = 1/12$. We can then approximate this sum by integrating the function $1200e^{-0.09t}$ over the half-year interval $[0, 0.5]$. Now

$$\int_0^{0.5} 1200e^{-0.09t}\, dt = \left[\frac{1200}{-0.09} e^{-0.09t} \right]_0^{0.5}$$

$$= -13{,}333.33(e^{-0.09(0.5)} - e^{-0.09 \cdot 0})$$

$$= -13{,}333.33(0.955997 - 1)$$

$$\approx \$586.71.$$

This is very close to the actual capital value.

Suppose a rental property has annual rent of R dollars paid in n equal payments per year. The capital value of the property over some time T is given by

$$\sum_{i=1}^{nT} \frac{R}{n} e^{-kt_i},$$

where the payment R/n is made at time t_i and the current interest rate is k, compounded continuously. This can be expressed as

$$\sum_{i=1}^{nT} Re^{-kt_i}\, \Delta t,$$

where $\Delta t = 1/n$. We can approximate this sum by the definite integral

$$\int_0^T Re^{-kt}\, dt.$$

Evaluating this integral, we get

$$\int_0^T Re^{-kt}\, dt = \frac{R}{-k}(e^{-kT} - e^{-k \cdot 0}) = \frac{R}{k}(1 - e^{-kT}).$$

3. Find the capital value of a rental property over a 20-year period where the annual rent is $1800 and the current interest rate is 10%.

CSS

THEOREM 3

The *capital value* V_T of a property over T years is *approximated* by

$$V_T = \frac{R}{k}(1 - e^{-kT}),$$

where R is the annual rent or income and k is the current interest rate.

Example 2 Find the capital value of a rental property over a 5-year period where the annual rent is $2400 and the current interest rate is 14%.

Solution

$$V_5 = \frac{2400}{0.14}(1 - e^{-0.14 \cdot 5}) = 17{,}142.86(1 - e^{-0.7})$$

$$= 17{,}142.86(1 - 0.496585)$$

$$\approx \$8629.97$$

DO EXERCISE 3.

The previous example is an application of the model

$$\int_0^T Pe^{-kt}\, dt = \frac{P}{k}(1 - e^{-kT}).$$

This model can be applied to a calculation of the buildup of a specific amount of radioactive material released into the atmosphere annually. Some of the material decays, but more keeps being released. The amount present at time T is given by the integral above.

EXERCISE SET 6.3

1. A parent, following the birth of a child, wants to make an initial investment P_0 that will grow to $5000 by the child's 20th birthday. Interest is compounded continuously at 9%. What should the initial investment be?

3. Find the present value of $60,000 due 8 years later at 12% compounded continuously.

2. A parent, following the birth of a child, wants to make an initial investment P_0 that will grow to $5000 by the child's 20th birthday. Interest is compounded continuously at 10%. What should this initial investment be?

4. Find the present value of $50,000 due 16 years later at 14% compounded continuously.

5. Find the capital value of a rental property over a 10-year period where the annual rent is $2700 and the current interest rate is 9%.

6. Find the capital value of a rental property over a 10-year period where the annual rent is $2700 and the current interest rate is 10%.

7. An MBA accepts the position of president of a company at age 35. Assuming retirement at age 65 and an annual salary of $45,000, what is the president's capital value? The current interest rate is 8%

8. A college dropout takes a job as a truck driver at age 25. Assuming retirement at age 65 and an annual salary of $14,000, what is the truck driver's capital value? The current interest rate is 7%.

9. ▦ *Radioactive buildup.* Plutonium has a decay rate of 0.003% per year. Suppose 1 lb of plutonium is released into the atmosphere each year for 20 years. What is the total amount of radioactive buildup?

10. ▦ *Radioactive buildup.* Cesium-137 has a decay rate of 2.3% per year. Suppose 1 lb of cesium-137 is released into the atmosphere each year for 20 years. What is the total amount of radioactive buildup?

▶ *Accumulated present values of a continuous cash flow.* Suppose we know that money will flow into an investment at the rate of $R(t)$ dollars per year, from now until some time T in the future. If an infinitesimal amount of time dt passes, $R(t)\,dt$ dollars will have accumulated. The present value of that amount is $[R(t)\,dt]e^{-kt}$, where k is the current interest rate. The accumulation of all the present values is given by

$$V(T) = \int_0^T R(t)e^{-kt}\,dt$$

and is called the *accumulated present value.*

11. ▦ Find $V(T)$ when $R(t) = t$, $k = 8\%$, and $T = 20$ years.

12. ▦ Find $V(t)$ when $R(t) = e^t$, $k = 7\%$, and $T = 10$ years.

Accumulated present values of dividends. Suppose $d(t)$ represents the instantaneous dividend payment of a stock at time t. Then $d(t)e^{-mt}$ is the present value of that payment, where m is the current interest rate. The accumulation of all present values from time 0 to time T is given by

$$D_p(T) = \int_0^T d(t)e^{-mt}\,dt.$$

13. Find $D_p(T)$ when $d(t) = \$10$, $m = 8\%$, and $T = 10$ years.

14. Find $D_p(T)$ when $d(t) = t$, $m = 7\%$, and $T = 20$ years.

15. *Capital value.* U-Rent-It, Inc., expects to get P dollars a day in rent on a certain tool. Find, as an integral, the capital value of the tool over a 1-year period. Evaluate the integral, in general, and when $P = \$14$. Assume the current interest rate is 12%.

16. *Capital value.* Repeat Exercise 15, assuming a 2-year period.

OBJECTIVES

You should be able to determine whether an improper integral is convergent or divergent, and calculate its value if it is convergent.

6.4 IMPROPER INTEGRALS

Let us try to find the area of the region under the graph of $y = 1/x^2$ on the inteval $[1,\infty)$.

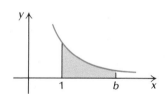
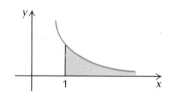

1. Complete.

b	$1 - \dfrac{1}{b}$
2	$1 - \dfrac{1}{2}$, or $\dfrac{1}{2}$
3	
10	
100	
200	

2. Find the area of the region under the graph of

$$y = \frac{1}{x^2}$$

on the interval $[2, \infty)$.

Note that this region is of infinite extent. We have not yet considered how to find the area of such a region. Let us find the area under the curve on the interval from 1 to b, and then see what happens as b gets very large. The area on $[1, b]$ is

$$\int_1^b \frac{dx}{x^2} = \left[-\frac{1}{x}\right]_1^b = \left(-\frac{1}{b}\right) - \left(-\frac{1}{1}\right) = -\frac{1}{b} + 1 = 1 - \frac{1}{b}.$$

Then

$$\lim_{b \to \infty} [\text{area from 1 to } b] = \lim_{b \to \infty}\left(1 - \frac{1}{b}\right).$$

Let us investigate this limit.

DO EXERCISE 1.

Note that as $b \to \infty$, $1/b \to 0$, so $[1 - 1/b)] \to 1$. Thus

$$\lim_{b \to \infty} [\text{area from 1 to } b] = \lim_{b \to \infty}\left(1 - \frac{1}{b}\right) = 1.$$

We *define* the area from 1 to ∞ to be this limit. Here we have an example of an infinitely long region with a finite area.

DO EXERCISE 2.

Such areas may not always be finite. Let us try to find the area of the region under the graph of $y = 1/x$ on the interval $[1, \infty)$.

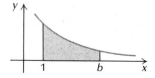 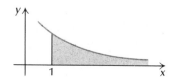

By definition, the area A from 1 to ∞ is the limit as $b \to \infty$ of the area from 1 to b, so

$$A = \lim_{b \to \infty} \int_1^b \frac{dx}{x} = \lim_{b \to \infty} [\ln x]_1^b = \lim_{b \to \infty} (\ln b - \ln 1) = \lim_{b \to \infty} \ln b.$$

In Section 4.2 we learned that $\ln b$ increases indefinitely as b increases. Therefore, the limit does not exist.

Thus we have an infinitely long region with an infinite area. Note that the graphs of $y = 1/x^2$ and $y = 1/x$ have similar shapes, but the region under one of them has a finite area and the other does not.

3. Find the area under the graph of

$$y = \frac{1}{x}$$

from $x = 2$ to $x = \infty$.

DO EXERCISE 3.

An integral such as

$$\int_a^\infty f(x)\,dx,$$

with an upper limit of ∞, is called an *improper integral*. Its value is defined to be the following limit.

DEFINITION

$$\int_a^\infty f(x)\,dx = \lim_{b \to \infty} \int_a^b f(x)\,dx$$

If the limit exists, then we say that the improper integral *converges*. If the limit does not exist, we say that the improper integral *diverges*. Thus

$$\int_1^\infty \frac{dx}{x^2} = 1 \quad \text{converges;} \qquad \text{and} \qquad \int_1^\infty \frac{dx}{x} = \infty \quad \text{diverges.}$$

Example 1

$$\int_0^\infty 2e^{-2x}\,dx = \lim_{b \to \infty} \int_0^b 2e^{-2x}\,dx = \lim_{b \to \infty} \left[2\left(-\frac{1}{2}\right)e^{-2x}\right]_0^b$$

$$= \lim_{b \to \infty} \left[-e^{-2x}\right]_0^b$$

$$= \lim_{b \to \infty} \left[-e^{-2b} - (-e^{-2\cdot 0})\right]$$

$$= \lim_{b \to \infty} (-e^{-2b} + 1)$$

$$= \lim_{b \to \infty} \left(1 - \frac{1}{e^{2b}}\right)$$

Now as $b \to \infty$, we know that $e^{2b} \to \infty$ (from Chapter 4), so

$$\frac{1}{e^{2b}} \to 0 \quad \text{and} \quad \left(1 - \frac{1}{e^{2b}}\right) \to 1.$$

Thus

$$\int_0^\infty 2e^{-2x}\,dx = 1.$$

(The integral is convergent.)

Determine whether each of the following improper integrals is convergent or divergent, and calculate its value if it is convergent.

4. $\displaystyle\int_{0}^{\infty} 5e^{-5x}\, dx$

5. $\displaystyle\int_{0}^{\infty} 2x\, dx$

DO EXERCISES 4 AND 5.

The following are definitions of two types of improper integrals.

DEFINITIONS

1. $\displaystyle\int_{-\infty}^{b} f(x)\, dx = \lim_{a \to -\infty} \int_{a}^{b} f(x)\, dx$

2. $\displaystyle\int_{-\infty}^{\infty} f(x)\, dx = \int_{-\infty}^{c} f(x)\, dx + \int_{c}^{\infty} f(x)\, dx$

For $\int_{-\infty}^{\infty} f(x)\, dx$ to converge, both integrals on the right above must converge.

Applications

In Section 6.3 we learned that the capital value of a rental property over T years is approximated by

$$A_T = \int_{0}^{T} R e^{-kt}\, dt$$

$$= \frac{R}{k}(1 - e^{-kT}),$$

where R is the annual rent and k is the current interest rate. Suppose that the rent is paid perpetually. Then under this assumption the capital value over this infinite time period would be

$$\lim_{T \to \infty} A_T = \int_{0}^{\infty} R e^{-kt}\, dt = \lim_{T \to \infty} \int_{0}^{T} R e^{-kt}\, dt$$

$$= \lim_{T \to \infty} \frac{R}{k}(1 - e^{-kT})$$

$$= \lim_{T \to \infty} \frac{R}{k}\left(1 - \frac{1}{e^{kT}}\right)$$

$$= \frac{R}{k}.$$

6. An annual rent of $2400 is being paid for a property for which there is a permanent lease. The current interest rate is 12%. Find the capital value.

THEOREM 4

> The *capital value* of a property for which the annual rent, or income, is being paid, or received, perpetually is
>
> $$\frac{R}{k},$$
>
> where k is the current interest rate compounded continuously.

Example 2 An annual rent of $2000 is being paid for a property for which there is a permanent lease. The current interest rate is 8%. Find the capital value.

Solution The capital value is 2000/0.08, or $25,000.

DO EXERCISE 6.

When an amount P of radioactive material is being released into the atmosphere annually, the amount present at time T is given by

$$A_T = \int_0^T Pe^{-kt}\,dt$$

$$= \frac{P}{k}(1 - e^{-kT}).$$

As $T \to \infty$ (the radioactive material is to be released forever),

$$A_T \to P/k.$$

That is, the buildup of radioactive material approaches a limiting value P/k.

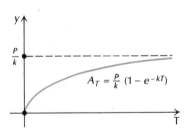

EXERCISE SET 6.4

Determine whether each of the following improper integrals is convergent or divergent, and calculate its value if it is convergent.

1. $\int_3^\infty \dfrac{dx}{x^2}$

2. $\int_4^\infty \dfrac{dx}{x^2}$

3. $\int_3^\infty \dfrac{dx}{x}$

4. $\int_4^\infty \dfrac{dx}{x}$

5. $\int_0^\infty 3e^{-3x}\,dx$

6. $\int_0^\infty 4e^{-4x}\,dx$

7. $\int_1^\infty \dfrac{dx}{x^3}$

8. $\int_1^\infty \dfrac{dx}{x^4}$

9. $\int_0^\infty \dfrac{dx}{1+x}$

10. $\int_0^\infty \dfrac{4\,dx}{1+x}$

11. $\int_1^\infty 5x^{-2}\,dx$

12. $\int_1^\infty 7x^{-2}\,dx$

13. $\int_0^\infty e^x\,dx$

14. $\int_0^\infty e^{2x}\,dx$

15. $\int_3^\infty x^2\,dx$

16. $\int_5^\infty x^4\,dx$

17. $\int_0^\infty xe^x\,dx$

18. $\int_0^\infty \ln x\,dx$

19. $\int_0^\infty me^{-mx}\,dx,\quad m>0$

20. $\int_0^\infty Qe^{-kt}\,dt,\quad k>0$

21. *Capital value.* An annual rent of $3600 is being paid for a property for which there is a permanent lease. The current interest rate is 10%. Find the capital value.

22. *Capital value.* An annual rent of $4500 is being paid for a property for which there is a permanent lease. The current interest rate is 9%. Find the capital value.

23. *Radioactive buildup.* Plutonium has a decay rate of 0.003% per year. Suppose 1 lb of plutonium is released into the atmosphere each year. What is the limiting value of the radioactive buildup?

24. *Radioactive buildup.* Cesium-137 has a decay rate of 2.3% per year. Suppose 1 lb of cesium-137 is released into the atmosphere each year. What is the limiting value of the radioactive buildup?

▶

Determine whether each of the following improper integrals is convergent or divergent, and calculate its value if it is convergent.

25. $\int_0^\infty \dfrac{dx}{x^{2/3}}$

26. $\int_1^\infty \dfrac{dx}{\sqrt{x}}$

27. $\int_0^\infty \dfrac{dx}{(x+1)^{3/2}}$

28. $\int_{-\infty}^0 e^{2x}\,dx$

29. $\int_0^\infty xe^{-x^2}\,dx$

30. $\int_{-\infty}^\infty xe^{-x^2}\,dx$

Accumulated present values of stock dividends paid perpetually. The accumulation of all present values of dividends that are assumed to be paid perpetually is given by

$$V = \int_0^\infty d(t)e^{-mt}\,dt,$$

where $d(t)$ is the instantaneous dividend payment and m is the current interest rate.

31. Find V when $d(t) = e^{-t}$ and $m = 7\%$.

32. Find V when $d(t) = \$1000$ and $m = 8\%$.

Biomedical: Drug dosage. Suppose an oral dose of a drug is taken. From that time, the drug is assimilated in the body and excreted through the urine. The total amount of the drug that has passed through the body in time T is given by

$$\int_0^T E(t)\,dt,$$

where E is the rate of excretion of the drug through the urine. A typical rate of excretion function is $E(t) = te^{-kt}$, where $k>0$ and t is time in hours.

33. Find $\int_0^\infty E(t)\,dt$ and interpret the answer. That is, what does the integral represent?

34. A physician prescribes a dosage of 100 mg. Find k.

OBJECTIVES

You should be able to
a) Verify that a given function satisfies the property

$$\int_a^b f(x)\,dx = 1$$

for being a probability density function.

b) Find k such that a function like

$$f(x) = kx^2$$

is a probability density function over an interval [a, b].

c) Solve applied problems involving probability density functions.

6.5 PROBABILITY

The definite integral plays a role in the theory of probability. Briefly, the *probability* of an event is a number from 0 to 1 that represents its chances of occurring. It is the "relative frequency" of occurrence—the percentage of times an event will occur in a large number of trials.

Example 1 What is the probability of drawing an ace from a well-shuffled deck of cards?

A desire to calculate odds in games of chance gave rise to the theory of probability. (*Marshall Henrichs*)

Solution Since there are 52 possible outcomes, each card has the same chance of being drawn, and there are 4 aces, the probability of drawing an ace is 4/52 or 1/13, or about 7.7%.

In practice we may not draw an ace 7.7% of the time, but in a large number of trials, after shuffling the cards and drawing one, shuffling the cards and drawing one, we would expect to get an ace about 7.7% of the time. That is, the more draws we make, the closer we get to the 7.7%.

Example 2 A bag contains 7 black balls, 6 yellow balls, 4 green balls, and 3 red balls. The bag is shaken well and you remove 1 ball without looking. What is the probability that it is red? white?

1. In Example 2, what is the probability that the ball you are holding is:

a) black?

b) yellow?

c) green?

d) purple?

2. Consider this dartboard.

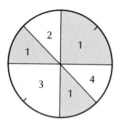

You throw a dart at the board without aiming at any particular region. What is the probability that you will score a:

a) 1?

b) 2?

c) 3?

d) 4?

Solution There are 20 balls altogether and of these 3 are red, so the probability of drawing a red ball is 3/20. There are no white balls, so the probability of drawing a white one is 0/20, or 0.

DO EXERCISES 1 AND 2.

Let us consider a table of probabilities from Example 2.

Color	Probability
Black (B)	$\frac{7}{20}$
Yellow (Y)	$\frac{6}{20}$
Green (G)	$\frac{4}{20}$
Red (R)	$\frac{3}{20}$

Note that the sum of these probabilities is 1. We are certain that we will draw either a black, yellow, green, or red ball. The probability of that event is 1. Let us arrange these data from the table into what is called a *frequency graph*. It shows the fraction of times each event occurs (the probability of each event).

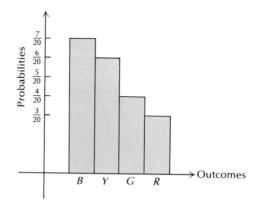

If we assign a width of 1 to each rectangle, then the sum of the areas of the rectangles is 1.

Continuous Random Variables

Suppose we throw a dart at a number line in such a way that is always lands in the interval [1, 3].

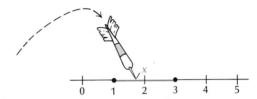

Let x be the number that the dart hits. Note that x is a quantity that can be observed (or measured) repeatedly and whose possible values consist of an entire interval of real numbers. Such a variable is called a *continuous random variable*. Suppose we throw the dart a large number of times and it lands in the subinterval [1.6, 2.8] 43% of the time; the probability, then, that the dart lands in that interval is 0.43.

Let us consider some other examples of continuous random variables.

Example 3 Suppose that x is the arrival time of buses at a bus stop in a three-hour period from 2 P.M. to 5 P.M. The interval is [2, 5].

Then x is a continuous random variable distributed over the interval [2, 5].

Example 4 Suppose that x is the corn acreage of each farm in the United States and Canada. The interval is [0, a], where a is the highest acreage. Or, not knowing what the highest acreage might be, the interval might be [0, ∞) to allow for all possibilities.

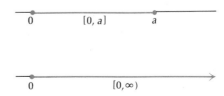

3. Suppose that dosage x of a drug is from 15 milligrams to 25 milligrams. What interval is determined?

(*Note:* It might be argued that there is a value in $[0, a]$ or $[0, \infty)$ for which no farm has that acreage, but for practical purposes these values are often disregarded.)

Then x is a continuous random variable distributed over the interval $[0, a]$ or $[0, \infty)$.

DO EXERCISES 3 AND 4.

Suppose, considering Example 3 on the arrival times of buses, that we wanted to know the probability that a bus will arrive between 4 P.M. and 5 P.M., as represented by

$$P([4, 5]), \quad \text{or} \quad P(4 \le x \le 5).$$

In some cases it is possible to find a function over $[2, 5]$ such that areas over subintervals give the probabilities that a bus will arrive during these subintervals. For example, suppose we had a constant function $f(x) = \frac{1}{3}$ that will give us these probabilities. Look at its graph.

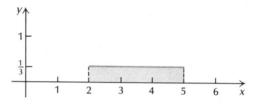

The area under the curve is $3 \cdot \frac{1}{3}$, or 1. The probability that a bus will arrive between 4 P.M. and 5 P.M. is that fraction of the large area that lies over the interval $[4, 5]$. That is,

$$P([4, 5]) = \frac{1}{3} = 33\frac{1}{3}\%.$$

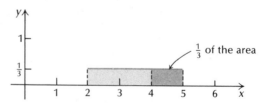

$\frac{1}{3}$ of the area

4. Suppose that distance x is the distance between successive cars on a highway. What interval is determined?

The probability that a bus will arrive between 2:00 P.M. and 4:30 P.M. is $\frac{5}{6}$, or $83\frac{1}{3}\%$.

5. Find the probability that a bus will arrive between 2:30 P.M. and 4:30 P.M.

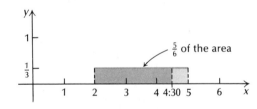

DO EXERCISE 5.

Note that any interval of length 1 has probability $\frac{1}{3}$. This may not always happen. Suppose we have a function

$$f(x) = \frac{3}{117}x^2$$

whose definite integral over the interval [4, 5] would yield the probability that a bus will arrive between 4 P.M. and 5 P.M. Then

$$P([4,\ 5]) = \int_4^5 f(x)\ dx = \int_4^5 \frac{3}{117}x^2\ dx$$

$$= \left[\frac{3}{117}\cdot\frac{1}{3}x^3\right]_4^5$$

$$= \frac{1}{117}(5^3 - 4^3)$$

$$= \frac{61}{117} \approx 0.52.$$

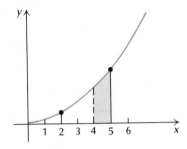

Thus 52% of the time you will be able to catch a bus between 4 P.M. and 5 P.M. The function f is called a *probability density function*. Its integral over *any* subinterval gives the probability that x "lands" in that subinterval.

6. Verify Property 3 of the definition of a probability density function for

$$f(x) = \frac{2}{3}x \quad \text{over } [1, 2].$$

DEFINITION

Let x be a **continuous random variable distributed over some interval** $[a, b]$. A function f is said to be a *probability density function* for x if

1. f is **nonnegative over** $[a, b]$, that is, $f(x) \geq 0$ for all x in $[a, b]$;

2. for any subinterval $[c, d]$ of $[a, b]$, the probability $P([c, d])$, or $P(c \leq x \leq d)$, that x **lands in that subinterval** is given by

$$P([c, d]) = \int_c^d f(x) \, dx;$$

3. the probability that x **lands in** $[a, b]$ **is 1**:

$$\int_a^b f(x) \, dx = 1.$$

That is, we are "certain" that x is in the interval $[a, b]$.

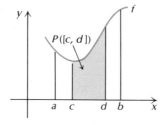 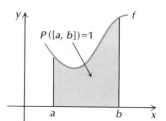

Example 5 Verify Property 3 of the above definition for

$$f(x) = \frac{3}{117}x^2.$$

Solution The "big" interval under consideration is $[2, 5]$. So

$$\int_2^5 \frac{3}{117}x^2 \, dx = \left[\frac{3}{117} \cdot \frac{1}{3}x^3 \right]_2^5 = \frac{1}{117}(5^3 - 2^3) = \frac{117}{117} = 1.$$

DO EXERCISE 6.

Example 6 A company produces transistors. It determines that the life t of a transistor is from 3 to 6 years and that the probability density function for t is given by

$$f(t) = \frac{24}{t^3}, \quad \text{for } 3 \leq t \leq 6.$$

7. In reference to Example 6:

a) Verify Property 3 of the definition of a probability density function.

b) Find the probability that a transistor will last no more than 5 years.

c) Find the probability that a transistor will last from 4 to 6 years.

a) Find the probability that a transistor will last no more than 4 years.
b) Find the probability that a transistor will last from 4 to 5 years.

Solution

a) The probability that a transistor will last no more than 4 years is

$$P(3 \leq t \leq 4) = \int_3^4 \frac{24}{t^3}\, dt = \left[24\left(-\frac{1}{2}t^{-2} \right) \right]_3^4$$

$$= \left[-\frac{12}{t^2} \right]_3^4 = -12\left(\frac{1}{4^2} - \frac{1}{3^2} \right)$$

$$= -12\left(\frac{1}{16} - \frac{1}{9} \right)$$

$$= -12\left(-\frac{7}{144} \right) = \frac{7}{12} \approx 0.58.$$

b) The probability that a transistor will last from 4 to 5 years is

$$P(4 \leq t \leq 5) = \int_4^5 \frac{24}{t^3}\, dt = \left[24\left(-\frac{1}{2}t^{-2} \right) \right]_4^5$$

$$= \left[-\frac{12}{t^2} \right]_4^5 = -12\left(\frac{1}{5^2} - \frac{1}{4^2} \right)$$

$$= -12\left(\frac{1}{25} - \frac{1}{16} \right)$$

$$= -12\left(-\frac{9}{400} \right) = \frac{27}{100} = 0.27.$$

DO EXERCISE 7.

Constructing Probability Density Functions

Suppose you have an arbitrary nonnegative function $f(x)$ whose definite integral over some interval $[a, b]$ is K. Then

$$\int_a^b f(x)\, dx = K.$$

Now multiply on both sides by $\frac{1}{K}$:

$$\frac{1}{K} \int_a^b f(x)\, dx = \frac{1}{K} \cdot K = 1 \quad \text{or} \quad \int_a^b \frac{1}{K} \cdot f(x)\, dx = 1.$$

8. Find k such that

$$f(x) = kx^2$$

is a probability density function over the interval [1, 3].

Thus when we multiply the function $f(x)$ by 1/K we have a function whose area over the given interval is 1.

Example 7 Find k such that

$$f(x) = kx^2$$

is a probability density function over the interval [2, 5].

Solution

$$\int_2^5 x^2 \, dx = \left[\frac{x^3}{3}\right]_2^5 = \frac{5^3}{3} - \frac{2^3}{3} = \frac{125}{3} - \frac{8}{3} = \frac{117}{3}$$

Thus

$$k = \frac{1}{(117/3)} = \frac{3}{117} \quad \text{and} \quad f(x) = \frac{3}{117}x^2.$$

DO EXERCISES 8 AND 9.

Uniform Distributions

Suppose the probability density function of a continuous random variable is constant. How is it described? Consider the following graph.

9. Find k such that

$$f(x) = kx^3$$

is a probability density function over the interval [0, 1].

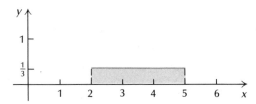

The length of the shaded rectangle is the length of the interval [2, 5], which is 3. For the shaded area to be 1, the height of the rectangle must be $\frac{1}{3}$. Thus $f(x) = \frac{1}{3}$.

For the general case, consider the following graph.

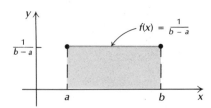

10. A number x is selected at random from the interval [7, 15]. The probability density function for x is given by

$$f(x) = \frac{1}{8}, \quad \text{for } 7 \leq x \leq 15.$$

Find the probability that a number selected is in the subinterval [11, 13].

The length of the shaded rectangle is the length of the interval $[a, b]$, which is $b - a$. For the shaded area to be 1, the height of the rectangle must be $1/(b - a)$. Thus $f(x) = 1/(b - a)$.

DEFINITION

> A continuous random variable x is said to be *uniformly distributed* over an interval $[a, b]$ if it has a probability density function f given by
>
> $$f(x) = \frac{1}{b - a}, \quad \text{for } a \leq x \leq b.$$

Example 8 A number x is selected at random from the interval [40, 50]. The probability density function for x is given by

$$f(x) = \frac{1}{10}, \quad \text{for } 40 \leq x \leq 50.$$

Find the probability that a number selected is in the subinterval [42, 48].

Solution The probability is

$$P(42 \leq x \leq 48) = \int_{42}^{48} \frac{1}{10}\, dx = \frac{1}{10}[x]_{42}^{48} = \frac{1}{10}(48 - 42) = \frac{6}{10} = 0.6.$$

DO EXERCISE 10.

Example 9 A company produces guitars for a rock concert. The maximum loudness L of the guitars ranges from 70 to 100 decibels. The probability density for L is

$$f(L) = \frac{1}{30}, \quad \text{for } 70 \leq L \leq 100.$$

A guitar is selected at random off the assembly line. Find the probability that its maximum loudness is from 70 to 92 decibels.

Solution The probability is

$$P(70 \leq L \leq 92) = \int_{70}^{92} \frac{1}{30}\, dL = \frac{1}{30}[L]_{70}^{92}$$

$$= \frac{1}{30}(92 - 70) = \frac{22}{30} = \frac{11}{15} \approx 0.73.$$

11. A person arrives at a bus stop. The waiting time t for a bus is 0 to 20 minutes. The probability density function for t is

$$f(t) = \frac{1}{20} \quad \text{for } 0 \le t \le 20.$$

What is the probability that the person will have to wait no more than 5 minutes for a bus?

DO EXERCISE 11.

Exponential Distributions

The duration of a phone call, the distance between successive cars on a highway, and the amount of time required to learn a task are all examples of exponentially distributed random variables. That is, their probability density functions are exponential.

DEFINITION

> **A continuous random variable is *exponentially distributed* if it has a probability density function given by**
>
> $$f(x) = ke^{-kx} \quad \text{over the interval } [0, \infty).$$

The function $f(x) = 2e^{-2x}$ is such a probability density function. That

$$\int_0^\infty 2e^{-2x}\, dx = 1$$

is shown in Section 6.4. The general case

$$\int_0^\infty ke^{-kx}\, dx = 1$$

can be verified in a similar way.

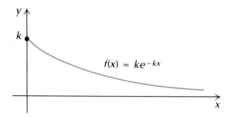

Why is it reasonable to assume that distance between cars is exponentially distributed? This is because there are many more cases in which distances are small. The same argument holds for the duration of a phone call. That is, there are more short calls than long ones.

A transportation planner can determine probabilities that cars are certain distances apart. (*Ellis Herwig: Stock, Boston*)

Example 10 *Transportation planning.* The distance x, in feet, between successive cars on a certain stretch of highway has a probability density function

$$f(x) = ke^{-kx}, \quad \text{for } 0 \leq x < \infty,$$

where $k = 1/a$ and a = the average distance between successive cars over some period of time.

A transportation planner determines that the average distance between cars on a certain stretch of highway is 166 ft. What is the probability that the distance between cars is 50 ft or less?

Solution We first determine k:

$$k = \frac{1}{166} \approx 0.006.$$

The probability density function for x is

$$f(x) = 0.006e^{-0.006x}, \quad \text{for } 0 \leq x < \infty.$$

The probability that the distance between cars is 50 ft or less is

$$P(0 \leq x \leq 50) = \int_0^{50} 0.006e^{-0.006x} \, dx$$

$$= \left[\frac{0.006}{-0.006} e^{-0.006x} \right]_0^{50}$$

$$= \left[-e^{-0.006x} \right]_0^{50}$$

$$= (-e^{-0.006 \cdot 50}) - (-e^{-0.006 \cdot 0})$$

$$= -e^{-0.3} + 1$$

$$= 1 - e^{-0.3}$$

$$= 1 - 0.740818$$

$$\approx 0.2592.$$

12. A transportation planner determines that the average distance between cars on a certain stretch of highway is 125 ft. What is the probability that the distance between cars is 50 ft or less?

DO EXERCISE 12.

EXERCISE SET 6.5

Verify Property 3 of the definition of a probability density function over the given interval.

1. $f(x) = 2x$, $[0, 1]$

2. $f(x) = \frac{1}{4}x$, $[1, 3]$

3. $f(x) = \frac{1}{3}$, $[4, 7]$

4. $f(x) = \frac{1}{4}$, $[9, 13]$

5. $f(x) = \frac{3}{26}x^2$, $[1, 3]$

6. $f(x) = \frac{3}{64}x^2$, $[0, 4]$

7. $f(x) = \frac{1}{x}$, $[1, e]$

8. $f(x) = \frac{1}{e-1}e^x$, $[0, 1]$

9. $f(x) = \frac{3}{2}x^2$, $[-1, 1]$

10. $f(x) = \frac{1}{3}x^2$, $[-2, 1]$

11. $f(x) = 3e^{-3x}$, $[0, \infty)$

12. $f(x) = 4e^{-4x}$, $[0, \infty)$

Find k such that each function is a probability density function over the given integral.

13. $f(x) = kx$, $[1, 3]$

14. $f(x) = kx$, $[1, 4]$

15. $f(x) = kx^2$, $[-1, 1]$

16. $f(x) = kx^2$, $[-2, 2]$

17. $f(x) = k$, $[2, 7]$

18. $f(x) = k$, $[3, 9]$

19. $f(x) = k(2 - x)$, $[0, 2]$

20. $f(x) = k(4 - x)$, $[0, 4]$

21. $f(x) = \frac{k}{x}$, $[1, 3]$

22. $f(x) = \frac{k}{x}$, $[1, 2]$

23. $f(x) = ke^x$, $[0, 3]$

24. $f(x) = ke^x$, $[0, 2]$

25. A dart is thrown at a number line in such a way that it always lands in the interval $[0, 10]$. Let $x =$ the number the dart hits. Suppose the probability density function for x is given by

$$f(x) = \frac{1}{50}x, \quad \text{for } 0 \le x \le 10.$$

Find $P(2 \le x \le 6)$, the probability that it lands in $[2, 6]$.

26. Suppose the situation of Exercise 25, but that the dart always lands in the interval $[0, 5]$, and that the probability density function for x is given by

$$f(x) = \frac{3}{125}x^2, \quad \text{for } 0 \le x \le 5.$$

Find $P(1 \le x \le 4)$, the probability that it lands in $[1, 4]$.

27. A number x is selected at random from the interval $[4, 20]$. The probability density function for x is given by

$$f(x) = \frac{1}{16}, \quad \text{for } 4 \le x \le 20.$$

Find the probability that a number selected is in the subinterval $[9, 17]$.

28. A number x is selected at random from the interval $[5, 29]$. The probability density function for x is given by

$$f(x) = \frac{1}{24}, \quad \text{for } 5 \le x \le 29.$$

Find the probability that a number selected is in the subinterval $[13, 29]$.

29. A transportation planner determines that the average distance between cars on a certain highway is 100 ft. What is the probability that the distance between cars is 40 ft or less?

30. A transportation planner determines that the average distance between cars on a certain highway is 200 ft. What is the probability that the distance between cars is 10 ft or less?

31. A telephone company determines that the duration t of a phone call is an exponentially distributed random variable with probability density function

$$f(t) = 2e^{-2t}, \quad 0 \le t < \infty.$$

Find the probability that a phone call will last no more than 5 minutes.

32. Referring to the data in Exercise 31, find the probability that a phone call will last no more than 2 minutes.

33. In a psychology experiment, the time t, in seconds, that it takes a rat to learn its way through a maze is an exponentially distributed random variable with probability density function

$$f(t) = 0.02e^{-0.02t}, \qquad 0 \le t < \infty.$$

Find the probability that a rat will learn its way through a maze in 150 seconds, or less.

34. Assume the situation and equation in Exercise 33, but find the probability that a rat will learn its way through a maze in 50 seconds or less.

The time it takes a rat to learn its way through a maze is an exponentially distributed random variable. (*Sol Schwartz from Monkmeyer*)

35. The *time to failure* t, in hours, of a certain machine can often be assumed to be exponentially distributed with probability density function

$$f(t) = ke^{-kt}, \qquad 0 \le t < \infty$$

where $k = 1/a$ and a = the average time that will pass before a failure occurs. Suppose the average time that will pass before a failure occurs is 100 hours. What is the probability that a failure will occur in 50 hours or less?

36. The *reliability* of the machine (probability that it will work) in Exercise 35 is defined as

$$R(T) = 1 - \int_0^T 0.01e^{-0.01t} \, dt,$$

where $R(T)$ is the reliability at time T. Find $R(T)$.

37. The function $f(x) = x^3$ is a probability density on $[0, b]$. What is b?

38. The function $f(x) = 12x^2$ is a probability density on $[-a, a]$. What is a?

6.6 PROBABILITY: EXPECTED VALUE; THE NORMAL DISTRIBUTION

Expected Value

Let us again consider throwing a dart at a number line in such a way that it always lands in the interval $[1, 3]$.

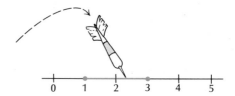

Suppose we throw the dart at the line 100 times and keep track of the numbers it hits. Then suppose we calculate the arithmetic mean (or average) $\overline{x}$ of all these numbers:

$$\overline{x} = \frac{x_1 + x_2 + x_3 + \cdots + x_{100}}{100} = \frac{\sum\limits_{i=1}^{100} x_i}{100} = \sum\limits_{i=1}^{100} x_i \cdot \frac{1}{100}.$$

The expression

$$\sum\limits_{i=1}^{n} x_i \cdot \frac{1}{n}$$

is analogous to the integral

$$\int_1^3 x \cdot f(x)\, dx,$$

where f is the probability density function for x. That is, $1/n$ gives a weight to x_i, and similarly $f(x)$ gives weight to x. We add all the $x_i \cdot (1/n)$ values when we find $\sum_{i=1}^{n} x_i \cdot (1/n)$; and similarly we add all the $x \cdot f(x)$ values when we find $\int_1^3 x \cdot f(x)\, dx$. Suppose $f(x) = \frac{1}{4}x$. Then

$$\int_1^3 x \cdot f(x)\, dx = \int_1^3 x \cdot \frac{1}{4}x\, dx = \left[\frac{1}{4} \cdot \frac{x^3}{3}\right]_1^3 = \left[\frac{x^3}{12}\right]_1^3$$

$$= \frac{1}{12}(3^3 - 1^3) = \frac{26}{12} \approx 2.17.$$

Suppose we keep throwing the dart and computing averages. The more times we throw the dart, the closer we expect the averages to come to 2.17.

Let x be a continuous random variable over the interval $[a, b]$ with probability density function f.

DEFINITION

The expected value of x is defined by

$$E(x) = \int_a^b x \cdot f(x) \, dx.$$

The notion of expected value generalizes to other functions of x. Suppose $y = g(x)$. Then we have the following.

DEFINITION

The expected value of g(x) is defined by

$$E(g(x)) = \int_a^b g(x) \cdot f(x) \, dx.$$

For example,

$$E(x) = \int_a^b xf(x) \, dx,$$

$$E(x^2) = \int_a^b x^2 f(x) \, dx,$$

$$E(e^x) = \int_a^b e^x f(x) \, dx,$$

and

$$E(2x + 3) = \int_a^b (2x + 3)f(x) \, dx.$$

Example 1 Given the probability density function

$$f(x) = \frac{1}{2}x, \quad \text{over } [0, 2],$$

find $E(x)$ and $E(x^2)$.

1. Given the probability density function

$$f(x) = 2x, \quad \text{over } [0, 1],$$

find $E(x)$ and $E(x^2)$.

Solution

$$E(x) = \int_0^2 x \cdot \frac{1}{2} x \, dx = \int_0^2 \frac{1}{2} x^2 \, dx = \frac{1}{2} \left[\frac{x^3}{3} \right]_0^2$$

$$= \frac{1}{2} \left(\frac{2^3}{3} - \frac{0^3}{3} \right) = \frac{1}{2} \cdot \frac{8}{3} = \frac{4}{3};$$

$$E(x^2) = \int_0^2 x^2 \cdot \frac{1}{2} x \, dx = \int_0^2 \frac{1}{2} x^3 \, dx = \frac{1}{2} \left[\frac{x^4}{4} \right]_0^2$$

$$= \frac{1}{2} \left(\frac{2^4}{4} - \frac{0^4}{4} \right) = \frac{1}{2} \cdot \frac{16}{4} = 2$$

DO EXERCISE 1.

DEFINITION

The *mean* μ of a continuous random variable is defined to be $E(x)$. That is,

$$\mu = E(x) = \int_a^b x f(x) \, dx.$$

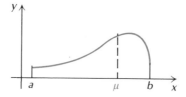 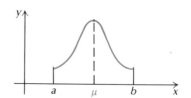

If we could imagine cutting out the region under the curve, the mean is the balance point. (The symbol μ is the lower-case Greek letter "mu.") Note that the mean can be thought of as an average on the x-axis in contrast to the "average value of a function" that lies on the y-axis.

DEFINITION

The *variance* σ^2 of a continuous random variable is defined as

$$\sigma^2 = E(x^2) - \mu^2 = E(x^2) - [E(x)]^2$$

$$= \int_a^b x^2 f(x) \, dx - \left[\int_a^b x f(x) \, dx \right]^2.$$

The *standard deviation* σ of a continuous random variable is defined as

$$\sigma = \sqrt{\text{variance}}.$$

2. Given the probability density function

$$f(x) = 2x, \quad \text{over } [0, 1],$$

find the mean, the variance, and the standard deviation.

The symbol σ is the lower-case Greek letter "sigma."

Example 2 Given the probability density function

$$f(x) = \frac{1}{2}x, \quad \text{over } [0, 2],$$

find the mean, variance, and standard deviation.

Solution From Example 1, we have

$$E(x) = \frac{4}{3} \quad \text{and} \quad E(x^2) = 2.$$

Then

$$the\ mean = \mu = E(x) = \frac{4}{3};$$

$$the\ variance = \sigma^2 = E(x^2) - [E(x)]^2$$

$$= 2 - \left(\frac{4}{3}\right)^2 = 2 - \frac{16}{9}$$

$$= \frac{18}{9} - \frac{16}{9} = \frac{2}{9};$$

$$the\ standard\ deviation = \sigma = \sqrt{\frac{2}{9}} = \frac{1}{3}\sqrt{2} \approx 0.47.$$

Loosely speaking, we say that the standard deviation is a measure of how close the graph of f is to the mean. Note these examples.

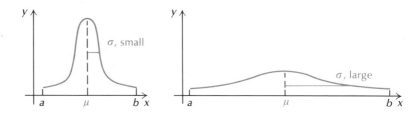

DO EXERCISE 2.

The Normal Distribution

Suppose the average on a test is 70. Usually there are about as many scores above the average as there are below; and the further away from the average, the fewer people there are who get a given score. For example, more people would score in the 80s than in the 90s; and more people would score in the 60s than in the 50s. Test scores, heights of human beings, and weights of human beings are all examples of random variables that may be *normally* distributed.

Consider the function

$$g(x) = e^{-x^2/2}, \quad \text{over the interval } (-\infty, \infty).$$

This function has the entire set of real numbers as its domain. Its graph is the bell-shaped curve below. We can find function values by using a calculator or Table 4:

$$y = e^{-x^2/2}.$$

x	0	1	2	3	−1	−2	−3
y	1	0.6	0.1	0.01	0.6	0.1	0.01

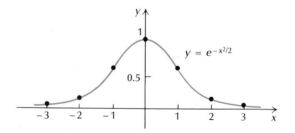

This function has an antiderivative, but that antiderivative has no elementary formula. Nevertheless, it has been shown that its improper integral converges over the interval $(-\infty, \infty)$, and

$$\int_{-\infty}^{\infty} e^{-x^2/2} \, dx = \sqrt{2\pi}.$$

That is, while an expression for the antiderivative cannot be found, there is a numerical value for the improper integral evaluated over the set of real numbers. Note that since the area is not 1, the function g is not a probability density function, but the following is:

$$\frac{1}{\sqrt{2\pi}} e^{-x^2/2}.$$

DEFINITION

A continuous random variable *x* has a *standard normal distribution* if its probability density function is

$$f(x) = \frac{1}{\sqrt{2\pi}} e^{-x^2/2}, \quad \text{over } (-\infty, \infty).$$

This distribution has a mean of 0 and standard deviation 1. Its graph is shown below.

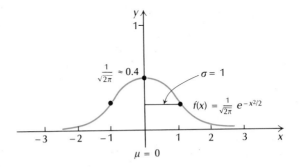

The general case is defined as follows.

DEFINITION

A continuous random variable *x* is *normally distributed* with mean μ and standard deviation σ if its probability density function is given by

$$f(x) = \frac{1}{\sigma\sqrt{2\pi}} \cdot e^{-(1/2)[(x-\mu)/\sigma]^2}, \quad \text{over } (-\infty, \infty).$$

The graph is a transformation of the graph of the standard density. This is done by translating the graph along the x-axis and changing the way the graph is clustered about the mean. Some examples are shown below.

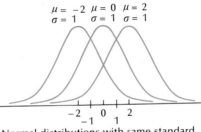

Normal distributions with same standard deviations but different means

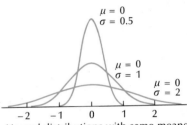

Normal distributions with same means but different standard deviations

The normal distribution is extremely important in statistics; it underlies much of the research in the behavioral and social sciences. Because of this, tables of approximate values of the definite integral of the standard density functions have been prepared. Table 6, at the back of the book, is such a table. It contains values of

$$P(0 \le x \le t) = \int_0^t \frac{1}{\sqrt{2\pi}} e^{-x^2/2}\, dx.$$

The symmetry of the graph about the mean allows many types of probabilities to be computed from the table.

Example 3 Let x be a continuous random variable with standard normal density. Using Table 6, find:

a) $P(0' \le x \le 1.68)$; **b)** $P(-0.97 \le x \le 0)$;
c) $P(-2.43 \le x \le 1.01)$; **d)** $P(1.90 \le x \le 2.74)$;
e) $P(-2.98 \le x \le -0.42)$; **f)** $P(x \ge 0.61)$.

Solution

a) $P(0 \le x \le 1.68)$ is the area bounded by the standard normal curve and the lines $x = 0$ and $x = 1.68$. We look this up in Table 6 by going down the left column to 1.6, then moving to the right to the column headed 0.08. There we read 0.4535. Thus

$$P(0 \le x \le 1.68) = 0.4535.$$

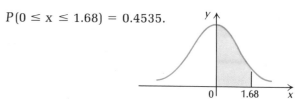

b) Because of the symmetry of the graph,

$P(-0.97 \le x \le 0)$

$= P(0 \le x \le 0.97) = 0.3340.$

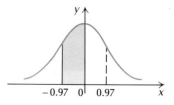

c) $P(-2.43 \le x \le 1.01)$

$= P(-2.43 \le x \le 0) + P(0 \le x \le 1.01)$

$= P(0 \le x \le 2.43) + P(0 \le x \le 1.01)$

$= 0.4925 + 0.3438$

$= 0.8363$

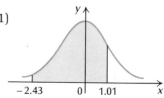

3. Let x be a continuous random variable with standard normal density. Using Table 6, find each of the following.  CSS

a) $P(0 \leq x \leq 2.17)$

b) $P(-1.76 \leq x \leq 0)$

c) $P(-1.77 \leq x \leq 2.53)$

d) $P(0.49 \leq x \leq 1.75)$

e) $P(-1.66 \leq x \leq -1.00)$

f) $P(x \geq 1.87)$

d) $P(1.90 \leq x \leq 2.74)$

$= P(0 \leq x \leq 2.74) - P(0 \leq x \leq 1.90)$

$= 0.4969 - 0.4713$

$= 0.0256$

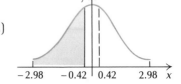

e) $P(-2.98 \leq x \leq -0.42)$

$= P(0.42 \leq x \leq 2.98)$

$= P(0 \leq x \leq 2.98) - P(0 \leq x \leq 0.42)$

$= 0.4986 - 0.1628$

$= 0.3358$

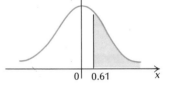

f) $P(x \geq 0.61)$

$= P(x \geq 0) - P(0 \leq x \leq 0.61)$

$= 0.5000 - 0.2291$

(Because of the symmetry about the line $x = 0$, half the area is on each side of the line, and since the entire area is 1, $P(x \geq 0) = 0.5000$.)

$= 0.2709$

DO EXERCISE 3.

In many applications, a normal distribution is not standard. It would be a hopeless task to make tables for all values of the mean μ and the standard deviation σ. In such cases, the transformation

$$X = \frac{x - \mu}{\sigma}$$

standardizes the distribution, permitting the use of Table 6 at the back of the book. That is,

$$P(a \leq x \leq b) = P\left(\frac{a - \mu}{\sigma} \leq X \leq \frac{b - \mu}{\sigma}\right),$$

and the probability on the right can be found using Table 6. To see this, consider

$$P(a \leq x \leq b) = \int_a^b \frac{1}{\sigma\sqrt{2\pi}} e^{-(1/2)[(x - \mu)/\sigma]^2} \, dx,$$

4. The daily profits p of a small firm are normally distributed with mean $200 and standard deviation $40. Find the probability that the daily profit will be from $230 to $250.

and make the substitution

$$X = \frac{x - \mu}{\sigma} = \frac{x}{\sigma} - \frac{\mu}{\sigma}.$$

Then

$$dX = \frac{1}{\sigma}\,dx.$$

When $x = a$, $X = (a - \mu)/\sigma$; and when $x = b$, $X = (b - \mu)/\sigma$. Then

$$P(a \le x \le b) = \int_a^b \frac{1}{\sigma\sqrt{2\pi}} e^{-(1/2)[(x - \mu)/\sigma]^2}\,dx$$

$$= \int_{(a-\mu)/\sigma}^{(b-\mu)/\sigma} \frac{1}{\sqrt{2\pi}} e^{-(1/2)X^2}\,dX$$

$$= P\left(\frac{a - \mu}{\sigma} \le X \le \frac{b - \mu}{\sigma}\right).$$

The integrand is now in the form of the standard density. We can look this up in Table 6.

Example 4 The weights w of the students in a calculus class are normally distributed with mean 150 lb and standard deviation 25 lb. Find the probability that a student's weight is from 160 lb to 180 lb.

Solution We first standardize the weights:

$$180 \text{ is standardized to } \frac{b - \mu}{\sigma} = \frac{180 - 150}{25} = 1.2;$$

$$160 \text{ is standardized to } \frac{a - \mu}{\sigma} = \frac{160 - 150}{25} = 0.4.$$

Then

$$P(160 \le w \le 180) = P(0.4 \le X \le 1.2) \quad \text{Now we can use Table 6.}$$

$$= P(0 \le X \le 1.2) - P(0 \le X \le 0.4)$$

$$= 0.3849 - 0.1554$$

$$= 0.2295.$$

Thus the probability that a student's weight is from 160 lb to 180 lb is 0.2295. That is, about 23% of the students have weights from 160 lb to 180 lb.

DO EXERCISE 4.

EXERCISE SET 6.6

For each probability density function, over the given interval, find $E(x)$, $E(x^2)$, the mean, the variance, and the standard deviation.

CSS

1. $f(x) = \frac{1}{3}$, [2, 5]

2. $f(x) = \frac{1}{4}$, [3, 7]

3. $f(x) = \frac{2}{9}x$, [0, 3]

4. $f(x) = \frac{1}{8}x$, [0, 4]

5. $f(x) = \frac{2}{3}x$, [1, 2]

6. $f(x) = \frac{1}{4}x$, [1, 3]

7. $f(x) = \frac{1}{3}x^2$, [−2, 1]

8. $f(x) = \frac{3}{2}x^2$, [−1, 1]

9. $f(x) = \frac{1}{\ln 3} \cdot \frac{1}{x}$, [1, 3]

10. $f(x) = \frac{1}{\ln 2} \cdot \frac{1}{x}$, [1, 2]

Let x be a continuous random variable with standard normal density. Using Table 6, find each of the following.

11. $P(0 \le x \le 2.69)$

12. $P(0 \le x \le 0.04)$

13. $P(-1.11 \le x \le 0)$

14. $P(-2.61 \le x \le 0)$

15. $P(-1.89 \le x \le 0.45)$

16. $P(-2.94 \le x \le 2.00)$

17. $P(1.76 \le x \le 1.86)$

18. $P(0.76 \le x \le 1.45)$

19. $P(-1.45 \le x \le -0.69)$

20. $P(-2.45 \le x \le -1.69)$

21. $P(x \ge 3.01)$

22. $P(x \ge 1.01)$

23. a) $P(-1 \le x \le 1)$
   b) What percentage of the area is from −1 to 1?

24. a) $P(-2 \le x \le 2)$
   b) What percentage of the area is from −2 to 2?

Let x be a continuous random variable that is normally distributed with mean $\mu = 22$ and standard deviation $\sigma = 5$. Using Table 6, find each of the following.

25. $P(24 \le x \le 30)$

26. $P(22 \le x \le 27)$

27. $P(19 \le x \le 25)$

28. $P(18 \le x \le 26)$

29. At the time of this writing, the bowling scores S of the author of this text were normally distributed with mean 188 and standard deviation 23.

a) Find the probability that a score is from 190 to 210.

b) Find the probability that a score is from 160 to 175.

c) Find the probability that a score is more than 200.

30. The daily production N of stereos by a recording company is normally distributed with mean 1000 and standard deviation 50. The company promises to pay bonuses to its employees on those days when the production of stereos is 1100 or more. What percentage of the days will the company have to pay a bonus?

31. The number of daily orders N received by a mail order firm is normally distributed with mean 250 and standard deviation 20. The company has to hire extra help or pay overtime on those days when the number of orders received is 300 or higher. What percentage of the days will the company have to hire extra help or pay overtime?

32. The scores S on a psychology test are normally distributed with mean 65 and standard deviation 20. A score of 80 to 89 is a B. What is the probability of getting a B?

For each probability density function over the given interval, find $E(x)$, $E(x^2)$, the mean, the variance, and the standard deviation.

33. The uniform probability density

$$f(x) = \frac{1}{b - a}, \quad \text{over } [a, b]$$

34. The exponential probability density

$$f(x) = ke^{-kx}, \quad \text{over } [0, \infty).$$

Median. Let x be a continuous random variable over $[a, b]$ with probability density function f. Then the *median* of x is that number m for which

$$\int_a^m f(x) \, dx = \frac{1}{2}.$$

Find the median.

35. $f(x) = \frac{1}{2}x, [0, 2]$

36. $f(x) = \frac{3}{2}x^2, [-1, 1]$

37. $f(x) = ke^{-kx}, [0, \infty).$

OBJECTIVE

You should be able to use integration to find the volume of a solid of revolution.

6.7 VOLUME

Consider the graph of $y = f(x)$. If the upper half-plane is rotated about the x-axis, then each point on the graph has a circular path, and the whole graph sweeps out a certain surface, called a *surface of revolution.*

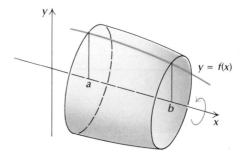

The plane region between the graph, the x-axis, and the interval $[a, b]$ sweeps out a *solid of revolution.* To calculate the volume of this solid, we first approximate it by a finite sum of thin right circular cylinders. We divide the interval $[a, b]$ into equal subintervals, each of length Δx. Thus the height of each cylinder is Δx. The radius of each cylinder is $f(x_i)$, where x_i is the right-hand endpoint of the subinterval that gives that cylinder.

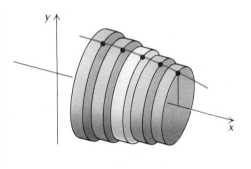
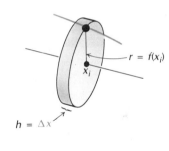

Since the volume of a right circular cylinder is

$$V = \pi r^2 h,$$

each of the approximating cylinders has volume

$$\pi[f(x_i)]^2 \Delta x.$$

The volume of the solid of revolution is approximated by the sum of all the cylinders:

$$V \approx \sum_{i=1}^{n} \pi[f(x_i)]^2 \Delta x.$$

The actual volume is the limit as the thickness of the cylinders approaches 0 or the number of them approaches infinity:

$$V = \lim_{n \to \infty} \sum_{i=1}^{n} \pi[f(x_i)]^2 \Delta x.$$

This is just the definite integral of the function $y = \pi[f(x)]^2$.

THEOREM 5

$$V = \int_a^b \pi[f(x)]^2 \, dx \qquad \textit{Volume of a solid of revolution}$$

Example 1 Find the volume of the solid of revolution generated by rotating the region under the graph of

$$y = \sqrt{x}$$

from $x = 0$ to $x = 1$.

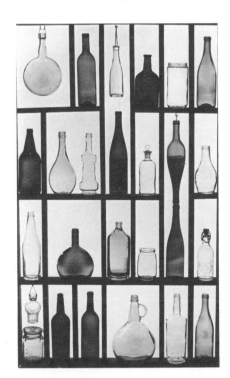

Select a bottle. Find a curve that can be rotated to form the solid of revolution. (*Cary Wolinsky: Stock, Boston*)

1. Find the volume of the solid of revolution generated by rotating the region under the graph of

$$y = x$$

from $x = 0$ to $x = 1$.

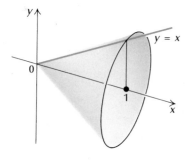

2. Find the volume of the solid of revolution generated by rotating the region under the graph of

$$y = e^x$$

from $x = -2$ to $x = 1$.

Explain how this could be interpreted as a solid of revolution. (*George Gerster from Rapho/Photo Researchers, Inc.*)

Solution

$$V = \int_0^1 \pi[f(x)]^2 \, dx$$

$$= \int_0^1 \pi[\sqrt{x}]^2 \, dx$$

$$= \int_0^1 \pi x \, dx$$

$$= \left[\frac{\pi x^2}{2}\right]_0^1$$

$$= \pi\left(\frac{1^2}{2} - \frac{0^2}{2}\right)$$

$$= \frac{\pi}{2}$$

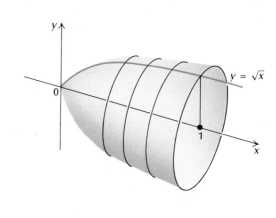

DO EXERCISE 1.

Example 2 Find the volume of the solid of revolution generated by rotating the region under the graph of

$$y = e^x$$

from $x = -1$ to $x = 2$.

Solution

$$V = \int_{-1}^2 \pi[f(x)]^2 \, dx$$

$$= \int_{-1}^2 [\pi e^x]^2 \, dx$$

$$= \int_{-1}^2 \pi e^{2x} \, dx$$

$$= \left[\frac{\pi}{2}e^{2x}\right]_{-1}^2$$

$$= \frac{\pi}{2}(e^{2\cdot 2} - e^{2(-1)})$$

$$= \frac{\pi}{2}(e^4 - e^{-2})$$

DO EXERCISE 2.

EXERCISE SET 6.7

Find the volume generated by revolving about the x-axis the regions bounded by the following graphs.

1. $y = \sqrt{x}$, $x = 0$, $x = 3$

2. $y = \sqrt{x}$, $x = 0$, $x = 2$

3. $y = x$, $x = 1$, $x = 2$

4. $y = x$, $x = 1$, $x = 3$

5. $y = e^x$, $x = -2$, $x = 5$

6. $y = e^x$, $x = -3$, $x = 2$

7. $y = \dfrac{1}{x}$, $x = 1$, $x = 3$

8. $y = \dfrac{1}{x}$, $x = 1$, $x = 4$

9. $y = \dfrac{1}{\sqrt{x}}$, $x = 1$, $x = 3$

10. $y = \dfrac{1}{\sqrt{x}}$, $x = 1$, $x = 4$

11. $y = 4$, $x = 1$, $x = 3$

12. $y = 5$, $x = 1$, $x = 3$

13. $y = x^2$, $x = 0$, $x = 2$

14. $y = x + 1$, $x = -1$, $x = 2$

15. $y = \sqrt{1 + x}$, $x = 2$, $x = 10$

16. $y = 2\sqrt{x}$, $x = 1$, $x = 2$

17. $y = \sqrt{4 - x^2}$, $x = -2$, $x = 2$

18. $y = \sqrt{r^2 - x^2}$, $x = -r$, $x = r$. Make a drawing for Exercise 17. Here you will derive a general formula for the volume of a sphere.

19. $y = \sqrt{\ln x}$, $x = e$, $x = e^3$

20. $y = \sqrt{xe^{-x}}$, $x = 1$, $x = 2$

21. Consider $y = \dfrac{1}{x}$.

a) Find $\displaystyle\int_1^\infty \dfrac{1}{x}\,dx$.

b) Find the volume generated by revolving the region under the graph of $y = 1/x$ for $x \geq 1$.

Note that the area under the graph does not exist, although the volume does. This is like a can of paint that has a finite volume, but not enough paint to cover a cross section of the can.

CHAPTER 6 TEST

1. Find the amount of an annuity where $1200 per year is being invested at 6% compounded continuously for 15 years.

2. What annual payment should be made so that the amount of an annuity over 25 years, at interest rate 6% compounded continuously, will be $20,000?

3. *The demand for iron ore.* In 1980 ($t = 0$) the world use of iron ore was 1,017,500 thousand tons, and the demand for it was growing exponentially at the rate of 6% per year. If the demand continues to grow at this rate, how many tons of iron ore will the world use from 1980 to 1990?

4. *The depletion of iron ore.* The world reserves of iron ore are 103,000,000 thousand tons. Assuming that the growth rate of 6% per year continues and that no new reserves are discovered, when will the world reserves of iron ore be exhausted?

5. A parent, following the birth of a child, wants to make an initial investment P_0 that will grow to $10,000 by the child's 20th birthday. Interest is compounded continuously at 7%. What should the initial investment be?

6. Find the capital value of a rental property over a 20-year period where the annual rent is $3800 and the current interest rate is 11%.

7. Find the capital value of the rental property in Exercise 6 if the rent is to be paid perpetually.

Determine if each of the following improper integrals is convergent or divergent, and calculate its value if convergent.

8. $\displaystyle\int_1^\infty \frac{dx}{x^5}$

9. $\displaystyle\int_0^\infty \frac{3}{1+x}\,dx$

10. Find k such that $f(x) = kx^3$ is a probability density function over the interval $[0, 2]$.

11. A telephone company determines that the length of time t of a phone call is an exponentially distributed random variable with probability density function

$$f(t) = 2e^{-2t}, \quad 0 \le t < \infty.$$

Find the probability that a phone call will last no more than 1 minute.

Given the probability density function $f(x) = 3x^2$, over $[0, 1]$, find each of the following.

12. $E(x)$ **13.** $E(x^2)$ **14.** The mean   **15.** The variance   **16.** The standard deviation

Let x be a continuous random variable with standard normal density. Using Table 6, find each of the following.

17. $P(0 \le x \le 1.5)$ **18.** $P(0.12 \le x \le 2.32)$ **19.** $P(-1.61 \le x \le 1.76)$

20. The price per pound p of T-bone steak at various stores in a certain city is normally distributed with mean $4.75 and standard deviation $0.25. What is the probability that the price per pound is $4.80 or more?

Given the demand and supply functions $D(x) = (x - 7)^2$, $S(x) = x^2 + x + 4$, find each of the following.

21. The equilibrium point   **22.** The consumer's surplus   **23.** The producer's surplus

Find the volume generated by revolving about the x-axis the regions bounded by the following graphs.

24. $y = \dfrac{1}{\sqrt{x}}$, $x = 1$, $x = 5$

25. $y = \sqrt{2 + x}$, $x = 0$, $x = 1$

26. The function $f(x) = x^5$ is a probability density on $[0, b]$. What is b?

27. Determine whether the following improper integral is convergent or divergent, and calculate its value if convergent.

$$\int_{-\infty}^0 x^3 e^{-x^4}\,dx$$

FUNCTIONS OF SEVERAL VARIABLES

(© 1983 Cheryl Shugars)

7

This shows a graph of a function of two variables. It has been generated by computer graphics.

OBJECTIVES

You should be able to

a) **Find a function value for a function of several variables.**

b) **Find the partial derivatives of a given function.**

c) **Evaluate the partial derivatives of a function at a given point.**

1. For $P(x, y) = 4x + 6y$:

 a) find $P(14, 12)$ and interpret its meaning;

 b) find $P(0, 8)$ and interpret its meaning.

2. A pitcher's *earned-run average* is given by

$$A(n, i) = 9 \cdot (n/i),$$

where n is the total number of earned runs given up in i innings of pitching. Find:

 a) find $A(4, 5)$;

 b) $A\left(7, \dfrac{2}{3}\right)$;

 c) $A(1, 15)$.

3. For $V(x, y, z) = xyz$, find $V(5, 10, 40)$.

7.1 PARTIAL DERIVATIVES

Functions of Several Variables

Suppose a one-product firm produces x items of its product at a profit of $4 per item. Then its total profit $P(x)$ is given by

$$P(x) = 4x.$$

This is a function of *one* variable.

Suppose a two-product firm produces x items of one product at a profit of $4 per item, and y items of a second at a profit of $6 per item. Then its total profit P is a function of the *two* variables, x and y, and is given by

$$P(x, y) = 4x + 6y.$$

This function assigns to the input pair (x, y) a unique output number $4x + 6y$.

Example 1 For $P(x, y) = 4x + 6y$, find $P(25, 10)$.

Solution $P(25, 10)$ is defined to be the value of the function found by substituting 25 for x and 10 for y:

$$P(25, 10) = 4 \cdot 25 + 6 \cdot 10$$

$$= 100 + 60$$

$$= 160.$$

This means that by selling 25 items of the first product and 10 of the second, the two-product firm will make a profit of $160.

DO EXERCISES 1–3.

The following are further examples of functions of several variables, that is, functions of two or more variables.

Example 2 The total revenue of a company in thousands of dollars, is given by

$$R(x, y, z, w) = 4x^2 + 5y + z - \ln(w + 1),$$

where x dollars are spent for labor, y dollars for raw materials, z dollars for advertising, and w dollars for machinery. This is a function of four variables.

Find $R(3, 2, 0, 10)$.

4. For the function R of Example 2, find:

a) $R(9, 5, 3, 0)$;

b) $R(1, 2, 3, 4)$.

5. Repeat Example 3 given that the original tank cost $65,000 and the new tank is triple the capacity of the original.

Solution We substitute 3 for x, 2 for y, 0 for z, and 10 for w:

$$R(3, 2, 0, 10) = 4 \cdot 3^2 + 5 \cdot 2 + 0 - \ln(10 + 1)$$

$$\approx 4 \cdot 9 + 10 + 0 - 2.4$$

$$= \$43.6 \text{ thousand.}$$

DO EXERCISE 4.

Example 3 A business purchases a piece of storage equipment that costs C_1 dollars and has capacity V_1. Later it wishes to replace the original with a new piece of equipment that costs C_2 dollars and has capacity V_2. The ratio of the new capacity to the original is

$$\frac{V_2}{V_1} = k,$$

so

$$V_2 = kV_1.$$

It has been found in industrial economics that the cost of the new piece of equipment can be estimated by the function of three variables:

$$C_2 = \left(\frac{V_2}{V_1}\right)^{0.6} C_1$$

$$= k^{0.6} C_1. \tag{1}$$

A beverage company buys a manufacturing tank for $45,000. It has a capacity of 10,000 gallons. Later it decides to buy a tank with double the capacity of the original. Estimate the cost of the new tank.

Solution We substitute 20,000 for V_2, 10,000 for V_1, and 45,000 for C_1 in Eq. (1):

$$C_2 = \left(\frac{20,000}{10,000}\right)^{0.6}(45,000) = 2^{0.6}(45,000) \approx (1.515712)(45,000)$$

$$= \$68,207.25.$$

Note that a 100% increase in capacity was achieved by about a 52% increase in cost. This is independent of any increase in labor, management, or the cost of other equipment resulting from the purchase of the tank.

DO EXERCISE 5.

6. The constant function g is given by

$$g(x, y) = 4$$

for all inputs x and y. Find:

a) $g(-9, 10)$;

b) $g(560, 43)$.

Example 4 *The gravity model.* The number of telephone calls between two cities is given by

$$N(d, P_1, P_2) = \frac{2.8P_1P_2}{d^{2.4}},$$

where d is the distance between the cities, and P_1 and P_2 are their populations.

Sociologists say that as two cities merge, the communication between them increases. (*USGS EROS Data Center*)

A constant can also be thought of as a function of several variables.

Example 5 The constant function f is given by

$$f(x, y) = -3, \quad \text{for all inputs } x \text{ and } y.$$

Find $f(5, 7)$ and $f(-2, 0)$.

Solution Since this is a constant function, it has the value -3 for any x and y. So

$$f(5, 7) = -3 \quad \text{and} \quad f(-2, 0) = -3.$$

DO EXERCISE 6.

Partial Derivatives

Consider the function f given by

$$z = f(x, y) = x^2y^3 + xy + 4y^2.$$

7. Consider

$$f(x, y) = 1 - x^2 - y^2.$$

a) Fix y at 4 and find $f(x, 4)$.

b) The answer to (a) could be interpreted as a function of one variable x. Find the first derivative.

Suppose for the moment that we fix y at 3. Then

$$f(x, 3) = x^2(3^3) + x(3) + 4(3^2) = 27x^2 + 3x + 36.$$

Note that we now have a function of only one variable. Taking the first derivative with respect to x, we have

$$54x + 3.$$

DO EXERCISE 7.

Now, without replacing y by a specific number, let us consider y fixed. Then f becomes a function of x alone and we can calculate its derivative with respect to x. This derivative is called the *partial derivative of f with respect to x*. Notation for this partial derivative is

$$\frac{\partial f}{\partial x} \quad \text{or} \quad \frac{\partial z}{\partial x}.$$

Thus let us consider again the function

$$z = f(x, y) = \boxed{x}^2 y^3 + \boxed{x}\, y + 4y^2.$$

8. For $f(x, y) = 1 - x^2 - y^2$, find $\partial f/\partial x$.

The color screens indicate the variable x when we fix y and treat it as a constant. The expressions y^3, y, and y^2 are then constants. We have

$$\frac{\partial f}{\partial x} = \frac{\partial z}{\partial x}$$

$$= 2xy^3 + y.$$

DO EXERCISE 8.

9. For $z = 3x^2y + 5x^3$, find (a) $\partial z/\partial x$; (b) $\partial z/\partial y$.

Similarly, we find $\partial f/\partial y$ or $\partial z/\partial y$ by fixing x (treating it as a constant) and calculating the derivative with respect to y. From

$$z = f(x, y) = x^2 \boxed{y}^3 + x\boxed{y} + 4\boxed{y}^2 \quad \text{The screens indicate the variable.}$$

we get

$$\frac{\partial f}{\partial y} = \frac{\partial z}{\partial y}$$

$$= 3x^2y^2 + x + 8y.$$

DO EXERCISE 9.

A definition of partial derivatives is as follows.

10. For $t = xy + xz + x^2 + y^3$, find
(a) $\partial t/\partial x$; (b) $\partial t/\partial y$; (c) $\partial t/\partial z$.

DEFINITION

For $z = f(x, y)$,

$$\frac{\partial z}{\partial x} = \lim_{h \to 0} \frac{f(x + h, y) - f(x, y)}{h},$$

$$\frac{\partial z}{\partial y} = \lim_{h \to 0} \frac{f(x, y + h) - f(x, y)}{h}.$$

Partial differentiation can be done for any number of variables.

Example 6 For $w = x^2 - xy + y^2 + 2yz + 2z^2 + z$, find

$$\frac{\partial w}{\partial x}, \quad \frac{\partial w}{\partial y}, \quad \text{and} \quad \frac{\partial w}{\partial z}.$$

Solution To find $\partial w/\partial x$, we consider x the variable and the other letters the constants. From

$$w = x^2 - xy + y^2 + 2yz + 2z^2 + z,$$

we get

$$\frac{\partial w}{\partial x} = 2x - y.$$

From

$$w = x^2 - xy + y^2 + 2yz + 2z^2 + z,$$

we get

$$\frac{\partial w}{\partial y} = -x + 2y + 2z.$$

From

$$w = x^2 - xy + y^2 + 2yz + 2z^2 + z,$$

we get

$$\frac{\partial w}{\partial z} = 2y + 4z + 1.$$

DO EXERCISE 10.

We will often make use of a simpler notation f_x for the partial derivative of f with respect to x, and f_y for the partial derivative of f with respect to y.

11. For $f(x, y) = 3x^3y + 2xy$, find:

a) f_x;

b) $f_x(-4, 1)$;

c) f_y;

d) $f_y(2, 6)$.

Example 7 For $f(x, y) = 3x^2y + xy$, find f_x and f_y.

Solution

$$f_x = 6xy + y,$$

$$f_y = 3x^2 + x$$

For the function in the preceding example, let us evaluate f_x at $(2, -3)$:

$$f_x(2, -3) = 6 \cdot 2 \cdot (-3) + (-3) = -39.$$

If we use the notation $\partial z/\partial x = 6xy + y$, where $z = 3x^2y + xy$, the value of the partial derivative at $(2, -3)$ is given by

$$\left. \frac{\partial z}{\partial x} \right|_{(2, -3)} = 6 \cdot 2 \cdot (-3) + (-3) = -39,$$

but this notation is not as convenient as $f_x(2, -3)$.

DO EXERCISE 11.

Example 8 For $f(x, y) = e^{xy} + y \ln x$, find f_x and f_y.

Solution

$$f_x = y \cdot e^{xy} + y \cdot \frac{1}{x} = ye^{xy} + \frac{y}{x},$$

$$f_y = x \cdot e^{xy} + 1 \cdot \ln x = xe^{xy} + \ln x$$

12. For $f(x, y) = \ln (xy) + ye^x$, find f_x and f_y.

DO EXERCISE 12.

Geometric Interpretations

Consider a function of two variables

$$z = f(x, y).$$

Recall the mapping interpretation of function that we considered in Chapter 1. As a mapping, a function of two variables can be thought of as mapping a point (x_1, y_1) in an xy-plane onto a point z_1 on a number line.

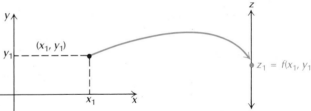

To graph a function of two variables, we need a three-dimensional coordinate system. The axes are usually placed as follows. The line z, called the z-axis, is placed perpendicular to the xy-plane at the origin.

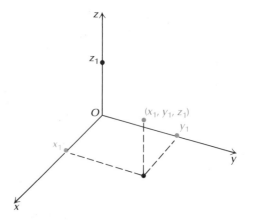

To help visualize this, think of looking into the corner of a room, where the floor is the xy-plane and the z-axis is the intersection of two walls. To plot a point (x_1, y_1, z_1) we locate the point (x_1, y_1) in the xy-plane, and move up or down in space according to the value of z_1.

Example 9 Plot these points: $P_1(2, 3, 5)$, $P_2(2, -2, -4)$, $P_3(0, 5, 2)$, and $P_4(2, 3, 0)$.

Solution

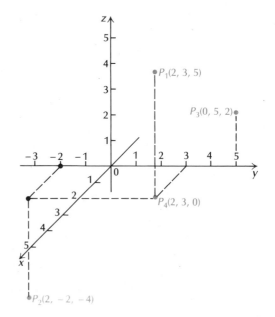

13. Using the axes shown below, graph $P_1(3, 2, 5)$, $P_2(2, 3, 1)$, and $P_3(-3, 2, 0)$.

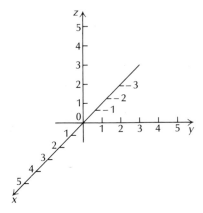

DO EXERCISE 13.

The *graph* of a function of two variables

$$z = f(x, y)$$

consists of ordered triples (x_1, y_1, z_1), where $z_1 = f(x_1, y_1)$. The domain of f is a region D in the xy-plane, and the graph of f is a surface S, as shown below.

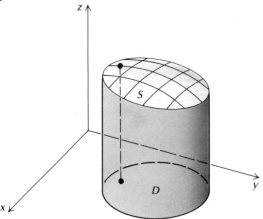

Here are some equations and their graphs.

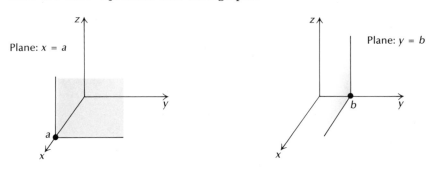

Plane: $x = a$

Plane: $y = b$

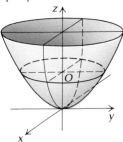

Elliptic paraboloid: $z = x^2 + y^2$

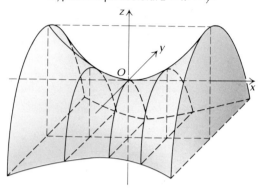

Hyperbolic paraboloid: $z = x^2 - y^2$

The following graphs have been generated by computer graphics.

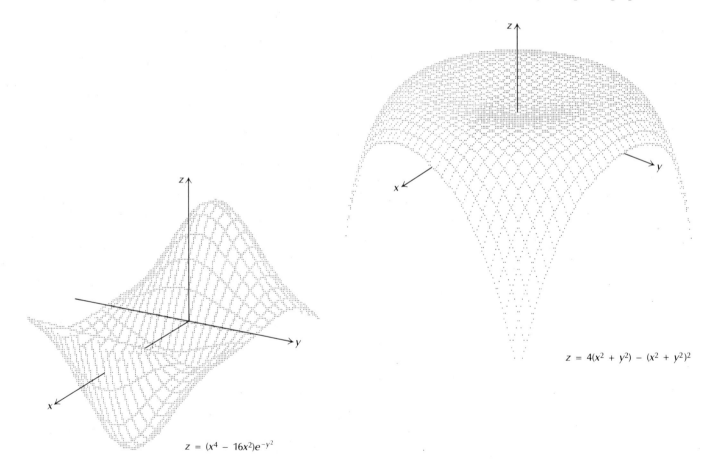

$z = (x^4 - 16x^2)e^{-y^2}$

$z = 4(x^2 + y^2) - (x^2 + y^2)^2$

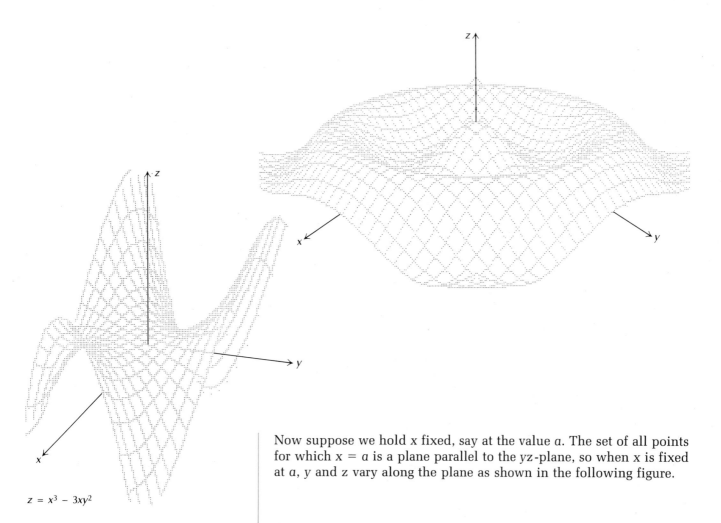

$z = x^3 - 3xy^2$

Now suppose we hold x fixed, say at the value a. The set of all points for which $x = a$ is a plane parallel to the yz-plane, so when x is fixed at a, y and z vary along the plane as shown in the following figure.

Slope is $\frac{\partial z}{\partial y}$

The plane shown cuts the surface in some curve C as shown. The partial derivative f_y gives the slopes of tangent lines to this curve. Similarly, if we hold y fixed, say at the value b, we obtain a curve C' as shown in the following figure. The partial derivative f_x gives the slopes of tangent lines to this curve.

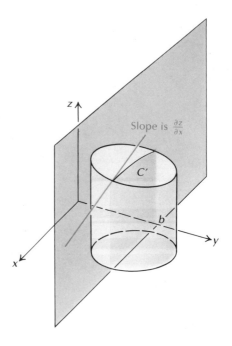

Economic Application: The Cobb–Douglas Production Function

One model of production that is frequently considered in business and economics is the *Cobb–Douglas production function*:

$$p(x, y) = Ax^a y^{1-a}, \qquad A > 0 \text{ and } 0 < a < 1,$$

where p is the number of units produced with x units of labor and y units of capital. Capital is the cost of machinery, buildings, tools, and other supplies. The partial derivatives

$$\frac{\partial p}{\partial x} \quad \text{and} \quad \frac{\partial p}{\partial y}$$

are called, respectively, the *marginal productivity of labor* and the *marginal productivity of capital*.

14. A company has the following production function for a certain product

$$p(x, y) = 800x^{3/4}y^{1/4}.$$

a) Find the production from 81 units of labor and 625 units of capital.

b) Find the marginal productivities.

c) Evaluate the marginal productivities at $x = 81$ and $y = 625$.

Example 10 A company has the following production function for a certain product:

$$p(x, y) = 50x^{2/3}y^{1/3}.$$

a) Find the production from 125 units of labor and 64 units of capital.

b) Find the marginal productivities.

c) Evaluate the marginal productivities at $x = 125$ and $y = 64$.

Solution

a) $p(125, 64) = 50(125)^{2/3}(64)^{1/3} = 50(25)(4) = 5000$ units

b) $\dfrac{\partial p}{\partial x} = 50\left(\dfrac{2}{3}\right)x^{-1/3}y^{1/3} = \dfrac{100y^{1/3}}{3x^{1/3}}$, or $\dfrac{100}{3}\left(\dfrac{y}{x}\right)^{1/3}$,

$\dfrac{\partial p}{\partial y} = 50\left(\dfrac{1}{3}\right)x^{2/3}y^{-2/3} = \dfrac{50x^{2/3}}{3y^{2/3}}$, or $\dfrac{50}{3}\left(\dfrac{x}{y}\right)^{2/3}$

c) $\dfrac{\partial p}{\partial x}\bigg|_{(125, 64)} = \dfrac{100(64)^{1/3}}{3(125)^{1/3}} = \dfrac{100(4)}{3(5)} = 26\dfrac{2}{3}$,

$\dfrac{\partial p}{\partial y}\bigg|_{(125, 64)} = \dfrac{50(125)^{2/3}}{3(64)^{2/3}} = \dfrac{50(25)}{3(16)} = 26\dfrac{1}{24}$

How can we interpret marginal productivities? Suppose the amount spent on capital is fixed at, say, $y = 64$. Then if the amount of labor changes by a small amount, production will change by about 27 units. Suppose the amount of labor is held fixed at, say, $x = 125$. Then if the amount of capital spent changes slightly, this will produce a change of about 26 units of production.

A Cobb–Douglas production function is consistent with the law of diminishing returns. That is, if one input (of either labor or capital) is held fixed while the other increases infinitely, then production will eventually increase at a decreasing rate. With such functions it also turns out that if a certain maximum production is possible, then the expense of more labor, for example, will not prevent that maximum output from still being attainable.

DO EXERCISE 14.

EXERCISE SET 7.1

1. For $f(x, y) = x^2 - 2xy$, find $f(0, -2)$, $f(2, 3)$, and $f(10, -5)$.

2. For $f(x, y) = (y^2 + 3xy)^3$, find $f(-2, 0)$, $f(3, 2)$, and $f(-5, 10)$.

3. For $f(x, y) = 3^x + 7xy$, find $f(0, -2)$, $f(-2, 1)$, and $f(2, 1)$.

4. For $f(x, y) = \log_{10} x - 5y^2$, find $f(10, 2)$, $f(1, -3)$, and $f(100, 4)$.

5. For $f(x, y) = \ln x + y^3$, find $f(e, 2)$, $f(e^2, 4)$, and $f(e^3, 5)$.

6. For $f(x, y) = 2^x - 3^y$, find $f(0, 0)$, $f(1, 1)$, and $f(2, 2)$.

7. For $f(x, y, z) = x^2 - y^2 + z^2$, find $f(-1, 2, 3)$ and $f(2, -1, 3)$.

8. For $f(x, y, z) = 2^x + 5zy - x$, find $f(0, 1, -3)$ and $f(1, 0, -3)$.

9. *Psychology: Intelligence quotient.* The intelligence quotient in psychology is given by

$$Q(m, c) = 100 \cdot \frac{m}{c},$$

where m is the mental age of a person and c is the chronological, or actual age. Find $Q(21, 20)$ and $Q(19, 20)$.

10. *Business: Price-earnings ratio.* The price-earnings ratio of a stock is given by

$$R(P, E) = \frac{P}{E},$$

where P is the price per share of the stock and E is the earnings per share. Recently, the price per share of IBM stock was $\$287\frac{3}{8}$ and the earnings per share were $\$23.30$. Find the price-earnings ratio. Give decimal notation to the nearest tenth.

11. *Business: Yield of a stock.* The yield of a stock is given by

$$Y(D, P) = \frac{D}{P},$$

where D is the dividends per share of a stock and P is the price per share. Recently, the price per share of Goodyear stock was $\$16\frac{7}{8}$ and the dividends per share were $\$1.30$. Find the yield. Give percent notation to the nearest tenth of a percent.

12. *Biomedical: Poiseuille's Law.* The speed of blood in a vessel is given by

$$V(L, p, R, r, v) = \frac{p}{4Lv}(R^2 - r^2),$$

where R is the radius of the vessel, r is the distance of the blood from the center of the vessel, L is the length of the blood vessel, p is pressure, and v is viscosity. Find $V(1, 100, 0.0075, 0.0025, 0.05)$.

Find $\dfrac{\partial z}{\partial x}, \dfrac{\partial z}{\partial y}, \dfrac{\partial z}{\partial x}\bigg|_{(-2, -3)}$ and $\dfrac{\partial z}{\partial y}\bigg|_{(0, -5)}$

13. $z = 2x - 3xy$ **14.** $z = (x - y)^3$ **15.** $z = 3x^2 - 2xy + y$ **16.** $z = 2x^3 + 3xy - x$

Find $f_x, f_y, f_x(-2, 4)$, and $f_y(4, -3)$.

17. $f(x, y) = 2x - 3y$ **18.** $f(x, y) = 5x + 7y$

Find $f_x, f_y, f_x(-2, 1)$, and $f_y(-3, -2)$.

19. $f(x, y) = \sqrt{x^2 + y^2}$ **20.** $f(x, y) = \sqrt{x^2 - y^2}$

Find f_x and f_y.

21. $f(x, y) = e^{2x+3y}$ **22.** $f(x, y) = e^{3x-2y}$ **23.** $f(x, y) = e^{xy}$ **24.** $f(x, y) = e^{2xy}$

25. $f(x, y) = y \ln(x + y)$ **26.** $f(x, y) = x \ln(x + y)$ **27.** $f(x, y) = x \ln(xy)$ **28.** $f(x, y) = y \ln(xy)$

29. $f(x, y) = \dfrac{x}{y} - \dfrac{y}{x}$ **30.** $f(x, y) = \dfrac{x}{y} + \dfrac{y}{x}$ **31.** $f(x, y) = 3(2x + y - 5)^2$ **32.** $f(x, y) = 4(3x + y - 8)^2$

Find $\dfrac{\partial f}{\partial b}$ and $\dfrac{\partial f}{\partial m}$.

33. $f(b, m)$

$$= (m + b - 4)^2 + (2m + b - 5)^2 + (3m + b - 6)^2$$

Find f_x, f_y, and f_λ.

35. $f(x, y, \lambda) = 3xy - \lambda(2x + y - 8)$

37. $f(x, y, \lambda) = x^2 + y^2 - \lambda(10x + 2y - 4)$

39. A company has the following production function for a certain product:

$$p(x, y) = 1800x^{0.621}y^{0.379},$$

where p is the number of units produced with x units of labor and y units of capital.

a)   Find the production from 2500 units of labor and 1700 units of capital.

b) Find the marginal productivities.

c)   Evaluate the marginal productivities at $x = 2500$ and $y = 1700$.

▶

41. ▦ *Wind chill temperature.* Wind speed affects the actual temperature, making a person colder due to extra heat loss from the skin. The *wind chill temperature* is what the temperature would have to be with no wind to give the same chilling effect. The wind chill temperature W is given by

$$W(v, T) = 91.4 - \frac{(10.45 + 6.68\sqrt{v} - 0.447v)(457 - 5T)}{110},$$

where T is the actual temperature as given by a thermometer, in degrees Fahrenheit, and v is the speed of the wind, in mph. Find the wind chill temperature in each case. Round to the nearest one degree.   **CSS**

a) $T = 30°F$, $v = 25$ mph   b) $T = 20°F$, $v = 20$ mph

c) $T = 20°F$, $v = 40$ mph   d) $T = -10°F$, $v = 30$ mph

34. $f(b, m)$

$$= (m + b - 6)^2 + (2m + b - 8)^2 + (3m + b - 9)^2$$

36. $f(x, y, \lambda) = 4xy - \lambda(3x - y + 7)$

38. $f(x, y, \lambda) = x^2 - y^2 - \lambda(4x - 7y - 10)$

40. A company has the following production function for a certain product:

$$p(x, y) = 2400x^{2/5}y^{3/5},$$

where p is the number of units produced with x units of labor and y units of capital.

a) Find the production from 32 units of labor and 1024 units of capital.

b) Find the marginal productivities.

c) Evaluate the marginal productivities at $x = 32$ and $y = 1024$.

42. ▦ For W in Exercise 41, find W_v and W_T.

Find f_x and f_t.

43. $f(x, t) = \dfrac{x^2 + t^2}{x^2 - t^2}$ **44.** $f(x, t) = \dfrac{x^2 - t}{x^3 + t}$

45. $f(x, t) = \dfrac{2\sqrt{x} - 2\sqrt{t}}{1 + 2\sqrt{t}}$ **46.** $f(x, t) = \sqrt[4]{x^3 t^5}$

47. $f(x, t) = 6x^{2/3} - 8x^{1/4}t^{1/2} - 12x^{-1/2}t^{3/2}$

48. $f(x, t) = \left(\dfrac{x^2 + t^2}{x^2 - t^2}\right)^5$

OBJECTIVE

You should be able to find the four second partial derivatives of a function.

7.2 HIGHER-ORDER PARTIAL DERIVATIVES

Consider

$$z = f(x, y) = 3xy^2 + 2xy + x^2. \tag{1}$$

Then

$$\frac{\partial z}{\partial x} = \frac{\partial f}{\partial x} = 3y^2 + 2y + 2x. \tag{2}$$

1. Consider

$$z = 3xy^2 + 2xy + x^2.$$

a) Find $\partial z/\partial y$.

b) For the function in (a), find the first partial derivative with respect to x.

c) For the function in (a), find the first partial derivative with respect to y; that is, differentiate "twice" with respect to y.

2. Consider

$$f(x, y) = 3xy^2 + 2xy + x^2.$$

a) Find f_y.

b) For the function in (a), find the first partial derivative with respect to x. Denote this f_{yx}.

c) For the function in (a), find the first partial derivative with respect to y. Denote this f_{yy}.

Suppose we find the first partial derivative of function (2) with respect to y. This will be a *second-order partial derivative*. Notation for it is as follows:

$$\frac{\partial}{\partial y}\left(\frac{\partial z}{\partial x}\right) = \frac{\partial}{\partial y}\left(\frac{\partial f}{\partial x}\right)$$

$$= \frac{\partial^2 z}{\partial y\,\partial x} = \frac{\partial^2 f}{\partial y\,\partial x} = 6y + 2.$$

DO EXERCISE 1.

We could also denote the preceding partial derivative using the notation f_{xy}. Then

$$f_{xy} = 6y + 2.$$

Note that in the notation f_{xy}, x and y are in the order (left to right) in which the differentiation is done. In the other symbolisms that order is reversed, but the meaning is not.

DO EXERCISE 2.

Notation for the four second-order partial derivatives is as follows.

DEFINITION

1. $\dfrac{\partial^2 z}{\partial x\,\partial x} = \dfrac{\partial^2 f}{\partial x\,\partial x} = \dfrac{\partial^2 z}{\partial x^2} = \dfrac{\partial^2 f}{\partial x^2} = f_{xx}$ Take the partial with respect to x, and then with respect to x again.

2. $\dfrac{\partial^2 z}{\partial y\,\partial x} = \dfrac{\partial^2 f}{\partial y\,\partial x} = f_{xy}$ Take the partial with respect to x, and then with respect to y.

3. $\dfrac{\partial^2 z}{\partial x\,\partial y} = \dfrac{\partial^2 f}{\partial x\,\partial y} = f_{yx}$ Take the partial with respect to y, and then with respect to x.

4. $\dfrac{\partial^2 z}{\partial y\,\partial y} = \dfrac{\partial^2 f}{\partial y\,\partial y} = \dfrac{\partial^2 z}{\partial y^2} = \dfrac{\partial^2 f}{\partial y^2} = f_{yy}$ Take the partial with respect to y, and then with respect to y again.

Example 1 For

$$z = f(x, y) = x^2y^3 + x^4y + xe^y,$$

find the four second-order partial derivatives.

3. For

$$z = f(x, y)$$
$$= 3xy^2 + 2xy + x^2 + x \ln y,$$

find the four second-order partial derivatives.

Solution

a) $\dfrac{\partial^2 f}{\partial x^2} = f_{xx} = \dfrac{\partial}{\partial x}(2xy^3 + 4x^3y + e^y)$ $\qquad$ Differentiate twice with respect to x.

$$= 2y^3 + 12x^2y$$

b) $\dfrac{\partial^2 f}{\partial y\, \partial x} = f_{xy} = \dfrac{\partial}{\partial y}(2xy^3 + 4x^3y + e^y)$ $\qquad$ Differentiate with respect to x, then with respect to y.

$$= 6xy^2 + 4x^3 + e^y$$

c) $\dfrac{\partial^2 f}{\partial x\, \partial y} = f_{yx} = \dfrac{\partial}{\partial x}(3x^2y^2 + x^4 + xe^y)$ $\qquad$ Differentiate with respect to y, then with respect to x.

$$= 6xy^2 + 4x^3 + e^y$$

d) $\dfrac{\partial^2 f}{\partial y^2} = f_{yy} = \dfrac{\partial}{\partial y}(3x^2y^2 + x^4 + xe^y)$ $\qquad$ Differentiate twice with respect to y.

$$= 6x^2y + xe^y$$

DO EXERCISE 3.

Note by comparing (b) and (c) above that

$$\frac{\partial^2 f}{\partial y\, \partial x} = \frac{\partial^2 f}{\partial x\, \partial y}. \qquad \text{And similarly, } f_{xy} = f_{yx}$$

This will be true for virtually all functions that we consider in this text, but is *not* true for all functions. Such an example is given in Exercise 19 of Exercise Set 7.2.

EXERCISE SET 7.2

Find the four second-order partial derivatives.

1. $f(x, y) = 3x^2 - xy + y$ $\qquad$ **2.** $f(x, y) = 5x^2 + xy - x$ $\qquad$ **3.** $f(x, y) = 3xy$

4. $f(x, y) = 4xy$ $\qquad$ **5.** $f(x, y) = x^5y^4 + x^3y^2$ $\qquad$ **6.** $f(x, y) = x^4y^3 - x^2y^3$

Find f_{xx}, f_{xy}, f_{yx}, and f_{yy}. (Remember, f_{yx} means to differentiate with respect to y, then x.)

7. $f(x, y) = 2x - 3y$ $\qquad$ **8.** $f(x, y) = 3x + 5y$ $\qquad$ **9.** $f(x, y) = e^{2xy}$

10. $f(x, y) = e^{xy}$ $\qquad$ **11.** $f(x, y) = x + e^y$ $\qquad$ **12.** $f(x, y) = y - e^x$

13. $f(x, y) = y \ln x$ $\qquad$ **14.** $f(x, y) = x \ln y$

Find f_{xx}, f_{xy}, f_{yx}, and f_{yy}.

15. $f(x, y) = \dfrac{x}{y^2} - \dfrac{y}{x^2}$

16. $f(x, y) = \dfrac{xy}{x - y}$

17. Consider $f(x, y) = \ln(x^2 + y^2)$. Show that f is a solution of the partial differential equation

$$\frac{\partial^2 f}{\partial x^2} + \frac{\partial^2 f}{\partial y^2} = 0.$$

18. Consider $f(x, y) = x^3 - 5xy^2$. Show that f is a solution of the partial differential equation

$$xf_{xy} - f_y = 0.$$

19. Consider the function f defined as follows:

$$f(x, y) = \begin{cases} \dfrac{xy(x^2 - y^2)}{x^2 + y^2} & \text{for } (x, y) \neq (0, 0), \\ 0 & \text{for } (x, y) = (0, 0). \end{cases}$$

a) Find $f_x(0, y)$ by evaluating the limit $\lim\limits_{h \to 0} \dfrac{f(h, y) - f(0, y)}{h}$.

b) Find $f_y(x, 0)$ by evaluating the limit $\lim\limits_{h \to 0} \dfrac{f(x, h) - f(x, 0)}{h}$.

c) Then find and compare $f_{yx}(0, 0)$ and $f_{xy}(0, 0)$.

OBJECTIVES

You should be able to find maximum and minimum values of functions of two variables.

7.3 MAXIMUM–MINIMUM PROBLEMS

We will now find maximum and minimum values of functions of two variables.

DEFINITION

A function f of two variables:

i) has a _relative maximum_ at (a, b) if

$$f(x, y) \leq f(a, b)$$

for all points in a circular region containing (a, b);

ii) has a _relative minimum_ at (a, b) if

$$f(x, y) \geq f(a, b)$$

for all points in a circular region containing (a, b).

This definition is illustrated in Fig. 1.

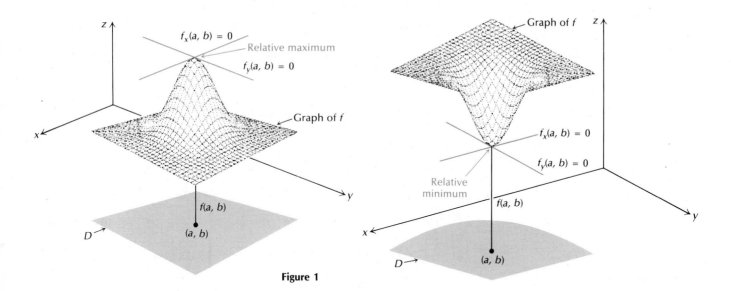

Figure 1

A relative maximum (minimum) may not be an "absolute" maximum (minimum), as illustrated below in Fig. 2.

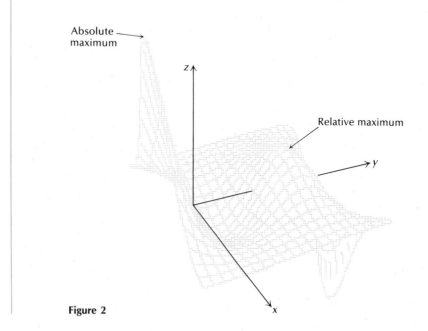

Figure 2

Determining Maximum and Minimum Values

Suppose a function f assumes a relative maximum (or minimum) value at some point (a, b) inside its domain. If we hold y constant at the value b, then $f(x, b)$ is a function of one variable x having its relative maximum value at $x = a$, so its derivative must be 0 there. That is, $f_x = 0$ at the point (a, b). Similarly, $f_y = 0$ at (a, b). The equations

$$f_x = 0, \qquad f_y = 0$$

are thus satisfied by the point (a, b) at which the relative maximum occurs. We call a point (a, b) where both partial derivatives are 0 a *critical point*. This is comparable to the earlier definition for functions of one variable. Thus one strategy for finding relative maximum or minimum values is to solve the above system of equations to find critical points. Just as for functions of one variable, this strategy does *not* guarantee that we will have a relative maximum or minimum value. We have argued only that *if* f has a maximum or minimum value at (a, b), *then* both its partial derivatives must be 0 at that point. Look at Fig. 1. That this does not hold in all cases is shown in Fig. 3.

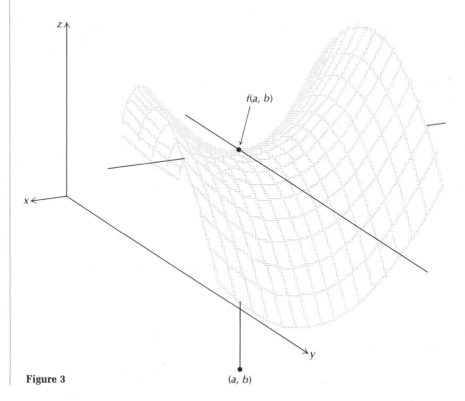

Figure 3

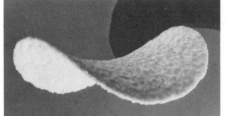

Where is the saddle point? (*Marshall Henrichs*)

Now suppose we fix y at a point b. Then $f(x, b)$, considered as a function of one variable, has a minimum at a, but f does not. Similarly, if we fix x at a, then $f(a, y)$, considered as a function of one variable, has a maximum at b, but f does not. The point $f(a, b)$ is called a *saddle point*. In other words, $f_x(a, b) = 0$ and $f_y(a, b) = 0$ (the point (a, b) is a critical point), but f does not attain a relative maximum or minimum at (a, b).

A test for finding relative maximum and minimum values that involves the use of first- and second-order partial derivatives is stated below. We shall not prove this theorem.

THEOREM 1   The D-test

To find the relative maximum and minimum values of f, we do the following.

1. **Find f_x, f_y, f_{xx}, f_{yy}, and f_{xy}.**
2. **Solve the system of equations $f_x = 0$, $f_y = 0$. Let (a, b) represent a solution.**
3. **Evaluate D where $D = f_{xx}(a, b) \cdot f_{yy}(a, b) - [f_{xy}(a, b)]^2$.**
4. **Then:**

    i) **f has a maximum at (a, b) if $D > 0$ and $f_{xx}(a, b) < 0$;**

    ii) **f has a minimum at (a, b) if $D > 0$ and $f_{xx}(a, b) > 0$;**

    iii) **f has neither a maximum nor a minimum at (a, b) if $D < 0$. The function has a *saddle point* at (a, b). See Fig. 3.**

    iv) **This test is not applicable if $D = 0$.**

A relative maximum or minimum *may not be an absolute maximum or minimum*. Tests for absolute maximum or minimum values are rather complicated. We shall restrict our attention to finding *relative* maximum or minimum values. Fortunately, in most applications, relative maximum and minimum values turn out to be absolute maximum and minimum values.

Example 1 Find the relative maximum and minimum values of

$$f(x, y) = x^2 + xy + y^2 - 3x.$$

Solution

1. Find f_x, f_y, f_{xx}, f_{yy}, and f_{xy}:

$$f_x = 2x + y - 3, \qquad f_y = x + 2y,$$

$$f_{xx} = 2, \qquad\qquad f_{yy} = 2,$$

$$f_{xy} = 1.$$

1. Find the relative maximum and minimum values of

$$f(x, y) = x^2 + 2xy + 2y^2 - 6y.$$

2. Solve the system of equations $f_x = 0$, $f_y = 0$:

$$2x + y - 3 = 0, \tag{1}$$

$$x + 2y = 0. \tag{2}$$

Solving Eq. (2) for x, we get $x = -2y$. Substituting $-2y$ for x in Eq. (1) and solving, we get

$$2(-2y) + y - 3 = 0$$

$$-4y + y - 3 = 0$$

$$-3y - 3 = 0$$

$$y = -1.$$

To find x when $y = -1$, we substitute -1 for y in either Eq. (1) or Eq. (2). We use Eq. (2):

$$x + 2(-1) = 0$$

$$x = 2.$$

Thus $(2, -1)$ is our candidate for a maximum or minimum.

3. We have to check to see if $f(2, -1)$ is a maximum or minimum:

$$D = f_{xx}(2, -1) \cdot f_{yy}(2, -1) - [f_{xy}(2, -1)]^2$$

$$= 2 \cdot 2 - [1]^2$$

$$= 3.$$

4. Thus $D = 3$ and $f_{xx}(2, -1) = 2$. Since $D > 0$ and $f_{xx}(2, -1) > 0$, it follows that f has a relative minimum at $(2, -1)$ and that the minimum is found as follows:

$$f(2, -1) = 2^2 + 2(-1) + (-1)^2 - 3 \cdot 2$$

$$= 4 - 2 + 1 - 6 = -3.$$

DO EXERCISE 1.

Example 2 Find the relative maximum and minimum values of

$$f(x, y) = xy - x^3 - y^2.$$

Solution

1. Find f_x, f_y, f_{xx}, f_{yy}, and f_{xy}:

$$f_x = y - 3x^2, \qquad f_y = x - 2y, \qquad f_{xx} = -6x,$$

$$f_{yy} = -2, \qquad f_{xy} = 1.$$

2. Solve the system of equations $f_x = 0$, $f_y = 0$:

$$y - 3x^2 = 0, \tag{1}$$

$$x - 2y = 0. \tag{2}$$

Solving Eq. (1) for y, we get $y = 3x^2$. Substituting $3x^2$ for y in Eq. (2) and solving, we get

$$x - 2(3x^2) = 0$$

$$x - 6x^2 = 0$$

$$x(1 - 6x) = 0. \quad \text{Factoring}$$

Setting each factor equal to 0 and solving, we have

$$x = 0 \quad \text{or} \quad 1 - 6x = 0$$

$$x = 0 \quad \text{or} \quad x = \frac{1}{6}.$$

To find y when $x = 0$ we substitute 0 for x in either Eq. (1) or Eq. (2). We use Eq. (2):

$$0 - 2y = 0$$

$$-2y = 0$$

$$y = 0.$$

Thus $(0, 0)$ is one critical value (candidate for a maximum or minimum). To find the other critical value, we substitute $\frac{1}{6}$ for x in either Eq. (1) or (2). We use Eq. (2):

$$\frac{1}{6} - 2y = 0$$

$$-2y = -\frac{1}{6}$$

$$y = \frac{1}{12}.$$

Thus $(\frac{1}{6}, \frac{1}{12})$ is another critical point.

3. We have to check both $(0, 0)$ and $(\frac{1}{6}, \frac{1}{12})$ as to whether they yield maximum or minimum values:

For $(0, 0)$: $\quad D = f_{xx}(0, 0) \cdot f_{yy}(0, 0) - [f_{xy}(0, 0)]^2$

$$= [-6 \cdot 0] \cdot [-2] - [1]^2$$

$$= -1,$$

2. Find the relative maximum and minimum values of

$$f(x, y) = 2xy - 4x^3 - y^2.$$

Since $D < 0$, it follows that $f(0, 0)$ is neither a maximum nor a minimum, but a saddle point.

For $\left(\dfrac{1}{6}, \dfrac{1}{12}\right)$: $D = f_{xx}\left(\dfrac{1}{6}, \dfrac{1}{12}\right)f_{yy}\left(\dfrac{1}{6}, \dfrac{1}{12}\right) - \left[f_{xy}\left(\dfrac{1}{6}, \dfrac{1}{12}\right)\right]^2$

$$= \left[-6 \cdot \dfrac{1}{6}\right] \cdot [-2] - [1]^2$$

$$= -1(-2) - 1$$

$$= 1.$$

4. Thus $D = 1$ and $f_{xx}(\frac{1}{6}, \frac{1}{12}) = -1$. Since $D > 0$ and $f_{xx}(\frac{1}{6}, \frac{1}{12}) < 0$, it follows that f has a relative maximum at $(\frac{1}{6}, \frac{1}{12})$ and that maximum is found as follows:

$$f\left(\dfrac{1}{6}, \dfrac{1}{12}\right) = \dfrac{1}{6} \cdot \dfrac{1}{12} - \left(\dfrac{1}{6}\right)^3 - \left(\dfrac{1}{12}\right)^2$$

$$= \dfrac{1}{72} - \dfrac{1}{216} - \dfrac{1}{144} = \dfrac{1}{432}.$$

DO EXERCISE 2.

Example 3 *Business: Maximizing profit.* A firm produces two kinds of golf balls, one that sells for \$3 each and the other for \$2 each. The total revenue from the sale of x thousand balls at \$3 each and y thousand at \$2 each is given by

$$R(x, y) = 3x + 2y.$$

The company determines that the total cost, in thousands of dollars, of producing x thousand of the \$3 ball and y thousand of the \$2 ball is given by

$$C(x, y) = 2x^2 - 2xy + y^2 - 9x + 6y + 7.$$

Find the amount of each type of ball that must be produced and sold in order to maximize profit.

Solution Total profit $P(x, y)$ is given by

$$P(x, y) = R(x, y) - C(x, y)$$

$$= 3x + 2y - (2x^2 - 2xy + y^2 - 9x + 6y + 7)$$

$$P(x, y) = -2x^2 + 2xy - y^2 + 12x - 4y - 7.$$

3. *Business: Maximizing profit.* A firm produces two kinds of calculators, one that sells for $15 each and the other for $20 each. The total revenue from the sale of x thousand calculators at $15 each and y thousand at $20 each is given by

$$R(x, y) = 15x + 20y.$$

The company determines that the total cost, in thousands of dollars, of producing x thousand of the $15 calculator and y thousand of the $20 calculator is given by

$C(x, y) =$

$3x^2 - 3xy + \dfrac{3}{2}y^2 + 6x + 14y - 50.$

Find the amount of each type of calculator that must be produced and sold in order to maximize profit.

1. Find P_x, P_y, P_{xx}, P_{yy}, and P_{xy}:

$$P_x = -4x + 2y + 12, \qquad P_y = 2x - 2y - 4,$$

$$P_{xx} = -4, \qquad\qquad P_{yy} = -2,$$

$$P_{xy} = 2.$$

2. Solve the system of equations $P_x = 0$, $P_y = 0$:

$$-4x + 2y + 12 = 0, \tag{1}$$

$$2x - 2y - 4 = 0. \tag{2}$$

Adding these equations, we get

$$-2x + 8 = 0.$$

Then

$$-2x = -8$$

$$x = 4.$$

To find y when x = 4, we substitute 4 for x in either Eq. (1) or Eq. (2). We use Eq. (2):

$$2 \cdot 4 - 2y - 4 = 0$$

$$-2y + 4 = 0$$

$$-2y = -4$$

$$y = 2.$$

Thus (4, 2) is our candidate for a maximum or minimum.

3. We have to check to see if $P(4, 2)$ is a maximum or minimum:

$$D = P_{xx}(4, 2)P_{yy}(4, 2) - [P_{xy}(4, 2)]^2$$

$$= (-4)(-2) - 2^2$$

$$= 4.$$

4. Thus $D = 4$ and $P_{xx} = -4$. Since $D > 0$ and $P_{xx}(4, 2) < 0$, it follows that P has a relative maximum at (4, 2). Thus to maximize profit, the company must produce and sell 4 thousand of the $3 golf balls and 2 thousand of the $2 golf balls.

DO EXERCISE 3.

EXERCISE SET 7.3

Find the relative maximum and minimum values.

1. $f(x, y) = x^2 + xy + y^2 - y$

2. $f(x, y) = x^2 + xy + y^2 - 5y$

3. $f(x, y) = 2xy - x^3 - y^2$

4. $f(x, y) = 4xy - x^3 - y^2$

5. $f(x, y) = x^3 + y^3 - 3xy$

6. $f(x, y) = x^3 + y^3 - 6xy$

7. $f(x, y) = x^2 + y^2 - 2x + 4y - 2$

8. $f(x, y) = x^2 + 2xy + 2y^2 - 6y + 2$

9. $f(x, y) = x^2 + y^2 + 2x - 4y$

10. $f(x, y) = 4y + 6x - x^2 - y^2$

11. $f(x, y) = 4x^2 - y^2$

12. $f(x, y) = x^2 - y^2$

In these problems assume that relative maximum and minimum values are absolute maximum and minimum values.

13. *Business: Maximizing profit.* A firm produces two kinds of radios, one that sells for $17 each and the other for $21 each. The total revenue from the sale of x thousand radios at $17 each and y thousand at $21 each is given by

$$R(x, y) = 17x + 21y.$$

The company determines that the total cost, in thousands of dollars, of producing x thousand of the $17 radio and y thousand of the $21 radio is given by

$$C(x, y) = 4x^2 - 4xy + 2y^2 - 11x + 25y - 3.$$

Find the amount of each type of radio that must be produced and sold in order to maximize profit.

15. A one-product company found that its profit in millions of dollars is a function P given by

$$P(a, p) = 2ap + 80p - 15p^2 - \frac{1}{10}a^2p - 100,$$

where $a =$ the amount spent on advertising, in millions of dollars, and $p =$ the price charged per item of the product, in dollars. Find the maximum value of P and the values of a and p at which it is attained.

▶

17. *Two-variable revenue maximization.* Boxowitz, Inc., a computer firm, markets two kinds of electronic calculators that compete with one another. Their demand functions are expressed by the following relationships

$$q_1 = 78 - 6p_1 - 3p_2, \tag{1}$$

$$q_2 = 66 - 3p_1 - 6p_2, \tag{2}$$

where $p_1, p_2 =$ the price of each calculator in multiples of $10, and $q_1, q_2 =$ the quantity of each calculator demanded in hundreds of units.

14. *Business: Maximizing profit.* A firm produces two kinds of baseball gloves, one that sells for $18 each and the other for $25 each. The total revenue from the sale of x thousand gloves at $18 each and y thousand at $25 each is given by

$$R(x, y) = 18x + 25y.$$

The company determines that the total cost, in thousands of dollars, of producing x thousand of the $18 glove and y thousand of the $25 glove is given by

$$C(x, y) = 4x^2 - 6xy + 3y^2 + 20x + 19y - 12.$$

Find the amount of each type of glove that must be produced and sold in order to maximize profit.

16. A one-product company finds that its profit in millions of dollars is a function P given by

$$P(a, n) = -5a^2 - 3n^2 + 48a - 4n + 2an + 300,$$

where $a =$ the amount spent on advertising, in millions of dollars, and $n =$ the number of items sold, in thousands. Find the maximum value of P and the values of a and n at which it is attained.

a) Find a formula for the total revenue function R in terms of the variables p_1 and p_2. (*Hint:* $R = p_1q_1 + p_2q_2$; then substitute expressions from Eqs. (1) and (2) to find $R(p_1, p_2)$.)

b) What prices p_1 and p_2 should be charged for each product in order to maximize total revenue?

c) How many units will be demanded?

d) What is the maximum total revenue?

18. Repeat Exercise 17, where

$$q_1 = 64 - 4p_1 - 2p_2 \quad \text{and} \quad q_2 = 56 - 2p_1 - 4p_2.$$

Find the relative maximum and minimum values.

19. $f(x, y) = e^x + e^y - e^{x+y}$

20. $f(x, y) = xy + \dfrac{2}{x} + \dfrac{4}{y}$

21. $S(b, m) = (m + b - 72)^2 + (2m + b - 73)^2 + (3m + b - 75)^2$

22. An open-top rectangular box with a square base is to be made with a 20-m² surface area. Find the dimensions that will yield the maximum volume.

OBJECTIVES

You should be able to

a) **Compute the total differential of a function of two or more variables.**

b) **Use the differential _dw_ of a function _w_ to approximate the actual change Δw in that function.**

1. Find dz if $z = xe^y$.

2. Find dz if $z = \ln(xy)$.

7.4 TOTAL DIFFERENTIAL

If $y = f(x)$, then the *differential of y, dy,* is defined by

$$dy = f'(x)\,dx,$$

where dx is the *differential of x*. Note that dy is a function of two variables, x and dx.

In Section 3.6, we saw that linear approximation is a useful tool for certain problems involving functions of one variable, and so we'd like to extend the techniques developed there to functions of two or more variables. To do so, we'll use differentials, beginning with the following definition.

DEFINITION

Let $z = f(x, y)$. Then the *total differential dz* is given by

$$dz = f_x\,dx + f_y\,dy.$$

Example 1 Find dz if $z = x^2y^3$.

Solution Here, $f(x, y) = x^2y^3$, so that $f_x = 2xy^3$ and $f_y = 3x^2y^2$. According to the definition,

$$dz = f_x\,dx + f_y\,dy$$
$$= 2xy^3\,dx + 3x^2y^2\,dy.$$

DO EXERCISES 1 AND 2.

Example 2 Evaluate dz for $x = 2$, $y = -1$ if $z = xy - 2x^2$.

Solution Since $f(x, y) = xy - 2x^2$, it follows that $f_x = y - 4x$ and $f_y = x$, so $dz = f_x\,dx + f_y\,dy = (y - 4x)\,dx + x\,dy$ is valid for all val-

3. Evaluate dz for $x = 2$, $y = 3$ if $z = xy^2$.

ues of x and y. If we substitute in $x = 2$ and $y = -1$, we find that

$$dz = (-1 - 4 \cdot 2) \, dx + 2 \, dy$$

$$= (-1 - 8) \, dx + 2 \, dy$$

$$= -9 \, dx + 2 \, dy.$$

DO EXERCISE 3.

Note in Example 2 that after specific values have been substituted for x and y, dz is still a function of the variables dx and dy. In general, if $z = f(x, y)$, then dz is a function of four variables: x, y, dx, and dy.

Example 3 Evaluate dz for $z = x^2 - y^2$ if $x = 2$, $y = 1$, $dx = 0.1$, and $dy = 0.15$.

Solution Since $f(x, y) = x^2 - y^2$, we have $f_x = 2x$ and $f_y = -2y$. From the definition,

$$dz = 2x \, dx + \,-2y \, dy.$$

If we let $x = 2$, $y = 1$, $dx = 0.1$, and $dy = 0.15$ in the last equation, we get

4. Evaluate dz for $z = x^3y + 5y^2$ if $x = 1$, $y = -1$, $dx = 0.25$, and $dy = -0.01$.

$$dz = 2 \cdot 2 \cdot (0.1) - 2 \cdot 1 \cdot (0.15)$$

$$= 4(0.1) - 2(0.15)$$

$$= 0.4 - 0.3 = 0.1.$$

DO EXERCISE 4.

Our definition may be easily extended to functions of three or more variables.

DEFINITION

Let $w = f(x, y, z)$. Then the *total differential dw* is given by

$$dw = f_x \, dx + f_y \, dy + f_z \, dz.$$

If w is a function of n variables, say $w = f(x_1, x_2, \ldots, x_n)$, then we define the *total differential dw* by

$$dw = f_{x_1} \, dx_1 + f_{x_2} \, dx_2 + \ldots + f_{x_n} \, dx_n$$

$$= \sum_{i=1}^{n} f_{x_i} \, dx_i.$$

5. Find dw if $w = \dfrac{1}{x} - xy + 2z$.

Example 4 Find dw given that $w = x^2 + y^3 + z^4$.

Solution Here, $f(x, y, z) = x^2 + y^3 + z^4$. Thus $f_x = 2x$, $f_y = 3y^2$, and $f_z = 4z^3$. Substituting these quantities into the definition gives

$$dw = f_x\, dx + f_y\, dy + f_z\, dz$$
$$= 2x\, dx + 3y^2\, dy + 4z^3\, dz.$$

DO EXERCISES 5 AND 6.

6. Find dw if $w = (x^2 + y^2 + z^2)^{1/2}$.

Example 5 Find dw if $w = x_1^2 + 2x_2x_3 - x_4$.

Solution Here, $f(x_1, x_2, x_3, x_4) = x_1^2 + 2x_2x_3 - x_4$, so that $f_{x_1} = 2x_1$, $f_{x_2} = 2x_3$, $f_{x_3} = 2x_2$, and $f_{x_4} = -1$. Substitution yields

$$dw = f_{x_1}\, dx_1 + f_{x_2}\, dx_2 + f_{x_3}\, dx_3 + f_{x_4}\, dx_4$$
$$= 2x_1\, dx_1 + 2x_3\, dx_2 + 2x_2\, dx_3 - 1\, dx_4$$
$$= 2x_1\, dx_1 + 2x_3\, dx_2 + 2x_2\, dx_3 - dx_4$$

DO EXERCISE 7.

7. Find dw if $w = x_1^2 + x_2^2 + x_3^2 + x_4^2$.

Example 6 Evaluate dw for $x = 1, y = 2, z = 3$, if $w = xy - xz + 2yz$.

Solution Since $f(x, y, z) = xy - xz + 2yz$ and $f_x = y - z, f_y = x + 2z$, and $f_z = -x + 2y$, we get

$$dw = (y - z)\, dx + (x + 2z)\, dy + (2y - x)\, dz$$

So, letting $x = 1, y = 2$, and $z = 3$, we have

$$dw = (2 - 3)\, dx + (1 + 2 \cdot 3)\, dy + (2 \cdot 2 - 1)\, dz$$

or

8. Evaluate dw for $x = -1, y = 1$, and $z = 1$, if $w = x^2y^3 - z^2$.

$$dw = -dx + 7\, dy + 3\, dz.$$

DO EXERCISE 8.

Note that after the three variables x, y, and z are assigned specific values in Example 6 above, w is still a function of the three variables dx, dy, and dz. If x, y, and z are allowed to vary, then we see that dw is actually a function of the six independent variables x, y, z, dx, dy, and dz. If w is a function of n variables, say $w = f(x_1, x_2, \ldots, x_n)$, then dw is a function of the $2n$ variables $x_1, x_2, \ldots, x_n, dx_1, dx_2, \ldots, dx_n$.

9. Evaluate dw for $x = 1$, $y = -1$, $z = 3$, $dx = 0.1$, $dy = -0.2$, and $dz = 0$ if $w = \ln(xyz)$.

10. Calculate Δz for
$$f(x, y) = 3x^2 + 4y$$
if $x = 2$, $y = 1$, $\Delta x = 0.07$, and $\Delta y = -0.015$.

Example 7 Evaluate dw for $x = 2$, $y = 5$, $z = -3$, $dx = -0.01$, $dy = 0.02$, and $dz = -0.03$ if $w = 4x^2 - 2y^2 + 3z^2$.

Solution Since $w = f(x, y, z) = 4x^2 - 2y^2 + 3z^2$, we have $f_x = 8x$, $f_y = -4y$, and $f_z = 6z$, so that $dw = 8x\,dx - 4y\,dy + 6z\,dz$. Substituting gives us

$$dw = 8 \cdot 2 \cdot (-0.01) - 4 \cdot 5 \cdot (0.02) + 6 \cdot (-3) \cdot (-0.03)$$

$$= 16(-0.01) - 20(0.02) - 18(-0.03)$$

$$= -0.16 - 0.40 + 0.54$$

$$= -0.02.$$

DO EXERCISE 9.

If $z = f(x, y)$, then the change Δz in z corresponding to changes Δx and Δy in x and y, respectively, is defined to be

$$\Delta z = f(x + \Delta x, y + \Delta y) - f(x, y).$$

Example 8 Calculate Δz for $f(x, y) = x^2 - y^2$ if $x = 1$, $y = 2$, $\Delta x = 0.01$, and $\Delta y = -0.03$.

Solution Since $x = 1$, $y = 2$, $\Delta x = 0.01$, and $\Delta y = -0.03$, we have $x + \Delta x = 1 + 0.01 = 1.01$ and $y + \Delta y = 2 + -0.03 = 1.97$. So,

$$\Delta z = f(x + \Delta x, y + \Delta y) - f(x, y)$$

$$= f(1.01, 1.97) - f(1, 2)$$

$$= ((1.01)^2 - (1.97)^2) - (1^2 - 2^2)$$

$$= (1.0201 - 3.8809) - (1 - 4)$$

$$= (-2.8608) - (-3) = 0.1392.$$

DO EXERCISE 10.

In Example 8 above, we calculated the exact change in $z = f(x, y)$ for certain values of x, y, Δx, and Δy. Just as we used the differential dy to approximate the exact change Δy when y is a function of x, we now want to use the differential dz to approximate the exact change Δz in $z = f(x, y)$.

Example 9 Use the differential dz to approximate Δz in Example 8 above.

11. Use the differential dz to approximate Δz in Margin Exercise 10.

Solution We had $z = f(x, y) = x^2 - y^2$ in Example 8, so that $f_x = 2x$, $f_y = -2y$, and $dz = 2x \, dx - 2y \, dy$. If we use the values that were given in Example 8, setting $\Delta x = dx = 0.01$ and $\Delta y = dy = -0.03$, then we get

$$dz = 2 \cdot 1 \cdot (0.01) - 2 \cdot (2) \cdot (-0.03)$$

$$= 2(0.01) - 4(-0.03)$$

$$= 0.02 + 0.12$$

$$= 0.14.$$

Comparing this with the exact value $\Delta z = 0.1392$ from Example 8, we see that $dz \approx \Delta z$.

DO EXERCISE 11.

These ideas extend immediately to functions of three or more variables.

Example 10 If $w = f(x, y, z) = 2x - y^2 + xz$, calculate both the exact change Δw and the approximate change dw for $x = 1$, $y = 1$, $z = 2$, $\Delta x = 0.01$, $\Delta y = -0.02$, and $\Delta z = 0.04$.

Solution Note that $x + \Delta x = 1.01$, $y + \Delta y = 0.98$, and $z + \Delta z = 2.04$, so that the exact change Δw is given by

$$\Delta w = f(x + \Delta x, y + \Delta y, z + \Delta z) - f(x, y, z)$$

$$\Delta w = f(1.01, 0.98, 2.04) - f(1, 1, 2)$$

$$\Delta w = [2(1.01) - (0.98)^2 + (1.01)(2.04)] - [2 \cdot 1 - 1^2 + 1 \cdot 2]$$

$$\Delta w = [2.02 - 0.9604 + 2.0604] - [2 - 1 + 2]$$

or

$$\Delta w = 3.12 - 3 = 0.12.$$

To compute dw, note that $f_x = 2 + z$, $f_y = -2y$, and $f_z = x$, so that

$$dw = (2 + z) \, dx - 2y \, dy + x \, dz.$$

Setting $dx = \Delta x = 0.01$, $dy = \Delta y = -0.02$, and $dz = \Delta z = 0.04$, we get

$$dw = (2 + 2) \cdot (0.01) - 2 \cdot 1 \cdot (-0.02) + 1 \cdot (0.04)$$

$$dw = 4(0.01) - 2(-0.02) + 0.04$$

$$dw = 0.04 + 0.04 + 0.04$$

$$dw = 0.12.$$

12. If $w = x^2 + y^2 + z^2$, calculate both the exact change Δw and the approximate change dw for $x = 4$, $y = -1$, $z = 5$, $\Delta x = 0.03$, $\Delta y = -0.04$, and $\Delta z = -0.01$.

In this case, coincidentally, $dw = \Delta w$.

DO EXERCISE 12.

If $z = f(x, y)$, then we had defined $\Delta z = f(x + \Delta x, y + \Delta y) - f(x, y)$. Rewriting, we get

$$f(x + \Delta x, y + \Delta y) = f(x, y) + \Delta z.$$

Now, if we use the approximation $dz \approx \Delta z$, we obtain

$$f(x + \Delta x, y + \Delta y) \approx f(x, y) + dz.$$

In a similar way, if $w = f(x, y, z)$, then we can show that

$$f(x + \Delta x, y + \Delta y, z + \Delta z) \approx f(x, y, z) + dw.$$

Example 11 If $z = f(x, y) = x^2y^3$, use differentials to calculate an approximate value for $(1.01)^2(0.98)^3$.

Solution We are asked to calculate $f(1.01, 0.98)$. If we set $x = 1$, $y = 1$, $\Delta x = 0.01$, and $\Delta y = -0.02$, then we must calculate $f(x + \Delta x, y + \Delta y)$. From above, we already know that $f(x + \Delta x, y + \Delta y) \approx f(x, y) + dz$. But $f(x, y) = f(1, 1) = 1^2(1)^3 = 1$, while

$$dz = f_x \, dx + f_y \, dy$$

$$= 2xy^3 \, dx + 3x^2y^2 \, dy$$

$$= 2 \cdot 1 \cdot (1)^3 \, dx + 3(1)^2(1)^2 \, dy$$

$$= 2 \, dx + 3 \, dy.$$

Setting $dx = \Delta x = 0.01$ and $dy = \Delta y = -0.02$, we find that

$$dz = 2(0.01) + 3(-0.02)$$

$$= 0.02 - 0.06 = -0.04.$$

So $f(1.01, 0.98) \approx f(1, 1) + dz$ and

$$\approx 1 - 0.04 = 0.96.$$

In this case, we can calculate the actual value of $f(1.01, 0.98)$ with a calculator. We find that

$$f(1.01, 0.98) = (1.01)^2(0.98)^3 \approx 0.9601.$$

In this case, our approximation was rather good.

13. If $z = \ln(xy)$, use differentials to calculate an approximate value for $\ln((1.01)(0.07))$.

14. If $z = \ln(x/y)$, use differentials to calculate an approximate value for $\ln(1.01/0.97)$.

15. The pressure P at the point (x, y, z) in a certain region of space is given (in pounds per square inch) by

$$P(x, y, z) = x^2 + 3xy + z^2.$$

Use differentials to calculate an approximate value for the pressure at $(1.99, 1.52, 3.98)$.

DO EXERCISES 13 AND 14.

Example 12 The temperature T at the point (x, y, z) of a certain region of space is given by $T(x, y, z) = (x^2 + y^2 + z^2)^{3/2}$. Use differentials to calculate the approximate temperature at $(2.014, 2.15, 1.06)$.

Solution We use

$$T(x + \Delta x, y + \Delta y, z + \Delta z) \approx T(x, y, z) + dT.$$

Since $T(2, 2, 1)$ is easy to evaluate, we let $x = 2$, $y = 2$, and $z = 1$; this forces us to choose $\Delta x = 0.014$, $\Delta y = 0.15$, and $\Delta z = 0.06$. Proceeding, we get

$$T(2, 2, 1) = (2^2 + 2^2 + 1^2)^{3/2}$$
$$= (4 + 4 + 1)^{3/2} = 9^{3/2} = 27.$$

Now,

$$dT = T_x\, dx + T_y\, dy + T_z\, dz$$

$$= \frac{3}{2}(x^2 + y^2 + z^2)^{1/2}\, 2x\, dx + \frac{3}{2}(x^2 + y^2 + z^2)^{1/2}\, 2y\, dy$$

$$+ \frac{3}{2}(x^2 + y^2 + z^2)^{1/2}\, 2z\, dz$$

$$= (x^2 + y^2 + z^2)^{1/2}(3x\, dx + 3y\, dy + 3z\, dz).$$

If we let $dx = \Delta x = 0.014$, $dy = \Delta y = 0.15$, and $dz = \Delta z = 0.06$, then we get

$$dT = (2^2 + 2^2 + 1^2)^{1/2}[3 \cdot 2 \cdot (0.014) + 3 \cdot 2 \cdot (0.15) + 3 \cdot 1 \cdot (0.06)]$$

$$= (9)^{1/2}[6(0.014) + 6(0.15) + 3(0.06)]$$

$$= 3[1.164] = 3.492.$$

Thus

$$T(2.014, 2.15, 1.06) \approx T(2, 2, 1) + dT$$

$$= 27 + 3.492$$

$$= 30.492.$$

DO EXERCISE 15.

EXERCISE SET 7.4

Find dz for each of the following.

1. $z = x^2 + 4y^2$

2. $z = 3x - 4y + xy$

3. $z = xe^{4y}$

4. $z = e^{xy}$

5. $z = x^4 + \dfrac{x}{y} + y^3$

6. $z = x \ln y$

Find dw for each of the following.

7. $w = x^2 - y^2 + 3z^2$

8. $w = xy + 2xz + yz$

9. $w = xy^2z^4$

10. $w = \ln (xyz)$

11. $w = x_1^2 + x_2x_3 - x_4^2$

12. $w = x_1x_2x_3x_4$

Evaluate both Δz and dz in each of the following cases.

13. $z = x^2 + y^2$, $x = 5$, $y = 12$; $dx = \Delta x = 0.01$, $dy = \Delta y = 0.02$.

14. $z = 4x + 3y$, $x = 1$, $y = -1$; $dx = \Delta x = 0.03$, $dy = \Delta y = -0.02$.

15. $z = e^{x+y}$, $x = y = 0$; $dx = \Delta x = 0.01$, $dy = \Delta y = 0.0005$.

16. $z = \ln (xy)$, $x = 2$, $y = 0.5$; $dx = \Delta x = 0.02$, $dy = \Delta y = 0.01$.

Evaluate both Δw and dw in each of the following cases.

17. $w = x^2 + y^2 - z^2$, $x = 2$, $y = 2$, $z = 1$, $dx = \Delta x = 0.01$, $dy = \Delta y = 0.02$, $dz = \Delta z = 0.02$

18. $w = 4x - 2y + 3z$, $x = 1$, $y = 1$, $z = 2$, $dx = \Delta x = 0.06$, $dy = \Delta y = 0.13$, $dz = \Delta z = -0.04$

19. $w = xyz$, $x = 1$, $y = 0$, $z = 3$, $dx = \Delta x = 0.04$, $dy = \Delta y = -0.02$, $dz = \Delta z = 0.01$

20. $w = \ln \left(\frac{xz}{y}\right)$, $x = 2$, $y = 6$, $z = 3$, $dx = \Delta x = 0.05$, $dy = \Delta y = 0.08$, $dz = \Delta z = 0.01$

Use differentials to calculate approximate values for each of the following quantities.

21. $(5.01)^2 + (11.94)^2$

22. $(1.06)(0.99)^3 + (2.03)^2$

23. $(1.04)^2 + (0.97)^2 + (1.02)^2 + (1.015)^2$

24. $(1.05)^3 - (0.93)^2$

25. Use differentials to find an approximate value for the area of a right triangle whose base and altitude are 2.013 ft and 4.501 ft, respectively.

26. A rectangular box has dimensions 1.54 ft by 1.01 ft by 7.05 ft. Use differentials to approximate its volume.

27. Use differentials to approximate the surface area of the box in Exercise 26.

28. Calculate dz if $z = x^y$, $x > 0$.

▶ ──

29. If $w = x_1x_2, \ldots, x_n$, show that

$$dw = w\left[\frac{1}{x_1}\, dx_1 + \frac{1}{x_2}\, dx_2 + \cdots + \frac{1}{x_n}\, dx_n\right].$$

30. *Business.* The price P per share of a certain stock is given by

$$P = D\left(1 + \frac{1}{i}\right),$$

where D is the annual dividend per share and i is the annual interest rate. Use differentials to estimate what happens to the price of the stock on the day its dividend is cut from \$5 to \$4 per share if interest rates increase from 10% to 11% on the same day.

You should be able to find the regression line for a given set of data points and use the regression line to make predictions regarding further data.

7.5 APPLICATION: THE LEAST SQUARES TECHNIQUE

The problem of fitting an equation to a set of data occurs frequently. We considered one procedure for doing this in Section 1.6. Such an equation provides a model of the phenomena from which predictions can be made. For example, in business one might want to predict future sales on the basis of past data. In ecology, one might want to predict future demands for natural gas on the basis of past need. Suppose we are trying to determine a linear equation

$$y = mx + b$$

to fit the data. To determine this equation is to determine the values of m and b. But how? Let us consider some factual data.

The graph shown in Fig. 4 appeared in a newspaper advertisement of the *Indianapolis Life Insurance Company*. It pertains to the total life insurance in force in various years. The same data are compiled in the following table.

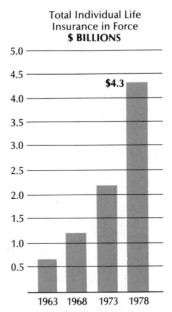

Figure 4

1. a) Use Fig. 5 to predict life insurance in force in 1983.

  b) Use Fig. 6 to predict life insurance in force in 1983.

  c) Compare your answers.

Year, x	1. 1963	2. 1968	3. 1973	4. 1978	5. 1983
Total individual life insurance in force (in billions), y	$0.6	$1.2	$2.2	$4.3	?

Suppose we plot these points and try to draw a line that fits. Note that there are several ways this might be done (see Figs. 5 and 6). Each would give a different estimate of the total insurance in force in 1983.

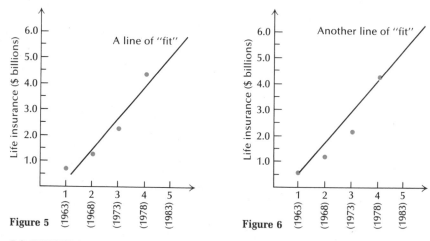

Figure 5   Figure 6

DO EXERCISE 1.

Note that time is incremented in fives of years, making computations easier. Consider the data points (1, 0.6), (2. 1.2), (3, 2.2), and (4, 4.3) as plotted in Fig. 7.

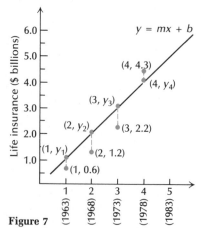

Figure 7

We will try to fit this data with a line

$$y = mx + b$$

by determining the values of m and b. Note the y-errors, or y-deviations, $y_1 - 0.6$, $y_2 - 1.2$, $y_3 - 2.2$, and $y_4 - 4.3$ between the observed points $(1, 0.6)$, $(2, 1.2)$, $(3, 2.2)$, and $(4, 4.3)$ and the points $(1, y_1)$, $(2, y_2)$, $(3, y_3)$, and $(4, y_4)$ on the line. We would like, somehow, to minimize these deviations in order to have a good fit. One way of minimizing the deviations is based on the so-called *least squares assumption*.

THE LEAST SQUARES ASSUMPTION

The line of best fit is the line for which the sum of squares of the y-deviations is a minimum. This is called the *regression line*.

Using the least squares assumption for the life insurance data, we would minimize

$$(y_1 - 0.6)^2 + (y_2 - 1.2)^2 + (y_3 - 2.2)^2 + (y_4 - 4.3)^2, \tag{1}$$

and since the points $(1, y_1)$, $(2, y_2)$, $(3, y_3)$, and $(4, y_4)$ must be solutions of $y = mx + b$, it follows that

$$y_1 = m \cdot 1 + b = m + b,$$

$$y_2 = m \cdot 2 + b = 2m + b,$$

$$y_3 = m \cdot 3 + b = 3m + b,$$

$$y_4 = m \cdot 4 + b = 4m + b.$$

Substituting $m + b$ for y_1, $2m + b$ for y_2, $3m + b$ for y_3, and $4m + b$ for y_4 in Eq. (1), we have

$$(m + b - 0.6)^2 + (2m + b - 1.2)^2 + (3m + b - 2.2)^2$$
$$+ (4m + b - 4.3)^2. \tag{2}$$

Thus, to find the regression line for the given set of data, we must find the values of m and b that minimize the function S given by the sum in Eq. (2).

2. Consider the following factual data on the cost of a 30-second commercial during the broadcast of the Super Bowl.

Year, x	Cost of a 30-Second Commercial During the Super Bowl, y
(1979) 1	$180,000
(1980) 2	$234,000
(1981) 3	$275,000
(1982) 4	$345,000

a) Find the regression line.

b) Use the regression line to predict the cost in 1985; in 1990.

To apply the D-test, we first find the partial derivatives $\partial S/\partial b$ and $\partial S/\partial m$:

$$\frac{\partial S}{\partial b} = 2(m + b - 0.6) + 2(2m + b - 1.2) + 2(3m + b - 2.2)$$
$$+ 2(4m + b - 4.3)$$
$$= 20m + 8b - 16.6;$$

$$\frac{\partial S}{\partial m} = 2(m + b - 0.6) + 2(2m + b - 1.2)2 + 2(3m + b - 2.2)3$$
$$+ 2(4m + b - 4.3)4$$
$$= 60m + 20b - 53.6.$$

We set these derivatives equal to 0 and solve the resulting system:

$$20m + 8b - 16.6 = 0, \qquad 5m + 2b = 4.15,$$
$$60m + 20b - 53.6 = 0; \quad \text{or} \quad 15m + 5b = 13.4.$$

The solution of this system is

$$b = -0.95, \qquad m = 1.21.$$

We leave it to the reader to complete the D-test to verify that $(-0.95, 1.21)$ does, in fact, yield the minimum of S. We need not bother to compute $S(-0.95, 1.21)$.

The values of m and b are all we need to determine $y = mx + b$. The regression line is

$$y = 1.21x - 0.95.$$

We can extrapolate from the data to predict the total life insurance in force in 1983:

$$y = 1.21(5) - 0.95 = 5.1.$$

Thus total life insurance in force in 1983 will be about $5.1 billion.

The method of least squares is a statistical process illustrated here with only four data points to ease the explanation. Most statistical researchers would warn that many more than four data points should be used to get a "good" regression line. Furthermore, making predictions too far in the future from any linear model may be suspect. It can be done, but the further into the future the prediction is made, the more dubious one should be about the prediction.

DO EXERCISE 2.

*The Regression Line for an Arbitrary Collection of Data Points $(c_1, d_1), (c_2, d_2), \ldots, (c_n, d_n)$

Look again at the regression line

$$y = 1.21x - 0.95$$

for the data points $(1, 0.6)$, $(2, 1.2)$, $(3, 2.2)$ and $(4, 4.3)$. Let us consider the arithmetic averages, or means, of the x-coordinates, denoted $\bar{x}$, and the y-coordinates, denoted $\bar{y}$:

$$\bar{x} = \frac{1 + 2 + 3 + 4}{4} = 2.5,$$

$$\bar{y} = \frac{0.6 + 1.2 + 2.2 + 4.3}{4} = 2.075.$$

It turns out that the point $(\bar{x}, \bar{y})$, or $(2.5, 2.075)$, is on the regression line for

$$2.075 = 1.21(2.5) - 0.95.$$

Thus the regression line is

$$y - \bar{y} = m(x - \bar{x}), \quad \text{or} \quad y - 2.075 = m(x - 2.5).$$

All that remains, in general, is to determine m. Suppose we wanted to find the regression line for an arbitrary number of points (c_1, d_1), (c_2, d_2), $\ldots$, (c_n, d_n).

To do so, we find the values m and b that minimize the function S given by

$$S(b, m) = (y_1 - d_1)^2 + (y_2 - d_2)^2 + \cdots + (y_n - d_n)^2 = \sum_{i=1}^{n} (y_i - d_i)^2,$$

where $y_i = mc_i + b$.

Using a procedure like the one we used earlier to minimize S, we can show that $y = mx + b$ takes the form

$$y - \bar{y} = m(x - \bar{x}),$$

where

$$\bar{x} = \frac{\sum_{i=1}^{n} c_i}{n}, \qquad \bar{y} = \frac{\sum_{i=1}^{n} d_i}{n}, \qquad \text{and} \quad m = \frac{\sum_{i=1}^{n} (c_i - \bar{x})(d_i - \bar{y})}{\sum_{i=1}^{n} (c_i - \bar{x})^2}.$$

Let us see how this works out for the life expectancy example done previously.

---

*This part is considered optional and can be omitted without loss of continuity.

3. Repeat Margin Exercise 2(a) using the procedure just outlined in the optional part of this section.

c_i	d_i	$c_i - \bar{x}$	$(c_i - \bar{x})^2$	$(d_i - \bar{y})$	$(c_i - \bar{x})(d_i - \bar{y})$
1	0.6	−1.5	2.25	−1.475	2.2125
2	1.2	−0.5	0.25	−0.875	0.4375
3	2.2	0.5	0.25	0.125	0.0625
4	4.3	1.5	2.25	2.225	3.3375

$$\sum_{i=1}^{4} c_i = 10 \qquad \sum_{i=1}^{4} d_i = 8.3 \qquad \sum_{i=1}^{4} (c_i - \bar{x})^2 = 5 \qquad \sum_{i=1}^{4} (c_i - \bar{x})(d_i - \bar{y}) = 6.05$$

$$\bar{x} = 2.5 \qquad \bar{y} = 2.075 \qquad\qquad\qquad m = \frac{6.05}{5} = 1.21$$

Thus the regression line is

$$y - 2.075 = 1.21(x - 2.5),$$

which simplifies to

$$y = 1.21x - 0.95.$$

DO EXERCISE 3.

*Nonlinear Regression

It can happen that data do not seem to fit a linear equation, but when logarithms of either the x-values or the y-values (or both) are taken, a linear relationship will exist. Indeed, on considering the graph in Fig. 4 it is not unreasonable to expect this data to fit an exponential function.

Example 1 Use logarithms and regression to find an equation

$$y = Be^{kx}$$

that fits the data. Then estimate the total life insurance in force in 1983.

Year, x	1. 1963	2. 1968	3. 1973	4. 1978
Total individual life insurance in force (in billions), y	$0.6	$1.2	$2.2	$4.3

---

*This part is considered optional and can be omitted without loss of continuity.

Solution If we take the natural logarithm of both sides of

$$y = Be^{kx},$$

we get

$$\ln y = \ln B + kx.$$

Note that $\ln B$ and k are constants. So, if we replace $\ln y$ by a new variable Y, the equation takes the form of a linear function

$$Y = mx + b,$$

where $m = k$ and $b = \ln B$.

We are going to find this regression line, but before starting we need to find the logarithms of the y-values.

x	1	2	3	4
$Y = \ln y$	−0.5108	0.1823	0.7885	1.4586

To find the regression line we use the abbreviated procedure described in the preceding part of this section.

c_i	d_i	$c_i - \bar{x}$	$(c_i - \bar{x})^2$	$d_i - \bar{Y}$	$(c_i - \bar{x})(d_i - \bar{Y})$
1	−0.5108	−1.5	2.25	−0.9905	1.4858
2	0.1823	−0.5	0.25	−0.2974	0.1487
3	0.7885	0.5	0.25	0.3088	0.1544
4	1.4586	1.5	2.25	0.9789	1.4684

$$\sum_{i=1}^{4} = 10 \qquad \sum_{i=1}^{4} d_i \qquad \sum_{i=1}^{4} (c_i - \bar{x})^2 \qquad \sum_{i=1}^{4} (c_i - \bar{x})(d_i - \bar{Y})$$

$$\bar{x} = \frac{10}{4} = 2.5 \qquad = 1.9186 \qquad = 5 \qquad = 3.2573$$

$$\bar{Y} = \frac{1.9186}{4} \qquad\qquad m = \frac{3.2573}{5}$$

$$= 0.4797 \qquad\qquad = 0.65146$$

Thus the regression line is

$$Y - 0.4797 = 0.65146(x - 2.5),$$

which simplifies to

$$Y = 0.65146x - 1.14895.$$

4. a) Use natural logarithms and regression to find an equation

$$y = Be^{kx}$$

that fits the data in Margin Exercise 2 regarding the cost of Super Bowl commercials.

b) Estimate the cost of a Super Bowl commerical in 1985; in 1990.

Recall that we were to find k and B. From this equation we know that

$$m = k = 0.65146$$

and

$$b = \ln B = -1.14895.$$

To find B we use the definition of logarithms (or take the antilog) and get

$$B = e^{-1.14895} = 0.3170.$$

Then the desired equation is

$$y = 0.317e^{0.65146x}.$$

Then, using this equation, we find that the total life insurance in force will be

$$y = 0.317e^{0.65146(5)} = \$8.2 \text{ billion}.$$

DO EXERCISE 4.

In conclusion, there are other kinds of nonlinear regression besides logarithmic. For example, a set of data might fit a quadratic equation

$$y = ax^2 + bx + c.$$

In such a case, one can still use regression to find the numbers a, b, and c that minimize the sums of squares of deviations.

EXERCISE SET 7.5

1. The factual data in the table below shows the total sales of Anacomp, Inc., over several years.

Year, x	Revenue of Anacomp, Inc. (in millions), y
1. 1978	$23.4
2. 1979	41.7
3. 1980	71.6
4. 1981	106.4
5. 1982	109.6

a) Find the regression line.

b) Use the regression line to predict sales in 1984; in 1989.

2. Consider the following factual data on natural gas demand.

Year, x	1. 1950	2. 1960	3. 1970
Demand (in quadrillion BTU)	19	21	22

a) Find the regression line $y = mx + b$.

b) Use the regression line to predict gas demand in 1990; in 2000.

3. A professor wanted to predict students' final examination scores on the basis of their midterm test scores. An equation was determined based on data (see below) on scores of three students who took the same course with the same instructor the previous semester.

Midterm Score (%), x	70	60	85
Final Exam Score (%), y	75	62	89

a) Find the regression line $y = mx + b$. (*Hint:* The y-deviations are $70m + b - 75, 60m + b - 62$, and so on.)

b) The midterm score of a student was 81. Use the regression line to predict the student's final exam score.

4. Consider the following total sales data of a company during the first 4 years of operation.

Year, x	1	2	3	4
Sales (in millions), y	$22	$34	$44	$60

a) Find the regression line $y = mx + b$.

b) Use the regression line to predict sales in the 5th year.

---

5. ▦ a) Find the regression line $y = mx + b$ that fits the set of data in the table.

b) Use the regression line to predict the world record in the mile in 1984.

c) In July 1980 Steve Ovett set a new world record of 3:48.8 for the mile. How does this compare with what can be predicted by the regression?
(*Hint:* Convert each time to decimal notation; for example, $4:24.5 = 4\frac{24.5}{60} = 4.4083$.)

Year, x (Use the Actual Year for x.)	World Record in mile, y (min:sec)
1875 (Walter Slade)	4:24.5
1894 (Fred Bacon)	4:18.2
1923 (Paavo Nurmi)	4:10.4
1937 (Sidney Wooderson)	4:06.4
1942 (Gunder Haegg)	4:06.2
1945 (Gunder Haegg)	4:01.4
1954 (Roger Bannister)	3:59.4
1964 (Peter Snell)	3:54.4
1967 (Jim Ryun)	3:51.1
1975 (John Walker)	3:49.4
1979 (Sebastian Coe)	3:49.0

6. ▦ a) Use logarithms and regression to find an equation

$$y = Be^{kx}$$

that fits this set of data.

Year (from 1976), x	0	1	2	3
Population of U.S. (in millions), y	216	218	219	221

b) Use the regression equation to find the population of the United States in 1987.

7. ▦ a) Use logarithms and regression to find an equation

$$y = Be^{kx}$$

that fits the set of data in Exercise 1.

b) Use the regression equation to predict sales in 1984; in 1989.
Compare your answers with those of Exercise 1.

8. ▦ a) Use regression (but not logarithms) to find an equation

$$y = ax^2 + bx + c$$

that fits this set of data.*

Average Number of hours of sleep, x	Death rate in one year (per 100,000 males), y
5	1121
6	805
7	626
8	813
9	967

b) Find the death rate of those who average 4 hours sleep; 10 hours sleep; 7.5 hours sleep.

*The set of data in Exercise 8 comes from a study by Dr. Harold J. Morowitz.

OBJECTIVES

You should be able to

a) **Find a maximum or minimum value of a given function subject to a given constraint, using the method of LaGrange multipliers.**

b) **Solve applied problems involving LaGrange multipliers.**

7.6 CONSTRAINED MAXIMUM AND MINIMUM VALUES: LAGRANGE MULTIPLIERS

Before we get into detail, let us look again at a problem we considered in Chapter 3.

Example 1 A hobby store has 20 ft of fencing to fence off a rectangular electric-train area in one corner of its display room. The two sides up against the wall require no fence. What dimensions of the rectangle will maximize the area?

We maximize the function

$$A = xy$$

subject to the condition or *constraint* $x + y = 20$. Note that A is a function of two variables:

$$A(x, y) = xy.$$

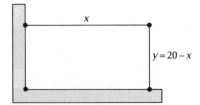

When we solved this earlier, we first solved the constraint for y:

$$y = 20 - x.$$

We then substituted $20 - x$ for y to obtain

$$A(x, 20 - x) = x(20 - x) = 20x - x^2,$$

which is a function of one variable. We then found a maximum value using Maximum–Minimum Principle 1 (see Section 3.3). By itself, the function of two variables

$$A(x, y) = xy$$

has no maximum value. This can be checked using the D-test. But, with the constraint $x + y = 20$, the function does have a maximum. We see this in the following figure.

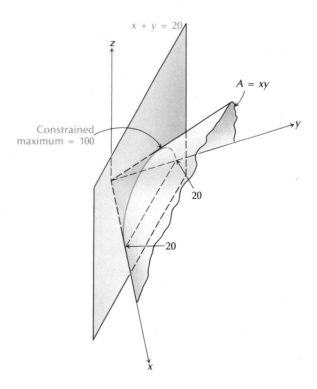

It may be quite difficult to solve a constraint for one variable. The procedure outlined below allows us to proceed without solving a constraint for one variable.

THEOREM 2   Method of LaGrange Multipliers

To find a maximum or minimum value of a function $f(x, y)$ subject to the constraint $g(x, y) = 0$, we do the following.

1. Form a new function:

$$F(x, y, \lambda) = f(x, y) - \lambda g(x, y).$$

2. Find the partial derivatives F_x, F_y, and F_λ.
3. Solve the system

$$F_x = 0, \quad F_y = 0, \quad \text{and} \quad F_\lambda = 0.$$

Let (a, b) represent a solution. We still must determine whether (a, b) yields a maximum or minimum, but we will assume one or the other in the problems considered here.

The variable λ (lambda) is called a *LaGrange multiplier*. We first illustrate the method of LaGrange multipliers by resolving the problem in Example 1.

Example 2   Find the maximum value of

$$A(x, y) = xy$$

subject to the constraint $x + y = 20$.

Solution

1. We form the new function F given by

$$F(x, y, \lambda) = xy - \lambda \cdot (x + y - 20).$$

Note that we first had to express $x + y = 20$ as $x + y - 20 = 0$.

2. We find the first partial derivatives:

$$F_x = y - \lambda,$$
$$F_y = x - \lambda,$$
$$F_\lambda = -(x + y - 20).$$

3. We set these derivatives equal to 0 and solve the resulting system:

$$y - \lambda = 0, \tag{1}$$
$$x - \lambda = 0, \tag{2}$$
$$x + y - 20 = 0. \tag{3}$$

1. A rancher has 50 ft of fencing to fence off a rectangular animal pen in the corner of a barn. What dimensions of the rectangle will yield the maximum area?

a) Express the area as a function of two variables with a constraint.

b) Find the maximum value of the function in (a) above, using the method of LaGrange multipliers.

(If $-(x + y - 20) = 0$, then $x + y - 20 = 0$.) From Eqs. (1) and (2) it follows that

$$x = y = \lambda.$$

Substituting λ for x and y in Eq. (3), we get

$$\lambda + \lambda - 20 = 0$$

$$2\lambda = 20$$

$$\lambda = 10.$$

Thus $x = \lambda = 10$ and $y = \lambda = 10$. The maximum occurs at $(10, 10)$ and is

$$A(10, 10) = 10 \cdot 10 = 100.$$

DO EXERCISE 1.

Example 3 Find the maximum value of

$$f(x, y) = 3xy$$

subject to the constraint

$$2x + y = 8.$$

(*Note:* f could be interpreted as a production function with budget constraint $2x + y = 8$.)

Solution

1. We form the new function F given by

$$F(x, y, \lambda) = 3xy - \lambda(2x + y - 8).$$

Note that we first had to express $2x + y = 8$ as $2x + y - 8 = 0$.

2. We find the first partial derivatives:

$$F_x = 3y - 2\lambda,$$

$$F_y = 3x - \lambda,$$

$$F_\lambda = -(2x + y - 8).$$

3. We set these derivatives equal to 0 and solve the resulting system:

$$3y - 2\lambda = 0, \qquad (1)$$

$$3x - \lambda = 0, \qquad (2)$$

$$-(2x + y - 8) = 0, \quad \text{or} \quad 2x + y - 8 = 0. \qquad (3)$$

2. Find the maximum value of

$$f(x, y) = 5xy$$

subject to the constraint.

$$4x + y = 20.$$

Solving Eq. (1) for y, we get

$$y = \frac{2}{3}\lambda.$$

Solving Eq. (2) for x, we get

$$x = \frac{\lambda}{3}.$$

Substituting $(2/3)\lambda$ for y and $(\lambda/3)$ for x in Eq. (3), we get

$$2\left(\frac{\lambda}{3}\right) + \left(\frac{2}{3}\lambda\right) - 8 = 0$$

$$\frac{4}{3}\lambda = 8$$

$$\lambda = \frac{3}{4} \cdot 8 = 6.$$

Then

$$x = \frac{\lambda}{3} = \frac{6}{3} = 2 \quad \text{and} \quad y = \frac{2}{3}\lambda = \frac{2}{3} \cdot 6 = 4.$$

The maximum of f subject to the constraint occurs at $(2, 4)$ and is

$$f(2, 4) = 3 \cdot 2 \cdot 4 = 24.$$

DO EXERCISE 2.

Example 4 *The beverage can problem.* The standard beverage can has a volume of 12 oz, or 26 in³. What dimensions yield the minimum surface area? Find the minimum surface area.

Solution We want to minimize the function s given by

$$s(h, r) = 2\pi rh + 2\pi r^2$$

subject to the volume constraint

$$\pi r^2 h = 26,$$

or

$$\pi r^2 h - 26 = 0.$$

Note that s does not have a minimum without the constraint.

1. We form the new function S given by

$$S(h, r, \lambda) = 2\pi rh + 2\pi r^2 - \lambda(\pi r^2 h - 26).$$

2. We find the first partial derivatives:

$$\frac{\partial S}{\partial h} = 2\pi r - \lambda\pi r^2,$$

$$\frac{\partial S}{\partial r} = 2\pi h + 4\pi r - 2\lambda\pi rh,$$

$$\frac{\partial S}{\partial \lambda} = -(\pi r^2 h - 26).$$

3. We set these derivatives equal to 0 and solve the resulting system:

$$2\pi r - \lambda\pi r^2 = 0, \qquad (1)$$

$$2\pi h + 4\pi r - 2\lambda\pi rh = 0, \qquad (2)$$

$$-(\pi r^2 h - 26) = 0, \quad \text{or} \quad \pi r^2 h - 26 = 0. \qquad (3)$$

Note that we can solve Eq. (1) for r:

$$\pi r(2 - \lambda r) = 0$$

$$\pi r = 0 \quad \text{or} \quad 2 - \lambda r = 0$$

$$r = 0 \quad \text{or} \qquad r = \frac{2}{\lambda}.$$

Note r = 0 cannot be a solution to the original problem, so we continue

3. Repeat Example 4 for a can of 16 oz, or 35 in³.

by substituting $2/\lambda$ for r in Eq. (2):

$$2\pi h + 4\pi \cdot \frac{2}{\lambda} - 2\lambda\pi \cdot \frac{2}{\lambda} \cdot h = 0$$

$$2\pi h + \frac{8\pi}{\lambda} - 4\pi h = 0$$

$$\frac{8\pi}{\lambda} - 2\pi h = 0$$

$$-2\pi h = -\frac{8\pi}{\lambda},$$

so

$$h = \frac{4}{\lambda}.$$

Since $h = 4/\lambda$ and $r = 2/\lambda$, it follows that $h = 2r$. Substituting $2r$ for h in Eq. (3) yields

$$\pi r^2 (2r) - 26 = 0$$

$$2\pi r^3 - 26 = 0$$

$$2\pi r^3 = 26$$

$$\pi r^3 = 13$$

$$r^3 = \frac{13}{\pi}$$

$$r = \sqrt[3]{\frac{13}{\pi}} \approx 1.6 \text{ in.} \quad \text{▤ or Table 1}$$

So when $r = 1.6$ in., $h = 3.2$ in., the surface area is a minimum and is about $2\pi(1.6)(3.2) + 2\pi(1.6)^2$, or 48.3 in².

DO EXERCISE 3.

The actual dimensions of a standard-sized 12-oz beverage can are $r = 1.25$ in. and $h = 4.875$ in. A natural question after studying Example 4 is, "Why don't beverage companies make cans using the dimensions found in that example?" To do this at this time would mean a monumental cost in retooling. New can-making machines would have to be purchased at a cost of millions. New beverage-filling machines would have to be purchased. Vending machines would no longer be the correct size. A partial response to the desire to save aluminum has been found in recycling and in manufacturing cans with a rippled effect at the top. These cans require less aluminum. As a result of many engineering ideas, the amount of aluminum required to make 1000 cans has been reduced from 36.5 lb to 28.1 lb. The consumer is actually a very important factor in the shape of the can. Market research has shown that a can with the dimensions found in Example 4 is not as comfortable to hold and might not be accepted by consumers.*

---

*Many thanks to Don Hauser of the Pepsi-Cola Co. and Bobby Ryals of the Continental Can Co. for the ideas in this paragraph.

EXERCISE SET 7.6

Find the maximum value of f, subject to the given constraint.

1. $f(x, y) = xy$; $2x + y = 8$

2. $f(x, y) = 2xy$; $4x + y = 16$

3. $f(x, y) = 4 - x^2 - y^2$; $x + 2y = 10$

4. $f(x, y) = 3 - x^2 - y^2$; $x + 6y = 37$

Find the minimum value of f subject to the given constraint.

5. $f(x, y) = x^2 + y^2$; $2x + y = 10$

6. $f(x, y) = x^2 + y^2$; $x + 4y = 17$

7. $f(x, y) = 2y^2 - 6x^2$; $2x + y = 4$

8. $f(x, y) = 2x^2 + y^2 - xy$; $x + y = 8$

9. $f(x, y, z) = x^2 + y^2 + z^2$; $y + 2x - z = 3$

10. $f(x, y, z) = x^2 + y^2 + z^2$; $x + y + z = 1$

Use the method of LaGrange multipliers to solve these problems.

11. Of all numbers whose sum is 70, find the two that have the maximum product.

12. Of all numbers whose sum is 50, find the two that have the maximum product.

13. Of all numbers whose difference is 6, find the two that have the minimum product.

14. Of all numbers whose difference is 4, find the two that have the minimum product.

15. A standard piece of typing paper has a perimeter of 39 in. Find the dimensions of the paper that will give the most typing area, subject to the perimeter constraint of 39 in. What is its area? Does the standard $8\frac{1}{2} \times 11$ paper have maximum area?

16. A carpenter is building a room with a fixed perimeter of 80 ft. What are the dimensions of the largest room that can be built? What is its area?

17. An oil drum of standard size has a volume of 200 gal, or 27 ft³. What dimensions yield the minimum surface area? Find the minimum surface area.

Do these drums appear to be made in such a way as to minimize surface area? (*Marshall Henrichs*)

18. A juice can of standard size has a volume of 99 in³. What dimensions yield the minimum surface area? Find the minimum surface area.

20. The total sales S of a one-product firm is given by

$$S(L, M) = 2ML - L^2,$$

where M = the cost of materials and L = the cost of labor. Find the maximum value of this function subject to the budget constraint

$$M + L = 60.$$

22. A container company is going to construct a shipping container of volume 12 cubic feet with a square bottom and top. The cost of the top and sides is $2 per square foot and $3 per square foot for the bottom. What dimensions will minimize the cost of the container?

19. The total sales S of a one-product firm is given by

$$S(L, M) = ML - L^2,$$

where M = the cost of materials and L = the cost of labor. Find the maximum value of this function subject to the budget constraint.

$$M + L = 80.$$

21. A company is planning to construct a warehouse whose cubic footage is to be 252,000 ft³. Construction costs are estimated to be as follows.

Walls: $3.00 per ft²

Floor: $4.00 per ft²

Ceiling: $3.00 per ft²

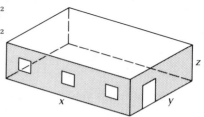

a) The total cost of the building is a function $C(x, y, z)$, where x is the length, y is the width, and z is the height. Find a formula for $C(x, y, z)$.

b) What dimensions of the building will minimize total cost? What is the minimum cost?

23. A product can be made entirely on machine A or machine B, or it can be made on both. The nature of the machines make their cost functions differ:

$$\text{Machine A:} \quad C(x) = 10 + \frac{x^2}{6};$$

$$\text{Machine B:} \quad C(y) = 200 + \frac{y^3}{9}.$$

Total cost is given by $C(x, y) = C(x) + C(y)$. How many units should be made on each machine in order to minimize total costs if $x + y = 10{,}100$ units are required?

Find the indicated maximum or minimum values of f, subject to the given constraint.

24. Minimum: $f(x, y) = xy; \ x^2 + y^2 = 4$

25. Minimum: $f(x, y) = 2x^2 + y^2 + 2xy + 3x + 2y;$
$y^2 = x + 1$

26. Maximum: $f(x, y, z) = x + y + z; \ x^2 + y^2 + z^2 = 1$

27. Maximum: $f(x, y, z) = x^2y^2z^2; \ x^2 + y^2 + z^2 = 1$

28. Maximum: $f(x, y, z) = x + 2y - 2z;$
$x^2 + y^2 + z^2 = 4$

29. Maximum: $f(x, y, z, t) = x + y + z + t;$
$x^2 + y^2 + z^2 + t^2 = 1$

30. Suppose $p(x, y)$ represents the production of a two-product firm. We give no formula for p. The company produces x items of the first product at a cost c_1 of each and y items of the second product at a cost c_2 of each. The budget constraint B is given by

$$B = c_1 x + c_2 y.$$

Find the value of λ in the LaGrange multiplier method in terms of p_x, p_y, c_1 and c_2. The resulting equation is called the *Law of Equimarginal Productivity*.

31. Minimum: $f(x, y, z) = x^2 + y^2 + z^2; x - 2y + 5z = 1$

OBJECTIVE

You should be able to find the dimensions of a building of fixed floor area, with a square base, which will minimize travel time from remotest points in the building.

7.7 APPLICATION: MINIMIZING TRAVEL TIME IN A BUILDING

In multilevel building design, one consideration is travel time between the most remote points in a rectangular building with a square base. Let us suppose each floor has a square grid of hallways, as shown below.

Suppose you are standing at the most remote point P in the top northeast corner of such a building with 12 floors. How long will it take to reach the southwest corner on the first floor? You will be going from point P to point Q in the illustration.

Let us call the time t. We find a formula for t in two steps:

1. You are to go from the twelfth floor to the first floor. This is a move in a *vertical direction*; and

2. You need to go across the first floor. This is a move in a *horizontal direction*.

This is the type of building we are considering. (*Marshall Henrichs*)

The vertical time is h, the height of the top floor from the ground, divided by a, the speed at which you can travel in a vertical direction (elevator speed). So, vertical time is h/a.

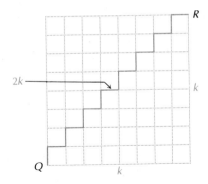

The horizontal time is the time it takes to go across the first level, by way of the square grid of hallways (from R to Q above). If each floor is a square with side of length k, then the distance from R to Q is $2k$. If the walking speed is b, then the horizontal time is $2k/b$. Thus the time it will take to go from P to Q is a function of two variables, h and k, given by

$$t(h, k) = \text{vertical time} + \text{horizontal time} = \frac{h}{a} + \frac{2k}{b}.$$

Now, what happens if we have to choose between two (or more) building plans with the same total floor area, but with different dimensions? Will the travel time be the same? Or will it be be different for the two buildings? First of all, what is the total floor area of a given building? Suppose the building has n floors, each a square of side k. Then the total floor area is given by

$$A = nk^2.$$

Note that the area of the roof is not included.

If h is the height of the top floor from the ground and c is the height of each floor, then $n = h/c$. So

$$A = \frac{h}{c}k^2.$$

Let us return to the problem of the two buildings with the same total floor area, but with different dimensions, and see what happens to $t(h, k)$.

EXPLORATORY EXERCISES

For each case, let the elevator speed $a = 10$ ft/sec, the walking speed $b = 4$ ft/sec, and the height of each floor $c = 15$ ft. Complete the following table.

Case	Building	Number of levels, n	Floor length (in feet), k	Total area (in square feet), A	Height (in feet), $h = cn$	Travel time (in seconds), $t(h, k)$
1	B_1	2	40.0	3200	30	23.00
	B_2	3	32.66	3200	45	20.83
2	B_1	2	60.0			
	B_2	3	48.99			
3	B_1	4	40.0			
	B_2	5	35.777			
4	B_1	5	60.0			
	B_2	10	42.246			
5	B_1	5	150.0			
	B_2	10	106.066			
6	B_1	10	40.0			
	B_2	17	30.679			
7	B_1	10	80.0			
	B_2	17	61.357			
8	B_1	17	40.0			
	B_2	26	32.344			
9	B_1	17	50.0			
	B_2	26	40.43			
10	B_1	26	77.0			
	B_2	50	55.525			

a) What did you notice about $t(h, k)$ when h and k were very nearly the same?

b) What about when h and k were very large?

c) Do you think there are values of h and k for a building with a given floor area that will minimize travel time?

Now let us use calculus to solve a building planning problem.

Example *Minimizing travel time.* The objective is to find the dimensions of a rectangular building with a square base that will minimize travel time t between the most remote points in the building. Each floor has a square grid of hallways. The height of the top floor from the ground is h, and the length of a side of each floor is k. The elevator speed is 10 ft/sec and the average speed of a person walking is 4 ft/sec. The total floor area of the building is 40,000 ft². The height of each floor is 8 ft.

Solution We want to find values of h and k that will minimize the function t given by

$$t(h, k) = \frac{h}{10} + \frac{2k}{4} = \frac{1}{10}h + \frac{1}{2}k,$$

subject to the constraint

$$\frac{h}{8}k^2 = 40,000, \quad \text{or} \quad hk^2 = 320,000.$$

We first form the new function T given by

$$T(h, k, \lambda) = \frac{1}{10}h + \frac{1}{2}k - \lambda(hk^2 - 320,000).$$

We take the first partial derivatives and set them equal to 0:

$$T_h = \frac{1}{10} - \lambda k^2 = 0, \tag{1}$$

$$T_k = \frac{1}{2} - 2\lambda hk = 0, \tag{2}$$

$$T_\lambda = -(hk^2 - 320,000) = 0 \quad \text{or} \quad hk^2 - 320,000 = 0. \tag{3}$$

To clear of fractions, we multiply the first equation by 10 and the second by 2:

$$1 - 10\lambda k^2 = 0, \tag{4}$$

$$1 - 4\lambda hk = 0, \tag{5}$$

$$hk^2 - 320,000 = 0. \tag{6}$$

Solving Eq. (4) for λ, we get

$$\lambda = \frac{1}{10k^2}.$$

1. Solve the example when the total floor area is 100,000 ft^2 and the height of each floor is 10 ft.

Solving Eq. (5) for λ, we get

$$\lambda = \frac{1}{4hk}.$$

Thus

$$\frac{1}{10k^2} = \frac{1}{4hk}. \tag{7}$$

Now k must be nonzero in the original problem. Assuming $k \neq 0$, we multiply Eq. (7) by k to simplify it and get

$$\frac{1}{10k} = \frac{1}{4h}.$$

Solving this equation for k, we get

$$k = \frac{2}{5}h. \tag{8}$$

Substituting $\frac{2}{5}h$ for k in Eq. (6), we get

$$h\left(\frac{2}{5}h\right)^2 - 320{,}000 = 0$$

$$\frac{4}{25}h^3 - 320{,}000 = 0$$

$$\frac{4}{25}h^3 = 320{,}000$$

$$h^3 = 320{,}000 \cdot \frac{25}{4} = 2{,}000{,}000$$

$$h = \sqrt[3]{2{,}000{,}000}$$

$$h \approx 126 \text{ ft.} \quad \blacksquare \text{ or Table 1}$$

Then from Eq. (8) we find k:

$$k = \frac{2}{5} \cdot 126$$

$$\approx 50 \text{ ft.}$$

The height of the building is 126 ft plus 8 ft. The height of the top floor is added on. Thus the dimensions of the building are 50 ft by 50 ft by 134 ft.

DO EXERCISE 1.

EXERCISE SET 7.7

Given the conditions in each of Exercises 1 and 2, find the values of h and k that minimize travel time

$$t(h, k) = \frac{h}{a} + \frac{2k}{b}$$

subject to the floor-area constraint A given by

$$A = \frac{h}{c}k^2,$$

where h = the height of the top floor from the ground (ft), k = the length of the side of the base (ft), a = the elevator speed (ft/sec), b = the average speed of humans walking in the building (ft/sec), and c = the height of each floor (ft). Then find the dimensions of the building.

1. $a = 20, b = 5, c = 8, A = 80,000 \text{ ft}^2$ **2.** $a = 20, b = 5, c = 10, A = 60,000 \text{ ft}^2$

3. Find a general solution in terms of a, b, c, and A.

OBJECTIVE

You should be able to evaluate the double integral of $f(x, y)$ over a rectangle R whose sides are parallel to the axes by evaluating two appropriately chosen iterated integrals.

7.8 DOUBLE INTEGRATION: RECTANGULAR REGIONS

Consider a rectangle R in the plane whose sides are parallel to the x- and y-axes. Such a rectangle is shown below.

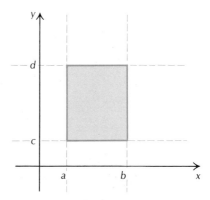

Now, let $z = f(x, y)$ be a continuous function* whose domain contains R. For the moment, assume that $f(x, y)$ is a positive function on R

---

*Intuitively speaking, a function of two variables is continuous if the surface formed by its graph has no "breaks." Formally a function f is continuous at (a, b) if

1. $f(a, b)$ exists;

2. $\lim_{(x, y) \to (a, b)} f(x, y)$ exists; and

3. $\lim_{(x, y) \to (a, b)} f(x, y) = f(a, b)$.

and consider the problem of calculating the volume of the solid figure (see the figure below) whose base is R and whose "top" is the portion of $z = f(x, y)$ that lies directly above R. Since the top is a curved surface, the volume cannot be computed using the standard formula $V = A \cdot h$, where V is the volume, A is the area of the base, and h is the height.

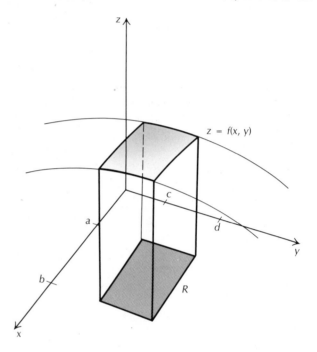

To calculate the volume, we proceed as follows. First, divide the interval $[a, b]$ along the x-axis into n subintervals, all with the same length Δx. This means that $\Delta x = (b - a)/n$. It also means that the endpoints of the subintervals, $x_1 = a$, x_2, x_3, . . . , $x_{n+1} = b$ are given by $x_{k+1} = a + k \cdot \Delta x$ for $k = 0, 1, 2, . . . , n$.

Next, divide the interval $[c, d]$ along the y-axis into m subintervals, all having the same length Δy. Then $\Delta y = (d - c)/m$ and $y_{k+1} = c + k \, \Delta y$ for $k = 0, 1, 2, . . . , m$.

Note in the figure below that if we draw the lines passing through x_2, x_3, . . . , x_n that are parallel to the y-axis and the lines through y_2, y_3, . . . , y_m that are parallel to the x-axis, then these lines will divide R into small rectangles. There are exactly $m \cdot n$ of these smaller rectangles, each having area $\Delta x \cdot \Delta y$. To keep track of the intervals, we label the rectangle that has x_i and y_j for its "smallest" x- and y-endpoints R_{ij}. So, the shaded rectangle whose x-endpoints are x_1 and x_2 and whose y-endpoints are y_1 and y_2 is called R_{11}, since x_1 and y_1 are the smaller x- and y-endpoints. Similarly, the colored rectangle is R_{22}.

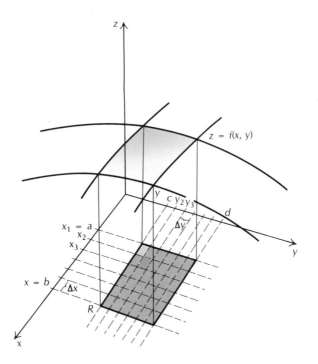

Continuing our calculation, look at the following figure.

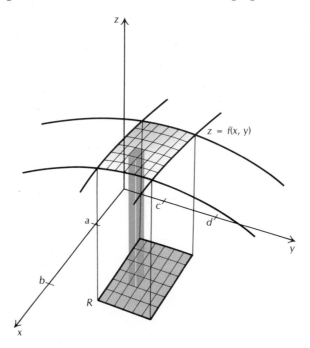

The portion of the volume that is colored has the rectangle R_{43} for its base. The pictured volume looks like a tall, thin box, but it is not a (rectangular) box, since its top is slightly curved. Nevertheless, there is so little variation in its altitude that we can approximate its volume using the volume of the box whose base is R_{43} and whose altitude or height is $f(x_4, y_3)$, the value of $f(x, y)$ at the corner point of R_{43} nearest the origin in Figure 7.35. Thus the volume of the colored portion in Figure 7.35 is approximately $f(x_4, y_3) \cdot \Delta x \cdot \Delta y$ since the area of the base rectangle R_{43} is $\Delta x \cdot \Delta y$. We can obtain a good approximation to the volume V of the original solid by constructing similar rectangular boxes for each base rectangle R_{ij} and then adding the volume of these boxes. The volume of the box whose base is R_{ij} will be approximately equal to $f(x_i, y_j) \cdot \Delta x \cdot \Delta y$ for all $i = 1, \ldots, n$ and all $j = 1, 2, \ldots, m$. Adding these numbers gives

$$V \approx \sum_{i=1}^{n} \sum_{j=1}^{m} f(x_i, y_j) \, \Delta x \, \Delta y.$$

We can increase the accuracy of the approximation by increasing m or n (or both). We expect that if we take the limit of these sums as m and n approach infinity, then we would obtain the exact volume. That is true.

DEFINITION

Let $z = f(x, y)$ be a continuous function on the rectangle R, where the sides of R are parallel to the x- and y-axes. Then the *double integral of f(x, y) over R*, denoted

$$\int_R \int f(x, y) \, dx \, dy,$$

is the limit

$$\lim_{m, n \to \infty} \sum_{i=1}^{n} \sum_{j=1}^{m} f(x_i, y_j) \, \Delta x \, \Delta y.$$

We can show that the limit above always exists. If $f(x, y) > 0$ on R, then we define the volume of the solid whose base is R and whose altitude at each point (x, y) of R is $f(x, y)$ to be the double integral defined above.

Some double integrals may be calculated directly from the definition, but the calculations are tedious. We'd like to have a method of evaluating double integrals that is fairly easy to apply. Recall that

1. Evaluate $\int_1^3 xy\ dx$ by factoring out the y-term and evaluating the remaining integral in the usual way.

although the integral

$$\int_a^b f(x)\ dx,$$

may be defined as a limit of sums (see Section 5.7), it is usually evaluated using antiderivatives. There is a corresponding method for handling double integrals. The next few examples lay the groundwork for this method.

Example 1 Evaluate $\int_1^3 xy\ dx$.

Solution First, we consider the key to the evaluation. The dx in the integral tells us that we are to carry out the integration with respect to x. This means that any other independent variables (like y above) that occur in the integral are to be treated as constants (much as you do when you calculate partial derivatives). Here, if we think of y as being a fixed number, then the antiderivative of xy is $y \cdot (x^2/2)$. Proceeding as usual, we get

$$\int_1^3 xy\ dx = \left[\frac{y \cdot x^2}{2} \right]_1^3 = \frac{y \cdot (3)^2}{2} - \frac{y \cdot (1)^2}{2}$$

$$= \frac{y \cdot 9}{2} - \frac{y \cdot 1}{2} = \frac{8y}{2} = 4y.$$

2. Evaluate $\int_0^1 xy^2\ dx$.

Note that the answer is a function of y alone.

DO EXERCISES 1 AND 2.

Example 2 Evaluate $\int_0^1 (xy + t)\ dy$.

Solution Here, the dy tells us that we are to integrate with respect to y, so that x and t are to be regarded as constants. Just as the antiderivative of $3y + 5$ is $3 \cdot (y^2/2) + 5y$, the antiderivative of $xy + t$ is $x \cdot (y^2/2) + ty$. Using the limits of integration, we get

$$\left[x \cdot \frac{y^2}{2} + ty \right]_0^1 = \left(x \cdot \frac{(1)^2}{2} + t \cdot 1 \right) - \left(x \cdot \frac{0^2}{2} + t \cdot 0 \right)$$

$$= \left(\frac{x}{2} + t \right) - (0 + 0)$$

$$= \frac{x}{2} + t.$$

3. Evaluate $\int_0^1 xy^2\,dy$.

4. Evaluate $\int_{-1}^2 (xy + x^2)\,dy$.

5. Evaluate $\int_0^2 (rx^2 + ys)\,ds$.

6. Evaluate

$$\int_0^1 \left\{ \int_0^1 xy^2\,dx \right\}\,dy.$$

(See Margin Exercise 2.)

7. Evaluate

$$\int_0^5 \left\{ \int_0^1 xy^2\,dy \right\}\,dx.$$

Note again that there is no y in the answer. The expression is a function of x and t only.

DO EXERCISES 3–5.

Example 3 Evaluate $\int_0^2 \{\int_1^3 xy\,dx\}\,dy$.

Solution Note that there are two integral signs here. This is our first example of a *repeated,* or *iterated, integral.* In order to evaluate it, we work from the inside out. We first attack the integral in brackets,

$$\int_1^3 xy\,dx,$$

which we recognize as the integral from Example 1 above. Using the answer from that exercise, we get

$$\int_0^2 \left\{ \int_1^3 xy\,dx \right\}\,dy = \int_0^2 4y\,dy.$$

Since the antiderivative of $4y$ (we are integrating with respect to y here) is $4 \cdot (y^2/2) = 2y^2$, we get

$$\int_0^2 \left\{ \int_1^3 xy\,dx \right\}\,dy = \int_0^2 4y\,dy$$

$$= [2y^2]_0^2$$

$$= 2(2)^2 - 2(0)^2$$

$$= 2 \cdot 4 - 2 \cdot 0 = 8.$$

DO EXERCISES 6 AND 7.

Example 4 Evaluate

$$\int_1^3 \left\{ \int_0^2 xy\,dy \right\}\,dx.$$

Solution As before, we work from the inside out. So, we look first at

$$\int_0^2 xy\,dy.$$

The dy tells us to integrate with respect to y. The antiderivative will then be $x \cdot (y^2/2)$. So,

$$\int_0^2 xy\,dy = \left[x \cdot \frac{y^2}{2} \right]_0^2 = x \cdot \frac{(2)^2}{2} - x \cdot \frac{(0)^2}{2} = 2x.$$

8. Evaluate

$$\int_0^1 \left\{ \int_0^1 xy^2 \, dy \right\} dx$$

and compare your answer with that of Margin Exercise 6.

9. Evaluate

$$\int_0^1 \left\{ \int_0^5 xy^2 \, dx \right\} dy$$

and compare your answer with that of Margin Exercise 7.

Then,

$$\int_1^3 \left\{ \int_0^2 xy \, dy \right\} dx = \int_1^3 2x \, dx$$
$$= [x^2]_1^3$$
$$= (3)^2 - (1)^2$$
$$= 8.$$

That is,

$$\int_1^3 \left\{ \int_0^2 xy \, dy \right\} dx = 8.$$

Note now that the integrals in Examples 3 and 4 are the same except that the order of integration has been reversed.

DO EXERCISES 8 AND 9.

The relationship between the double integral

$$\int_R \int f(x, y) \, dx \, dy$$

and the *iterated* (or *repeated*) *integral*, such as

$$\int_a^b \left\{ \int_c^d f(x, y) \, dy \right\} dx,$$

is given in the following theorem.

THEOREM 3

Let $f(x, y)$ be continuous on the rectangle R (R as in the definition of double integral above). Then

$$\int_R \int f(x, y) \, dx \, dy = \int_c^d \left[\int_a^b f(x, y) \, dx \right] dy$$

$$= \int_a^b \left[\int_c^d f(x, y) \, dy \right] dx.$$

That is, the double integral of $f(x, y)$ over R can be evaluated as an *iterated integral* (in two different ways). The order of integration is *irrelevant*.

Example 5 Evaluate the double integral of $f(x, y) = x^2y$ over the rectangle shown below.

10. a) Evaluate $\int_1^3 x^2y\ dy$.

b) Evaluate the double integral I from Example 5 as

$$\int_2^4 \left\{ \int_1^3 x^2y\ dy \right\} dx.$$

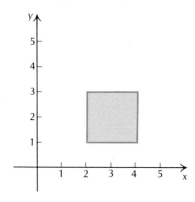

11. Evaluate $\int_R\int 3xy\ dx\ dy$, where R is the rectangular region in Example 5.

Solution We apply the theorem (x^2y is certainly continuous) to obtain

$$I = \int_R \int x^2y\ dx\ dy = \int_1^3 \left\{ \int_2^4 x^2y\ dx \right\} dy.$$

We evaluate the inside integral first, getting

$$I = \int_1^3 \left\{ \left[\frac{x^3}{3} \cdot y \right]_2^4 \right\} dy.$$

As before, we treat y as a constant in finding the antiderivative above, since we are integrating with respect to x. So,

$$I = \int_1^3 \left\{ \frac{4^3}{3}y - \frac{2^3}{3}y \right\} dy$$

$$= \int_1^3 \left\{ \frac{64}{3}y - \frac{8}{3}y \right\} dy$$

$$= \int_1^3 \frac{56}{3}y\ dy = \frac{56}{3} \int_1^3 y\ dy$$

$$= \frac{56}{3} \left[\frac{y^2}{2} \right]_1^3 = \frac{56}{3} \left[\frac{3^2}{2} - \frac{1^2}{2} \right]$$

$$= \frac{56}{3} \left[\frac{9}{2} - \frac{1}{2} \right] = \frac{56}{3} \cdot 4$$

$$I = \frac{224}{3}.$$

DO EXERCISES 10 AND 11.

Example 6 Evaluate

$$I = \int_R \int \frac{x}{y} \, dx \, dy$$

(over the rectangle $R = \{(x, y) \mid 0 \leq x \leq 1, 1 \leq y \leq 2\}$ shown below) in *two* ways. This notation means: R is the set of all points (x, y) such that $0 \leq x \leq 1$ and $1 \leq y \leq 2$.

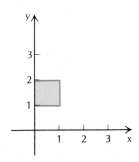

Solution If we integrate first with respect to x and then with respect to y, we get

$$I = \int_1^2 \left\{ \int_0^1 \frac{x}{y} \, dx \right\} dy$$

$$= \int_1^2 \frac{1}{y} \left\{ \int_0^1 x \, dx \right\} dy$$

$$= \int_1^2 \frac{1}{y} \cdot \left\{ \left[\frac{x^2}{2} \right] \right\}_0^1 \right\} dy$$

$$= \int_1^2 \frac{1}{y} \left\{ \frac{1^2}{2} - \frac{0^2}{2} \right\} dy$$

$$= \int_1^2 \frac{1}{y} \cdot \frac{1}{2} \, dy = \left[\frac{1}{2} \ln |y| \right]_1^2$$

$$= \frac{1}{2} \ln |2| - \frac{1}{2} \ln |1|$$

$$= \frac{1}{2} \ln 2 - 0$$

$$= \frac{1}{2} \ln 2.$$

12. Evaluate $I = \int_R \int ye^x \, dx \, dy$ in two ways, where R is the rectangle in Example 6.

If we integrate first with respect to y and then with respect to x, we get

$$I = \int_0^1 \left\{ \int_1^2 \frac{x}{y} \, dy \right\} dx$$

$$= \int_0^1 x \left\{ \int_1^2 \frac{1}{y} \, dy \right\} dx$$

$$= \int_0^1 x \{ [\ln |y|]_1^2 \} \, dx$$

$$= \int_0^1 x \{ \ln 2 \} \, dx$$

$$= \ln 2 \cdot \left[\frac{x^2}{2} \right]_0^1 = \frac{1}{2} \ln 2.$$

DO EXERCISES 12 AND 13.

Example 7 Find the volume of the solid (shown below) whose base is the rectangle $R = \{(x, y) \,|\, 0 \leq x \leq 1, 0 \leq y \leq 1\}$ and whose height at (x, y) is given by $f(x, y) = x + y$.

13. Evaluate $I = \int_R \int xy^3 \, dx \, dy$ in two ways, where R is the rectangle in Example 6.

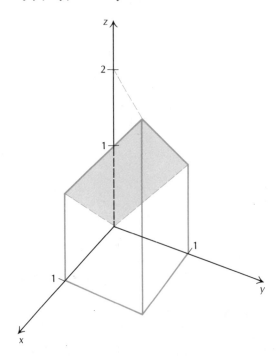

14. Find the volume of the solid whose base is the rectangle R in Example 7 and whose height at (x, y) is $f(x, y) = x^2 + y$.

Solution According to the definition, the volume V is given by the double integral

$$V = \int_R \int (x + y)\, dx\, dy.$$

Writing the double integral as two iterated integrals, we get

$$V = \int_0^1 \left\{ \int_0^1 (x + y)\, dx \right\} dy$$

$$= \int_0^1 \left\{ \left[\frac{x^2}{2} + xy \right]_0^1 \right\} dy$$

$$= \int_0^1 \left\{ \left(\frac{1^2}{2} + 1 \cdot y \right) - \left(\frac{(0)^2}{2} + 0 \cdot y \right) \right\} dy$$

$$= \int_0^1 \left(\frac{1}{2} + y \right) dy$$

$$= \left[\frac{1}{2}\, y + \frac{y^2}{2} \right]_0^1 = \left(\frac{1}{2} \cdot 1 + \frac{1^2}{2} \right) - \left(\frac{1}{2} \cdot 0 - \frac{(0)^2}{2} \right)$$

$$= \left(\frac{1}{2} + \frac{1}{2} \right) - (0 - 0) = 1.$$

Geometrically, we can see that this is correct by noting that our solid comprises one-half of the rectangular box whose base is R and whose altitude is 2. This box has volume $1 \cdot 1 \cdot 2 = 2$. Thus $1 = \frac{1}{2} \cdot 2$ is correct.

DO EXERCISE 14.

EXERCISE SET 7.8

Evaluate each of the following integrals.

1. $\displaystyle\int_0^1 x^2 y\, dx$

2. $\displaystyle\int_{-1}^1 (x^2 + xy + 1)\, dy$

3. $\displaystyle\int_{-1}^1 (x^2 + xy + 1)\, dx$

4. $\displaystyle\int_0^3 (ax^2 + bx + y)\, dx$

Evaluate each of the following iterated integrals.

5. $\displaystyle\int_0^1 \left\{ \int_0^2 x^2 y^3\, dy \right\} dx$

6. $\displaystyle\int_{-1}^0 \left\{ \int_0^1 (3y^2 - 1 + 2x)\, dx \right\} dy$

7. $\displaystyle\int_0^1 \left\{ \int_0^{\ln 2} 2xe^y\, dy \right\} dx$

8. $\displaystyle\int_0^1 \left\{ \int_0^1 (2axy + x^2)\, dy \right\} dx$

Evaluate the double integral $\int_R \int f(x, y)\, dx\, dy$ over the indicated rectangle by reducing to an appropriate iterated integral.

9. $f(x, y) = x + y$, $R = \{(x, y) \mid 0 \leq x \leq 2, 1 \leq y \leq 3\}$

10. $f(x, y) = x$, $R = \{(x, y) \mid -1 \leq x \leq 1, 0 \leq y \leq 1\}$

11. $f(x, y) = \dfrac{x^2}{y}$, $R = \{(x, y) \mid 0 \leq x \leq 1, 2 \leq y \leq 4\}$

12. $f(x, y) = 1$, $R = \{(x, y) \mid a \leq x \leq b, c \leq y \leq d\}$

13. Find the volume of the solid whose base is the rectangle

$$R = \{(x, y) \mid 0 \leq x \leq 1, 0 \leq y \leq 1\}$$

and whose height at (x, y) is $f(x, y) = 3x^2 + 3y^2$.

14. Find the volume of the solid whose base is the rectangle

$$R = \{(x, y) \mid 0 \leq x \leq 1, 0 \leq y \leq 1\}$$

and whose height at (x, y) is $f(x, y) = e^x$.

15. Find the volume of the solid whose base is the rectangle

$$\{(x, y) \mid 1 \leq x \leq 2, 0 \leq y \leq 4\}$$

and whose height at (x, y) is $f(x, y) = \dfrac{y}{x}$.

DEFINITION

If $f(x, y)$ is a continuous function on the rectangle R, then f_{av}, **the average value of $f(x, y)$ on R,** is

$$f_{av} = \frac{1}{\text{area } R} \int_R \int f(x, y)\, dx\, dy.$$

Calculate f_{av} for each of the following choices of $f(x, y)$ and R.

16. $f(x, y) = x^2 + y^2$, $R = \{(x, y) : 0 \leq x \leq 1, 0 \leq y \leq 1\}$

17. $f(x, y) = xy$, $R = \{(x, y) : 0 \leq x \leq 2, 0 \leq y \leq 2\}$

18. $f(x, y) = xy$, $R = \{(x, y) : -1 \leq x \leq 1, 0 \leq y \leq 1\}$

▶

19. Let $R = \{(x, y) \mid a \leq x \leq b, c \leq y \leq d\}$ and let $g(x)$ be **continuous on** $[a, b]$ and $h(y)$ be continuous on $[c, d]$. Let $f(x, y) = g(x)h(y)$ on R. Show that

$$f_{av} = g_{av} h_{av},$$

where f_{av} is the average value of $f(x, y)$ on R and g_{av} and h_{av} are the average values of $g(x)$ on $[a, b]$ and $h(y)$ on $[c, d]$, respectively.

OBJECTIVE

You should be able to evaluate the double integral $f(x, y)$ over a suitable region G by evaluating two appropriately chosen iterated integrals.

7.9 DOUBLE INTEGRATION: MORE GENERAL REGIONS

In Section 7.8, we saw how to define and evaluate the double integral of $f(x, y)$ over R, where R is a rectangle in the plane whose sides are parallel to the x- and y-axes. It is useful, however, to be able to define and evaluate the double integral of $f(x, y)$ over more general regions. To avoid technical problems, we will restrict ourselves for the time being to integration over regions G that look like those shown below. In these cases, the region G lies between the graphs of $y = g(x)$, $y = h(x)$, $x = a$, and $x = b$, where $g(x)$ and $h(x)$ are smooth functions with $g(x) < h(x)$ for $a < x < b$.

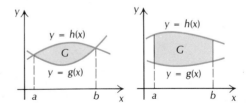

For such a region G, we can cover G with rectangles as shown below, form the sum $\sum_{i=1}^{n} \sum_{j=1}^{m} f(x_i, y_j) \, \Delta x \, \Delta y$ as before, and take the limit as m and n both tend to infinity. If $f(x, y)$ is continuous on G and if $g(x)$ and $h(x)$ have a continuous derivative on $[a, b]$, then this limit always exists. We denote it by

$$\int_G \int f(x, y) \, dx \, dy.$$

We will soon see that we can evaluate the double integral by an appropriate iterated integral.

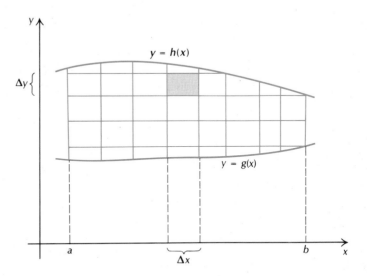

Example 1 Evaluate

$$\int_{x^2}^{x} xy \, dy.$$

1. Evaluate

$$\int_x^1 3y^2 \, dy.$$

2. Evaluate

$$\int_{-x}^x (2y + 4) \, dy.$$

3. Evaluate the iterated integral

$$\int_0^1 \left\{ \int_x^1 3y^2 \, dy \right\} dx.$$

4. Evaluate the iterated integral

$$\int_0^1 \left\{ \int_{-x}^x (2y + 4) \, dy \right\} dx.$$

Solution We proceed just as we did in Section 7.8. The dy indicates that we are to integrate with respect to y. The antiderivative of $f(x, y) = xy$ (with x regarded as a constant) is $x \cdot (y^2/2)$. So,

$$\int_{x^2}^x xy \, dy = \left[x \cdot \frac{y^2}{2} \right]_{x^2}^x$$

$$= x \cdot \frac{(x)^2}{2} - \frac{x \cdot (x^2)^2}{2}$$

$$= \frac{x^3}{2} - \frac{x^5}{2}$$

$$= \frac{x^3 - x^5}{2}.$$

Note that we use the upper and lower limits of integration (to evaluate the integral) in the usual way.

DO EXERCISES 1 AND 2.

Example 2 Evaluate the iterated integral

$$\int_0^1 \left\{ \int_{x^2}^x xy \, dy \right\} dx.$$

Solution We proceed as usual, evaluating the inside integral first. Using the result of Example 1, we get

$$\int_0^1 \left\{ \int_{x^2}^x xy \, dy \right\} dx = \int_0^1 \frac{x^3 - x^5}{2} \, dx$$

$$= \int_0^1 \left(\frac{1}{2} x^3 - \frac{1}{2} x^5 \right) dx$$

$$= \left[\frac{1}{8} x^4 - \frac{1}{12} x^6 \right]_0^1$$

$$= \left[\frac{1}{8} (1)^4 - \frac{1}{12} (1)^6 \right] - \left[\frac{1}{8} (0)^4 - \frac{1}{12} (0)^6 \right]$$

$$= \left(\frac{1}{8} - \frac{1}{12} \right) - (0) = \frac{3}{24} - \frac{2}{24}$$

$$= \frac{1}{24}.$$

DO EXERCISES 3 AND 4.

To see why we can evaluate the double integral by iterated integrals, we consider the volume of the solid based on G and capped above by the piece of the surface $z = f(x, y)$ lying over G, where f is a positive (continuous) function of two variables.

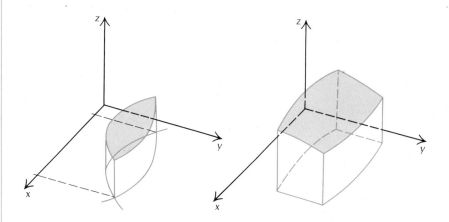

The cross section of this solid in the plane $x = x_0$ is the plane region under the graph of $z = f(x_0, y)$ from $y_1 = g(x_0)$ to $y_2 = h(x_0)$. Its area is

$$A(x_0) = \int_{y_1}^{y_2} f(x_0, y)\, dy$$

$$= \int_{g(x_0)}^{h(x_0)} f(x_0, y)\, dy.$$

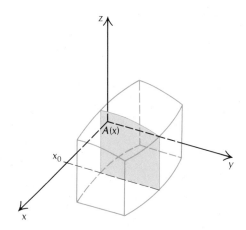

As we vary the slicing plane, the cross section also changes. It follows that the volume in question is given by

$$V = \int_a^b A(x) \, dx$$

$$V = \int_a^b \left\{ \int_{g(x)}^{h(x)} f(x, y) \, dy \right\} dx.$$

We give the relationship between the double integral $\int_G \int f(x, y) \, dx \, dy$ and an iterated integral in the following theorem.

THEOREM 4

Let $g(x)$ and $h(x)$ have continuous derivatives and satisfy $g(x) < h(x)$ for $a \le x \le b$. Let G be either of the regions shown at the top of page 459, so that

$$G = \{(x, y) \mid a \le x \le b \quad \text{and} \quad g(x) \le y \le h(x)\}.$$

Let $f(x, y)$ be continuous at every point of G. Then the double integral

$$\int_G \int f(x, y) \, dx \, dy$$

exists and is equal to the iterated integral

$$\int_a^b \left\{ \int_{g(x)}^{h(x)} f(x, y) \, dy \right\} dx.$$

Note that the iterated integral is taken from the lower boundary of G, $y = g(x)$, to the upper boundary of G, $y = h(x)$. In particular, the first (inside) integration takes place with respect to y. The second (outside) integration is taken from the leftmost value of x in G, $x = a$, to the rightmost value of x in G, $x = b$. The second integration, then, is taken with respect to the variable x.

Example 3 Evaluate $\int_G \int f(x, y) \, dx \, dy$, where $G = \{(x, y) \mid 0 \le x \le 1$ and $x^2 \le y \le x\}$ and $f(x, y) = xy$.

Solution We apply Theorem 4 to reduce the double integral to an iterated integral. The region G is shown below. As in the last section, the notation in the statement of the example is read: G is the set of all points (x, y) such that $0 \le x \le 1$ and $x^2 \le y \le x$. The first, or inside, integral will extend from the lower boundary $y = x^2$ to the upper boundary $y = x$. So it will be

$$\int_{x^2}^x f(x, y) \, dy = \int_{x^2}^x xy \, dy.$$

5. a) Sketch the region
   $G = \{(x, y) \mid 0 \le x \le 1$ and
   $x \le y \le 1\}$.

   b) Then evaluate

   $$\int_G\!\!\int 3y^2 \, dx \, dy.$$

6. a) Sketch the region
   $G = \{(x, y) \mid 0 \le x \le 1$ and
   $-x \le y \le x\}$.

   b) Evaluate

   $$\int_G\!\!\int (2y + 4) \, dx \, dy.$$

7. a) Sketch the region
   $G = \{(x, y) \mid 0 \le x \le 2$ and
   $-x \le y \le 1\}$.

   b) Evaluate

   $$\int_G\!\!\int 6x \, dx \, dy.$$

The second, or outside, integral will be taken from $x = 0$ to $x = 1$. So

$$\int_G\!\!\int xy \, dx \, dy = \int_0^1 \left\{ \int_{x^2}^x xy \, dy \right\} dx.$$

For the record, note that the order of integration is important in this case.

To complete the problem, note that the iterated integral above is the integral from Example 2. Using the answer from that example, we get

$$\int_G\!\!\int xy \, dx \, dy = \frac{1}{24}.$$

DO EXERCISES 5–7.

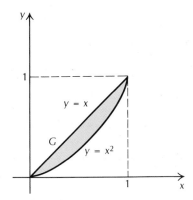

Example 4 Find the volume of the solid whose base is the triangular region $G = \{(x, y) \mid 0 \le x \le 1, \ 0 \le y \le x\}$ and whose height at each point of G is $z = f(x, y) = x - y$.

Solution The solid is shown in the figure below. As in Section 7.8, we define the volume V of the solid to be the double integral of $z = f(x, y) = x - y$ over G. So

$$V = \int_G\!\!\int f(x, y) \, dx \, dy.$$

We can evaluate the double integral by reducing it to an iterated integral. The first, or inside, integral will extend from the x-axis ($y = 0$) to the upper boundary $y = x$. The first integral will be

$$\int_0^x (x - y) \, dy.$$

8. Find the volume of the solid whose base is the triangular region
$G = \{(x, y) \mid 0 \leq x \leq 1, x \leq y \leq 1\}$
and whose height at (x, y) is
$z = f(x, y) = 3y^2$.

9. Find the volume of the solid whose base is the region
$G = \{(x, y) \mid 0 \leq x \leq 1, 0 \leq y \leq e^x$
and whose height is given by
$z = 2y$.

The outer integral is taken from $x = 0$ to $x = 1$. So

$$V = \int_0^1 \left\{ \int_0^x (x - y) \, dy \right\} dx.$$

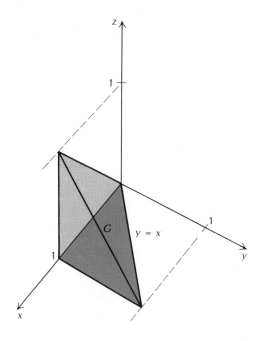

Evaluating, we get

$$V = \int_0^1 \left\{ \left[xy - \frac{y^2}{2} \right]_0^x \right\} dx$$

$$= \int_0^1 \left\{ \left(x \cdot x - \frac{(x)^2}{2} \right) - \left(x \cdot 0 - \frac{(0)^2}{2} \right) \right\} dx$$

$$= \int_0^1 \left\{ \left(x^2 - \frac{x^2}{2} \right) - 0 \right\} dx = \int_0^1 \frac{1}{2} x^2 \, dx$$

$$= \left[\frac{1}{2} \cdot \frac{x^3}{3} \right]_0^1 = \left(\frac{1}{2} \cdot \frac{1^3}{3} \right) - \left(\frac{1}{2} \cdot \frac{0^3}{3} \right)$$

$$= \frac{1}{2} \cdot \frac{1}{3} - \frac{1}{2} \cdot 0 = \frac{1}{6}.$$

DO EXERCISES 8 AND 9.

10. Calculate the average value of
$f(x, y) = 3y^2$ over
$G = \{(x, y) \mid 0 \le x \le 1, x \le y \le 1\}$.

11. Calculate the average value of
$f(x, y) = x$ over
$G = \{(x, y) \mid 0 \le x \le 1, x^2 \le y \le 1\}$.

DEFINITION

Let $f(x, y)$ be continuous on G, where G satisfies the condition of the previous theorem. Then *the average value of f(x, y) on G*, denoted f_{av}, is defined by

$$f_{av} = \frac{1}{\text{area }(G)} \int_G\!\!\int f(x, y) \, dx \, dy.$$

Example 5 Calculate the average value of $f(x, y) = x - y$ over the region $G = \{(x, y) \mid 0 \le x \le 1, 0 \le y \le x\}$.

Solution The region G is triangular. The triangle has base and height 1 so that its area is $\frac{1}{2}$. So

$$f_{av} = \frac{1}{\text{area }(G)} \int_G\!\!\int f(x, y) \, dx \, dy$$

$$= \frac{1}{\frac{1}{2}} \int_G\!\!\int (x - y) \, dx \, dy.$$

The double integral that appears here was evaluated in Example 4. Using that result, we have

$$f_{av} = 2 \cdot \frac{1}{6}$$

$$= \frac{1}{3}.$$

Note that in most cases, we need to use calculus to compute the area of G.

DO EXERCISES 10 AND 11.

Example 6 Evaluate $\int_G\!\!\int x \, dx \, dy$, where G is the portion of the plane above $y = x^2$ and below $y = x + 2$.

Solution This problem differs from previous ones only in that the x-limits of integration are not given. The region G is shown below.

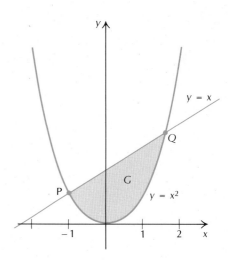

We must find the points P and Q of intersection. Setting $x^2 = x + 2$ and solving gives

$$x^2 - x - 2 = 0$$

or

$$(x - 2)(x + 1) = 0.$$

We see that P is $(-1, 1)$ and Q is $(2, 4)$. So we obtain

$$\iint_G x\,dx\,dy = \int_{-1}^{2}\left\{\int_{x^2}^{x+2} x\,dy\right\}dx$$

$$= \int_{-1}^{2}\left\{\left[xy\right]_{x^2}^{x+2}\right\}dx$$

$$= \int_{-1}^{2}(x^2 + 2x - x^3)\,dx$$

$$= \left[\frac{x^3}{3} + x^2 - \frac{x^4}{4}\right]_{-1}^{2}$$

$$= \left(\frac{8}{3} + 4 - \frac{16}{4}\right) - \left(\frac{-1}{3} + 1 - \frac{1}{4}\right)$$

$$= \frac{9}{3} + 3 - \frac{15}{4}$$

$$= \frac{9}{4}.$$

12. Calculate $\int_G\int 2y\ dx\ dy$ where G is the portion of the plane above the parabola $y = x^2 + x$ and below the parabola $y = 1 - x^2$.

DO EXERCISE 12.

In the iterated integrals that have appeared so far in this section, the first, or inside, integral has always been taken with respect to y. If the region G looks like the ones below, however, it is more convenient to integrate first with respect to x.

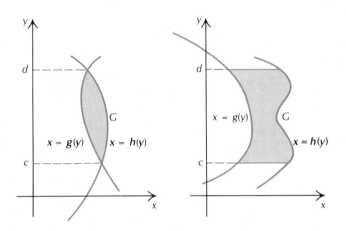

In the case shown, we integrate from the left-hand boundary $x = g(y)$ to the right-hand boundary $x = h(y)$. So

$$\int_{g(y)}^{h(y)} f(x, y)\ dx$$

is the inside, or first, integral. Then we integrate with respect to y, integrating from the smallest y-value in G, $y = c$, to the largest, $y = d$. The result is the equality

$$\int_G\int f(x, y)\ dx\ dy = \int_c^d\left\{\int_{g(y)}^{h(y)} f(x, y)\ dx\right\}dy.$$

Example 7 Evaluate $\int_G\int 6xy\ dx\ dy$, where G is the region between $x = y^2 - 2$ and $y = x$.

Solution Solving as in Example 6, we find that the region G looks like the following figure.

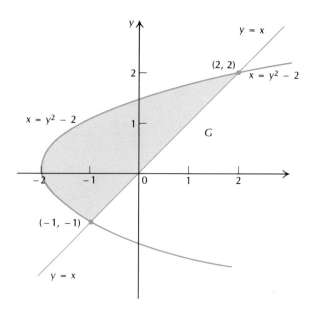

In this case, we integrate from the left-hand boundary, $x = y^2 - 2$, to the right-hand boundary, $x = y$, getting

$$\int_{y^2-2}^{y} 6xy \ dx.$$

Then we integrate from the smallest y-value in G, -1, to the largest, 2, obtaining

$$\int_{G}\int 6xy \ dx \ dy = \int_{-1}^{2}\left\{\int_{y^2-2}^{y} 6xy \ dx\right\} dy.$$

Evaluating, we get

$$\int_{-1}^{2}\{[3x^2y]_{y^2-2}^{y}\} \ dy = \int_{-1}^{2}\{3(y)^2y - 3(y^2 - 2)^2y\} \ dy$$

$$= \int_{-1}^{2}\{3y^3 - 3(y^4 - 4y^2 + 4)y\} \ dy$$

$$= \int_{-1}^{2}\{3y^3 - 3y^5 + 12y^3 - 12y\} \ dy$$

$$= \int_{-1}^{2}\{-3y^5 + 15y^3 - 12y\} \ dy$$

13. Evaluate $\int_G\int 2y\ dx\ dy$, where G is the region in the first quadrant to the left of $x = 2y - y^2$.

$$= \left[-\frac{1}{2}y^6 + \frac{15}{4}y^4 - 6y^2 \right]_{-1}^{2}$$

$$= \left\{ -\frac{1}{2}(2)^6 + \frac{15}{4}(2)^4 - 6(2)^2 \right\}$$

$$- \left\{ -\frac{1}{2}(-1)^6 + \frac{15}{4}(-1)^4 - 6(-1)^2 \right\}$$

$$= \{-32 + 60 - 24\} - \left\{ -\frac{1}{2} + \frac{15}{4} - 6 \right\}$$

$$= 4 + \frac{1}{2} - \frac{15}{4} + 6 = 10 + \frac{2}{4} - \frac{15}{4} = \frac{27}{4}.$$

DO EXERCISE 13.

In Example 7 and Margin Exercise 13, it is much more convenient to integrate first with respect to x. In some cases, we can apply both methods. In these cases, we obtain the same answer, as the following example shows.

Example 8 Rework Example 3, taking the inside integral with respect to x.

Solution We integrate from the (see Figure 7.43) left-hand boundary, $x = y$, to the right-hand boundary, $x = \sqrt{y}$, getting

$$\int_{y}^{\sqrt{y}} xy\ dx.$$

14. Rework Example 4 integrating first with respect to x.

Integrating next from $y = 0$ to $y = 1$, we find that

$$\int_G\int xy\ dx\ dy = \int_0^1 \left\{ \int_y^{\sqrt{y}} xy\ dx \right\} dy = \int_0^1 \left\{ \left[\frac{1}{2}x^2 y \right]_y^{\sqrt{y}} \right\} dy$$

$$= \int_0^1 \left\{ \frac{1}{2}(\sqrt{y})^2 y - \frac{1}{2}(y)^2 y \right\} dy$$

15. Rework Example 6 by first breaking up the given region into two smaller regions, integrating first with respect to x in both cases.

$$= \int_0^1 \left\{ \frac{1}{2}y^2 - \frac{1}{2}y^3 \right\} dy$$

$$= \left[\frac{1}{6}y^3 - \frac{1}{8}y^4 \right]_0^1 = \frac{1}{6} - \frac{1}{8} = \frac{1}{24}.$$

DO EXERCISES 14 AND 15.

Application to Probability

Suppose we throw a dart at a region G in a plane. It lands on a point (x, y). We can think of (x, y) as a continuous random variable that assumes all values in some region G. A function f is said to be a *joint probability density* if

$$f(x, y) \geq 0 \qquad \text{for all } (x, y) \text{ in } G$$

and

$$\int_G \int f(x, y) \, dx \, dy = 1,$$

the double integral of $f(x, y)$ over G. Now if G_0 is a subregion of G and is the sort of region we discussed earlier, then the probability that the dart "lands in G_0" is given by

$$\int_{G_0} \int f(x, y) \, dx \, dy.$$

For appropriate regions, we can evaluate this double integral as an iterated integral.

EXERCISE SET 7.9

Evaluate each of the following.

1. $\int_0^1 \int_0^1 2y \, dx \, dy$

2. $\int_0^1 \int_0^1 2x \, dx \, dy$

3. $\int_{-1}^1 \int_x^1 xy \, dy \, dx$

4. $\int_{-1}^1 \int_x^2 (x + y) \, dy \, dx$

5. $\int_0^1 \int_{-1}^3 (x + y) \, dy \, dx$

6. $\int_0^1 \int_{-1}^1 (x + y) \, dy \, dx$

7. $\int_0^1 \int_{x^2}^x (x + y) \, dy \, dx$

8. $\int_0^1 \int_{-1}^x (x^2 + y^2) \, dy \, dx$

9. $\int_0^2 \int_0^x (x + y^2) \, dy \, dx$

10. $\int_1^3 \int_0^x 2e^{x^2} \, dy \, dx$

Evaluate $\int_G \int f(x, y) \, dx \, dy$ for the given function $f(x, y)$ and region G. Sketch G.

11. $f(x, y) = 1$, $G = \{(x, y) \mid 0 \leq x \leq 2, 0 \leq y \leq x\}$.

12. $f(x, y) = x$, $G = \{(x, y) \mid 0 \leq y \leq 1 - x^2\}$.

13. $f(x, y) = 1 + 2x$, $G = \{(x, y) \mid \frac{1}{2}y \leq x \leq y, y \leq 1\}$.

14. $f(x, y) = 4x$, $G = \{(x, y) \mid y \geq 0, x^3 \leq y \leq x\}$.

15. Find the volume of the solid capped by the surface $z = 1 - y - x^2$ over the region bounded above and below by $y = 0$ and $y = 1 - x^2$, and left and right by $x = 0$ and $x = 1$, by evaluating the integral

$$\int_0^1 \int_0^{1-x^2} (1 - y - x^2) \, dy \, dx.$$

16. Find the volume of the solid capped by the surface $z = x + y$ over the region bounded above and below by $y = 0$ and $y = 1 - x$, and left and right by $x = 0$ and $x = 1$, by evaluating the integral

$$\int_0^1 \int_0^{1-x} (x + y) \, dy \, dx.$$

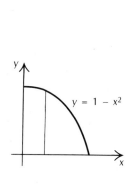

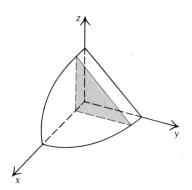

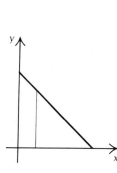

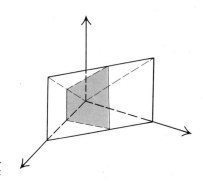

17. Find the volume of the solid whose base is the region in the xy-plane that is bounded by the parabola $y = 4 - x^2$ and the line $y = 3x$, while the top of the solid is capped by $z = x + 6$.

18. Find the volume of the solid whose base is the region in the xy-plane above the x-axis and below $y = 2 - x^2$ and which is capped by $z = 3y^2$.

Find the average value f_{av} of the given function over the region R.

19. $f(x, y) = x^2 + y^2$, $G = \{(x, y) \mid 0 \le x \le 2, 0 \le y \le 2x\}$ **20.** $f(x, y) = e^{x+y}$, $G = \{(x, y) \mid 0 \le y \le x, 0 \le x \le 1\}$

In each of the following, write an equivalent iterated integral with the order of integration reversed. As a check, evaluate *both* iterated integrals.

21. $\int_0^2 \int_1^{e^x} dy \, dx$

22. $\int_0^2 \int_0^y dx \, dy$

In Exercises 23–26, suppose that a continuous random variable has a joint probability function given by

$$f(x, y) = x^2 + \frac{1}{3} xy, \quad 0 \le x \le 1, \quad 0 \le y \le 2.$$

23. Find $\int_0^2 \int_0^1 f(x, y) \, dx \, dy$.

24. Find the probability that a point (x, y) is in the region bounded by $0 \le x \le \frac{1}{2}$, $1 \le y \le 2$, by evaluating the integral

$$\int_1^2 \int_0^{1/2} f(x, y) \, dx \, dy.$$

25. Find the probability that a point (x, y) is in the region $\{(x, y): 0 \le x \le 1, 0 \le y \le 1\}$, that is, the lower half of the rectangle.

26. Find the probability that a point (x, y) is in the part of the rectangle above the line $y = 2 - 2x$.

A *triple integral* of a continuous function $f(x, y, z)$ over an appropriate region G in three dimensions, say

$$\iiint_G f(x, y, z)\, dx\, dy\, dz,$$

can be defined in a manner similar to that in which the double integral was defined. Triple integrals can frequently be evaluated by the iterated integral

$$\int_a^b \left\{ \int_{G(x)}^{H(x)} \left\{ \int_{g(x, y)}^{h(x, y)} f(x, y, z)\, dz \right\} dy \right\} dx,$$

where the region G is bounded above and below by the surfaces $z = g(x, y)$ and $z = h(x, y)$ and where the y- and x-limits of integration are obtained from the boundaries of the region bounded by $g(x, y) = h(x, y)$ in the xy-plane. Evaluate each of the following triple integrals.

27. $\displaystyle\int_0^1 \int_1^3 \int_{-1}^2 (2x + 3y - z)\, dx\, dy\, dz$

28. $\displaystyle\int_0^2 \int_1^4 \int_{-1}^6 (8x - 2y + z)\, dx\, dy\, dz$

29. $\displaystyle\int_0^1 \int_0^{1-x} \int_0^{2-x} xyz\, dz\, dy\, dx$

30. $\displaystyle\int_0^2 \int_{2-y}^{6-2y} \int_0^{\sqrt{4-y^2}} z\, dz\, dx\, dy$

If $f(x, y, z)$ is continuous on a region G in three dimensions, define the *average value* f_{av} of $f(x, y, z)$ on G by

$$f_{av} = \frac{1}{\text{vol }(G)} \iiint_G f(x, y, z)\, dx\, dy\, dz,$$

where vol (G) is the volume of G.

31. Find the average value of $f(x, y, z) = xyz$ on the unit cube

$$\{(x, y, z) \mid 0 \le x \le 1, 0 \le y \le 1 \text{ and } 0 \le z \le 1\}.$$

CHAPTER 7 TEST

Given $f(x, y) = e^x + 2x^3y + y$, find each of the following.

1. $\dfrac{\partial f}{\partial x}$

2. $\dfrac{\partial f}{\partial y}$

3. $\dfrac{\partial^2 f}{\partial x^2}$

4. $\dfrac{\partial^2 f}{\partial x\, \partial y}$

5. $\dfrac{\partial^2 f}{\partial y\, \partial x}$

6. $\dfrac{\partial^2 f}{\partial y^2}$

7. Find the relative maximum and minimum values.

$$f(x, y) = x^2 - xy + y^3 - x$$

8. Find the relative maximum and minimum values.

$$f(x, y) = y^2 - x^2$$

9. Calculate the total differential dz given that z $x^3 + e^y$.

10. Calculate the total differential dz given that $z = 3xy^2$.

11. A rectangular box is determined to measure 2 ft by 3 ft by 5 ft. If errors of 0.01 ft, 0.02 ft, and 0.04 ft are possible in the three measurements, respectively, use differentials to calculate the maximum possible error in measuring the volume of the box.

12. Consider this data regarding the total sales of a company during the first three years of operation.

Year, x	1	2	3
Sales (in millions), y	$10	$15	$19

a) Find the regression line $y = mx + b$.

b) Use the regression line to predict sales in the fourth year.

13. Find the maximum value of

$$f(x, y) = 6xy - 4x^2 - 3y^2$$

subject to the constraint $x + 3y = 19$.

15. Find the volume of the solid whose base is the semi-circular region

$$\{(x, y) \mid 0 \le y, x^2 + y^2 \le 1\}$$

and which is capped by the plane $z = 2y$.

14. Evaluate $\int_0^2 \int_1^x (x^2 - y) \, dy \, dx$.

16. Find the average value f_{av} of $f(x, y) = xy$ on the region of the plane bounded by the positive x- and y-axes and the curve $y = 1 - x^3$.

▶

17. A company has the following Cobb–Douglas production function for a certain product

$$p(x, y) = 50x^{2/3}y^{1/3},$$

where $x =$ labor, measured in dollars, and $y =$ capital, measured in dollars. Suppose a company can make a total investment in labor and capital of $600,000. How should it allocate the investment between labor and capital in order to maximize production?

18. Suppose beverages could be packaged in either a cylindrical container or a rectangular container with a square top and bottom. If we assume a volume of 26 in³, which container would have the minimum surface area?

TRIGONOMETRIC
FUNCTIONS

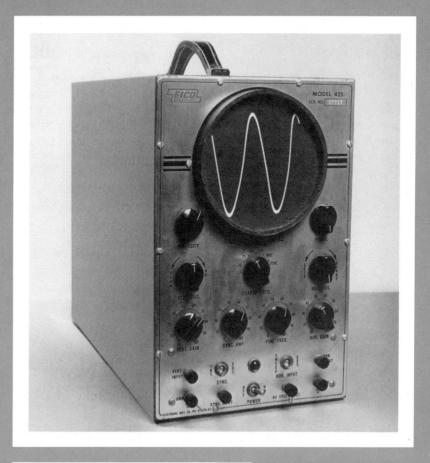

This oscilloscope shows the graph of a trigonometric function. (*Lester V. Bergman & Assoc., Inc.*)

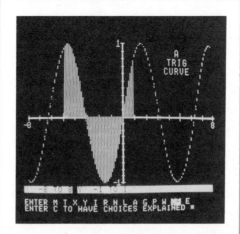

This graph shows a trigonometric curve and some area under the curve over the interval [−5, 1].

8

OBJECTIVES

You should be able to

a) **Tell in which quadrant the terminal side of an angle lies.**

b) **Convert from radian measure to degree measure and from degree measure to radian measure.**

c) **Verify certain identities.**

8.1 INTRODUCTION TO TRIGONOMETRY

Angles and Rotations

Our goal here is to introduce the trigonometric functions together with their derivatives and integrals.

We will consider a rotating ray, with its endpoint at the origin of an xy-plane. The ray starts in position along the positive half of the x-axis. A counterclockwise rotation is called *positive*, and a clockwise rotation is called *negative*.

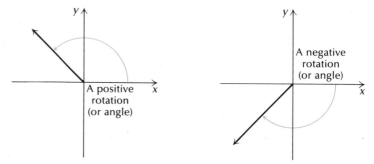

Note that the rotating ray and the positive half of the x-axis form an angle. Thus we often speak of "rotations" and "angles" interchangeably. The rotating ray is often called the *terminal side* of the angle, and the positive half of the x-axis is called the *initial side*.

Measures of Rotations or Angles

The size, or *measure*, of an angle, or rotation, may be given in degrees. Thus a complete revolution has a measure of 360°, half a revolution has a measure of 180°, and so on. We can also speak of an *angle* of 90° or 720° or −240°.

An angle between 0° and 90° has its terminal side in the first quadrant. An angle between 90° and 180° has its terminal side in the second

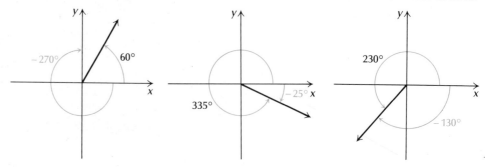

1. In which quadrant does the terminal side of each angle lie?

   a) 47°

   b) 212°

   c) −43°

   d) −135°

   e) 365°

   f) −365°

   g) 740°

quadrant. An angle between 180° and 270° has its terminal side in the third quadrant. An angle between 0° and −90° has its terminal side in the fourth quadrant.

Note that angles with measure 0°, 360°, and 720° have the same terminal side, as do 270° and −90°.

DO EXERCISE 1.

Radian Measure

A unit of angle or rotation measure other than the degree is very useful for many purposes. This unit is called the *radian*. Consider a circle with radius of length 1, centered at the origin. The distance around this circle, from the initial side to the terminal side, is the measure of the angle, or rotation, in *radians*.

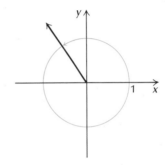

Since the circumference of the circle is $2\pi \cdot 1$, or 2π, a complete revolution (360°) has a measure of 2π radians. Half of this (180°) is π radians, and a fourth of this (90°) is $\pi/2$ radians. In general, we can convert from one measure to the other using this proportion.

THEOREM 1

$$\frac{\textbf{Radian measure}}{\boldsymbol{\pi}} = \frac{\textbf{Degree measure}}{\textbf{180}}$$

Example 1 Convert 270° to radians.

Solution

$$\frac{Radian\ measure}{\pi} = \frac{270}{180}$$

$$Radian\ measure = \frac{270}{180} \cdot \pi, \quad \text{or} \quad \frac{3}{2}\pi.$$

2. Convert to radian measure. Leave answers in terms of π.

a) $135°$

b) $315°$

c) $-90°$

d) $720°$

e) $-225°$

f) $-315°$

g) $405°$

h) $480°$

3. Convert these radian measures to degrees.

a) $\dfrac{\pi}{3}$

b) $\dfrac{3}{4}\pi$

c) $\dfrac{5}{2}\pi$

d) 10π

e) $-\dfrac{7\pi}{6}$

f) 300π

g) -270π

h) $\dfrac{25\pi}{4}$

When no unit is specified for an angle measure, it is understood to be given in radians.

Example 2 Convert $\pi/4$ radians to degrees.

Solution

$$\frac{\pi/4}{\pi} = \frac{Degree\ measure}{180}$$

$$Degree\ measure = 180 \cdot \frac{\pi/4}{\pi}, \quad or \quad 45°.$$

DO EXERCISES 2 AND 3.

Trigonometric Functions

The concept of rotation or angle is important to functions called *trigonometric*, or *circular*, functions.

Consider an angle t, measured in radians, shown as follows on a circle with radius 1.

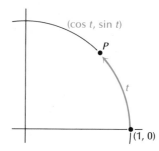

The terminal side of the angle intersects this circle at point P. The distance *around the circle* from $(1, 0)$ to P is t. We define cos t (cosine t) and sin t (sine t) as the first and second coordinates of P:

$$\cos t = \text{the first coordinate of } P,$$

$$\sin t = \text{the second coordinate of } P.$$

Certain values of these functions are easy to determine. When $t = \pi$, for example, the terminal side of the angle is on the horizontal axis, and P is one unit to the left; hence the first coordinate is -1 and the second coordinate is 0. Thus,

$$\cos \pi = -1 \quad \text{and} \quad \sin \pi = 0.$$

Similarly, when $t = \pi/2$, the point P is one unit up on the vertical axis; hence the first coordinate is 0 and the second coordinate is 1. Thus,

$$\cos \frac{\pi}{2} = 0 \quad \text{and} \quad \sin \frac{\pi}{2} = 1.$$

When $t = 0$, the terminal side is on the horizontal axis and P is one unit to the right; hence the first coordinate is 1 and the second coordinate is 0. Thus,

$$\cos 0 = 1 \quad \text{and} \quad \sin 0 = 0.$$

Using properties of right triangles, we can develop these other values.

For $t = \dfrac{\pi}{4}$,

$$\cos \frac{\pi}{4} = \frac{\sqrt{2}}{2} \approx 0.707 \text{ and } \sin \frac{\pi}{4} = \frac{\sqrt{2}}{2} \approx 0.707.$$

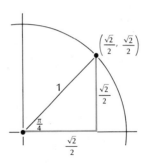

For $t = \dfrac{\pi}{3}$,

$$\cos \frac{\pi}{3} = \frac{1}{2} = 0.5 \text{ and } \sin \frac{\pi}{3} = \frac{\sqrt{3}}{2} \approx 0.866.$$

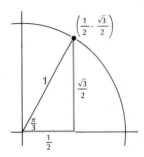

For $t = \dfrac{\pi}{6}$, $\cos \dfrac{\pi}{6} = \dfrac{\sqrt{3}}{2} \approx 0.866$ and $\sin \dfrac{\pi}{6} = \dfrac{1}{2} = 0.5$.

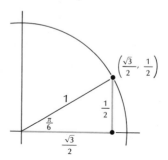

The following table summarizes these important values of the sine and cosine functions. It should be memorized.

t	$\sin t$	$\cos t$
0	0	1
$\dfrac{\pi}{6}$	$\dfrac{1}{2}$	$\dfrac{\sqrt{3}}{2}$
$\dfrac{\pi}{4}$	$\dfrac{\sqrt{2}}{2}$	$\dfrac{\sqrt{2}}{2}$
$\dfrac{\pi}{3}$	$\dfrac{\sqrt{3}}{2}$	$\dfrac{1}{2}$

Other function values follow from certain symmetries on the unit circle. Some are shown below.

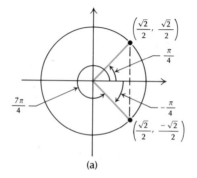

(a)

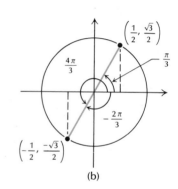

(b)

4. Using the following,

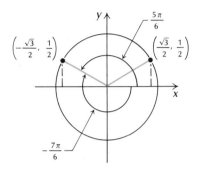

find:

a) $\cos\left(\dfrac{5\pi}{6}\right)$;

b) $\sin\left(\dfrac{5\pi}{6}\right)$;

c) $\cos\left(-\dfrac{7\pi}{6}\right)$;

d) $\sin\left(-\dfrac{7\pi}{6}\right)$.

From (a) we get

$$\cos\left(-\frac{\pi}{4}\right) = \frac{\sqrt{2}}{2} \approx 0.707,$$

$$\sin\left(-\frac{\pi}{4}\right) = -\frac{\sqrt{2}}{2} \approx -0.707,$$

and

$$\cos\left(\frac{7\pi}{4}\right) = \frac{\sqrt{2}}{2} \approx 0.707,$$

$$\sin\left(\frac{7\pi}{4}\right) = -\frac{\sqrt{2}}{2} \approx -0.707.$$

From (b) we get

$$\cos\left(-\frac{2\pi}{3}\right) = -\frac{1}{2} = -0.5,$$

$$\sin\left(-\frac{2\pi}{3}\right) = -\frac{\sqrt{3}}{2} \approx -0.866,$$

and

$$\cos\left(\frac{4\pi}{3}\right) = -\frac{1}{2} = -0.5,$$

$$\sin\left(\frac{4\pi}{3}\right) = -\frac{\sqrt{3}}{2} \approx -0.866.$$

Table 7 at the back of the book shows many values of the trigonometric functions. Many calculators also have trigonometric keys.

DO EXERCISE 4.

Graphs of cos *t* and sin *t*   `CSS`

Note that cos *t* and sin *t* are functions of *t* defined for all real numbers *t*. For very large $|t|$ we may "wrap around" the circle several times before coming to the terminal point *P*. Nevertheless, *P* still has one first coordinate, cos *t*, and one second coordinate, sin *t*. For example,

$$\cos 3\pi = \cos \pi = -1 \quad \text{and} \quad \sin 3\pi = \sin \pi = 0.$$

Also,

$$\cos\left(\frac{15\pi}{4}\right) = \cos\left(\frac{7\pi}{4}\right) = \frac{\sqrt{2}}{2} \quad \text{and} \quad \sin\left(\frac{15\pi}{4}\right) = \sin\left(\frac{7\pi}{4}\right) = -\frac{\sqrt{2}}{2}.$$

Plotting points previously obtained, we graph the cosine and sine functions as follows.

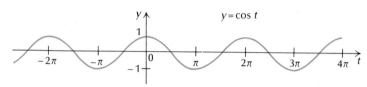

The cosine function

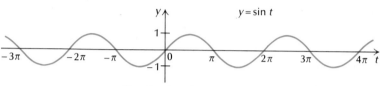

The sine function

At the origin, t is of course 0. Hence the point P is on the horizontal axis, so $\cos 0 = 1$ and $\sin 0 = 0$. Moving to the right on the graphs corresponds to having the terminal side of the angle rotate counterclockwise. Moving to the left corresponds to having the terminal side move clockwise. Note in particular that $\sin t = 0$ has solutions $t = 0$, $\pm \pi$, $\pm 2\pi$, . . . ; and $\cos t = 0$ has solutions $t = \pm(\pi/2)$, $\pm(3\pi/2)$, $\pm(5\pi/2)$, ... Note that each curve repeats itself, as the terminal side makes successive revolutions. From 0 to 2π is one complete revolution, or *cycle*. The cycle repeats itself from there on. We say that the *period* of each function is 2π. Algebraically, this means that for any t,

$$\cos (t + 2\pi) = \cos t \quad \text{and} \quad \sin (t + 2\pi) = \sin t. \qquad \boxed{\textit{CSS}}$$

DEFINITION

A function f is *periodic* if there exists a positive number p such that

$$f(x + p) = f(x).$$

This means that adding p to an input does not change the output. The smallest such number p is called the *period*.

A printout from an electrocardiogram may, like the one on the next page, form a periodic function.

5. Find each of the following.

a) $\tan \dfrac{\pi}{4}$

b) $\tan \dfrac{\pi}{3}$

c) $\tan 0$

d) $\sec \dfrac{\pi}{4}$

e) $\sec \dfrac{\pi}{3}$

f) $\sec 0$

6. Find each of the following.

a) $\cot \dfrac{\pi}{6}$

b) $\cot 0$

c) $\cot \dfrac{\pi}{3}$

d) $\csc \dfrac{\pi}{6}$

e) $\csc 0$

f) $\csc \dfrac{\pi}{4}$

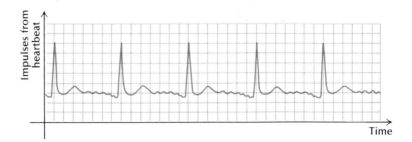

Other Trigonometric Functions

The functions sin x and cos x are the basic trigonometric functions, but there are four others—the tangent, cotangent, secant, and cosecant functions—defined as follows. **CSS**

DEFINITION

$$\tan x = \frac{\sin x}{\cos x},$$

$$\cot x = \frac{\cos x}{\sin x} = \frac{1}{\tan x},$$

$$\sec x = \frac{1}{\cos x},$$

$$\csc x = \frac{1}{\sin x},$$

provided the denominators are not equal to 0.

Let us find some values of the tangent function.

Example 3 Find $\tan \pi/6$ and $\tan \pi/2$.

Solution

$$\tan \frac{\pi}{6} = \frac{\sin (\pi/6)}{\cos (\pi/6)} = \frac{1/2}{\sqrt{3}/2} = \frac{1}{\sqrt{3}} = \frac{1}{\sqrt{3}} \cdot \frac{\sqrt{3}}{\sqrt{3}} = \frac{\sqrt{3}}{3} \approx 0.577,$$

$$\tan \frac{\pi}{2} = \frac{\sin (\pi/2)}{\cos (\pi/2)} = \frac{1}{0} = \text{undefined}$$

DO EXERCISES 5 AND 6.

The graph of tan x is as follows.

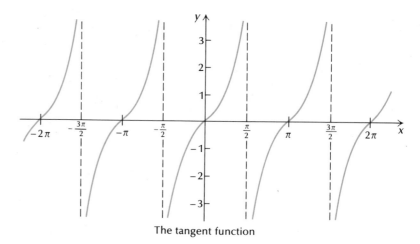

The tangent function

Identities

The properties

$$\cos (x + 2\pi) = \cos x \quad \text{and} \quad \sin (x + 2\pi) = \sin x$$

hold for all real numbers x. They are examples of *trigonometric identities*. Another identity that holds for all real numbers is the following:

THEOREM 2

$$\sin^2 t + \cos^2 t = 1, \tag{1}$$

where $\sin^2 t$ means $(\sin t)^2$ and $\cos^2 t$ means $(\cos t)^2$.

To see why this holds, note the right triangle inside the unit circle of the figure below.

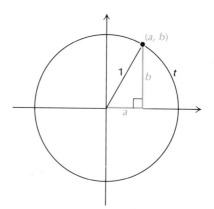

7. Multiply Identity (1) by

$$\frac{1}{\sin^2 t}$$

to develop another identity.

Since the length of the radius is 1, we know from the Pythagorean Theorem that

$$a^2 + b^2 = 1^2, \quad \text{or} \quad a^2 + b^2 = 1;$$

and, since for any point (a, b) on the unit circle, $\cos t = a$ and $\sin t = b$, the identity follows.

If we multiply Identity (1) by $1/(\cos^2 t)$, we get another identity:

$$\frac{\sin^2 t}{\cos^2 t} + \frac{\cos^2 t}{\cos^2 t} = \frac{1}{\cos^2 t}$$

THEOREM 3

$$\tan^2 t + 1 = \sec^2 t \qquad\qquad (2)$$

DO EXERCISE 7.

The following are *sum and difference* identities.

THEOREM 4

$$\cos (u + v) = \cos u \cos v - \sin u \sin v, \qquad (3)$$

$$\cos (u - v) = \cos u \cos v + \sin u \sin v, \qquad (4)$$

$$\sin (u + v) = \sin u \cos v + \cos u \sin v \qquad (5)$$

$$\sin (u - v) = \sin u \cos v - \cos u \sin v \qquad (6)$$

8. Use Identity (6) to find $\sin 15°$.

9. Use Identity (5) to find

$$\sin \left(\frac{\pi}{4} + \frac{\pi}{3}\right).$$

Example 4 Use Identity (4) to find $\cos 15°$.

Solution If we think of $15°$ as $45° - 30°$, then

$$\cos 15° = \cos (45° - 30°)$$

$$= \cos 45° \cos 30° + \sin 45° \sin 30°$$

$$= \frac{\sqrt{2}}{2} \cdot \frac{\sqrt{3}}{2} + \frac{\sqrt{2}}{2} \cdot \frac{1}{2}$$

$$= \frac{\sqrt{6} + \sqrt{2}}{4}.$$

DO EXERCISES 8 AND 9.

10. Let u = v in Identity (5) to find an identity for sin 2u.

If we let u = v in Identity (3), we obtain a *double-angle* identity:

$$\cos 2u = \cos (u + u)$$

$$= \cos u \cos u - \sin u \sin u;$$

or the following:

CSS

THEOREM 5

$$\cos 2u = \cos^2 u - \sin^2 u$$

DO EXERCISE 10.

EXERCISE SET 8.1

In what quadrant does the terminal side of each angle lie?

1. 34°   **2.** 320°   **3.** −120°   **4.** −205°

Convert to radian measure. Leave answers in terms of π.

5. 30°   **6.** 15°   **7.** 60°   **8.** 200°   **9.** 75°   **10.** 300°

Convert these radian measures to degrees.

11. $\frac{3}{2}\pi$ **12.** $\frac{5}{4}\pi$ **13.** $-\frac{\pi}{4}$ **14.** $-\frac{\pi}{6}$ **15.** 8π **16.** -12π

17. 1 radian   **18.** 2 radians

Find each of the following.

19. $\sin \frac{\pi}{3}$ **20.** $\cos \frac{\pi}{4}$ **21.** $\cos \frac{3\pi}{2}$ **22.** $\sin \frac{5\pi}{4}$ **23.** $\sin \frac{\pi}{6}$ **24.** $\cos \pi$

25. $\tan \frac{\pi}{2}$ **26.** $\tan \frac{\pi}{6}$ **27.** $\cot \frac{\pi}{3}$ **28.** $\cot \frac{\pi}{6}$ **29.** $\sec \pi$ **30.** $\csc \frac{\pi}{4}$

Verify the following identities:

31. $\tan (u + v) = \dfrac{\tan u + \tan v}{1 - \tan u \tan v}$ **32.** $\tan 2u = \dfrac{2 \tan u}{1 - \tan^2 u}$

▦ If your calculator has trigonometric keys, find answers to the following problems to four decimal places. Check to see if your calculator is using degree or radian measure for angles.

33. $\sin 31.4°$ **34.** $\cos 1.07°$ **35.** $\tan 139.2°$ **36.** $\cot 153.5°$

37. $\cos (-1.91)$ **38.** $\sin (-11.2\pi)$ **39.** $\cot 49\pi$ **40.** $\tan (-17.4)$

41. ▦*Biomedical: Temperature during an illness.* The temperature of a patient during a 12-day illness is given by

$$T(t) = 101.6° + 3 \sin\left(\frac{\pi}{8}t\right).$$

The graph is shown at the right. Find $T(0)$, $T(1)$, $T(2)$, $T(4)$, and $T(12)$. Round to the nearest tenth of a degree.

CSS

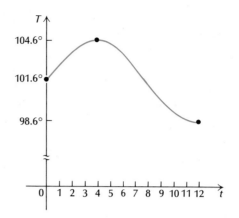

42. ▦*Business: Sales.* Sales of certain products fluctuate in cycles, as seen in the graph at the right. A company in a northern climate has total sales of skis as given by

$$S(t) = 7\left(1 - \cos\frac{\pi}{6}t\right).$$

where S is sales in thousands of dollars during the tth month. Find $S(0)$, $S(1)$, $S(2)$, $S(3)$, $S(6)$, $S(12)$, and $S(15)$. Round to the nearest tenth.

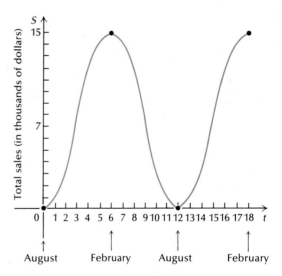

OBJECTIVES

You should be able to

a) **Differentiate trigonometric functions.**

b) **Solve problems involving derivatives of trigonometric functions.**

8.2 DERIVATIVES OF THE TRIGONOMETRIC FUNCTIONS

The development of the derivatives of $\sin x$ and $\cos x$ is comparable to the development of the derivative of e^x. We first find the value of the derivative at 0, and then extend this to the general formula. The derivatives at 0 are given by the following limits:

$$\sin'(0) = \lim_{h \to 0} \frac{\sin(0 + h) - \sin 0}{h} = \lim_{h \to 0} \frac{\sin h}{h};$$

$$\cos'(0) = \lim_{h \to 0} \frac{\cos(0 + h) - \cos 0}{h} = \lim_{h \to 0} \frac{\cos h - 1}{h}.$$

We find the limits using the following input–output table.

h (in radians) with decimal approximation	$\sin h$	$\cos h$	$\dfrac{\sin h}{h}$	$\dfrac{\cos h - 1}{h}$
$\dfrac{\pi}{2}(1.5708)$	1	0	0.6366	−0.6366
$\dfrac{\pi}{3}(1.0472)$	$\dfrac{\sqrt{3}}{2}(0.8660)$	$\dfrac{1}{2}(0.5000)$	0.8270	−0.4775
$\dfrac{\pi}{4}(0.7854)$	$\dfrac{\sqrt{2}}{2}(0.7071)$	$\dfrac{\sqrt{2}}{2}(0.7071)$	0.9003	−0.3729
$\dfrac{\pi}{6}(0.5236)$	$\dfrac{1}{2}(0.5000)$	$\dfrac{\sqrt{3}}{2}(0.8660)$	0.9549	−0.2559
$\dfrac{\pi}{20}(0.1571)$	0.1564	0.9877	0.9955	−0.0783
$\dfrac{\pi}{50}(0.0628)$	0.0628	0.9980	1.0000	−0.0318
$\dfrac{\pi}{4000}(0.0008)$	0.0008	0.99997	1.0000	−0.0038

Values approaching 0 from the left can also be checked. It follows that

$$\lim_{h \to 0} \frac{\sin h}{h} = 1 \quad \text{and} \quad \lim_{h \to 0} \frac{\cos h - 1}{h} = 0.$$

Now let us consider the general derivatives:

$$\frac{d}{dx} \sin x = \lim_{h \to 0} \frac{\sin (x + h) - \sin x}{h}.$$

Using Identity (5), we get

$$\frac{\sin (x + h) - \sin x}{h} = \frac{\sin x \cos h + \cos x \sin h - \sin x}{h}$$

$$= \frac{\sin x \cos h - \sin x}{h} + \frac{\cos x \sin h}{h}$$

$$= \sin x \left(\frac{\cos h - 1}{h} \right) + \cos x \left(\frac{\sin h}{h} \right).$$

Then using the limits just developed, we have

$$\frac{d}{dx} \sin x = \lim_{h \to 0} \frac{\sin (x + h) - \sin x}{h}$$

$$= (\sin x) \cdot 0 + (\cos x) \cdot 1$$

$$= \cos x.$$

A development for the derivative of cos x is similar but uses Identity (3). The result is −sin x.

In summary, we have the following.   CSS

THEOREM 6

$$\frac{d}{dx} \sin x = \cos x \quad \text{and} \quad \frac{d}{dx} \cos x = -\sin x.$$

The derivatives of the remaining trigonometric functions are computed from their definitions in terms of sin x and cos x, together with the Quotient Rule and/or the Extended Power Rule. The remaining formulas are as follows.

THEOREM 7

$$\frac{d}{dx} \tan x = \sec^2 x,$$

$$\frac{d}{dx} \cot x = -\csc^2 x,$$

$$\frac{d}{dx} \sec x = \tan x \sec x,$$

$$\frac{d}{dx} \csc x = -\cot x \csc x$$

Example 1 Prove that $\frac{d}{dx} \tan x = \sec^2 x$.

Solution By definition,

$$\tan x = \frac{\sin x}{\cos x}.$$

1. Prove that

$$\frac{d}{dx} \cot x = -\csc^2 x.$$

We can therefore find its derivative using the Quotient Rule:

$$\frac{d}{dx} \tan x = \frac{\cos x (\cos x) - \sin x (-\sin x)}{\cos^2 x}$$

$$= \frac{\cos^2 x + \sin^2 x}{\cos^2 x}$$

$$= \frac{1}{\cos^2 x} = \sec^2 x.$$

DO EXERCISE 1.

Example 2 Find the derivative of $y = \sec^3 x$.

Solution We use the Extended Power Rule:

$$\frac{dy}{dx} = 3 \sec^2 x \cdot \left(\frac{d}{dx} \sec x \right)$$

$$= 3 \sec^2 x \cdot \tan x \cdot \sec x.$$

Replacing the factors by the definitions in terms of sin x and cos x, we can simplify this as follows:

2. Differentiate

$$y = \tan^3 x.$$

$$3 \sec^2 x \cdot \tan x \cdot \sec x = 3 \cdot \frac{1}{\cos^2 x} \cdot \frac{\sin x}{\cos x} \cdot \frac{1}{\cos x}$$

$$= \frac{3 \sin x}{\cos^4 x}.$$

DO EXERCISE 2.

Using the Chain Rule, we can find other derivatives.

Example 3 Differentiate $f(x) = \sin (x^3 - 5x)$.

Solution

$$f'(x) = \cos (x^3 - 5x) \cdot (3x^2 - 5)$$

$$= (3x^2 - 5) \cos (x^3 - 5x)$$

Example 4 Differentiate $y = \cos (e^{4x}) \sin x^2$.

Differentiate.

3. $f(x) = \sin(x^4 + 3x^2)$

Solution We use the Chain Rule and the Product Rule:

$$\frac{dy}{dx} = -\sin(e^{4x}) \cdot (4e^{4x}) \sin x^2 + \cos(e^{4x}) \cdot 2x \cdot \cos x^2$$

$$= -4e^{4x} \sin(e^{4x}) \sin x^2 + 2x \cos(e^{4x}) \cos x^2.$$

DO EXERCISES 3 AND 4.

Applications

Equations of the type

$$y = A \sin(Bx - C) + D$$

have many applications. We can also express this equation as

$$y = A \sin\left[B\left(x - \frac{C}{B}\right)\right] + D.$$

The numbers A, B, and C play an important role in graphing such an equation. The number A corresponds to a vertical stretching or shrinking of the graph of $y = \sin x$, B corresponds to a horizontal stretching or shrinking, and C/B corresponds to a shift to the left or right of the entire graph. These names are attached to the numbers:

$$Amplitude = |A|,$$

$$Period = \frac{2\pi}{B},$$

$$Phase\ shift = \frac{C}{B}.$$

4. $y = \cos(e^{x^2}) \sin x^3$

Example 5 A weight is attached to the end of a spring. When the weight is disturbed, it bobs up and down with a definite frequency. If the motion were to occur in a perfect vacuum, and if the spring were perfectly elastic, then the oscillatory motion would continue undiminished forever. Suppose a spring oscillates in such a way that its vertical position at time t, from its position at rest, is given by

$$y = 3 \sin\left(2t + \frac{\pi}{2}\right) = 3 \sin\left[2\left(t - \left(-\frac{\pi}{4}\right)\right)\right].$$

a) Find the amplitude, period, and phase shift.

b) Graph the equation.

c) Find dy/dt.

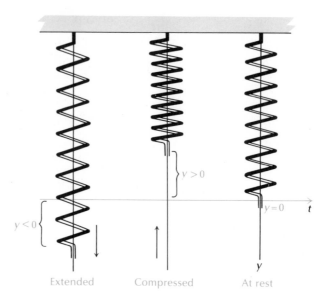

Extended   Compressed   At rest

Solution

a) The amplitude $= |3| = 3$. The period $= 2\pi/2 = \pi$. The phase shift $= -(\pi/4)$.

b) We plot the various equations needed to get the final graph:

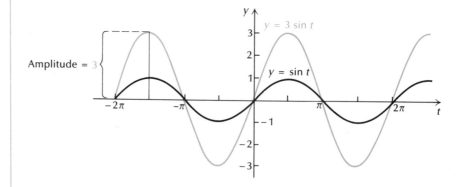

The graph of $y = 3 \sin t$ is a vertical stretching, by a factor of 3, of the graph of $y = \sin t$. The period of $y = 3 \sin t$ is still 2π. The graph of $y = 3 \sin 2t$ is a horizontal shrinking, by a factor of $\frac{1}{2}$, of the graph of $y = 3 \sin t$. The period of $y = 3 \sin 2t$ is π.

5. Suppose a spring oscillates in such a way that its vertical position at time t, from its position at rest, is given by

$$y = 4 \sin \left(2t - \frac{\pi}{2} \right).$$

a) Find the amplitude, period, and phase shift.

b) Graph the equation.

c) Find dy/dt.   CSS

6. A sound wave is given by

$$y = 0.03 \sin 1.198\pi x.$$

Find dy/dx.

7. A light wave is given by

$$L = A \sin \left(2\pi f T - \frac{d}{\omega} \right).$$

Find dL/dT.

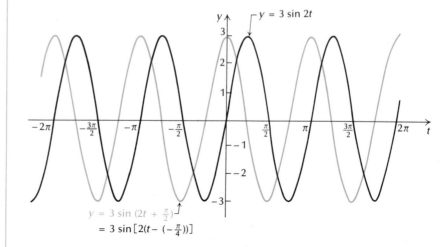

$$y = 3 \sin \left(2t + \frac{\pi}{2} \right)$$
$$= 3 \sin \left[2 \left(t - \left(-\frac{\pi}{4} \right) \right) \right]$$

The graph of

$$y = 3 \sin \left[2 \left(t - \left(-\frac{\pi}{4} \right) \right) \right]$$

is shifted $\pi/4$ units to the left of the graph of $y = 3 \sin 2t$. If the phase shift were positive, the shift would be to the right.

c) $\dfrac{dy}{dt} = 3 \cos \left(2t + \dfrac{\pi}{2} \right) \cdot 2 = 6 \cos \left(2t + \dfrac{\pi}{2} \right)$

DO EXERCISE 5.

The motion of the spring in the preceding example is called *simple harmonic motion*. Other types of simple harmonic motion are sound waves and light waves. For sound waves the amplitude is the loudness. For light waves the amplitude is the brightness. The *frequency*, which is the reciprocal of the period, is the tone of a sound wave and the color of a light wave.

DO EXERCISES 6 AND 7.

Biological Application: Biorhythms

Some people conjecture that a person's life has cycles of good and bad days that begin at birth. There are supposedly three such cycles, or *biorhythms*.

1. *Physical.* A cycle represented by a sine function with a period of 23 days:

$$y = \sin \frac{2\pi}{23} t.$$

This cycle is related to physical qualities such as vitality, strength, and energy.

2. *Emotional (sensitivity).* A cycle represented by a sine function with a period of 28 days:

$$y = \sin \frac{2\pi}{28} t.$$

This cycle is related to creativity, moodiness, intuition, and cheerfulness.

3. *Mental (intellectual).* A cycle represented by a sine function with a period of 33 days:

$$y = \sin \frac{2\pi}{33} t.$$

This cycle is related to the ability to study, think, react, and remember. (The positive part of *this* cycle would be a good time to study calculus.) Because these cycles have periods of different lengths, there are varying combinations of highs and lows throughout one's life. This is illustrated in the following graph.*

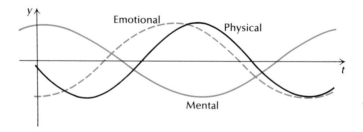

Example 6 *Biology: Minimizing the surface area of a bee's cell.* Did you know that a honey bee constructs the cells in its comb in such a way

*For more on these "biorhythms," write to Biorhythm Computers, Inc., 298 Fifth Ave, New York, NY 10001.

that the minimum amount of wax is used?* Let us see why. One cell of a honey comb is shown here. It is a prism whose base, the open part at the top, is a regular hexagon. The bottom comes together at point A. The surface area is given by

$$S(\theta) = 6ab + \frac{3}{2}a^2\left(\frac{\sqrt{3} - \cos\theta}{\sin\theta}\right),$$

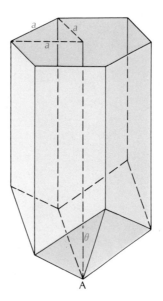

where θ is the measure of what is called the angle of inclination. Find the value of θ between $0°$ and $90°$ that minimizes S.

Solution We first find $dS/d\theta$:

$$\frac{dS}{d\theta} = \frac{3}{2}a^2\left[\frac{\sin\theta(\sin\theta) - \cos\theta(\sqrt{3} - \cos\theta)}{\sin^2\theta}\right]$$

$$= \frac{3}{2}a^2\left[\frac{\sin^2\theta - \sqrt{3}\cos\theta + \cos^2\theta}{\sin^2\theta}\right]$$

$$= \frac{3}{2}a^2\left[\frac{1 - \sqrt{3}\cos\theta}{\sin^2\theta}\right].$$

---

*This discovery was published in a study by Sir D'Arcy Wentworth Thompson, *On Growth and Form*, Cambridge University Press, 1917.

A honeycomb. (*Marshall Henrichs*)

8. A corridor of width a meets a corridor of width b at right angles. Workers wish to push a heavy beam of length L on dollies around the corner. Before starting, however, they want to be sure they can make the turn. How long a beam will go around the corner? (Disregard the width of the beam.)

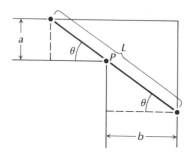

The length L of the beam is given by

$$L = \frac{a}{\sin \theta} + \frac{b}{\cos \theta},$$

where θ is the angle shown above.

a) Find L in terms of a and b, such that L is of maximum length.

b) Use the formula in (a) to find how long a beam will go around the corner when $a = 8$ and $b = 5\sqrt{5}$.   CSS

Critical points occur where $\sin \theta$ is 0; but if we check such values, we see that they are multiples of 90°, and that at such values the bottom of the cell would be flat, which is not the way it is constructed. Thus we set $dS/d\theta = 0$ and solve for θ with the assumption that $\sin \theta \neq 0$:

$$\frac{dS}{d\theta} = \frac{3}{2}a^2 \left[\frac{1 - \sqrt{3} \cos \theta}{\sin^2 \theta} \right] = 0$$

$$\frac{1 - \sqrt{3} \cos \theta}{\sin^2 \theta} = 0$$

$$1 - \sqrt{3} \cos \theta = 0 \qquad \text{Multiplying by } \sin^2 \theta$$

$$\cos \theta = \frac{1}{\sqrt{3}} \approx 0.5774.$$

We need to find θ such that $\cos \theta \approx 0.5774$. We can work backward in Table 7, or we can use $\boxed{\text{Arc}}$ $\boxed{\cos}$ key(s) on a calculator. By either method it follows that

$$\theta \approx 54.7°.$$

The study showed that bees tend to use this angle. That they do may be explained by a genetic theory that those who survive to pass on their genetic traits are those who are more "successful" at living. Perhaps bees that waste less time and wax by making cells with the minimum amount of wax were, at one time, such successful genetic survivors.

DO EXERCISE 8.

EXERCISE SET 8.2

Prove the following derivative formulas.

1. $\dfrac{d}{dx} \sec x = \tan x \sec x$

2. $\dfrac{d}{dx} \csc x = -\cot x \csc x$

Differentiate.

3. $y = x \sin x$ **4.** $y = x \cos x$ **5.** $f(x) = e^x \sin x$ **6.** $f(x) = e^x \cos x$

7. $y = \dfrac{\sin x}{x}$ **8.** $y = \dfrac{\cos x}{x}$ **9.** $f(x) = \sin^2 x$ **10.** $f(x) = \cos^2 x$

11. $y = \sin x \cos x$ **12.** $y = \cos^2 x + \sin^2 x$ **13.** $f(x) = \dfrac{\sin x}{1 + \cos x}$ **14.** $f(x) = \dfrac{1 - \cos x}{\sin x}$

15. $y = \tan^2 x$ **16.** $y = \sec^2 x$ **17.** $f(x) = \sqrt{1 + \cos x}$ **18.** $f(x) = \sqrt{1 - \sin x}$

19. $y = x^2 \cos x - 2x \sin x - 2 \cos x$ **20.** $y = x^2 \sin x - 2x \cos x + 2 \sin x$

21. $y = e^{\sin x}$ **22.** $y = e^{\cos x}$

23. Find d^2y/dx^2 if $y = \sin x$.   **24.** Find d^2y/dx^2 if $y = \cos x$.

25. $y = \sin (x^2 + x^3)$ **26.** $y = \sin (x^5 - x^4)$ **27.** $f(x) = \cos (x^5 - x^4)$ **28.** $f(x) = \cos (x^2 + x^3)$

29. $f(x) = \cos \sqrt{x}$ **30.** $f(x) = \sin \sqrt{x}$ **31.** $y = \sin (\cos x)$ **32.** $y = \cos (\sin x)$

33. $y = \sqrt{\cos 4x}$ **34.** $y = \sin^2 5x$ **35.** $f(x) = \cot \sqrt[3]{5 - 2x}$ **36.** $f(x) = \sec (\tan 7x)$

37. $y = \tan^4 3x - \sec^4 3x$ **38.** $y = \sqrt[5]{\cot 5x - \cos 5x}$

Differentiate. Use the formula $\dfrac{d}{du} \ln |u| = \dfrac{1}{u} du$.

39. $y = \ln |\sin x|$

40. $f(x) = \ln |x - \cos x|$

41. *Biomedical: Temperature during an illness.* The temperature of a patient during a 12-day illness is given by

$$T(t) = 101.6° + 3 \sin \frac{\pi}{8} t.$$

Find $T'(t)$.

42. *Business: Sales.* A company in a northern climate has sales of skis as given by

$$S(t) = 7\left(1 - \cos \frac{\pi}{6} t\right).$$

Find $S'(t)$.

43. *Satellite location.* A satellite circles the earth in such a manner that it is y miles from the equator (north or south, height not considered) t minutes after its launch, where

$$y = 5000\left[\cos \frac{\pi}{45} (t - 10)\right].$$

Find dy/dt.

44. *Rollercoaster layout.* A rollercoaster is constructed in such a way that it is y meters above ground x meters from the starting point, where

$$y = 15 + 15 \sin \frac{\pi}{50} x.$$

Find dy/dx.

45. A spring oscillates in such a way that its vertical position at time t, from its position at rest, is given by

$$y = 5 \sin (4t + \pi).$$

a) Find the amplitude, period, and phase shift.

b) Find dy/dt.

46. The current i at time t of a wire passing through a magnetic field is given by

$$i = I \sin (\omega t + a).$$

a) Find the amplitude, period, and phase shift.

b) Find di/dt.

47. A company determines that sales during the tth month are given by

$$S(t) = 40{,}000 (\sin t + \cos t).$$

The sales are seasonal and fluctuate. Find $S'(t)$.

48. A piston, connected to a crankshaft (see the figure to the right) moves up and down in such a way that its second coordinate after time t is given by

$$y(t) = \sin t + \sqrt{25 - \cos^2 t}.$$

Find the rate of change $y'(t)$.

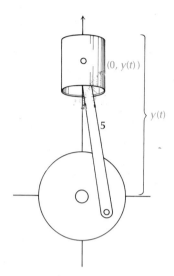

49. Referring to Margin Exercise 8, determine how long a beam will go around the corner when $a = 8$ and $b = 8$.

50. Referring to Margin Exercise 8, determine how long a beam will go around the corner when $a = 3\sqrt{3}$ and $b = 5\sqrt{5}$.

51. Two towers 40 ft apart are 30 and 20 ft high, respectively. A wire fastened to the top of each tower is guyed to the ground at a point between the towers, and is tightened so there is no sag. How far from the tallest tower will the wire touch the ground if the length of the wire is a minimum?

52. The illumination from a light source is inversely proportional to the square of the distance from the light and directly proportional to the sine of the angle of incidence. How high should a light be placed on a pole in order to maximize the illumination on the ground along the circumference of a circle of radius 25 feet?

Find the partial derivatives $f_x, f_y, f_{xx}, f_{yx}, f_{xy},$ and f_{yy}.

53. $f(x, y) = \sin 2y - ye^x$

54. $f(x, y) = e^{-x} \cos y$

55. $f(x, y) = \cos(2x + 3y)$

56. $f(x, y) = y \ln(\sin x)$

57. $f(x, y) = x^3 \tan(5xy)$

58. $f(x, y) = \sin x \cos y$

Differentiate implicitly to find y'.

59. $y = x \cos y$

60. $xy = e^x \sin y$

OBJECTIVE

You should be able to integrate trigonometric functions.

8.3 INTEGRATION OF THE TRIGONOMETRIC FUNCTIONS

Each of the previously developed differentiation formulas yields an integration formula. For example, we have the following.

THEOREM 8

$$\int \sin x \, dx = -\cos x + C, \qquad \int \cos x = \sin x + C,$$

$$\int \sec^2 x \, dx = \tan x + C, \qquad \int \csc^2 x \, dx = -\cot x + C$$

1. Find the area under the graph of $y = \sin x$ on the interval $[0, \pi]$.

Example 1 Find the area under $y = \cos x$ on the interval $[0, \pi/2]$.

Solution

$$\int_0^{\pi/2} \cos x \, dx = [\sin x]_0^{\pi/2} = \left(\sin \frac{\pi}{2}\right) - (\sin 0) = 1 \quad \boxed{CSS}$$

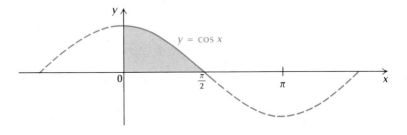

DO EXERCISE 1.

We use substitution in the following examples.

2. Integrate

$$\int (\sin x)^3 \cos x \, dx.$$

Example 2

$$\int (\sin x)^2 \cos x \, dx = \int u^2 \, du \qquad \underline{\text{Substitution}} \quad \boxed{\begin{array}{l} u = \sin x, \\ du = \cos x \, dx \end{array}}$$

$$= \frac{u^3}{3} + C$$

$$= \frac{(\sin x)^3}{3} + C$$

DO EXERCISE 2.

3. Integrate

$$\int \cos 4x \, dx.$$

Example 3

$$\int \sin 3x \, dx = \frac{1}{3} \int \sin u \, du \qquad \underline{\text{Substitution}} \quad \boxed{\begin{array}{l} u = 3x, \\ du = 3 \, dx \end{array}}$$

$$= -\frac{1}{3} \cos u + C$$

$$= -\frac{1}{3} \cos 3x + C$$

DO EXERCISE 3.

4. Integrate

$$\int \frac{2x - \sin x}{x^2 + \cos x} \, dx.$$

In an expression where natural logarithms of trigonometric functions are involved, absolute value signs are usually necessary because the trigonometric functions alternate periodically between positive and negative values. Accordingly, we use the integration formula

$$\int \frac{du}{u} = \ln |u| + C.$$

Example 4

$$\int \frac{1 + \cos x}{x + \sin x} \, dx = \int \frac{du}{u} \qquad \underline{\text{Substitution}} \qquad \boxed{\begin{array}{l} u = x + \sin x, \\ du = (1 + \cos x) \, dx \end{array}}$$

$$= \ln |u| + C$$

$$= \ln |x + \sin x| + C$$

DO EXERCISE 4.

We use integration by parts in the following example.

Example 5 Integrate $\int x \cos x \, dx$.

5. Integrate

$$\int x \sin x \, dx.$$

Solution Let

$$u = x \quad \text{and} \quad dv = \cos x \, dx.$$

Then

$$du = dx \quad \text{and} \quad v = \sin x.$$

Using the integration-by-parts formula, we get

$$\int \overset{u}{(x)} \overset{dv}{(\cos x \, dx)} = \overset{u}{(x)} \overset{v}{(\sin x)} - \int \overset{v}{(\sin x)} \overset{du}{(dx)}$$

$$= x \sin x - (-\cos x) + C$$

$$= x \sin x + \cos x + C.$$

DO EXERCISE 5.

To find an integral such as $\int \sec x \, dx$, we first have to use some algebra to rename the expression. Then we use substitution.

Example 6 Integrate $\int \sec x \, dx$.

6. Integrate

$$\int \csc x \, dx.$$

(*Hint:* Multiply by 1 using

$$\frac{\csc x + \cot x}{\csc x + \cot x}.)$$

Solution

$$\int \sec x \, dx = \int (\sec x) \cdot 1 \, dx \qquad \text{Multiplying by 1}$$

$$= \int \sec x \cdot \frac{\sec x + \tan x}{\sec x + \tan x} \, dx$$

$$\text{Substituting } \frac{\sec x + \tan x}{\sec x + \tan x} \text{ for 1}$$

$$= \int \frac{\sec x (\sec x + \tan x)}{\sec x + \tan x} \, dx$$

$$= \int \frac{\sec x \tan x + \sec^2 x}{\sec x + \tan x} \, dx$$

$$= \int \frac{du}{u} \qquad \underline{\text{Substitution}} \quad \boxed{\begin{array}{l} u = \sec x + \tan x, \\ du = (\sec x \tan x + \sec^2 x) \, dx \end{array}}$$

$$= \ln |u| + C$$

$$= \ln |\sec x + \tan x| + C$$

DO EXERCISE 6.

EXERCISE SET 8.3

1. Find the area under $y = \sin x$ on the interval $[0, \pi/3]$. **2.** Find the area under $y = \cos x$ on the interval $[0, \pi/3]$.

Integrate, using substitution.

3. $\int (\sin x)^4 \cos x \, dx$

4. $\int (\sin x)^5 \cos x \, dx$

5. $\int -(\cos x)^2 \sin x \, dx$

6. $\int (\cos x)^3 (-\sin x) \, dx$

7. $\int \cos (x + 3) \, dx$

8. $\int \sin (x + 4) \, dx$

9. $\int \sin 2x \, dx$

10. $\int \cos 3x \, dx$

11. $\int x \cos x^2 \, dx$

12. $\int x \sin x^2 \, dx$

13. $\int e^x \sin (e^x) \, dx$

14. $\int e^x \cos (e^x) \, dx$

15. $\int \tan x \, dx$ (*Hint:* $\tan x = (\sin x)/(\cos x)$.)

16. $\int \cot x \, dx$ (*Hint:* $\cot x = (\cos x)/(\sin x)$.)

Integrate by parts.

17. $\int x \cos 4x \, dx$

18. $\int x \sin 3x \, dx$

19. $\int 3x \cos x \, dx$

20. $\int 2x \sin x \, dx$

21. $\int x^2 \sin x \, dx$ (*Hint*: Let $u = x$ and $dv = x \sin dx$ and use the result of Margin Exercise 5.)

22. $\int x^2 \cos x \, dx$ (*Hint*: Let $u = x$ and $dv = x \cos x \, dx$ and use the result of Example 5.)

23. $\int \tan^2 x \, dx$

24. $\int \cot^2 x \, dx$

25. *Business: Total sales.* A company in a northern climate has sales of skis as given by

$$S(t) = 7\left(1 - \cos\frac{\pi}{6}t\right),$$

where S is sales in thousands of dollars during the tth month. Total sales for the first year are given by

$$\int_0^{12} S(t) \, dt.$$

Find the total sales.

26. *Total area under a rollercoaster.* A rollercoaster is made in such a way that it is y meters above ground x meters from the starting point, where

$$y = 15 + 15 \sin\frac{\pi}{50}x.$$

The area under the rollercoaster, from the starting point to 100 meters away, is given by

$$\int_0^{100} y \, dx.$$

Find this area.

Integrate by parts.

27. $\int \sin(\ln x) \, dx$

28. $\int \cos(\ln x) \, dx$

29. $\int e^x \cos x \, dx$

30. $\int e^x \sin x \, dx$

Integrate.

31. $\int \sec^2 7x \, dx$

32. $\int e^x \csc^2(e^x) \, dx$

33. $\int \sec x(\sec x + \tan x) \, dx$

34. $\int \sec u \tan u \, du$

35. $\int \csc u \cot u \, du$

36. $\int e^{2x} \sec(e^{2x}) \, dx$

37. $\int \frac{\cos^2 x}{\sin x} \, dx$

38. $\int \cot 7x \sin 7x \, dx$

39. $\int (1 + \sec x)^2 \, dx$

40. $\int (1 - \csc x)^2 \, dx$

OBJECTIVES

You should be able to

a) **Find certain values of inverse trig-onometric functions.**

b) **Integrate certain functions whose antiderivatives involve inverse trig-onometric functions.**

Find each of the following.   CSS

1. $\sin^{-1}\left(\dfrac{1}{2}\right)$

2. $\sin^{-1}(0)$

3. $\sin^{-1}\left(-\dfrac{\sqrt{3}}{2}\right)$

Find each of the following.

4. $\tan^{-1}(\sqrt{3})$

5. $\cos^{-1}\left(\dfrac{\sqrt{3}}{2}\right)$

6. $\cos^{-1}\left(-\dfrac{1}{2}\right)$

*8.4 INVERSE TRIGONOMETRIC FUNCTIONS

Look back at the graph of sin x. Note that outputs ranged from -1 to 1. Suppose we wanted to work backward from an output to an input. More specifically, suppose x is some number such that $-1 \le x \le 1$ and that we wanted to find a number y such that

$$-\frac{\pi}{2} \le y \le \frac{\pi}{2} \quad \text{and} \quad \sin y = x.$$

This determines a function, called the inverse sine function, given by

$$y = \sin^{-1} x \quad \text{or} \quad \arcsin x.$$

Such a function is an *inverse trigonometric function*.

Example 1 Find $\sin^{-1}(\sqrt{3}/2)$.

Solution *Think:* The number from $-(\pi/2)$ to $\pi/2$ whose sine is $\sqrt{3}/2$ is $\pi/3$. Thus

$$\sin^{-1}\left(\frac{\sqrt{3}}{2}\right) = \frac{\pi}{3}.$$

DO EXERCISES 1–3.

The inverse cosine function is defined as

$$y = \cos^{-1} x,$$

where

$$x = \cos y \quad \text{and} \quad 0 \le y \le \pi.$$

The inverse tangent function is defined as

$$y = \tan^{-1} x,$$

where

$$x = \tan y \quad \text{and} \quad -\frac{\pi}{2} < y < \frac{\pi}{2}.$$

DO EXERCISES 4–6.

---

*This section can be omitted without loss of continuity.

7. Find

$$\int \frac{1}{1 + 25x^2} \, dx$$

by using the substitution $u = 5x$ in the integral

$$\int \frac{1}{1 + x^2} \, dx = \tan^{-1} x + C.$$

Graphs of these functions are as follows.   **CSS**

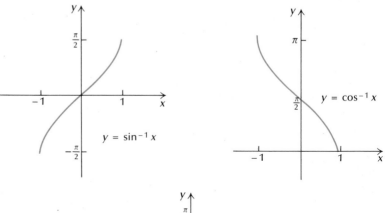

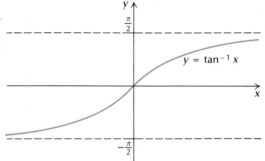

Inverse trigonometric functions are integrals of certain functions. You will often see them in tables of integrals.

Example 2 Prove that

$$\int \frac{1}{1 + x^2} \, dx = \tan^{-1} x + C.$$

Solution We use substitution:

$$\int \frac{1}{1 + x^2} \, dx = \int \frac{1}{1 + \tan^2 u} \sec^2 u \, du \quad \underline{\text{Substitution}} \quad \boxed{\begin{array}{l} u = \tan^{-1} x, \\ x = \tan u, \\ dx = \sec^2 u \, du \end{array}}$$

$$= \int \frac{1}{\sec^2 u} \sec^2 u \, du \qquad \text{By Identity (2)}$$

$$= \int du = u + C = \tan^{-1} x + C.$$

DO EXERCISE 7.

8. Find

$$\int -\frac{1}{\sqrt{1 - 4x^2}}\, dx$$

by using substitution $u = 2x$ in the integral

$$\int -\frac{1}{\sqrt{1 - x^2}}\, dx = \cos^{-1} x + C.$$

Example 3 Prove that

$$\int -\frac{1}{\sqrt{1 - x^2}}\, dx = \cos^{-1} x + C.$$

Solution Again we use substitution:

$$\int -\frac{1}{\sqrt{1 - x^2}}\, dx = \int \frac{1}{\sqrt{1 - \cos^2 u}}\, \sin u\, du.$$

Substitution
| $u = \cos^{-1} x,$ |
| $x = \cos u,$ |
| $dx = -\sin u\, du$ |

By Identity (1), $\sin^2 u + \cos^2 u = 1$, so

$$1 - \cos^2 u = \sin^2 u.$$

Then, since u is such that $0 \le u \le \pi$, $\sin u \ge 0$, so $\sqrt{\sin^2 u} = \sin u$ and the integral becomes

$$\int \frac{1}{\sqrt{\sin^2 u}} \cdot \sin u\, du \qquad \text{By Identity (1)}$$

$$= \int \frac{1}{\sin u} \cdot \sin u\, du = \int du = u + C = \cos^{-1} x + C.$$

DO EXERCISE 8.

EXERCISE SET 8.4

Find each of the following.

1. $\sin^{-1}\left(\dfrac{\sqrt{2}}{2}\right)$

2. $\sin^{-1}\left(-\dfrac{\sqrt{2}}{2}\right)$

3. $\cos^{-1}(0)$

4. $\cos^{-1}\left(\dfrac{\sqrt{2}}{2}\right)$

5. $\tan^{-1}(1)$

6. $\tan^{-1}(-1)$

7. $\sin^{-1}\left(-\dfrac{1}{2}\right)$

8. $\cos^{-1}\left(-\dfrac{\sqrt{3}}{2}\right)$

Integrate. Use substitution.

9. $\displaystyle\int \frac{e^t}{1 + e^{2t}}\, dt$

10. $\displaystyle\int \frac{1}{\sqrt{1 - 25x^2}}\, dx$

11. Show that

$$\int \frac{1}{\sqrt{1 - x^2}}\, dx = \sin^{-1} x + C.$$

12. Integrate. Use the formula in Exercise 11.

$$\int \frac{1}{\sqrt{1 - 49x^2}}\, dx$$

⊞ If your calculator has an inverse function key, find each of the following.

13. $\sin^{-1}(0.9874)$ **14.** $\cos^{-1}(-0.3487)$ **15.** $\cos^{-1}(0.9988)$ **16.** $\tan^{-1}(2000)$

17. Since

$$\int \frac{1}{1+x^2}\, dx = \tan^{-1} x + C,$$

it follows that if $u = \tan^{-1} x$, then

$$\frac{du}{dx} = \frac{1}{1+x^2}.$$

Use this result and integration by parts to find

$$\int \tan^{-1} x\, dx.$$

A summary of the important formulas for this chapter is given on the inside back cover.

CHAPTER 8 TEST

1. Convert $120°$ to radian measure. (Leave the answer in terms of π.)

2. Convert $5\pi/6$ radian measure to degrees.

Find each of the following.

3. $\sin \dfrac{\pi}{6}$

4. $\cos \pi$

Differentiate.

5. $y = \cos t$

6. $y = \sin(3x^2 - 5x)$

7. $f(x) = \dfrac{x}{\sin x}$

8. $f(t) = \tan^2 t$

9. $f(x) = \sqrt{\sin x + \cos x}$

10. $y = \dfrac{\sin x + \cos x}{\sin x - \cos x}$

11. A company has sales as given by

$$S(t) = 20\left(1 + \sin \frac{\pi}{8}t\right).$$

Find $S'(t)$.

12. Find the area under $y = \sin x$ on the interval $[0, \pi/6]$.

Integrate, using substitution.

13. $\displaystyle\int (\sin x)^9 \cos x\, dx$

14. $\displaystyle\int \cos 5t\, dt$

15. Integrate by parts.

$$\int 5x \sin 5x \, dx$$

16. Integrate. Use substitution.

$$\int \frac{1}{1 + 16t^2} \, dt$$

17. Integrate.

$$\int \frac{2 - \cos x}{2x - \sin x} \, dx$$

18. Find

$$\frac{dy}{dx} \quad \text{and} \quad \frac{d^2y}{dx^2},$$

if $y = 2e^{\cos x}$.

19. Find f_x and f_{xy}.

$$f(x, y) = \frac{\cos x}{\cos y}$$

DIFFERENTIAL EQUATIONS

The population of this island is probably pretty close to its limiting value. (*Georg Gerster: Rapho/Photo Researchers, Inc.*)

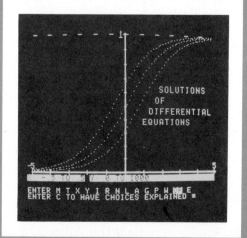

These graphs are solutions to the differential equation

$$\frac{dP}{dt} = kP(L - P).$$

9

OBJECTIVES

You should be able to

a) **Solve certain differential equations, giving both general and particular solutions.**

b) **Solve certain differential equations given a condition $f(a) = b$.**

c) **Verify that a given function is a solution of a given differential equation.**

1. Solve $y' = 3x^2$.

At the 1968 Olympic Games in Mexico City, Bob Beamon made what was believed to be a miracle long jump of 29 ft. $2\frac{1}{2}$ in. Many believed this was due to the altitude, which was 7400 ft. Using differential equations for analysis, M. N. Bearley refuted the altitude theory in "The Long Jump Miracle of Mexico City," (*Mathematics Magazine*, vol. 45, November 1972, pp. 241–246). Bearley argues that the world record jump was a result of Beamon's exceptional speed (9.5 seconds in the 100-yd dash) and the fact that he hit the take-off board in perfect position. (*United Press International Photo*)

9.1 DIFFERENTIAL EQUATIONS

A *differential equation* is an equation that involves derivatives or differentials. In Chapter 4 we studied one very important differential equation

$$\frac{dP}{dt} = kP,$$

where P, or $P(t)$, is the population at time t. This equation is a model of uninhibited population growth. Its solution is

$$P = P_0 e^{kt},$$

where the constant P_0 is the size of the initial population; that is, at $t = 0$. As this one equation illustrated, differential equations are rich in application.

Solving Certain Differential Equations

In this chapter we will frequently use the notation y' for a derivative—mainly because it is simple. Thus, if $y = f(x)$, then

$$y' = \frac{dy}{dx} = f'(x).$$

We have already found solutions of certain differential equations when we found antiderivatives or indefinite integrals. The differential equation

$$\frac{dy}{dx} = g(x) \quad \text{or} \quad y' = g(x)$$

has the solution

$$y = \int g(x) \, dx + C.$$

Example 1 Solve $y' = 2x$.

Solution

$$y = \int 2x \, dx + C = x^2 + C$$

DO EXERCISE 1.

Look again at the solution to Example 1. Note the constant of integration. This solution is called a *general solution* because taking all values of C gives *all* the solutions. Taking specific values of C gives particular

2. Given

$$y' = 3x^2.$$

a) Write the general solution.

b) Write three particular solutions.

solutions. For example, the following are particular solutions of $y' = 2x$:

$$y = x^2 + 3,$$

$$y = x^2,$$

$$y = x^2 - 3.$$

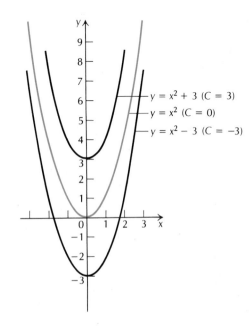

This graph shows the curves of a few particular solutions. The general solution can be envisioned as the set of all particular solutions, a *family* of curves.

DO EXERCISE 2.

Knowing the value of a function at a particular point may allow us to select a particular solution from the general solution.

Example 2 Solve $f'(x) = e^x + 5x - x^{1/2}$, given that $f(0) = 8$.

Solution

a) First find the general solution:

$$f(x) = \int f'(x)\, dx + C = e^x + \frac{5}{2}x^2 - \frac{2}{3}x^{3/2} + C.$$

3. Solve

$$y' = 2x$$

given that $y = 7$ when $x = 1$.

b) Since $f(0) = 8$, we substitute to find C:

$$8 = e^0 + \frac{5}{2} \cdot 0^2 - \frac{2}{3} \cdot 0^{3/2} + C$$

$$8 = 1 + C$$

$$7 = C.$$

Thus the solution is $f(x) = e^x + \frac{5}{2}x^2 - \frac{2}{3}x^{3/2} + 7$.

DO EXERCISES 3 AND 4.

4. Solve

$$f'(x) = \frac{1}{x} - 2x + x^{1/2}$$

given that $f(1) = 4$.

Verifying Solutions

To verify that a function is a solution of a differential equation, we find the necessary derivatives and substitute.

Example 3 Show that $y = 4e^x + 5e^{3x}$ is a solution of

$$y'' - 4y' + 3y = 0.$$

Solution

a) We first find y' and y'':

$$y' = 4e^x + 15e^{3x},$$

$$y'' = 4e^x + 45e^{3x}.$$

5. Show that

$$y = 2e^x - 7e^{3x}$$

is a solution of

$$y'' - 4y' + 3y = 0.$$

b) Then we substitute as follows in the differential equation.

$$y'' - 4y' + 3y = 0$$

$(4e^x + 45e^{3x}) - 4(4e^x + 15e^{3x}) + 3(4e^x + 5e^{3x})$	0
$4e^x + 45e^{3x} - 16e^x - 60e^{3x} + 12e^x + 15e^{3x}$	
	0

6. Show that

$$y = xe^{2x}$$

is a solution of

$$\frac{dy}{dx} - 2y = e^{2x}.$$

DO EXERCISES 5 AND 6.

EXERCISE SET 9.1

Find the general solution and three particular solutions.

1. $y' = 4x^3$

2. $y' = 6x^5$

3. $y' = e^{2x} + x$

4. $y' = e^{3x} - x$

5. $y' = \dfrac{3}{x} - x^2 + x^5$

6. $y' = \dfrac{5}{x} + x^2 - x^4$

Find the particular solution determined by the given condition.

7. $y' = x^2 + 2x - 3$;　　$y = 4$ when $x = 0$

8. $y' = 3x^2 - x + 5$;　　$y = 6$ when $x = 0$

9. $f'(x) = x^{2/3} - x$; $f(1) = -6$

10. $f'(x) = x^{2/5} + x$; $f(1) = -7$

11. Show that $y = x \ln x + 3x - 2$ is a solution of

$$y'' - \frac{1}{x} = 0 \quad \text{for } x > 0.$$

12. Show that $y = x \ln x - 5x + 7$ is a solution of

$$y'' - \frac{1}{x} = 0 \quad \text{for } x > 0.$$

13. Show that $y = e^x + 3xe^x$ is a solution of

$$y'' - 2y' + y = 0.$$

14. Show that $y = -2e^x + xe^x$ is a solution of

$$y'' - 2y' + y = 0.$$

15. Marginal cost for a certain product is $C'(x) = 2.6 - 0.02x$. Find the total cost function $C(x)$ and the average cost $A(x)$, assuming fixed costs are \$120; that is, $C(0) = \$120$.

16. Marginal revenue for a certain product is $R'(x) = 300 - 2x$. Find the total revenue function $R(x)$ assuming $R(0) = 0$.

17. A firm's marginal profit P as a function of total cost C is given by

$$\frac{dP}{dC} = \frac{-200}{(C + 3)^{3/2}}.$$

a) Find the profit function $P(C)$, if $P = \$10$ when $C = \$61$.

b) At what cost will the firm break even $(P = 0)$?

OBJECTIVE

You should be able to solve certain differential equations using separation of variables.

9.2 SEPARATION OF VARIABLES

Consider the differential equation

$$\frac{dy}{dx} = 2xy. \tag{1}$$

We treat dy/dx as a quotient, as we did in Chapter 5. We multiply Eq. (1) by dx and then by $1/y$ to get

$$\frac{dy}{y} = 2x \, dx. \tag{2}$$

1. Use separation of variables to solve

$$\frac{dy}{dx} = 3x^2 y.$$

We say that we have *separated the variables*, meaning that all the expressions involving y are on one side, and those involving x are on the other. We then integrate both sides of Eq. (2):

$$\int \frac{dy}{y} = \int 2x\, dx + C.$$

We use only one constant because any two antiderivatives differ by a constant:

$$\ln y = x^2 + C. \qquad y > 0$$

Recall that the definition of logarithms says that if $\log_a b = t$, then $b = a^t$. Now, $\ln y = \log_e y = x^2 + C$, so by the definition of logarithms, we have

$$y = e^{x^2 + C} = e^{x^2} \cdot e^C.$$

Thus the solution of differential equation (1) is

$$y = C_1 e^{x^2},$$

where

$$C_1 = e^C.$$

DO EXERCISE 1.

In the exercises you will be asked to show, using separation of variables, that the solution of

$$\frac{dP}{dt} = kP$$

is

$$P = P_0 e^{kt}.$$

Example 1 Solve

$$3y^2 \frac{dy}{dx} + x = 0.$$

Solution We first separate the variables as follows:

$$3y^2 \frac{dy}{dx} = -x$$

$$3y^2 \, dy = -x\, dx.$$

2. Solve

$$3y^2\frac{dy}{dx} - 2x = 0.$$

3. Solve

$$\frac{dy}{dx} = \frac{5}{y}.$$

We integrate both sides:

$$\int 3y^2 \, dy = \int -x \, dx + C$$

$$y^3 = -\frac{x^2}{2} + C = C - \frac{x^2}{2}$$

$$y = \sqrt[3]{C - \frac{x^2}{2}}. \qquad \text{Taking the cube root}$$

DO EXERCISE 2.

Example 2 Solve

$$\frac{dy}{dx} = \frac{x}{y}.$$

Solution We first separate variables as follows:

$$y\frac{dy}{dx} = x$$

$$y \, dy = x \, dx.$$

We then integrate both sides:

$$\int y \, dy = \int x \, dx + C$$

$$\frac{y^2}{2} = \frac{x^2}{2} + C$$

$$y^2 = x^2 + 2C$$

$$y^2 = x^2 + C_1,$$

where $C_1 = 2C$. We make this substitution to simplify the equation. We then obtain the solutions

$$y = \sqrt{x^2 + C_1} \quad \text{and} \quad y = -\sqrt{x^2 + C_1}.$$

DO EXERCISE 3.

Example 3 Solve $y' = x - xy$.

4. Solve

$$y' = 2x + xy.$$

Solution Before we separate variables we replace y' by dy/dx:

$$\frac{dy}{dx} = x - xy.$$

Now we separate variables:

$$dy = (x - xy)\, dx$$

$$dy = x(1 - y)\, dx$$

$$\frac{dy}{1 - y} = x\, dx.$$

Then we integrate both sides:

$$\int \frac{dy}{1 - y} = \int x\, dx + C$$

$$-\ln(1 - y) = -\frac{x^2}{2} + C \qquad 1 - y > 0$$

$$\ln(1 - y) = -\frac{x^2}{2} - C$$

$$1 - y = e^{-x^2/2 - C} \qquad \text{Definition of logarithms}$$

$$-y = e^{-x^2/2 - C} - 1$$

$$y = -e^{-x^2/2 - C} + 1$$

$$= -e^{-x^2/2} \cdot e^{-C} + 1.$$

Thus

$$y = 1 + C_1 e^{-x^2/2} \qquad \text{where} \quad C_1 = -e^{-C}.$$

DO EXERCISE 4.

Application: Elasticity

Example 4 The elasticity of demand for a product is 1 for all $x > 0$. That is, $E(x) = 1$ for all $x > 0$. Find the demand function $p = D(x)$ (see Section 4.6).

Solution Since $E(x) = 1$ for all $x > 0$,

$$1 = E(x) = -\frac{p}{x} \cdot \frac{1}{dp/dx}. \tag{1}$$

5. Find the demand function $p = D(x)$ given the elasticity condition

$$E(x) = 3 \quad \text{for all } x > 0.$$

Then

$$\frac{dp}{dx} = -\frac{p}{x}.$$

Separating variables, we get

$$\frac{dp}{p} = -\frac{dx}{x}.$$

Now we integrate both sides:

$$\int \frac{dp}{p} = -\int \frac{dx}{x} + C$$

$$\ln p = -\ln x + C \qquad p > 0 \text{ since } E(x) = 1 \text{ and } x > 0. \text{ See Eq. (1)}.$$

We can express $C = \ln C_1$, since any real number C is the natural logarithm of some number C_1. Then

$$\ln p = \ln C_1 - \ln x = \ln \frac{C_1}{x},$$

so

$$p = \frac{C_1}{x}.$$

This characterizes those demand functions for which the elasticity is always 1.

DO EXERCISE 5.

A Psychological Application: Reaction to a Stimulus

THE WEBER–FECHNER LAW

In psychology, one model of stimulus-response asserts that the rate of change dR/dS of the reaction R with respect to a stimulus S is inversely proportional to the stimulus. That is,

$$\frac{dR}{dS} = \frac{k}{S},$$

where k is some positive constant.

To solve this equation we first separate the variables:

$$dR = k \cdot \frac{dS}{S}.$$

We then integrate both sides:

$$\int dR = \int k \cdot \frac{dS}{S} + C$$

$$R = k \ln S + C. \tag{1}$$

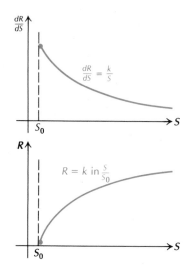

Now suppose we let S_0 be the lowest level of the stimulus that can be detected consistently. This is the *threshold value* or *detection threshold*. For example, the lowest level of sound that can be consistently detected is the tick of a watch at 20 feet, under very quiet conditions. If S_0 is the lowest level of sound that can be detected, it seems reasonable that the reaction to it would be 0. That is, $R(S_0) = 0$. Substituting this condition in Eq. (1) we get

$$0 = k \ln S_0 + C, \quad \text{or} \quad -k \ln S_0 = C.$$

Replacing C in Eq. (1) by $-k \ln S_0$, we get

$$R = k \cdot \ln S - k \cdot \ln S_0$$

$$R = k(\ln S - \ln S_0).$$

Using a property of logarithms we now have

$$\boldsymbol{R = k \cdot \ln \frac{S}{S_0}.}$$

Compare this equation with that for the loudness of sound and the magnitude of an earthquake given on p. 267. Look at the graphs of dR/dS and

6. *The Brentano–Stevens Law.* The Weber–Fechner Law has been the subject of great debate among psychologists as to its validity. The model

$$\frac{dR}{dS} = k \cdot \frac{R}{S},$$

where k is a positive constant, has also been conjectured and experimented with. Find the general solution of this equation. (This has also been referred to as the *Power Law of Stimulus-Response.*)

R given previously. Note that as the stimulus gets larger the rate of change decreases; that is, the reaction becomes smaller as the stimulation you receive gets stronger. For example, suppose you are in a room with one lamp and that lamp has a 50-watt bulb in it. If the bulb were suddenly changed to 100 watts, you would probably be very aware of the difference. That is, your reaction would be strong. If the bulb were then changed to 150 watts, your reaction would not be as great as it was to the change from 50 to 100 watts. A change from a 150- to a 200-watt bulb would cause even less reaction and so on.

For your interest, here are some other detection thresholds.

STIMULUS	DETECTION THRESHOLD
Light	The flame of a candle 30 miles away on a dark night.
Taste	Sugar dissolved in water in the ratio of 1 teaspoon to two gallons.
Smell	One drop of perfume diffused into the volume of three average size rooms.
Touch	The wing of a bee dropped on your cheek at a distance of 1 centimeter (about $\frac{3}{8}$ of an inch).

DO EXERCISE 6.

EXERCISE SET 9.2

Solve.

1. $\dfrac{dy}{dx} = 4x^3 y$

2. $\dfrac{dy}{dx} = 5x^4 y$

3. $3y^2 \dfrac{dy}{dx} = 5x$

4. $3y^2 \dfrac{dy}{dx} = 7x$

5. $\dfrac{dy}{dx} = \dfrac{2x}{y}$

6. $\dfrac{dy}{dx} = \dfrac{x}{2y}$

7. $\dfrac{dy}{dx} = \dfrac{3}{y}$

8. $\dfrac{dy}{dx} = \dfrac{4}{y}$

9. $y' = 3x + xy$

10. $y' = 2x - xy$

11. $y' = 5y^{-2}$

12. $y' = 7y^{-2}$

13. $\dfrac{dy}{dx} = 3y$

14. $\dfrac{dy}{dx} = 4y$

15. $\dfrac{dP}{dt} = 2P$

16. $\dfrac{dP}{dt} = 4P$

17. a) Use separation of variables to solve the differential equation model of uninhibited growth,

$$\frac{dP}{dt} = kP.$$

b) Rewrite the solution in terms of the condition $P_0 = P(0)$.

18. *Domar's Capital Expansion Model* is

$$\frac{dI}{dt} = hkI,$$

where I = investment, h = investment productivity (constant), k = marginal productivity to consume (constant), and t = time.

a) Use separation of variables to solve the differential equation.

b) Rewrite the solution in terms of the condition $I_0 = I(0)$.

19. *Economics: Utility.* The reaction R in pleasure units, by a consumer receiving S units of a product can be modeled by the differential equation

$$\frac{dR}{dS} = \frac{k}{S + 1},$$

where k is a positive constant.

a) Use separation of variables to solve the differential equation.

b) Rewrite the solution in terms of the initial condition $R(0) = 0$.

c) Explain why the condition $R(0) = 0$ is reasonable.

20. *Newton's Law of Cooling.* The temperature T of a cooling object drops at a rate that is proportional to the difference $T - M$, where M is the constant temperature of the surrounding medium. Thus

$$\frac{dT}{dt} = -k(T - M),$$

`CSS`

where k is a positive constant and t is time.

a) Solve the differential equation.

b) Rewrite the solution in terms of the condition $T(0) = 200°$.

Elasticity. Find the demand function $p = D(x)$ given the following elasticity conditions.

21. $E(x) = \frac{200 - x}{x}$; $p = 190$ when $x = 10$

22. $E(x) = \frac{4}{x}$; $p = e^{-1}$ when $x = 4$

23. $E(x) = 2$ for all $x > 0$

24. $E(x) = k$ for some constant k and all $x > 0$

OBJECTIVES

You should be able to

a) State the differential equation

$$\frac{dP}{dt} = kP(L - P)$$

for the inhibited growth model.

b) State the solution to the inhibited growth model

$$P(t) = \frac{P_0 L}{P_0 + e^{-Lkt}(L - P_0)}.$$

c) Given the point of inflection, find the limiting value.

d) Given the values of the constants for the equation in (b),
   i) write the equation for $P(t)$ in terms of the given constant;
   ii) sketch a graph of $P(t)$, given the time t at which the point of inflection occurs;
   iii) find t such that $P(t) = M$, for some number M.

9.3 APPLICATIONS: THE INHIBITED GROWTH MODEL $\frac{dP}{dt} = kP(L - P)$

Recall the model of uninhibited growth

$$\frac{dP}{dt} = kP,$$

with solution (by separation of variables)

$$P = P_0 e^{kt}.$$

A more realistic model of population growth will take into account factors other than the size of the population. For example, there may be some compelling reason why the population P can never grow beyond a limiting value L, due perhaps to a limitation on food, living space, or other natural resources. In such cases we expect *growth rate* to lessen as the population size increases, and to approach 0 as P approaches L.

Example 1 Buy an ant colony at a pet shop. Place a fixed amount of food and water in the colony (enough to last a long period of time) and close it up. The growth rate of the ants is inhibited by the fixed amount of food and water.

The simplest model for such an inhibition to express itself is for the growth rate dP/dt to be directly proportional to *both* the population

e) Given certain constants and function values, find the value of k in the formula in (b).

size P and to its remaining possible room for growth $L - P$. Then we have the following.

DEFINITION

> **The Model of Inhibited Growth is**
>
> $$\frac{dP}{dt} = kP(L - P),$$
>
> **where L = the *limiting value* and $k > 0$.**

Work through the following exploratory exercise set, either on your own or in class with your instructor.

Now let us solve the differential equation

$$\frac{dP}{dt} = kP(L - P).$$

Note that this equation defines P as a function of t implicitly. That is, it is assumed that P is a function of t even though t does not appear on the right side of the equation. We first separate variables:

$$dP = kP(L - P)\, dt$$

$$\frac{dP}{P(L - P)} = k\, dt.$$

Now we integrate both sides. We use Table 5 to integrate the left side. A similar example appears on p. 324.

$$\int \frac{dP}{P(L - P)} = \int k\, dt + C$$

$$\frac{1}{L} \ln \frac{P}{L - P} = kt + C$$

To solve for P we first find an expression for $P/(L - P)$:

$$\ln \frac{P}{L - P} = Lkt + LC$$

$$\frac{P}{L - P} = e^{Lkt + LC} = e^{Lkt} \cdot e^{LC}.$$

Now, letting $C_1 = e^{LC}$, we have

$$\frac{P}{L - P} = C_1 e^{Lkt}.$$

EXPLORATORY EXERCISES: SHIPWRECKED ON AN ISLAND

A ship carrying 100 passengers wrecks on an island never to be rescued. The population grows over the next 120 years as indicated in the table.

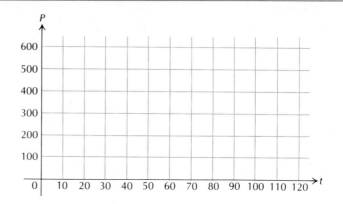

t (IN YEARS)	P (POPULATION)	$\Delta P/\Delta t$ (GROWTH RATE)
0	100	0 (assumed)
10	120	2
20	150	3
30	200	
40	270	
50	350	
60	420	
70	470	
80	500	
90	520	
100	530	
110	535	
120	538	

a) Complete the table.

b) Plot P versus t. Use solid dots and connect them with a smooth curve.

c) Plot $\Delta P/\Delta t$ versus t. Use open dots and connect them with a smooth curve.

d) Does P resemble the graph of an exponential function of the type $P(t) = P_0 e^{kt}$?

e) Between what values of t is $\Delta P/\Delta t$ increasing?

f) Between what values of t is $\Delta P/\Delta t$ decreasing?

g) Is there a point of inflection? Where?

h) Does there appear to be a limiting value to the population growth? If so, what is it?

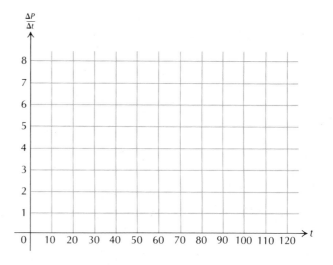

1. Let $C = 1$, $L = 1$, and $k = 1$. Then Eq. (2) becomes

$$P(t) = \frac{e^t}{1 + e^t}.$$

We want to graph $P(t)$, but to ease computations we multiply by 1 as follows:

$$P(t) = \frac{e^t}{1 + e^t} \cdot \frac{e^{-t}}{e^{-t}} = \frac{e^t \cdot e^{-t}}{1 \cdot e^{-t} + e^t \cdot e^{-t}}$$

$$P(t) = \frac{1}{e^{-t} + 1}$$

a) (▦ or Table 4). Complete this table of function values.

t	-3	-2	-1	0	1	2	3
$P(t)$							0.95

For example,

$$P(3) = \frac{1}{e^{-3} + 1} \approx \frac{1}{0.05 + 1}$$

$$= \frac{1}{1.05} \approx 0.95.$$

b) Graph $P(t)$.

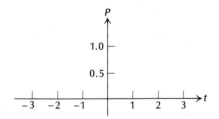

We complete the solution for P as follows, dropping the subscript from C_1 for simplicity:

$$P = Ce^{Lkt}(L - P)$$

$$P = CLe^{Lkt} - PCe^{Lkt}$$

$$P + PCe^{Lkt} = CLe^{Lkt}$$

$$P(1 + Ce^{Lkt}) = CLe^{Lkt}$$

$$P = \frac{CLe^{Lkt}}{1 + Ce^{Lkt}}.$$

THEOREM 1

The **Model of Inhibited Growth**

$$\frac{dP}{dt} = kP(L - P) \qquad (1)$$

has the solution

$$P(t) = \frac{CLe^{Lkt}}{1 + Ce^{Lkt}}. \qquad (2)$$

DO EXERCISE 1.

For each value of the constant C we get an S-shaped curve. These curves fill up the strip $0 < P < L$.

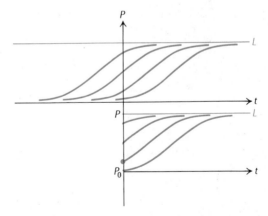

If we don't project these curves into the past (that is, if we don't use *negative time*), then the solution curves are confined to the part of the strip for which $t \geq 0$. For any initial population P_0, there is a growth

2. Given

$$P = \frac{CLe^{Lkt}}{1 + Ce^{Lkt}} \qquad (1)$$

and

$$C = \frac{P_0}{L - P_0}. \qquad (2)$$

a) Replace C in Eq. (1) by the right side of Eq. (2) and show your work in simplifying to get

$$P = \frac{P_0 Le^{Lkt}}{P_0 e^{Lkt} + (L - P_0)}. \qquad (3)$$

b) Simplify Eq. (3) by multiplying by e^{-Lkt}/e^{-Lkt}.

curve emanating from $(0, P_0)$, showing the evolution of that population. That is, $P = P_0$ at $t = 0$ is a condition that picks out a unique solution. We can eliminate the constant C and express P in terms of P_0 as follows:

$$P_0 = P(0) = \frac{CLe^{Lk0}}{1 + Ce^{Lk0}} = \frac{CL}{1 + C}. \qquad (3)$$

Solving for C, we get

$$C = \frac{P_0}{L - P_0},$$

$$P(t) = \frac{P_0 Le^{Lkt}}{P_0 e^{Lkt} + (L - P_0)}. \qquad (4)$$

We can simplify this further by multiplying by e^{-Lkt}/e^{-Lkt}.

DO EXERCISE 2.

THEOREM 2

The *Model of Inhibited Growth* (often called the *logistic function*)

$$\frac{dP}{dt} = kP(L - P)$$

has the solution

$$P(t) = \frac{P_0 L}{P_0 + e^{-Lkt}(L - P_0)}.$$

Recall that an inflection point is a point across which a curve changes concavity (see p. 161). The inflection point on the graph of $P(t)$ is important in application. It gives the time and population size at which the *rate* of population growth reaches a maximum and changes from an increasing function to a decreasing function. The following tells us where the inflection point of P occurs.

THEOREM 3

P has an inflection point at that value of t where

$$P = \frac{1}{2}L.$$

The following example is based on an actual experiment.*

A ring-necked pheasant. (©*Russ Kinne, Photo Researchers, Inc.*)

---

* A. S. Einarsen, "Some Factors Affecting Ring-necked Pheasant Population Density," *Murrelet*, **26** (1945): 2–9, 39–44.

3. The following shows the growth of an ant colony with a point of inflection at P_1. What is the limiting value?

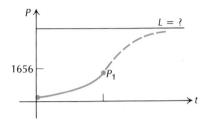

Example 2 Prior to 1937 there were no pheasants on Protection Island off the coast of Washington State. In 1937 eight pheasants, two cocks, and six hens were released on the island. The following curve shows the growth of the pheasant population over the next six years.

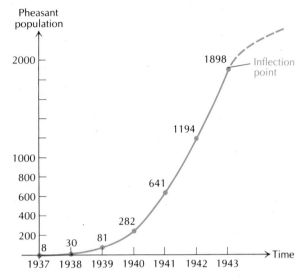

In 1943 the army arrived and started shooting the pheasants. It was theorized on the basis of the data that the population had reached its point of inflection. That is, the growth would stop turning upward and start turning downward as it approached its limiting value. What was the limiting value of the pheasant population?

Solution At the inflection point

$$P = \frac{1}{2}L.$$

Now $P = 1898$ from the graph, so

$$1898 = \frac{1}{2}L.$$

$$3796 = L.$$

DO EXERCISE 3.

The spread of an epidemic can also be described by the model of inhibited growth.

Example 3 *Spread of an epidemic.* In a town whose total population is 2000, the disease *Rottenich* creates an epidemic. The initial number of people infected is 10. The rate of spread of the infection follows the inhibited growth model

$$\frac{dP}{dt} = kP(L - P),$$

where

      P = the number of people infected after time t (in weeks),

      L = 2000, the total population of the town,

      k = 0.003, and

      P_0 = the number of people initially infected = 10.

a) Write the equation for $P(t)$ in terms of the given constants.

b) Sketch a graph of $P(t)$ given that the point of inflection occurs at t = 0.882 week.

c) At what time t are 1600 people affected?

Solution

a) $P(t) = \dfrac{P_0 L}{P_0 + e^{-Lkt}(L - P_0)} = \dfrac{10 \cdot 2000}{10 + e^{-2000(.003)t}(2000 - 10)}$

    $P(t) = \dfrac{20{,}000}{10 + 1990e^{-6t}}$

b) We first mark the limiting value 2000 on the vertical axis in the figure and draw a horizontal line through it. Then we mark P_0, or 10, on the vertical axis and the inflection point (0.882, 1000). Finally, we draw an S-shaped curve from P_0 through the inflection point, and upward approaching the limiting value.

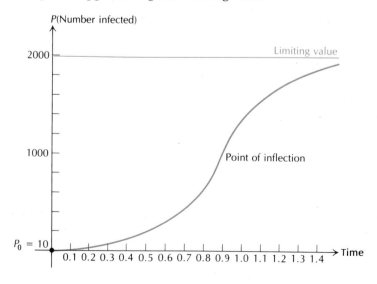

4. A town whose population is 4000 is struck by the disease *Andromedamama*. The initial number of people infected is 20. The rate of spread of the infection follows the inhibited growth model

$$\frac{dP}{dt} = kP(L - P), \quad \boxed{CSS}$$

where

$P =$ the number infected after t (months),

$L = 4000$, the total population of the town,

$k = 0.0005$, and

$P_0 = 20$, the number of people initially infected.

a) Write the equation for $P(t)$ in terms of the given constants.

b) Sketch a graph of $P(t)$, given that the point of inflection occurs at $t = 2.9$ mo. Use graph paper.

c) At what time t are 800 people infected?

5. State the solution, $Q(y)$, of

$$\frac{dQ}{dy} = rQ(M - Q).$$

c) We solve the following equation for t:

$$1600 = \frac{20,000}{10 + 1990e^{-6t}}$$

$$1600(10 + 1990e^{-6t}) = 20,000$$

$$10 + 1990e^{-6t} = \frac{20,000}{1600} = 12.5$$

$$1990e^{-6t} = 2.5$$

$$e^{-6t} = \frac{2.5}{1990}.$$

We use natural logarithms to solve this equation:

$$\ln e^{-6t} = \ln \frac{2.5}{1990}$$

$$-6t = \ln 2.5 - \ln 1990$$

$$-6t = 0.916291 - 7.595890 \quad \text{⊞ or Table 3}$$

$$t = \frac{0.916291 - 7.595890}{-6}$$

$$t \approx 1.113 \text{ wk}.$$

DO EXERCISE 4.

As you will see better after you do the exercises in this section, and as you have already seen in the preceding example, the differential equation $dP/dt = kP(L - P)$ has many applications. The solution of the equation, however, can be used in all of the applications. Of course, different variables or constants are used in different applications. For example, the solution of

$$\frac{dN}{dw} = cN(K - N)$$

is

$$N(w) = \frac{N_0 K}{N_0 + e^{-Kcw}(K - N_0)}.$$

DO EXERCISE 5.

Population Growth: Some Comments

Can a J-shaped curve (graph of the equation $P = P_0 e^{kt}$) or an S-shaped curve be used to model world population growth? To find out let us look at a log–log graph* of population growth over the last one million years.

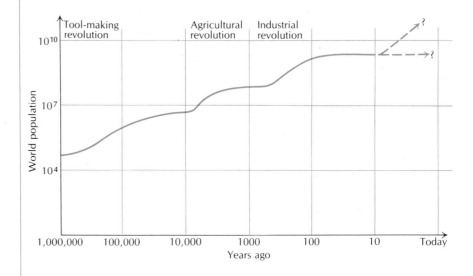

What happens is that we get a series of S-shaped curves. At three intervals there are surges of growth. These smaller intervals could be modeled by J-shaped curves. Following each surge there seems to be a leveling off, but an advance in technology causes another surge. One comes after the *tool-making revolution*. Another comes after the *agricultural revolution*, and another comes after the *industrial revolution*. Judging from the graph, and from the current concerns about availability of food, energy, and other resources, we might speculate that we are in the midst of a leveling-off period when birth rate and death rate are about the same, or will be by the year 2000.

---

* The axes are scaled using logarithms of the numbers in question. This condenses the graph.

EXERCISE SET 9.3

For the inhibited growth model:

1. $P = 450$ at the point of inflection. What is the limiting value?

2. $P = 670$ at the point of inflection. What is the limiting value?

3. In an experiment* it was found that the rate of growth of yeast cells in a laboratory satisfied the inhibited growth model

$$\frac{dP}{dt} = kP(L - P), \qquad \boxed{CSS}$$

where

$P =$ the number of cells present after time t (in hours),

$L = 700$, the limiting value,

$k = 0.0008$, and

$P_0 = 10$.

a) Write the equation for $P(t)$ in terms of the given constants.

b) Sketch a graph of $P(t)$ given that the point of inflection occurs at $t = 8$.

c) At what time t are 500 cells present?

4. A company introduced a new product on a trial run in a city. They advertised the product on television and gathered data concerning the percentage P of people who bought the product in relation to the number of times the product was advertised on television. The graph of the data resembled an S-shaped curve, so they hypothesized that it satisfied the inhibited growth model

$$\frac{dP}{dt} = kP(L - P),$$

where

$P =$ the percentage of people who buy the product,

$t =$ the number of times the product is advertised on TV,

$L = 100\%$, or 1, the highest percentage of people who could buy the product,

$k = 0.13$, and

$P_0 = 2\%$, or 0.02, the percentage who bought the product without having seen an ad.

a) Write the equation for $P(t)$ in terms of the given constants.

b) Sketch a graph of $P(t)$ given that the point of inflection (50%) occurs at $t = 30$.

c) How many times t would the company have to advertise its product so that 80% of the people would buy?

In Exercises 5 and 6 you will study two problems concerning the *diffusion of information*. This is a sociological term for the spread of information such as a rumor, or news of some new fashion rage, or news of some new product or medicine. In a limited population L the rate of diffusion of the information is proportional to the number who have heard the information N, and to the number who have not heard the information $L - N$.

---

*G. F. Gause, *The Struggle For Existence*, Baltimore: Williams and Wilkins, 1934.

5. *Spread of a rumor.* In a college with a student population of 800, a group of 6 students spread the rumor "Men go for women who study calculus, and women go for men who study calculus!" The rate at which the rumor spreads satisfies the differential equation

$$\frac{dN}{dt} = kN(L - N),$$

CSS

where

> N = the number of people who have heard the rumor after t minutes,
>
> L = 800, and
>
> N_0 = 6, the number of people who started the rumor.

a) Write the equation for $N(t)$, the solution of the differential equation, in terms of the given constants. Note that we do not yet know k.

b) The point of inflection occurs at $t = 12.2$; that is, $N(12.2) = 400$. Use this to find k,

c) After what time t have 700 students heard the rumor?

6. *Acceptance of hybrid corn.* The rate of acceptance of hybrid corn in Kentucky* followed the differential equation

$$\frac{dC}{dt} = kC(L - C),$$

where

> C = the percentage of corn acreage in hybrid at time t in years (after 1938),
>
> L = 100%, or 1, and
>
> C_0 = 3%, the percent of corn acreage at $t = 0$ (1938).

a) Write the equation for $C(t)$, the solution of the differential equation, in terms of the given constants. Note that we do not yet know k.

b) The point of inflection occurs at $t = 7$; that is, $C(7) = 50\%$, or 0.5. Use this to find k.

c) After what time t is 80% of the acreage in hybrid corn?

The acceptance of hybrid corn in Kentucky followed the uninhibited growth model. (©*John Urban, Stock, Boston*)

---

* *Agricultural Statistics*, U.S. Department of Agriculture, Washington, D.C., 1961.

EXPLORATORY EXERCISES: A PSYCHOLOGY EXPERIMENT ON PERCEPTION

Preparation

a) Find a 100-gram ("gram" is abbreviated "g") weight to be used as a *standard*.

b) Find rocks (or other objects) that weigh, respectively, about 85 g, 90 g, 95 g, 105 g, 110 g, and 115 g.

Carrying Out the Experiment   The experiment can be performed by the instructor or a student on the rest of the class.

a) Arrange the rocks in random order. Pick one. Hand the standard and the rock to each member of the class and ask the question "Is the rock heavier than the standard?"

b) Repeat (a) with each rock and keep track of the responses in a table such as that to the right.

c) Plot the data points using the axes below and connect them with a smooth curve.

ROCK WGT. (g)	STUDENTS						% Yes
	1	**2**	**3**	**4**	**5**	$\cdots$	
105	Yes	No	Yes				80
85	No	No	No				5
110							
100							
90							
95							
115							

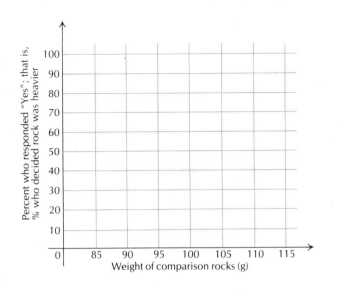

d)   Answer the following discussion questions.

1. Does the graph resemble an S-shaped curve?

2. Does a small percent of the class respond "yes" when handed the 85-g rock? Does this seem reasonable?

3. Does a large percent of the class respond "yes" when handed the 115-g rock? Does this seem reasonable?

4. What percent responds "yes" to the 100-g rock? Does this seem reasonable? Why?

5. How could the experiment be repeated to improve the shape of the curve?

EXERCISE SET 9.3 (Continued)

7. A *perception* experiment in psychology is done as in the preceding exploratory exercises. If you have not done the experiment, you should read it over. The percentage P who respond "yes" to a rock of weight w satisfies the differential equation

$$\frac{dP}{dw} = kP(L - P),$$

where

P = % who respond "yes" to a rock of weight w,

L = 100%, or 1, and

P_{85} = 4%, percent who respond "yes" to the weight of w = 85 g (gram).

Look the graph over. Doesn't it seem reasonable that only 4% would have judged the 85-g rock to be heavier than the 100-g rock? As the weight of the comparison rocks gets to the standard, the percent increases. When the comparison weights 100 g, the same as the standard, 50% say it is heavier. That is, half would say it is heavier and half would say it isn't. Then as the comparison rock gets heavier and heavier, the percent of "yes" responses increases toward 100%.

a) Write the equation for $P(w)$, the solution of the differential equation, in terms of the given constants. Note that w = 0 means 85 g, w = 1 means 90 g, w = 2 means 95 g, and so on. Note that we have not yet determined k.

b) Use the fact that P = 50% = 0.5 at w = 3 (100 g) to find k.

c) Rewrite the equation for $P(w)$ in terms of k and the other constants.

d) At what weight w does 80% respond "yes"?

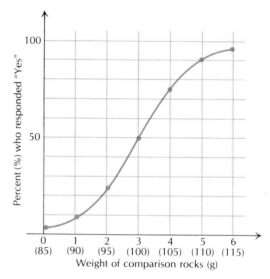

OBJECTIVES

You should be able to

a) State the solution to the growth model

$$\frac{dP}{dt} = k(L - P).$$

b) Use Table 4 to graph equations of the type

$$P(t) = L(1 - e^{-kt}).$$

c) Given L and a function value for the equation in (b), find k.

d) Given L and k for the equation in (b), find a certain function value $P(t)$.

e) Given L and k for the equation in (b), find t such that $P(t) = M$, for some given number M.

1. Exploratory exercises: Advertising. A company introduces a new product on a trial run in a city. It advertises the product on TV and gathers data concerning the percentage P of the people in the city who bought the product after it was advertised a certain number of times.

a) The following table contains the data. Complete the calculations in the table.

t (NUMBER OF TIMES AD RAN)	P (% WHO BOUGHT PRODUCT)	$\Delta P/\Delta t$ (% RATE OF CHANGE)
0	0	undefined
10	50	5
20	75	
30	87	
40	94	
50	97	

(Exercise continued on next page.)

9.4 APPLICATIONS: THE GROWTH MODEL $\frac{dP}{dt} = k(L - P)$

We have considered two models of growth:

$$\frac{dP}{dt} = kP, \qquad \text{Uninhibited growth model}$$

$$\frac{dP}{dt} = kP(L - P). \qquad \text{Inhibited growth model}$$

A third kind of growth assumes a limited population L, but assumes that the growth rate dP/dt is directly proportional only to its remaining possible room for growth $L - P$, independent of the number who already exist. The model is

$$\frac{dP}{dt} = k(L - P), \qquad \text{where} \quad P = 0 \text{ when } t = 0.$$

This model has not proved suitable to describe biological population growth. Thus we will not consider that here. It is applicable, however, to other areas. Before solving the differential equation we consider one such application in Margin Exercise 1.

DO EXERCISE 1.

We have considered a problem similar to this before in Exercise 4 of Exercise Set 9.3 where we encountered an S-shaped curve. In Margin Exercise 1 the graph did not resemble an S-shaped curve. In most advertising applications a curve like that in Margin Exercise 1 is what occurs. Models can vary, of course.

Now let us solve the differential equation

$$\frac{dP}{dt} = k(L - P).$$

We first separate the variables

$$\frac{dP}{L - P} = k \, dt.$$

Now we integrate both sides:

$$\int \frac{dP}{L - P} = \int k \, dt + C$$

$$-\ln (L - P) = kt + C$$

$$\ln (L - P) = -kt - C$$

$$L - P = e^{-kt-C} = e^{-kt} \cdot e^{-C}.$$

b) Plot P versus t. Connect the points with a smooth curve.

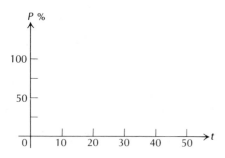

c) Plot $\Delta P/\Delta t$ versus t. Connect the points with a smooth curve.

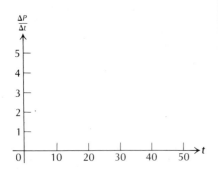

d) Does P resemble an S-shaped curve?

e) Does there seem to be a point of inflection for P?

f) What appears to be the limiting value for P?

g) Does the slope of the graph in (b) appear to be increasing? decreasing? neither?

Letting $C_1 = e^{-C}$, we have

$$L - P = C_1 e^{-kt}$$
$$-P = -L + C_1 e^{-kt}$$
$$P = L - C_1 e^{-kt}.$$

Now $P = 0$ when $t = 0$, so

$$0 = L - C_1 e^{-k \cdot 0}$$
$$= L - C_1 \cdot 1$$
$$= L - C_1.$$

Thus $L = C_1$, and

$$P = L - L e^{-kt}$$
$$= L(1 - e^{-kt}).$$

THEOREM 4

The solution of

$$\frac{dP}{dt} = k(L - P) \qquad P(0) = 0 \qquad (1)$$

is

$$P(t) = L(1 - e^{-kt}). \qquad (2)$$

DO EXERCISE 2 (on the next page).

As Margin Exercise 2 illustrates, the following is a graph of $P(t)$. Note that the graph starts at $(0, 0)$ and increases upward toward the limiting value L. The slope is decreasing and there is no point of inflection.

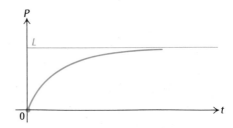

2. Let $L = 3$ and $k = 1$, so   `CSS`

$$P(t) = 3(1 - e^{-t}).$$

a) (▦ or Table 4). Complete this table of function values.

t	0	1	2	3	4	5
$P(t)$			2.6			

For example,

$$P(2) = 3(1 - e^{-2})$$
$$= 3(1 - 0.135335)$$
$$\approx 2.6.$$

b) Graph $P(t)$.

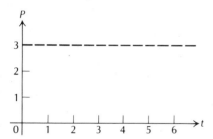

3. a) In another advertising experiment it was determined that $L = 100\%$, so

$$P(t) = 1 - e^{-kt}.$$

One of the data points was

$$P(10) = 40\%, \quad \text{or} \quad 0.4.$$

Use this value to determine k. Round to the nearest hundredth.

b) Rewrite $P(t)$ in terms of k.

c) Use the equation in (b) to find $P(60)$.

d) How many times must the product be advertised so that 99% of the people will buy it?

Now reconsider the problem on advertising in Margin Exercise 1.

Example Reread Margin Exercise 1.

a) Suppose we want to "fit" Eq. (2) to this data. we know that $L = 100\%$, or 1, so

$$P(t) = 1 - e^{-kt}.$$

Thus we must determine a value for k. A way to do this is to arbitrarily pick a data point, substitute, and solve for k. We use $(10, 50\%)$; that is $P(10) = 50\%$, or 0.5. Substituting and solving, we have

$$0.5 = 1 - e^{-k \cdot 10}$$
$$-0.5 = -e^{-10k}$$
$$0.5 = e^{-10k}.$$

We use natural logarithms to solve this equation:

$$\ln 0.5 = \ln e^{-10k}$$
$$-0.693147 = -10k \qquad ▦ \text{ or Table 3}$$
$$0.0693147 = k$$
$$0.07 \approx k.$$

b) We then rewrite $P(t)$ in terms of k:

$$P(t) = 1 - e^{-0.07t}.$$

c) Use the equation in (b) to find $P(50)$:

$$P(50) = 1 - e^{-0.07 \cdot 50} = 1 - e^{-3.5}$$
$$= 1 - 0.030197$$
$$\approx 0.9698.$$

Note that this is close to the data value of 97%. This may not always happen in practice.

d) How many times must the product be advertised so that 99% of the people will buy it? We solve

$$0.99 = 1 - e^{-0.07t}$$

for t and get $t \approx 66$.

DO EXERCISE 3.

EXERCISE SET 9.4

Graph.   CSS

1. $P(t) = 4(1 - e^{-t})$ **2.** $P(t) = 2(1 - e^{-t})$ **3.** $P(t) = 1 - e^{-2t}$ **4.** $P(t) = 1 - e^{-3t}$

5. In a different advertising experiment it was determined that $L = 50\%$, or 0.5; that is, no matter how much a company advertised, the percentage of people who bought the product would never reach or exceed 50%.

a) Express $P(t)$ in terms of $L = 0.5$.

b) It was determined that $P(10) = 25\%$, or 0.25. Use this to determine k. Round to the nearest hundredth.

c) Rewrite $P(t)$ in terms of k.

d) Use the equation in (c) to find $P(30)$.

e) How many times must the product be advertised so that 49% of the people will buy it?

6. *Biomedical: Acceptance of a new medicine.* A different model for the diffusion of information—for example, awareness of a new medicine by doctors—is the differential equation

$$\frac{dP}{dt} = k(L - P),$$

where

$P =$ the percentage of doctors who are aware of the new medicine after time t (in months),

$L = 100\%$, or 1, and

$k =$ a constant.

a) Write the equation for $P(t)$, the solution of the differential equation, in terms of L and k.

b) Suppose $P(4) = 50\%$, or 0.5. Use this to determine k. Round to the nearest tenth.

c) Rewrite $P(t)$ in terms of k.

d) Use the equation in (c) to find $P(6)$.

e) How many months will it take for 90% of the doctors to become aware of the new medicine?

7. *Psychology: Hullian Model of Learning.* This model is expressed by

$$\frac{dP}{dt} = k(L - P),$$

where

$P =$ the probability of mastery of a certain concept after t learning trials,

$L = 1$, the limiting value of mastery of the learning, and

$k =$ a constant.

a) Write the solution $P(t)$ in terms of L and k.

b) Suppose $P(5) = 0.6$. Use this to determine k. Round to the nearest hundredth.

c) Rewrite $P(t)$ in terms of k.

d) Use the equation in (c) to find $P(10)$.

e) Find t such that $P(t) = 0.90$. That is, how many trials are necessary for the probability of mastery of learning to be 0.90?

8. *Biology: Yield due to the spread of fertilizer.* A farmer is growing soybeans in a field. The more fertilizer he spreads, the greater his yield, up to some limiting value L. A model for this, applied to any given specific farm, is

$$\frac{dY}{dn} = k(L - Y),$$

where

> Y = the yield, in bushels per acre, upon spreading n pounds of fertilizer per acre,
>
> L = 60 bushels per acre, and
>
> k = a constant.

a) Write the solution $Y(n)$ in terms of L and k.

b) Suppose $P(10) = 21$. Use this to determine k. Round to the nearest hundredth.

c) Rewrite $Y(n)$ in terms of k.

d) Use the equation in (c) to find $Y(5)$.

e) How many pounds of fertilizer, per acre, must be spread to yield 42 bushels per acre?

(*Note:* This model is not completely valid in the sense that there usually is a certain amount of fertilizer that will kill the crop.)

The yield due to fertilizer can be modeled by $Y' = k(L - Y)$. (*Charles O'Rear, EPA Documerical*)

9. Inside nuclear reactors, radioactive material is produced that satisfies

$$\frac{dM}{dt} = k\left(\frac{p}{k} - M\right), \qquad M = 0 \text{ when } t = 0;$$

where

> M = the mass of radioactive material inside the reactor after time t (in years),
>
> p = the rate at which the radioactive material is produced, and
>
> k = the annual decay rate of the radioactive material.

Find the solution, $M(t)$, of the differential equation in terms of p and k.

Nuclear power station at Rowe, Massachusetts, in operation since 1960. (*Photo courtesy of Yankee Atomic Electric Company*)

10. The *growth rate of a certain stock* is modeled by

$$\frac{dV}{dt} = k(L - V), \qquad V = \$20 \text{ when } t = 0;$$

where

> V = the value of the stock, per share, after time t (in months),
>
> L = \$24.81, the limiting value of the stock, and
>
> k = a constant.

Find the solution of the differential equation in terms of L and k.

11. *Fick's Law in Biology* is given by

$$\frac{dC}{dt} = \frac{kA}{V}(C_T - C),$$

where

> C = the amount of a substance that passes through a tube with cross-sectional area A and volume V at time t,
>
> k = a constant, and
>
> C_T = the total amount of substance.

Find the solution of the differential equation assuming $C(0) = C_0$.

OBJECTIVES

You should be able to

a) **Determine whether or not a given first-order differential equation is exact.**

b) **Solve an exact first-order differential equation.**

9.5 EXACT EQUATIONS

Consider the differential equation

$$y' = \frac{dy}{dx} = \frac{-3x^2 - y}{x + 1}.$$

Although this equation appears to be a likely candidate for separation of variables, a little experimentation will convince you that this cannot be done.

Instead, we try the following approach. Write the equation in *differential form*. This means that we clear the equation of "fractions" by formally multiplying both sides by $(x + 1) \, dx$. This gives us

$$(x + 1) \, dy = (-3x^2 - y) \, dx,$$

or, rewriting,

$$(3x^2 + y) \, dx + (x + 1) \, dy = 0.$$

Now consider the function $F(x, y) = x^3 + xy + y$. Its partial derivatives are $F_x = 3x^2 + y$ and $F_y = x + 1$. The equation above is just

$$F_x \, dx + F_y \, dy = 0.$$

If this last expression looks familiar, it's because we studied the total differential in Chapter 7. We can rewrite this equation as

$$dF = 0,$$

where we have used the fact that

$$dF = F_x \, dx + F_y \, dy.$$

1. Check by substitution that

$$y = \frac{C - x^3}{x + 1}$$

satisfies the differential equation

$$y' = \frac{dy}{dx} = \frac{-3x^2 - y}{x + 1}.$$

Since $dF = 0$ when $F(x, y) = C$, the last equation suggests that the solution is $F(x, y) = C$. That turns out to be correct. We get

$$x^3 + xy + y = C$$

or, solving for y,

$$y(x + 1) = C - x^3,$$

or

$$y = \frac{C - x^3}{x + 1}.$$

Now that we have the solution, we can check it by direct substitution.

DO EXERCISES 1 AND 2.

Unfortunately, it's not always possible to take a first-order differential equation, write it in differential form, and then find a function F such that the equation becomes $dF = 0$.

2. Write the differential equation

$$y' = \frac{dy}{dx} = \frac{-x}{y}$$

in differential form.

DEFINITION
___

The first-order differential equation

$$P(x, y)\, dx + Q(x, y)\, dy = 0$$

is *exact* if there is a function $F(x, y)$ such that

$$F_x = \frac{\partial F}{\partial x} = P(x, y) \quad \text{and} \quad F_y = \frac{\partial F}{\partial y} = Q(x, y).$$

___

In other words, $P(x, y)\, dx + Q(x, y)\, dy$, or simply $P\, dx + Q\, dy = 0$ is exact if we can solve it just as we solved the example above. The following theorem tells us when an equation is exact.

THEOREM 5
___

If $P(x, y)$ and $Q(x, y)$ have continuous partial derivatives for all (x, y) in a rectangle R, then

$$P(x, y)\, dx + Q(x, y)\, dy = 0$$

is *exact* if and only if

$$P_y = Q_x.$$
___

3. Show that

$$x \, dx + y \, dy = 0$$

is exact.

4. Show that

$$(6xy - 2y) \, dx + (3x^2 - 2x + 1) \, dy = 0$$

is exact.

5. a) "Simplify" the equation

$$2xy \, dx + x^2 dy = 0$$

by dividing by x.
b) Show that the resulting equation is *not* exact.

Example 1 Show that the equation

$$2xy \, dx + x^2 \, dy = 0$$

is exact.

Solution Here $P(x, y) = 2xy$ and $Q(x, y) = x^2$. So,

$$\frac{\partial P}{\partial y} = P_y = 2x$$

and

$$\frac{\partial Q}{\partial x} = Q_x = 2x.$$

Since $P_y = Q_x$, the equation is exact.

DO EXERCISES 3–5.

Example 2 Solve $x \, dx + y \, dy = 0$.

Solution We know from Margin Exercise 3 that this equation is exact. The problem here is to find $F(x, y)$. Now, $F(x, y)$ must satisfy $F_x = x$ and $F_y = y$. Consider the first equation. We can antidifferentiate it to get $F(x, y) = \frac{1}{2} x^2 + C$. To be correct, however, recall that since F is a function of two variables and since we are integrating with respect to x (and thinking of y as a constant), the constant C above may involve y-terms. We write

$$F(x, y) = \frac{1}{2} x^2 + g(y),$$

where g is an unknown function. To find g, recall that F must also satisfy $F_y = y$. Here, $F_y = \partial F/\partial y = g'(y)$. Thus $g'(y) = y$, and, antidifferentiating, we have $g(y) = \frac{1}{2} y^2 + C_1$, where C_1 is a constant. So we find that

$$F(x, y) = \frac{1}{2} x^2 + \frac{1}{2} y^2 + C_1.$$

The solution to

$$x \, dx + y \, dy = 0,$$

then, is given by $\frac{1}{2} x^2 + \frac{1}{2} y^2 + C_1 = C$, where C is a constant. We can simplify this result by subtracting C_1 on both sides and multiplying by

6. Solve the exact equation in Example 1.

7. Solve $y\, dx + x\, dy = 0$.

8. Solve the equation in Margin Exercise 4.

2. We then get

$$x^2 + y^2 = K,$$

where $K = 2(C - C_1)$ is a constant. The solution curves are a family of concentric circles, each centered at (0, 0).

Notice, in retrospect, that the constant C_1 was irrelevant and could have been left out. We will do so in the future.

DO EXERCISES 6–8.

Example 3 Solve the initial-value problem

$$(2xy^2 - y + 5)\, dx + (2x^2y - x)\, dy = 0, \qquad y(1) = 2.$$

Solution We first solve the equation just as we would if $y(1) = 2$ were not given. After solving the equation, we'll use the initial condition to eliminate the constant of integration.

Here, $P(x, y) = 2xy^2 - y + 5$ and $Q(x, y) = 2x^2y - x$. Testing for exactness, we find that $P_y = 4xy - 1$ and $Q_x = 4xy - 1$. The equation is exact, since $P_y = Q_x$.

To find $F(x, y)$, we recall that $F_x = P = 2xy^2 - y + 5$ and integrate with respect to x. We get

$$F(x, y) = \int P(x, y)\, dx$$

$$= \int (2xy^2 - y + 5)\, dx$$

$$= x^2y^2 - xy + 5x + g(y).$$

So $\partial F/\partial y = F_y = 2x^2y - x + g'(y)$. Since we also must have $F_y = Q = 2x^2y - x$, we see that $g'(y) = 0$. So, $g(y) = C_0$, a constant, and may be taken to be 0. Thus $F(x, y) = x^2y^2 - xy + 5x$ and $x^2y^2 - xy + 5x = C$ gives the solution.

To evaluate C, note that $y = 2$ when $x = 1$. Direct substitution above yields

$$(1)^2(2)^2 - 1 \cdot 2 + 5 \cdot 1 = C$$

or

$$C = 1 \cdot 4 - 2 + 5$$

$$= 9 - 2 = 7.$$

9. Use the quadratic formula to solve

$$x^2 y^2 - xy + 5x = 7$$

for y.

10. Solve the initial-value problem

$$2xy^3 \, dx + 3x^2 y^2 \, dy, \qquad y(1) = 2.$$

11. Solve the initial-value problem

$(1 + 2y) \, dx + 2x \, dy = 0, \; y(1) = -1.$

Therefore,

$$x^2 y^2 - xy + 5x = 7.$$

It isn't necessary to do so, but you may solve the last equation for y by regarding $a = x^2$, $b = -x$, and $c = 5x - 7$ and applying the quadratic formula to $ay^2 + by + c = 0$.

DO EXERCISES 9–11.

EXERCISE SET 9.5

Determine whether the given equation is exact.

1. $(2x + 3) \, dx + 6y \, dy = 0$

2. $(2y + 3) \, dx + 5x \, dy = 0$

3. $(6y + 3) \, dx + 6x \, dy = 0$

4. $\sin y \, dx + \cos x \, dy = 0$

5. $\sin y \, dx + x \cos y \, dy = 0$

6. $\sin x \, dx + y \cos x \, dy = 0$

7. $\left(y^3 e^x + \dfrac{y}{x} \right) dx + (3y^2 e^x + 2y + \ln x) \, dy = 0$

8. $(4y^5 x^3 - 6xy^2) \, dx + (5x^4 y^4 - 6x^3 y) \, dy = 0$

9. Solve the equation in Exercise 1.

10. Solve the equation in Exercise 3.

11. Solve the equation in Exercise 5.

12. Solve the equation in Exercise 7.

Show that the following equations are exact. Then solve the initial-value problem.

13. $(3x^2 + 3y) \, dx + (3x + 3y^2) \, dy = 0, \qquad y(1) = 1$

14. $2xy \, dx + (1 + x^2) \, dy = 0, \qquad y(1) = \dfrac{1}{2}$

15. $(\cos y - y \sin x) \, dx + (\cos x - x \sin y) = 0,$
$y(0) = 1$

16. $(x^3 + 2xy) \, dx + (x^2 + y^3) \, dy = 0, \qquad y(2) = 2$

17. Find the equation of the curve that satisfies the differential equation

$$\frac{dy}{dx} = \frac{y - x}{2y - x}$$

and passes through $(1, 1)$.

18. *Business.* A company finds that if p is the price per unit (in dollars) of its goods and if x represents the quantity of goods sold, then the elasticity of demand is

$$E(x) = \frac{1 + x^2}{1 + 3x^2}$$

Use the fact that

$$E = -\frac{p}{x} \cdot \frac{dx}{dp}$$

(see Section 4.7) to set up and solve a differential equation for p, given that $p(2) = \$0.10$.

19. Show that the substitution $y = vx$ in the differential equation in Exercise 17 changes it to a problem that can be solved by separation of the variables.

You should be able to solve any first-order linear differential equation.

9.6 FIRST-ORDER LINEAR EQUATIONS

We begin this section with the following definition.

DEFINITION

A first-order differential equation is called *linear* if it can be written in the form

$$a_0(x)y' + a_1(x)y = b(x). \tag{1}$$

Note that $a_0(x)$ and $a_1(x)$, which are called the *coefficients* of the equation, are functions of x alone. So too is the function $b(x)$ on the right-hand side of the equation. So, for example, the equation

$$(x^2 + 1)y' - xy = e^x$$

is a first-order linear differential equation with $a_0(x) = x^2 + 1$, $a_1(x) = -x$, and $b(x) = e^x$. On the other hand, $(y')^3(y) - e^y = 0$ is *not* a linear equation.

In this text, we will always assume that the coefficients $a_0(x)$ and $a_1(x)$ and the function $b(x)$ (sometimes called the *forcing function*) are all continuous on some closed interval $[c, d]$. This guarantees that Eq. (1) has a solution on $[c, d]$.

Example 1 Solve $xy' - 3x^3y = 0$.

Solution Comparing this equation with the definition above, we have $a_0(x) = x$, $a_1(x) = -3x^3$, and $b(x) = 0$. (Equations in which $b(x) = 0$ are called *homogeneous*.) Dividing through by x, we have

$$y' - 3x^2y = 0.$$

Using $y' = dy/dx$, we have

$$\frac{dy}{dx} = 3x^2y.$$

We see now that we can separate the variables in this equation. Doing so, we obtain

$$\frac{dy}{y} = 3x^2 \, dx.$$

Integrating both sides, we have

$$\int \frac{dy}{y} = \int 3x^2 \, dx + C$$

$$\ln y = x^3 + C.$$

1. Solve $xy' - y = 0$.

Applying the definition of logarithms to $\ln y = \log_e y = x^3 + C$, we get

$$y = e^{x^3 + C}$$

or

$$y = e^{x^3} e^C$$

or

$$y = C_1 e^{x^3} \qquad \text{where } C_1 = e^C.$$

DO EXERCISE 1.

It turns out that solving an arbitrary *homogeneous* first-order differential equation is no harder than solving Example 1 or Margin Exercise 1. That is, every homogeneous linear first-order equation can be solved by separating the variables (see Exercise Set 9.6). Unfortunately, though, separation of variables usually doesn't work in the nonhomogeneous case. Nor is such an equation exact. Consider, for example,

$$a_0(x)y' + a_1(x)y = b(x).$$

Dividing through by $a_0(x)$ yields

$$y' + \frac{a_1(x)}{a_0(x)} y = \frac{b(x)}{a_0(x)},$$

where we assume that $a_0(x) \neq 0$. For simplicity, let

$$p(x) = \frac{a_1(x)}{a_0(x)} \quad \text{and} \quad q(x) = \frac{b(x)}{a_0(x)}.$$

Then our equation becomes

$$y' + p(x)y = q(x).$$

This is sometimes called the *standard form* of the equation. Now, letting $y' = dy/dx$ and rearranging gives

$$p(x)y - q(x) + \frac{dy}{dx} = 0$$

or

$$[p(x)y - q(x)]\, dx + dy = 0 \qquad (2)$$

in differential form. Here, to consider exactness, let $P(x, y) = p(x)y - q(x)$ and $Q(x, y) = 1$. Testing for exactness, we find that $\partial P/\partial y = p(x)$ and $\partial Q/\partial x = 0$, so that the equation isn't exact except in the uninteresting case $p(x) = 0$.

In Margin Exercise 5 of Section 9.5 we saw that exactness could be destroyed by dividing by a term. If this is the case, then we can restore it by multiplying by that term. Here, we can make the equation exact by multiplying by $e^{\int p(x)\,dx}$. We'll check this out in a moment, but first, let's calculate the derivative (with respect to x) of $e^{\int p(x)\,dx}$. Recall that (by the Chain Rule) the derivative of $e^{f(x)}$ is $e^{f(x)} \cdot f'(x)$ for any function $f(x)$. If $v = e^{\int p(x)\,dx}$, then

$$\frac{dv}{dx} = e^{\int p(x)\,dx} \cdot \frac{d}{dx}\left[\int p(x)\,dx\right].$$

But the derivative of $\int p(x)\,dx$ is just $p(x)$. So,

$$\frac{dv}{dx} = p(x) \cdot e^{\int p(x)\,dx}.$$

In differential form, our original equation was

$$[p(x)y - q(x)]\,dx + dy = 0.$$

Multiplying by $e^{\int p(x)\,dx}$ gives

$$(p(x)e^{\int p(x)\,dx}y - q(x)e^{\int p(x)\,dx})\,dx + e^{\int p(x)\,dx}\,dy = 0.$$

The partial derivative with respect to y of the left-hand term is $p(x)e^{\int p(x)\,dx}$. The partial derivative with respect to x of $e^{\int p(x)\,dx}$ is just its derivative with respect to x, and, from above, that's also $p(x)e^{\int p(x)\,dx}$. The new form of the equation is exact!

We'll illustrate with an example.

Example 2 Solve $xy' - 3x^3y = 0$.

Solution Dividing through by x, we have $y' - 3x^2y = 0$. Here, $p(x) = -3x^2$ so that $\int p(x)\,dx = -x^3$ (to simplify matters, we omit the constant of integration). So we'll multiply the equation by e^{-x^3}. Doing so yields

$$e^{-x^3}y' - 3x^2e^{-x^3}y = 0 \cdot e^{-x^3}$$

or

$$e^{-x^3}y' - 3x^2e^{-x^3}y = 0.$$

Now the left-hand side of the equation is just the derivative of $e^{-x^3} \cdot y$ (you should check this yourself using the Product Rule). Thus we have

$$(e^{-x^3}y)' = 0.$$

Integrating, we get

$$e^{-x^3}y = C, \quad \text{a constant.}$$

2. By direct substitution, show that $y = Ce^{x^3}$ is a solution to $xy' - 3x^3y = 0$.

Multiplying both sides by e^{x^3} gives

$$e^{x^3}e^{-x^3}y = Ce^{x^3}$$

or

$$e^0 \cdot y = Ce^{x^3}$$

or

$$y = Ce^{x^3} \qquad \text{since } e^0 = 1.$$

DO EXERCISES 2 AND 3.

Example 3 Solve $xy' + 2y = 5x^3$.

Solution We divide through by x (valid if we restrict ourselves to positive values of x), obtaining

$$y' + \frac{2}{x}y = 5x^2, \qquad x > 0.$$

Here,

3. Solve $y' + 2xy = 0$ in two different ways.

$$p(x) = \frac{2}{x} \quad \text{and} \quad \int p(x)\,dx = 2\int \frac{dx}{x} = 2\ln x.$$

Note that we can write $\int p(x)\,dx = 2\ln x = \ln x^2$, using a property of logarithms. Now we'll multiply the equation by $e^{\int p(x)\,dx} = e^{\ln x^2}$ (this function is sometimes called the *integrating factor* for the equation). Note that $e^{\ln x^2} = x^2$, using Property 7 of Section 4.5. Multiplying through by x^2 gives

$$x^2y' + x^2 \cdot \left(\frac{2}{x}\right)y = 5x^2 \cdot x^2,$$

or

$$x^2y' + 2xy = 5x^4.$$

Now the left-hand side is just the derivative of x^2y. That is, we have

$$(x^2y)' = 5x^4.$$

Integrating, we have $x^2y = x^5 + C$, where C is a constant of integration. So, solving for y, we now have

$$y = x^3 + \frac{C}{x^2} = x^3 + Cx^{-2},$$

valid for $x > 0$. Note that the solution is undefined at $x = 0$. We knew

4. Solve $x^2y' + 4xy = 8x$.

that our solution might not be valid there, since we had to divide through by x in the first step of the solution.

DO EXERCISES 4 AND 5.

Example 4 *Mixing.* A tank initially contains a brine solution consisting of 400 pounds of salt dissolved in 1200 gallons of water. Salt water containing 1 pound of salt per gallon of water is pumped into the tank at the rate of 5 gallons per minute. The brine is pumped out at the rate of 5 gallons per minute. Assuming that the mixture is kept uniform by stirring, find the amount of salt in the tank after 5 hours.

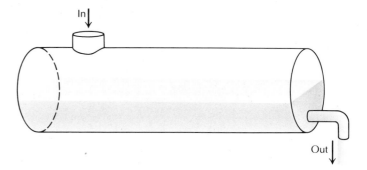

5. Solve $y' + 2y = e^{3x}$.

Solution Let $A(t)$ be the amount of salt (in pounds) that is in the tank after t minutes. We are given that $A(0) = 400$ and we want to find $A(300)$. We will find this by setting up and then solving a differential equation for $A(t)$. Note that the expression $A'(t) = dA/dt$ represents the rate of change (the rate of increase or decrease) of the amount of salt in the tank at time t.

Since salt water containing 1 pound per gallon is being pumped into the tank at a rate of 5 gallons per minute, we see that salt is entering the tank at the rate of 5 pounds per minute.

The mixture is being pumped out of the tank at the rate of 5 gallons per minute. How much salt does each gallon of the mixture contain? At time t, there are $A(t)$ pounds of salt dissolved in 1200 gallons of water. The concentration of salt (in pounds per gallon) at time t is $A(t)/1200$. Since 5 gallons per minute are being pumped out, this means that

$$\frac{A(t)}{1200} \cdot 5 = \frac{A(t)}{240}$$

pounds per minute are being pumped out.

Now the rate of change of salt in the tank, $dA/dt = A'(t)$, is just the difference between the rate at which it enters the tank and the rate at

6. Show that the equation

$$A'(t) = 5 - \frac{A(t)}{240}$$

from Example 4 can be solved by separating the variables.

which it leaves the tank. That is,

$$\frac{dA}{dt} = \text{(rate entering)} - \text{(rate leaving)}$$

$$\frac{dA}{dt} = 5 - \frac{A(t)}{240}.$$

Rearranging, we have

$$A'(t) + \frac{A(t)}{240} = 5,$$

a first-order linear equation. To solve it, note that the integrating factor is $e^{\int (1/240)\,dt} = e^{t/240}$. The equation becomes

$$(e^{t/240}A)' = 5e^{t/240},$$

or, after integrating,

$$e^{t/240}A = 1200e^{t/240} + C$$

or

$$A(t) = 1200 + Ce^{-t/240}.$$

Recall that $A(0) = 400$. Letting $t = 0$ and $A = 400$ above, we get

$$400 = 1200 + Ce^{-0/240}$$

7. Find the answer to the question in Example 4 if the solution pumped into the tanks contains only 0.6 pound of salt per gallon.

or

$$400 = 1200 + Ce^{0}$$

and

$$C = -800.$$

So

$$A(t) = 1200 - 800e^{-t/240}.$$

In particular,

$$A(300) = 1200 - 800e^{-300/240}$$

$$= 1200 - 800e^{-1.25}$$

$$= 1200 - 800(0.2865048)$$

$$\approx 1200 - 229.204$$

$$\approx 970.796 \text{ pounds}$$

DO EXERCISES 6 AND 7.

EXERCISE SET 9.6

Solve each of the following by separating the variables.

1. $y' - 3y = 0$

2. $y' - 2xy = 0$

3. $y' + \dfrac{3}{x}y = 0$

4. $y' - 2xy = 2x$

In Exercises 5–8, show that the given equation is not exact. Then find an integrating factor $m(x)$ for the equation. Don't actually solve the equation.

5. $y' - 3y = e^{5x}$

6. $y' + \dfrac{4}{x}y = \sin x$

7. $y' - \dfrac{5}{x^2}y = x^3$

8. $y' + \dfrac{2x}{x^2 + 1}y = 3x + 2$

Solve each of the following by finding an appropriate integrating factor.

9. $y' - 3y = 0$

10. $y' + 4y = 0$

11. $y' - 2xy = 0$

12. $y' + \dfrac{3}{x}y = 0$

13. $y' + 5y = 0,\ y(0) = 7$

14. $xy' + y = 0,\ y(1) = 3$

15. $y' - 2xy = 2x$

16. $y' + \dfrac{3}{x}y = 6x^2 - 4$

17. $y' + \dfrac{2x}{1 + x^2}y = x$

18. $y' - \dfrac{\sin x}{\cos x}y = \dfrac{1}{\cos x}$

19. $x^2y' + xy = x^4,\ y(2) = 5$

20. $y' + \dfrac{3x^2}{(1 + x^3)}y = 1,\ y(0) = 2$

21. A tank initially contains 180 pounds of salt dissolved in 500 gallons of water. Salt water containing 0.5 pound of salt per gallon enters the tank at the rate of 3 gallons per minute. The mixture (kept uniform by stirring) is removed at the same rate. How many pounds of salt are in the tank after an hour? a day?

22. A 2000-gallon tank initially contains 200 pounds of salt dissolved in 500 gallons of water. Pure water is pumped into the tank at the rate of 5 gallons per minute, while the (well-stirred) mixture is drawn off at the rate of 2 gallons per minute. The process stops when the tank is full. How much salt is left in the tank when the tank is full?

23. Let $f(x)$ satisfy $f'(x) = p(x)$, where $p(x)$ is continuous on $[a, b]$.

a) Show that if $v = ye^{f(x)}$, then $v' = e^{f(x)}y' + p(x)e^{f(x)}y$.

b) Then show that under the substitution $v = ye^{f(x)}$, equation $y' + y \cdot p(x) = q(x)$ becomes $v' = e^{-\int p(x)\,dx} \cdot q(x)$, which may be solved by an integration.

24. In each of the following, show that the given equation is not exact. Then show that the given function $m(x, y)$ is an integrating factor for the equation, and solve the equation.

a) $3y^4\,dx + 4xy^3\,dy = 0,\ m(x, y) = x^2$

b) $5xy\,dx + 3x^2\,dy = 0;\ m(x, y) = x^3y^2$

c) $(2y - 15xy^2)\,dx + (3x - 20x^2y)\,dy = 0;\ m(x, y) = xy^2$

d) $2xy\,dx + 5x^2\,dy = 0;\ m(x, y) = y^4$

25. Consider the first-order homogeneous equation $a_0y' + a_1y = 0$, where the coefficients are assumed to be constant. Show that $y = Ce^{-a_1x/a_0}$ is the solution to this equation.

9.7 HIGHER-ORDER DIFFERENTIAL EQUATIONS

In the preceding section, we saw that we could solve any first-order linear differential equation. Unfortunately, not every higher-order linear differential equation can be solved explicitly. For the moment, we'll consider second-order linear homogeneous equations with constant coefficients, that is, equations of the form

$$a_0 y'' + a_1 y' + a_2 y = 0, \tag{1}$$

where a_0, a_1, and a_2, the coefficients of the equation, are constants.

The corresponding first-order equation was solved explicitly in Exercise 25 of Exercise Set 9.6. The solution is an exponential function. This suggests that the solution to Eq. (1) may be an exponential function. This turns out to be correct. If we let $y = e^{rx}$ (r a constant) in Eq. (1) above, then, since $y' = re^{rx}$ and $y'' = r^2 e^{rx}$, we'll get

$$a_0(r^2 e^{rx}) + a_1(re^{rx}) + a_2(e^{rx}) = 0.$$

Noticing that each term in the equation above has the common factor e^{rx}, we write

$$e^{rx}(a_0 r^2 + a_1 r + a_2 r) = 0.$$

Now e^{rx} is a positive number for all choices of real numbers r and x. Dividing by e^{rx}, we get

$$a_0 r^2 + a_1 r + a_2 = 0.$$

The quadratic formula yields two values of r.

Example 1 Solve $y'' - 5y' + 6y = 0$.

Solution We try $y = e^{rx}$. Then $y' = re^{rx}$ and $y'' = r^2 e^{rx}$. Substituting these quantities into the equation gives

$$r^2 e^{rx} - 5re^{rx} + 6e^{rx} = 0$$

or

$$e^{rx}(r^2 - 5r + 6) = 0$$

or

$$r^2 - 5r + 6 = 0.$$

This last equation factors as $(r - 2)(r - 3) = 0$, and we get solutions, or roots, 2 and 3. This gives us the two solutions e^{2x} and e^{3x}.

It turns out that the general solution to the equation is given by $y = C_1 e^{2x} + C_2 e^{3x}$, where C_1 and C_2 are arbitrary constants. We can

1. a) By substituting directly, show that e^{2x} is a solution to $y'' - 5y' + 6y = 0$.

b) Repeat part (a) for e^{3x}.

c) Repeat part (a) for $2e^{3x} - 7e^{2x}$.

2. Solve $y'' - 6y' + 8y = 0$.

3. Solve $2y'' - 7y' + 3y = 0$.

verify this by finding y' and y'' and substituting y, y', and y'' into the original equation. We get

$$y = C_1 e^{2x} + C_2 e^{3x},$$

$$y' = 2C_1 e^{2x} + 3C_2 e^{3x},$$

$$y'' = 4C_1 e^{2x} + 9C_2 e^{3x}.$$

Then, substituting, we get

$$y'' - 5y' + 6y$$

$$= (4C_1 e^{2x} + 9C_2 e^{3x}) - 5(2C_1 e^{2x} + 3C_2 e^{3x}) + 6(C_1 e^{2x} + C_2 e^{3x})$$

$$= 4C_1 e^{2x} + 9C_2 e^{3x} - 10C_1 e^{2x} - 15C_2 e^{3x} + 6C_1 e^{2x} + 6C_2 e^{3x}$$

$$= 0.$$

DO EXERCISES 1–3.

In Example 1 we used the fact that if $f(x)$ and $g(x)$ are solutions of a linear homogeneous differential equation, then so too is $C_1 f(x) + C_2 g(x)$. We state this important principle in the form of a theorem.

THEOREM 6   The Principle of Superposition

If $f_1(x), f_2(x), \ldots, f_n(x)$ are solutions to a linear homogeneous differential equation, then so too is

$$\sum_{i=1}^{n} C_i f_i(x) = C_1 f_1(x) + C_2 f_2(x) + \cdots + C_n f_n(x),$$

where $C_1, C_2, \ldots, C_n$ are arbitrary constants.

Note that the equation need not have constant coefficients. Nor must it be a second-order equation. The only requirements are that it be linear and homogeneous. We leave the proof of this theorem for the exercises.

In Example 1, we obtained the expression $y = C_1 e^{2x} + C_1 e^{3x}$ for the solution to $y'' - 5y' + 6y = 0$. This expression is called the *general solution* to the equation. This means that every function of this form is a solution to the equation and that every solution to the equation can be written in this form.

For the equation $y'' - 5y' + 6y = 0$, the general solution contained two arbitrary constants. In all of the equations that we'll consider, the number of constants in the general solution will be equal to the order of the equation.

4. Solve $y''' + y'' - 2y' = 0$.

Example 2 Find the general solution to $y''' + y'' - 6y' = 0$.

Solution As in Example 1, we let $y = e^{rx}$. We obtain $y' = re^{rx}$, $y'' = r^2 e^{rx}$, and $y''' = r^3 e^{rx}$. Substituting into the equation, we obtain

$$r^3 e^{rx} + r^2 e^{rx} - 6re^{rx} = 0,$$

or

$$e^{rx}(r^3 + r^2 - 6r) = 0.$$

Since e^{rx} is never 0, we get $r^3 + r^2 - 6r = 0$, or

$$r(r^2 + r - 6) = 0.$$

Factoring yields $r(r + 3)(r - 2) = 0$, so that 0, -3, and 2 are its roots. It is easy to check that e^{-3x}, e^{2x}, and e^{0x}, or 1, are solutions to $y''' + y'' - 6y' = 0$. According to the Principle of Superposition, the general solution is

$$y = C_1 e^{-3x} + C_2 e^{2x} + C_3 \cdot 1,$$

or

$$y = C_1 e^{-3x} + C_2 e^{2x} + C_3.$$

DO EXERCISE 4.

In each of Examples 1 and 2, we made the substitution $y = e^{rx}$, reduced the given differential equation to a polynomial equation in the variable r, and then solved this polynomial equation for r. The resulting values of r determined the solution to the differential equation.

But there's no need to repeat the substitution $y = e^{rx}$ and the subsequent algebra each time. We can bypass this work, going directly from the given differential equation to the associated polynomial equation (called the *characteristic equation*).

5. Find the characteristic equation of $y'' - 7y' + 12y = 0$.

6. Find the characteristic equation of $y''' - 6y'' + 11y' - 6y = 0$.

Example 3 Find the characteristic equation of $y'' - 7y' + 10y = 0$.

Solution Replacing y'' by r^2, y' by r, and $y(= y^{(0)})$ by $r^0 = 1$ gives

$$r^2 - 7r + 10 = 0.$$

DO EXERCISES 5 AND 6.

Example 4 Solve $y'' - 7y' + 10y = 0$.

Solution From Example 3, the characteristic equation is $r^2 - 7r + 10 = 0$. Factoring, we get $(r - 2)(r - 5) = 0$, so that its solutions are 2

7. Solve $y'' - 7y' + 12y = 0$.

and 5. The general solution to the differential equation is

$$y = C_1 e^x + C_2 e^x.$$

DO EXERCISE 7.

The technique we have developed works well so long as the characteristic equation has real roots, all of which are distinct. It must be modified if the real roots are repeated or if the roots are complex.

Example 5 Solve $y'' - 2y' + y = 0$.

Solution The characteristic equation is $r^2 - 2r + 1 = 0$. Factoring yields $(r - 1)^2 = 0$. The roots are 1 and 1, or simply 1. (This is the case of repeated real roots.) Sure enough, $r = 1$ yields the solution $y = e^x$. We might try $y = C_1 e^x + C_2 e^x$ as the general solution. But

$$C_1 e^x + C_2 e^x = (C_1 + C_2)e^x = Ke^x,$$

where $K = C_1 + C_2$. In other words, simply repeating the e^x gets us nowhere.

Let's substitute $y = ve^x$ into the original equation. Then, by the Product Rule,

$$y' = ve^x + v'e^x$$

8. Solve $y'' - 4y' + 4y = 0$.

and

$$y'' = (ve^x + v'e^x) + (v'e^x + v''e^x) = ve^x + 2v'e^x + v''e^x$$

and substituting all of this into

$$y'' - 2y' + y = 0$$

gives

$$(ve^x + 2v'e^x + v''e^x) - 2(ve^x + v'e^x) + ve^x = 0.$$

Simplifying, we get $v''e^x = 0$, or, since e^x is never 0, $v'' = 0$. Integrating twice yields $v = C_1 + C_2x$. Replacing v in the expression $y = ve^x$ gives

$$y = (C_1 + C_2x)e^x$$

or

$$y = C_1 e^x + C_2 xe^x.$$

This is the general solution to the original equation.

DO EXERCISE 8.

9. Solve $y'' - 6y' + 9y = 0$.

The calculations in Example 5 weren't easy. Rather than repeat them each time repeated roots occur, we simply remember that if the root r is repeated, the corresponding solutions of the differential equation are e^{rx} and xe^{rx}.

Example 6 Solve $y'' + 4y' + 4y = 0$.

Solution The characteristic equation is $r^2 + 4r + 4 = 0$, or $(r + 2)^2 = 0$. So -2 and -2 are the roots. The corresponding solutions of the differential equation are e^{-2x} and xe^{-2x}. The general solution is

$$y = C_1 e^{-2x} + C_2 x e^{-2x}.$$

DO EXERCISES 9 AND 10.

10. Solve $4y'' + 4y' + y = 0$.

If a given number, say r, appears as a root three times, we use e^{rx}, xe^{rx}, and $x^2 e^{rx}$ for the three solutions.

Example 7 Solve $y''' - 3y'' + 3y' - y = 0$.

Solution The characteristic equation is $r^3 - 3r^2 + 3r - 1 = 0$, or $(r - 1)^3 = 0$. So the roots are 1, 1, and 1. The three solutions are e^x, xe^x, and $x^2 e^x$. The general solution is

$$y = C_1 e^x + C_2 x e^x + C_3 x^2 e^x.$$

DO EXERCISE 11.

11. Solve $y''' + 3y'' + 3y' + y = 0$.

We have considered the cases in which the roots of the characteristic equation are real numbers. Consider the equation $y'' + y = 0$. The characteristic equation is $r^2 + 1 = 0$. The roots of this equation are $r = \pm\sqrt{-1}$, or $r = \pm i$, where i denotes $\sqrt{-1}$ (which doesn't exist as a real number). On the other hand, it's easy to solve $y'' + y = 0$. If $y = \sin x$, for example, then $y' = \cos x$ and $y'' = -\sin x$. In other words, $y = \sin x$ is one solution. A similar calculation (do it!) reveals that $y = \cos x$ is another solution. The general solution is $y = C_1 \sin x + C_2 \cos x$.

This example illustrates the general case. If the roots of the characteristic equation are the complex numbers $r_1 \pm r_2 i$, then the two corresponding solutions are $e^{r_1 x} \cos (r_2 x)$ and $e^{r_1 x} \sin (r_2 x)$.

Example 8 Solve $y'' + 16y = 0$.

12. Solve $y'' + 9y = 0$.

13. Solve $y'' - 4y' + 29y = 0$.

14. Solve $y'' + 2y' + 5y = 0$.

15. Solve $y''' - 2y'' + 8y' = 0$.

Solution The characteristic equation is $r^2 + 16 = 0$. Solving directly yields $r^2 = -16$, or $r = \pm\sqrt{-16}$, or $r = \pm 4\sqrt{-1} = \pm 4i = 0 \pm 4i$. This is the case where $r_1 = 0$ and $r_2 = 4$. The two solutions are $e^{0x} \sin 4x = \sin 4x$ and $e^{0x} \cos(4x) = \cos 4x$. The general solution is $y = C_1 \sin 4x + C_2 \cos 4x$.

DO EXERCISE 12.

Example 9 Solve $y'' - 2y' + 10y = 0$.

Solution The characteristic equation is $r^2 - 2r + 10 = 0$. Applying the quadratic formula yields

$$r = \frac{-(-2) \pm \sqrt{(-2)^2 - 4(1)(10)}}{2 \cdot 1}$$

$$= \frac{2 \pm \sqrt{4 - 40}}{2} = \frac{2 \pm \sqrt{-36}}{2}$$

$$= \frac{2 \pm 6\sqrt{-1}}{2} = 1 \pm 3i.$$

So $r_1 = 1$ and $r_2 = 3$. The corresponding solutions are $e^x \sin 3x$ and $e^x \cos 3x$ and the general solution is $y = C_1 e^x \sin 3x + C_2 e^x \cos 3x$.

DO EXERCISES 13 AND 14.

If some of the roots are real and some are complex, we simply combine our earlier techniques.

Example 10 Solve $y''' - 4y'' + 8y' - 8y = 0$.

Solution The characteristic equation is $r^3 - 4r^2 + 8r - 8 = 0$. This factors as $(r - 2)(r^2 - 2r + 4) = 0$. The roots turn out to be $r = 2$, $1 \pm \sqrt{3}i$. The corresponding solutions are e^{2x}, $e^x \sin(\sqrt{3}x)$, and $e^x \cos(\sqrt{3}x)$. The general solution is

$$y = C_1 e^{2x} + C_2 e^x \sin(\sqrt{3}x) + C_3 e^x \cos(\sqrt{3}x).$$

DO EXERCISE 15.

If complex roots are repeated, then we modify the solutions just as we did in the case of repeated real roots.

Example 11 Solve $y'''' + 2y'' + y = 0$.

16. Solve $y'''' + 8y'' + 16y = 0$.

Solution The characteristic equation is $r^4 + 2r^2 + 1 = 0$, or $(r^2 + 1)^2 = 0$. The roots are $\pm i$ and $\pm i$. Two solutions corresponding to $\pm i$ are sin x and cos x. Two more corresponding to the second pair $\pm i$ are x sin x and x cos x. The general solution is

$$y = C_1 \sin x + C_2 \cos x + C_3 x \sin x + C_4 x \cos x.$$

DO EXERCISE 16.

Nonhomogeneous Equations

So far in this section, the equations considered have been linear homogeneous equations with constant coefficients. We'll relax that now, considering equations of the form

17. Check by direct calculation that

$$y = C_1 \sin x + C_2 \cos x + 5$$

is a solution to $y'' + y = 5$ for all choices of C_1 and C_2.   `CSS`

$$a_0 y^{(n)} + a_1 y^{(n-1)} + \cdots + a_{n-1} y' + a_n y = b(x); \tag{2}$$

that is, we no longer require that the equation be homogeneous. We'll only consider the case in which $b(x)$ is a polynomial.

It turns out to be easy to solve such an equation. To do so, we first find the general solution (call it y_h) to the *associated homogeneous equation*, that is, to Eq. (2) with $b(x) = 0$. Next, we find a *particular solution* (call it y_p) to Eq. (2). Finally, we add y_p to y_h to obtain the general solution to Eq. (2).

Example 12 Solve $y'' + y = 5$.

18. Solve $4y'' + 4y' + y = 3$ (see Margin Exercise 10).

Solution We must first solve the associated homogeneous equation, $y'' + y = 0$. From above, the general solution, y_h, is $y_h = C_1 \sin x + C_2 \cos x$. It remains to find a particular solution y_p. To do so, we make an educated guess. Here, we guess that $y_p = C$ must be a constant (more about how to make this "guess" later). Since $y_p' = 0$, we get $y_p'' = 0$, and, since y_p must satisfy $y'' + y = 5$, we get

$$(0) + (C) = 5,$$

or

$$C = 5.$$

In other words, $y_p = 5$ is a particular solution to $y'' + y = 5$. In retrospect, we see that this is correct.

To complete the problem, we add y_p and y_h, getting

$$y = C_1 \sin x + C_2 \cos x + 5.$$

DO EXERCISES 17 AND 18.

In Example 12 and Margin Exercise 18, it turned out that y_p was equal to the forcing term $b(x)$. That was a coincidence.

Example 13 Solve $y'' - 6y' + 8y = 2x^2 - 1$.

Solution We first solve $y'' - 6y' + 8y = 0$. From Margin Exercise 2, we know that its solution is $y_h = C_1 e^{4x} + C_2 e^{2x}$. It remains only to find a particular solution y_p. To do this, we try $y_p = Ax^2 + Bx + C$, where A, B, and C are numbers to be determined. Then $y_p' = 2Ax + B$ and $y_p'' = 2A$. Since y_p must satisfy the given equation, we substitute, getting

$$(2A) - 6(2Ax + B) + 8(Ax^2 + Bx + C) = 2x^2 - 1,$$

or

$$2A - 12Ax - 6B + 8Ax^2 + 8Bx + 8C = 2x^2 - 1.$$

Regrouping yields

$$8Ax^2 + (-12A + 8B)x + (2A - 6B + 8C) = 2x^2 - 1.$$

In order for these two expressions to be equal, the coefficients of corresponding terms must be equal. So,

$$8A = 2, \qquad x^2\text{-coefficients are equal}$$

$$8B - 12A = 0, \qquad x\text{-coefficients are equal}$$

$$2A - 6B + 8C = -1 \qquad \text{Constant terms are equal}$$

The second equation gives $B = \frac{3}{2}A$. Since $8A = 2$, we get $A = \frac{1}{4}$ and $B = \frac{3}{2} \cdot \frac{1}{4} = \frac{3}{8}$. Solving the third equation for C yields

$$8C = -2A + 6B - 1 = -2\left(\frac{1}{4}\right) + 6\left(\frac{3}{8}\right) - 1,$$

or

$$8C = -\frac{1}{2} + \frac{18}{8} - 1 = -\frac{2}{4} + \frac{9}{4} - \frac{4}{4} = \frac{3}{4},$$

and

$$C = \frac{3}{32}.$$

So,

$$y_p = \frac{1}{4}x^2 + \frac{3}{8}x + \frac{3}{32}.$$

The general solution to the given equation is

$$y = y_h + y_p,$$

19. Solve $y'' - 9y = x^2 + 2x$.

or

$$y = C_1 e^{4x} + C_2 e^{2x} + \frac{1}{4}x^2 + \frac{3}{8}x + \frac{3}{32}.$$

DO EXERCISE 19.

The two examples above show that making the correct guess for the form of the particular solution y_p is the crucial step in solving these equations. If the forcing term is a polynomial, then we will usually let y_p be a polynomial of the same degree. This works in almost all cases. But consider the equation below.

Example 14 Solve $y'' - 2y' = 5$.

Solution We first solve $y'' - 2y' = 0$. The characteristic equation is $r^2 - 2r = 0$, or $r(r - 2) = 0$. So the roots are 0 and 2 and

$$y_h = C_1 e^{0 \cdot x} + C_2 e^{2x} = C_1 + C_2 e^{2x}.$$

To complete the problem, we need only find a particular solution y_p. Since the forcing term is a constant, 5, we might try $y_p = C$. Then $y_p' = 0$ and $y_p'' = 0$. Substitution into the original equation gives $(0) - 2(0) = 5$, or $0 = 5$, which is not true. We conclude that $y_p = C$ was the wrong trial solution. The trouble is that 0 was a root of the characteristic equation. Whenever 0 is a root of the characteristic equation, we must remember to multiply the trial solution we would generally use by x^m (m is the number of times that 0 occurs as a root of the characteristic equation). Above, we would ordinarily have used $y_p = C$. Since 0 is a root of the characteristic equation, we must multiply by x. So we use $y_p = Cx$. Then $y_p' = C$ and $y_p'' = 0$. Substituting, we obtain

$$(0) - 2(C) = 5,$$

or

$$-2C = 5$$

or

$$C = -\frac{5}{2}.$$

20. Solve $y'' + y' = 7$.

Thus $y_p = -\frac{5}{2}x$ is a particular solution to $y'' - 2y' = 5$. The general solution is then

$$y = y_h + y_p$$

$$y = C_1 + C_2 e^{2x} - \frac{5}{2}x.$$

DO EXERCISE 20.

Example 15 Solve $y''' - y'' = 2x - 3$, $y(0) = 1$, $y'(0) = 0$, $y''(0) = 3$.

Solution As before, we first find the general solution to the differential equation. Then we use the initial conditions $y(0) = 1$, $y'(0) = 0$, and $y''(0) = 3$ to eliminate the constants.

To solve $y''' - y'' = 2x - 3$, we first solve the associated homogeneous equation $y''' - y'' = 0$. The characteristic equation is $r^3 - r^2 = 0$, or $r^2(r - 1) = 0$. The roots are 0, 0, and 1. So $y_h = C_1 + C_2 x + C_3 e^x$.

Next, we must find y_p. Since the forcing term is $2x - 3$, a polynomial of degree one, we would ordinarily use $y_p = Ax + B$. In this case, that won't work (as you would discover by trying it), since 0 is a root (here, a double root) of the characteristic equation. Since 0 occurs twice as a root of $r^3 - r^2 = 0$, we multiply our usual trial solution, $Ax + B$, by x^2, obtaining $y_p = Ax^3 + Bx^2$. Then $y_p' = 3Ax^2 + 2Bx$ and $y_p'' = 6Ax + 2B$. Finally, $y_p''' = 6A$. Substituting into the original equation, we get

$$(6A) - (6Ax + 2B) = 2x - 3, \qquad \boxed{\text{CSS}}$$

or

$$-6Ax + (6A - 2B) = 2x - 3.$$

Equating corresponding coefficients yields $-6A = 2$ and $6A - 2B = -3$. Solving the first one gives $A = -\frac{1}{3}$. Substituting this into the second equation yields $6(-\frac{1}{3}) - 2B = -3$, or $-2 - 2B = -3$. Solving gives $-2B = -1$, or $B = \frac{1}{2}$. Thus $y_p = -\frac{1}{3}x^3 + \frac{1}{2}x^2$. The general solution is then

$$y = y_h + y_p = C_1 + C_2 x + C_3 e^x - \frac{1}{3}x^3 + \frac{1}{2}x^2.$$

It remains to use the initial conditions to eliminate C_1, C_2, and C_3. Letting $x = 0$ above, we get

$$y(0) = C_1 + C_2(0) + C_3 e^0 - \frac{1}{3}(0)^3 + \frac{1}{2}(0)^2,$$

or

$$y(0) = C_1 + C_3.$$

But $y(0) = 1$, so $C_1 + C_3 = 1$. Now, differentiating the solution gives

$$y' = C_2 + C_3 e^x - x^2 + x.$$

Letting $x = 0$, we have

$$y'(0) = C_2 + C_3 e^0 - (0)^2 + 0,$$

21. Solve $y''' + 2y'' = 6x - 4$.

or

$$y'(0) = C_2 + C_3.$$

So, since $y'(0) = 0$, we get $C_2 + C_3 = 0$. Finally, differentiating again gives

$$y'' = C_3 e^x - 2x + 1,$$

so that

$$y''(0) = C_3 e^0 - 2(0) + 1 = C_3 + 1.$$

22. Solve $y'' - y = -5$, $y(0) = 1$, $y'(0) = 13$.

Since $y''(0) = 3$, we have $C_3 + 1 = 3$, or $C_3 = 2$. This enables us to find C_1 and C_2, for $C_2 + C_3 = 0$ gives $C_2 = -C_3 = -2$ and $C_1 + C_3 = 1$ yields $C_1 = 1 - C_3 = 1 - 2 = -1$. So,

$$y = -1 - 2x + 2e^x - \frac{1}{3}x^3 + \frac{1}{2}x^2.$$

Observe that three initial conditions are needed to eliminate the three constants in the general solution.

DO EXERCISES 21 AND 22.

EXERCISE SET 9.7

Solve.

1. $y'' - 6y' + 5y = 0$

2. $y'' - 8y' + 15y = 0$

3. $y'' - y' - 2y = 0$

4. $y'' + 2y' - 3y = 0$

5. $y'' + 3y' + 2y = 0$

6. $y'' + 5y' + 6y = 0$

7. $2y'' - 5y' + 2y = 0$

8. $2y'' - y' - y = 0$

9. $y'' - 9y = 0$

10. $y'' - 3y = 0$

11. $y'' + 10y' + 25y = 0$

12. $y'' - 8y' + 16y = 0$

13. $4y'' + 12y' + 9y = 0$

14. $y''' + 8y'' + 16y' = 0$

15. $y''' - y'' - 8y' + 12y = 0$

16. $y''' - 7y'' + 11y' - 5y = 0$

17. $y'''' + 3y''' + 3y'' + y' = 0$

18. $y'''' - 4y''' = 0$

19. $y'''' - 5y''' + 4y'' = 0$

20. $y'''' + 3y''' - y'' - 3y' = 0$

21. $y'' + 36y = 0$

22. $y'' + 4y' + 13y = 0$

23. $y'' + 8y' + 41y = 0$

24. $y'' + y' + y = 0$

25. $y''' + 2y'' + 5y' = 0$

26. $y''' - 2y'' + 4y' - 8y = 0$

27. $y''' + y'' + 15y' - 17y = 0$

28. $y'''' + 4y'' + 4y = 0$

29. $y'' + y = 7$

30. $y'' - 3y' + 2y = 4$

31. $y'' - 2y' + y = 3$

32. $y'' - 7y' + 10y = 10x - 27$

33. $y'' + 4y' + 4y = 8 - 12x$

34. $y'' - 4y' + 5y = 5x + 1$

35. $y'' + 4y' + 3y = 6x^2 - 4$

36. $y'' + 4y = 8x^2 - 12x$

37. $y'' - 3y' = 4$

38. $y''' + y' = -2$

39. $y''' + y'' = -2$

40. $y'' - 2y' = -4x$

41. $y''' + 4y'' + 20y' = 40x - 12$

42. $y'' - y' = 3x^2 - 8x + 5$

43. $y'' - y' = 0, \; y(0) = 0, \; y'(0) = -1$

44. $y'' + y = 0, \; y(0) = 4, \; y'(0) = 3$

45. $y'' - y = 0, \; y(0) = 1, \; y'(0) = 3$

46. $y'' - 2y' + y = 0, \; y(0) = 1, \; y'(0) = -2$

47. $y''' - y' = 0, \; y(0) = 1, \; y'(0) = 8, \; y''(0) = 4$

48. $y'' - y' - 2y = 2x - 1, \; y(0) = 6, \; y'(0) = 0$

49. $y'' + 9y = 9 - 9x, \; y(0) = 3, \; y'(0) = 2$

50. $y''' + 3y'' = 18x - 18, \; y(0) = 3, \; y'(0) = -1,$
$y''(0) = 10$

51. Let y_1 and y_2 both be solutions to the second-order linear homogeneous differential equation

$$a_0(x)y'' + a_1(x)y' + a_2(x)y = 0.$$

(Note that we are *not* assuming that the coefficients are constants.) Show that $C_1y_1 + C_2y_2$ is also a solution to the equation, where C_1 and C_2 are arbitrary constants.

53. a) Let y_p be a solution to

$$a_0(x)y'' + a_1(x)y' + a_2(x)y = b(x)$$

and let y_h be a solution to

$$a_0(x)y'' + a_1(x)y' + a_2(x)y = 0.$$

Show that $y_p + y_h$ is a solution to the nonhomogeneous equation.

b) Let y_1 and y_2 *both* be solutions to the nonhomogeneous equation

$$a_0(x)y'' + a_1(x)y' + a_2(x)y = b(x).$$

Show that $y_1 - y_2$ is a solution to the homogeneous equation

$$a_0(x)y'' + a_1(x)y' + a_2(x)y = 0.$$

c) Use part (b) to show that if y_1 is a given solution to

$$a_0(x)y'' + a_1(x)y' + a_2(x)y = b(x),$$

then every other solution to it is of the form $y_1 + y_2$, where y_2 is a solution to the associated homogeneous equation.

52. Show that the conclusion of Exercise 51 holds for the nth-order homogeneous linear equation

$$a_0(x)y^{(n)} + a_1(x)y^{(n-1)} + \cdots + a_{n-1}(x)y' + a_n(x)y = 0.$$

54. You are told that the roots of the characteristic equation of a certain sixth-order linear homogeneous differential equation with constant coefficients are 2, 2, 3, 3, 3, and 3. Find the general solution to the equation.

55. Repeat Exercise 54 if the roots are 7, 9, $2 \pm 4i$, and $2 \pm 4i$.

You should be able to

a) Solve a system of first-order linear homogeneous differential equations.

b) Solve the corresponding non-homogeneous systems if the forcing functions are polynomials.

1. Solve

$$\frac{dx}{dt} = 4x,$$

$$\frac{dy}{dt} = y.$$

9.8 SYSTEMS OF DIFFERENTIAL EQUATIONS

In many applications we are interested in finding the values of two (or more) quantities, say x and y, that are functions of yet another variable t (usually thought of as being time), whose rates of change are given as a system of equations.

Example 1 Solve

$$\frac{dx}{dt} = 5x,$$

$$\frac{dy}{dt} = 3y.$$

Solution We want to find x and y in terms of t. In this case, that's particularly easy to do because the equations in the system aren't inter-related. That is, the first equation doesn't involve y and the second one doesn't involve x.

The first equation is a first-order linear equation (with variables separable) whose solution is $x = C_1 e^{5t}$. Similarly, the solution to the second equation is $y = C_2 e^{3t}$.

Note that there will be two constants in the solution.

DO EXERCISE 1.

Example 2 Solve

$$\frac{dx}{dt} = y,$$

$$\frac{dy}{dt} = 4x.$$

Solution In this system the variables *are* interrelated, or coupled. That is, x and y both occur in the first equation as well as the second equation. We can "uncouple" them as follows. We differentiate the first equation to get

$$\frac{d^2x}{dt^2} = x'' = \frac{dy}{dt}$$

From the second equation, we have

$$\frac{dy}{dt} = 4x.$$

2. Solve

$$\frac{dx}{dt} = 2y,$$

$$\frac{dy}{dt} = 8x.$$

3. Solve

$$\frac{dx}{dt} = y,$$

$$\frac{dy}{dt} = -4x.$$

Combining the equations gives us

$$x'' = \frac{dy}{dt} = 4x,$$

or

$$x'' - 4x = 0.$$

This is a second-order linear homogeneous equation with constant coefficients. Its characteristic equation is $r^2 - 4 = 0$, or $(r + 2)(r - 2) = 0$. Since 2 and -2 are the roots, we find that $x(t) = C_1 e^{2t} + C_2 e^{-2t}$. Note that since $y = dx/dt$, we know immediately that we have

$$y(t) = 2C_1 e^{2t} - 2C_2 e^{-2t} = \frac{dx}{dt}.$$

DO EXERCISES 2 AND 3.

We can solve any first-order linear system of the form

$$\frac{dx}{dt} = ax + by,$$

$$\frac{dy}{dt} = cx + dy,$$

where a, b, c, and d are constants, by combining the technique shown in Example 2 with a little algebra. To see how, consider the following example.

Example 3 Solve

$$\frac{dx}{dt} = x + 2y,$$

$$\frac{dy}{dt} = 4x - y.$$

Solution Our objective is to combine the two equations so as to eliminate one of the dependent variables, say y. We write the first equation as $x' = x + 2y$, and then differentiate it to get

$$x'' = x' + 2y'.$$

Now, we'll use the second equation, $y' = 4x - y$, to eliminate y' above. We get

$$x'' = x' + 2(4x - y),$$

4. Check the solutions obtained in Example 3 by direct substitution in the original system of equations.

or

$$x'' = x' + 8x - 2y.$$

Unfortunately, a y-term remains. But we can eliminate it by using the original version of the first equation. Solving $x' = x + 2y$ for y yields $2y = x' - x$. Substituting this into the last equation yields

$$x'' = x' + 8x - (x' - x),$$

or

$$x'' = x' + 8x - x' + x,$$

or

$$x'' - 9x = 0.$$

We have eliminated y and what remains is a second-order linear homogeneous equation in x. Its characteristic equation is $r^2 - 9 = 0$. The roots here are 3 and -3, so that $x = C_1 e^{3t} + C_2 e^{-3t}$. It still remains to find y. Recall though that in the process of eliminating y, we had obtained $2y = x' - x$. Now using our solution for x gives

$$2y = (3C_1 e^{3t} - 3C_2 e^{-3t}) - (C_1 e^{3t} + C_2 e^{-3t}),$$

5. Solve

$$\frac{dx}{dt} = 4x + 3y,$$

$$\frac{dy}{dt} = -4x - 4y.$$

or

$$2y = 2C_1 e^{3t} - 4C_2 e^{-3t},$$

or

$$y = C_1 e^{3t} - 2C_2 e^{-3t}.$$

So the solution is

$$x = C_1 e^{3t} + C_2 e^{-3t},$$
$$y = C_1 e^{3t} - 2C_2 e^{-3t}.$$

DO EXERCISES 4 AND 5.

Example 4 Solve

$$\frac{dx}{dt} = y, \qquad x(0) = 1,$$

$$\frac{dy}{dt} = -x + 2y, \qquad y(0) = 3.$$

Solution Just as in previous sections, we'll first find the general solution to the problem. Then we'll use the initial conditions to evaluate the constants in the general solution.

6. Solve

$$\frac{dx}{dt} = 3x + y, \qquad x(0) = 1,$$

$$\frac{dy}{dt} = -x + y, \qquad y(0) = 3.$$

Differentiating the first equation gives $x'' = y'$. Using the second equation to replace y', we get $x'' = -x + 2y$. From the first equation, we have $y = x'$. Combining the last two equations gives $x'' = -x + 2x'$, or $x'' - 2x' + x = 0$.

Again, we have a second-order linear equation with constant coefficients. The characteristic equation is $r^2 - 2r + 1 = 0$, or $(r - 1)^2 = 0$, so that the roots are 1 and 1. Since the roots are repeated, we have

$$x = C_1 e^t + C_2 t e^t.$$

Since $y = dx/dt = x'$, we get

$$y = C_1 e^t + C_2 t e^t + C_2 e^t,$$

or

$$y = (C_1 + C_2) e^t + C_2 t e^t.$$

Letting $t = 0$ in the equation for x gives

$$x(0) = C_1 e^0 + C_2 \cdot 0 \cdot e^0 = C_1.$$

Since $x(0) = 1$ was one of the initial conditions, we have $x(0) = C_1 = 1$. Letting $t = 0$ in the equation for y gives $y(0) = C_1 + C_2$. Since $y(0) = 3$, we have $C_1 + C_2 = 3$. Since $C_1 = 1$, we must have $C_2 = 2$. So, $x = e^t + 2t e^t$ and $y = 3e^t + 2t e^t$.

It is easy to check that these solutions satisfy both the original equations and the initial conditions.

DO EXERCISES 6 AND 7.

7. Solve

$$\frac{dx}{dt} = 3x - y, \qquad x(0) = 3,$$

$$\frac{dy}{dt} = 2x, \qquad y(0) = 5.$$

If $f(t)$ and $g(t)$ are polynomials in t, then the preceding techniques can be used to solve systems of the form

$$\frac{dx}{dt} = ax + by + f(t),$$

$$\frac{dy}{dt} = cx + dy + g(t),$$

where, as above, a, b, c, and d are constants.

Example 5 Solve

$$\frac{dx}{dt} = x + y + 4,$$

$$\frac{dy}{dt} = 4x + y + 3.$$

8. Solve

$$\frac{dx}{dt} = 4x + y,$$

$$\frac{dy}{dt} = 2x + 3y + 50.$$

Solution We differentiate the first equation to obtain $x'' = x' + y'$. Using the second equation to replace y' yields

$$x'' = x' + (4x + y + 3),$$

or

$$x'' = x' + 4x + y + 3.$$

Solving the first equation for y yields $y = x' - x - 4$, so that combining the last two equations gives us

$$x'' = x' + 4x + (x' - x - 4) + 3,$$

or

$$x'' = 2x' + 3x - 1,$$

or

$$x'' - 2x' - 3x = -1.$$

Note that the only difference between this example and our previous ones is that we end up with a nonhomogeneous second-order equation for x.

Here, the characteristic equation of the associated homogeneous equation $(x'' - 2x' - 3x = 0)$ is $r^2 - 2r - 3 = 0$, which factors as $(r - 3)(r + 1) = 0$. The roots are 3 and -1, and $x_h = C_1 e^{3t} + C_2 e^{-t}$. Note that the constant function $x_p = \frac{1}{3}$ is a particular solution, so that

$$x = C_1 e^{3t} + C_2 e^{-t} + \frac{1}{3}$$

is the general solution for x. Since $y = x' - x - 4$, we get

$$y = (3C_1 e^{3t} - C_2 e^{-t}) - \left(C_1 e^{3t} + C_2 e^{-t} + \frac{1}{3}\right) - 4$$

$$y = 3C_1 e^{3t} - C_2 e^{-t} - C_1 e^{3t} - C_2 e^{-t} - \frac{1}{3} - \frac{12}{3},$$

or

$$y = 2C_1 e^{3t} - 2C_2 e^{-t} - \frac{13}{3}.$$

DO EXERCISE 8.

Example 6 Tank A contains 2000 pounds of salt dissolved in 1000 gallons of water. Tank B contains 1000 pounds of salt dissolved in 1000 gallons of water. The mixture from tank A is pumped to tank B at the rate of 500 gallons per day, while that from tank B is pumped to tank A at the same rate. Assuming that the mixture in each tank is kept uniform by stirring, how much salt is in each tank after t days?

Solution The initial concentration of salt in tank A is 2000 lb/1000 gal = 2 lb/gal. The initial concentration in tank B is 1000 lb/1000 gal = 1 lb/gal. As we pump the saltier mixture from tank A to tank B, we'd expect the mixture in Tank B to get saltier. At the same time, the mixture being pumped from tank B to tank A should gradually dilute the mixture in tank A. We might guess that the concentrations in each tank might approach an equilibrium concentration.

To see what actually happens, let

$A(t)$ = the number of pounds of salt in tank A after t days, and

$B(t)$ = the number of pounds of salt in tank B after t days.

Note that $A(0) = 2000$ and $B(0) = 1000$. The concentration of salt in tank A after t days is $(A(t)/1000)$ lb/gal, since tank A always contains exactly 1000 gallons of water. Similarly, the concentration in tank B after t days is $B(t)/1000$. To set up our differential equation we observe that

Change in salt in tank A

$$= \text{(amt. of salt flowing in)} - \text{(amt. of salt flowing out)}.$$

The amount of salt flowing into tank A is $(500) \cdot B(t)/1000$ pounds per day, the product of the flow into tank A, 500 gallons per day, with the concentration in tank B (since the flow *into* tank A comes *from* tank B), $B(t)/1000$ pounds per gallon.

Similarly, the flow out of tank A is $(500) \cdot A(t)/1000$. So the change in the amount of salt in tank A, dA/dt, is

$$\frac{dA}{dt} = \frac{500B(t)}{1000} - \frac{500A(t)}{1000},$$

or

$$\frac{dA}{dt} = \frac{1}{2}B(t) - \frac{1}{2}A(t).$$

The amount of salt flowing into tank B is just the negative of that flowing into tank A (equivalently, what flows out of tank A goes into tank B and vice versa). So

$$\frac{dB}{dt} = -\frac{dA}{dt} = \frac{1}{2}A(t) - \frac{1}{2}B(t).$$

We have a system of equations, namely,

$$\frac{dA}{dt} = -\frac{1}{2}A + \frac{1}{2}B,$$

$$\frac{dB}{dt} = \frac{1}{2}A - \frac{1}{2}B$$

with the initial conditions $A(0) = 2000$ and $B(0) = 1000$.

Differentiating the first equation, $A' = -\frac{1}{2}A + \frac{1}{2}B$, gives $A'' = -\frac{1}{2}A' + \frac{1}{2}B'$. Using the second equation, we get

$$A'' = -\frac{1}{2}A' + \frac{1}{2} \cdot \left(\frac{1}{2}A - \frac{1}{2}B\right),$$

or

$$A'' = -\frac{1}{2}A' + \frac{1}{4}A - \frac{1}{4}B.$$

From the first equation, we have $\frac{1}{2}B = A' + \frac{1}{2}A$ or, substituting above,

$$A'' = -\frac{1}{2}A' + \frac{1}{4}A - \frac{1}{4}(2A' + A),$$

or

$$A'' = -A'.$$

The characteristic equation of $A'' + A' = 0$ is $r^2 + r = 0$, or $r(r + 1) = 0$. Since its roots are 0 and -1, we get

$$A(t) = C_1 + C_2 e^{-t}.$$

Since $B = A + 2A'$ (solving the first equation for B), we get

$$B(t) = (C_1 + C_2 e^{-t}) + 2(-C_2 e^{-t}),$$

or

$$B(t) = C_1 - C_2 e^{-t}.$$

To evaluate the constants, recall that $A(0) = 2000$, whereas if we let $t = 0$ in our formula for $A(t)$, we get

$$A(0) = C_1 + C_2 e^{-0} = C_1 + C_2.$$

So $C_1 + C_2 = 2000$. Similarly, $B(0) = 1000$ while $B(0) = C_1 - C_2 e^{-0} = C_1 - C_2$. Thus $C_1 - C_2 = 1000$. Adding, we get

$$\begin{array}{r} C_1 + C_2 = 2000 \\ C_1 - C_2 = 1000 \\ \hline 2C_1 = 3000 \end{array}$$

9. Find the answer to Example 6 if tank A initially contains 1600 pounds of salt, while tank B contains 800 pounds.

10. Find the answer to Example 6 if both tanks initially contain 1100 pounds of salt. (*Warning:* Before computing, stop and *think!*).

or $C_1 = 1500$. From the first equation, $1500 + C_2 = 2000$, or $C_2 = 500$. We have

$$A(t) = 1500 + 500e^{-t},$$ CSS

$$B(t) = 1500 - 500e^{-t}.$$

Note that $A(t) + B(t) = 3000$, which is correct, since salt neither enters nor leaves the system.

As for our intuitive feeling that the concentration of salt in each tank should approach an equilibrium concentration, note that if t is a large positive number, then e^{-t} is a very small (positive) number. If you have a calculator handy, you can check that

$$e^{-3} \approx 0.04978707,$$

$$e^{-14} \approx 0.00000083,$$

and

$$e^{-30} \approx 1 \times 10^{-13} \quad (0.0000000000001).$$

Then we can see that as t increases, the amount of salt in each tank will approach 1500 pounds, as we might have expected.

DO EXERCISES 9 AND 10.

Example 7 *Commodities prices.* The price of two commodities after t days is $P(t)$ and $Q(t)$, respectively. Traders notice that over a period of time the prices obey the system

$$\frac{dP}{dt} = 3000 - 5Q$$

and

$$\frac{dQ}{dt} = 20P - 5000.$$

If the initial prices are $P(0) = \$260$ and $Q(0) = \$560$, find the prices at any later time (assume t is in days and P and Q are in dollars per unit).

Solution Differentiating the first equation yields $P'' = -5Q'$. Since $Q' = 20P - 5000$ (the second equation), we get

$$P'' = -5(20P - 5000),$$

or

$$P'' = -100P + 25{,}000,$$

or

$$P'' + 100P = 25{,}000.$$

We have a nonhomogeneous second-order equation with constant coefficients. Note that the constant function $P_p = 250$ is a particular solution to the equation. Since the characteristic equation of the associated homogeneous equation $(P'' + 100P = 0)$ is $r^2 + 100 = 0$, the roots are $r = \pm 10i$ and we get

$$P_h = C_1 \sin(10t) + C_2 \cos(10t).$$

So

$$P = C_1 \sin(10t) + C_2 \cos(10t) + 250.$$

Solving the first equation for Q gives $5Q = -dP/dt + 3000$, or

$$Q = -\frac{1}{5} \cdot \frac{dP}{dt} + 600.$$

So

$$Q = -\frac{1}{5}(10C_1 \cos(10t) - 10C_2 \sin(10t)) + 600$$

$$Q = -2C_1 \cos(10t) + 2C_2 \sin(10t) + 600,$$

or

$$Q = 2C_2 \sin(10t) - 2C_1 \cos(10t) + 600.$$

We were given that $P(0) = 260$. From our solution,

$$P(0) = C_1 \sin(10 \cdot 0) + C_2 \cos(10 \cdot 0) + 250$$

$$P(0) = C_1 \sin(0) + C_2 \cos(0) + 250,$$

or

$$P(0) = C_2 + 250.$$

So $260 = C_2 + 250$, or $C_2 = 10$. Similarly, $Q(0) = 560$, while, from the solution,

$$Q(0) = 2C_2 \sin(10 \cdot 0) - 2C_1 \cos(10 \cdot 0) + 600,$$

or

$$Q(0) = 2C_2 \sin 0 - 2C_1 \cos 0 + 600,$$

or

$$Q(0) = 600 - 2C_1.$$

So $560 = 600 - 2C_1$ and we get $C_1 = 20$. Thus after t days, the prices of the commodities are

$$P(t) = 20 \sin(10t) + 10 \cos(10t) + 250$$

11. If $P(0) = \$24$, $Q(0) = \$38$, and P and Q are related by

$$\frac{dP}{dt} = 160 - 4Q,$$

$$\frac{dQ}{dt} = 16P - 320,$$

find $P(t)$ and $Q(t)$.

and

$$Q(t) = 20 \sin (10t) - 40 \cos (10t) + 600.$$

Table 9.1 gives the values of P and Q after the first few days.

TABLE 9.1

t	P	Q
0	260.00	560.00
1	230.73	622.68
2	272.34	601.94
3	231.78	574.07
4	258.23	641.58
5	254.40	556.15

The prices fluctuate back and forth about the values $250 and $600, respectively. It is *not* true, however, that the greatest price for P coincides with the least for Q (there is a time lag involved).

DO EXERCISE 11.

Example 8 *Population growth.* The populations of two species at time t are $X(t)$ and $Y(t)$, respectively. If it is assumed that the species interact according to

$$\frac{dX}{dt} = 3X - 5Y,$$

$$\frac{dY}{dt} = 5X - 3Y - 480,$$

and if $X(0) = 160$ and $Y(0) = 92$, find $X(t)$ and $Y(t)$. Describe what happens to the populations over a long period of time.

Solution The usual sequence of steps (we omit the details) leads to the second-order equation $X'' + 16X = 2400$. The general solution to this is $X = C_1 \sin 4t + C_2 \cos 4t + 150$. Solving for Y in the first equation yields

$$Y = \left(\frac{3}{5}C_1 + \frac{4}{5}C_2\right) \sin 4t + \left(-\frac{4}{5}C_1 + \frac{3}{5}C_2\right) \cos 4t + 90.$$

Using the initial conditions yields $160 = C_2 + 150$, or $C_2 = 10$, and $92 = -\frac{4}{5}C_1 + \frac{3}{5}C_2 + 90$, or $2 = -\frac{4}{5}C_1 + 6$, or $C_1 = 5$. Substitution and simplification result in

$$X = 5 \sin 4t + 10 \cos 4t + 150,$$ [CSS]

$$Y = 11 \sin 4t + 2 \cos 4t + 90.$$

The populations fluctuate. If we use the fact that $A \sin (\lambda t) + B \cos (\lambda t)$ always remains between $\pm \sqrt{A^2 + B^2}$, then we see that the X-population stays between $150 \pm \sqrt{5^2 + 10^2}$, or between 139 and 161, while the Y-population remains between 79 and 101 (approximately).

EXERCISE SET 9.8

Find the general solution to each of the following systems of equations.

1. $x' = 2x$, $y' = 3y$

2. $x' = -x$, $y' = 4y$

3. $x' = y$, $y' = -2x + 3y$

4. $x' = x + 2y$, $y' = x + 2y$

5. $x' = 2x + y$, $y' = 3x + 4y$

6. $x' = 3x - 4y$, $y' = 2x - 3y$

7. $x' = -\frac{1}{2}x + y$, $y' = \frac{1}{2}x$

8. $x' = -x - 3y$, $y' = -y$

9. $x' = 4x + y$, $y' = -x + 2y$

10. $x' = 2x - y$, $y' = 4x - 2y$

11. $x' = 2y$, $y' = -18x$

12. $x' = x - y$, $y' = 5x - y$

13. $x' = 3x - 5y$, $y' = x - y$

14. $x' = -3x + 5y$, $y' = -x + y$

Find the functions that satisfy both the system of equations and the given initial conditions.

15. $x' = 2x - y$, $y' = 2x + 5y$, $x(0) = 3$, $y(0) = -5$

16. $x' = x + y$, $y' = x + y$, $x(0) = 2$, $y(0) = 0$

17. $x' = 2x + 3y$, $y' = -3x + 8y$, $x(0) = 1$, $y(0) = 1$

18. $x' = 2x + 3y$, $y' = -3x + 8y$, $x(0) = 1$, $y(0) = 2$

19. $x' = y$, $y' = -4x$, $x(0) = 1$, $y(0) = 2$

20. $x' = x + 5y$, $y' = -x - 3y$, $x(0) = 10$, $y(0) = -3$

Find the solution to each of the following systems of equations, finding the particular solution if initial conditions are given.

21. $x' = 5x - y + 5$, $y' = 2x + 2y + 2$

22. $x' = 2x + y + 1$, $y' = -2y - 22$

23. $x' = y + 2t + 3$, $y' = -x + 4t - 2$

24. $x' = -x + y + 2t + 2$, $y' = 2x + 2t - 1$

25. $x' = 2x + y + 3$, $y' = 5x - 2y + 12$, $x(0) = 0$, $y(0) = -3$

26. $x' = y + 5$, $y' = -x + 2y + 10$, $x(0) = 1$, $y(0) = -2$

27. *Mixing chemicals.* Tank A initially contains 200 pounds of salt dissolved in 500 gallons of water, while tank B contains 100 pounds of salt dissolved in 500 gallons of water. The mixture in tank A is pumped to tank B at the rate of 100 gallons per day and the mixture from B is pumped to A at the same rate. Set up the system of differential equations for $A(t)$ and $B(t)$, the number of pounds of salt in tanks A and B, after t days, respectively. Then solve them. What is the "limiting concentration" of salt in tank A? in tank B?   [CSS]

28. *Population growth.* The numbers $X(t)$ and $Y(t)$ of two species in a certain area (in hundreds) satisfy the system of equations

$$\frac{dX}{dt} = X + Y - 100, \qquad X(0) = 40,$$

$$\frac{dY}{dt} = -8X - 3Y + 450, \qquad Y(0) = 60,$$

where t is measured in months.

a) Find $X(t)$ and $Y(t)$.

b) Find $X(1)$, $Y(1)$, $X(10)$, and $Y(10)$.

c) What are $X(\infty)$ and $Y(\infty)$?

29. *Commodities prices.* The prices P and Q of two inter-related commodities are determined by the following system of equations:

$$P' = P + 13Q - 400, \qquad P(0) = 49,$$

$$Q' = -2P - Q + 100, \qquad Q(0) = 32.$$

Find $P(t)$ and $Q(t)$.

30. *Projected sales.* A securities analyst finds that the projected sales $X(t)$ and $Y(t)$ of two competing firms (in millions of dollars) is given by

$$X' = 4Y - 100t, \qquad X(0) = 377,$$

$$Y' = -X + 40t + 400, \qquad Y(0) = 12.$$

Find $X(t)$ and $Y(t)$ (t is in years).

31. Solve

$$x' = -2x + 2y + 2e^t,$$

$$y' = 2x + y.$$

32. Solve

$$\frac{dx}{dt} = -2x + y - z,$$

$$\frac{dy}{dt} = y + z,$$

$$\frac{dz}{dt} = 3z.$$

CHAPTER 9 TEST

1. Solve the differential equation $y' = x^2 + 3x - 5$, given the condition $y = 7$ when $x = 0$.

Solve the following differential equations.

2. $\dfrac{dy}{dx} = 8x^7 y$

3. $\dfrac{dy}{dx} = \dfrac{9}{y}$

4. $\dfrac{dy}{dt} = 6y$

5. For the inhibited growth model, $P = 680$ at the point of inflection. What is the limiting value?

6. The growth rate for a certain population of bacteria cells is given by

$$\frac{dP}{dt} = kP(L - P),$$

where

$P =$ the number of cells present after time t (in hours);
$L = 800$, the limiting value;
$k = 0.0009$; and $P_0 = 20$.

a) Write the solution $P(t)$ in terms of the given constraints.

b) Sketch a graph of $P(t)$ given that the point of inflection occurs at $t = 5$.

c) At what time t are 500 cells present?

7. Graph $P(t) = 5(1 - e^{-t})$.

8. The growth rate of a company is modeled by

$$\frac{dS}{dt} = k(L - S), \qquad S = \$0 \text{ when } t = 0;$$

where

  $S =$ the total sales, in millions of dollars, in the
    tth year of operation,
  $L = \$20$ million, the limiting value, and
  $k =$ a constant.

a) Write the solution $S(t)$ in terms of L and k.

b) Suppose $S(2) = \$4$ million; that is, sales during the second year are $4 million. Use this to determine k. Round to the nearest hundredth.

c) Rewrite $S(t)$ in terms of k.

d) Use the equation in (c) to find $S(10)$, the sales in the tenth year.

e) In what year will the sales be $15 million?

Solve each of the following.

9. $3x^2 \, dx + 2y \, dy = 0, \; y(0) = 1$

10. $(3x^2 + 4xy) \, dx + (2x^2 - 5y^4) \, dy = 0, \; y(0) = 1$

11. $xy' + 3y = 0, \; y(1) = 4$

12. $y' - 3y = e^{4x}$

13. $y'' - 5y' + 6y = 0$

14. $y'' - 12y' + 36y = 0$

15. $y'' + 36y = 72$

16. $x' = 2x + 5y, \; y' = 3x$

17. $x' = 6y, \; y' = -24x$

18. $x' = -6x + 3y, \; y' = -3x$

19. $x' = -6x + y, \; y' = -2x - 3y, \; x(0) = 0, \; y(0) = -1$

TAYLOR SERIES AND L'HOPITAL'S RULE

10

A "series" of deposits (*Cheryl Shugars*).

This graph shows the effect of $1000 placed into the economy in a bank when 90% of it is loaned out and redeposited elsewhere, then 90% of that is loaned out, and so on.

OBJECTIVE

You should be able to expand a given polynomial in powers of $x - a$ (for any given real number a).

10.1 TAYLOR'S FORMULA FOR POLYNOMIALS

The function f given by

$$f(x) = 2x^2 - 16x + 35$$

can also be expressed in the form

$$f(x) = 2(x - 3)^2 - 4(x - 3) + 5.$$ `CSS`

This can be checked by multiplying and simplifying the right-hand side of the second equation. In the second form, $f(x)$ is expressed as a sum of powers of $x - 3$. In general, the formulas for many functions can be written in the form*

$$f(x) = \sum_{n=0}^{\infty} c_n (x - a)^n,$$

where a is a fixed number and $c_1, c_2, \ldots, c_n, \ldots$ is a sequence of real numbers. In this chapter we will develop expansions like this. Each is called a *Taylor series expansion*. This tool is useful in many areas of mathematics, in particular when solving differential equations.

To introduce the subject, we first consider the Taylor expansion for polynomial functions. Recall the following definition from Section 1.5.

DEFINITION

A *polynomial function* (in *x*) is a function that can be written as

$$f(x) = c_n x^n + c_{n-1} x^{n-1} + \cdots + c_1 x + c_0,$$

where *n* is a nonnegative integer and $c_n, c_{n-1}, \ldots c_0$ are real numbers, called *coefficients*.

Thus, each of the following is a polynomial function:

$$f(x) = -5 \qquad \text{(a constant function)},$$
$$f(x) = x + 3 \qquad \text{(a linear function)},$$

and

$$f(x) = 2x^3 - 4x^2 + x + 1.$$

The degree of a polynomial in x is the exponent of the highest order x-term present that has a nonzero coefficient, with the understanding that nonzero constant functions have degree 0 and the polynomial that is 0 has no degree.

---

*You may need to review summation notation in Section 5.7.

1. Simplify

$$\frac{2(x - 3)^2 + 4(x - 3) + 1}{x - 3}$$

for $x \neq 3$.

For example, the polynomial $2x^3 - 4x^2 + x + 1$ has degree 3. The polynomial $x + 3$ has degree 1 (since $x^1 = x$). The polynomial $0 \cdot x^5 + 3x^4$ has degree 4. The polynomial 6, or $6x^0$, has degree 0.

Consider the following example.

Example 1 Simplify

$$\frac{x^2 - 3x + 1}{x}, \qquad x \neq 0.$$

Solution We write this expression in the form

$$\frac{x^2}{x} - \frac{3x}{x} + \frac{1}{x},$$

and, performing the indicated divisions, we obtain

$$x - 3 + \frac{1}{x} \qquad \text{for } x \neq 0.$$

DO EXERCISE 1.

2. Use the identity

$$x^3 - 6x^2 + 14x - 8$$
$$= (x - 2)^3 + 2(x - 2) + 4$$

to simplify

$$\frac{x^3 - 6x^2 + 14x - 8}{(x - 2)^2}.$$

Example 2 Simplify

$$\frac{x^3 - 3x^2 + 3x - 4}{x - 1}, \qquad x \neq 1.$$

Solution This looks harder. We can write this expression as

$$\frac{x^3}{x - 1} - \frac{3x^2}{x - 1} + \frac{3x}{x - 1} - \frac{4}{x - 1},$$

but this doesn't simplify as in Example 1. Suppose, however, that we know that

$$x^3 - 3x^2 + 3x - 4 = (x - 1)^3 - 3.$$

This is true, although you would not know it offhand. Then we can simplify as follows:

$$\frac{x^3 - 3x^2 + 3x - 4}{x - 1} = \frac{(x - 1)^3 - 3}{x - 1} = \frac{(x - 1)^3}{x - 1} - \frac{3}{x - 1}$$

$$= (x - 1)^2 - \frac{3}{x - 1},$$

whenever $x \neq 1$.

DO EXERCISE 2.

In Example 2, the key to the solution was the identity $x^3 - 3x^2 + 3x - 4 = (x - 1)^3 - 3$. We were given the expansion of $x^3 - 3x^2 + 3x - 4$ in powers of $x - 1$. We want to be able to write down identities of this kind quickly and easily. Suppose that we want to write the polynomial $2x^2 - 16x + 35$ in powers of $x - 3$. That is, we want to find real numbers c_2, c_1, and c_0 so that

$$2x^2 - 16x + 35 = c_2(x - 3)^2 + c_1(x - 3) + c_0.$$

One possible solution is to multiply out the right-hand side of the equation, grouping the x^2, x, and constant terms together. Then we can equate corresponding coefficients to obtain three linear equations in the three unknowns c_2, c_1, and c_0, each of which can then be found. But here's another way.

Example 3 Let $f(x) = 2x^2 - 16x + 35$. Find c_2, c_1, and c_0 so that

$$f(x) = c_2(x - 3)^2 + c_1(x - 3) + c_0. \tag{1}$$

Solution The equation above must hold for all real numbers x. So it must be true if we let $x = 3$. We obtain

$$f(3) = c_2(3 - 3)^3 + c_1(3 - 3) + c_0$$

$$= c_2 \cdot 0 + c_1 \cdot 0 + c_0$$

$$= c_0.$$

Since $f(3) = 2(3)^2 - 16(3) + 35 = 18 - 48 + 35 = 5$, we get

$$c_0 = f(3) = 5.$$

Next, we apply some calculus. We differentiate both sides of Eq. (1) with respect to x to get

$$f'(x) = 2c_2(x - 3) + c_1 + 0 \tag{2}$$

$$= 2c_2(x - 3) + c_1.$$

Letting $x = 3$, we get

$$f'(3) = 2c_2(3 - 3) + c_1,$$

or

$$c_1 = f'(3).$$

Since $f(x) = 2x^2 - 16x + 35$, then $f'(x) = 4x - 16$, and $f'(3) = 4(3) - 16 = -4$. So we get

$$c_1 = f'(3) = -4.$$

To find c_2, we differentiate Eq. (2). We get $f''(x) = 2c_2$, a constant function. Since $f'(x) = 4x - 16$, we also have $f''(x) = 4$, so that $2c_2 = 4$, or

3. Let $f(x) = x^2 - 4x + 2$. Find $c_2, c_1,$ and c_0 so that
$$f(x) = c_2(x - 3)^2 + c_1(x - 3) + c_0.$$

$c_2 = 2$. Thus $c_2 = 2$, $c_1 = -4$, $c_0 = 5$, and
$$2x^2 - 16x + 35 = 2(x - 3)^2 - 4(x - 3) + 5.$$

DO EXERCISE 3.

Example 4 Write $f(x) = x^3 - 7x + 3$ in powers of $x - 2$.

Solution We want to write
$$f(x) = c_3(x - 2)^3 + c_2(x - 2)^2 + c_1(x - 2) + c_0. \tag{3}$$
We let $x = 2$ in Eq. (3) to get
$$f(2) = c_3(2 - 2)^3 + c_2(2 - 2)^2 + c_1(2 - 2) + c_0$$
$$= c_0.$$

Differentiating, we obtain
$$f'(x) = 3c_3(x - 2)^2 + 2c_2(x - 2) + c_1. \tag{4}$$
Letting $x = 2$, we get
$$f'(2) = c_1.$$

Differentiating Eq. (4) yields
$$f''(x) = 6c_3(x - 2) + 2c_2,$$
and, letting $x = 2$, we get
$$f''(2) = 2c_2.$$

Another differentiation yields
$$f'''(x) = 6c_3.$$

Since this equation is true for all values of x—that is, since $f'''(x)$ is a constant function—we have, in particular,
$$f'''(2) = 6c_3.$$

Before we compute $c_3, c_2, c_1,$ and c_0, note that there seems to be a pattern to the c_n's. That is,
$$c_0 = f(2),$$
$$c_1 = f'(2),$$
$$c_2 = \frac{1}{2}f''(2),$$
$$c_3 = \frac{1}{6}f'''(2).$$

4. Compute. a) $4!$

  b) $\dfrac{16}{2!}$

  c) $\dfrac{-5}{0!}$

  d) $\dfrac{120}{6!}$

Since $f(x) = x^3 - 7x + 3$, we have

$$c_0 = f(2) = 2^3 - 7 \cdot 2 + 3$$
$$= 8 - 14 + 3$$
$$= -3.$$

Differentiating, we have $f'(x) = 3x^2 - 7$, so that

$$c_1 = f'(2) = 3(2)^2 - 7$$
$$= 12 - 7 = 5.$$

Also, $f''(x) = 6x$, so that $f''(2) = 6 \cdot 2 = 12$, and $c_2 = \frac{1}{2}f''(2) = \frac{1}{2} \cdot 12 = 6$. Finally, $f'''(x) = 6$, so that $c_3 = \frac{1}{6}f'''(2) = \frac{1}{6} \cdot 6 = 1$. Thus

$$x^3 - 7x + 3 = (x - 2)^3 + 6(x - 2)^2 + 5(x - 2) - 3.$$

Factorial Notation

Before continuing, we introduce some notation. Products like $3 \cdot 2 \cdot 1$ or $6 \cdot 5 \cdot 4 \cdot 3 \cdot 2 \cdot 1$ will occur so often that it is convenient to adopt a notation for them. We define $5 \cdot 4 \cdot 3 \cdot 2 \cdot 1 = 5!$, read "5 factorial."

DEFINITION

We define $n!$, read "n factorial," as follows:

$$n! = n(n - 1)(n - 2) \ldots 3 \cdot 2 \cdot 1,$$

for any integer n, $n > 1$. For convenience in later notation, we define $1!$ (*one factorial*) to be 1 and $0!$ (*zero factorial*) to be 1.

Thus we have

$$3! = 3 \cdot 2 \cdot 1 = 6,$$
$$5! = 5 \cdot 4 \cdot 3 \cdot 2 \cdot 1 = 120,$$

and

$$6! = 6 \cdot 5 \cdot 4 \cdot 3 \cdot 2 \cdot 1$$
$$= 6 \cdot (5 \cdot 4 \cdot 3 \cdot 2 \cdot 1)$$
$$= 6 \cdot 5! = 6 \cdot 120 = 720.$$

DO EXERCISE 4.

For a moment, look back at Example 4, where we listed the coefficients c_3, c_2, c_1, and c_0 in the expansion of $f(x) = x^3 - 7x + 3$ in powers of x − 2. Note now that we can rewrite them as follows:

$$c_0 = f(2) = \frac{f(2)}{0!} \qquad \text{since } 0! = 1,$$

$$c_1 = f'(2) = \frac{f'(2)}{1!} \qquad \text{since } 1! = 1,$$

$$c_2 = \frac{1}{2}f''(2) = \frac{f''(2)}{2} = \frac{f''(2)}{2!},$$

and

$$c_3 = \frac{1}{6}f'''(2) = \frac{f'''(2)}{6} = \frac{f'''(2)}{3!}.$$

In words, c_3 is the value of the third derivative of $f(x)$ at $x = 2$ (since we were expanding in powers of $x − 2$) divided by 3!.

To make the pattern stand out more clearly, recall (Section 2.8) that $f^{(3)}(x)$ is sometimes used for the third derivative of $f(x)$. Thus $f^{(1)}(x) = f'(x)$ and $f^{(6)}(x)$ denotes the sixth derivative of $f(x)$. Finally, we let $f^{(0)}(x) = f(x)$. Then we can rewrite the preceding formulas as follows:

$$c_0 = \frac{f(2)}{0!} = \frac{f^{(0)}(2)}{0!},$$

$$c_1 = \frac{f'(2)}{1!} = \frac{f^{(1)}(2)}{1!},$$

$$c_2 = \frac{f''(2)}{2!} = \frac{f^{(2)}(2)}{2!},$$

and

$$c_3 = \frac{f'''(2)}{3!} = \frac{f^{(3)}(2)}{3!}.$$

Examples 3 and 4 illustrate the following theorem.

THEOREM 1   Taylor's Formula For Polynomials

Let $f(x)$ be a polynomial of degree n and let a be any number. Then $f(x)$ has a unique expansion in powers of $x − a$:

$$f(x) = c_0 + c_1(x − a) + c_2(x − a)^2 + \cdots + c_{n-1}(x − a)^{n-1} + c_n(x − a)^n.$$

Here, the jth coefficient, c_j, can be computed from the formula

$$c_j = \frac{f^{(j)}(a)}{j!}, \qquad j = 0, 1, \ldots, n.$$

5. Write $f(x) = 2x^2 - x + 3$ in powers of $x - 1$.

Example 5 Write $f(x) = x^3 - 7x^2 + 2x + 3$ in powers of $x - 1$.

Solution We apply Taylor's formula with $a = 1$. Thus

$$c_0 = \frac{f^{(0)}(1)}{0!} = \frac{(1)^3 - 7(1)^2 + 2 \cdot 1 + 3}{1}$$

$$= 1 - 7 + 2 + 3$$

$$= -1.$$

To compute c_1, we need a formula for $f'(x)$. Differentiating $f(x) = x^3 - 7x^2 + 2x + 3$, we get

$$f^{(1)}(x) = f'(x)$$

$$= 3x^2 - 14x + 2.$$

Hence,

$$c_1 = \frac{f^{(1)}(1)}{1!} = \frac{3(1)^2 - 14 \cdot 1 + 2}{1}$$

$$= -9.$$

Continuing, we have $f^{(2)}(x) = f''(x) = 6x - 14$, so that

$$c_2 = \frac{f^{(2)}(1)}{2!} = \frac{6 \cdot 1 - 14}{2} = \frac{-8}{2} = -4.$$

6. Write $f(x) = x^3 - x$ in powers of $x - 3$.

Finally, $f^{(3)}(x) = f'''(x) = 6$, so that

$$c_3 = \frac{f^{(3)}(1)}{3!} = \frac{6}{3!}$$

$$= \frac{6}{6} = 1.$$

Note that $f^{(4)}(x) = 0$ for all x. In particular, c_4 must be 0. Since the same is true for all higher derivatives of $f(x)$, we can stop here—that is, we have found all the nonzero c_j's. So the expansion of $f(x) = x^3 - 7x^2 + 2x + 3$ in powers of $x - 1$ is

$$f(x) = -1 - 9(x - 1) - 4(x - 1)^2 + (x - 1)^3,$$

or, equivalently,

$$f(x) = (x - 1)^3 - 4(x - 1)^2 - 9(x - 1) - 1.$$

DO EXERCISES 5 AND 6.

Example 6 Expand $f(x) = x^3 - 7x^2 + 2x + 3$ in powers of $x - 2$.

7. Write $f(x) = x^2 - 3x + 1$ in powers of $x - 4$.

Solution We have

$$c_0 = \frac{f^{(0)}(2)}{0!} = \frac{f(2)}{0!}$$

$$= (2)^3 - 7(2)^2 + 2 \cdot 2 + 3$$

$$= 8 - 28 + 4 + 3 = -13.$$

Since $f'(x) = 3x^2 - 14x + 2$ (from Example 5), then

$$c_1 = \frac{f^{(1)}(2)}{1!} = \frac{3 \cdot 2^2 - 14 \cdot 2 + 2}{1}$$

$$= \frac{12 - 28 + 2}{1}$$

$$= -14.$$

Continuing, we have $f^{(2)}(x) = 6x - 14$ and

$$c_2 = \frac{f^{(2)}(2)}{2!} = \frac{6 \cdot 2 - 14}{2} = \frac{-2}{2} = -1.$$

Since $f^{(3)}(x) = 6$,

$$c_3 = \frac{f^{(3)}(2)}{3!} = \frac{6}{6} = 1.$$

8. Write $f(x) = x^2 - 3x + 1$ in powers of $x - 2$.

So

$$f(x) = (x - 2)^3 - 1(x - 2)^2 - 14(x - 2) - 13.$$

Note that even though the function being expanded is the same in Examples 5 and 6, the coefficients of the two expansions are different.

DO EXERCISES 7 AND 8.

Example 7 Expand $f(x) = x^2 + 3x + 1$ in powers of $x + 1$.

Solution Since $x + 1 = x - (-1)$, we have $a = -1$. So

$$c_0 = \frac{f^{(0)}(-1)}{0!} = \frac{(-1)^2 + 3(-1) + 1}{1}$$

$$= 2 - 3 = -1.$$

Next, $f^{(1)}(x) = f'(x) = 2x + 3$, so that

$$c_1 = \frac{f^{(1)}(-1)}{1!} = \frac{2(-1) + 3}{1} = 1.$$

Finally, $f^{(2)}(x) = 2$ and

$$c_2 = \frac{f^{(2)}(-1)}{2!} = \frac{2}{2} = 1.$$

So

$$f(x) = 1 \cdot (x + 1)^2 + 1 \cdot (x + 1) - 1$$

or

$$f(x) = (x + 1)^2 + (x + 1) - 1. \tag{5}$$

Note that if we multiply this last expression out, we obtain

$$f(x) = (x^2 + 2x + 1) + (x + 1) - 1$$
$$= x^2 + (2x + x) + (1 + 1 - 1)$$
$$= x^2 + 3x + 1.$$

Our check shows that formula (5) is correct.

So far, we have regarded Taylor's theorem for polynomials only as a way to write a given polynomial $f(x)$ in powers of $x - a$ (a given). There is a slightly different point of view that will be helpful to us later in the chapter. According to this view, Taylor's theorem says that if $f(x)$ is a polynomial of degree n, and if we know the value of $f(x)$ and its first n derivatives at $x = a$, then we can find the value of $f(x)$ at any point. To illustrate this, we consider the following example.

Example 8 Find a polynomial of degree 3 such that $f(5) = 2, f'(5) = 0$, $f''(5) = -3$, and $f'''(5) = 12$.

Solution If we use the given data to expand $f(x)$ about $x = 5$—that is, in powers of $x - 5$—we obtain

$$c_0 = \frac{f^{(0)}(5)}{0!} = \frac{f(5)}{1} = 2,$$

$$c_1 = \frac{f^{(1)}(5)}{1!} = \frac{0}{1} = 0,$$

$$c_2 = \frac{f^{(2)}(5)}{2!} = \frac{-3}{2} = -\frac{3}{2},$$

and

$$c_3 = \frac{f^{(3)}(5)}{3!} = \frac{12}{6} = 2.$$

9. Find a polynomial $f(x)$ of degree 2 that satisfies $f(1) = 2$, $f'(1) = -1$, and $f''(1) = 2$.

10. Find a polynomial $f(x)$ of degree 3 such that $f(-1) = 1$, $f'(-1) = -\frac{1}{2}$, $f''(-1) = 4$, and $f'''(-1) = 18$.

11. Find a polynomial of degree 5 that satisfies $f(0) = 1$, $f'(0) = 2$, $f''(0) = -1$, $f^{(3)}(0) = 42$, $f^{(4)}(0) = 12$, and $f^{(5)}(0) = -60$.

Thus

$$f(x) = 2(x - 5)^3 - \frac{3}{2}(x - 5)^2 + 0(x - 5) + 2$$

$$= 2(x - 5)^3 - \frac{3}{2}(x - 5)^2 + 2.$$

From Theorem 1, we know that $f(x)$ is the *only* polynomial of degree 3 that satisfies the given conditions.

DO EXERCISES 9–11.

EXERCISE SET 10.1

Simplify.

1. $\dfrac{x^2 - 4x + 3}{x}$, $x \neq 0$

2. $\dfrac{(x + 4)^2 - 3(x + 4) + 1}{x + 4}$, $x \neq -4$

3. $\dfrac{2(x - 3)^2 - 6(x - 3) + 7}{x - 3}$, $x \neq 3$

4. $\dfrac{x^3 - 5x^2 + 7x - 3}{x}$, $x \neq 0$

5. $\dfrac{(x - 2)^3 + 7(x - 2) + 4}{x - 2}$, $x \neq 2$

6. $\dfrac{3(x + 4)^3 + 3(x + 4)^2 + 5(x + 4)}{x + 4}$, $x \neq -4$

Expand each of the given polynomials in powers of $x - a$.

7. $x^2 - 2x + 3$, $a = 2$

8. $x^2 - 2x + 3$, $a = 1$

9. $x^2 - 5x + 7$, $a = 2$

10. $x^2 - 5x + 7$, $a = 1$

11. $x^2 - 5x + 7$, $a = -1$

12. $x^2 - 5x + 7$, $a = 0$

13. $x^2 - 3x + 4$, $a = -2$

14. $x^2 - 3x + 4$, $a = 2$

15. $x^3 - 2x^2 + 4x + 1$, $a = 1$

16. $x^3 - 2x^2 + 4x + 1$, $a = 2$

17. $x^3 - 2x^2 + 4x + 1$, $a = -1$

18. $x^3 + 5$, $a = 1$

19. $2x^2 - 6x + 3$, $a = 3$

20. $x^3 - x - 1$, $a = 2$

21. $3x^4 - 7x^2 + 1$, $a = 1$

22. $x^4 - 5x^3 + 3$, $a = 1$

23. $x^5 - 5x^4 + 5x - 1$, $a = 1$

24. $x^5 - 3x + 1$, $a = 1$

25. Check the answers to Exercises 11 and 13 by multiplying them out and comparing your results with the given polynomial.

26. Repeat Exercise 25 using the solutions to Exercises 15 and 21.

Simplify using Taylor's formula.

27. $\dfrac{x^2 - 5}{x - 1}$, $x \neq 1$

28. $\dfrac{x^2 - 3x + 1}{x - 2}$, $x \neq 2$

29. $\dfrac{x^2 + 1}{x + 1}$, $x \neq -1$

30. $\dfrac{x^2 - 7x - 1}{x + 3}$, $x \neq -3$

▶

31. Calculate $f'(1)$ and $f''(1)$ for $f(x) = x^4 - 4x^3 + 6x^2 - 4x + 4$.

Conclude that $x = 1$ is a critical point for $f(x)$ but that Maximum–Minimum Principle 2 (Section 3.2) fails. Next, write $f(x)$ in powers of $x - 1$. Use your answers to show that $f(x)$ has a minimum point at $x = 1$ and that $x = 1$ is the only maximum or minimum point for $f(x)$.

32. Calculate $f'(1)$ and $f''(1)$ for $f(x) = x^3 - 3x^2 + 3x - 3$. Conclude, as in Exercise 31, that Maximum–Minimum Principle 2 fails. Write $f(x)$ in powers of $x - 1$. Use your answers to conclude that $f(x)$ has no maximum or minimum points and that $f(x)$ has exactly one point of inflection, at $x = 1$.

OBJECTIVES

You should be able to

a) **Compute the Taylor polynomial of degree n for $f(x)$ at the point $x = a$.**

b) **Find a Taylor polynomial and use it to approximate a function value.**

10.2 TAYLOR POLYNOMIALS

Finding Taylor Polynomials

We now extend the ideas discussed in Section 10.1 to a wide class of functions.

DEFINITION

Let $f(x)$ have derivatives of all orders. Then the *Taylor polynomial* $p_n(x)$ *of degree n at $x = a$* is defined to be the polynomial given by

$$p_n(x) = c_0 + c_1(x - a) + \cdots + c_n(x - a)^n$$

$$= \sum_{j=0}^{n} c_j(x - a)^j,$$

where

$$c_j = \frac{f^{(j)}(a)}{j!}, \qquad \text{for } j = 0, 1, \ldots, n.$$

Compare this with the formula in Theorem 1 (Section 10.1).

Example 1 Find the Taylor polynomial of degree 2 at $x = 0$ if $f(x) = e^x$.

Solution We must apply the definition with $n = 2$ and $a = 0$. Doing so, we obtain $p_2(x) = c_0 + c_1 x + c_2 x^2$, where

$$c_j = \frac{f^{(j)}(0)}{j!}.$$

`CSS`

But $f^{(0)}(x) = f(x) = e^x$, so that

$$c_0 = \frac{f^{(0)}(0)}{0!} = \frac{e^0}{0!} = 1.$$

1. Find the Taylor polynomial of degree 2 at $x = 0$ if $f(x) = 2e^{-3x}$.

Since $f^{(1)}(x) = f'(x) = e^x$ and $f^{(2)}(x) = e^x$, we have

$$c_1 = \frac{f^{(1)}(0)}{1!} = \frac{e^0}{1} = 1$$

and

$$c_2 = \frac{f^{(2)}(0)}{2!} = \frac{e^0}{2} = \frac{1}{2}.$$

So

$$p_2(x) = 1 + x + \frac{1}{2}x^2.$$

DO EXERCISE 1.

Example 2 Find the Taylor polynomial of degree 2 at $x = -1$ if $f(x) = x^2 + 3x + 1$.

Solution We have $p_2(x) = c_0 + c_1(x + 1) + c_2(x + 1)^2$, where

$$c_j = \frac{f^{(j)}(-1)}{j!}.$$

Since $f(x) = f^{(0)}(x) = x^2 + 3x + 1$, we get

$$c_0 = \frac{f^{(0)}(-1)}{0!}$$

$$= \frac{(-1)^2 + 3(-1) + 1}{1}$$

$$= \frac{1 - 3 + 1}{1} = -1.$$

Similarly, $f'(x) = f^{(1)}(x) = 2x + 3$ and

$$c_1 = \frac{f^{(1)}(-1)}{1!} = \frac{2(-1) + 3}{1} = \frac{-2 + 3}{1} = 1,$$

and since $f''(x) = f^{(2)}(x) = 2$, we have

$$c_2 = \frac{f^{(2)}(-1)}{2!} = \frac{2}{2} = 1.$$

So

$$p_2(x) = -1 + (x + 1) + (x + 1)^2.$$

2. a) Find the Taylor polynomial of degree 2 at $x = 4$ if $f(x) = x^2 - 3x + 1$.

b) Compare Example 2 and Margin Exercise 2 with Example 7 and Margin Exercise 7 of Section 10.1.

DO EXERCISE 2.

Example 3 Find the Taylor polynomial of degrees 2 and 3 at $x = 0$ if $f(x) = \cos x$.

Solution Recalling that $f'(x) = f^{(1)}(x) = -\sin x$ and that $f''(x) = f^{(2)}(x) = -\cos x$, we have

$$c_0 = \frac{f^{(0)}(0)}{0!} = \frac{\cos (0)}{1} = 1,$$

$$c_1 = \frac{f^{(1)}(0)}{1!} = \frac{-\sin (0)}{1} = 0,$$

and

$$c_2 = \frac{f^{(2)}(0)}{2!} = \frac{-\cos (0)}{2!} = \frac{-1}{2} = -\frac{1}{2}.$$

We find that

$$p_2(x) = 1 + 0 \cdot x - \frac{1}{2} x^2 = 1 - \frac{1}{2} x^2.$$

To compute $p_3(x)$, we need only calculate

$$c_3 = \frac{f^{(3)}(0)}{3!}.$$

Since $f''(x) = -\cos x$, differentiation yields $f'''(x) = f^{(3)}(x) = \sin x$ and $f^{(3)}(0) = \sin (0) = 0$. So

$$c_3 = \frac{f^{(3)}(0)}{3!} = 0,$$

and we have

$$p_3(x) = 1 + 0 \cdot x - \frac{1}{2} x^2 + 0 \cdot x^3 = 1 - \frac{1}{2} x^2.$$

3. Find the Taylor polynomial of degrees 2 and 3 at $x = 0$ if $f(x) = \sin x$.

DO EXERCISE 3.

Example 4 Compute the Taylor polynomial of degree 2 at $x = 1$ for

$$f(x) = \frac{1}{1 + x^2}.$$

4. Find the Taylor polynomial of degree 2 at $x = 0$ if $f(x) = 1/(1 - x^2)$.

Solution Since $f^{(0)}(1) = \frac{1}{2}$, we have $c_0 = \frac{1}{2}$. Applying the Quotient Rule, we get

$$f'(x) = \frac{(1 + x^2) \cdot 0 - 1(2x)}{(1 + x^2)^2} = \frac{-2x}{(1 + x^2)^2}.$$

So

$$f'(1) = f^{(1)}(1) = \frac{-2(1)}{(1 + 1^2)^2} = \frac{-2}{4} = -\frac{1}{2}$$

and

$$c_1 = \frac{f^{(1)}(1)}{1!} = -\frac{1}{2}.$$

Finally, differentiating $f'(x)$ yields

$$f''(x) = \frac{(1 + x^2)^2(-2) - (-2x) \cdot 2(1 + x^2) \cdot 2x}{(1 + x^2)^4}$$

$$= \frac{-2(1 + x^2)^2 + 8x^2(1 + x^2)}{(1 + x^2)^4}$$

$$= \frac{-2(1 + x^2) + 8x^2}{(1 + x^2)^3}$$

$$= \frac{6x^2 - 2}{(1 + x^2)^3}.$$

So

$$c_2 = \frac{f^{(2)}(1)}{2!} = \frac{1}{2!} \cdot \frac{6(1)^2 - 2}{(1 + 1^2)^3} = \frac{1}{2} \cdot \frac{4}{8} = \frac{1}{4}.$$

Thus

$$p_2(x) = \frac{1}{2} - \frac{1}{2}(x - 1) + \frac{1}{4}(x - 1)^2.$$

DO EXERCISE 4.

Approximating Using Taylor Polynomials

We turn now to the geometrical or graphical meaning of Taylor polynomials. Let $f(x) = e^x$, the graph that appears in color in the figure below. We'll consider the Taylor polynomials for e^x at $x = 0$.

The Taylor polynomial of degree 0 at $x = 0$ is the constant function $p_0(x) = f(0) = e^0 = 1$. This function is the horizontal (black) line passing through $(0, 1)$ in the figure below.

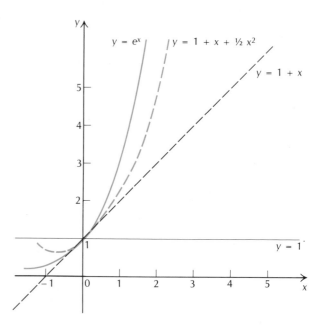

The Taylor polynomial of degree 1 at $x = 0$ is $p_1(x) = f(0) + f'(0)x = 1 + x$. This function is also linear and its graph is the dashed line in the figure above. Note that it, too, passes through $(0, 1)$. Since the slope of $1 + x$ is 1 and the slope of e^x at $(0, 1)$ is also $e^0 = 1$, we see that $1 + x$ is the tangent line to the graph of $y = e^x$ at $(0, 1)$.

In general, the Taylor polynomial of degree 1 for $f(x)$ at $x = a$ is always the tangent line to the graph of $y = f(x)$ at the point $(a, f(a))$ on the graph.

Next, consider the Taylor polynomial of degree 2 for e^x at $x = 0$. From Example 1, we have $p_2(x) = 1 + x + \frac{1}{2}x^2$. This polynomial is the colored dashed curve in the figure above. Since $p_2(0) = 1$, it passes through $(0, 1)$. By the way in which it was constructed, $p_2'(0) = 1 = f'(0)$ and $p_2''(0) = 1 = f''(0)$. We might, therefore, think of $p_2(x)$ as being the "tangent quadratic" to the graph of $f(x) = e^x$ at $x = 0$.

Note that although $p_2(x)$ agrees with e^x only at $x = 0$, it is "close to e^x" for values of x that are close to 0. This suggests that we might use the Taylor polynomials of e^x to approximate e^x.

Example 5 Use $p_2(x) = 1 + x + \frac{1}{2}x^2$ to calculate an approximate value for $e^{0.1}$.

5. Find the Taylor polynomial of degree 3 to e^x at $x = 0$ and use it to calculate an approximate value for $e^{0.1}$. CSS

Solution We know that

$$p_2(0.1) = 1 + 0.1 + \frac{1}{2}(0.1)^2 = 1.105.$$

Thus $e^{0.1}$ is approximately equal to 1.105. A hand-held calculator gives $e^{0.1} = 1.1051709$, so that the error involved here is about 0.0001709. Generally speaking, we can obtain a more accurate answer by using Taylor polynomials of higher degree.

DO EXERCISE 5.

Example 6 Use Taylor polynomials to calculate $\sin\left(\frac{1}{5}\right)$.

Solution We calculate $p_3(x)$ for $f(x) = \sin x$ at $x = 0$ and then approximate $\sin\left(\frac{1}{5}\right)$ by $p_3\left(\frac{1}{5}\right)$. Since $f'(x) = \cos x$, $f''(x) = -\sin x$, and $f'''(x) = -\cos x$, we have

$$c_0 = \frac{f(0)}{0!} = \frac{\sin(0)}{1} = 0,$$

$$c_1 = \frac{f^{(1)}(0)}{1!} = \frac{\cos(0)}{1} = 1,$$

$$c_2 = \frac{f^{(2)}(0)}{2!} = \frac{-\sin(0)}{2} = 0,$$

and

$$c_3 = \frac{f^{(3)}(0)}{3!} = \frac{-\cos(0)}{6} = \frac{-1}{6},$$

so that $p_3(x) = x - \frac{1}{6}x^3$. Now CSS

$$p_3\left(\frac{1}{5}\right) = \frac{1}{5} - \frac{1}{6}\left(\frac{1}{5}\right)^3 = 0.2 - 0.0013333 = 0.1986667.$$

Thus

$$\sin\frac{1}{5} \approx 0.1986667.$$

(Recall that $\approx$ means *approximately equal to*.) A hand-held calculator (make sure that yours is set to handle radians before you compute) yields

$$\sin\left(\frac{1}{5}\right) \approx 0.19866933.$$

6. Use Taylor polynomials to approximate $\tan\left(\frac{1}{10}\right)$.

DO EXERCISE 6.

EXERCISE SET 10.2

Find $p_2(x)$ for the given function and the specified value of x.

1. $y = e^{-x}$; $x = 0$

2. $y = \dfrac{1}{x}$; $x = 1$

3. $y = \sin(x^2)$; $x = 0$

4. $y = 5x^2 - 6x + 3$; $x = 0$

5. $f(x) = 2x^3 - 4x^2 + x - 1$; $x = 0$

6. $f(x) = \sqrt{x}$; $x = 1$

7. $f(x) = \sec x$; $x = 0$

8. $f(x) = \ln x$; $x = 1$

9. $y = x \sin x$; $x = 0$

10. $y = x^2 \cos x$; $x = 0$

Find $p_3(x)$ for the given function and the specified value of x.

11. $y = e^{(1/2)x}$; $x = 0$

12. $y = xe^x$; $x = 0$

13. $f(x) = x^3$; $x = 1$

14. $f(x) = \ln(1 + x)$; $x = 0$

15. $y = \ln(1 - x)$; $x = 0$

16. $y = \dfrac{1}{1 - x}$; $x = 0$

Find both $p_4(x)$ and $p_5(x)$ for the given function and the specified value of x.

17. $f(x) = \sin x$; $x = 0$

18. $f(x) = e^x$; $x = 0$

19. $y = \cos x$; $x = 0$

20. $y = \ln(1 + x)$; $x = 0$

Sketch the graphs of $f(x)$, $p_0(x)$, $p_1(x)$, and $p_2(x)$ near the given value of x for each of the following.

21. $f(x) = \sin x$; $x = 0$

22. $f(x) = \cos x$; $x = 0$

23. $f(x) = e^{-x}$; $x = 0$

24. $f(x) = \sqrt{x}$; $x = 1$

Use $p_3(x)$ from above to calculate an approximate value for each of the following.

25. $\sin(0.13)$

26. $\cos(1/8)$

27. $e^{-0.05}$

28. $\ln(1.05)$

29. $\ln(0.95)$

30. $\sin(1/5)$

31. *Psychology: The Hullian model of learning.* It is found that the probability P of mastery of a certain concept after t trials is given by

$$P(t) = 1 - e^{-0.18t}.$$

a) Compute $P(1)$, $P(2)$, and $P(3)$.

b) Find the Taylor polynomial $p_3(t)$ of order 3 for $P(t)$ at $t = 0$.

c) Use the polynomial in (b) to calculate approximate values for $P(1)$, $P(2)$, and $P(3)$.

32. *Newton's law of cooling.* An object heated to 200°F is placed in a room whose temperature is kept constant at 50°F. The temperature of the object after t hours in the room is determined to be given by

$$T(t) = 50 + 150e^{-0.5t}.$$

a) Find the temperature of the object after 1 and 2 hours.

b) Use the Taylor polynomial $p_3(t)$ of degree 3 at $t = 0$ to compute approximate values for $T(1)$ and $T(2)$.

33. *Ecology: Population growth.* The population of fish (in thousands) in a certain lake is given by

$$P(t) = \sqrt{22 + t}$$

over a five-year period beginning at $t = 0$.

a) Calculate the Taylor polynomial $p_2(t)$ of order 2 about $t = 3$ for $P(t)$.

b) Use $p_2(3.5)$ to find an approximate value for $P(3.5)$. Do the same for $P(4)$.

34. *Science: Motion of a spring.* The displacement y of a spring after t seconds is given by

$$y(t) = 6.5 \sin\left(t + \frac{\pi}{4}\right).$$

a) Find the Taylor polynomial $p_2(t)$ for $y(t)$ at $t = \frac{\pi}{4}$.

b) Use $p_2\left(\frac{\pi}{3}\right)$ to compute an approximate value for the displacement after $\frac{\pi}{3}$ seconds.

OBJECTIVES

You should be able to

a) **Tell if a given geometric series converges or diverges by calculating its ratio. If it converges, you should be able to compute its sum.**

b) **Tell if an alternating series converges or diverges. If it converges, you should be able to approximate its sum by the *n*th partial sum and to determine the error involved in this approximation.**

10.3 INFINITE SERIES

An *infinite series* is a special kind of sum

$$a_1 + a_2 + \cdots + a_n + \cdots, \tag{1}$$

where each number of the sequence $a_1, a_2, \ldots a_n$ is a real number. The numbers $a_1, a_2, \ldots, a_n$ are called the *terms* of the infinite series.

Two examples of infinite series are

$$\frac{1}{2} + \frac{1}{4} + \cdots + \frac{1}{2^n} + \cdots \tag{2}$$

and

$$1 + 2 + \cdots + n + \cdots.$$

In the first series, the nth term of the series is $1/2^n$. The fourth term, for instance, is

$$\frac{1}{2^4} = \frac{1}{16}$$

and the first term is

$$\frac{1}{2^1} = \frac{1}{2}.$$

In the second series, the nth term is n.

In ordinary arithmetic, numbers are added two at a time. To add $2 + 4 + 5 + 8$, for example, requires three additions. But an infinite sum like

$$\frac{1}{2} + \frac{1}{4} + \cdots + \frac{1}{2^n} + \cdots$$

requires an infinite number of additions, so that it cannot be carried out in a finite length of time. So, in order to give meaning to an expression like (1) or (2), we use the limit concept.

If the series (1) is given, then

$$s_n = a_1 + a_2 + \cdots + a_n$$

is called the nth *partial sum* of the series.

DEFINITION

The infinite series (1) *converges to s*, or *has the sum s*, written

$$s = a_1 + a_2 + \cdots + a_n + \cdots,$$

if the limit of the partial sums s_n of the series is s. Here s is a (finite) real number. In other words,

$$s = \lim_{n \to \infty} (a_1 + a_2 + \cdots + a_n)$$

$$= \lim_{n \to \infty} s_n.$$

If the partial sums do not have a limit, then we say that the series *diverges*.

Example 1 Find the sum of (2), that is, of

$$\frac{1}{2} + \frac{1}{4} + \frac{1}{8} + \cdots + \frac{1}{2^n} + \cdots.$$

Solution Let's consider s_n for a few values of n:

$$s_1 = \frac{1}{2} \qquad\qquad\qquad = \frac{1}{2} = 0.5,$$

$$s_2 = \frac{1}{2} + \frac{1}{4} \qquad\qquad\qquad = \frac{3}{4} = 0.75,$$

$$s_3 = \frac{1}{2} + \frac{1}{4} + \frac{1}{8} \qquad\qquad = \frac{7}{8} = 0.875,$$

$$s_4 = \frac{1}{2} + \frac{1}{4} + \frac{1}{8} + \frac{1}{16} \qquad = \frac{15}{16} = 0.9375,$$

$$s_5 = \frac{1}{2} + \frac{1}{4} + \frac{1}{8} + \frac{1}{16} + \frac{1}{32} = \frac{31}{32} = 0.96875.$$

Note that when s_n is written as a fraction, as in the next-to-last column, the numerator is one less than the denominator. Since the denominator is 2^n, this suggests that

$$s_n = \frac{2^n - 1}{2^n} = 1 - \frac{1}{2^n}$$

1. a) Find an expression for the nth term a_n of the infinite series

$$\frac{2}{3} + \frac{2}{9} + \frac{2}{27} + \frac{2}{81} + \cdots.$$

b) Calculate the first six partial sums of the series above. (Round your answers to seven decimal places.)

for all n. This formula is true. Note that as n increases, the sequence $1/2^n$ decreases rapidly toward 0. So

$$\lim_{n \to \infty} \left(1 - \frac{1}{2^n} \right) = 1.$$

Then

$$s = \lim_{n \to \infty} s_n = 1.$$

DO EXERCISE 1.

Example 2 Find the sum of $1 + 1 + \cdots + 1 + \cdots$,

Solution The nth partial sum of this series is

$$\underbrace{1 + 1 + \cdots + 1}_{n \text{ times}} = n.$$

So the partial sums tend to infinity as n increases. Since $+\infty$ is not a real number, the series diverges.

DO EXERCISE 2.

2. a) Find an expression for the nth term of

$$1 + 3 + 5 + 7 + 9 + \cdots.$$

b) Calculate the first five partial sums of this series.

c) Find an expression for the nth partial sum s_n of this series.

d) Does the series converge?

A special case in which we can always determine convergence or divergence is that of a geometric series.

DEFINITION

The infinite series

$$a + ar + ar^2 + \cdots + ar^{n-1} + \cdots \qquad (3)$$

is called a *geometric series with initial term a and ratio r.* Here $a \neq 0$.

The series in Example 1 is a geometric series with initial term $\frac{1}{2}$ and ratio $\frac{1}{2}$.

If $a + ar + \cdots + ar^{n-1} + \cdots$ is a geometric series, then the ratio of the second term, ar, to the first term, a, is

$$\frac{ar}{a} = r.$$

The ratio of the third term, ar^2, to the second term, ar, is

$$\frac{ar^2}{ar} = r.$$

3. a) Find an expression for the nth term a_n of the geometric series

$$2 + \frac{2}{5} + \frac{2}{25} + \cdots .$$

b) Find the fifth and sixth terms of the series above.

In general, the ratio of any term of a geometric series to the preceding term is r. In other words, $a_{n+1}/a_n = r$, or $a_{n+1} = ra_n$ for all n.

Example 3 Find the fifth and sixth terms of the geometric series $2 + 6 + 18 + \cdots$.

Solution The initial term a is 2. The ratio of the second term to the first is $6/2 = 3$. So $r = 3$. This means that the nth term of the series is $ar^{n-1} = 2 \cdot 3^{n-1}$. Thus the fifth term is $2 \cdot 3^{5-1} = 2 \cdot 3^4 = 2 \cdot 81 = 162$. The sixth term is three times the fifth term, or $a_6 = 3 \cdot 162 = 486$.

DO EXERCISES 3 AND 4.

The Partial Sums of a Geometric Series

We want to find a formula for the nth partial sum s_n of a geometric series:

$$s_n = a + ar + ar^2 + \cdots + ar^{n-1}.$$

If we multiply both sides of this equation by the ratio r, we get

$$rs_n = ar + ar^2 + \cdots + ar^{n-1} + ar^n.$$

If we multiply the formula for s_n by -1, we have

$$-s_n = -a - ar - ar^2 - \cdots - ar^{n-1}.$$

Adding these expressions for rs_n and $-s_n$, we get

$$rs_n - s_n = ar^n - a,$$

or

$$(r - 1)s_n = a(r^n - 1).$$

To solve this equation for s_n, we can divide both sides by $r - 1$ (if $r \neq 1$), getting the following formula.

4. Find the initial term and the ratio of the geometric series

$$\frac{1}{5} + \frac{1}{15} + \frac{1}{45} + \cdots .$$

THEOREM 2

The **n**th partial sum of a geometric series is given by

$$s_n = \frac{a(r^n - 1)}{r - 1} \tag{4}$$

for any $r \neq 1$.

5. Find the nth partial sum s_n of the geometric series

$$2 + \frac{2}{3} + \frac{2}{9} + \frac{2}{27} + \cdots.$$

Example 4 Find the sum of the first n terms of the geometric series

$$\frac{1}{2} + \frac{1}{4} + \cdots + \frac{1}{2^n} + \cdots. \qquad \boxed{CSS}$$

Solution This is a geometric series with $a = \frac{1}{2}$ and $r = \frac{1}{2}$. Using Eq. (4), we have

$$s_n = \frac{\frac{1}{2}\left[\left(\frac{1}{2}\right)^n - 1\right]}{\frac{1}{2} - 1} = \frac{\frac{1}{2}\left[\left(\frac{1}{2}\right)^n - 1\right]}{-\frac{1}{2}}$$

$$s_n = -\left[\left(\frac{1}{2}\right)^n - 1\right]$$

or

$$s_n = 1 - \frac{1}{2^n}.$$

Note that this justifies our use of this formula in Example 1.

DO EXERCISE 5.

The Sum of a Geometric Series

We want to be able to tell when a given geometric series converges and when it diverges. If it converges, we want to be able to find its sum. Formula (4) enables us to do this. For if r is a number between -1 and 1, then it follows that r^n gets closer and closer to 0 as n gets larger. (Pick a number r between -1 and 1 and check this by finding larger and larger powers on your calculator.) As r^n gets closer and closer to 0, so does ar^n. Since

$$s_n = \frac{a(r^n - 1)}{r - 1} = \frac{ar^n - a}{r - 1},$$

we see that s_n gets closer and closer to

$$\frac{-a}{r - 1} = \frac{a}{1 - r}$$

if r is between -1 and 1.

6. Find the sum of the geometric series

$$2 + \frac{2}{5} + \frac{2}{25} + \cdots.$$

If $|r|$ is greater than 1, then $|r|^n$ gets larger and larger as n increases. (▤ Check this on your calculator.) In this case, s_n does not have a limit.

If $|r| = 1$, the series also diverges. This case is considered in exercises 53 and 54.

THEOREM 3

The geometric series $a + ar + ar^2 + \cdots + ar^n + \cdots$ converges to

$$s = \frac{a}{1 - r}$$

if $|r| < 1$, that is, if r is between -1 and 1. Otherwise it diverges.

Example 5 Find the sum of

$$2 + \frac{2}{3} + \frac{2}{9} + \frac{2}{27} + \cdots + \frac{2}{3^{n-1}} + \cdots.$$

7. Find the sum of the geometric series

$$\frac{1}{5} + \frac{1}{15} + \frac{1}{45} + \cdots.$$

Solution This is a geometric series with initial term $a = 2$ and ratio $r = \frac{1}{3}$. Since r is between -1 and 1, the series converges and the sum is

$$s = \frac{a}{1 - r} = \frac{2}{1 - \frac{1}{3}} = \frac{2}{\frac{2}{3}} = 2 \cdot \frac{3}{2} = 3.$$

DO EXERCISES 6 AND 7.

Example 6 Find the sum of

$$1 - 1 + 1 - 1 + \cdots + (-1)^{n+1} + \cdots.$$

8. Find the sum of

$$1 - 2 + 4 - 8 + \cdots.$$

Solution This is a geometric series with initial term $a = 1$ and ratio $r = -1$. Since r is not between -1 and 1, this series diverges. Note that the partial sums of this series are 1, 0, 1, 0, 1, 0, 1, . . . , which has no limit.

DO EXERCISE 8.

Example 7 Find the sum of

$$1 - \frac{1}{2} + \frac{1}{4} - \frac{1}{8} + \cdots + \left(-\frac{1}{2}\right)^{n-1} + \cdots.$$

Solution This is a geometric series with $a = 1$ and $r = -\frac{1}{2}$. Thus it

9. Find the sum of

$$1 - \frac{2}{5} + \frac{4}{25} - \frac{8}{125} + \cdots .$$

converges and its sum is

$$\frac{a}{1-r} = \frac{1}{1 - \left(-\frac{1}{2}\right)} = \frac{1}{1 + \frac{1}{2}} = \frac{1}{\frac{3}{2}} = \frac{2}{3}.$$

DO EXERCISE 9.

Example 8 *Economic multiplier.* The United States banking laws require banks to maintain a reserve equivalent to a certain proportion of their outstanding deposits. This enables such banks, when they wish and when they can find borrowers, to loan out a certain proportion of the funds that have been deposited in them. Let us assume that this proportion is 0.90 (or 90%). Now suppose a corporation deposits $1000 in a bank that subsequently is able to loan the maximum amount legally possible, and this loan is redeposited elsewhere, and so on. What is the total effect of the $1000 on the economy?

10. Find *s* if the proportion in Example 8 is changed to 0.95 (or 95%).

Solution The total effect can be modeled as the sum of the infinite geometric series

$$\$1000 + \$1000(0.90) + \$1000(0.90)^2 + \$1000(0.90)^3 + \cdots,$$

which is given by

$$s = \frac{\$1000}{1 - 0.90} = \$10,000.$$

The sum $10,000 is the result of what is referred to in economics as the *multiplier effect*.

11. Find *s* if the proportion in Example 8 is changed to 85%.

DO EXERCISES 10 AND 11.

Example 9 Write 0.6363636 . . . , or $0.\overline{63}$, as a fraction. (The bar above sets off the portion that repeats.)

Solution The key to the problem is to note that we can write this decimal as a geometric series. Thus

$$0.63636363 \ldots = 0.63 + 0.0063 + 0.000063 + \cdots$$

$$= \frac{63}{100} + \frac{63}{10,000} + \frac{63}{1,000,000} + \cdots .$$

The right-hand side is a geometric series with initial term $a = 63/100 = 0.63$ and ratio $r = 1/100 = 0.01$. Since the ratio is be-

12. Write 0.4444 . . . as a fraction.

tween -1 and 1, the sum of the series can be expressed as

$$\frac{a}{1-r} = \frac{63/100}{1 - 1/100} = \frac{63/100}{99/100} = \frac{63}{99} = \frac{7}{11}.$$

Thus

$$\frac{7}{11} = 0.63636363 \ldots , \quad \text{or} \quad 0.\overline{63}.$$

DO EXERCISES 12 AND 13.

Alternating Series

We reconsider the series

$$1 - \frac{1}{2} + \frac{1}{4} - \frac{1}{8} + \cdots + \left(-\frac{1}{2}\right)^{n-1} + \cdots = \frac{2}{3}.$$

13. Write 0.4594594 . . . as a fraction.

The first six partial sums are as follows:

$$s_1 = 1 \qquad\qquad\qquad = 1 \ = 1,$$

$$s_2 = 1 - \frac{1}{2} \qquad\qquad\qquad = \frac{1}{2} \ = 0.5,$$

$$s_3 = 1 - \frac{1}{2} + \frac{1}{4} \qquad\qquad = \frac{3}{4} \ = 0.75,$$

$$s_4 = 1 - \frac{1}{2} + \frac{1}{4} - \frac{1}{8} \qquad = \frac{5}{8} \ = 0.625,$$

14. Calculate the first six partial sums of

$$2 - \frac{2}{3} + \frac{2}{9} - \frac{2}{27} + \frac{2}{81} - \cdots .$$

$$s_5 = 1 - \frac{1}{2} + \frac{1}{4} - \frac{1}{8} + \frac{1}{16} \qquad = \frac{11}{16} = 0.6875,$$

$$s_6 = 1 - \frac{1}{2} + \frac{1}{4} - \frac{1}{8} + \frac{1}{16} - \frac{1}{32} = \frac{21}{32} = 0.65625.$$

Note that the odd-numbered partial sums $(1, 0.75, 0.6875, \ldots)$ are all greater than $\frac{2}{3}$, the sum of the given series. The even-numbered partial sums are all less than $\frac{2}{3}$.

Furthermore, the odd-numbered partial sums $(1, 0.75, 0.6875, \ldots)$ are decreasing. The even-numbered partial sums $(0.5, 0.625, 0.65625, \ldots)$ are increasing.

DO EXERCISE 14.

The discussion above and margin Exercise 14 illustrate the following theorem.

15. Determine whether or not

$$1 - \frac{1}{4} + \frac{1}{9} - \frac{1}{16} + \cdots$$
$$+ (-1)^{n+1} \frac{1}{n^2} + \cdots$$

converges or diverges.

THEOREM 4   Alternating Series Test

Let

$$a_1 \geq a_2 \geq \cdots \geq a_n \geq \cdots$$

be positive numbers with $\lim_{n \to \infty} a_n = 0$. Then the *alternating series*

$$a_1 - a_2 + a_3 - a_4 + \cdots + (-1)^{n-1} a_n + \cdots$$

converges. Furthermore, the odd-numbered partial sums are all greater than s, the sum of the series. In fact, they decrease to s. The even-numbered partial sums are all less than s (and increase to s).

Example 10 Determine whether the series

$$1 - \frac{1}{2} + \frac{1}{3} - \frac{1}{4} + \cdots + (-1)^{n-1} \frac{1}{n} + \cdots$$

converges or diverges.

Solution The given series is an alternating series with $a_n = 1/n$ for all positive integers n. Since these numbers decrease to 0, the series converges.

DO EXERCISE 15.

Example 11 Show that

$$1 - \frac{1}{8} + \frac{1}{27} + \cdots + (-1)^{n-1} \frac{1}{n^3} + \cdots$$

converges and use the fourth and fifth partial sums to estimate the sum s of the series.

Solution The series is an alternating series with $a_n = 1/n^3$ for each positive integer n. Since these numbers decrease to 0, the series converges. The fourth partial sum is

$$1 - \frac{1}{8} + \frac{1}{27} - \frac{1}{64} \approx 0.896412.$$

The fifth partial sum is

$$1 - \frac{1}{8} + \frac{1}{27} - \frac{1}{64} + \frac{1}{125} \approx 0.904412.$$

According to the test, the sum s of the series is between these two

16. Use the third and fourth partial sums to estimate the sum of

$$1 - \frac{1}{16} + \frac{1}{81} - \frac{1}{256} + \cdots$$

$$+ (-1)^{n+1} \frac{1}{n^4} + \cdots.$$

numbers. Thus

$$0.896412 < s < 0.904412.$$

DO EXERCISE 16.

In Example 11 we saw that the sum s of the series

$$1 - \frac{1}{8} + \frac{1}{27} + \cdots + (-1)^{n-1} \frac{1}{n^3} + \cdots$$

was between 0.896412 and 0.904412, the fourth and fifth partial sums of the series. The difference between the fifth partial sum and the fourth partial sum of this series is the fifth term of the series, $\frac{1}{125}$. If we want to use 0.896412 as an approximate value for s, then the error involved is no more than $\frac{1}{125} = 0.008$. This fact is so important that we state it separately.

THEOREM 5

If $a_1 \geq a_2 \geq \cdots \geq a_n \geq \cdots$ are positive numbers with $\lim_{n \to \infty} a_n = 0$ and if s is the sum of the (convergent) alternating series

$$a_1 - a_2 + a_3 + \cdots + (-1)^{n-1} a_n + \cdots,$$

then the absolute value of the difference between s and the nth partial sum is less than or equal to a_{n+1}. That is,

$$|s - s_n| \leq a_{n+1}.$$

Example 12 Let

$$s = 1 - \frac{1}{7} + \frac{1}{49} + \cdots + \frac{(-1)^{n-1}}{7^{n-1}} + \cdots.$$

Find the error involved in approximating s by the fourth partial sum

$$s_4 = 1 - \frac{1}{7} + \frac{1}{49} - \frac{1}{343} \approx 0.87463557.$$

Check by finding s explicitly.

Solution By the alternating series test, we know that the series converges. According to Theorem 5 above, the error involved in approximating s by 0.87463557 is less than $1/7^4 = 1/2401 \approx 0.0004165$. Since the given series happens to be a geometric series with $a = 1$ and

17. a) Use the third and fourth partial sums to estimate the sum of

$$1 - \frac{1}{9} + \frac{1}{81} - \frac{1}{729} + \cdots.$$

b) Then check your answer by finding the sum of this geometric series.

$r = -\frac{1}{7}$, we see that the sum is

$$s = \frac{a}{1 - r} = \frac{1}{1 - \left(-\frac{1}{7}\right)} = \frac{1}{\frac{8}{7}} = \frac{7}{8} = 0.875.$$

Thus the error involved in approximating s by 0.87463557 is 0.00036443, so that our error estimate above is correct.

DO EXERCISE 17.

It is important to keep in mind that the alternating series test applies only if the series is indeed alternating.

Example 13 Use the alternating series test to determine whether

$$1 + \frac{1}{2} + \frac{1}{3} + \cdots + \frac{1}{n} + \cdots$$

converges or diverges.

Solution This series is not an alternating series! Thus the alternating series test cannot be applied. Contrary to what you might expect, this series diverges, as we will show in Exercise Set 10.3.

EXERCISE SET 10.3

Find an expression for the nth term a_n of each of the following series.

1. $\frac{1}{5} + \frac{1}{25} + \frac{1}{125} + \cdots$

2. $1 + 4 + 9 + 16 + 25 + \cdots$

3. $\frac{1}{2} + \frac{2}{3} + \frac{3}{4} + \frac{4}{5} + \frac{5}{6} + \cdots$

4. $1 - \frac{1}{2} + \frac{1}{3} - \frac{1}{4} + \cdots$

5. $\frac{2}{3} + \frac{4}{9} + \frac{8}{27} + \frac{16}{81} + \frac{32}{243} + \cdots$

6. $\frac{1}{4} - \frac{1}{16} + \frac{1}{64} - \frac{1}{256} + \cdots$

7. $1 - 2 + 3 - 4 + \cdots$

8. $1 + \frac{1}{6} + \frac{1}{36} + \frac{1}{216} + \cdots$

Find the initial term, the ratio, and the fifteenth partial sum for each of the following geometric series.

9. $1 + \frac{1}{3} + \frac{1}{9} + \cdots$

10. $1 + \frac{1}{5} + \frac{1}{25} + \cdots$

11. $1 - \frac{1}{3} + \frac{1}{9} - \cdots$

12. $1 - \frac{1}{5} + \frac{1}{25} - \cdots$

13. $\dfrac{1}{8} - \dfrac{1}{12} + \dfrac{1}{18} - \cdots$

14. $\dfrac{1}{2} + \dfrac{2}{3} + \dfrac{8}{9} + \cdots$

15. $\dfrac{5}{3} - \dfrac{2}{3} + \dfrac{4}{15} - \cdots$

16. $\dfrac{1}{3} - \dfrac{1}{2} + \dfrac{3}{4} - \dfrac{9}{8} + \cdots$

Determine whether each of the geometric series below converges or diverges. If it converges, find its sum.

17. $1 + \dfrac{1}{8} + \dfrac{1}{64} + \cdots$

18. $1 + 3 + 9 + 27 + \cdots$

19. $1 + \dfrac{1}{9} + \dfrac{1}{81} + \dfrac{1}{729} + \cdots$

20. $1 - 2 + 4 - 8 + \cdots$

21. $1 - \dfrac{2}{3} + \dfrac{4}{9} - \cdots$

22. $\dfrac{1}{2} - \dfrac{1}{2} + \dfrac{1}{2} - \dfrac{1}{2} + \cdots$

23. $\dfrac{1}{4} + \dfrac{1}{4} + \dfrac{1}{4} + \cdots$

24. $\dfrac{1}{2} - \dfrac{1}{6} + \dfrac{1}{18} - \cdots$

Write each of the following decimals as a rational fraction (the bar above sets off the portion that repeats).

25. $0.7\overline{777}\ldots$

26. $0.41\overline{4141}\ldots$

27. $2.38\overline{3838}\ldots$

28. $0.036\overline{666}\ldots$

29. $0.142857\overline{142857}\ldots$

30. $0.1234567890\overline{1234567890}$

Determine whether each of the following infinite series converges or diverges.

31. $\dfrac{1}{7} + \dfrac{2}{7} + \dfrac{4}{7} + \dfrac{8}{7} + \cdots$

32. $1 - \dfrac{1}{3} + \dfrac{1}{5} - \dfrac{1}{7} + \cdots$

33. $1 - \dfrac{9}{10} + \dfrac{81}{100} - \dfrac{729}{1000} + \cdots$

34. $1 - \dfrac{1}{9} + \dfrac{1}{25} - \dfrac{1}{49} + \dfrac{1}{81} - \cdots$

35. $1 - \dfrac{1}{4} + \dfrac{1}{7} - \dfrac{1}{10} + \cdots$

36. $\dfrac{1}{9} - \dfrac{1}{6} + \dfrac{1}{4} - \dfrac{3}{8} + \cdots$

37. $1 - \dfrac{3}{4} + \dfrac{9}{16} - \dfrac{27}{64} + \cdots$

38. $-\dfrac{1}{5} + \dfrac{1}{6} - \dfrac{1}{7} + \dfrac{1}{8} - \cdots$

39. $\dfrac{1}{5} - \dfrac{1}{7} + \dfrac{1}{9} - \dfrac{1}{11} + \cdots$

40. $\dfrac{1}{36} + \dfrac{1}{24} + \dfrac{1}{16} + \dfrac{3}{32} + \cdots$

Use the fourth and fifth partial sums of the (convergent) series below to estimate the sum of the series.

41. $1 - \dfrac{1}{3} + \dfrac{1}{9} - \dfrac{1}{27} + \cdots$

42. $1 - \dfrac{1}{6} + \dfrac{1}{36} - \dfrac{1}{216} + \cdots$

43. $1 - \dfrac{1}{16} + \dfrac{1}{81} - \dfrac{1}{256} + \cdots + (-1)^{n+1} \cdot \dfrac{1}{n^4} + \cdots$

44. $\dfrac{1}{2} - \dfrac{1}{8} + \dfrac{1}{24} - \dfrac{1}{64} + \cdots + (-1)^{n-1} \cdot \dfrac{1}{n\,2^n} + \cdots$

45. $\dfrac{1}{9} - \dfrac{1}{25} + \dfrac{1}{49} - \dfrac{1}{81} + \cdots + (-1)^{n+1} \cdot \dfrac{1}{(2n+1)^2} + \cdots$

46. $-\dfrac{1}{4} + \dfrac{1}{16} - \dfrac{1}{36} + \dfrac{1}{64} - \dfrac{1}{100} + \cdots$

47. *Economics.* The government makes an $8,000,000,000 expenditure for a new type of aircraft. If 75% of this gets spent again, and 75% of that gets spent, and so on, what is the total effect on the economy?

49. If, in Problem 48, the daily dosage is 5 mg and the percentage eliminated daily is 60%, find:

a) the amount of the drug in the patient's system immediately after the first, second, and the third doses are administered;

b) the amount of the drug in the patient's system immediately after the ninth dose is administered;

c) the "limiting value" of the amount of the drug in the patient's system, that is, the sum of the geometric series in question.

51. *Investments.* A certain stock pays a dividend of $5 per year at present. If the dividend grows 2% each year and if the interest rate is 12.5%, then the present value V of the next twelve payments (beginning today) is given by

$$V = 5 + 5\left(\frac{1.02}{1.125}\right) + 5\left(\frac{1.02}{1.125}\right)^2 + \cdots + 5\left(\frac{1.02}{1.125}\right)^{11}.$$

Find the sum.

53. Show, by computing the nth partial sum s_n, that a geometric series with initial term $a \neq 0$ and ratio $r = 1$ diverges.

55. A patient is able to eliminate p $(0 < p < 1)$ of the concentration of a certain drug between successive doses. If the daily dosage of the drug is d mg, find the limiting value of the amount of the drug in the patient's system.

48. *Medicine.* A patient is to receive 1 mg (milligram) of a certain drug once every day. By the time the next dose is administered, it is expected that 50% of the drug will have been eliminated from the patient's system.

a) Find the amount of the drug in the patient's system immediately after the first four doses are administered.

b) If we assume that the pattern in (a) holds, how much of the drug will be present after the fifteenth dose is administered?

c) When will the amount of the drug in the patient's system exceed 3 mg?

50. *Finance.* A certain *perpetuity* is to pay $100 each year (starting today). If the current interest rate is i, then the present value of a payment of $100 n years from today is defined to be $100(1 + i)^{-n}$, so that the *total* present value P of the payments is given by

$$P = 100 + \frac{100}{1 + i} + \frac{100}{(1 + i)^2} + \cdots.$$

a) Find the present value of the perpetuity if $i = 0.10$ (10% interest).

b) What happens to the present value if the interest rates drop from 10% to 8%?

52. *Chemistry.* A factory deposits waste material containing 1 kg of a certain radioactive isotope on a slag heap each day. It is known that after one day, 3% of the isotope will decay.

a) How much of the isotope will be present after the first, second, and third deposits?

b) How much of the isotope will be present after the fiftieth deposit?

c) Find the limiting value of the amount of isotope in the slag heap.

54. Show, by computing the nth partial sum s_n, that a geometric series with initial term $a \neq 0$ and ratio $r = -1$ diverges.

56. A perpetuity pays r annually beginning today. If the interest rate is i, then the present value of the annuity, denoted P, is given by

$$P = r + \frac{r}{1 + i} + \frac{r}{(1 + i)^2} + \cdots.$$

Find the sum of the series.

57. Find the value of the perpetuity in Problem 56 if the first payment is to be made one year from today.

58. We consider the harmonic series

$$1 + \frac{1}{2} + \frac{1}{3} + \frac{1}{4} + \frac{1}{5} + \cdots + \frac{1}{n} + \cdots .$$

Note that

$$\frac{1}{3} + \frac{1}{4} > \frac{1}{4} + \frac{1}{4} = \frac{1}{2},$$

that

$$\frac{1}{5} + \frac{1}{6} + \frac{1}{7} + \frac{1}{8} > \frac{1}{8} + \frac{1}{8} + \frac{1}{8} + \frac{1}{8} = \frac{1}{2},$$

and so forth.

a) Show that $\dfrac{1}{2^n + 1} + \dfrac{1}{2^n + 2} + \cdots + \dfrac{1}{2^{n+1}} > \dfrac{1}{2}$.

b) Use the result from (a) to show that $s_4 > 2$, $s_8 > \dfrac{5}{2}$,

$s_{16} > 3$; then show that $s_{2^{n+1}} > \dfrac{n + 3}{2}$ for all integers n.

c) Conclude that the harmonic series diverges.

OBJECTIVES

You should be able to

a) **Compute the Taylor series expansion for a given function.**

b) **Use Taylor series and the alternating series test to approximate certain integrals.**

10.4 TAYLOR SERIES

In Section 10.2, we saw that we can approximate many functions by using an appropriately constructed polynomial, called the Taylor polynomial (of degree n) of $f(x)$ at $x = a$. Although exact equality occurs only at $x = a$, the approximation is excellent near $x = a$ and the accuracy can be improved by using a polynomial of increased degree. This raises the question of whether we can obtain exact equality by taking the limit (as n tends to infinity) of the Taylor polynomial of degree n. When this occurs, we say that $f(x)$ *is represented by its Taylor series at* $x = a$.

DEFINITION

If $f(x)$ has derivatives of all orders at $x = a$, then *the Taylor series of* $f(x)$ *at* $x = a$ *is the series*

$$\sum_{j=0}^{\infty} c_j(x - a)^j,$$

where

$$c_j = \frac{f^{(j)}(a)}{j!} \quad \text{for } j = 0, 1, \ldots .$$

1. Compute the Taylor series of

$$f(x) = e^{-x} \text{ at } x = 0.$$

Note that the *n*th partial sum of the series is given by

$$p_n(x) = \sum_{j=0}^{n} c_j(x - a)^j,$$

the Taylor polynomial of $f(x)$ (of degree *n*) at $x = a$. We say that $f(x)$ is *represented by its Taylor series at $x = a$* if

$$f(x) = \sum_{j=0}^{\infty} c_j(x - a)^j$$

$$= \lim_{n \to \infty} \sum_{j=0}^{n} c_j(x - a)^j$$

$$= \lim_{n \to \infty} p_n(x).$$

Example 1 Compute the Taylor series of $f(x) = e^x$ at $x = 0$.

Solution In this problem, $a = 0$. We must compute the coefficients, given by $c_j = f^{(j)}(0)/j!$. Since $f(x) = e^x$, the required derivatives are easy to compute in this case. We have $f^{(j)}(x) = e^x$ for all j. So $f^{(j)}(0) = e^0 = 1$ for all values of j and $c_j = f^{(j)}(0)/j! = 1/j!$ for all j. Thus the Taylor series for e^x at $x = 0$ is

$$1 + x + \frac{x^2}{2!} + \frac{x^3}{3!} + \cdots + \frac{x^n}{n!} + \cdots = \sum_{j=0}^{\infty} \frac{x^j}{j!}.$$

DO EXERCISE 1.

Example 2 Compute the Taylor series of $f(x) = \sin x$ at $x = 0$.

Solution We have $f'(x) = \cos x$ and $f''(x) = f^{(2)}(x) = -\sin x$. Differentiating again, we obtain $f^{(3)}(x) = -\cos x$, and, differentiating again, $f^{(4)}(x) = \sin x$. Note that $f^{(4)}(x) = f(x)$, the function with which we started. This means that $f^{(5)}(x) = f'(x) = \cos x$, that $f^{(6)}(x) = f^{(2)}(x) = -\sin x$, and so forth, repeating every four terms. If we let $x = 0$, we get

$$f(0) = \sin (0) = 0,$$

$$f'(0) = \cos(0) = 1,$$

$$f''(0) = f^{(2)}(0) = -\sin (0) = 0,$$

$$f^{(3)}(0) = -\cos (0) = -1,$$

$$f^{(4)}(0) = \sin (0) = 0,$$

$$f^{(5)}(0) = f^{(1)}(0) = 1,$$

2. Compute the Taylor series of

$$f(x) = \cos x \text{ at } x = 0.$$

and so on. In other words, the derivatives, when evaluated at $x = 0$, produce the sequence $0, 1, 0, -1, 0, 1, 0, -1, 0, 1, \ldots$, in which the four numbers $0, 1, 0, -1$ (in that order) are continually repeated.

It is helpful to be able to describe this sequence concisely. The even-numbered terms are all equal to 0. This is equivalent to $f^{(j)}(0) = 0$ for $j = 2k$, or $f^{(2k)}(0) = 0$ for all k. If we let $j = 2k + 1$, then, as k takes on the values $0, 1, 2, 3, \ldots$, j takes on the values $1, 3, 5, 7, \ldots$. Using this, we can write $f^{(j)}(0) = (-1)^k$ for $j = 2k + 1$, or $f^{(2k+1)}(0) = (-1)^k$. Note that as k takes on the values $0, 1, 2, \ldots$, the quantity $(-1)^k$ is equal to $1, -1, 1, -1, \ldots$.

Finally, since $c_j = f^{(j)}(0)/j!$, we see that for j even (that is, $j = 2k$), we have $c_{2k} = 0$, whereas for j odd ($j = 2k + 1$), we have $c_{2k+1} = (-1)^k/(2k+1)!$. Thus the Taylor series for $\sin x$ at $x = 0$ is

$$\sum_{k=0}^{\infty} \frac{(-1)^k x^{2k+1}}{(2k+1)!},$$

or

$$x - \frac{x^3}{3!} + \frac{x^5}{5!} - \frac{x^7}{7!} + \frac{x^9}{9!} - \cdots.$$

DO EXERCISE 2.

Example 3 Compute the Taylor series of $f(x) = 2x^2 - 3x + 5$ at $x = 0$.

Solution We have $f(0) = 2(0)^2 - 3(0) + 5 = 5$. Since $f'(x) = 4x - 3$, we have $f'(0) = 4(0) - 3 = -3$. Next, $f^{(2)}(x) = 4$, so that $f^{(2)}(0) = 4$. Observe that $f^{(3)}(x)$ and all higher-order derivatives of $f(x)$ are equal to 0. So

$$c_0 = \frac{f(0)}{0!} = \frac{5}{1} = 5,$$

$$c_1 = \frac{f^{(1)}(0)}{1!} = \frac{-3}{1} = -3,$$

and

$$c_2 = \frac{f^{(2)}(0)}{2!} = \frac{4}{2} = 2,$$

with all other coefficients equal to 0. Thus the Taylor series for $f(x) = 2x^2 - 3x + 5$ at $x = 0$ is $5 - 3x + 2x^2 + 0 \cdot x^3 + 0 \cdot x^4 + \ldots$, or $5 - 3x + 2x^2$, which is what we started with.

3. Compute the Taylor series of

$$f(x) = 7x^3 - 5x^2 + 2x + 4$$

at $x = 0$.

DO EXERCISE 3.

Example 4 Compute the Taylor series of $f(x) = \ln x$ at $x = 1$.

Solution We have $f'(x) = 1/x = x^{-1}$. So, differentiating again, we get $f^{(2)}(x) = f''(x) = (-1)x^{-2} = -1/x^2$. Continuing, we have

$$f^{(3)}(x) = (-1)(-2x^{-3}) = 2x^{-3} = 2/x^3,$$

$$f^{(4)}(x) = 2(-3x^{-4}) = -(3!)x^{-4},$$

and

$$f^{(5)}(x) = -(3!)(-4x^{-5}) = (4!)x^{-5} = \frac{4!}{x^5}.$$

To see the pattern, consider the following:

$$f^{(1)}(x) = x^{-1},$$

$$f^{(2)}(x) = -1x^{-2},$$

$$f^{(3)}(x) = (2!)x^{-3},$$

$$f^{(4)}(x) = -(3!)x^{-4},$$

$$f^{(5)}(x) = (4!)x^{-5}.$$

We see that $f^{(j)}(x) = (-1)^{j+1}(j-1)! \, x^{-j}$ (here, the $(-1)^{j+1}$ term takes care of the alternating signs of the derivatives). We have then

$$f^{(j)}(1) = (-1)^{j+1}(j-1)!(1)^{-j} = (-1)^{j+1}(j-1)! \quad \text{for } j > 0,$$

so that

$$c_j = \frac{f^{(j)}(1)}{j!} = \frac{(-1)^{j+1}(j-1)!}{j!} = \frac{(-1)^{j+1}(j-1)!}{j(j-1)!} = \frac{(-1)^{j+1}}{j} \quad \text{for } j > 0.$$

Recalling that

$$c_0 = \frac{f^{(0)}(1)}{0!} = \frac{\ln(1)}{1} = 0,$$

we see that the Taylor series of $\ln x$ at $x = 1$ is

$$(x-1) - \frac{1}{2}(x-1)^2 + \frac{1}{3}(x-1)^3 - \frac{1}{4}(x-1)^4 + \cdots$$

$$= \sum_{j=1}^{\infty} \frac{(-1)^{j+1}(x-1)^j}{j}.$$

Example 5 Compute the Taylor series of $f(x) = 1/(1-x)$ at $x = 0$.

4. Compute the Taylor series for $1/(1 + x^2)$ at $x = 0$.

Solution   Writing $f(x) = (1 - x)^{-1}$, we see that

$$f'(x) = -1(1 - x)^{-2}(-1) = (1 - x)^{-2}.$$

So

$$f^{(2)}(x) = f''(x) = -2(1 - x)^{-3}(-1) = 2(1 - x)^{-3}$$

and

$$f^{(3)}(x) = 2(-3)(1 - x)^{-4}(-1) = 3!(1 - x)^{-4}.$$

We see that

$$f^{(j)}(x) = (j!)(1 - x)^{-(j+1)} \quad \text{for all } j > 0.$$

Thus

$$f^{(j)}(0) = (j!)(1 - 0)^{-(j+1)} = (j!)(1)^{-(j+1)} = j! \, .$$

We get

$$c_j = \frac{f^{(j)}(0)}{j!} = \frac{j!}{j!} = 1 \quad \text{for all } j \geq 0$$

and the Taylor series for $1/(1 - x)$ at $x = 0$ is

$$1 + x + x^2 + \cdots + x^n + \cdots = \sum_{j=0}^{\infty} x^j.$$

Note that this is a geometric series.

DO EXERCISE 4.

In Example 5, we showed that the Taylor series for $1/(1 - x)$ at $x = 0$ is given by $1 + x + x^2 + \cdots + x^n + \cdots$. In Section 10.3, we showed that the geometric series $1 + r + r^2 + \cdots r^n + \cdots$ converges to $1/(1 - r)$ if $-1 < r < 1$ and diverges otherwise. This shows that the Taylor series for $f(x) = 1/(1 - x)$ *is equal to $f(x)$* for values of x between -1 and 1, that is,

$$\frac{1}{1 - x} = 1 + x + x^2 + \cdots + x^n + \cdots, \qquad -1 < x < 1,$$

and that this formula makes no sense for other values of x.

    It turns out that this example illustrates the general case. The series $\sum_{j=0}^{\infty} c_j(x - a)^j$ always converges at $x = a$, since all the terms except the first are 0 in this case. Also, there is a nonnegative number R so that the series converges on the interval $(a - R, a + R)$ and diverges if $|x - a| > R$ (where R may be 0 or $+\infty$). If R is positive and finite, then

5. Compute the Taylor series for $1/(1 + x)^2$ at $x = 0$.

the series converges on the following interval

and diverges for $x > a + R$ and $x < a - R$ (convergence at the end-points $a + R$ and $a - R$ must be decided on a case-by-case basis). If $R = 0$, then the series converges only at $x = a$ (where it always converges) and diverges for all other values of x. If $R = +\infty$, then the series converges for all real numbers x. Fortunately, for a large class of functions, it turns out that R (the *radius of convergence*) is positive and $f(x)$ is equal to its Taylor series on $(a - R, a + R)$, *the interval of convergence* of the series.

If $f(x) = \sum_{j=0}^{\infty} c_j(x - a)^j$ on $(a - R, a + R)$ for R positive, then this formula can be differentiated or integrated termwise within the given interval.

We can also make appropriate substitutions. These techniques may make the computation of Taylor series much less tedious. We illustrate with several examples.

Example 6 Compute the Taylor series of $f(x) = 1/(1 - x)^2$ at $x = 0$.

Solution We use

$$(1 - x)^{-1} = \frac{1}{1 - x} = 1 + x + x^2 + \cdots, \qquad \text{valid for } -1 < x < 1,$$

6. Compute the Taylor series for $1/(1 - x)^3$ at $x = 0$.

from Example 5. Differentiating $(1 - x)^{-1}$ gives $(-1)(1 - x)^{-2}(-1) = (1 - x)^{-2}$. Differentiating the series $1 + x + x^2 + x^3 + \cdots$ term by term gives $0 + 1 + 2x + 3x^2 + \cdots$. Since the formula with which we started was correct for all x between -1 and 1, so too is the new formula. That is,

$$(1 - x)^{-2} = \frac{1}{(1 - x)^2} = 1 + 2x + 3x^2 + \cdots,$$

for all x with $-1 < x < 1$.

DO EXERCISES 5 AND 6.

Example 7 Compute the Taylor series of $1/(1 + x^2)$ at $x = 0$.

Solution We could compute the Taylor series directly, but those calculations are lengthy (see Margin Exercise 4). For a shortcut, we use

$$\frac{1}{1 - t} = 1 + t + t^2 + t^3 + \cdots, \qquad -1 < t < 1.$$

7. Compute the Taylor series of $1/(1 - x^3)$ at $x = 0$.

Letting $t = -x^2$, we get

$$\frac{1}{1 - (-x^2)} = 1 + (-x^2) + (-x^2)^2 + (-x^2)^3 + \cdots$$

or,

$$\frac{1}{1 + x^2} = 1 - x^2 + x^4 - x^6 + \cdots .$$

This formula is valid for $-1 < x^2 < 1$ or, in other words, for $-1 < x < 1$.

DO EXERCISE 7.

Example 8 Compute the Taylor series for $f(x) = e^{-x^2}$ at $x = 0$.

Solution We use the fact that e^x is equal to its Taylor series at $x = 0$ for *all* values of x. Recall that

8. Compute the Taylor series of $f(x) = \sin(x^2)$ at $x = 0$.

$$e^t = 1 + t + \frac{t^2}{2!} + \frac{t^3}{3!} + \cdots \quad \text{for all } t.$$

Letting $t = -x^2$, we get

$$e^{-x^2} = 1 - x^2 + \frac{x^4}{2!} - \frac{x^6}{3!} + \frac{x^8}{4!} - \frac{x^{10}}{5!} + \cdots \quad \text{for all } x.$$

DO EXERCISES 8 AND 9.

Example 9 Compute the Taylor series for $f(x) = \ln(1 + x)$ at $x = 0$.

9. Compute the Taylor series of $f(x) = e^{-5x}$ at $x = 0$.

Solution Letting $x = -t$ in the series for $1/(1 - x)$ yields

$$\frac{1}{1 + t} = 1 - t + t^2 - t^3 + \cdots , \qquad -1 < t < 1.$$

Integrating both sides from 0 to x, we obtain

$$\int_0^x \frac{1}{1 + t} \, dt = \int_0^x (1 - t + t^2 - t^3 + \cdots) \, dt$$

$$= \int_0^x dt - \int_0^x t \, dt + \int_0^x t^2 \, dt - \int_0^x t^3 \, dt + \cdots ,$$

or

$$[\ln(1 + t)]_0^x = [t]_0^x - \left[\frac{t^2}{2}\right]_0^x + \left[\frac{t^3}{3}\right]_0^x - \left[\frac{t^4}{4}\right]_0^x + \cdots ,$$

10. Compute the Taylor series of $f(x) = \ln(1 - x)$ at $x = 0$.

or

$$\ln(1 + x) - \ln(1 + 0) = (x - 0) - \left(\frac{x^2}{2} - 0\right)$$

$$+ \left(\frac{x^3}{3} - 0\right) - \left(\frac{x^4}{4} - 0\right) + \cdots .$$

Thus

$$\ln(1 + x) = x - \frac{x^2}{2} + \frac{x^3}{3} - \frac{x^4}{4} + \cdots , \quad -1 < x < 1.$$

DO EXERCISE 10.

Example 10 Compute the Taylor series for $f(x) = \tan^{-1} x$ at $x = 0$.

Solution Recall that the derivative of $\tan^{-1} x$ is $1/(1 + x^2)$ (see Section 8.4). In Example 7, we found that the Taylor series at $x = 0$ of $1/(1 + x^2)$ is

$$\frac{1}{1 + x^2} = 1 - x^2 + x^4 - x^6 + \cdots , \quad -1 < x < 1.$$

Replacing x by t and integrating from $t = 0$ to $t = x$ yields

$$\int_0^x \frac{1}{1 + t^2} \, dt = \int_0^x \{1 - t^2 + t^4 - t^6 + \cdots\} \, dt, \quad -1 < x < 1,$$

or

$$[\tan^{-1} t]_0^x = \left[t - \frac{t^3}{3} + \frac{t^5}{5} - \frac{t^7}{7} + \cdots\right]_0^x, \quad -1 < x < 1$$

and

$$\tan^{-1} x - \tan^{-1} 0 = \left(x - \frac{x^3}{3} + \frac{x^5}{5} - \frac{x^7}{7} + \cdots\right) - 0.$$

Since $\tan^{-1} 0 = 0$, we have

$$\tan^{-1} x = x - \frac{x^3}{3} + \frac{x^5}{5} - \frac{x^7}{7} + \cdots , \quad -1 < x < 1.$$

Example 11 Use Taylor series to estimate

$$\int_0^{1/2} \frac{1}{1 + x^3} \, dx$$

with an error less than 0.001.

11. Use Taylor series to compute

$$\int_0^{1/2} \frac{1}{1 + x^4} \, dx$$

with an error less than 0.001.

Solution We use a combination of several techniques from this chapter. First, we compute the Taylor series for $1/(1 + x^3)$ at $x = 0$ by letting $t = x^3$ in the expression for $1/(1 + t)$ to get

$$\frac{1}{1 + x^3} = 1 - x^3 + x^6 - x^9 + \cdots, \quad -1 < x < 1.$$

Integrating both sides from 0 to $\frac{1}{2}$ yields

$$\int_0^{1/2} \frac{1}{1 + x^3} \, dx = \left[x - \frac{x^4}{4} + \frac{x^7}{7} - \frac{x^{10}}{10} + \cdots \right]_0^{1/2},$$

or

$$\int_0^{1/2} \frac{1}{1 + x^3} \, dx = \frac{1}{2} - \frac{1}{4}\left(\frac{1}{2}\right)^4 + \frac{1}{7}\left(\frac{1}{2}\right)^7 - \frac{1}{10}\left(\frac{1}{2}\right)^{10} + \cdots.$$

Before proceeding, note that the series on the right is an alternating series. Thus if we were to approximate the integral by using, say, $\frac{1}{2} - \frac{1}{4}\left(\frac{1}{2}\right)^4$, the resulting approximation would be smaller than the actual value of the integral and the error would be less than

$$\frac{1}{7}\left(\frac{1}{2}\right)^7 = \frac{1}{7 \cdot 2^7} = \frac{1}{7 \cdot 128} = \frac{1}{896}.$$

If we use the sums of the first three terms to approximate the integral, then we will have an overestimate of the integral and the error will be less than

$$\frac{1}{10} \cdot \left(\frac{1}{2}\right)^{10} = \frac{1}{10 \cdot 2^{10}} = \frac{1}{10,240}.$$

Since we want an error of less than $0.001 = \frac{1}{1000}$, and since $\frac{1}{896} > \frac{1}{1000}$, we must use three terms. We get

$$\int_0^{1/2} \frac{1}{1 + x^3} \, dx \approx \frac{1}{2} - \frac{1}{4}\left(\frac{1}{2}\right)^4 + \frac{1}{7}\left(\frac{1}{2}\right)^7$$

$$= \frac{1}{2} - \frac{1}{64} + \frac{1}{896},$$

or

$$\int_0^{1/2} \frac{1}{1 + x^3} \, dx \approx 0.485491$$

with an error of less than $1/10,000 = 0.0001$.

DO EXERCISE 11.

12. Compute the Taylor series of $f(x) = x^2 \cos x$ at $x = 0$.

Example 12 Compute the Taylor series expansion of $f(x) = x^2 e^x$ at $x = 0$.

Solution We could calculate the coefficients of the Taylor series of $x^2 e^x$ by directly computing the derivatives of $f(x)$ (see Exercise Set 10.4). There is, however, an easier way. From Example 1, we have

$$e^x = 1 + x + \frac{1}{2}x + \frac{1}{3!}x^3 + \cdots,$$

valid for all values of x. Simply multiplying both sides of this equation yields

$$x^2 e^x = x^2\left(1 + x + \frac{1}{2}x^2 + \frac{1}{3!}x^3 + \cdots\right)$$

$$= x^2 + x^3 + \frac{1}{2}x^4 + \frac{1}{3!}x^5 + \cdots,$$

valid for all real numbers x.

DO EXERCISES 12 AND 13.

13. Compute the Taylor series of $f(x) = x \tan^{-1} x$ at $x = 0$.

Example 13 Calculate the Taylor series expansion of $f(x)$ at $x = 0$, where $f(x)$ is given by

$$f(x) = \begin{cases} \dfrac{\sin x}{x}, & x \neq 0, \\ 1, & x = 0. \end{cases}$$

Solution The formula $(\sin x)/x$ makes sense only for $x \neq 0$. From Example 2, we get

$$\sin x = x - \frac{1}{3!}x^3 + \frac{1}{5!}x^5 - \frac{1}{7!}x^7 + \cdots,$$

valid for all real numbers x. If x is not equal to 0, then we can divide both sides of this equation by x, getting

$$\frac{\sin x}{x} = 1 - \frac{1}{3!}x^2 + \frac{1}{5!}x^4 - \frac{1}{7!}x^6 + \cdots,$$

valid for all x (except $x = 0$, of course). Note that as x approaches 0 in the formula above, the right-hand side will approach 1. That is,

$$\lim_{x \to 0} \frac{\sin x}{x} = 1,$$

14. Compute the Taylor series of
$f(x) =$ at $x = 0$ if

$$f(x) = \begin{cases} \dfrac{e^x - 1}{x}, & x \neq 0 \\ 1, & x = 0 \end{cases}$$

a fact that we will use in Section 10.5. For the present, note that we have shown that

$$f(x) = 1 - \frac{x^2}{3!} + \frac{x^4}{5!} - \frac{x^6}{7!} + \cdots$$

for *all* values of x.

DO EXERCISE 14.

EXERCISE SET 10.4

In each of the following cases, compute the Taylor series expansion of $f(x)$ at the prescribed value of x by directly calculating the coefficients.

1. $f(x) = e^{4x}$; $x = 0$

2. $f(x) = \sin(5x)$; $x = 0$

3. $f(x) = e^{1-x}$; $x = 1$

4. $f(x) = \cos(x - 2)$; $x = 2$

5. $f(x) = x^3 - 1$; $x = -1$

6. $f(x) = x^2 - 5x + 1$; $x = 3$

7. $f(x) = \dfrac{1}{1 - 2x}$; $x = 0$

8. $f(x) = \dfrac{1}{2 - x}$; $x = 0$

Using any method you wish, compute the Taylor series expansion of $f(x)$ for the prescribed value of x.

9. $f(x) = e^{6x}$; $x = 0$

10. $f(x) = \cos(x^2)$; $x = 0$

11. $f(x) = \dfrac{1}{1 - x^5}$; $x = 0$

12. $f(x) = \dfrac{1}{1 - bx}$ $(b \neq 0)$; $x = 0$

13. $f(x) = \ln(1 + 4x)$; $x = 0$

14. $f(x) = \dfrac{1}{(1 - bx)^2}$ $(b \neq 0)$; $x = 0$

15. $f(x) = \dfrac{1}{2 + x}$; $x = -1$ $\left(\text{Hint: } \dfrac{1}{2 + x} = \dfrac{1}{2} \cdot \dfrac{1}{1 + (1 + x)}.\right)$

16. $f(x) = \dfrac{1}{(2 + x)^2}$; $x = -1$

17. $f(x) = \ln(2 + x)$; $x = -1$

18. $f(x) = \dfrac{1}{(1 + x)^3}$; $x = 0$

19. $f(x) = \dfrac{5x^4}{1 - x^5}$; $x = 0$ (See Exercise 11 above.)

20. $f(x) = \ln(1 - x^5)$; $x = 0$

21. $f(x) = \begin{cases} \dfrac{\cos x - 1}{x^2}, & x \neq 0, \\ -\dfrac{1}{2}, & x = 0; \end{cases}$

22. $f(x) = x^3 e^{-x}$; $x = 0$

23. $f(x) = x^5 \sin(x^3)$; $x = 0$

24. $f(x) = \begin{cases} \dfrac{\sin(x^3)}{x}, & x \neq 0, \\ 0, & x = 0; \end{cases}$

Use the Taylor series and the alternating series test to find an approximate value for each of the following integrals (use three terms of the series and give an error estimate).

25. $\displaystyle\int_0^1 \cos(x^2)\, dx$ (See Exercise 10 above.)

26. $\displaystyle\int_0^1 x^5 \sin(x^3)\, dx$ (See Exercise 23 above.)

27. $\int_0^1 e^{-x^3} \, dx$

28. $\int_0^{0.1} \ln(1+x) \, dx$

29. $\int_0^{0.5} x^5 \sin(x^3) \, dx$

30. $\int_0^{0.25} \cos(x^2) \, dx$

31. *Probability theory.* Let

$$F(t) = \int_0^t e^{(-x^2)/2} \, dx.$$

a) Find the Taylor series expansion of $F(t)$ at $x = 0$.
b) Use your answer to (a) to calculate

$$P(0 \le t \le 0.1) = \frac{1}{\sqrt{2\pi}} F(0.1)$$

for the normal distribution.
c) Check your answer to (b) in Table 6.

What we have given here is a method for constructing that table.

OBJECTIVE

You should be able to determine whether or not a given limit expression is an indeterminate form. If it is, you should be able to use a version of L'Hôpital's Rule to evaluate it.

1. Evaluate

$$\lim_{x \to 3} \frac{x+1}{x^2+2}.$$

10.5 INDETERMINATE FORMS AND L'HÔPITAL'S RULE

We will lay the groundwork for this section with three preliminary examples. You may wish to review the definition of continuous function given in Section 2.1.

Example 1 Evaluate

$$\lim_{x \to 0} \frac{x^2+1}{x+3}.$$

Solution Let $f(x) = x^2 + 1$ and $g(x) = x + 3$. Then $f(x)$ and $g(x)$ are continuous functions for all x. It follows that

$$\frac{f(x)}{g(x)} = \frac{x^2+1}{x+3}$$

is continuous except when $g(x) = 0$ (that is, when $x = -3$). So we can evaluate the given limit as follows:

$$\lim_{x \to 0} \frac{x^2+1}{x+3} = \frac{\lim_{x \to 0}(x^2+1)}{\lim_{x \to 0}(x+3)} = \frac{(0)^2+1}{0+3} = \frac{1}{3}.$$

DO EXERCISE 1.

2. Evaluate

$$\lim_{x \to 0} \frac{1 - \cos x}{x^2 + 2}.$$

Example 2 Evaluate

$$\lim_{x \to 0} \frac{\sin x}{x + 3}.$$

Solution Let $f(x) = \sin x$ and $g(x) = x + 3$. Then $f(x)$ and $g(x)$ are continuous functions. It follows that $f(x)/g(x)$ is continuous except at $x = -3$. So, just as in Example 1, we have

$$\lim_{x \to 0} \frac{\sin x}{x + 3} = \frac{\lim\limits_{x \to 0} \sin x}{\lim\limits_{x \to 0} x + 3} = \frac{\sin (0)}{0 + 3} = \frac{0}{3} = 0.$$

Note that the fact that the *numerator* of the resulting fraction is 0 does not affect the validity of the steps that were used.

DO EXERCISES 2 AND 3.

Example 3 Evaluate

$$\lim_{x \to 1} \frac{x^2}{x - 1}.$$

3. $\lim\limits_{x \to 1} \dfrac{\ln x}{x}$

Solution Let $f(x) = x^2$ and $g(x) = x - 1$. Then, just as before, $f(x)$ and $g(x)$ are continuous functions. Just as before,

$$\frac{f(x)}{g(x)} = \frac{x^2}{x - 1}$$

is continuous for all values of x such that $g(x) \neq 0$ (in this case, except for $x = 1$). But this is exactly the point at which we wish to evaluate the limit. We cannot use the techniques of Examples 1 and 2 since the expression $1/0$ is meaningless.

To try to understand what is happening, we compute a few values of $x^2/(x - 1)$ for choices of x near 1 (but different from 1). For $x = 1.1$, we obtain

$$\frac{(1.1)^2}{1.1 - 1} = \frac{1.21}{0.1} = 12.1.$$

For $x = 1.01$, we get

$$\frac{(1.01)^2}{1.01 - 1} = \frac{1.0201}{0.01} = 102.01.$$

For $x = 1.001$, we get

$$\frac{(1.001)^2}{1.001 - 1} = \frac{1.002001}{0.001} = 1002.001.$$

4. Compute the ratio $x^2/(x-1)$ for $x = 0.93$, $x = 0.991$, and $x = 0.999$.

CSS

5. Evaluate

$$\lim_{x \to -1} \frac{x}{x+1}.$$

CSS

Note that as we take values of x that are closer and closer to 1, the numerator of $x^2/(x-1)$ gets closer and closer to 1 while the denominator gets closer and closer to 0, so that the resulting quotients are larger and larger positive numbers.

The values of x used above were all larger than 1. If we had used values of x smaller than 1 (and getting closer and closer to 1), the quotients would be negative numbers with large absolute values.

In any event, the quotients do not get closer to a real number. In this case, we say that the limit *does not exist*.

DO EXERCISES 4 AND 5.

In investigating

$$\lim_{x \to a} \frac{f(x)}{g(x)},$$

where $f(x)$ and $g(x)$ are continuous functions, we have seen that we can compute the limit by substitution so long as $g(a) \neq 0$. If the denominator approaches 0 as x approaches a, while the numerator approaches a nonzero number, then the limit does not exist. This leaves only the case in which both the numerator and the denominator are 0 at $x = a$. This is the most interesting case.

The Indeterminate Form 0/0

Example 4 Evaluate

$$\lim_{x \to 0} \frac{x^2 + 3x}{5x}.$$

Solution This limit is of the form

$$\lim_{x \to a} \frac{f(x)}{g(x)},$$

where $a = 0$, $f(x) = x^2 + 3x$, and $g(x) = 5x$. Substitution in the numerator yields $f(0) = (0)^2 + 3(0) = 0$. Substitution in the denominator yields $g(0) = 5(0) = 0$. Since substitution of $x = a$ in both numerator and denominator yields 0, we say that

$$\lim_{x \to 0} \frac{x^2 + 3x}{5x}$$

is an indeterminate form of the type $0/0$.

6. Evaluate

$$\lim_{x \to 0} \frac{x^2 - 7x}{x}.$$

In this case, we cannot evaluate the limit by direct substitution. Note that we can factor an x out of $x^2 + 3x$, obtaining

$$\lim_{x \to 0} \frac{x^2 + 3x}{5x} = \lim_{x \to 0} \frac{x(x + 3)}{5x}.$$

If x is not zero, $\frac{x(x + 3)}{5x} = \frac{x + 3}{5}$. Thus we have

$$\lim_{x \to 0} \frac{x^2 + 3x}{5x} = \lim_{x \to 0} \frac{x + 3}{5}.$$

Note that the limit on the right-hand side of the last equation is *not* an indeterminate form. In fact, direct substitution yields

$$\lim_{x \to 0} \frac{x + 3}{5} = \frac{(0) + 3}{5} = \frac{3}{5}.$$

So

$$\lim_{x \to 0} \frac{x^2 + 3x}{5x} = \frac{3}{5}.$$

DO EXERCISE 6.

7. Evaluate

$$\lim_{x \to -3} \frac{x^2 + 4x + 3}{x + 3}.$$

Example 5 Evaluate

$$\lim_{x \to -1} \frac{x^2 + 4x + 3}{x + 1}.$$

Solution Substitution in the numerator yields

$$(-1)^2 + 4(-1) + 3 = 1 - 4 + 3 = 0.$$

Since substitution of $x = -1$ in the denominator also yields 0, this is an indeterminate form. We can write the numerator in the form $x^2 + 4x + 3 = (x + 1)(x + 3)$. For values of x that are not equal to -1, we have

$$\frac{x + 4x + 3}{x + 1} = \frac{(x + 1)(x + 3)}{x + 1} = x + 3.$$

So

$$\lim_{x \to -1} \frac{x^2 + 4x + 3}{x + 1} = \lim_{x \to -1} (x + 3) = (-1) + 3 = 2.$$

DO EXERCISE 7.

8. Evaluate

$$\lim_{x \to 0} \frac{1 - \cos x}{x^2}.$$

Example 6 Evaluate

$$\lim_{x \to 0} \frac{\sin x}{x}.$$

Solution Both the numerator and the denominator are continuous functions of x, but substitution of x = 0 yields sin (0) = 0 in the numerator and 0 in the denominator. This limit is an indeterminate form. In this case, we recall from Section 10.4 that $f(x) = \sin x$ can be represented as

$$\sin x = x - \frac{x^3}{3!} + \frac{x^5}{5!} - \frac{x^7}{7!} + \cdots.$$

This formula is valid for all values of x. Dividing both sides of x, we see that if x is not equal to 0, then

$$\frac{\sin x}{x} = 1 - \frac{x^2}{3!} + \frac{x^4}{5!} - \frac{x^6}{7!} + \cdots.$$

So

$$\lim_{x \to 0} \frac{\sin x}{x} = \lim_{x \to 0} \left(1 - \frac{x^2}{3!} + \frac{x^4}{5!} - \frac{x^6}{7!} + \cdots\right)$$

$$= 1 - \frac{(0)^2}{3!} + \frac{(0)^4}{5!} - \frac{(0)^6}{7!} + \cdots = 1.$$

DO EXERCISE 8.

So far, we have evaluated several indeterminate forms using algebraic techniques that are often tedious and time-consuming. Fortunately, there is an easier way to do many of these problems.

Consider

$$\lim_{x \to a} \frac{f(x)}{g(x)},$$

where we assume that $f(a) = 0$ and $g(a) = 0$, so that this limit is an indeterminate form. Suppose that $f(x)$ and $g(x)$ are both represented by their Taylor series expansion (at x = a) on an interval containing x = a. Then

$$f(x) = f(a) + f'(a)(x - a) + \frac{f''(a)}{2!}(x - a)^2 + \cdots.$$

Since $f(a) = 0$, we have

$$f(x) = f'(a)(x - a) + \frac{f''(a)}{2!}(x - a)^2 + \frac{f'''(a)}{3!}(x - a)^3 + \cdots.$$

Factoring the common term $x - a$ from the right-hand side yields

$$f(x) = (x - a)\left[f'(a) + \frac{f''(a)}{2!}(x - a) + \frac{f'''(a)}{3!}(x - a)^2 + \cdots \right].$$

Likewise,

$$g(x) = (x - a)\left[g'(a) + \frac{g''(a)}{2!}(x - a) + \frac{g'''(a)}{3!}(x - a)^2 + \cdots \right],$$

since $g(a) = 0$. So, for values of x that are not equal to a, we have

$$\frac{f(x)}{g(x)} = \frac{(x - a)\left[f'(a) + \frac{f''(a)}{2!}(x - a) + \cdots \right]}{(x - a)\left[g'(a) + \frac{g''(a)}{2!}(x - a) + \cdots \right]}$$

$$= \frac{f'(a) + \frac{f''(a)}{2!}(x - a) + \cdots}{g'(a) + \frac{g''(a)}{2!}(x - a) + \cdots}.$$

Note that substituting $x = a$ in the right-hand side yields $f'(a)/g'(a)$. This suggests the following rule.

THEOREM 6   L'Hôpital's Rule

If

$$\lim_{x \to a} \frac{f(x)}{g(x)}$$

is an indeterminate form of the type 0/0 and if

$$\lim_{x \to a} \frac{f'(x)}{g'(x)}$$

exists, then

$$\lim_{x \to a} \frac{f(x)}{g(x)} = \lim_{x \to a} \frac{f'(x)}{g'(x)}.$$

Example 7 Use L'Hôpital's Rule to evaluate

$$\lim_{x \to 0} \frac{x^2 + 3x}{5x}.$$

Solution This is an indeterminate form of the type 0/0, with

9. Use L'Hôpital's Rule to evaluate

$$\lim_{x \to -1} \frac{x^2 + 4x + 3}{x + 1}.$$

10. Use L'Hôpital's Rule to evaluate

$$\lim_{x \to -3} \frac{x^2 + 4x + 3}{x + 3}.$$

11. Use L'Hôpital's Rule to evaluate

$$\lim_{x \to 0} \frac{\sin x}{x}.$$

$f(x) = x^2 + 3x$ and $g(x) = 5x$. Here, $f'(x) = 2x + 3$ and $g'(x) = 5$. So

$$\lim_{x \to 0} \frac{x^2 + 3x}{5x} = \lim_{x \to 0} \frac{2x + 3}{5}$$

$$= \frac{2(0) + 3}{5} = \frac{3}{5}.$$

CSS

This is the correct answer, as we saw in Example 4.

DO EXERCISES 9 AND 10.

Example 8 Evaluate

$$\lim_{x \to 0} \frac{e^{2x} - 1}{x}.$$

Solution Substitution of $x = 0$ in $f(x) = e^{2x} - 1$ yields $f(0) = e^{2(0)} - 1 = e^0 - 1 = 1 - 1 = 0$. Since the same substitution into $g(x) = x$ also yields 0, this is an indeterminate form. Since $f'(x) = 2e^{2x}$ and $g'(x) = 1$, we have

$$\lim_{x \to 0} \frac{e^{2x} - 1}{x} = \lim_{x \to 0} \frac{2e^{2x}}{1} = \frac{2e^{2(0)}}{1} = 2.$$

DO EXERCISE 11.

Example 9 Evaluate

$$\lim_{x \to 0} \frac{1 - \cos x}{x^2}.$$

Solution This is an indeterminate form of the type 0/0. Since $f(x) = 1 - \cos x$ and $g(x) = x^2$, we have $f'(x) = \sin x$ and $g'(x) = 2x$. Now, both $\sin x$ and $2x$ tend to 0 as x approaches 0. In other words,

$$\lim_{x \to 0} \frac{f'(x)}{g'(x)} = \lim_{x \to 0} \frac{\sin x}{2x}$$

is *another* indeterminate form. But we may apply L'Hôpital's Rule again and again, stopping as soon as we obtain a limit that is not an indeterminate form. In this case, we obtain

$$\lim_{x \to 0} \frac{f''(x)}{g''(x)} = \lim_{x \to 0} \frac{\cos x}{2}$$

$$= \frac{\cos (0)}{2} = \frac{1}{2}.$$

12. Use L'Hôpital's Rule to evaluate

$$\lim_{x \to 1} \frac{1 + \ln x - x}{(x - 1)^2}.$$

Thus

$$\lim_{x \to 0} \frac{f(x)}{g(x)} = \lim_{x \to 0} \frac{f'(x)}{g'(x)} = \lim_{x \to 0} \frac{f''(x)}{g''(x)} = \frac{1}{2}.$$

DO EXERCISE 12.

Example 10 Use L'Hôpital's Rule to evaluate

$$\lim_{x \to 0} \frac{e^x}{\sin x}.$$

Solution As x approaches 0, substitution shows that the numerator approaches $e^0 = 1$, while the denominator approaches $\sin(0) = 0$. This is *not* an indeterminate form of the type $0/0$ and L'Hôpital's Rule does *not* apply. In fact, this limit is similar to the limit in Example 3. So

$$\lim_{x \to 0} \frac{e^x}{\sin x} \quad \text{does not exist.}$$

DO EXERCISE 13.

13. Use L'Hôpital's Rule, if possible, to evaluate

$$\lim_{x \to 2} \frac{e^x}{x - 1}.$$

Example 11 Evaluate

$$\lim_{x \to 0} \frac{\sin x}{x^2}.$$

Solution This is an indeterminate form of the type $0/0$. Here, $f(x) = \sin x$ and $g(x) = x^2$, so that $f'(x) = \cos x$ and $g'(x) = 2x$. But

$$\lim_{x \to 0} \frac{\cos x}{2x}$$

does not exist since as x approaches 0, the numerator approaches $\cos(0) = 1$ while the denominator approaches 0. L'Hôpital's Rule, as stated earlier, does not apply. It turns out, though, that if the numerator $f'(x)$ has a nonzero limit and the denominator $g'(x)$ approaches 0 as x approaches a, then the original limit,

$$\lim_{x \to a} \frac{f(x)}{g(x)},$$

does not exist either. Thus

$$\lim_{x \to 0} \frac{\sin x}{x^2}$$

does not exist.

14. Evaluate

$$\lim_{x \to 0} \frac{e^x - 1}{x^3}.$$

DO EXERCISE 14.

Indeterminate Forms of the Form ∞ / ∞

In applications, one encounters limits of the form

$$\lim_{x \to a} \frac{f(x)}{g(x)},$$

where both $f(x)$ and $g(x)$ tend to infinity as x approaches a. In this case, the following version of L'Hôpital's Rule applies.

THEOREM 7

If $f(x)$ and $g(x)$ are continuous and if $f(x)$ and $g(x)$ approach infinity as x approaches a, then

$$\lim_{x \to a} \frac{f(x)}{g(x)} = \lim_{x \to a} \frac{f'(x)}{g'(x)}$$

if this latter limit exists. If $f'(x)/g'(x)$ tends to infinity as x approaches a, then so does $f(x)/g(x)$.

15. Evaluate

$$\lim_{x \to +\infty} \frac{5x - 1}{4 - 3x}.$$

Example 12 Evaluate

$$\lim_{x \to +\infty} \frac{2x - 3}{x + 1}.$$

Solution Both numerator and denominator tend to infinity as x does. Applying L'Hôpital's Rule, we have

$$\lim_{x \to +\infty} \frac{2x - 3}{x + 1} = \lim_{x \to +\infty} \frac{2}{1} = 2.$$

Thus

$$\lim_{x \to +\infty} \frac{2x - 3}{x + 1} = 2.$$

DO EXERCISE 15.

Example 13 Evaluate

$$\lim_{x \to +\infty} \frac{e^x}{x}.$$

Solution Again, both numerator and denominator tend to infinity as x

16. Evaluate

$$\lim_{x \to +\infty} \frac{7e^{3x}}{5x}.$$

does. By L'Hôpital's Rule,

$$\lim_{x \to +\infty} \frac{e^x}{x} = \lim_{x \to +\infty} \frac{e^x}{1} = +\infty,$$

That is, the limit does not exist.

DO EXERCISE 16.

Example 14 Evaluate

$$\lim_{x \to +\infty} \frac{3x^2 - 5x + 1}{x^2 - 3x + 7}.$$

17. Evaluate

$$\lim_{x \to +\infty} \frac{3x^2 - 5x + 1}{x^2 + 1}.$$ CSS

Solution One application of L'Hôpital's Rule yields

$$\lim_{x \to +\infty} \frac{3x^2 - 5x + 1}{x^2 - 3x + 7} = \lim_{x \to +\infty} \frac{6x - 5}{2x - 3},$$

which is again an indeterminate form of the type ∞/∞. A second application, however, gives

$$\lim_{x \to +\infty} \frac{6x - 5}{2x - 3} = \lim_{x \to +\infty} \frac{6}{2} = 3.$$

Thus

$$\lim_{x \to +\infty} \frac{3x^2 - 5x + 1}{x^2 - 3x + 7} = 3.$$

DO EXERCISE 17.

EXERCISE SET 10.5

Find each of the following limits, if possible, *without* using L'Hôpital's Rule.

1. $\lim_{x \to 1} \dfrac{1}{x}$

2. $\lim_{x \to 2} \dfrac{x-2}{x^2}$

3. $\lim_{x \to -1} \dfrac{3}{x+1}$

4. $\lim_{x \to -5} \dfrac{7}{x+5}$

5. $\lim_{x \to -1} \dfrac{x+1}{3}$

6. $\lim_{x \to -5} \dfrac{x+3}{7}$

7. $\lim_{x \to 0} \dfrac{7x+3}{x^2}$

8. $\lim_{x \to 0} \dfrac{x^2}{7x+3}$

9. $\lim_{x \to 0} \dfrac{7x+3x^2}{x}$

10. $\lim_{x \to -1} \dfrac{x+1}{x^2+6x+5}$

11. $\lim_{x \to 2} \dfrac{x-2}{x^2-4}$

12. $\lim_{x \to 1} \dfrac{x-1}{x^3-1}$

13. $\lim_{x \to +\infty} \dfrac{5}{x}$

14. $\lim_{x \to 1} \dfrac{x^3}{x-1}$

15. $\lim_{x \to +\infty} \dfrac{2x+1}{x-6}$

16. $\lim_{x \to 0} \dfrac{7+3x}{x}$

Evaluate each of the following limits, if it exists, using L'Hôpital's Rule, if it applies.   `CSS`

17. $\lim\limits_{x\to 1} \dfrac{x-1}{x^3-1}$

18. $\lim\limits_{x\to 2} \dfrac{x-2}{x^2-4}$

19. $\lim\limits_{x\to 0} \dfrac{1-e^x}{x}$

20. $\lim\limits_{x\to 0} \dfrac{\sin (ax)}{x}$

21. $\lim\limits_{x\to -1} \dfrac{x+1}{x^3+3x^2-x-3}$

22. $\lim\limits_{x\to 1} \dfrac{\ln x}{x-1}$

23. $\lim\limits_{x\to \pi} \dfrac{\sin x}{x-\pi}$

24. $\lim\limits_{x\to 0} \dfrac{x \cos x}{\sin (2x)}$

25. $\lim\limits_{x\to 0} \dfrac{e^x}{\cos x}$

26. $\lim\limits_{x\to +\infty} \dfrac{2x+1}{x-6}$

27. $\lim\limits_{x\to 1} \dfrac{\ln x-x+1}{x-1}$

28. $\lim\limits_{x\to 0} \dfrac{e^x-x-1}{x^2}$

29. $\lim\limits_{x\to +\infty} \dfrac{x^2}{7}$

30. $\lim\limits_{x\to +\infty} \dfrac{x^2}{e^x}$

31. $\lim\limits_{x\to 0} \dfrac{x^2-2x+2}{3x+4}$

32. $\lim\limits_{x\to 1} \dfrac{\ln x}{(x-1)^3}$

33. $\lim\limits_{x\to +\infty} \dfrac{x^2-5x+1}{3x^2-7x+2}$

34. $\lim\limits_{x\to +\infty} \dfrac{x^3}{x^3-1}$

35. $\lim\limits_{x\to +\infty} \dfrac{e^{5x}}{x^3}$

36. $\lim\limits_{x\to +\infty} \dfrac{\tan^{-1} x}{x}$

37. $\lim\limits_{x\to 0} \dfrac{e^{5x}}{x^3}$

38. $\lim\limits_{x\to 0} \dfrac{e^{3x}-1}{x}$

▶

39. In a certain chemical reaction, the amount of the product at time t is given by

$$x(t)=ab\left\{\dfrac{r-pe^{kt}}{pa-be^{kt}}\right\},$$

here a, b, k, p, and r are positive constants. Use L'Hôpital's Rule to find the limiting amount of the product present, that is, to evaluate

$$\lim\limits_{t\to +\infty} x(t).$$

40. The velocity v of a parachutist t seconds after ejection from a plane is given by

$$v(t)=a\left[\dfrac{e^{bt}-e^{-bt}}{e^{bt}+e^{-bt}}\right],$$

where a and b are positive constants. Determine the limiting velocity of the parachutist,

$$\lim\limits_{t\to +\infty} v(t).$$

41. If f is continuous on $(-\infty, +\infty)$, find

$$\lim\limits_{x\to 0} \dfrac{1}{x} \int_0^x f(t)\, dt.$$

(Franklin Wing: Stock, Boston).

CHAPTER 10 TEST

Find $p_2(x)$ *and* $p_3(x)$ for the given function and the specified value of x.

1. $y = x^3$; $x = -2$ **2.** $y = e^{x^3}$; $x = 0$ **3.** $y = \sin (x - 1)$; $x = 1$

Use a Taylor polynomial of degree 3 to find an approximate value for each of the following.

4. $\sin (0.04)$ **5.** $e^{0.07}$

6. Find the sum of the infinite series   **7.** Evaluate

$$\frac{3}{5} - \frac{2}{5} + \frac{4}{15} - \frac{8}{45} + \frac{16}{135} - \frac{32}{405} + \cdots .$$

$$\frac{1}{2} + \frac{1}{7} + \frac{2}{49} + \cdots + \frac{1}{2}\left(\frac{2}{7}\right)^{12} .$$

8. Write $2.431\overline{3131}$ as a rational fraction.

Determine whether the given infinite series converges or diverges. If it converges, estimate its sum with an error of less than 0.01.

9. $1 - \dfrac{1}{3 \cdot 4} + \dfrac{1}{5 \cdot 4^2} - \dfrac{1}{7 \cdot 4^3} + \dfrac{1}{9 \cdot 4^4} - \cdots$ **10.** $1 - 2 + 4 - 8 + 16 - 32 + \cdots$

11. $\dfrac{2}{3} - \dfrac{3}{4} + \dfrac{4}{5} - \dfrac{5}{6} + \dfrac{6}{7} - \dfrac{7}{8} + \cdots$ **12.** $1 - \dfrac{1}{16} + \dfrac{1}{256} - \dfrac{1}{4^6} + \dfrac{1}{4^8} - \cdots$

Expand each of the following functions in a Taylor series using the prescribed value of x.

13. $y = x^2 \sin x$; $x = 0$ **14.** $y = e^{(x - 1)^2}$; $x = 1$

15. $\dfrac{1}{1 - 5x}$; $x = 0$ **16.** Use a Taylor series to evaluate

$$\lim_{x \to 0} \frac{1 - \cos (x^2)}{x^4} .$$

Evaluate each of the following limits.

17. $\lim\limits_{x \to 0} \dfrac{x \sin x}{\cos x - 1}$ **18.** $\lim\limits_{x \to 1} \dfrac{3x \ln x}{e^x}$

19. $\lim\limits_{x \to \pi/2} \dfrac{\cos x}{x - \pi/2}$ **20.** $\lim\limits_{x \to +\infty} \dfrac{x^n}{e^x}$ (n is a positive integer)

21. $\lim\limits_{x \to +\infty} \dfrac{x^3 - 5x + 6}{2x^3 - 7x^2 + 1}$

22. A certain stock pays a dividend of $2 per year. Assuming that the dividend remains constant forever and that interest rates are constant at 10% so that the present value of the dividend payment n years from today is $2(1.10)^{-n}$, find the sum of the present value of all future dividend payments, assuming that the first payment is one year from today.

23. In Problem 22, if the dividend payment is $D and the interest rate is i, find the present value of all future dividend payments.

24. Find a function $f(x)$ (if there is one) so that $f(0) = 0$ and $f^{(n)}(0) = 1/n$ for all positive integers n.

NUMERICAL TECHNIQUES

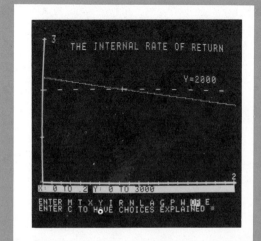

An internal rate of return key occurs on some business calculators and computers. (*Cheryl Shugars*)

Your computer software supplement uses Newton's Method to solve equations. Shown here are the two graphs related to solving an internal rate of return problem.

11

OBJECTIVE

You should be able to solve equations using the method of iteration.

11.1 SOLVING EQUATIONS NUMERICALLY: ITERATION

The solutions of many equations that arise in practical applications cannot be found exactly using standard algebraic techniques. An example of an equation of this sort is $\cos x = x$, which we consider below. We use the following general procedure.

To solve a given equation, we start by making an *initial approximation*, or *initial guess* (denoted x_1), of the solution to the equation. Then we use our initial guess x_1 to construct a new value x_2 (which we expect to be closer to the solution than x_1 is). Continuing, we use x_2 to construct a new value x_3, and so forth. In theory, we construct an infinite sequence $x_1, x_2, \ldots, x_n, \ldots$, which converges to a solution, say $\overline{x}$, of the given equation. In practice, we must stop the process after a finite number of steps N, using the approximate solution x_N instead of the actual solution $\overline{x}$.

Example 1 Solve $\cos x = x$.

Solution The graphs of $y = \cos x$ and $y = x$ are plotted as shown below.

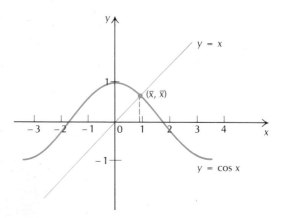

They appear to intersect at one point only. The x-coordinate of that point, $\overline{x}$, is the solution to our equation. Judging from the graph, $x_1 = \frac{1}{2} = 0.5000000$ is not an unreasonable initial guess.

We'll construct x_2 using what is called the *method of iteration*. In this case, it means that $x_2 = \cos(x_1) = \cos(0.5)$. Computing (using a calculator in radian mode), we get $x_2 \approx 0.8775826$. We continue by computing $x_3 = \cos(x_2) = \cos(0.8775826)$. We get $x_3 \approx 0.6390125$. Then, with $x_4 = \cos(x_3)$, we find that $x_4 \approx 0.8026851$. We continue

1. Using $x_1 = 1.8$ and $x_{n+1} = \sqrt{x_n}$ for $n = 1, 2, \ldots$, solve $\sqrt{x} = x$.

with $x_{n+1} = \cos(x_n)$, for all n. We list a few values (check them on your calculator):

$$x_5 \approx 0.6947780,$$

$$x_6 \approx 0.7681958,$$

$$x_7 \approx 0.7191654,$$

$$x_8 \approx 0.7523558,$$

$$x_9 \approx 0.7300811,$$

$$x_{10} \approx 0.7451203,$$

.

.

.

$$x_{17} \approx 0.7387045,$$

$$x_{18} \approx 0.7393415,$$

$$x_{19} \approx 0.7389124,$$

.

.

.

$$x_{43} \approx 0.7390851,$$

$$x_{44} \approx 0.7390851,$$

with the readings unchanged after that. We conclude that $\bar{x} \approx 0.7390851$. As a check, we substitute this value into the original equation to get

$$\cos(\bar{x}) \approx 0.7390851 \approx \bar{x}.$$

DO EXERCISE 1.

We will apply the method of iteration only to equations of the form $g(x) = x$, where $g(x)$ is a continuous function. Remember that $x_{n+1} = g(x_n)$ for all n. If the sequence $x_1, x_2, \ldots x_n, \ldots$, has a limit, say $\bar{x}$, then

$$\bar{x} = \lim_{x \to \infty} x_n.$$

2. Try to solve $x^2 = x$ using $x_1 = 1.1$.

For continuous functions g,

$$g(\bar{x}) = \lim_{n \to \infty} g(x_n) = \lim_{n \to \infty} x_{n+1} = \bar{x},$$

or

$$g(\bar{x}) = \bar{x}.$$

Thus if the sequence of iterates $x_1, x_2, \ldots x_n, \ldots$, has a limit $\bar{x}$ and if g is continuous, then $\bar{x}$ is a solution to $g(x) = x$. The catch, of course, is that the sequence of iterates may not converge.

Example 2 Solve $x^2 = x$ using $x_1 = 1.5$.

Solution Here, $g(x) = x^2$. So $x_{n+1} = x_n^2$ for $n = 1, 2, \ldots$. Using $x_1 = 1.5$, we obtain $x_2 = (1.5)^2 = 2.25, x_3 = 5.0625, x_4 \approx 25.62891$, and so on. The sequence of iterates diverges. Note that $x = 0$ and $x = 1$ are the only solutions to $x^2 = x$.

DO EXERCISES 2 AND 3.

3. Try to solve $x^2 = x$ using each of the following.

a) $x_1 = \dfrac{1}{3}$

b) $x_1 = -\dfrac{1}{2}$

c) $x_1 = \dfrac{2}{3}$

Example 3 Solve $x^2 - 5x + 6 = 0$.

Solution We write $x^2 + 6 = 5x$, or $\frac{1}{5}x^2 + \frac{6}{5} = x$. Letting $g(x) = \frac{1}{5}x^2 + \frac{6}{5}$, we have $g(x) = x$.

The original equation can be factored as $(x - 2)(x - 3) = 0$, so that $\bar{x} = 2$ and $\bar{x} = 3$ are the solutions. Alternatively, we could have used the quadratic formula to obtain the solutions.

Ignoring the fact that we can easily find the solutions in this case, we start with $x_1 = 1$. Then

$$x_2 = g(x_1) = \frac{1}{5}(1)^2 + \frac{6}{5} = 1.4.$$

Continuing, we have

$$x_3 = g(x_2) = \frac{1}{5}(1.4)^2 + \frac{6}{5} = 1.592$$

and

$$x_4 = g(x_3) = \frac{1}{5}(1.592)^2 + \frac{6}{5} \approx 1.7068928.$$

Similarly, $x_5 \approx 1.782697$, $x_6 \approx 1.835601$, $x_7 \approx 1.873887$, and $x_8 \approx 1.902290$. We find that $x_{14} \approx 1.976536$ and, after a few more iterations, $x_{20} \approx 1.99398$. It seems clear that these iterates have the limit $\bar{x} = 2$.

4. What happens in Example 3 if we use $x_1 = 2.9$?

Now, what happens if we start with the initial guess $x_1 = 3.2$? Then $x_2 = \frac{1}{5}(3.2)^2 + \frac{6}{5} = 3.248$. Continuing, we have $x_3 = g(x_2) = \frac{1}{5}(3.248)^2 + \frac{6}{5} \approx 3.30990$, a bad sign, since $x_3 - x_2$ is larger than $x_2 - x_1$. Continuing, we find that $x_4 \approx 3.391089$, $x_5 \approx 3.499896$, $x_6 \approx 3.649855$, and $x_7 \approx 3.864288$, with the sequence of iterates diverging.

DO EXERCISES 4 AND 5.

Examples 2 and 3 and the Margin Exercises 2, 3, and 4 show that the sequence of iterates may diverge for one choice of x_1 and converge for another choice. The following theorem clarifies the situation.

THEOREM 1

> If $g(x)$ and $g'(x)$ are continuous and $|g'(x)| < 1$ on an interval containing *both* x_1 and an $\overline{x}$ satisfying $g(\overline{x}) = \overline{x}$, and if we put
>
> $$x_{n+1} = g(x_n) \quad \text{for} \quad n = 1, 2, \ldots,$$
>
> then the sequence $x_1, x_2, \ldots, x_n, \ldots$, converges to $\overline{x}$.

5. Solve $x^2 + 1 = 4x$ using $x_1 = 2$.

We reconsider Examples 2 and 3. In Example 2, we had $g(x) = x^2$, so that $g'(x) = 2x$. If x is between $-\frac{1}{2}$ and $\frac{1}{2}$, then $g'(x) = 2x$ will be between -1 and 1. Since the interval contains $\overline{x} = 0$, the theorem guarantees us that if we pick x_1 between $-\frac{1}{2}$ and $\frac{1}{2}$ and define $x_{n+1} = x_n^2$ for $n = 1, 2, \ldots$, then the iterates will converge to $\overline{x} = 0$.

In Example 3, we had $g(x) = \frac{1}{5}x^2 + \frac{6}{5}$. Thus $g'(x) = \frac{2}{5}x$. In order to guarantee that $-1 < \frac{2}{5}x < 1$, we must choose x so that $-\frac{5}{2} < x < \frac{5}{2}$ (multiply the original inequality by $\frac{5}{2}$). This interval contains $\overline{x} = 2$. It follows that if we select x_1 between $-\frac{5}{2}$ and $\frac{5}{2}$, then the iterates will converge to 2. For this function, we can show that the iterates will converge to 2 if we select any x_1 between -3 and 3.

Example 4 Use iteration to solve $x^2 - 5x + 6 = 0$.

Solution Yes! This is the same equation that appeared in Example 2 above. But there may be many ways to rewrite $x^2 - 5x + 6 = 0$ in the form $g(x) = x$. For instance, let's write $x^2 = 5x - 6$. Then, taking the principal square roots of both sides (assuming that x and $5x - 6$ are nonnegative), we get $x = \sqrt{5x - 6}$. Let $g(x) = \sqrt{5x - 6}$. If we define

$$x_{n+1} = g(x_n) = \sqrt{5x_n - 6},$$

then we obtain a new and different iteration formula for this equation.

6. Solve $x^2 + 1 = 4x$ using $x_1 = 2$ and $g(x) = \sqrt{4x - 1}$.

If we pick $x_1 = 4$, for example, then $x_2 = \sqrt{5x_1 - 6} = \sqrt{5(4) - 6} = \sqrt{14} \approx 3.741657$. Continuing, we get $x_3 = \sqrt{5x_2 - 6} \approx \sqrt{12.708287} \approx 3.564868$, $x_4 \approx 3.438654$, $x_5 \approx 3.345635$, $x_6 \approx 3.275389, \ldots$, $x_{24} \approx 3.008502$, $x_{25} \approx 3.007077$, $x_{26} \approx 3.005892$, and so forth. Theorem 1 can be applied to show that the iterates converge to $\overline{x} = 3$, although the convergence is very slow. Examples 2 and 4 show that in solving an equation by iteration, some experimentation may be required to determine a workable iteration formula.

DO EXERCISE 6.

EXERCISE SET 11.1

Solve the given equation by iteration, using the indicated initial guess. Check your answers.   **CSS**

1. $x = \frac{1}{3}x^2 + \frac{1}{3}$; $x_1 = 0$

2. $x = \frac{1}{3}x^2 + \frac{1}{3}$; $x_1 = 3$

3. $x = \sin x + \frac{1}{2}$; $x_1 = 1$ (x in radians)

4. $x = \frac{1}{2}\cos x$; $x_1 = 1$

5. $x = 2\cos x$; $x_1 = 1$

6. $x = \sqrt{3x - 1}$; $x_1 = 2$

7. $x = \frac{1}{3}x^3 + 1$; $x_1 = 0$

8. $x = \sqrt[3]{3x - 3}$; $x_1 = 0$

Use iteration to find the solution to each of the following. Check your answers.

9. $x - e^{-x} = 0$ **10.** $x^5 - 5x + 5 = 0$ **11.** $3\cos x - 7x = 0$ **12.** $\cos x - 2x^2 = 0$

13. Define $x_{n+1} = \frac{1}{3}x_n^2 + \frac{1}{3}$ for all n, as in Exercise 2 above.

a) Show that if $x_n > 3$, then $x_{n+1} > \frac{10}{3} > 3$. Conclude that if $x_1 > 3$, then all the succeeding x_n's are greater than $\frac{10}{3}$.

b) By writing

$$\frac{x_n + 1}{x_n} = \frac{1}{3}x_n + \frac{1}{3x_n},$$

show that if $x_n > \frac{10}{3}$, then

$$\frac{x_{n+1}}{x_n} > \frac{10}{9}.$$

c) By combining (a) and (b), show that if $x_1 > 3$, then $x_2 > \frac{10}{9}$ and $x_{k+2} > \left(\frac{10}{9}\right)^k \cdot x_2$, which in turn shows that the iterates in Exercise 2 diverge.

You should be able to solve equations using Newton's method.

11.2 NEWTON'S METHOD

We introduce Newton's method, another technique for solving equations numerically. This method is fairly easy to apply and usually yields more rapid convergence than the method of iteration. To apply this method, we write the given equation in the form $f(x) = 0$. Let x_1 be our initial approximation, as shown in the following figure.

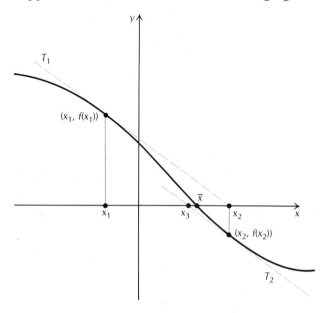

Assuming that $f'(x_1) \neq 0$, the tangent line T_1 to the graph of $y = f(x)$ at the point $(x_1, f(x_1))$ will intersect the x-axis. We call this point $(x_2, 0)$. The x-coordinate of this point, x_2, will be our second approximation. Next we locate $(x_2, f(x_2))$ on the graph of $y = f(x)$ and construct the tangent line T_2 to the graph at this new point. Then the tangent line T_2 will intersect the x-axis at $(x_3, 0)$ (assuming that $f'(x_2) \neq 0$). The x-coordinate of this point, x_3, is our third approximation to the solution $\overline{x}$. Continuing, we can generate as many approximations as we need.

We now have a geometric method for constructing x_{n+1} from x_n. To obtain a formula relating x_n and x_{n+1}, we return to the tangent line T_1. The slope of T_1 is $f'(x_1)$, the value of the derivative of $f(x)$ at the point at which $x = x_1$. Now, T_1 passes through the point $(x_1, f(x_1))$. Applying the point-slope formula from Section 1.4, we see that the equation of T_1 is

$$y - f(x_1) = f'(x_1)(x - x_1).$$

Now, the point $(x_2, 0)$ is on T_1 and so its coordinates must satisfy the equation of T_1. Thus, setting $x = x_2$ and $y = 0$, we have

$$0 - f(x_1) = f'(x_1)(x_2 - x_1),$$

or

$$-f(x_1) = f'(x_1)(x_2 - x_1).$$

Dividing both sides by $f'(x_1)$ (recall that $f'(x_1) \neq 0$), we get

$$\frac{-f(x_1)}{f'(x_1)} = x_2 - x_1.$$

Solving for x_2, we get

$$x_2 = x_1 - \frac{f(x_1)}{f'(x_1)}. \qquad \boxed{CSS}$$

In general, we have the following:

$$x_{n+1} = x_n - \frac{f(x_n)}{f'(x_n)}. \qquad (1)$$

Example 1 Use Newton's method with $x_1 = 1.5$ to solve $x^2 = x$.

Solution To apply Newton's method, we must write the equation in the form $f(x) = 0$. So we write $x^2 - x = 0$. If we put $f(x) = x^2 - x$, then the equation takes the form $f(x) = 0$ and we can apply formula (1). Since $f(x) = x^2 - x$, we get $f'(x) = 2x - 1$. So formula (1) becomes

$$x_{n+1} = x_n - \frac{f(x_n)}{f'(x_n)} = x_n - \frac{x_n^2 - x_n}{2x_n - 1}.$$

If $n = 1$, we get

$$x_2 = x_1 - \frac{x_1^2 - x_1}{2x_1 - 1}$$

and, with $x_1 = 1.5$,

$$x_2 = 1.5 - \frac{(1.5)^2 - 1.5}{2(1.5) - 1} = 1.5 - \frac{2.25 - 1.5}{3 - 1}$$

$$= 1.5 - \frac{0.75}{2} = 1.5 - 0.375 = 1.125.$$

Then

$$x_3 = x_2 - \frac{x_2^2 - x_2}{-2x_2 - 1} = 1.125 - \frac{(1.125)^2 - (1.125)}{2(1.125) - 1}$$

$$= 1.125 - \frac{1.265625 - 1.125}{2.25 - 1} = 1.125 - \frac{0.140625}{1.25}$$

or

$$x_3 = 1.125 - 0.1125 = 1.0125.$$

Then

$$x_4 = 1.0125 - \frac{(1.0125)^2 - 1.0125}{2(1.0125) - 1}$$

$$\approx 1.0125 - 0.1234756$$

$$\approx 1.000152,$$

and

$$x_5 \approx 1.000152 - \frac{(1.000152)^2 - (1.000152)}{2(1.000152) - 1}$$

$$\approx 1.0000000.$$

With $x_5 = 1$ (correct to seven decimal places), we get $x_6 = 1, \ldots$, and we see that 1 is a solution to $f(x) = 0$.

Note the graph of $f(x) = x^2 - x$ below. It's a parabola and the initial guess x_1, the tangent line T_1, and x_2 are all shown. If we pick x_1 greater than 0.5, then the iterates given by Newton's method will converge to 1. If x_1 is less than 0.5, then the iterates will converge to 0. Note that we may not choose $x_1 = 0.5$ since $f'(x_1) = 2x_1 - 1 = 2(0.5) - 1 = 0$ and the tangent line at $x = 0.5$ is horizontal.

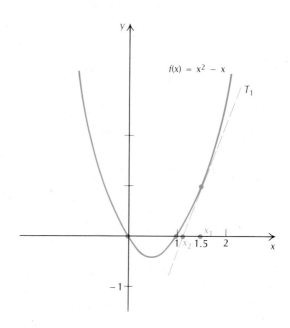

1. Use Newton's method with $x_1 = \frac{1}{3}$ to solve $x^2 = x$.

DO EXERCISE 1.

Example 2 Solve $\cos x = x$.

Solution For convenience, we write $x - \cos x = 0$ and let $f(x) = x - \cos x$. Then $f'(x) = 1 + \sin x$ and formula (1) becomes

$$x_{n+1} = x_n - \frac{x_n - \cos(x_n)}{1 + \sin(x_n)}.$$

If $x_1 = 0.5$, then

$$x_2 = x_1 - \frac{x_1 - \cos(x_1)}{1 + \sin(x_1)}$$

$$= 0.5 - \frac{(0.5) - \cos(0.5)}{1 + \sin(0.5)}$$

$$= 0.5 - \frac{0.5 - (0.8775826)}{1 + 0.4794255}$$

$$= 0.5 - \frac{-0.3775826}{1.4794255}$$

$$\approx 0.5 + 0.2552224$$

$$\approx 0.7552224.$$

So,

$$x_3 \approx 0.7552224 - \frac{(0.7552224) - \cos(0.7552224)}{1 + \sin(0.7552224)}$$

$$\approx 0.7552224 - \frac{0.0271033}{1.685451}$$

$$\approx 0.7391417.$$

Continuing, we have

$$x_4 \approx 0.7391417 - \frac{(0.7391417) - \cos(0.7391417)}{1 + \sin(0.7391417)}$$

$$\approx 0.7391417 - \frac{0.000094672}{1.6736538}$$

$$\approx 0.7391417 - 0.0000565661$$

$$\approx 0.7390851,$$

which is correct to seven decimal places, as we saw in Example 1 of Section 11.1.

2. Use Newton's method to solve $\cos x = 2x$. Check your answer.

Be sure to note the difference in the speed of convergence of the two methods. Newton's method generally gives much faster convergence than simple iteration does.

DO EXERCISE 2.

Newton's method doesn't always work. One equation on which it must fail is $x^2 + 1 = 0$, because the equation has no real solutions.

Even if the equation has a solution, though, the iterates generated by Newton's method may not converge to it. Consider the graph below. The function has no particular formula—call it $f(x)$.

The equation $f(x) = 0$ has exactly one solution. But if x_1 is chosen as shown, then x_2 will be farther from $\bar{x}$ than x_1 is. In fact, it appears from the graph that successive estimates will get worse and worse. Note that if the initial guess had been a_1 instead of x_1, the successive approximations would seem to converge rapidly to $\bar{x}$.

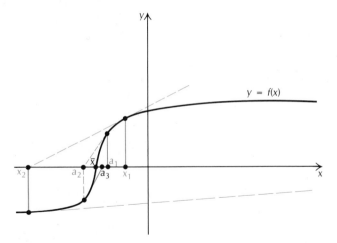

The following theorem gives us a way to make a correct choice.

THEOREM 2 Newton's Method

If there exists an interval containing a solution $\bar{x}$ to $f(x) = 0$ on which $f(x)$, $f'(x)$, and $f''(x)$ are continuous, $f'(x) \neq 0$, and

$$\left| \frac{f(x)f''(x)}{(f'(x))^2} \right| < 1,$$

and if x_1 is chosen in this interval, then the sequence given by

$$x_{n+1} = x_n - \frac{f(x_n)}{f'(x_n)}$$

converges to $\bar{x}$.

We omit the proof. To see how to apply it, consider the following example.

Example 3 Use Newton's method to solve $x^2 - 3 = 0$.

Solution We set $f(x) = x^2 - 3$. Then $f(1) = (1)^2 - 3 = -2$ and $f(2) = (2)^2 - 3 = 1$, so that the equation $f(x) = 0$ has a solution $\overline{x}$ in the interval from 1 to 2.

In this case, $f'(x) = 2x$ and $f''(x) = 2$ so that

$$\frac{f(x)f''(x)}{(f'(x))^2} = \frac{(x^2 - 3)(2)}{(2x)^2}$$

$$= \frac{2x^2 - 6}{4x^2} = \frac{1}{2} - \frac{3}{2x^2}.$$

Note that this expression is always less than 1. To keep this ratio between -1 and 1, as the theorem requires, we need only choose x so that

$$-1 < \frac{2x^2 - 6}{4x^2}$$

or

$$-4x^2 < 2x^2 - 6.$$

Rearranging terms, we get

$$6 < 2x^2 + 4x^2,$$

or

$$6 < 6x^2,$$

which is satisfied if x is greater than 1 (or less than -1). Now $f'(x) = 2x$ is 0 only when $x = 0$. So, if we choose x_1 to be greater than 1, then all the requirements of the theorem are met and the iterates will converge to the solution $\overline{x}$, which lies between 1 and 2.

If $x_1 = 2$, for example, then we have

$$x_{n+1} = x_n - \frac{x_n^2 - 3}{2x_n}$$

$$= \frac{2x_n^2}{2x_n} - \frac{x_n^2 - 3}{2x_n}$$

$$= \frac{2x_n^2 - x_n^2 + 3}{2x_n} = \frac{x_n^2 + 3}{2x_n}.$$

3. Using Newton's method with $x_1 = -2$, solve $x^3 - 3x + 3 = 0$. Check your answer.

So

$$x_2 = \frac{(2)^2 + 3}{2 \cdot 2} = \frac{4 + 3}{4} = 1.75.$$

Then

$$x_3 = \frac{(1.75)^2 + 3}{2 \cdot (1.75)} = \frac{3.0625 + 3}{3.5},$$

or

$$x_3 \approx 1.732143.$$

Continuing, we have

$$x_4 \approx \frac{(1.732143)^2 + 3}{2 \cdot (1.732143)} \approx 1.732051.$$

Another iteration yields $x_5 \approx x_4$. If we put $\overline{x} = 1.732051$, then $(\overline{x})^2 \approx 3$.

For this particular function, the iterates given by Newton's method will converge for any initial choice x_1 except $x_1 = 0$ (the one value of x for which $f'(x) = 0$). Note that if $x_1 > 0$, the iterates will converge to $\sqrt{3}$, whereas if $x_1 < 0$, they will converge to $-\sqrt{3}$ (see the following figure).

4. Use Newton's method to solve $x^3 = 10$. Check your answer.

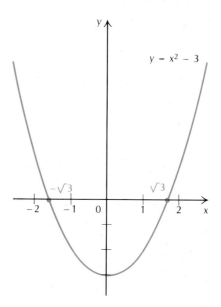

DO EXERCISES 3 AND 4.

EXERCISE SET 11.2

Use the indicated choice of x_1 and Newton's method to solve the given equation. Check your answers.

1. $x = \frac{1}{3}x^2 + \frac{1}{3}$; $x_1 = 0$

2. $x = \frac{1}{3}x^2 + \frac{1}{3}$; $x_1 = 3$ **CSS**

3. $x = \sin x + \frac{1}{2}$ (x in radians); $x_1 = 1$

4. $x = 2 \cos x$; $x_1 = 1$

5. $x^3 - 3x + 3 = 0$; $x_1 = -3$

6. $x^4 - 4x + 1 = 0$; $x_1 = 0$

Use Newton's method to find all real solutions to the given equation.

7. $x^3 = e^{-x}$

8. $x^4 - x - 2 = 0$

9. $2x^3 + 4x^2 - 6x - 1 = 0$

10. $x^4 + 4x^2 + 1 = 0$

11. $x^7 - 7x + 8 = 0$

12. $3 \cos x - 7x = 0$

13. $\cos x - 2x^2 = 0$

14. $x + \ln x = 0$

15. a) Show that applying Newton's method to
$$x^2 - a = 0$$
yields the iteration scheme
$$x_{n+1} = \frac{x_n}{2} + \frac{a}{2x_n},$$
provided $x_n \neq 0$.

    b) Use the formula above to approximate $\sqrt{19}$.

16. a) Sketch the graph of $y = f(x) = \tan^{-1} x$ to show that the equation $\tan^{-1} x = 0$ has only the solution $\bar{x} = 0$.

    b) Calculate
$$\left| \frac{f(x)f''(x)}{(f'(x))^2} \right|.$$

    c) Use Newton's method and $x_1 = 4$ to try to find the solution.

    d) Repeat (c) using $x_1 = 0.5$.

    e) Explain the different results in parts (c) and (d) by referring to Theorem 2 and (b) above.

17. a) Use Newton's method to show that the iterates
$$x_{n+1} = \frac{k-1}{k} x_n + \frac{a}{kx_n^{k-1}} \qquad (x_1, a > 0)$$
converge to $a^{1/k}$.

    b) Use the algorithm above to calculate $\sqrt[10]{20}$.

OBJECTIVES

You should be able to

a) **Set up the equation for the yield rate for a (finite) sequence of investments and returns on those investments.**

b) **Solve the equations resulting from (a).**

11.3 APPLICATION: YIELD RATE (INTERNAL RATE OF RETURN)

Which is the better deal—a payment of $1000 immediately or a payment of $1100 one year from today? (It's understood here that inflation and taxes are not to be taken into account in our analysis.) Consider the following. If we deposit $1000 in a savings account (or money-market fund) at 12% interest, then, at the end of one year, we will have $1000 plus 12% of $1000, that is, $1120. In this situation (12% interest), we would be better off taking the $1000 now and depositing it at 12% interest, since, at the end of one year, we would have $1120

1. If $8000 is deposited in a savings account at $8\frac{1}{4}\%$ interest, find its accumulated value after three years.

instead of $1100. On the other hand, if the interest rate is 8%, then our $1000 would earn $(0.08) \times (\$1000) = \$80$ interest. At the end of one year, we would have only $1080. In this case (8% interest), we would be better off taking $1100 at the end of one year.

A little thought shows that if the interest rate i is greater than 10%, then we should take the $1000 immediately. If i is less than 10%, we should take the $1100 at the end of the year. If i equals 10%, then the choices are equivalent.

If the interest rate is 8%, that is, if $i = 0.08$, then $1000 will accumulate to $1080. A deposit of P will accumulate to $P + \$(0.08)P = \$(1.08)P$. In general, P deposited at a rate of interest i (compounded annually) will accumulate to $P(1 + i)$ at the end of one year and $P(1 + i)^n$ at the end of n years.

Example 1 If $1400 is deposited in a money-market fund at $11\frac{1}{2}\%$ interest (compounded annually), find its accumulated value after two years.

Solution Here, $P = 1400$, $i = 0.115$, and $n = 2$. If A denotes the accumulated value, then

$$A = (1400)(1 + 0.115)^2$$

$$= (1400)(1.115)^2$$

$$= \$1740.52.$$

DO EXERCISE 1.

To find the *present value* of an amount A payable n years from today at a rate of interest i, we solve the equation $P(1 + i)^n = A$ for P. In other words, we find the amount P, which, if deposited at a rate of interest i, would accumulate to A in n years. Dividing both sides by $(1 + i)^n$, we have

$$P = \frac{A}{(1 + i)^n},$$

or

$$P = A(1 + i)^{-n}.$$

DEFINITION

The *present value* *P of a payment *of A to be made *n* years from today at a rate of interest *i* (compounded annually) is

$$P = A(1 + i)^{-n}.$$

2. Find the present value of $1100, payable one year from today, at a rate of interest of 12%.

Unless otherwise indicated, assume in this section that interest is compounded annually.

Example 2 Find the present value of $1100, payable one year from today, at a rate of interest of 8%.

Solution Here, $A = 1100$, $i = 0.08$, and $n = 1$. Thus

$$P = (1100)(1 + 0.08)^{-1} = (1100)(1.08)^{-1}$$
$$= (1100)(0.9259259) \approx \$1018.52.$$

DO EXERCISE 2.

If we revisit the problem of the present payment of $1000 versus the payment of $1100 one year from today, we see, in light of Example 2 and Margin Exercise 2, that the present value of the future payment of $1100 one year from today will be less than $1000 if the rate of interest is greater than 10% and more than $1000 if the rate of interest is less than 10%. Although the actual numbers are different, the results of the analysis are identical.

We define the *present value* of a series of future payments to be the *sum of the present values of the payments.*

3. An annuity will pay $250 at the end of each year for the next four years. Find the present value of the annuity if the interest rate is 12%.

Example 3 An annuity will pay $1000 at the end of each year for the next three years. Find the present value of the annuity if the interest rate is 8%.

Solution The present value of the first payment (one year from today) is $(1000)(1.08)^{-1}$. The present value of the second payment is $(1000)(1.08)^{-2}$ and the present value of the third payment is $(1000)(1.08)^{-3}$. The present value P of the annuity is the sum

$$P = (1000)(1.08)^{-1} + (1000)(1.08)^{-2} + (1000)(1.08)^{-3}$$
$$= 925.93 + 857.34 + 793.83 = \$2577.10.$$

DO EXERCISE 3.

In the preceding problems, we were given the dollar amount of a series of future payments and the interest rate and were asked to find the *present value* of the future payments. In practice, we may be given the dollar amount of the future payments and the present value of the future payments and asked to find an interest rate i. In this case, we can call i the *yield rate*, or the *internal rate of return*, of the initial investment.

Example 4　An investor makes two investments of $1000 each. The first returns $1050 at the end of one year, while the second returns $1210 at the end of two years. Find the yield rate i.

Solution　Note what happens if we consider the investments separately. The first investment of $1000 returns $1050 at the end of one year. Solving the equation

$$1000 = 1050(1 + i_1)^{-1}$$

for i_1 gives us $i_1 = 5\%$. The second investment of $1000 returns $1210 at the end of two years. Solving

$$1000 = 1210(1 + i_2)^{-2}$$

for i_2 gives us $i_2 = 10\%$. We have two interest rates, each coming as a part of the investment. What we want is one interest rate that represents the entire investment. This is the *yield rate*.

　　We define the yield rate i as the value of i at which the sum of the present value of the returns equals the sum of the present value of the investments. In this case, the present value of the investments is $2000. The present value of the returns is $(1050)(1 + i)^{-1}$ and $(1210)(1 + i)^{-2}$, respectively. Therefore, the value of i that we are looking for is the solution to

$$2000 = (1050)(1 + i)^{-1} + (1210)(1 + i)^{-2}.$$

　　To simplify the equation, let

$$v = \frac{1}{1 + i} = (1 + i)^{-1}.$$

Then we get

$$2000 = 1050v + 1210v^2,$$

a quadratic equation in v. Writing it in the form

$$1210v^2 + 1050v - 2000 = 0$$

and applying the quadratic formula gives

$$v = \frac{-1050 \pm \sqrt{(1050)^2 - 4(1210)(-2000)}}{2(1210)}$$

$$= \frac{-1050 \pm \sqrt{1{,}102{,}500 + 9{,}680{,}000}}{2420}$$

$$= \frac{-1050 \pm 3283.67}{2420}.$$

4. An investor invests a total of $4000 in two projects. The first returns $2000 at the end of one year, while the second returns $2500 at the end of two years. Find the investor's yield rate on this package.

We get $v_1 \approx 0.923005$ and $v_2 \approx -1.790773$. Note that even if i is the worst case possible (-100%, or $i = -1$, at which point we lose all our money), $1 + i$ is a nonnegative number. In this situation, then, negative values of v are irrelevant and v_2 can be discarded. To complete the problem, observe that since

$$v = \frac{1}{1 + i},$$

CSS

we can take reciprocals to get

$$\frac{1}{v} = 1 + i.$$

Then,

$$i = \frac{1}{v} - 1 = \frac{1}{v} - \frac{v}{v} = \frac{1 - v}{v}.$$

Here,

$$i \approx \frac{1 - 0.923005}{0.923005} \approx 0.083418.$$

The yield rate of this investment "package" is approximately 8.34%. We now have one interest rate that represents the entire investment.

DO EXERCISE 4.

Example 5 An investor buys an annuity for $2700. The annuity will return $1000 for three years, beginning one year from today. Find the yield rate on this investment.

Solution The present value of the three payments, $2700, is given by

$$2700 = 1000(1 + i)^{-1} + 1000(1 + i)^{-2} + 1000(1 + i)^{-3}.$$

Letting

$$v = \frac{1}{1 + i} = (1 + i)^{-1}$$

(v is sometimes called *the discount factor*), we see that the equation to be solved is

$$2700 = 1000v + 1000v^2 + 1000v^3,$$

or, if we rearrange and divide by 1000,

$$v^3 + v^2 + v - 2.7 = 0.$$

High Yields- Triple Tax-free!

Exempt from Federal, State and NY City taxes for New York residents:

High triple tax-free income—exempt from Federal, NY State and NY City taxes.

• Actively managed, widely diversified portfolio.
• No sales charge.
• No penalties on redemption.
• Complete liquidity.
• Free exchange between funds…even by phone.
• Initial investment: $2,500. Additions as low as $100.
• For information and a free tax rate chart, send coupon or call toll free number.

Dreyfus New York
Tax Exempt Bond Fund, Inc.

For more complete information, including management fee charges and expenses, obtain a Prospectus by sending this coupon. Read it carefully before you invest or send money.

Post Office Box 600
Middlesex, NJ 08846

FOR A PROSPECTUS
Call toll free, day or night, 7 days a week.
1-800-USA-LION

Name
Address
City
State   Zip
013M027   NY124

Does this yield rate conform to the notion of yield rate introduced in this section?

5. An investor pays $5000 for three annual payments of $2000 per year beginning one year from today. Find the yield rate on this annuity.

To solve this, we'll use Newton's method. First, though, observe that since i is generally a small positive number, $v = \frac{1}{1+i}$ will be close to 1 (and usually less than 1). So, $v_1 = 1$ is a good first approximation. Newton's method yields

$$v_{i+1} = v_i - \frac{f(v_i)}{f'(v_i)}, \qquad i = 1, 2, \ldots.$$

Since $f(v) = v^3 + v^2 + v - 2.7$, differentiation yields $f'(v) = 3v^2 + 2v + 1$. Thus

$$v_{i+1} = v_i - \frac{v^3 + v^2 + v - 2.7}{3v^2 + 2v + 1}.$$

If we let $v_1 = 1$, we have

$$v_2 = 1 - \frac{(1)^3 + (1)^2 + 1 - 2.7}{3(1)^2 + 2(1) + 1}$$

$$= 1 - \frac{1 + 1 + 1 - 2.7}{3 + 2 + 1}$$

$$= 1 - \frac{0.3}{6} = 1 - 0.05 = 0.95.$$

Using $v_2 = 0.95$ gives us

$$v_3 = 0.95 - \frac{(0.95)^3 + (0.95)^2 + (0.95) - 2.7}{3(0.95)^2 + 2(0.95) + 1}$$

or

$$v_3 = 0.95 - \frac{0.857375 + 0.9025 + 0.95 - 2.7}{2.7075 + 1.90 + 1}$$

$$= 0.95 - \frac{0.009875}{5.6075}$$

$$= 0.95 - 0.00176,$$

or

$$v_3 \approx 0.948239.$$

Using this value for v yields

$$i = \frac{1 - v}{v} \approx \frac{1 - 0.948239}{0.948239} \approx 0.0545865.$$

Therefore, $i \approx 5.46\%$ on this annuity.

DO EXERCISE 5.

It is interesting to compare the results of Examples 3 and 5. The series of payments is identical. Note that as the price (present value) of the annuity increased from $2577.10 to $2700, the yield rate *dropped* from 8% to roughly 5.46%. This illustrates the general principle that as interest rates rise, the prices of annuities (and bonds) fall, while as interest rates fall, the prices of annuities (and bonds) rise.

Note too that in Example 4, an initial investment with return payments after one and two years led to a quadratic equation in v, whereas in Example 5 an investment with return payments in one, two, and three years led to a cubic equation. The general situation of an initial investment followed by return payments at the end of the year for a period of n years will lead to a polynomial equation of degree n. If payments are made at times other than the end of the year, fractional exponents may be required (see Exercise Set 11.3). In any case, a technique such as Newton's method (interpolation in tables is another way) is generally needed.

The Method of Bisection

We revisit the equation $f(x) = x^3 - 3x + 3 = 0$, which we solved using Newton's method (in Margin Exercise 3 of Section 11.2). We note that the solution $\overline{x}$ to $f(x) = 0$ lies between -2 and -3. In fact,

$$f(-3) = (-3)^3 - 3(-3) + 3 = -15 \qquad \boxed{\text{CSS}}$$

and

$$f(-2) = (-2)^3 - 3(-2) + 3 = 1,$$

so that, by continuity, a solution to $f(x) = 0$ must lie between -2 and -3. Now, we bisect the interval from -3 to -2. The midpoint is -2.5. Thus,

$$f(-2.5) = (-2.5)^3 - 3(-2.5) + 3 = -5.125.$$

Since $f(-2.5)$ is negative, while $f(-2)$ is positive, there must be a root between -2.5 and -2 (there *may* also be a root (or even several roots) between -3 and -2.5, but the argument *guarantees* one between -2.5 and -2). We repeat the procedure. Bisecting the interval yields -2.25. We compute

$$f(-2.25) = (-2.25)^3 - 3(-2.25) + 3 = -1.640625,$$

which is negative. Arguing as before, we see that a solution to $f(x) = 0$ must lie between -2.25 and -2. Bisecting again, we arrive at -2.125. Calculating, we now have

$$f(-2.125) = (-2.125)^3 - 3(-2.125) + 3 \approx -0.22070313.$$

A solution lies between -2.125 and -2. Note that had we wanted to stop at this step, we could have used the midpoint of the interval, -2.0625, for $\bar{x}$ with the assurance that the error involved here is less than one-half the length of the interval (in this case, 0.0625).

The method of bisection is simple, but usually much too laborious for hand calculations. It is, however, ideal for high-speed computing machinery since it is easy to program and gives an error estimate at each stage of the calculation.

EXERCISE SET 11.3

Find the accumulated value of an initial amount P for n years at a rate of interest $i\%$ (compounded annually).

1. $P = 500, n = 7, i = 10\%$

2. $P = 500, n = 8, i = 10\%$

3. $P = 1000, n = 30, i = 12\%$

4. $P = 1000, n = 30, i = 10\%$

5. $P = 1000, n = 30, i = 8\%$

6. $P = 1000, n = 30, i = 6\%$

Find the present value of a payment of $\$A$ to be made n years from today at a rate of interest $i\%$ (compounded annually).

7. $A = \$1000, n = 3, i = 10\%$

8. $A = \$1000, n = 5, i = 10\%$

9. $A = \$1000, n = 20, i = 10\%$

10. $A = \$1000, n = 10, i = 12\%$

11. $A = \$1000, n = 10, i = 9\%$

12. $A = \$1000, n = 10, i = 6\%$

13. $A = \$10{,}062.66, n = 30, i = 8\%$

14. $A = \$974.36, n = 7, i = 10\%$

An annuity pays $\$R$ per year for n years. Find the present value of the annuity if the rate of interest is $i\%$ (compounded annually).

15. $R = \$1000, n = 2, i = 12\%$

16. $R = \$1000, n = 2, i = 9\%$

17. $R = \$1000, n = 2, i = 6\%$

18. $R = \$1000, n = 2, i = 19\%$

19. $R = \$200, n = 5, i = 10\%$

20. $R = \$5000, n = 6, i = 11\frac{1}{2}\%$

21. An investor makes two different investments of $5000, and receives a payment of $6000 after one year on the first and $5800 after two years on the second. Find the equation that determines the yield rate i (in terms of the discount factor $v = (1 + i)^{-1}$) for this pair of investments. Then calculate i.

22. An investor makes two investments, one of $3000, the second of $5000. The first returns $3500 after one year, the second $6500 after four years. Find the equation that determines the yield rate and use Newton's method to solve it.

CSS

23. An investor invests a total of $9000 in two projects, receiving return payments of $4000 after two years and $7000 after five years. Find the internal rate of return on the investment.

24. An investor pays $3000 for four annual payments of $1000, beginning one year from today. Find the yield rate of this annuity. Check your answer.

25. An investor must choose between two annuities. The first pays $1000 a year for three years and sells for $2400. The second pays $800 a year for five years and sells for $2900. Which should the investor buy?

26. An investor makes an initial investment of $2000 in a company, receives a payment of $4300 one year later, and makes another investment of $2310 in the company two years after the initial investment. Set up the equation for the investor's yield rate and show that the equation has two solutions, that is, that the yield rate is not well defined.

28. An investor makes an initial investment of P and receives payment of $R_1, R_2, \ldots, R_k$ dollars after $1, 2, \ldots, k$ years. Find the equation that determines the investor's yield rate and use calculus to show that there is *exactly one* positive number v that solves the equation.

27. Repeat Exercise 26 given a second investment of $2500 (instead of $2310).

OBJECTIVE

You should be able to find a numerical estimate for $\int_a^b f(x)\, dx$ using:

a) **Riemann sums;**

b) **the Trapezoidal Rule; and**

c) **Simpson's Rule.**

11.4 NUMERICAL INTEGRATION

In Chapters 5 and 6, we saw that many interesting problems in business, medicine, probability, and the physical sciences lead to the evaluation of definite integrals, that is, expressions of the form

$$\int_a^b f(x)\, dx.$$

In many cases, $f(x)$ has an antiderivative that we can find and evaluate. Thus, for example, $\int_0^5 3x^2\, dx = [x^3]_0^5 = 5^3 - 0^3 = 125$. In many other cases, though, this won't be so. As we saw in Chapter 6, this happens with the function $f(x) = e^{-(1/2)x^2}$, a function that occurs naturally in many probability applications, but has no elementary antiderivative.

In cases where $f(x)$ has no elementary antiderivative, we may have to evaluate $\int_a^b f(x)\, dx$ approximately. In this section, we will consider several approximation schemes for estimating integrals.

Let $f(x)$ be continuous on $[a, b]$ and consider $\int_a^b f(x)\, dx$. If we subdivide $[a, b]$ into n equal pieces (see the figure below) of length Δx, then

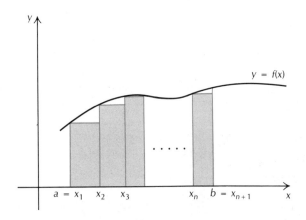

$\Delta x = (b - a)/n$. Let $a = x_1$ and choose $x_2, x_3, \ldots, x_n, x_{n+1}$ so that $x_{i+1} - x_i = \Delta x$ for $i = 1, 2, \ldots, n$, that is, so that the points are equally spaced. Just as in Section 5.7, we see that the area of the first rectangle is $f(x_1) \cdot \Delta x$, and so forth, so that the sum of the areas of the n inscribed rectangles indicated in the figure above is

$$\sum_{i=1}^{n} f(x_i) \, \Delta x.$$

In Section 5.7, we saw that

$$\int_{a}^{b} f(x) \, dx = \lim_{n \to \infty} \sum_{i=1}^{n} f(x_i) \, \Delta x.$$

For large values of n, then, we expect that $\sum_{i=1}^{n} f(x_i) \, \Delta x$ can yield a good approximation to $\int_{a}^{b} f(x) \, dx$.

Example 1 Approximate $\int_{0}^{1} (1 + x^2) \, dx$ using $\sum_{i=1}^{n} f(x_i) \, \Delta x$ with $n = 5$ and compare your approximation with the exact answer.

Solution The integral represents the area shown below.

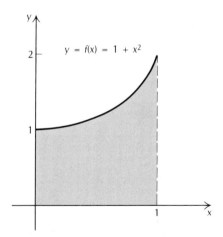

We can find its exact value as follows:

$$\int_{0}^{1} (1 + x^2) \, dx = \left[x + \frac{1}{3} x^3 \right]_{0}^{1}$$

$$= \left(1 + \frac{1}{3}(1)^3 \right) - \left(0 + \frac{1}{3}(0)^3 \right)$$

$$= 1 + \frac{1}{3} = \frac{4}{3} \approx 1.333333.$$

If n = 5, then $\Delta x = (1 - 0)/5 = 1/5 = 0.2$. The points of subdivision are $x_1 = 0$, $x_2 = 0.2$, $x_3 = 0.4$, $x_4 = 0.6$, $x_5 = 0.8$, and $x_6 = 1$. Since $f(x) = 1 + x^2$, we have

$$f(x_1) = f(0) = 1 + (0)^2 = 1,$$

$$f(x_2) = f(0.2) = 1 + (0.2)^2 = 1.04,$$

$$f(x_3) = f(0.4) = 1 + (0.4)^2 = 1.16,$$

$$f(x_4) = f(0.6) = 1 + (0.6)^2 = 1.36,$$

$$f(x_5) = f(0.8) = 1 + (0.8)^2 = 1.64.$$

Therefore,

$$\sum_{i=1}^{n} f(x_i)\, \Delta x = f(x_1)\, \Delta x + f(x_2)\, \Delta x + \cdots + f(x_5)\, \Delta x$$

$$= [f(x_1) + f(x_2) + \cdots + f(x_5)]\, \Delta x$$

$$= [1 + (1.04) + (1.16) + (1.36) + (1.64)] \cdot \frac{1}{5}$$

$$= [6.2](0.2)$$

$$= 1.24.$$

This sum equals the sum of the areas of the rectangles shown in the following figure.

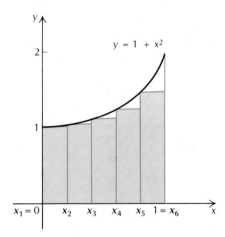

The error involved in our approximation is about $(1.333333) - (1.24) \approx 0.09$.

1. Approximate $\int_0^1 (1 + x)\, dx$ using $\sum_{i=1}^{n} f(x_i)\, \Delta x$ with $n = 4$ and compare your approximation with the exact answer.

2. Approximate $\int_0^1 (1 + x)\, dx$ using $\sum_{i=1}^{n} f(x_{i+1})\, \Delta x$ with $n = 4$ and compare your approximation with the exact answer.

DO EXERCISE 1.

In Example 1, we used the values of $f(x)$ at the left-hand endpoints of the subintervals, $f(x_1), f(x_2), \ldots, f(x_5)$, for the altitudes of the approximating rectangles. If, instead, we had used the values of $f(x)$ at the right-hand endpoints of the subintervals, $f(x_2), f(x_3), \ldots, f(x_6)$, we would have obtained the rectangles shown below.

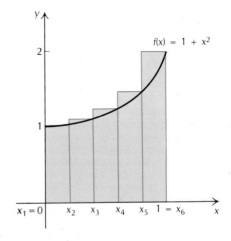

Example 2 Approximate $\int_0^1 (1 + x^2)\, dx$ by using $\sum_{i=1}^{n} f(x_{i+1})\, \Delta x$ with $n = 5$, as shown above. Compare your result with the exact answer.

Solution We have

$$\sum_{i=1}^{5} f(x_{i+1})\, \Delta x = [f(x_2) + f(x_3) + \cdots + f(x_6)]\, \Delta x$$

$$= [(1.04) + (1.16) + (1.36) + (1.64) + 2] \cdot \frac{1}{5}$$

$$= [7.2](0.2)$$

$$= 1.44.$$

The error involved is about $(1.44) - (1.333333) \approx 0.106667$.

DO EXERCISE 2.

In each of Examples 1 and 2, we approximated $\int_0^1 (1 + x^2)\, dx$ by using the sum of areas of rectangles. In Example 1, we used the values of $f(x)$

at the left-hand endpoints for the altitudes of the rectangles, while in Example 2 we used the right-hand endpoints. Both cases are examples of *approximation by Riemann sums*. In the general case, we pick any number c_1 in $[x_1, x_2]$ and use $f(c_1)$ for the altitude of the first rectangle (see the following figure).

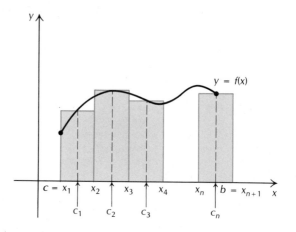

We pick c_2 between x_2 and x_3 and use $f(c_2) \, \Delta x$ for the area of the second rectangle. Continuing, we pick $c_3, c_4, \ldots, c_n$ so that c_i is between x_i and x_{i+1} for $i = 1, 2, \ldots, n$. Then we can approximate $\int_a^b f(x) \, dx$ by the sum $\sum_{i=1}^n f(c_i) \, \Delta x$. This sum is the sum of the areas of the rectangles shown in the figure above. Any sum of the form

$$\sum_{i=1}^n f(c_i) \, \Delta x$$

is called a *Riemann sum* for $\int_a^b f(x) \, dx$.

The sums $\sum_{i=1}^n f(x_i) \, \Delta x$ (using the left-hand endpoints of the subintervals for c_i) and $\sum_{i=1}^n f(x_{i+1}) \, \Delta x$ (using the right-hand endpoints) are both special cases of Riemann sums. We might use the midpoints of the subintervals or any other convenient choice of $c_1, c_2, \ldots, c_n$.

In general, though, Riemann sums are not an efficient way to approximate integrals. For example, suppose that we were to approximate $\int_0^1 (1 + x^2) \, dx$ using left-hand endpoints and $n = 100$ (think of the computations required!). It turns out that we would obtain

$$\sum_{i=1}^{100} f(x_i) \, \Delta x = 1.32835.$$

Note that even with 100 subintervals, the error (which is about $1.333333 - 1.32835 \approx 0.005$) is rather large.

The Trapezoidal Rule

There is an easy and obvious way of improving on approximation by Riemann sums. As before, we partition $[a, b]$ into n equal subintervals. Instead of using rectangles, however, we use trapezoids, as shown below.

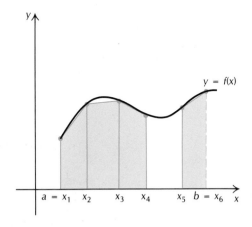

Note that the sum of the areas of the trapezoids in the figure, although not equal to $\int_a^b f(x)\, dx$, certainly appears to yield a better approximation than the inscribed rectangles would.

Note that the area A of a trapezoid whose base is b and whose altitudes are h_1 and h_2 (see the figure) is

$$\begin{pmatrix} \text{the area of} \\ \text{the rectangle} \end{pmatrix} + \begin{pmatrix} \text{the area of} \\ \text{the triangle} \end{pmatrix} = bh_1 + \frac{1}{2} b(h_2 - h_1).$$

Then,

$$A = bh_1 + \frac{1}{2} b(h_2 - h_1)$$

$$= b\left[h_1 + \frac{1}{2}(h_2 - h_1) \right]$$

$$= b\left[h_1 + \frac{1}{2}h_2 - \frac{1}{2}h_1 \right]$$

$$= b\left[\frac{1}{2}h_1 + \frac{1}{2}h_2 \right]$$

$$= \frac{1}{2}b[h_1 + h_2].$$

Applying these calculations to the situation shown in the figure, we see that each of the trapezoids will have base equal to Δx. The first trapezoid has altitudes $f(x_1)$ and $f(x_2)$, the values of $f(x)$ at the right- and left-hand endpoints. Thus the area of the first trapezoid will be $\frac{1}{2}(f(x_1) + f(x_2)) \cdot \Delta x$. The area of the second trapezoid will be $\frac{1}{2}(f(x_2) + f(x_3)) \cdot \Delta x$. The area of the last (nth) trapezoid will be $\frac{1}{2}(f(x_n) + f(x_{n+1})) \cdot \Delta x$. Let T_n denote the sum of the areas of all n trapezoids. Then,

$$T_n = \frac{1}{2}(f(x_1) + f(x_2)) \cdot \Delta x + \frac{1}{2}(f(x_2) + f(x_3)) \cdot \Delta x$$

$$+ \cdots + \frac{1}{2}(f(x_n) + f(x_{n+1})) \cdot \Delta x$$

$$= \frac{1}{2}[(f(x_1) + f(x_2)) + (f(x_2) + f(x_3)) + \cdots + (f(x_n) + f(x_{n+1}))] \cdot \Delta x$$

$$= \frac{1}{2}\left[f(x_1) + 2f(x_2) + 2f(x_3) + \cdots + 2f(x_n) + f(x_{n+1})\right] \cdot \Delta x$$

$$= \left[\frac{1}{2}f(a) + f(x_2) + f(x_3) + \cdots + f(x_n) + \frac{1}{2}f(b)\right] \cdot \Delta x.$$

Note that the endpoints a and b of the interval receive "special treatment" in the formula since they occur only once as endpoints of subintervals, whereas the other points occur twice.

Example 3 Approximate $\int_0^1 (1 + x^2)\, dx$ using the Trapezoidal Rule with $n = 5$, that is, T_5. Compare your answer with the exact value of the integral.

Solution As in the solution to Example 1, we have $x_1 = 0$, $x_2 = 0.2$, $x_3 = 0.4$, $x_4 = 0.6$, $x_5 = 0.8$, and $x_6 = 1$. Also,

$$f(x_1) = 1 + (x_1)^2 = 1 + (0)^2 \ \ = 1.00,$$

$$f(x_2) = 1 + (x_2)^2 = 1 + (0.2)^2 = 1.04,$$

$$f(x_3) = 1 + (x_3)^2 = 1 + (0.4)^2 = 1.16,$$

$$f(x_4) = 1 + (x_4)^2 = 1 + (0.6)^2 = 1.36,$$

$$f(x_5) = 1 + (x_5)^2 = 1 + (0.8)^2 = 1.64,$$

and

$$f(x_6) = 1 + (x_6)^2 = 1 + (1)^2 = 2.00.$$

3. Approximate $\int_0^1 (1 + x)\, dx$ using the Trapezoidal Rule with $n = 4$, that is, T_4. Compare your result with the exact answer.

Thus,

$$T_5 = \left[\frac{1}{2}f(x_1) + f(x_2) + \cdots + f(x_5) + \frac{1}{2}f(x_6)\right] \cdot \Delta x$$

$$= \left[\frac{1}{2}(1) + (1.04) + (1.16) + (1.36) + (1.64) + \frac{1}{2}(2)\right] \cdot \frac{1}{5}$$

$$= [6.70](0.2) = 1.34.$$

The error involved is about $(1.34) - (1.333333) = 0.006667$. Note that this compares with the error we get with $n = 100$ using Riemann sums with left-hand endpoints.

DO EXERCISE 3.

In Margin Exercise 3, the Trapezoidal Rule is exact. This is always the case if $f(x) = c_0 + c_1 x$ is a polynomial of degree 1. In fact, we have the following theorem.

THEOREM 3   Trapezoidal Rule

> **If $f(x)$ and its first and second derivatives are continuous on $[a, b]$, and if**
>
> $$T_n = \left[\frac{1}{2}f(x_1) + f(x_2) + \cdots + f(x_n) + \frac{1}{2}f(x_{n+1})\right] \cdot \Delta x,$$
>
> **then,**
>
> $$\int_a^b f(x)\, dx \approx T_n$$
>
> **and the error involved in this approximation is**
>
> $$E = \frac{b - a}{12} f''(c)\, (\Delta x)^2,$$
>
> **where $a < c < b$.**

If we let $K = \dfrac{b - a}{12} f''(c)$ and observe that K is a constant, we see that the error is $K(\Delta x)^2$. The error is proportional to $1/n^2$.

Example 4   Use the error estimate in Theorem 3 to calculate the error in approximating $\int_0^1 (1 + x^2)\, dx$ by T_5 in Example 3.

Solution   We have $a = 0$, $b = 1$, $\Delta x = \frac{1}{5}$, and $f(x) = 1 + x^2$. So

4. Use the error estimate in Theorem 3 to calculate the error in approximating $\int_0^1 (1 + x)\, dx$ by T_4 in Margin Exercise 3.

$f'(x) = 2x$ and $f''(x) = 2$. Thus $f''(c) = 2$, independent of c. The error E is then given by

$$\frac{b - a}{12} f''(c) (\Delta x)^2 = \frac{1 - 0}{12} \cdot 2 \cdot \left(\frac{1}{5}\right)^2$$

$$= \frac{1}{12} \cdot 2 \cdot \frac{1}{25} = \frac{1}{150} \approx 0.006667.$$

From Example 3, we know that this is correct.

DO EXERCISE 4.

Margin Exercise 3 and Example 3 suggest that the Trapezoidal Rule is simply the average of the Riemann sums corresponding to the right- and left-hand endpoints, and that is correct.

Example 5 Use T_{10} to approximate $\int_1^2 (1/x)\, dx$. Estimate the error involved.

Solution The points of subdivision are $x_1 = 1$, $x_2 = 1.1$, $x_3 = 1.2$, $,\ldots, x_{10} = 1.9$, and $x_{11} = 2.0$. Also, $\Delta x = 0.1$. Applying the Trapezoidal Rule gives us

$$T_{10} = \left[\frac{1}{2} \cdot \frac{1}{1} + \frac{1}{1.1} + \frac{1}{1.2} + \cdots + \frac{1}{1.9} + \frac{1}{2} \cdot \frac{1}{2.0}\right](0.1)$$

$$= [6.9377140](0.1) = 0.69377140.$$

According to Theorem 3, the error, call it E, is given by

$$E = \frac{b - a}{12} \cdot f''(c) \cdot (\Delta x)^2,$$

where $a < c < b$. Here, $a = 1$, $b = 2$, and $\Delta x = 0.1$. Since $f(x) = 1/x = x^{-1}$, we have $f'(x) = -1x^{-2}$ and $f''(x) = (-1)(-2)x^{-3} = 2/x^3$. So we get

$$E = \frac{2 - 1}{12} \cdot \frac{2}{c^3} \cdot (0.1)^2 = \frac{1}{12} \cdot \frac{2}{c^3} \cdot (0.01) = \frac{1}{600} \cdot \frac{1}{c^3},$$

where $1 < c < 2$. Since $1 < c < 2$, we have $1 < c^3 < 8$ and $1/8 < 1/c^3 < 1$. Thus $E < 1/600 \approx .0016667$. In fact,

$$\int_1^2 \frac{1}{x}\, dx = [\ln x]_1^2$$

$$= \ln 2 - \ln 1 = \ln 2.$$

5. Use T_{10} to approximate $\int_0^1 x^3 \, dx$. Estimate the error involved.

So, to seven decimal places,

$$\int_1^2 \frac{1}{x} \, dx = \ln 2 \approx 0.6931472.$$

The actual error is about 0.0006242, which, of course, is less than 0.00166667, the error estimate obtained above.

DO EXERCISE 5.

Simpson's Rule

In approximating $\int_a^b f(x) \, dx$ by Riemann sums, we partitioned $[a, b]$ into n equal subintervals and used $f(c_i)$ to approximate $f(x)$ on the ith subinterval. In approximating the same integral using the Trapezoidal Rule, we approximated $f(x)$ on the ith subinterval by means of the line segment joining $(x_i, f(x_i))$ to $(x_{i+1}, f(x_{i+1}))$. This line segment is the top of the ith trapezoid in the figure on page 655.

To obtain Simpson's Rule, we partition $[a, b]$ into n equal subintervals and fit a parabola to $f(x)$ on each subinterval. Since it takes three points to uniquely determine a parabola, we use $(x_i, f(x_i))$, $(x_{i+1}, f(x_{i+1}))$ and the point on $y = f(x)$ corresponding to the midpoint of the subinterval,

$$\left(\frac{x_i + x_{i+1}}{2}, \; f\left(\frac{x_i + x_{i+1}}{2} \right) \right),$$

as shown in the following figure.

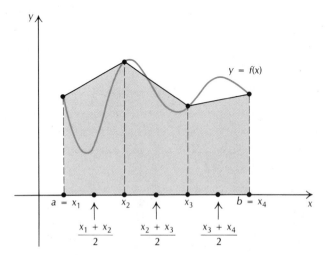

Calculating the area under each parabolic arch, summing the areas, and simplifying yields

$$S_n = \frac{1}{6}\left[f(x_1) + 4f\left(\frac{x_1 + x_2}{2}\right) + 2f(x_2) + 4f\left(\frac{x_2 + x_3}{2}\right) + 2f(x_3)\right.$$

$$\left. + \cdots + 2f(x_n) + 4f\left(\frac{x_n + x_{n+1}}{2}\right) + f(x_{n+1})\right] \cdot \Delta x.$$

We reiterate this result and the error estimate in the following theorem.

THEOREM 4   Simplson's Rule   `CSS`

If $f(x)$ and its first four derivatives are continuous on $[a, b]$, and if

$$S_n = \frac{1}{6}\left[f(x_1) + 4f\left(\frac{x_1 + x_2}{2}\right) + 2f(x_2)\right.$$

$$\left. + \cdots + 4f\left(\frac{x_n + x_{n+1}}{2}\right) + f(x_{x_n + 1})\right] \cdot \Delta x$$

then

$$\int_a^b f(x)\, dx \approx S_n,$$

and the error involved is

$$\frac{(b - a)}{180} \cdot f^{(4)}(c) \cdot (\Delta x)^4,$$

where $a < c < b$.

Example 6 Approximate $\int_1^2 (1/x)\, dx$ using Simpson's Rule with $n = 5$. Estimate the error involved.

Solution The points of subdivision are $x_1 = 1.0$, $x_2 = 1.2$, $x_3 = 1.4$, $x_4 = 1.6$, $x_5 = 1.8$, and $x_6 = 2.0$. Since $f(x) = 1/x$, we can arrange the calculations in the form of the following table.

x	$f(x) = \frac{1}{x}$	MULTIPLYING FACTOR	PRODUCT
1.0	1	1	1.000000
1.1	0.9090909	4	3.636364
1.2	0.8333333	2	1.666667
1.3	0.7692308	4	3.076923
1.4	0.7142857	2	1.428571
1.5	0.6666667	4	2.666667
1.6	0.6250000	2	1.250000
1.7	0.5882353	4	2.352941
1.8	0.5555556	2	1.111111
1.9	0.5263158	4	2.105263
2.0	0.5000000	1	0.500000

6. Use S_5 to approximate $\int_0^1 x^3\, dx$. Estimate the error involved.

Summing the products on the right, we get

$$S_5 \approx \frac{1}{6}(20.794506) \cdot \left(\frac{1}{5}\right)$$

$$\approx 0.6931502.$$

So

$$\int_1^2 \frac{1}{x}\, dx \approx 0.6931502.$$

According to Theorem 4, the error E is given by

$$\frac{(b-a)}{180} \cdot f^{(4)}(c) \cdot (\Delta x)^4,$$

where $a < c < b$. In this case, $a = 1$, $b = 2$, and $\Delta x = 1/5$. Since $f(x) = x^{-1}$ and, from above, $f''(x) = 2x^{-3}$, we get

$$f^{(3)}(x) = 2(-3)x^{-4}$$

$$= -6x^{-4}$$

and

$$f^{(4)}(x) = -6(-4)x^{-5}$$

$$= \frac{24}{x^5}.$$

Thus

$$E = \frac{(2-1)}{180} \cdot \frac{24}{c^5} \cdot \left(\frac{1}{5}\right)^4$$

$$= \frac{1}{180} \cdot \frac{24}{c^5} \cdot \frac{1}{625}$$

$$= \frac{2}{9375c^5}.$$

Here, since $1 < c < 2$, we have $1/32 < 1/c^5 < 1$. Therefore,

$$E < \frac{2}{9375} \approx 0.0002133.$$

In fact, the actual error is about 0.000003, which is less than the error estimate computed above.

DO EXERCISE 6.

EXERCISE SET 11.4

Approximate each of the given integrals by the Riemann sum for the given Δx and using (a) left-hand endpoints and (b) right-hand endpoints.

1. $\int_0^1 (2x + 1)\, dx;\ \Delta x = 0.1$

2. $\int_{-1}^2 x^2\, dx;\ \Delta x = 0.5$

3. $\int_0^1 \frac{1}{(1 + x^2)}\, dx;\ \Delta x = 0.2$

4. $\int_0^1 x^4\, dx;\ \Delta x = 0.25$

Use the Trapezoidal Rule with the indicated value of n to approximate the integrals below.

5. $\int_0^1 (2x + 1)\, dx;\ n = 10$

6. $\int_{-1}^2 x^2\, dx;\ n = 6$

7. $\int_0^1 \frac{1}{1 + x^2}\, dx;\ n = 5$

8. $\int_0^1 x^4\, dx;\ n = 4$

Use the Trapezoidal Rule with the indicated value of n to approximate each of the following integrals. Then use the result of Theorem 3 to estimate the error involved in your approximation.

9. $\int_0^1 x^5\, dx;\ n = 10$

10. $\int_1^2 \frac{1}{\sqrt{x}}\, dx;\ n = 10$

11. $\int_1^2 \ln x\, dx;\ n = 10$

12. $\int_0^1 (x - x^3)\, dx;\ n = 5$

Use Simpson's Rule with the indicated value of n to approximate each of the following integrals.

13. $\int_0^1 (2x + 1)\, dx;\ n = 5$

14. $\int_{-1}^2 x^2\, dx;\ n = 6$

15. $\int_0^1 \frac{1}{1 + x^2}\, dx;\ n = 5$

16. $\int_0^1 x^4\, dx;\ n = 4$

Use Simpson's Rule with the indicated value of n to approximate each of the following integrals. Then use the estimate in Theorem 4 to estimate the error involved in your approximation.

17. $\int_0^1 x^3\, dx;\ n = 4$

18. $\int_0^1 x^5\, dx;\ n = 5$

19. $\int_1^2 \ln x\, dx;\ n = 5$

20. $\int_0^1 (x - x^3)\, dx;\ n = 5$

*11.5 EXTRAPOLATION: SPEEDING UP CONVERGENCE

In Section 11.1 we saw that the method of iteration, although a convenient way to solve equations of the form $g(x) = x$, is often ineffective because the iterates converge so slowly. Similar difficulties arose in Section 11.4, where our extensive computations sometimes yielded approximations to integrals that were valid to only one or two decimal places. In this section, we will investigate two methods that will "speed up" convergence. Both methods are examples of *extrapolation* techniques.

We reconsider the equation $g(x) = x$. Let $\bar{x}$ be a solution to the equation and suppose that $x_1, x_2, x_3, \ldots, x_n$, the sequence of iterates generated by an initial guess x_1, converges to $\bar{x}$. Then for all n, $x_{n+1} = g(x_n)$. For the moment, let $e_n = x_n - \bar{x}$ denote the error involved in approximating $\bar{x}$ by x_n (for all n). It turns out that if both x_1 and $\bar{x}$ lie in an interval on which the derivative of $g(x)$ satisfies $|g'(x)| < K$ (where $K < 1$), then we can show that

$$|e_2| < K|e_1|,$$

$$|e_3| < K|e_2| < K(K|e_1|) = K^2|e_1|,$$

and so forth, yielding

$$|e_{n+1}| < K|e_n| < \cdots < K^n|e_1|.$$

In particular, e_n tends to 0 as n tends to infinity. Assume, for the moment, that equality holds above, so that $e_n = K^{n-1}e_1$. Then

$$x_n = \bar{x} + K^{n-1}e_1. \tag{1}$$

If we also assume that $e_{n+1} = K^n e_1$ and $e_{n+2} = K^{n+1}e_1$, then we get

$$x_{n+1} = \bar{x} + K^n e_1 \tag{2}$$

and

$$x_{n+2} = \bar{x} + K^{n+1}e_1. \tag{3}$$

We could solve Eqs. (1), (2), and (3) simultaneously to eliminate K and e_1. After some simplification (see Exercise Set 11.5), we would have

$$\bar{x} = \frac{x_n x_{n+2} - (x_{n+1})^2}{x_{n+2} - 2x_{n+1} + x_n}.$$

If we only assume that Eqs. (1), (2), and (3) are approximately true, then

---

*This section can be omitted without loss of continuity.

1. Using $x_1 = 4$ and $x_{n+1} = \sqrt{5x_n - 6}$, compute x_2 and x_3 by iteration and x_4 by Aitken acceleration. Then compute x_5 and x_6 (from x_4) by iteration and use them to compute x_7 by Aitken's method. Compare your results with those of Example 4 in Section 11.1.

we obtain

$$\overline{x} \approx \frac{x_n x_{n+2} - (x_{n+1})^2}{x_{n+2} - 2x_{n+1} + x_n}.$$

So we can use three values of the iterates to generate a new approximation. This technique is sometimes called *Aitken acceleration*.

Example 1 Use $x_1 = 0.5$, iteration, and Aitken acceleration to solve $\cos x = x$.

Solution As in Example 1 of Section 11.1, we obtain

$$x_1 = 0.5,$$

$$x_2 = \cos (x_1) = \cos (0.5) \approx 0.8775826,$$

and

$$x_3 = \cos (x_2) \approx 0.63901249.$$

Applying Aitken acceleration, we obtain a new estimate, call it x_4 (this is *not* the same x_4 as in Example 1 of Section 11.1), as follows:

$$x_4 = \frac{x_1 x_3 - (x_2)^2}{x_3 - 2x_2 + x_1}$$

$$\approx \frac{(0.5)(0.63901249) - (0.8775826)^2}{(0.63901249) - 2(0.8775826) + 0.5}$$

$$\approx 0.7313852.$$

Of course, we don't have to stop here. Using x_4 and iteration, we get

$$x_5 = \cos (x_4) \approx 0.7442499$$

and

$$x_6 = \cos (x_5) \approx 0.7355962.$$

Using Aitken acceleration to generate x_7, we get

$$x_7 = \frac{x_4 x_6 - (x_5)^2}{x_6 - 2x_5 + x_4}$$

$$\approx \frac{(0.7313852)(0.7355962) - (0.7442499)^2}{(0.7355962) - 2(0.7442499) + (0.7313852)}$$

$$\approx 0.7390763.$$

Compare this with the results of Example 1 of Section 11.1.

DO EXERCISE 1.

We obtained Aitken's acceleration formula above because we were able to estimate the error in successive applications of the method of iteration. The technique used above, called *extrapolation*, can be used in numerical integration, too.

Suppose that we would like to evaluate the integral $I = \int_a^b f(x)\,dx$ numerically. We might fix n, partition $[a, b]$ into n subintervals of equal length, and use the Trapezoidal Rule to obtain an estimate T_n for I. According to Theorem 3, the error here, $|I - T_n|$, is less than $K(\Delta x)^2$, where K is a constant and $\Delta x = (b - a)/n$ is the length of each subinterval. The key here is going to be this error estimate. To see what happens, suppose, for a moment, that we repeat the process above using twice as many points. We obtain a new estimate, T_{2n}, for I. How much better is our new estimate? Since we doubled the number of subintervals, it follows that the length of each of the new subintervals is exactly one-half the length of those used the first time. In other words, we simply replaced Δx by $\Delta x/2$. Our new error estimate will look like

$$K\left(\frac{\Delta x}{2}\right)^2 = K\frac{(\Delta x)^2}{4} = \frac{1}{4}K(\Delta x)^2.$$

This suggests that doubling the number of points used should improve the accuracy by a factor of 4. But it gives us a lot more than that.

If we pretend for a moment that our error estimates are exact, then we get

$$\frac{1}{4}(I - T_n) = I - T_{2n},$$

or

$$I - T_n = 4(I - T_{2n}).$$

Thus

$$I - T_n = 4I - 4T_{2n}.$$

Solving for I, we get

$$3I = 4T_{2n} - T_n,$$

or

$$I = \frac{4T_{2n} - T_n}{3}.$$

Of course, the error estimates are only approximations, so that we write

$$I \approx \frac{4T_{2n} - T_n}{3}.$$

2. Approximate $\int_1^2 (1/x)\, dx$ using T_5. Use T_5 and the value of T_{10} from Example 5 of Section 11.4 to compute a new approximation to the integral.

Example 2 Find an approximate value for $\int_0^1 x^4\, dx$.

Solution We know that the exact value of the integral is $\frac{1}{5}$. Using the Trapezoidal Rule with $n = 4$ and $f(x) = x^4$, we have

$$T_4 = \left[\frac{1}{2} f(0) + f\left(\frac{1}{4}\right) + f\left(\frac{1}{2}\right) + f\left(\frac{3}{4}\right) + \frac{1}{2} f(1) \right] \cdot \frac{1}{4}$$

$$= \left[\frac{1}{2} (0)^4 + \left(\frac{1}{4}\right)^4 + \left(\frac{1}{2}\right)^4 + \left(\frac{3}{4}\right)^4 + \frac{1}{2} (1)^4 \right] \cdot \frac{1}{4}$$

$$= \left[0 + \frac{1}{256} + \frac{1}{16} + \frac{81}{256} + \frac{1}{2} \right] \cdot \frac{1}{4}$$

$$= [0.8828125] \cdot \frac{1}{4} = 0.220703125.$$

Similarly,

$$T_8 = \left[\frac{1}{2} (0)^4 + \left(\frac{1}{8}\right)^4 + \left(\frac{2}{8}\right)^4 + \cdots + \left(\frac{7}{8}\right)^4 + \frac{1}{2} (1)^4 \right] \cdot \frac{1}{8}$$

$$\approx [1.6416016] \cdot \frac{1}{8} \approx 0.2052002.$$

Using the formula above, we get

$$I \approx \frac{4T_8 - T_4}{3}$$

$$= \frac{4(0.2052002) - (0.220703125)}{3}$$

$$= 0.2000325521.$$

We see that this is an excellent approximation.

DO EXERCISE 2.

Compare the answer to Margin Exercise 2 to that in Example 6 of Section 11.4. The answers are the same. This is no coincidence. The answer that we have obtained from T_n and T_{2n} by extrapolation (or, as it is usually called, *Romberg integration*) is S_n. In other words,

$$S_n = \frac{4T_{2n} - T_n}{3}.$$

This will be proved in Exercise Set 11.5. In the meantime, note that we can use our earlier argument, along with the error estimate from Theorem 4, to obtain even better numerical estimates.

If $I = \int_a^b f(x)\,dx$ and S_n is the Simpson's Rule approximation to I, then the absolute error $|I - S_n|$ is less than $M\,(\Delta x)^4$, where M is a constant and Δx is the length of the subintervals used. Here, if we double the number of points used, then we halve the length of each subinterval and, replacing Δx by $\Delta x/2$ in our error estimate, we get

$$M\left(\frac{\Delta x}{2}\right)^4 = M \cdot \frac{(\Delta x)^4}{16}.$$

Doubling the number of points should improve the accuracy by a factor of 16. Thus

$$I - S_{2n} \approx \frac{1}{16}(I - S_n)$$

or

$$16I - 16S_{2n} \approx I - S_n$$

or

$$15I \approx 16S_{2n} - S_n.$$

Solving, we get

$$I \approx \frac{16S_{2n} - S_n}{15}.$$

Define the Romberg sum R_n by

$$R_n = \frac{16S_{2n} - S_n}{15}$$

for all positive integers n.

Example 3 Estimate

$$\int_0^1 \frac{x}{1 + x^2}\,dx$$

by S_2, S_4, and the Romberg sum

$$R_2 = \frac{16S_4 - S_2}{15}.$$

Solution Here,

$$f(x) = \frac{x}{1 + x^2}.$$

3. Estimate

$$\int_0^1 \frac{1}{1+x}\, dx$$

using S_2, S_4, and the Romberg sum

$$R_2 = \frac{16S_4 - S_2}{15}.$$

Thus

$$S_2 = \frac{1}{6} \cdot \left[f(0) + 4f\left(\frac{1}{4}\right) + 2f\left(\frac{1}{2}\right) + 4f\left(\frac{3}{4}\right) + f(1) \right] \cdot \Delta x$$

$$= \frac{1}{6} \cdot \left[0 + 4\left(\frac{4}{17}\right) + 2\left(\frac{2}{5}\right) + 4\left(\frac{12}{25}\right) + \frac{1}{2} \right] \cdot \frac{1}{2}$$

$$\approx \frac{1}{6} \cdot [4.161176] \cdot \frac{1}{2}$$

$$\approx 0.3467647.$$

Similarly,

$$S_4 = \frac{1}{6}\left[f(0) + 4f\left(\frac{1}{8}\right) + 2f\left(\frac{1}{4}\right) + \cdots + 4f\left(\frac{7}{8}\right) + f(1) \right] \cdot \Delta x$$

$$= \frac{1}{6} \cdot \left[0 + 4 \cdot \frac{8}{65} + 2 \cdot \frac{4}{17} + \cdots + 4 \cdot \frac{56}{113} + \frac{1}{2} \right] \cdot \frac{1}{4}$$

$$\approx \frac{1}{24} \cdot [8.3180181]$$

$$\approx 0.3465841.$$

Now,

$$R_2 = \frac{16S_4 - S_2}{15}$$

$$\approx \frac{16(0.3465841) - (0.3467647)}{15}$$

$$\approx \frac{5.1985809}{15}$$

$$\approx 0.34657206.$$

To get an idea of the accuracy, we write

$$I = \int_0^1 \frac{x}{1+x^2}\, dx$$

$$= \left[\frac{1}{2} \ln (1 + x^2) \right]_0^1 = \frac{1}{2} \ln 2.$$

So, using a calculator, we get $I = \frac{1}{2} \ln 2 \approx 0.34657359.$

DO EXERCISE 3.

EXERCISE SET 11.5

Use the initial guess x_1 and iteration to compute x_2 and x_3; then calculate x_4 using Aitken's method.

1. $x = \dfrac{1}{3}x^2 + \dfrac{1}{3}; x_1 = 0$

2. $x = \dfrac{1}{3}x^2 + \dfrac{1}{3}; x_1 = \dfrac{1}{2}$

3. $x = \sin x + \dfrac{1}{2}$ (x in radians); $x_1 = 1$

4. $x = \dfrac{1}{2}\cos x; x_1 = 1$

5. $x = \sqrt[3]{3x - 3}; x_1 = -1$

6. $x = \sqrt{3x - 1}; x_1 = 2$

Calculate T_2, T_4, and T_8 for each of the following integrals. Then use Romberg integration to calculate S_2 and S_4. Finally, use Romberg integration to calculate R_2 from S_2 and S_4.

7. $\displaystyle\int_0^1 x^2\, dx$

8. $\displaystyle\int_0^1 x^4\, dx$

9. $\displaystyle\int_0^1 \dfrac{1}{1 + x^2}\, dx$

10. $\displaystyle\int_0^1 \dfrac{1}{1 + x^6}\, dx$

11. Let

$$G = \frac{x_n x_{n+2} - (x_{n+1})^2}{x_{n+2} - 2x_{n+1} + x_n}$$

for a fixed positive integer n. Also, let

$$x_n = \bar{x} + K^{n-1}e_1, \qquad (1)$$

$$x_{n+1} = \bar{x} + K^n e_1, \qquad (2)$$

and

$$x_{n+2} = \bar{x} + K^{n+1}e_1, \qquad (3)$$

where K, $\bar{x}$, and e_1 are constants. Substitute Eqs. (1), (2), and (3) into the expression for G, expand, and simplify to obtain $G = \bar{x}$.

12. Let $f(x)$ be a continuous function on $[a, b]$, and let T_n and S_n be as in Sections 11.4 and 11.5. Show that

$$\frac{4T_{2n} - T_n}{3} = S_n \quad \text{for all } n.$$

OBJECTIVES

You should be able to

a) **Find an approximate solution to a first-order differential equation with initial condition using either Euler's method or the three-term method.**

b) **Find an approximate solution to a system of two first-order differential equations with initial conditions using Euler's method.**

11.6 EULER'S METHOD

In Chapter 9, we saw that many differential equations can be solved directly using various methods that depend on the form of the equations. Unfortunately, many applications give rise to differential equations that cannot be solved directly. In these cases, it is necessary to have methods for generating accurate, but approximate, solutions to the given equation.

We first need to introduce some notation. In Example 3 of Section 9.2, we solved the differential equation

$$y' = x - xy.$$

Note that this equation involves variables x and y, with y being a function of x. It can then be thought of as a function of two variables and can be written

$$y' = x - xy = f(x, y).$$

The solution of the equation is

$$y = 1 + C_1 e^{-x^2/2}.$$

Now we know that y is a function of x. Though we have not done so in the past, we can express this as $y = y(x)$. With the initial condition $y(0) = 2$, the solution can be written as

$$y = 1 + e^{-x^2/2}.$$

In this section we will be considering first-order differential equations (no derivative will occur higher than the first)

$$y' = f(x, y), \tag{1}$$

with initial condition

$$y(x_0) = y_0. \tag{2}$$

If we assume (and we always will) that f and $\partial f / \partial y$ are continuous functions on some rectangle in the plane containing (x_0, y_0), then we can show that there is an interval containing x_0 on which there is a unique solution $y = g(x)$ to Eq. (1) that also satisfies Eq. (2). In the specific example mentioned above, these conditions are certainly satisfied.

The first numerical technique to be considered is Euler's method. Let's suppose that we must solve

$$y' = f(x, y), \tag{1}$$

$$y(x_0) = y_0, \tag{2}$$

where f and its partials are so smooth (have continuous partial derivatives) as to guarantee a unique solution to the given system on some interval $[x_0, b]$. We will construct an approximate solution as follows. First, select n. Partition $[x_0, b]$ into n equal subintervals (of length $(b - x_0)/n$, sometimes called the *step size*). Let $x_1, x_2, \ldots, x_{n-1}, x_n = b$ denote the points in the partition. Now, condition (2) tells us that the point (x_0, y_0) is on the graph of the solution (see the following figure). If we substitute $x = x_0$ and $y = y_0$ into Eq. (1), we can find the slope of the exact solution $y = g(x)$ at the point (x_0, y_0). The slope will be $f(x_0, y_0) = y'|_{(x_0, y_0)}$. With this information, we can compute the equation of the tangent line to $y = g(x)$ at (x_0, y_0). The equation will be

$$y - y_0 = f(x_0, y_0)(x - x_0).$$

We use this equation to find the y-value of our approximation when $x = x_1$. Graphically, we find the point of intersection of the tangent line T_0 with the vertical line $x = x_1$ (see the figure). The y-coordinate of this point, call it y_1, will be the value that we use. If we let $x = x_1$ in the equation of the tangent line and solve for y, we get

$$y = y_0 + f(x_0, y_0)(x_1 - x_0).$$

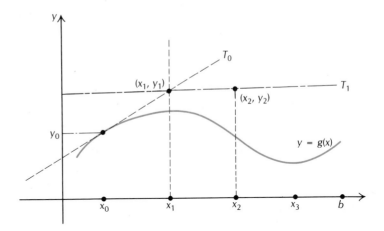

We now define y_1 by this formula, so that

$$y_1 = y_0 + f(x_0, y_0)(x_1 - x_0).$$

But now we can repeat the whole process. We substitute the coordinates (x_1, y_1) into $y' = f(x, y)$ to obtain $f(x_1, y_1)$. The equation of the line T_1 through (x_1, y_1) with slope $f(x_1, y_1)$ is

$$y - y_1 = f(x_1, y_1)(x - x_1).$$

Letting $x = x_2$ and solving for $y = y_2$ yields

$$y_2 = y_1 + f(x_1, y_1)(x_2 - x_1).$$

Note that if we let $\Delta x = x_2 - x_1$ denote the step size, then we get

$$y_2 = y_1 + f(x_1, y_1) \cdot \Delta x.$$

Continuing, we obtain the general formula

$$y_{n+1} = y_n + f(x_n, y_n) \cdot \Delta x$$

for all nonnegative integers n.

Example 1 Apply Euler's method to $y' = y$, $y(0) = 2$ on the interval $[0, 1]$ with $n = 5$.

Solution We have deliberately chosen a differential equation whose solution we know, so that we can compare our approximate solution with the exact solution, which is $y = g(x) = 2e^x$.

First, since $n = 5$ and our interval has length 1, note that the step-size is $\frac{1}{5} = 0.2 = \Delta x$. The points to be used as x-coordinates are $x_1 = 0.2$, $x_2 = 0.4$, $x_3 = 0.6$, $x_4 = 0.8$, and, of course, $x_5 = 1.0$.

1. **a)** By separating the variables, solve $y' = -2xy$, $y(0) = 1$ explicitly.

   b) Using Euler's method with $n = 5$ on $[0, 1]$, find an approximate solution to $y' = -2xy$, $y(0) = 1$.

   c) Repeat (b) with a step size of 0.1.

The initial condition is $y(0) = 2$. In other words, $x_0 = 0$ and $y_0 = 2$. So, using $\Delta x = 0.2$ and $f(x, y) = y$, we get

$$y_1 = y_0 + f(x_0, y_0) \cdot \Delta x$$
$$= 2 + (y_0) \cdot (0.2)$$
$$= 2 + 2 \cdot (0.2)$$
$$= 2 + 0.4 = 2.4.$$

Next,

$$y_2 = y_1 + f(x_1, y_1) \Delta x$$
$$= 2.4 + y_1 \cdot (0.2)$$
$$= 2.4 + (2.4) \cdot (0.2)$$
$$= 2.88.$$

Continuing, we have

$$y_3 = 2.88 + (2.88)(0.2)$$
$$= 3.456,$$
$$y_4 = 3.456 + (3.456)(0.2)$$
$$= 4.1472,$$

and

$$y_5 = 4.1472 + (4.1472)(0.2)$$
$$= 4.97664.$$

If these points were plotted against the graph of the exact solution $y = 2e^x = g(x)$, then we would find that on the left-hand side of the interval, the approximations aren't so bad, but when we reach $x_5 = 1$, the approximate value, $y_5 = 4.97664$, bears little resemblance to the actual value,

$$g(1) = 2e^1 = 2e \approx 5.4365637.$$

The great defect in Euler's method (sometimes called the *tangent-line method*) is that it isn't very accurate.

DO EXERCISE 1.

There are many ways to alter Euler's method so as to improve its accuracy. To obtain one such improvement, return to Figure 11.13. There we obtained our initial approximations by use of the tangent line T_1 to the graph of the exact solution $y = g(x)$. To obtain better accuracy, we could use the "tangent quadratic" to the graph of $y = g(x)$. Suppose for the moment that $y = g(x)$ is represented by its Taylor series at $x = x_0$. Then,

$$g(x) = g(x_0) + g'(x_0)(x - x_0) + \frac{g''(x_0)}{2!}(x - x_0)^2 + \cdots.$$

We want to use this formula to compute $y_1 = g(x_1)$. Setting $x = x_1$ and ignoring all terms after the third one, we obtain

$$y_1 = g(x_1) \approx g(x_0) + g'(x_0)(x_1 - x_0) + \frac{g''(x_0)}{2!}(x_1 - x_0)^2.$$

Now $x_1 - x_0 = \Delta x$, the step size. So,

$$y_1 \approx g(x_0) + g'(x_0) \cdot \Delta x + \frac{g''(x_0)}{2!}(\Delta x)^2.$$

But $g(x_0) = y_0$ and $g'(x_0)$ is just $y'(x_0) = f(x_0, y_0)$, which is given. Substituting, we get

$$y_1 \approx y_0 + f(x_0, y_0)\,\Delta x + \frac{1}{2}g''(x_0)(\Delta x)^2.$$

Every quantity on the right-hand side above is given except Δx, the step size, which we are free to choose, and $g''(x_0)$. The last quantity may or may not be easy to compute. We illustrate with two examples.

Example 2 Compute $g''(x_0) = y''(x_0)$ for the initial value problem

$$y' = xy^2,$$

$$y(1) = 2.$$

Solution In this problem, $f(x, y) = xy^2$, $x_0 = 1$, and $y_0 = 2$. Thus $y'(x_0) = f(x_0, y_0) = f(1, 2) = 1 \cdot (2)^2 = 4$. To get y'', we must differentiate $y' = g(x, y) = xy^2$ with respect to x. Using the Product Rule, we obtain

$$y'' = x(y^2)' + y^2(x)'$$
$$= x(y^2)' + y^2,$$

and, applying the Chain Rule, we get

$$y'' = x(2yy') + y^2.$$

So $y'' = 2xyy' + y^2$. Replacing y' by xy^2 gives

$$y'' = 2xy(xy^2) + y^2,$$

2. Compute y'' for the initial value problem $y' = y \sin(xy)$, $y(1) = 1$.

or

$$y'' = 2x^2y^3 + y^2.$$

Even a very simple choice of $f(x, y)$ results in a fairly complicated expression for y''.

DO EXERCISE 2.

When y'' is relatively simple, however, this method (sometimes called the *three-term method*) yields superior results. Note that the general iteration formula will be as follows:

$$y_{n+1} = y_n + f(x_n, y_n) \cdot \Delta x + \frac{1}{2} y''(x_n, y_n)(\Delta x)^2.$$

Example 3 Apply the three-term method to $y' = y$, $y(0) = 2$, on the interval $[0, 1]$ with $n = 5$.

Solution This is the problem from Example 1 given earlier. To apply the method, we must compute y''. Here, since $y' = y$, we get $y'' = y'$ by differentiation. Combining the two equations, we have $y'' = y$. Since $f(x, y) = y$ and $y'' = g'' = y$, the iteration formula becomes

$$y_{n+1} = y_n + y_n \cdot (\Delta x) + \frac{1}{2} y_n \cdot (\Delta x)^2,$$

or

$$y_{n+1} = y_n \left(1 + \Delta x + \frac{1}{2}(\Delta x)^2\right).$$

Here, $\Delta x = 0.2$. So,

$$y_{n+1} = y_n \left(1 + (0.2) + \frac{1}{2}(0.2)^2\right)$$

$$= y_n(1 + 0.2 + 0.02)$$

$$= 1.22y_n.$$

Using $y(x_0) = y(0) = 2$, we find that

$$y_1 = y(x_1) = 1.22y_0 = 1.22 \cdot (2)$$

$$= 2.44.$$

Next, we have

$$y_2 = 1.22y_1 = 1.22(2.44),$$

3. Use the three-term method to find an approximate solution to $y' = -2xy$, $y(0) = 1$, on the interval $[0, 1]$ using a step size of 0.2.

or

$$y_2 = 2.9768.$$

Similarly,

$$y_3 = 1.22(2.9768) = 3.631696,$$

$$y_4 = 1.22(3.631696) \approx 4.4306691,$$

and

$$y_5 \approx 1.22(4.4306691) \approx 5.4054163.$$

DO EXERCISE 3.

We compare the result of using Euler's method with the three-term method on the initial value problem $y' = y$, $y(0) = 2$, in the table below. The exact values in the right-hand column were calculated using $y = 2e^x$.

4. Compare the results of Margin Exercises 1 and 3 with the corresponding values of the exact solution $y = e^{-x^2}$ as in the table at right.

	EULER	THREE-TERM METHOD	EXACT
$y_0 = y(0)$	2	2.	2
$y_1 = y(0.2)$	2.4	2.44	2.4428055
$y_2 = y(0.4)$	2.88	2.9768	2.9836494
$y_3 = y(0.6)$	3.456	3.631696	3.6442376
$y_4 = y(0.8)$	4.1472	4.4306691	4.4510819
$y_5 = y(1)$	4.97664	5.4054163	5.4365637

DO EXERCISE 4.

As expected, the three-term method yields greater accuracy (although more computation is involved) than Euler's method.

We can also apply these techniques to systems of first-order differential equations. They could be used if, for instance, we have

$$x' = f(x, y, t),$$

$$y' = g(x, y, t)$$

with the variables x and y being thought of as functions of the third variable t. If the initial conditions are

$$x(t_0) = x_0,$$

$$y(t_0) = y_0.$$

Then Euler's method becomes

$$x_{n+1} = x_n + f(x_n, y_n, t_n) \cdot \Delta t$$

and

$$y_{n+1} = y_n + g(x_n, y_n, t_n) \cdot \Delta t$$

for all n.

Example 4 Use Euler's method to find approximate values for $x(t)$ and $y(t)$ on $[0, 1]$, where x and y satisfy

$$x' = x - 5y,$$
$$y' = 2x - 6y$$

with

$$x(0) = 6,$$
$$y(0) = 3$$

using $\Delta t = 0.2$.

Solution Here, $f(x, y, t) = x - 5y$ and $g(x, y, t) = 2x - 6y$. Our formulas become

$$x_{n+1} = x_n + (x_n - 5y_n) \cdot \Delta t$$

and

$$y_{n+1} = y_n + (2x_n - 6y_n) \cdot \Delta t,$$

or, if we use $\Delta t = 0.2$,

$$x_{n+1} = x_n + (x_n - 5y_n) \cdot (0.2)$$

and

$$y_{n+1} = y_n + (2x_n - 6y_n) \cdot (0.2).$$

Simplification yields

$$x_{n+1} = 1.2x_n - y_n$$

and

$$y_{n+1} = 0.4x_n - 0.2y_n.$$

Using $x_0 = x(0) = 6$ and $y_0 = y(0) = 3$, we find that

$$x_1 = 1.2x_0 - y_0$$
$$= (1.2)(6) - 3$$
$$= 4.2,$$

5. a) Show that $x(t) = 5e^{-t} + e^{-4t}$ and $y(t) = 2e^{-t} + e^{-4t}$ are the exact solutions to the initial value problem in Example 4.

b) Use (a) to compute the exact values of $y(0.4)$, $x(0.6)$, and $y(1)$.

6. Use Euler's method to find two approximate values for $t = 0.2$ and $t = 0.4$ for the system

$$x' = x - 2y$$
$$y' = -2x + y,$$

with initial conditions $x(0) = 2$, $y(0) = 0$. Use $\Delta t = 0.2$.

whereas

$$y_1 = 0.4x_0 - 0.2y_0$$
$$= (0.4)(6) - (0.2)(3)$$
$$= 1.8.$$

Continuing, we get

$$x_2 = (1.2)x_1 - y_1 = (1.2)(4.2) - (1.8) = 3.24,$$
$$y_2 = (0.4)x_1 - (0.2)y_1 = (0.4)(4.2) - (0.2)(1.8) = 1.32,$$
$$x_3 = (1.2)(3.24) - (1.32) = 2.568,$$
$$y_3 = (0.4)(3.24) - (0.2)(1.32) = 1.032,$$
$$x_4 = (1.2)(2.568) - (1.032) = 2.0496,$$
$$y_4 = (0.4)(2.568) - (0.2)(1.032) = 0.8208,$$

and

$$x_5 = (1.2)(2.0496) - (0.8208) = 1.63872,$$
$$y_5 = (0.4)(2.0496) - (0.2)(0.8208) = 0.65568.$$

DO EXERCISES 5 AND 6.

EXERCISE SET 11.6

Find the solution to each of the following initial value problems using Euler's method on $[0, 1]$ with the indicated step size Δx.

1. $y' = -1$, $y(0) = 1$, $\Delta x = 0.1$

2. $y' = 2x$, $y(0) = -1$, $\Delta x = 0.1$

3. $y' = 2x$, $y(0) = 0$, $\Delta x = 0.25$

4. $y' = 3x^2 - 1$, $y(0) = 1$, $\Delta x = 0.2$

5. $y' = 2xy$, $y(0) = 1$, $\Delta x = 0.2$

6. $y' = -y$, $y(0) = 1$, $\Delta x = 0.25$

7. $y' = x\sqrt{y}$, $y(0) = 1$, $\Delta x = 0.2$

Find the solutions to each of the following initial value problems using the three-term method on $[0, 1]$ with the indicated step size Δx.

8. $y' = -1$, $y(0) = 1$, $\Delta x = 0.1$

9. $y' = 2x$, $y(0) = -1$, $\Delta x = 0.1$

10. $y' = 2x$, $y(0) = 0$, $\Delta x = 0.25$

11. $y' = 3x^2 - 1$, $y(0) = 1$, $\Delta x = 0.2$

12. $y' = 2xy$, $y(0) = 1$, $\Delta x = 0.2$

13. $y' = -y$, $y(0) = 1$, $\Delta x = 0.25$

14. $y' = x\sqrt{y}$, $y(0) = 1$, $\Delta x = 0.2$

For each of the following initial value problems, find the exact solution $y = g(x)$. Then make a table comparing the exact values with the approximate values obtained using Euler's method and the three-term method.

15. $y' = -1$, $y(0) = 1$

16. $y' = 2x$, $y(0) = -1$

17. $y' = 2xy$, $y(0) = 1$

18. $y' = -y$, $y(0) = 1$

Use Euler's method with the indicated value of Δt to approximate the solution to the given system of differential equations on $[0, 1]$. Compare these with the exact solutions, where possible.

19. $x'(t) = x + y$,
$y'(t) = 3x - y$,
$x(0) = 3, y(0) = -1, \Delta t = 0.2$

21. $x'(t) = y$,
$y'(t) = -2x + 3y$,
$x(0) = 0, y(0) = 1, \Delta t = 0.2$

23. $x' = x^3$, $y' = -y$, $x(0) = 0, y(0) = 1, \Delta t = 0.1$

25. *Accumulated present value of a continuous cash flow.* If money flows continuously into an investment at the rate $R(t)$ (in thousands of dollars), then the *accumulated present value* $V(t)$ satisfies the initial value problem (see Exercise Set 6.3)

$$V' = R(t)e^{-kt}, \qquad V(0) = 0,$$

where k is the current interest rate. Using step size $\Delta t = 0.2$, calculate $V(1)$ using Euler's method if $R(t) = 1/(1 + t^2)$ and $k = 8\%$.

27. *Fick's law in biology* is given by

$$\frac{dC}{dt} = \frac{kA}{V}(C_T - C),$$

where $C(t)$ is the amount of a substance that passes through a tube with cross-sectional area A and volume V at time t (see Exercise 11 of Exercise Set 7.4), k is a constant, and C_T is the total amount of substance. Assuming that $k = 1$, $A = 3, V = 20, C_T = 1000$, and $C(0) = C_0 = 2$, use step size $\Delta t = 1$ to estimate $C(5)$ using (a) Euler's method; (b) the three-term method; and (c) the exact solution.

28. *Accumulated present value of dividends.* Suppose that $d(t)$ represents the instantaneous dividend payment of a stock at time t. Then $d(t)e^{-mt}$ is the present value of that payment where m is the current interest rate. Then the accumulated value $D_p(t)$ of the dividends at time t satisfies the initial value problem

$$D'_p(t) = d(t)e^{-mt}, \qquad D(0) = 0.$$

Use step size $\Delta t = 1$ to estimate $D_p(10)$ if $m = 8\%$ and $d(t) = \$10$ using (a) Euler's method; and (b) the three-term method.

20. $x'(t) = x + y$,
$y'(t) = 4x - 2y$,
$x(0) = 3, y(0) = -2, \Delta t = 0.25$

22. $x'(t) = -3x + 4y$,
$y'(t) = -2x + 3y$,
$x(0) = 4, y(0) = 3, \Delta t = 0.1$

24. $x' = x + ty$, $y' = xy$, $x(0) = 1, y(0) = 1, \Delta t = 0.2$

26. In an experiment (see Exercise 3 of Exercise Set 9.3) it was found that the rate of growth of yeast cells in a laboratory satisfied the initial value problem

$$\frac{dP}{dt} = 0.0008P(700 - P), \qquad P(0) = 10,$$

where $P(t)$ is the number of cells present after time t (in hours). Using step size 1, calculate $P(5)$ and $P(8)$ using (a) Euler's method; (b) the three-term method; and (c) the exact solution.

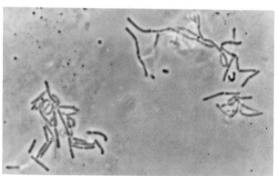

Yeast cells growing in a Petri dish (*Lester V. Bergman and Assoc.*)

29. *Population.* The population of two species satisfies the system

$$\frac{dx}{dt} = 0.10x - 1.5y,$$

$$\frac{dy}{dt} = 0.015x + 0.3y,$$

$$x(0) = 1600,$$

$$y(0) = 75,$$

where $x(t)$ and $y(t)$ are the populations of the two species at time t (in years), respectively. Use Euler's method with step size $\Delta t = 1$ to estimate $x(5)$ and $y(5)$.

30. *Commodities prices.* The price (in dollars) of two commodities at time t (in weeks) are $P(t)$ and $Q(t)$, respectively. Given that the prices satisfy the system

$$P' = -0.5P + 0.5Q + e^{-t^2},$$

$$Q' = 0.25P - 0.25Q - 3e^{-t^2},$$

$$P(0) = 225,$$

$$Q(0) = 100,$$

use Euler's method with step size $\Delta t = 1$ to estimate $P(10)$ and $Q(10)$.

31. Let $F(x)$ be defined by

$$F(x) = \int_0^x f(t)\, dt$$

where $f(t)$ is continuous on $[0, 10]$.

a) Show that $F(x)$ satisfies the initial value problem

$$F'(x) = f(x), \qquad F(0) = 0$$

for all x in $[0, 10]$.

b) Use Euler's method with step size $\Delta x = 1$ to estimate $F(10)$.

Application: Predator–Prey. Suppose that $x(t)$ and $y(t)$ denote the number of hares and the number of foxes, respectively, present at time t in a certain enclosed wilderness area.

If we assume that plenty of food is provided for the hares, then it seems reasonable that their birth rate should be a (large) constant that is independent of time. Their death rate, on the other hand, ought to be proportional to the number of encounters between foxes and hares. Since the rate of change of the hare population will be the difference between the birth and death rates, we assume that

$$\frac{dx}{dt} = ax - bxy,$$

where a and b are positive constants.

For the foxes, the birth rate tends to increase with the supply of food (hares). Here, we might take the birth rate to be proportional to xy. As the fox population increases, there will be shortages of food and the death rate of the foxes will be proportional to the number of foxes present. We get

$$\frac{dy}{dt} = cxy - dy,$$

where c and d are positive constants. Thus we have the (nonlinear) system of equations

$$\frac{dx}{dt} = ax - bxy,$$

$$\frac{dy}{dt} = cxy - dy.$$

We will omit the derivation but we can show that these equations lead to the equation

$$\frac{dy}{dx} = \frac{cxy - dy}{ax - bxy} = \frac{y(cx - d)}{x(a - by)}.$$

Commodities listing

32. a) Use separation of variables to solve the equation in Exercise 31.

b) Evaluate the constant of integration in part (a) in terms of a, b, c, d, and the initial populations $x_0 = x(0)$ and $y_0 = y(0)$.

Although the equation relating $x(t)$ and $y(t)$ that we obtained in Exercise 31 cannot generally be solved for y or x as a function of the other variable, the graph of the relation can be plotted, as shown below.

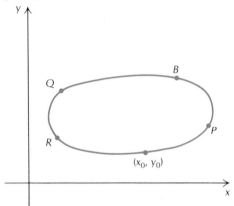

Note that if we start at (x_0, y_0) and proceed counterclockwise around the figure, then the number of hares increases at first, as does the number of foxes. This results in a dramatic increase in the number of foxes (starting at P). Eventually, the foxes begin to devour the hares and we have a long period in the cycle during which the hare population drops steadily, while the fox population remains constant (arc BQ). Finally, there are so few hares that the foxes begin to starve (at Q) and their population dwindles. As the number of foxes declines, the hare population begins to grow (R), and the cycle begins anew.

33. In a certain forest, the population of hares and foxes (in thousands) is determined by the system

$$\frac{dx}{dt} = 0.75x - 0.04xy, \quad x(0) = 60,$$

$$\frac{dy}{dt} = 0.004xy - 0.25y, \quad y(0) = 20.$$

Use Euler's method with step size $\Delta t = 1$ to find an approximation for:

a) $x(5)$ and $y(5)$;

b) $x(10)$ and $y(10)$.

CHAPTER 11 TEST

Use iteration and the suggested initial guesses to solve each of the given equations.

1. $x = \sqrt{3x - 2}$; $x_1 = 1.5$

2. $x = \sqrt[3]{3x - 2}$; $x_1 = 0$

3. $x = \sqrt[4]{\dfrac{x + 3}{6}}$; $x_1 = 1$

4. $x = \dfrac{2}{5} \cos x$; $x_1 = 0.75$

5. $x = \dfrac{5}{2} \cos x$; $x_1 = 0.75$

Use Newton's method to find all solutions in $[0, 2]$ of each of the given equations.

6. $2x^3 - 5x + 1 = 0$

7. $6x^4 - x - 3 = 0$

8. $x^9 - 4x + 2 = 0$

9. Find the equilibrium point for the supply and demand functions

$$S(x) = x^3 + 6x + 1 \quad \text{and} \quad D(x) = \frac{12}{x^2}.$$

11. An investor makes an initial investment of $3000 followed by an additional $1000 one year later. The investment returns $6000 four years after the original investment was made.

a) Set up the equation for the yield rate for this investment.

b) Find the investor's yield rate.

13. Let

$$f(x) = \begin{cases} \dfrac{\sin x}{x}, & \text{if } x > 0, \\ \\ 1, & \text{if } x = 0. \end{cases}$$

Use the Trapezoidal Rule with $n = 2$, 4, and 8 to approximate

$$\int_0^1 \frac{\sin x}{x}\, dx.$$

15. Use the initial guess $x_1 = 0.75$ and iteration to obtain x_2 and x_3 for $x = \frac{5}{2} \cos x$. Then use Aitken acceleration to calculate x_4, etc. Compare your answer here with your answer to Problem 5 above.

17. Use Romberg integration and the results of Problem 14 to find an approximate value for

$$\int_0^1 \frac{\sin x}{x}\, dx.$$

19. Use the three-term method with $\Delta x = 0.2$ to find an approximate solution to Problem 18.

21. Use Euler's method with step size $\Delta t = 0.2$ on the interval $[0, 1]$ to approximate a solution to

$$\frac{dx}{dt} = -3x + 2y, \quad x(0) = 1,$$

$$\frac{dy}{dt} = -x, \quad y(0) = 2.$$

22. Show that the present value V of a series of (equal) payments of R dollars per year for n years (starting one year from today) at a rate of interest i is

$$V = Rv \cdot \frac{1 - v^n}{1 - v},$$

where $v = (1 + i)^{-1}$.

10. If interest is compounded annually at 9.25%, find the present value of a payment of $100,000 to be made twenty years from today.

12. An investor must choose between two investment packages. The first requires an initial investment of $1000 and returns $1500 after three years. The second requires an initial investment of $1000, another $1000 after one year, and returns $10,000 12 years after the initial investment was made. Find the respective yield rates.

14. Use Simpson's Rule to calculate S_2 and S_4 for Problem 13.

16. Repeat Problem 15 with $x = \frac{2}{5} \cos x$.

18. Use Euler's method with step size $\Delta x = 0.2$ to find an approximate solution to $y' = -y^2$ with $y(0) = 1$ on the interval $[0, 1]$.

20. a) Separate the variables to find the exact solution to Problem 18.

b) Compare the exact values with the approximate values obtained in Problems 18 and 19.

23. Show that Aitken's formula can be written in the form

$$x_n - \frac{(x_{n+1} - x_n)^2}{x_{n+2} - 2x_{n+1} + x_n}.$$

CUMULATIVE
REVIEW

1. Write an equation of the line with slope -4 and containing the point $(-7, 1)$.

2. For $f(x) = x^2 - 5$, find $f(x + h)$.

3. a) Graph:

$$f(x) = \begin{cases} 5 - x & \text{for } x \neq 2, \\ -3 & \text{for } x = 2. \end{cases}$$

b) Find $\lim_{x \to 2} f(x)$.

c) Find $f(2)$.

d) Is f continuous at 2?

Find each limit, if it exists.

4. $\lim\limits_{x \to -4} \dfrac{x^2 - 16}{x + 4}$

5. $\lim\limits_{x \to 1} \sqrt{x^3 + 8}$

6. $\lim\limits_{x \to 3} \dfrac{4}{x - 3}$

7. $\lim\limits_{x \to \infty} \dfrac{12x - 7}{3x + 2}$

8. $\lim\limits_{h \to 0} \dfrac{3x^2 + h}{x - 2h}$

Differentiate.

9. $y = -9x + 3$

10. $y = x^2 - 7x + 3$

11. $y = x^{1/4}$

12. $f(x) = x^{-6}$

13. $f(x) = (x - 3)(x + 1)^5$

14. $f(x) = \dfrac{x^3 - 1}{x^5}$

15. $y = \dfrac{e^x + x}{e^x}$

16. $y = \ln (x^2 + 5)$

17. $y = e^{\ln x}$

18. $y = e^{3x} + x^2$

19. $y = e^{-x} \sin (x^2 + 3)$

20. $y = \ln |\cos x|$

21. For $y = \tan x$, find $\dfrac{d^2y}{dx^2}$.

22. Differentiate implicitly to find $\dfrac{dy}{dx}$ if $x^3 + \dfrac{x}{y} = 7$.

Find maximum and minimum values, if they exist, over the indicated interval. If no interval is indicated, use the real line.

23. $f(x) = 3x^2 - 6x - 4$

24. $f(x) = -5x + 1$

25. $f(x) = \frac{1}{3}x^3 - x^2 - 3x + 5; [-2, 0]$

26. For a certain product the total revenue and total cost functions are given by

$$R(x) = 4x^2 + 11x + 110,$$

$$C(x) = 4.2x^2 + 5x + 10.$$

Find the number of units that must be produced and sold in order to maximize profit.

28. A certain population of furbearing animals has the reproduction curve

$$f(P) = P(400 - P),$$

where P is measured in thousands. Find the population at which the maximum sustainable harvest occurs, and find the maximum sustainable harvest.

30. Sketch the graph of $y = x^3 - 3x + 1$.

31. The demand for oil in the world is increasing at the rate of 10% per year; that is,

$$\frac{dA}{dt} = 0.1A,$$

where A = the amount of oil used and t = the time in years from 1980.

a) Given that 66,164 million barrels were used in 1980, find the solution of the equation, assuming $A_0 = 66,164$ and $k = 0.1$.

b) How much oil will be used in 1990?

Integrate.

33. $\int 3x^5 \, dx$

36. $\int x^3 e^{x^4} \, dx$ (Use substitution. Do not use Table 5.)

37. $\int (x + 3) \ln x \, dx$

40. Find the area under the graph of $y = x^2 + 3x$ on the interval $[1, 5]$.

42. Find the capital value of a rental property over a 50-year period where the annual rent is $4500 and the current interest rate is 9%.

27. An appliance store sells 450 pocket radios each year. It costs $4 to store one radio for one year. To order radios there is a fixed cost of $1 plus $.75 for each radio. How many times per year should the store reorder radios, and in what lot size, in order to minimize inventory costs?

29. For $f(x) = 3x^2 - 7$, $x = 5$, and $\Delta x = 0.1$, find Δy and $f'(x) \, \Delta x$.

32. Find the elasticity: $p = D(x) = 80 - 30 \ln (x + 1)$.

34. $\int_{-1}^{0} (2e^x + 1) \, dx$

35. $\int \frac{x}{(7 - 3x)^2} \, dx$ (Use Table 5.)

38. $\int 2x \sin (x^2 + 3) \, dx$

39. $\int e^x \cos (e^x) \, dx$

41. What annual payment should be made so the amount of an annuity, after 30 years at 9% compounded continuously, will be $81,000?

43. Determine whether the following improper integral is convergent or divergent, and calculate its value if convergent:

$$\int_3^{\infty} \frac{1}{x^7} \, dx.$$

44. Given the probability density function

$$f(x) = \frac{3}{2x^2} \text{ over } [1, 3],$$

find $E(x)$.

46. Given the demand and supply functions

$$D(x) = (x - 20)^2 \quad \text{and} \quad S(x) = x^2 + 10x + 50,$$

find the equilibrium point and the consumer's surplus.

48. Solve the differential equation $dy/dx = xy$.

45. Let x be a continuous random variable that is normally distributed with mean $\mu = 3$ and standard deviation $\sigma = 5$. Using Table 6, find $P(-2 \le x \le 8)$.

47. Find the volume of the solid of revolution generated by rotating the region under the graph of

$$y = e^{-x} \quad \text{from } x = 0 \text{ to } x = 5,$$

about the x-axis.

49. In an advertising experiment it was determined that

$$\frac{dP}{dt} = k(L - P),$$

where P = the percentage of the people in the city who bought the product after the ad was run t times, and $L = 100\%$, or 1.

a) Express $P(t)$ in terms of $L = 1$.

b) It was determined that $P(10) = 70\%$. Use this to determine k. Round your answer to the nearest hundredth.

c) Rewrite $P(t)$ in terms of k.

d) Use the equation in (c) to find $P(20)$.

Given $f(x, y) = e^y + 4x^2y^3 + 3x$, find each of the following.

50. f_x

51. f_{yy}

52. Find the relative maximum and minimum values of $f(x, y) = 8x^2 - y^2$.

53. Maximize $f(x, y) = 4x + 2y - x^2 - y^2 + 4$, subject to the constraint $x + 2y = 9$.

54. Evaluate.

$$\int_0^3 \int_{-1}^2 e^x \, dy \, dx$$

Find each of the following.

55. $\sin \frac{\pi}{4}$

56. $\cos \frac{\pi}{3}$

57. $\tan \pi$

Integrate.

58. $\displaystyle\int \sec 8x \, dx$

59. $\displaystyle\int \frac{dx}{1 + 9x^2}$

Differentiate.

60. $y = \cot (\sin x^3)$

61. $f(x) = \sin^{-1} x$

Find.

62. dw when $w = 3x^5y^2$

63. dw when $w = 7e^x \sin(yz)$

64. A rectangular box has dimensions 7.01 feet by 2.03 feet by 0.97 feet. Use differentials to approximate the volume of the box.

66. Find the volume of the solid whose base is the triangular region

$$G = \{(x, y) \,|\, x \le y \le 2 - x, \quad 0 \le x \le 1\}$$

and whose height at each point is $2xy$.

Solve.

68. $(y - 3)dx + (x + 3y^2)dy = 0$

71. $y'' - 14y' + 49y = 0$

74. Expand $5x^3 - 13x^2 + 8x - 5$ in powers of $x - 1$.

76. Find the sum of $\dfrac{1}{5} - \dfrac{1}{35} + \dfrac{1}{245} - \dfrac{1}{1715} + \cdots$.

78. Does $\dfrac{1}{9} - \dfrac{1}{16} + \dfrac{1}{25} - \dfrac{1}{36} + \cdots$ converge or diverge?

80. Find an approximate value for $\displaystyle\int_0^1 x \sin(x^4)dx$.

Evaluate.

81. $\displaystyle\lim_{x \to 1} \dfrac{2x}{x - 1}$

83. $\displaystyle\lim_{x \to 0} \dfrac{e^{4x} - 1}{x}$

85. Use iteration to solve $5x - 2\cos x = 0$. Then check your answer.

87. Find the present value of a payment of $1000 to be made 8 years from today at a rate of interest of 10%.

89. Use the Trapezoidal rule with $n = 5$ to estimate

$$\int_0^1 \frac{dx}{x + 2}.$$

Compare your value with the exact answer.

91. a) Use Simpson's Rule with $n = 2$ and $n = 4$ to estimate

$$I = \int_0^1 x^8 \, dx.$$

   b) Then use S_2 and S_4 to compute R_2.

65. Evaluate

$$\int_0^1 \left\{ \int_0^1 x^3 y^2 dx \right\} dy.$$

67. Evaluate $\displaystyle\iint_G x^4 y$ where G is the unit disk,

$$G = \{(x, y) \,|\, x^2 + y^2 \le 1\}.$$

69. $y' + (\cos x)y = 0$

72. $y'' + 4y' + 3y = 15$

70. $y''' + 4y' = 0$

73. $x' = -4x + y$
     $y' = -15x + 4y$

75. Use the Taylor polynomial of degree 3 for $y = \sin x$ at $a = 0$ to find an approximate value for $\sin(0.07)$.

77. Write $0.679\overline{679}$ as a rational fraction.

79. Compute the Taylor series expansion of $f(x) = x \sin(x^4)$ at $x = 0$.

82. $\displaystyle\lim_{x \to 1} \dfrac{x^3 - 5x^2 + 3x + 1}{x^2 - 1}$

84. $\displaystyle\lim_{x \to 1} \dfrac{2x}{x + 1}$

86. Use Newton's Method to find the solution of $x^3 - 6x + 4 = 0$ that lies between 0 and 1.

88. An investor pays $7000 for three annual payments of $3000 each, beginning one year from today. Find the yield rate of this investment and check your answer.

90. Use Simpson's Rule with $n = 5$ to estimate

$$\int_0^1 \frac{dx}{x + 2}.$$

TABLES

TABLE 1 POWERS, ROOTS, AND RECIPROCALS

n	n^2	n^3	$\sqrt{n}$	$\sqrt[3]{n}$	$\sqrt{10n}$	$\dfrac{1}{n}$	n	n^2	n^3	$\sqrt{n}$	$\sqrt[3]{n}$	$\sqrt{10n}$	$\dfrac{1}{n}$
1	1	1	1.000	1.000	3.162	1.0000	51	2,601	132,651	7.141	3.708	22.583	.0196
2	4	8	1.414	1.260	4.472	.5000	52	2,704	140,608	7.211	3.733	22.804	.0192
3	9	27	1.732	1.442	5.477	.3333	53	2,809	148,877	7.280	3.756	23.022	.0189
4	16	64	2.000	1.587	6.325	.2500	54	2,916	157,464	7.348	3.780	23.238	.0185
5	25	125	2.236	1.710	7.071	.2000	55	3,025	166,375	7.416	3.803	23.452	.0182
6	36	216	2.449	1.817	7.746	.1667	56	3,136	175,616	7.483	3.826	23.664	.0179
7	49	343	2.646	1.913	8.367	.1429	57	3,249	185,193	7.550	3.849	23.875	.0175
8	64	512	2.828	2.000	8.944	.1250	58	3,364	195,112	7.616	3.871	24.083	.0172
9	81	729	3.000	2.080	9.487	.1111	59	3,481	205,379	7.681	3.893	24.290	.0169
10	100	1,000	3.162	2.154	10.000	.1000	60	3,600	216,000	7.746	3.915	24.495	.0167
11	121	1,331	3.317	2.224	10.488	.0909	61	3,721	226,981	7.810	3.936	24.698	.0164
12	144	1,728	3.464	2.289	10.954	.0833	62	3,844	238,328	7.874	3.958	24.900	.0161
13	169	2,197	3.606	2.351	11.402	.0769	63	3,969	250,047	7.937	3.979	25.100	.0159
14	196	2,744	3.742	2.410	11.832	.0714	64	4,096	262,144	8.000	4.000	25.298	.0156
15	225	3,375	3.873	2.466	12.247	.0667	65	4,225	274,625	8.062	4.021	25.495	.0154
16	256	4,096	4.000	2.520	12.648	.0625	66	4,356	287,496	8.124	4.041	25.690	.0152
17	289	4,913	4.123	2.571	13.038	.0588	67	4,489	300,763	8.185	4.062	25.884	.0149
18	324	5,832	4.243	2.621	13.416	.0556	68	4,624	314,432	8.246	4.082	26.077	.0147
19	361	6,859	4.359	2.668	13.784	.0526	69	4,761	328,509	8.307	4.102	26.268	.0145
20	400	8,000	4.472	2.714	14.142	.0500	70	4,900	343,000	8.367	4.121	26.458	.0143
21	441	9,261	4.583	2.759	14.491	.0476	71	5,041	357,911	8.426	4.141	26.646	.0141
22	484	10,648	4.690	2.802	14.832	.0455	72	5,184	373,248	8.485	4.160	26.833	.0139
23	529	12,167	4.796	2.844	15.166	.0435	73	5,329	389,017	8.544	4.179	27.019	.0137
24	576	13,824	4.899	2.884	15.492	.0417	74	5,476	405,224	8.602	4.198	27.203	.0135
25	625	15,625	5.000	2.924	15.811	.0400	75	5,625	421,875	8.660	4.217	27.386	.0133
26	676	17,576	5.099	2.962	16.125	.0385	76	5,776	438,976	8.718	4.236	27.568	.0132
27	729	19,683	5.196	3.000	16.432	.0370	77	5,929	456,533	8.775	4.254	27.749	.0130
28	784	21,952	5.292	3.037	16.733	.0357	78	6,084	474,552	8.832	4.273	27.928	.0128
29	841	24,389	5.385	3.072	17.029	.0345	79	6,241	493,039	8.888	4.291	28.107	.0127
30	900	27,000	5.477	3.107	17.321	.0333	80	6,400	512,000	8.944	4.309	28.284	.0125
31	961	29,791	5.568	3.141	17.607	.0323	81	6,561	531,441	9.000	4.327	28.460	.0123
32	1,024	32,768	5.657	3.175	17.889	.0312	82	6,724	551,368	9.055	4.344	28.636	.0122
33	1,089	35,937	5.745	3.208	18.166	.0303	83	6,889	571,787	9.110	4.362	28.810	.0120
34	1,156	39,304	5.831	3.240	18.439	.0294	84	7.056	592,704	9.165	4.380	28.983	.0119
35	1,225	42,875	5.916	3.271	18.708	.0286	85	7,225	614,125	9.220	4.397	28.155	.0118
36	1,296	46,656	6.000	3.302	18.974	.0278	86	7,396	636,056	9.274	4.414	29.326	.0116
37	1,369	50,653	6.083	3.332	19.235	.0270	87	7,569	658,503	9.327	4.431	29.496	.0115
38	1,444	54,872	6.164	3.362	19.494	.0263	88	7,744	681,472	9.381	4.448	29.665	.0114
39	1,521	59,319	6.245	3.391	19.748	.0256	89	7,921	704,969	9.434	4.465	29.833	.0112
40	1,600	64,000	6.325	3.420	20.000	.0250	90	8,100	729,000	9.487	4.481	30.000	.0111
41	1,681	68,921	6.403	3.448	20.248	.0244	91	8,281	753,571	9.539	4.498	30.166	.0110
42	1,764	74,088	6.481	3.476	20.494	.0238	92	8,464	778,688	9.592	4.514	30.332	.0109
43	1,849	79,507	6.557	3.503	20.736	.0233	93	8,649	804,357	9.644	4.531	30.496	.0108
44	1,936	85,184	6.633	3.530	20.976	.0227	94	8,836	830,584	9.695	4.547	30.659	.0106
45	2,025	91,125	6.708	3.557	21.213	.0222	95	9,025	857,375	9.747	4.563	30.822	.0105
46	2,116	97,336	6.782	3.583	21.448	.0217	96	9,216	884,736	9.798	4.579	30.984	.0104
47	2,209	103,823	6.856	3.609	21.679	.0213	97	9,409	912,673	9.849	4.595	31.145	.0103
48	2,304	110,592	6.928	3.634	21.909	.0208	98	9,604	941,192	9.899	4.610	31.305	.0102
49	2,401	117,649	7.000	3.659	22.136	.0204	99	9,801	970,299	9.950	4.626	31.464	.0101
50	2,500	125,000	7.071	3.684	22.361	.0200	100	10,000	1,000,000	10.000	4.642	31.623	.0100

TABLE 2 COMMON LOGARITHMS

x	0	1	2	3	4	5	6	7	8	9
1.0	.0000	.0043	.0086	.0128	.0170	.0212	.0253	.0294	.0334	.0374
1.1	.0414	.0453	.0492	.0531	.0569	.0607	.0645	.0682	.0719	.0755
1.2	.0792	.0828	.0864	.0899	.0934	.0969	.1004	.1038	.1072	.1106
1.3	.1139	.1173	.1206	.1239	.1271	.1303	.1335	.1367	.1399	.1430
1.4	.1461	.1492	.1523	.1553	.1584	.1614	.1644	.1673	.1703	.1732
1.5	.1761	.1790	.1818	.1847	.1875	.1903	.1931	.1959	.1987	.2014
1.6	.2041	.2068	.2095	.2122	.2148	.2175	.2201	.2227	.2253	.2279
1.7	.2304	.2330	.2355	.2380	.2405	.2430	.2455	.2480	.2504	.2529
1.8	.2553	.2577	.2601	.2625	.2648	.2672	.2695	.2718	.2742	.2765
1.9	.2788	.2810	.2833	.2856	.2878	.2900	.2923	.2945	.2967	.2989
2.0	.3010	.3032	.3054	.3075	.3096	.3118	.3139	.3160	.3181	.3201
2.1	.3222	.3243	.3263	.3284	.3304	.3324	.3345	.3365	.3385	.3404
2.2	.3424	.3444	.3464	.3483	.3502	.3522	.3541	.3560	.3579	.3598
2.3	.3617	.3636	.3655	.3674	.3692	.3711	.3729	.3747	.3766	.3784
2.4	.3802	.3820	.3838	.3856	.3874	.3892	.3909	.3927	.3945	.3962
2.5	.3979	.3997	.4014	.4031	.4048	.4065	.4082	.4099	.4116	.4133
2.6	.4150	.4166	.4183	.4200	.4216	.4232	.4249	.4265	.4281	.4298
2.7	.4314	.4330	.4346	.4362	.4378	.4393	.4409	.4425	.4440	.4456
2.8	.4472	.4487	.4502	.4518	.4533	.4548	.4564	.4579	.4594	.4609
2.9	.4624	.4639	.4654	.4669	.4683	.4698	.4713	.4728	.4742	.4757
3.0	.4771	.4786	.4800	.4814	.4829	.4843	.4857	.4871	.4886	.4900
3.1	.4914	.4928	.4942	.4955	.4969	.4983	.4997	.5011	.5024	.5038
3.2	.5051	.5065	.5079	.5092	.5105	.5119	.5132	.5145	.5159	.5172
3.3	.5185	.5198	.5211	.5224	.5237	.5250	.5263	.5276	.5289	.5307
3.4	.5315	.5328	.5340	.5353	.5366	.5378	.5391	.5403	.5416	.5428
3.5	.5441	.5453	.5465	.5478	.5490	.5502	.5514	.5527	.5539	.5551
3.6	.5563	.5575	.5587	.5599	.5611	.5623	.5635	.5647	.5658	.5670
3.7	.5682	.5694	.5705	.5717	.5729	.5740	.5752	.5763	.5775	.5786
3.8	.5798	.5809	.5821	.5832	.5843	.5855	.5866	.5877	.5888	.5899
3.9	.5911	.5922	.5933	.5944	.5955	.5966	.5977	.5988	.5999	.6010
4.0	.6021	.6031	.6042	.6053	.6064	.6075	.6085	.6096	.6107	.6117
4.1	.6128	.6138	.6149	.6160	.6170	.6180	.6191	.6201	.6212	.6222
4.2	.6232	.6243	.6253	.6263	.6274	.6284	.6294	.6304	.6314	.6325
4.3	.6335	.6345	.6355	.6365	.6375	.6385	.6395	.6405	.6415	.6425
4.4	.6435	.6444	.6454	.6464	.6474	.6484	.6493	.6503	.6513	.6522
4.5	.6532	.6542	.6551	.6561	.6571	.6580	.6590	.6599	.6609	.6618
4.6	.6628	.6637	.6646	.6656	.6665	.6675	.6684	.6693	.6702	.6712
4.7	.6721	.6730	.6739	.6749	.6758	.6767	.6776	.6785	.6794	.6803
4.8	.6812	.6821	.6830	.6839	.6848	.6857	.6866	.6875	.6884	.6893
4.9	.6902	.6911	.6920	.6928	.6937	.6946	.6955	.6964	.6972	.6981
5.0	.6990	.6998	.7007	.7016	.7024	.7033	.7042	.7050	.7059	.7067
5.1	.7076	.7084	.7093	.7101	.7110	.7118	.7126	.7135	.7143	.7152
5.2	.7160	.7168	.7177	.7185	.7193	.7202	.7210	.7218	.7226	.7235
5.3	.7243	.7251	.7259	.7267	.7275	.7284	.7292	.7300	.7308	.7316
5.4	.7324	.7332	.7340	.7348	.7356	.7364	.7372	.7380	.7388	.7396
x	0	1	2	3	4	5	6	7	8	9

TABLE 2—(*cont.*)

x	0	1	2	3	4	5	6	7	8	9
5.5	.7404	.7412	.7419	.7427	.7435	.7443	.7451	.7459	.7466	.7474
5.6	.7482	.7490	.7497	.7505	.7513	.7520	.7528	.7536	.7543	.7551
5.7	.7559	.7566	.7574	.7582	.7589	.7597	.7604	.7612	.7619	.7627
5.8	.7634	.7642	.7649	.7657	.7664	.7672	.7679	.7686	.7694	.7701
5.9	.7709	.7716	.7723	.7731	.7738	.7745	.7752	.7760	.7767	.7774
6.0	.7782	.7789	.7796	.7803	.7810	.7818	.7825	.7832	.7839	.7846
6.1	.7853	.7860	.7868	.7875	.7882	.7889	.7896	.7903	.7910	.7917
6.2	.7924	.7931	.7938	.7945	.7952	.7959	.7966	.7973	.7980	.7987
6.3	.7993	.8000	.8007	.8014	.8021	.8028	.8035	.8041	.8048	.8055
6.4	.8062	.8069	.8075	.8082	.8089	.8096	.8102	.8109	.8116	.8122
6.5	.8129	.8136	.8142	.8149	.8156	.8162	.8169	.8176	.8182	.8189
6.6	.8195	.8202	.8209	.8215	.8222	.8228	.8235	.8241	.8248	.8254
6.7	.8261	.8267	.8274	.8280	.8287	.8293	.8299	.8306	.8312	.8319
6.8	.8325	.8331	.8338	.8344	.8351	.8357	.8363	.8370	.8376	.8382
6.9	.8388	.8395	.8401	.8407	.8414	.8420	.8426	.8432	.8439	.8445
7.0	.8451	.8457	.8463	.8470	.8476	.8482	.8488	.8494	.8500	.8506
7.1	.8513	.8519	.8525	.8531	.8537	.8543	.8549	.8555	.8561	.8567
7.2	.8573	.8579	.8585	.8591	.8597	.8603	.8609	.8615	.8621	.8627
7.3	.8633	.8639	.8645	.8651	.8657	.8663	.8669	.8675	.8681	.8686
7.4	.8692	.8698	.8704	.8710	.8716	.8722	.8727	.8733	.8739	.8745
7.5	.8751	.8756	.8762	.8768	.8774	.8779	.8785	.8791	.8797	.8802
7.6	.8808	.8814	.8820	.8825	.8831	.8837	.8842	.8848	.8854	.8859
7.7	.8865	.8871	.8876	.8882	.8887	.8893	.8899	.8904	.8910	.8915
7.8	.8921	.8927	.8932	.8938	.8943	.8949	.8954	.8960	.8965	.8971
7.9	.8976	.8982	.8987	.8993	.8998	.9004	.9009	.9015	.9020	.9025
8.0	.9031	.9036	.9042	.9047	.9053	.9058	.9063	.9069	.9074	.9079
8.1	.9085	.9090	.9096	.9101	.9106	.9112	.9117	.9122	.9128	.9133
8.2	.9138	.9143	.9149	.9154	.9159	.9165	.9170	.9175	.9180	.9186
8.3	.9191	.9196	.9201	.9206	.9212	.9217	.9222	.9227	.9232	.9238
8.4	.9243	.9248	.9253	.9258	.9263	.9269	.9274	.9279	.9284	.9289
8.5	.9294	.9299	.9304	.9309	.9315	.9320	.9325	.9330	.9335	.9340
8.6	.9345	.9350	.9555	.9360	.9365	.9370	.9375	.9380	.9385	.9390
8.7	.9395	.9400	.9405	.9410	.9415	.9420	.9425	.9430	.9435	.9440
8.8	.9445	.9450	.9455	.9460	.9465	.9469	.9474	.9479	.9484	.9489
8.9	.9494	.9499	.9504	.9509	.9513	.9518	.9523	.9528	.9533	.9538
9.0	.9542	.9547	.9552	.9557	.9562	.9566	.9571	.9576	.9581	.9586
9.1	.9590	.9595	.9600	.9605	.9609	.9614	.9619	.9624	.9628	.9633
9.2	.9638	.9643	.9647	.9652	.9657	.9661	.9666	.9671	.9675	.9680
9.3	.9685	.9689	.9694	.9699	.9703	.9708	.9713	.9717	.9722	.9727
9.4	.9731	.9736	.9741	.9745	.9750	.9754	.9759	.9763	.9768	.9773
9.5	.9777	.9782	.9786	.9791	.9795	.9800	.9805	.9809	.9814	.9818
9.6	.9823	.9827	.9832	.9836	.9841	.9845	.9850	.9854	.9859	.9863
9.7	.9868	.9872	.9877	.9881	.9886	.9890	.9894	.9899	.9903	.9908
9.8	.9912	.9917	.9921	.9926	.9930	.9934	.9939	.9943	.9948	.9952
9.9	.9956	.9961	.9965	.9969	.9974	.9978	.9983	.9987	.9991	.9996
x	0	1	2	3	4	5	6	7	8	9

TABLE 3   NATURAL LOGARITHMS (ln x)

x	0.00	0.01	0.02	0.03	0.04	0.05	0.06	0.07	0.08	0.09
1.0	0.0000	0.0100	0.0198	0.0296	0.0392	0.0488	0.0583	0.0677	0.0770	0.0862
1.1	0.0953	0.1044	0.1133	0.1222	0.1310	0.1398	0.1484	0.1570	0.1655	0.1740
1.2	0.1823	0.1906	0.1989	0.2070	0.2151	0.2231	0.2311	0.2390	0.2469	0.2546
1.3	0.2624	0.2700	0.2776	0.2852	0.2927	0.3001	0.3075	0.3148	0.3221	0.3293
1.4	0.3365	0.3436	0.3507	0.3577	0.3646	0.3716	0.3784	0.3853	0.3920	0.3988
1.5	0.4055	0.4121	0.4187	0.4253	0.4318	0.4383	0.4447	0.4511	0.4574	0.4637
1.6	0.4700	0.4762	0.4824	0.4886	0.4947	0.5008	0.5068	0.5128	0.5188	0.5247
1.7	0.5306	0.5365	0.5423	0.5481	0.5539	0.5596	0.5653	0.5710	0.5766	0.5822
1.8	0.5878	0.5933	0.5988	0.6043	0.6098	0.6152	0.6206	0.6259	0.6313	0.6366
1.9	0.6419	0.6471	0.6523	0.6575	0.6627	0.6678	0.6729	0.6780	0.6831	0.6881
2.0	0.6931	0.6981	0.7031	0.7080	0.7130	0.7178	0.7227	0.7275	0.7324	0.7372
2.1	0.7419	0.7467	0.7514	0.7561	0.7608	0.7655	0.7701	0.7747	0.7793	0.7839
2.2	0.7885	0.7930	0.7975	0.8020	0.8065	0.8109	0.8154	0.8198	0.8242	0.8286
2.3	0.8329	0.8372	0.8416	0.8459	0.8502	0.8544	0.8587	0.8629	0.8671	0.8713
2.4	0.8755	0.8796	0.8838	0.8879	0.8920	0.8961	0.9002	0.9042	0.9083	0.9123
2.5	0.9163	0.9203	0.9243	0.9282	0.9322	0.9361	0.9400	0.9439	0.9478	0.9517
2.6	0.9555	0.9594	0.9632	0.9670	0.9708	0.9746	0.9783	0.9821	0.9858	0.9895
2.7	0.9933	0.9969	1.0006	1.0043	1.0080	1.0116	1.0152	0.0188	1.0225	1.0260
2.8	1.0296	1.0332	1.0367	1.0403	1.0438	1.0473	1.0508	1.0543	1.0578	1.0613
2.9	1.0647	1.0682	1.0716	1.0750	1.0784	1.0818	1.0852	1.0886	1.0919	1.0953
3.0	1.0986	1.1019	1.1053	1.1086	1.1119	1.1151	1.1184	1.1217	1.1249	1.1282
3.1	1.1314	1.1346	1.1378	1.1410	1.1442	1.1474	1.1506	1.1537	1.1569	1.1600
3.2	1.1632	1.1663	1.1694	1.1725	1.1756	1.1787	1.1817	1.1848	1.1878	1.1909
3.3	1.1939	1.1970	1.2000	1.2030	1.2060	1.2090	1.2119	1.2149	1.2179	1.2208
3.4	1.2238	1.2267	1.2296	1.2326	1.2355	1.2384	1.2413	1.2442	1.2470	1.2499
3.5	1.2528	1.2556	1.2585	1.2613	1.2641	1.2669	1.2698	1.2726	1.2754	1.2782
3.6	1.2809	1.2837	1.2865	1.2892	1.2920	1.2947	1.2975	1.3002	1.3029	1.3056
3.7	1.3083	1.3110	1.3137	1.3164	1.3191	1.3218	1.3244	1.3271	1.3297	1.3324
3.8	1.3350	1.3376	1.3403	1.3429	1.3455	1.3481	1.3507	1.3533	1.3558	1.3584
3.9	1.3610	1.3635	1.3661	1.3686	1.3712	1.3737	1.3762	1.3788	1.3813	1.3838
4.0	1.3863	1.3888	1.3913	1.3938	1.3962	1.3987	1.4012	1.4036	1.4061	1.4085
4.1	1.4110	1.4134	1.4159	1.4183	1.4207	1.4231	1.4255	1.4279	1.4303	1.4327
4.2	1.4351	1.4375	1.4398	1.4422	1.4446	1.4469	1.4493	1.4516	1.4540	1.4563
4.3	1.4586	1.4609	1.4633	1.4656	1.4679	1.4702	1.4725	1.4748	1.4770	1.4793
4.4	1.4816	1.4839	1.4861	1.4884	1.4907	1.4929	1.4952	1.4974	1.4996	1.5019
4.5	1.5041	1.5063	1.5085	1.5107	1.5129	1.5151	1.5173	1.5195	1.5217	1.5239
4.6	1.5261	1.5282	1.5304	1.5326	1.5347	1.5369	1.5390	1.5412	1.5433	1.5454
4.7	1.5476	1.5497	1.5518	1.5539	1.5560	1.5581	1.5602	1.5623	1.5644	1.5665
4.8	1.5686	1.5707	1.5728	1.5748	1.5769	1.5790	1.5810	1.5831	1.5851	1.5872
4.9	1.5892	1.5913	1.5933	1.5953	1.5974	1.5994	1.6014	1.6034	1.6054	1.6074
5.0	1.6094	1.6114	1.6134	1.6154	1.6174	1.6194	1.6214	1.6233	1.6253	1.6273
5.1	1.6292	1.6312	1.6332	1.6351	1.6371	1.6390	1.6409	1.6429	1.6448	1.6467
5.2	1.6487	1.6506	1.6525	1.6544	1.6563	1.6582	1.6601	1.6620	1.6639	1.6658
5.3	1.6677	1.6696	1.6715	1.6734	1.6752	1.6771	1.6790	1.6808	1.6827	1.6845
5.4	1.6864	1.6882	1.6901	1.6919	1.6938	1.6956	1.6974	1.6993	1.7011	1.7029
5.5	1.7047	1.7066	1.7084	1.7102	1.7120	1.7138	1.7156	1.7174	1.7192	1.7210
5.6	1.7228	1.7246	1.7263	1.7281	1.7299	1.7317	1.7334	1.7352	1.7370	1.7387
5.7	1.7405	1.7422	1.7440	1.7457	1.7475	1.7492	1.7509	1.7527	1.7544	1.7561
5.8	1.7579	1.7596	1.7613	1.7630	1.7647	1.7664	1.7682	1.7699	1.7716	1.7733
5.9	1.7750	1.7766	1.7783	1.7800	1.7817	1.7834	1.7851	1.7867	1.7884	1.7901

TABLE 3—(*cont.*)

x	0.00	0.01	0.02	0.03	0.04	0.05	0.06	0.07	0.08	0.09
6.0	1.7918	1.7934	1.7951	1.7967	1.7984	1.8001	1.8017	1.8034	1.8050	1.8066
6.1	1.8083	1.8099	1.8116	1.8132	1.8148	1.8165	1.8181	1.8197	1.8213	1.8229
6.2	1.8245	1.8262	1.8278	1.8294	1.8310	1.8326	1.8342	1.8358	1.8374	1.8390
6.3	1.8406	1.8421	1.8437	1.8453	1.8469	1.8485	1.8500	1.8516	1.8532	1.8547
6.4	1.8563	1.8579	1.8594	1.8610	1.8625	1.8641	1.8656	1.8672	1.8687	1.8703
6.5	1.8718	1.8733	1.8749	1.8764	1.8779	1.8795	1.8810	1.8825	1.8840	1.8856
6.6	1.8871	1.8886	1.8901	1.8916	1.8931	1.8946	1.8961	1.8976	1.8991	1.9006
6.7	1.9021	1.9036	1.9051	1.9066	1.9081	1.9095	1.9110	1.9125	1.9140	1.9155
6.8	1.9169	1.9184	1.9199	1.9213	1.9228	1.9242	1.9257	1.9272	1.9286	1.9301
6.9	1.9315	1.9330	1.9344	1.9359	1.9373	1.9387	1.9402	1.9416	1.9430	1.9445
7.0	1.9459	1.9473	1.9488	1.9502	1.9516	1.9530	1.9544	1.9559	1.9573	1.9587
7.1	1.9601	1.9615	1.9629	1.9643	1.9657	1.9671	1.9685	1.9699	1.9713	1.9727
7.2	1.9741	1.9755	1.9769	1.9782	1.9796	1.9810	1.9824	1.9838	1.9851	1.9865
7.3	1.9879	1.9892	1.9906	1.9920	1.9933	1.9947	1.9961	1.9974	1.9988	2.0001
7.4	2.0015	2.0028	2.0042	2.0055	2.0069	2.0082	2.0096	2.0109	2.0122	2.0136
7.5	2.0149	2.0162	2.0176	2.0189	2.0202	2.0215	2.0229	2.0242	2.0255	2.0268
7.6	2.0282	2.0295	2.0308	2.0321	2.0334	2.0347	2.0360	2.0373	2.0386	2.0399
7.7	2.0412	2.0425	2.0438	2.0451	2.0464	2.0477	2.0490	2.0503	2.0516	2.0528
7.8	2.0541	2.0554	2.0567	2.0580	2.0592	2.0605	2.0618	2.0631	2.0643	2.0665
7.9	2.0669	2.0681	2.0694	2.0707	2.0719	2.0732	2.0744	2.0757	2.0769	2.0782
8.0	2.0794	2.0807	2.0819	2.0832	2.0844	2.0857	2.0869	2.0882	2.0894	2.0906
8.1	2.0919	2.0931	2.0943	2.0956	2.0968	2.0980	2.0992	2.1005	2.1017	2.1029
8.2	2.1041	2.1054	2.1066	2.1078	2.1090	2.1102	2.1114	2.1126	2.1133	2.1150
8.3	2.1163	2.1175	2.1187	2.1199	2.1211	2.1223	2.1235	2.1247	2.1258	2.1270
8.4	2.1282	2.1294	2.1306	2.1318	2.1330	2.1342	2.1353	2.1365	2.1377	2.1389
8.5	2.1401	2.1412	2.1424	2.1436	2.1448	2.1459	2.1471	2.1483	2.1494	2.1506
8.6	2.1518	2.1529	2.1541	2.1552	2.1564	2.1576	2.1587	2.1599	2.1610	2.1622
8.7	2.1633	2.1645	2.1656	2.1668	2.1679	2.1691	2.1702	2.1713	2.1725	2.1736
8.8	2.1748	2.1759	2.1770	2.1782	2.1793	2.1804	2.1815	2.1827	2.1838	2.1849
8.9	2.1861	2.1872	2.1883	2.1894	2.1905	2.1917	2.1928	2.1939	2.1950	2.1961
9.0	2.1972	2.1983	2.1994	2.2006	2.2017	2.2028	2.2039	2.2050	2.2061	2.2072
9.1	2.2083	2.2094	2.2105	2.2116	2.2127	2.2138	2.2148	2.2159	2.2170	2.2181
9.2	2.2192	2.2203	2.2214	2.2225	2.2235	2.2246	2.2257	2.2268	2.2279	2.2289
9.3	2.2300	2.2311	2.2322	2.2332	2.2343	2.2354	2.2364	2.2375	2.2386	2.2396
9.4	2.2407	2.2418	2.2428	2.2439	2.2450	2.2460	2.2471	2.2481	2.2492	2.2502
9.5	2.2513	2.2523	2.2534	2.2544	2.2555	2.2565	2.2576	2.2586	2.2597	2.2607
9.6	2.2618	2.2628	2.2638	2.2649	2.2659	2.2670	2.2680	2.2690	2.2701	2.2711
9.7	2.2721	2.2732	2.2742	2.2752	2.2762	2.2773	2.2783	2.2793	2.2803	2.2814
9.8	2.2824	2.2834	2.2844	2.2854	2.2865	2.2875	2.2885	2.2895	2.2905	2.2915
9.9	2.2925	2.2935	2.2946	2.2956	2.2966	2.2976	2.2986	2.2996	2.3006	2.3016

Examples.

$$\ln 96{,}700 = \ln 9.67 + 4 \ln 10$$
$$= 2.2690 + 9.2103$$
$$= 11.4793.$$

$$\ln 0.00967 = \ln 9.67 - 3 \ln 10$$
$$= 2.2690 - 6.9078$$
$$= -4.6388.$$

$\ln 10$	$= 2.3026$	$7 \ln 10$	$= 16.1181$
$2 \ln 10$	$= 4.6052$	$8 \ln 10$	$= 18.4207$
$3 \ln 10$	$= 6.9078$	$9 \ln 10$	$= 20.7233$
$4 \ln 10$	$= 9.2103$	$10 \ln 10$	$= 23.0259$
$5 \ln 10$	$= 11.5129$	$11 \ln 10$	$= 25.3284$
$6 \ln 10$	$= 13.8155$	$12 \ln 10$	$= 27.6310$

Note: Adapted from *Functional Approach to Precalculus*, 2nd ed., Mustafa A. Munem and James P. Yizze (New York, NY: Worth Publishers, Inc., © 1974), pp. 500–501. Reproduced by permission of the publisher.

TABLE 4 EXPONENTIAL FUNCTIONS

x	e^x	e^{-x}	x	e^x	e^{-x}	x	e^x	e^{-x}
0.00	1.0000	1.0000	0.55	1.7333	0.5769	3.6	36.598	0.0273
0.01	1.0101	0.9900	0.60	1.8221	0.5488	3.7	40.447	0.0247
0.02	1.0202	0.9802	0.65	1.9155	0.5220	3.8	44.701	0.0224
0.03	1.0305	0.9704	0.70	2.0138	0.4966	3.9	49.402	0.0202
0.04	1.0408	0.9608	0.75	2.1170	0.4724	4.0	54.598	0.0183
0.05	1.0513	0.9512	0.80	2.2255	0.4493	4.1	60.340	0.0166
0.06	1.0618	0.9418	0.85	2.3396	0.4274	4.2	66.686	0.0150
0.07	1.0725	0.9324	0.90	2.4596	0.4066	4.3	73.700	0.0136
0.08	1.0833	0.9231	0.95	2.5857	0.3867	4.4	81.451	0.0123
0.09	1.0942	0.9139	1.0	2.7183	0.3679	4.5	90.017	0.0111
0.10	1.1052	0.9048	1.1	3.0042	0.3329	4.6	99.484	0.0101
0.11	1.1163	0.8958	1.2	3.3201	0.3012	4.7	109.95	0.0091
0.12	1.1275	0.8869	1.3	3.6693	0.2725	4.8	121.51	0.0082
0.13	1.1388	0.8781	1.4	4.0552	0.2466	4.9	134.29	0.0074
0.14	1.1503	0.8694	1.5	4.4817	0.2231	5	148.41	0.0067
0.15	1.1618	0.8607	1.6	4.9530	0.2019	6	403.43	0.0025
0.16	1.1735	0.8521	1.7	5.4739	0.1827	7	1096.6	0.0009
0.17	1.1853	0.8437	1.8	6.0496	0.1653	8	2981.0	0.0003
0.18	1.1972	0.8353	1.9	6.6859	0.1496	9	8103.1	0.0001
0.19	1.2092	0.8270	2.0	7.3891	0.1353	10	22026	0.00005
0.20	1.2214	0.8187	2.1	8.1662	0.1225	11	59874	0.00002
0.21	1.2337	0.8106	2.2	9.0250	0.1108	12	162,754	0.000006
0.22	1.2461	0.8025	2.3	9.9742	0.1003	13	442,413	0.000002
0.23	1.2586	0.7945	2.4	11.023	0.0907	14	1,202,604	0.0000008
0.24	1.2712	0.7866	2.5	12.182	0.0821	15	3,269,017	0.0000003
0.25	1.2840	0.7788	2.6	13.464	0.0743			
0.26	1.2969	0.7711	2.7	14.880	0.0672			
0.27	1.3100	0.7634	2.8	16.445	0.0608			
0.28	1.3231	0.7558	2.9	18.174	0.0550			
0.29	1.3364	0.7483	3.0	20.086	0.0498			
0.30	1.3499	0.7408	3.1	22.198	0.0450			
0.35	1.4191	0.7047	3.2	24.533	0.0408			
0.40	1.4918	0.6703	3.3	27.113	0.0369			
0.45	1.5683	0.6376	3.4	29.964	0.0334			
0.50	1.6487	0.6065	3.5	33.115	0.0302			

TABLE 5 INTEGRATION FORMULAS

(Whenever $\ln X$ is used it is assumed that $X > 0$.)

1. $\int x^n \, dx = \dfrac{x^{n+1}}{n+1} + C, n \neq -1$

2. $\int \dfrac{dx}{x} = \ln x + C$

3. $\int u \, dv = uv - \int v \, du$

4. $\int e^x \, dx = e^x + C$

5. $\int e^{ax} \, dx = \dfrac{1}{a} \cdot e^{ax} + C$

6. $\int x e^{ax} \, dx = \dfrac{1}{a^2} \cdot e^{ax}(ax - 1) + C$

7. $\int x^n e^{ax} \, dx = \dfrac{x^n e^{ax}}{a} - \dfrac{n}{a} \int x^{n-1} e^{ax} \, dx$

8. $\int \ln x \, dx = x \ln x - x + C$

9. $\int (\ln x)^n \, dx = x(\ln x)^n - n \int (\ln x)^{n-1} \, dx, n \neq -1$

10. $\int x^n \ln x \, dx = x^{n+1} \left[\dfrac{\ln x}{n+1} - \dfrac{1}{(n+1)^2} \right] + C, n \neq -1$

11. $\int a^x \, dx = \dfrac{a^x}{\ln a} + C, a > 0, a \neq 1$

12. $\int \dfrac{1}{\sqrt{x^2 + a^2}} \, dx = \ln(x + \sqrt{x^2 + a^2}) + C$

13. $\int \dfrac{1}{\sqrt{x^2 - a^2}} \, dx = \ln(x + \sqrt{x^2 - a^2}) + C$

14. $\int \dfrac{1}{x^2 - a^2} \, dx = \dfrac{1}{2a} \ln\left(\dfrac{x - a}{x + a} \right) + C$

15. $\int \dfrac{1}{a^2 - x^2} \, dx = \dfrac{1}{2a} \ln\left(\dfrac{a + x}{a - x} \right) + C$

16. $\int \dfrac{1}{x\sqrt{a^2 + x^2}} \, dx = -\dfrac{1}{a} \ln\left(\dfrac{a + \sqrt{a^2 + x^2}}{x} \right) + C$

17. $\int \dfrac{1}{x\sqrt{a^2 - x^2}} \, dx = -\dfrac{1}{a} \ln\left(\dfrac{a + \sqrt{a^2 - x^2}}{x} \right) + C, 0 < x < a$

18. $\int \dfrac{x}{ax + b} \, dx = \dfrac{b}{a^2} + \dfrac{x}{a} - \dfrac{b}{a^2} \ln(ax + b) + C$

19. $\int \dfrac{x}{(ax + b)^2} \, dx = \dfrac{b}{a^2(ax + b)} + \dfrac{1}{a^2} \ln(ax + b) + C$

20. $\int \dfrac{1}{x(ax + b)} \, dx = \dfrac{1}{b} \ln\left(\dfrac{x}{ax + b} \right) + C$

21. $\int \dfrac{1}{x(ax + b)^2} \, dx = \dfrac{1}{b(ax + b)} + \dfrac{1}{b^2} \ln\left(\dfrac{x}{ax + b} \right) + C$

22. $\int \sqrt{x^2 \pm a^2} \, dx$

$\qquad = \frac{1}{2}[x\sqrt{x^2 \pm a^2} \pm a^2 \ln(x + \sqrt{x^2 \pm a^2})] + C$

Area = Probability = $P(0 \leq x \leq t)$

$$= \int_0^t \frac{1}{\sqrt{2\pi}} e^{-x^2/2} \, dx$$

TABLE 6   AREAS FOR A STANDARD NORMAL DISTRIBUTION

Entries in the table represent area under the curve between $t = 0$ and a positive value of t. Because of the symmetry of the curve, area under the curve between $t = 0$ and a negative value of t would be found in a like manner.

t	0.00	0.01	0.02	0.03	0.04	0.05	0.06	0.07	0.08	0.09
0.0	.0000	.0040	.0080	.0120	.0160	.0199	.0239	.0279	.0319	.0359
0.1	.0398	.0438	.0478	.0517	.0557	.0596	.0636	.0675	.0714	.0753
0.2	.0793	.0832	.0871	.0910	.0948	.0987	.1026	.1064	.1103	.1141
0.3	.1179	.1217	.1255	.1293	.1331	.1368	.1406	.1443	.1480	.1517
0.4	.1554	.1591	.1628	.1664	.1700	.1736	.1772	.1808	.1844	.1879
0.5	.1915	.1950	.1985	.2019	.2054	.2088	.2123	.2157	.2190	.2224
0.6	.2257	.2291	.2324	.2357	.2389	.2422	.2454	.2486	.2517	.2549
0.7	.2580	.2611	.2642	.2673	.2704	.2734	.2764	.2794	.2823	.2852
0.8	.2881	.2910	.2939	.2967	.2995	.3023	.3051	.3078	.3106	.3133
0.9	.3159	.3186	.3212	.3238	.3264	.3289	.3315	.3340	.3365	.3389
1.0	.3413	.3438	.3461	.3485	.3508	.3531	.3554	.3577	.3599	.3621
1.1	.3643	.3665	.3686	.3708	.3729	.3749	.3770	.3790	.3810	.3830
1.2	.3849	.3869	.3888	.3907	.3925	.3944	.3962	.3980	.3997	.4015
1.3	.4032	.4049	.4066	.4082	.4099	.4115	.4131	.4147	.4162	.4177
1.4	.4192	.4207	.4222	.4236	.4251	.4265	.4279	.4292	.4306	.4319
1.5	.4332	.4345	.4357	.4370	.4382	.4394	.4406	.4418	.4429	.4441
1.6	.4452	.4463	.4474	.4484	.4495	.4505	.4515	.4525	.4535	.4545
1.7	.4554	.4564	.4573	.4582	.4591	.4599	.4608	.4616	.4625	.4633
1.8	.4641	.4649	.4656	.4664	.4671	.4678	.4686	.4693	.4699	.4706
1.9	.4713	.4719	.4726	.4732	.4738	.4744	.4750	.4756	.4761	.4767
2.0	.4772	.4778	.4783	.4788	.4793	.4798	.4803	.4808	.4812	.4817
2.1	.4821	.4826	.4830	.4834	.4838	.4842	.4846	.4850	.4854	.4857
2.2	.4861	.4864	.4868	.4871	.4875	.4878	.4881	.4884	.4887	.4890
2.3	.4893	.4896	.4898	.4901	.4904	.4906	.4909	.4911	.4913	.4916
2.4	.4918	.4920	.4922	.4925	.4927	.4929	.4931	.4932	.4934	.4936
2.5	.4938	.4940	.4941	.4943	.4945	.4946	.4948	.4949	.4951	.4952
2.6	.4953	.4955	.4956	.4957	.4959	.4960	.4961	.4962	.4963	.4964
2.7	.4965	.4966	.4967	.4968	.4969	.4970	.4971	.4972	.4973	.4974
2.8	.4974	.4975	.4976	.4977	.4977	.4978	.4979	.4979	.4980	.4981
2.9	.4981	.4982	.4982	.4983	.4984	.4984	.4985	.4985	.4986	.4986
3.0	.4987	.4987	.4987	.4988	.4988	.4989	.4989	.4989	.4990	.4990

TABLE 7 TRIGONOMETRIC VALUES

Angle	Sine	Cosine	Tan-gent	Angle	Sine	Cosine	Tan-gent
0°	0.000	1.000	0.000				
1°	0.017	1.000	0.017	46°	0.719	0.695	1.036
2°	0.035	0.999	0.035	47°	0.731	0.682	1.072
3°	0.052	0.999	0.052	48°	0.743	0.669	1.111
4°	0.070	0.998	0.070	49°	0.755	0.656	1.150
5°	0.087	0.996	0.087	50°	0.766	0.643	1.192
6°	0.105	0.995	0.105	51°	0.777	0.629	1.235
7°	0.122	0.993	0.123	52°	0.788	0.616	1.280
8°	0.139	0.990	0.141	53°	0.799	0.602	1.327
9°	0.156	0.988	0.158	54°	0.809	0.588	1.376
10°	0.174	0.985	0.176	55°	0.819	0.574	1.428
11°	0.191	0.982	0.194	56°	0.829	0.559	1.483
12°	0.208	0.978	0.213	57°	0.839	0.545	1.540
13°	0.225	0.974	0.231	58°	0.848	0.530	1.600
14°	0.242	0.970	0.249	59°	0.857	0.515	1.664
15°	0.259	0.966	0.268	60°	0.866	0.500	1.732
16°	0.276	0.961	0.287	61°	0.875	0.485	1.804
17°	0.292	0.956	0.306	62°	0.883	0.469	1.881
18°	0.309	0.951	0.325	63°	0.891	0.454	1.963
19°	0.326	0.946	0.344	64°	0.899	0.438	2.050
20°	0.342	0.940	0.364	65°	0.906	0.423	2.145
21°	0.358	0.934	0.384	66°	0.914	0.407	2.246
22°	0.375	0.927	0.404	67°	0.921	0.391	2.356
23°	0.391	0.921	0.424	68°	0.927	0.375	2.475
24°	0.407	0.914	0.445	69°	0.934	0.358	2.605
25°	0.423	0.906	0.466	70°	0.940	0.342	2.748
26°	0.438	0.899	0.488	71°	0.946	0.326	2.904
27°	0.454	0.891	0.510	72°	0.951	0.309	3.078
28°	0.469	0.883	0.532	73°	0.956	0.292	3.271
29°	0.485	0.875	0.554	74°	0.961	0.276	3.487
30°	0.500	0.866	0.577	75°	0.966	0.259	3.732
31°	0.515	0.857	0.601	76°	0.970	0.242	4.011
32°	0.530	0.848	0.625	77°	0.974	0.225	4.332
33°	0.545	0.839	0.649	78°	0.978	0.208	4.705
34°	0.559	0.829	0.675	79°	0.982	0.191	5.145
35°	0.574	0.819	0.700	80°	0.985	0.174	5.671
36°	0.588	0.809	0.727	81°	0.988	0.156	5.314
37°	0.602	0.799	0.754	82°	0.990	0.139	7.115
38°	0.616	0.788	0.781	83°	0.993	0.122	8.144
39°	0.629	0.777	0.810	84°	0.995	0.105	9.514
40°	0.643	0.766	0.839	85°	0.996	0.087	11.43
41°	0.656	0.755	0.869	86°	0.998	0.070	14.30
42°	0.669	0.743	0.900	87°	0.999	0.052	19.08
43°	0.682	0.731	0.933	88°	0.999	0.035	28.64
44°	0.695	0.719	0.966	89°	1.000	0.017	57.29
45°	0.707	0.707	1.000	90°	1.000	0.000	

ANSWERS

CHAPTER 1

Margin Exercises, Section 1.1, pp. 2–8

1. $3 \cdot 3 \cdot 3 \cdot 3$, or 81　**2.** $(-3)(-3)$, or 9　**3.** $1.02 \times 1.02 \times 1.02$, or 1.061208　**4.** $\frac{1}{4} \cdot \frac{1}{4}$, or $\frac{1}{16}$　**5.** 1　**6** $5t$　**7.** 1　**8.** m

9. $\frac{1}{4}$　**10.** 1　**11.** $\frac{1}{2 \cdot 2 \cdot 2 \cdot 2}$, or $\frac{1}{16}$　**12.** $\frac{1}{10 \cdot 10}$, or $\frac{1}{100}$, or 0.01　**13.** 64　**14.** $\frac{1}{t^7}$　**15.** $\frac{1}{e^t}$　**16.** $\frac{1}{M}$　**17.** $\frac{1}{(x+1)^2}$　**18.** t^9

19. t^{-3}　**20.** $50e^{-13}$　**21.** t^{-6}　**22.** $24b^3$　**23.** x^4　**24.** x^{-4}　**25.** 1　**26.** e^{2-k}　**27.** e^{12}　**28.** e^2　**29.** x^{-12}

30. e^4　**31.** e^{3x}　**32.** $25x^6y^{10}$　**33.** $\frac{1}{256}x^{20}y^{24}z^{-8}$, or $\frac{x^{20}y^{24}}{256z^8}$　**34.** $2x + 14$　**35.** $P - Pi$　**36.** $x^2 + 3x - 28$

37. $a^2 - 2ab + b^2$　**38.** $a^2 - b^2$　**39.** $x^2 - 2xh + h^2$　**40.** $9x^2 + 6xt + t^2$　**41.** $25t^2 - m^2$　**42.** $P(1 - i)$　**43.** $(x + 5y)^2$

44. $4(x + 5)(x + 2)$　**45.** $(5c - d)(5c + d)$　**46.** $h(3x^2 + 3xh + h^2)$　**47.** 1.01　**48.** \$1299.60　**49.** \$1378.84

Exercise Set 1.1, p. 9–10

1. $5 \cdot 5 \cdot 5$, or 125　**3.** $(-7)(-7)$, or 49　**5.** 1.0201　**7.** $\frac{1}{16}$　**9.** 1　**11.** t　**13.** 1　**15.** $\frac{1}{3^2}$, or $\frac{1}{9}$　**17.** 8　**19.** 0.1　**21.** $\frac{1}{e^b}$

23. $\frac{1}{b}$　**25.** x^5　**27.** x^{-6}, or $\frac{1}{x^6}$　**29.** $35x^5$　**31.** x^4　**33.** 1　**35.** x^3　**37.** x^{-3}, or $\frac{1}{x^3}$　**39.** 1　**41.** e^{t-4}　**43.** t^{14}　**45.** t^2

47. t^{-6}, or $\frac{1}{t^6}$　**49.** e^{4x}　**51.** $8x^6y^{12}$　**53.** $\frac{1}{81}x^8y^{20}z^{-16}$, or $\frac{x^8y^{20}}{81z^{16}}$　**55.** $9x^{-16}y^{14}z^4$, or $\frac{9y^{14}z^4}{x^{16}}$　**57.** $5x - 35$

59. $x - xt$　**61.** $x^2 - 7x + 10$　**63.** $a^3 - b^3$　**65.** $2x^2 + 3x - 5$　**67.** $a^2 - 4$　**69.** $25x^2 - 4$　**71.** $a^2 - 2ah + h^2$

73. $25x^2 + 10xt + t^2$　**75.** $5x^5 + 30x^3 + 45x$　**77.** $a^3 + 3a^2b + 3ab^2 + b^3$　**79.** $x^3 - 15x^2 + 75x - 125$

81. $x(1 - t)$　**83.** $(x + 3y)^2$　**85.** $(x - 5)(x + 3)$　**87.** $(x - 5)(x + 4)$　**89.** $(7x - t)(7x + t)$　**91.** $4(3t - 2m)(3t + 2m)$

93. $ab(a + 4b)(a - 4b)$　**95.** $(a^4 + b^4)(a^2 + b^2)(a + b)(a - b)$　**97.** $10x(a + 2b)(a - 2b)$　**99.** $2(1 + 4x^2)(1 + 2x)(1 - 2x)$

101. $(x + 2)(x^2 - 2x + 4)$　**103.** (a) 0.81; (b) 0.0801; (c) 0.008001　**105.** (a) 1.261; (b) 0.120601; (c) 0.012006001

107. (a) \$1160; (b) \$1166.40; (c) \$1169.86; (d) \$1173.47; assuming 365 days in a year; (e) \$1173.51

Margin Exercises, Section 1.2, pp. 11–16

1. $\frac{56}{9}$ **2.** \$725   **3.** $0, -2, \frac{3}{2}$ **4.** $-4, 3$ **5.** $0, -1, 1$ **6.** $x < \frac{11}{5}$ **7.** $\frac{20}{17} \le x$ **8.** More than 19,975 suits

9. (a) $(-1, 3)$; (b) $(1, 4)$ **10.** (a) $(-1, 4)$; (b) $\left(-\frac{1}{4}, \frac{1}{4}\right)$ **11.** (a) $[-1, 4]$; (b) $(-1, 4]$; (c) $[-1, 4)$; (d) $(-1, 4)$ **12.** (a) $(-\sqrt{2}, \sqrt{2})$;
(b) $[0, 1)$; (c) $(-6.7, -4.2]$; (d) $\left[3, 7\frac{1}{2}\right]$ **13.** (a) $(-\infty, 5]$; (b) $(4, \infty)$; (c) $(-\infty, 4.8)$; (d) $[3, \infty)$ **14.** (a) $[8, \infty)$; (b) $(-\infty, -7)$;
(c) $(10, \infty)$; (d) $(-\infty, -0.78]$

Exercise Set 1.2, p. 17–18

1. $\frac{7}{4}$ **3.** -8 **5.** 120 **7.** 200 **9.** 480 lb   **11.** \$650   **13.** 810,000   **15.** $0, -3, \frac{4}{5}$ **17.** $0, 2$ **19.** $0, 3$ **21.** $0, 7$
23. $0, \frac{1}{3}, -\frac{1}{3}$ **25.** 1 **27.** $-\frac{4}{5} \le x$ **29.** $x > -\frac{1}{12}$ **31.** $x > -\frac{4}{7}$ **33.** $x \le -3$ **35.** $x > \frac{2}{3}$ **37.** $x < -\frac{2}{5}$
39. $2 < x < 4$ **41.** $\frac{3}{2} \le x \le \frac{11}{2}$ **43.** $-1 \le x \le \frac{14}{5}$ **45.** More than 7000 units   **47.** $60\% \le x < 100\%$ **49.** $(0, 5)$
51. $[-9, -4)$ **53.** $[x, x + h]$ **55.** (p, ∞) **57.** $[-3, 3]$ **59.** $[-14, -11)$ **61.** $(-\infty, -4]$

Margin Exercises, Section 1.3, pp. 19–29

1. 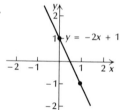   **2.** (a) Yes; (b) no   **3.** 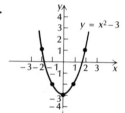   **4.**   **5.**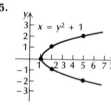

6.

Inputs	Outputs
5	$\frac{1}{5}$
$-\frac{2}{3}$	$-\frac{3}{2}$
$\frac{1}{4}$	4
$\frac{1}{a}$	a
k	$\frac{1}{k}$
$1 + t$	$\frac{1}{1 + t}$

7. $f(5) = \frac{1}{5}, f(-2) = -\frac{1}{2}, f\left(\frac{1}{4}\right) = 4, f\left(\frac{1}{a}\right) = a,$
$f(k) = \frac{1}{k}, f(1 + t) = \frac{1}{1 + t}, f(x + h) = \frac{1}{x + h}$
8. $t(5) = 30, t(-5) = 20, t(x + h) = x + h + x^2 + 2xh + h^2$ **9.** (a) All real numbers
except 3, since an input of 3 would result in
division by 0; (b) $f(5) = \frac{1}{2}, f(4) = 1, f(2.5) = -2, f(x + h) = \frac{1}{x + h - 3}$.   **10.** Same as
Margin Exercise 3, only labeled $f(x) = -2x + 1$
11. Same as Margin Exercise 4, only labeled
$g(x) = x^2 - 3$ **12.** c, d

13.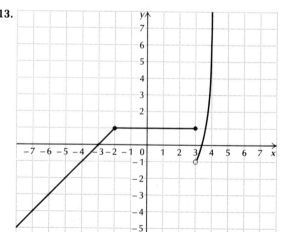

Exercise Set 1.3, p. 30–31

1. (a)

Inputs	Outputs
4.1	11.2
4.01	11.02
4.001	11.002
4	11

(b) $f(5) = 13, f(-1) = 1, f(k) = 2k + 3, f(1 + t) = 2t + 5, f(x + h) = 2x + 2h + 3$

3. $g(-1) = -2, g(0) = -3, g(1) = -2, g(5) = 22, g(u) = u^2 - 3, g(a + h) = a^2 + 2ah + h^2 - 3, g(1 - h) = h^2 - 2h - 2$

5. (a) $f(4) = 1, f(-2) = 25, f(0) = 9, f(a) = a^2 - 6a + 9, f(t + 1) = t^2 - 4t + 4, f(t + 3) = t^2, f(x + h) = x^2 + 2xh + h^2 - 6x - 6h + 9$; (b) Take an input, square it, subtract 6 times the input, add 9.

7.

9.

11.
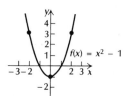

13.
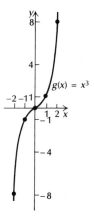

15. Yes   **17.** Yes   **19.** No   **21.** No   **23** (a)
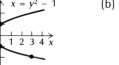
(b) no

25. $f(x + h) = x^2 + 2xh + h^2 - 3x - 3h$ **27.** $R(10) = \$70. R(100) = \$250$

29.
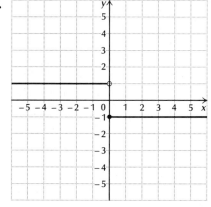

31.
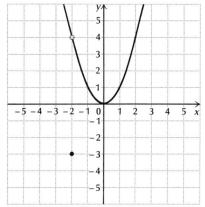

33. $y = 5$; a function   **35.** $y = \pm\sqrt{x}$; not a function

Margin Exercises, Section 1.4, pp. 32–44

1. (a) Horizontal line through $(0, 3)$; (b) yes   **2.** (a) Vertical line through $(1, 0)$; (b) no   **3.** (a)

(b) yes; (c) -2 **4.** (a) A, B; (b) C, D, E; (c) A; (d) E **5.** (a) $T = \frac{1}{36}h$; (b) 4.5

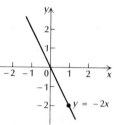

6. (a) 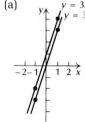   (b) by moving it upward 1 unit   **7.** $m = -\frac{2}{3}$, y-intercept: $(0, 2)$ **8.** $y = -4x + 1$

9. $y + 7 = -4(x - 2)$, or $y = -4x + 1$ **10.** 2   **11.** $\frac{1}{8}$ **12.** $-\frac{17}{2}$ **13.** 0   **14.** 0   **15.** No slope

16. (a)

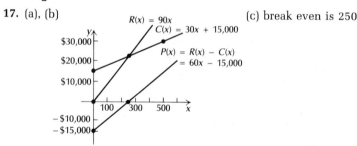

(b) $C(100) = \$18,000$, $C(400) = \$27,000$;
(c) $C(400) - C(100) = \$9000$

17. (a), (b)   (c) break even is 250

$P(x) = R(x) - C(x)$
$= 60x - 15,000$

Exercise Set 1.4, pp. 44–47

1. Horizontal line through $(0, -4)$ **3.** Vertical line through $(4.5, 0)$

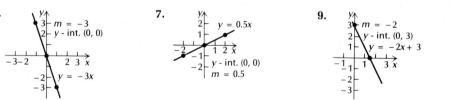

5. $m = -3$, y-int. $(0, 0)$, $y = -3x$ **7.** $y = 0.5x$, y-int. $(0, 0)$, $m = 0.5$ **9.** $m = -2$, y-int. $(0, 3)$, $y = -2x + 3$ **11.** $m = -1$, y-int. $(0, -2)$, $y = -x - 2$

13. $m = -2$, y-int.: $(0, 2)$ **15.** $m = -1$, y-int.: $\left(0, -\frac{5}{2}\right)$ **17.** $y + 5 = -5(x - 1)$, or $y = -5x$ **19.** $y - 3 = -2(x - 2)$, or

$y = -2x + 7$ **21.** $y = \frac{1}{2}x - 6$ **23.** $y = 3$ **25.** $\frac{3}{2}$ **27.** $\frac{1}{2}$ **29.** No slope   **31.** 0   **33.** 3   **35.** 2

37. $y - 1 = \frac{3}{2}(x + 2)$, or $y + 2 = \frac{3}{2}(x + 4)$, $y = \frac{3}{2}x + 4$ **39.** $y + 4 = \frac{1}{2}(x - 2)$, or $y = \frac{1}{2}x - 5$ **41.** $x = 3$

43. $y = 3$ **45.** $y = 3x$ **47.** $y = 2x + 3$ **49.** (a) $R = 4.17T$; (b) $R = 25.02$

51. (a) $B = 0.025W$; (b) $B = 2.5\%\ W$. The weight of the brain is 2.5% of the body weight. (c) 3 lb

53. (a) $A = P + 14\%P = P + 0.14P = 1.14P$; (b) \$114; (c) \$240

55. (a) $D(0°) = 115$ ft, $D(-20°) = 75$ ft, $D(10°) = 135$ ft, $D(32°) = 179$ ft;

(b) 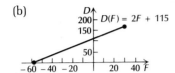   (c) Temperature below $-57.5°$ would yield a negative stopping distance, which has no meaning here. For temperatures above $32°$ there would be no ice.

57. (a) $A(0) = 2$, $A(1) = 3.1$, $A(4) = 6.4$, $A(10) = 13$; (b) straight line through $(0, 2)$ and $(10, 13)$; (c) The area is measured only from the time the organism is released. Thus only nonnegative values of t would be used as inputs.

59. (a) $C(x) = 20x + 100,000$; (b) $R(x) = 45x$; (c) $P(x) = R(x) - C(x) = 25x - 100,000$; (d) \$3,650,000, a profit; (e) 4000

61. (a) 200.69 cm; (b) 195.23 cm

Margin Exercises, Section 1.5, pp. 47–60

1.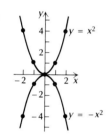

2. (a) 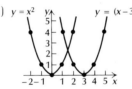   (b) by moving it to the right 3 units

3. $\dfrac{-1 \pm \sqrt{22}}{3}$

4.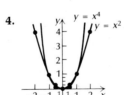

5. (a) All real numbers except -4; (b) all real numbers except -5, 1; (c) all real numbers except 5.

6.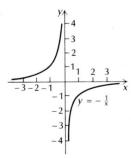

7. 160

8.

9.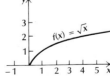

10. $\left[-\dfrac{3}{2}, \infty\right)$

11. $t^{3/4}$ **12.** $y^{1/5}$ **13.** $x^{-2/5}$ **14.** $t^{-1/3}$ **15.** x^3 **16.** $x^{7/2}$ **17.** $\sqrt[7]{y}$ **18.** $\sqrt{x^3}$ **19.** $\dfrac{1}{\sqrt{t^3}}$ **20.** $\dfrac{1}{\sqrt{b}}$ **21.** 32   **22.** 9
23. (2, $9)

Exercise Set 1.5, p. 60–62

1.

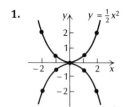

3. See Margin Exercise 1 for the graph of $y = x^2$. Move it to the right 1 unit to get the graph of $y = (x - 1)^2$.

5. See Margin Exercise 1 for $y = x^2$. Move it to the left 1 unit to get $y = (x + 1)^2$.   **7.** $y = |x + 3|$ $y = |x|$

9. See Exercise Set 1.3, Ex. 13 for $y = x^3$. Move it up 1 unit for $y = x^3 + 1$.   **11.** See Margin Exercise 9 for $y = \sqrt{x}$.
Move it to the left 1 unit for $y = \sqrt{x + 1}$.   **13.** $y = x^2 - 4x + 3$ **15.** $y = -x^2 + 2x - 1$

17.   **19.**   **21.**

x	-2	-1	$-\dfrac{1}{2}$	$\dfrac{1}{2}$	1	2
y	$\dfrac{1}{4}$	1	4	4	1	$\dfrac{1}{4}$

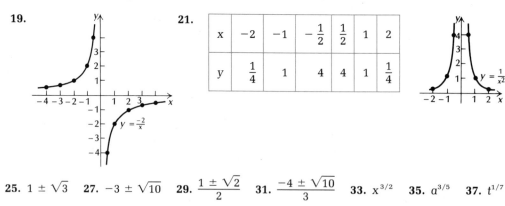

23.   **25.** $1 \pm \sqrt{3}$ **27.** $-3 \pm \sqrt{10}$ **29.** $\dfrac{1 \pm \sqrt{2}}{2}$ **31.** $\dfrac{-4 \pm \sqrt{10}}{3}$ **33.** $x^{3/2}$ **35.** $a^{3/5}$ **37.** $t^{1/7}$

39. $t^{-4/3}$ **41.** $t^{-1/2}$ **43.** $(x^2 + 7)^{-1/2}$ **45.** $\sqrt[5]{x}$ **47.** $\sqrt[3]{y^2}$ **49.** $\dfrac{1}{\sqrt[5]{t^2}}$ **51.** $\dfrac{1}{\sqrt[3]{b}}$ **53.** $\dfrac{1}{\sqrt[6]{e^{17}}}$ **55.** $\dfrac{1}{\sqrt{x^2 - 3}}$ **57.** 27

59. 16 **61.** 8 **63.** All real numbers except 5 **65.** All real numbers except 2, 3 **67.** $\left[-\dfrac{4}{5}, \infty\right)$ **69.** (2, \$4)

71. (1, \$4) **73.** (2, \$4) **75.** \$2.27

Margin Exercises, Section 1.6, pp. 62–68

1. $3.77744 \approx 3:46.6$ **2.** 1984 **3.** (a) $T = 0.5x - 946$; (b) 46.5¢; 1.5¢ more than in (b) of Example 1

4. (a) $A = \dfrac{131}{300}x^2 - 39\dfrac{7}{10}x + 1039\dfrac{1}{3}$, or $A = 0.437x^2 - 39.7x + 1039.333$; (b) 516

Exercise Set 1.6, pp. 68–69

1. (b) Linear; (c) $y = 32x + 9998$; (d) \$10,126; \$10,318; (e) $C = 29.5x + 10,000$; (f) \$10,118; \$10,295 **3.** (b) Quadratic; (c) $D = 0.9875x^2 - 121.5x + 3756.25$; (d) 21.25 **5.** (b) Constant, linear; (c) $S = -20x + 100,330$; (d) Let $S = b$, where b is the average of sales totals: \$100,296.25.

Chapter 1 Test, p. 69–70

1. $\dfrac{1}{e^k}$ **2.** e^{-13} or $\dfrac{1}{e^{13}}$ **3.** $x^2 + 2xh + h^2$ **4.** $(5x - t)(5x + t)$ **5.** \$920 **6.** $x > -4$ **7.** (a) $f(-3) = 5$;

(b) $x^2 + 2xh + h^2 - 4$ **8.** $m = -3$; y-int.: (0, 2) **9.** $y + 5 = \dfrac{1}{4}(x - 8)$, or $y = \dfrac{1}{4}x - 7$ **10.** $m = 6$

11. $F = \dfrac{2}{3}W$ **12.** (a) $C(x) = 0.5x + 10,000$; (b) $R(x) = 1.3x$; (c) $P(x) = R(x) - C(x) = 0.8x - 10,000$; (d) 12,500

13. (3, \$16) **14.**

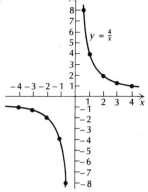

15. $t^{-1/2}$ **16.** $\dfrac{1}{\sqrt[5]{t^3}}$ **17.** All real numbers except 2, −7 **18.** $[-2, \infty)$

19. $y = 4x - 1$ **20.** $y = -4.5x^2 + 17.5x - 8$ **21.** $[c, d)$ **22.**

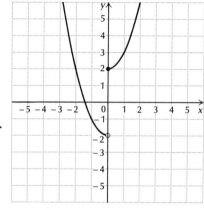

23. (a) \$9.25; (b) 525

CHAPTER 2

Margin Exercises, Section 2.1, pp. 72–84

1. a, b **2.** (a) No, yes;   (b) yes, no

3. (a) 14, 16.4, 16.7, 16.97, 16.997; 17.003, 17.03, 17.3, 18.2, 20; (b) 17;

(c)

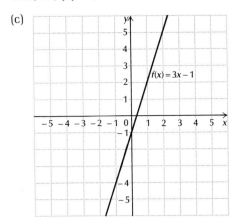

(d) −4; (e) 5; (f) −1

4. (a) −0.5, −1, −10, −100, −1000; 1000, 100, 10, 1.25, 1; (b) does not exist;

(c)

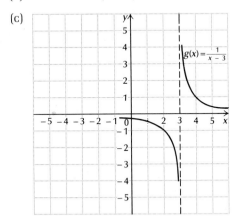

(d) does not exist; (e) −0.5; (f) 1

5. (a) Does not exist; (b) 0; (c) 3

6. (a)

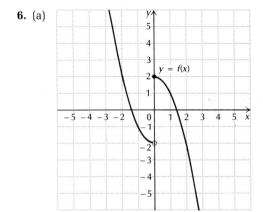

(b) does not exist; (c) 2

7. (a)

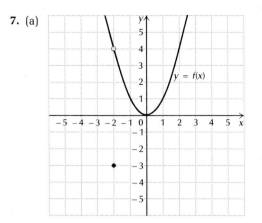

(b) 4; (c) 4; (d) no; (e) yes

8. (a) 2; (b) 3; (c) yes; (d) no   **9.** (a) 17; (b) 17; (c) yes; (d) yes   **10.** (a) No; (b) yes   **11.** No   **12.** No
13. $\sqrt[3]{x}$ is continuous by (ii); 7 is continuous by (i), and x^2 is continuous by (ii), so $7x^2$ is continuous by (iii). Then
$\sqrt[3]{x} - 7x^2$ is continuous by (iii). Now x is continuous by (ii), and 2 is continuous by (i), so x − 2 is continuous by (iii),
and $\dfrac{1}{x-2}$ is continuous by (iv). Thus $\dfrac{\sqrt[3]{x} - 7x^2}{x - 2}$ is continuous by (iii).   **14.** 53   **15** $\sqrt{8}$ **16.** −8.

17. $\dfrac{1}{6}$ **18.** (a) 2x + 1, 2x + 0.7, 2x + 0.4, 2x + 0.1, 2x + 0.01, 2x + 0.001; (b) 2x   **19.** (a) 4.00, 4.50, 5.14, 6.00, 7.20,

9.00, 12.00, 18.00, 36.00, 54.00, 108.00; (b) ∞; (c) ∞   **20.** (a) 3.25, 2.25, 2.0625, 2.025, 2.005, 2.0005; (b) 2   **21.** $\dfrac{2}{3}$

Exercise Set 2.1, pp. 85–87

1. No   **3.** Yes   **5.** (a) Does not exist; (b) −1; (c) no; (d) 3; (e) 3; (f) yes   **7.** (a) 2; (b) 2; (c) yes; (d) 0; (e) 0; (f) yes
9. No, yes, no, yes   **11.** Does not exist   **13.** 54¢   **15.** −2   **17.** Does not exist   **19.** 11   **21.** −10   **23.** $-\dfrac{5}{2}$ **25.** 5

27. 2   **29.** 3   **31.** Does not exist   **33.** $\dfrac{2}{7}$ **35.** $\dfrac{5}{4}$ **37.** $\dfrac{1}{2}$ **39.** $6x^2$ **41.** $\dfrac{-2}{x^3}$ **43.** $\dfrac{2}{5}$ **45.** 5   **47.** $\dfrac{1}{2}$ **49.** $\dfrac{2}{3}$
51. (a) 1; (b) 1; (c) yes, no   **53.** (a) $800, $736, $677.12, $622.95, $573.11; (b) $5656.12; (c) $10,000   **55.** 0, no
57. −0.25   **59.** 0

Margin Exercises, Section 2.2, pp. 88–94

1. (a) 20 $\dfrac{\text{suits}}{\text{hr}}$, 35 $\dfrac{\text{suits}}{\text{hr}}$, 9 $\dfrac{\text{suits}}{\text{hr}}$, 36 $\dfrac{\text{suits}}{\text{hr}}$; (b) 11 A.M. to 12 P.M.; (c) workers were anticipating the lunch break; answers
may vary; (d) 10 A.M. to 11 A.M.; (e) workers were fatigued or took longer breaks than they should have; answers may vary;
(f) 25 $\dfrac{\text{suits}}{\text{hr}}$ **2.** (a) 21; (b) 7; (c) 28; (d) 21   **3.** (a) $\dfrac{1}{2}$; (b) $\dfrac{1}{2}$; (c) $\dfrac{1}{2}$

4.

x	h	x + h	f(x)	f(x + h)	f(x + h) − f(x)	$\dfrac{f(x + h) - f(x)}{h}$
3	2	5	36	100	64	32
3	1	4	36	64	28	28
3	0.1	3.1	36	38.44	2.44	24.4
3	0.01	3.01	36	36.2404	0.2404	24.04
3	0.001	3.001	36	36.024004	0.024004	24.004

5. (a) $f(x + h) = 4x^2 + 8xh + 4h^2$; (b) $f(x + h) - f(x) = 8xh + 4h^2$; (c) $\dfrac{f(x + h) - f(x)}{h} = 4(2x + h)$; (d) 36, 40, 44, 47.6,

47.96, 47.996   **6.** (a) $4(3x^2 + 3xh + h^2)$; (b) 28, 45.64, 47.7604, 47.976004   **7.** (a) $\dfrac{-1}{x(x + h)}$; (b) −0.1; −0.1667, −0.2381,
−0.2488, −0.2499

Exercise Set 2.2, pp. 95–97

1. (a) 70, 39, 29, 23 pleasure units/unit of product; (b) the more you get, the less pleasure you get from each additional unit
3. (a) $1.25\dfrac{\text{words}}{\text{min}}$, $1.25\dfrac{\text{words}}{\text{min}}$, $0.625\dfrac{\text{words}}{\text{min}}$, $0\dfrac{\text{words}}{\text{min}}$, $0\dfrac{\text{words}}{\text{min}}$; (b) you have reached a saturation point; you cannot memorize any more   **5.** (a) $125\dfrac{\text{million people}}{\text{yr}}$ for each; (b) no; (c) A: $290\dfrac{\text{million people}}{\text{yr}}$, $-40\dfrac{\text{million people}}{\text{yr}}$, $-50\dfrac{\text{million people}}{\text{yr}}$, $300\dfrac{\text{million people}}{\text{yr}}$; B: $125\dfrac{\text{million people}}{\text{yr}}$ in all intervals; (d) A   **7.** (a) \$300,093.99; (b) \$299,100;
(c) \$993.99; (d) \$993.99 **9.** (a) 144 ft; (b) 400 ft; (c) $128\dfrac{\text{ft}}{\text{sec}}$ **11.** $72,500\dfrac{\text{marriages}}{\text{yr}}$ **13.** \$13,151,000 per yr
15. (a) $7(2x+h)$; (b) 70, 63, 56.7, 56.07   **17.** (a) $-7(2x+h)$; (b) $-70, -63, -56.7, -56.07$ **19.** (a) $7(3x^2+3xh+h^2)$;
(b) 532, 427, 344.47, 336.8407   **21.** (a) $\dfrac{-5}{x(x+h)}$; (b) $-0.2083, -0.25, -0.3049, -0.3117$ **23.** (a) -2; (b) all -2
25. (a) $2x+h-1$; (b) 9, 8, 7.1, 7.01   **27.** $2ax+b+ah$ **29.** $\dfrac{1}{\sqrt{x+h}+\sqrt{x}}$ **31.** $\dfrac{-(2x+h)}{x^2(x+h)^2}$ **33.** $\dfrac{1}{(x+1)(x+1+h)}$

Margin Exercises, Section 2.3, pp. 98–107

1. (a) L_2, L_3, L_4, L_6; (b) $-2, -1, 0, 1, 2$; (c) $m(x)=2x$ **2.** $f'(5)=10$ **3.** $f'(x)=8x, f'(5)=40=$ the slope of the tangent
line at $(5, f(5))$, or $(5, 100)$ **4.** $f'(x)=12x^2, f'(-5)=300, f'(0)=0$ **5.** $f'(x)=\dfrac{-1}{x^2}, f'(-10)=-\dfrac{1}{100}=-0.01,$
$f'(-2)=-\dfrac{1}{4}=-0.25$ **6.** x_2, x_4, x_5, x_6

Exercise Set 2.3, pp. 107–108

1. $f'(x)=10x, f'(-2)=-20, f'(-1)=-10, f'(0)=0, f'(1)=10, f'(2)=20$ **3.** $f'(x)=-10x, f'(-2)=20, f'(-1)=10,$
$f'(0)=0, f'(1)=-10, f'(2)=-20$ **5.** $f'(x)=15x^2, f'(-2)=60, f'(-1)=15, f'(0)=0, f'(1)=15, f'(2)=60$
7. $f'(x)=2,$ all 2   **9.** $f'(x)=-4,$ all -4 **11.** $f'(x)=2x+1, f'(-2)=-3, f'(-1)=-1, f'(0)=1, f'(1)=3, f'(2)=5$
13. $f'(x)=\dfrac{-4}{x^2}; f'(-2)=-1; f'(-1)=-4; f'(0)$ does not exist; $f'(1)=-4; f'(2)=-1$ **15.** $f'(x)=m,$ all m

17. $x_0, x_3, x_4, x_6, x_{12}$ **19.** $x=-3$ **21.** $\dfrac{-2}{x^3}$ **23.** $\dfrac{1}{(1+x)^2}$

Margin Exercises, Section 2.4, pp. 108–116

1. (a) $3x^2$; (b) $3x^2$; (c) 48   **2.** $\dfrac{dy}{dx}=6x^5$ **3.** $\dfrac{dy}{dx}=-7x^{-8}$, or $-\dfrac{7}{x^8}$ **4.** $\dfrac{dy}{dx}=\dfrac{1}{3}x^{-2/3}$, or $\dfrac{1}{3\sqrt[3]{x^2}}$ **5.** $\dfrac{dy}{dx}=-\dfrac{1}{4}x^{-5/4}$,
or $\dfrac{-1}{4\sqrt[4]{x^5}}$ **6.** $g'(x)=0$ **7.** $\dfrac{dy}{dx}=100x^{19}$ **8.** $\dfrac{dy}{dx}=\dfrac{3}{x^2}$ **9.** $\dfrac{dy}{dx}=-4x^{-1/2}$, or $\dfrac{-4}{\sqrt{x}}$ **10.** $\dfrac{dy}{dx}=x^{5.25}$ **11.** $\dfrac{dy}{dx}=-\dfrac{1}{4}$
12. $\dfrac{dy}{dx}=28x^3+12x$ **13.** $\dfrac{dy}{dx}=30x-\dfrac{4}{x^2}+\dfrac{1}{2\sqrt{x}}$ **14.** $\left(2, \dfrac{8}{3}\right)$ **15.** $\left(2+\sqrt{3}, \dfrac{8}{3}+\sqrt{3}\right), \left(2-\sqrt{3}, \dfrac{8}{3}-\sqrt{3}\right)$

Exercise Set 2.4, pp. 117–118

1. $7x^6$ **3.** 0   **5.** $600x^{149}$ **7.** $3x^2 + 6x$ **9.** $\dfrac{4}{\sqrt{x}}$ **11.** $0.07x^{-0.93}$ **13.** $\dfrac{2}{5 \cdot \sqrt[5]{x}}$ **15.** $\dfrac{-3}{x^4}$ **17.** $6x - 8$ **19.** $\dfrac{1}{4\sqrt[4]{x^3}} + \dfrac{1}{x^2}$

21. $1.6x^{1.5}$ **23.** $\dfrac{-5}{x^2} - 1$ **25.** 4   **27.** 4   **29.** x^3 **31.** $-0.02x - 0.5$ **33.** $-2x^{-5/3} + \dfrac{3}{4}x^{-1/4} + \dfrac{6}{5}x^{1/5} - 24x^{-4}$ **35.** $(0, 0)$

37. $(0, 0)$ **39.** $\left(\dfrac{5}{6}, \dfrac{23}{12}\right)$ **41.** $(-25, 76.25)$ **43.** There are none.   **45.** The tangent is horizontal at all points on graph.

47. $\left(\dfrac{5}{3}, \dfrac{148}{27}\right), (-1, -4)$ **49.** $(\sqrt{3}, 2 - 2\sqrt{3}), (-\sqrt{3}, 2 + 2\sqrt{3})$ **51.** $\left(\dfrac{19}{2}, \dfrac{399}{4}\right)$ **53.** $(60, 150)$ **55.** $\left(-2 + \sqrt{3}, \dfrac{4}{3} - \sqrt{3}\right),$

$\left(-2 - \sqrt{3}, \dfrac{4}{3} + \sqrt{3}\right)$ **57.** $(0, -4), \left(\sqrt{\dfrac{2}{3}}, -\dfrac{40}{9}\right), \left(-\sqrt{\dfrac{2}{3}}, -\dfrac{40}{9}\right)$ **59.** $2x - 1$ **61.** $2x + 1$ **63.** $3x^2 - \dfrac{1}{x^2}$ **65.** $-192x^2$

67. $\dfrac{2}{3 \cdot \sqrt[3]{x^2}}$ **69.** $3x^2 + 6x + 3$ **71.** $\dfrac{F(x + h) - F(x)}{h} = \dfrac{[f(x + h) - g(x + h)] - [f(x) - g(x)]}{h}$

$= \dfrac{f(x + h) - f(x)}{h} - \dfrac{g(x + h) - g(x)}{h}$. As $h \to 0$, the two terms on the right approach $f'(x)$ and $g'(x)$, respectively, so their difference approaches $f'(x) - g'(x)$. Thus $F'(x) = f'(x) - g'(x)$.

Margin Exercises, Section 2.5, pp. 118–123

1. (a) $70\dfrac{\text{mi}}{\text{hr}}$; (b) $100\dfrac{\text{mi}}{\text{hr}}$ **2.** (a) $v(t) = 32t$; (b) $v(2) = 64$ ft/sec; (c) $v(10) = 320$ ft/sec   **3.** $a(t) = 32$ ft/sec^2

4. (a) $V'(s) = 3s^2$; (b) $V'(10) = 300$ ft^2 **5.** (a) $P'(t) = 9700 + 20{,}000t$; (b) $308{,}500$; $109{,}700\dfrac{\text{bacteria}}{\text{hr}}$; (c) $428{,}200$;

$129{,}700\dfrac{\text{bacteria}}{\text{hr}}$ **6.** (a) $P(x) = 40x - 0.5x^2 - 3$; (b) $R(40) = \$1200$, $C(40) = \$403$, $P(40) = \$797$; (c) $R'(x) = 50 - x$,

$C'(x) = 10$, $P'(x) = 40 - x$; (d) $R'(40) = \$10$ per unit, $C'(40) = \$10$ per unit, $P'(40) = \$0$ per unit; (e) no

Exercise Set 2.5, pp. 124–125

1. (a) $v(t) = 3t^2 + 1$; (b) $a(t) = 6t$; (c) $v(4) = 49$ ft/sec, $a(4) = 24$ ft/sec^2 **3.** (a) $\dfrac{dV}{dh} = 0.61/\sqrt{h}$; (b) 244 miles; (c) 0.00305

miles per foot   **5.** $P'(t) = 1.25$ **7.** $\dfrac{dC}{dr} = 2\pi$ **9.** (a) $T'(t) = -0.2t + 1.2$; (b) $100.175°$; (c) 0.9 degrees/day   **11.** $\dfrac{dB}{dx} =$

$0.1x - 0.9x^2$ **13.** $\dfrac{dT}{dW} = 1.31W^{0.31}$ **15.** (a) $P(x) = -0.001x^2 + 3.8x - 60$; (b) $R(100) = \$500$, $C(100) = \$190$,

$P(100) = \$310$; (c) $R'(x) = 5$, $C'(x) = 0.002x + 1.2$, $P'(x) = -0.002x + 3.8$; (d) $R'(100) = \$5$ per unit, $C'(100) = \$1.4$ per unit, $P'(100) = \$3.6$ per unit

Margin Exercises, Section 2.6, pp. 126–129

1. $f'(x) = 54x^{17}$ **2.** $f'(x) = (9x^3 + 4x^2 + 10)(-14x + 4x^3) + (27x^2 + 8x)(-7x^2 + x^4)$ **3.** $f'(x) = 4x^3$ **4.** $\dfrac{3x^2 - 5}{x^6}$

5. $\dfrac{-x^4 + 3x^2 + 2x}{(x^3 + 1)^2}$ **6.** (a) $R(x) = x(200 - x) = 200x - x^2$; (b) $R'(x) = 200 - 2x$

Exercise Set 2.6, pp. 129–130

1. $11x^{10}$ **3.** $\dfrac{1}{x^2}$ **5.** $3x^2$ **7.** $(8x^5 - 3x^2 + 20)\left(32x^3 - \dfrac{3}{2\sqrt{x}}\right) + (40x^4 - 6x)(8x^4 - 3\sqrt{x})$ **9.** $300 - 2x$ **11.** $\dfrac{300}{(300 - x)^2}$

13. $\dfrac{17}{(2x + 5)^2}$ **15.** $\dfrac{-x^4 - 3x^2 - 2x}{(x^3 - 1)^2}$ **17.** $\dfrac{1}{(1 - x)^2}$ **19.** $\dfrac{2}{(x + 1)^2}$ **21.** $\dfrac{-1}{(x - 3)^2}$ **23.** $\dfrac{-2x^2 + 6x + 2}{(x^2 + 1)^2}$ **25.** $\dfrac{-18x + 35}{x^8}$

27. (a) $R(x) = x(400 - x) = 400x - x^2$; (b) $R'(x) = 400 - 2x$ **29.** (a) $R(x) = 4000 + 3x$; (b) $R'(x) = 3$

31. $A'(x) = \dfrac{xC'(x) - C(x)}{x^2}$ **33.** $\dfrac{5x^3 - 30x^2\sqrt{x}}{2\sqrt{x}(\sqrt{x} - 5)^2}$ **35.** $\dfrac{-3(1 + 2v)}{(1 + v + v^2)^2}$ **37.** $\dfrac{2t^3 - t^2 + 1}{(1 - t + t^2 - t^3)^2}$ **39.** $\dfrac{5x^3 + 15x^2 + 2}{2x\sqrt{x}}$

41. $[x(9x^2 + 6) + (3x^3 + 6x - 2)](3x^4 + 7) + 12x^4(3x^3 + 6x - 2)$ **43.** $\dfrac{6t^2(t^5 + 3)}{(t^3 + 1)^2} + \dfrac{5t^4(t^3 - 1)}{t^3 + 1}$

45. $\dfrac{(x^7 - 2x^6 + 9)[(2x^2 + 3)(12x^2 - 7) + 4x(4x^3 - 7x + 2)] - (7x^6 - 12x^5)(2x^2 + 3)(4x^3 - 7x + 2)}{(x^7 - 2x^6 + 9)^2}$

Margin Exercises, Section 2.7, pp. 130–134

1. $20x(1 + x^2)^9$ **2.** $\dfrac{-x}{\sqrt{1 - x^2}}$ **3.** $-2x(1 + x^2)(1 + 3x^2)$ **4.** $2(x - 4)^4(6 - x)^2(21 - 4x)$ **5.** $\left(\dfrac{x + 5}{x - 4}\right)^{-2/3} \cdot \dfrac{-3}{(x - 4)^2}$

6. $3(x^2 - 1), 9x^2 - 1$ **7.** $4 \cdot \sqrt[3]{x} + 5, \sqrt[3]{4x + 5}$

Exercise Set 2.7, p. 135

1. $-55(1 - x)^{54}$ **3.** $\dfrac{4}{\sqrt{1 + 8x}}$ **5.** $\dfrac{3x}{\sqrt{3x^2 - 4}}$ **7.** $-240x(3x^2 - 6)^{-41}$ **9.** $\sqrt{2x + 3} + \dfrac{x}{\sqrt{2x + 3}}$, or $\dfrac{3(x + 1)}{\sqrt{2x + 3}}$

11. $2x\sqrt{x - 1} + \dfrac{x^2}{2\sqrt{x - 1}}$, or $\dfrac{5x^2 - 4x}{2\sqrt{x - 1}}$ **13.** $\dfrac{-6}{(3x + 8)^3}$ **15.** $(1 + x^3)^2(-3x^2 - 12x^5)$, or $-3x^2(1 + x^3)^2(1 + 4x^3)$

17. $4x - 400$ **19.** $2(x + 6)^9(x - 5)^3(7x - 13)$ **21.** $4(x - 4)^7(3 - x)^3(10 - 3x)$ **23.** $4(2x - 3)^2(3 - 8x)$

25. $\left(\dfrac{1 - x}{1 + x}\right)^{-1/2} \cdot \dfrac{-1}{(x + 1)^2}$ **27.** (a) $\dfrac{2x - 3x^2}{(1 + x)^6}$; (b) $\dfrac{2x - 3x^2}{(1 + x)^6}$; (c) same   **29.** $C'(x) = \dfrac{1500x^2}{\sqrt{x^3 + 2}}$ **31.** $\$3000\,(1 + i)^2$

33. $\dfrac{x^2 - 2}{\sqrt[3]{(x^3 - 6x + 1)^2}}$ **35.** $\dfrac{x - 2}{2(x - 1)^{3/2}}$ **37.** $\dfrac{-4(1 + 2v)^3}{v^5}$ **39.** $\dfrac{1}{\sqrt{1 - x^2}(1 - x)}$ **41.** $3\left(\dfrac{x^2 - x - 1}{x^2 + 1}\right)^2 \cdot \dfrac{x^2 + 4x - 1}{(x^2 + 1)^2}$

43. $\dfrac{1}{\sqrt{t}(1 + \sqrt{t})^2}$

Margin Exercises, Section 2.8, pp. 136–138

1. $f'(x) = 12x^5 - 5x^4, f''(x) = 60x^4 - 20x^3, f'''(x) = 240x^3 - 60x^2, f^{(4)}(x) = 720x^2 - 120x, f^{(5)}(x) = 1440x - 120,$
$f^{(6)}(x) = 1440$ **2.** (a) $\dfrac{dy}{dx} = 7x^6 - 3x^2$; (b) $\dfrac{d^2y}{dx^2} = 42x^5 - 6x$; (c) $\dfrac{d^3y}{dx^3} = 210x^4 - 6$; (d) $\dfrac{d^4y}{dx^4} = 840x^3$ **3.** $\dfrac{d^2y}{dx^2} = \dfrac{4}{x^3}$

4. $y' = 60(x^2 - 12x)^{29}(x - 6); y'' = 60(x^2 - 12x)^{28}[59x^2 - 708x + 2088]$ **5.** $a(t) = 12t^2$

Exercise Set 2.8, pp. 138–139

1. 0 **3.** $-\dfrac{2}{x^3}$ **5.** $-\dfrac{3}{16}x^{-7/4}$ **7.** $12x^2 + \dfrac{8}{x^3}$ **9.** $\dfrac{12}{x^5}$ **11.** $n(n - 1)x^{n-2}$ **13.** $12x^2 - 2$ **15.** $-\dfrac{1}{4}(x - 1)^{-3/2}$, or

$\dfrac{-1}{4\sqrt{(x - 1)^3}}$ **17.** $2a$ **19.** 24 **21.** $720x$ **23.** $20(x^2 - 5)^8[19x^2 - 5]$ **25.** $a(t) = 6t + 2$ **27.** $P''(t) = 200,000$

29. $y' = -x^{-2} - 2x^{-3}$, $y'' = 2x^{-3} + 6x^{-4}$, $y''' = -6x^{-4} - 24x^{-5}$　**31.** $y' = \dfrac{1 + 2x^2}{\sqrt{1 + x^2}}$, $y'' = \dfrac{2x^3 + 3x}{(1 + x^2)^{3/2}}$, $y''' = \dfrac{3}{(1 + x^2)^{5/2}}$

33. $y' = \dfrac{11}{(2x + 3)^2}$, $y'' = \dfrac{-44}{(2x + 3)^3}$, $y''' = \dfrac{264}{(2x + 3)^4}$　**35.** $y' = \dfrac{x - 2}{2(x - 1)^{3/2}}$, $y'' = \dfrac{4 - x}{4(x - 1)^{5/2}}$, $y''' = \dfrac{3(x - 6)}{8(x - 1)^{7/2}}$

37. $\dfrac{2}{(x - 1)^3}$

Chapter 2 Test, pp. 139–140

1. Yes　**2.** No　**3.** Does not exist　**4.** 1　**5.** No　**6.** 3　**7.** 3　**8.** Yes　**9.** 6　**10.** $\dfrac{1}{2}$　**11.** Does not exist　**12.** 4

13. $3(2x + h)$　**14.** $(0, 0), (2, -4)$　**15.** $84x^{83}$　**16.** $\dfrac{5}{\sqrt{x}}$　**17.** $\dfrac{10}{x^2}$　**18.** $\dfrac{5}{4}x^{1/4}$, or $\dfrac{5}{4} \cdot \sqrt[4]{x}$　**19.** $-x + 0.61$

20. $x^2 - 2x + 2$　**21.** $\dfrac{-6x + 20}{x^5}$　**22.** $\dfrac{5}{(5 - x)^2}$　**23.** $(x + 3)^3(7 - x)^4(-9x + 13)$

24. $-5(x^5 - 4x^3 + x)^{-6}(5x^4 - 12x^2 + 1)$　**25.** $\sqrt{x^2 + 5} + \dfrac{x^2}{\sqrt{x^2 + 5}}$, or $\dfrac{2x^2 + 5}{\sqrt{x^2 + 5}}$　**26.** $24x$

27. (a) $P(x) = -0.001x^2 + 48.8x - 60$; (b) $R(10) = \$500, C(10) = \$72.10, P(10) = \$427.90$; (c) $R'(x) = 50$, $C'(x) = 0.002x + 1.2, P'(x) = -0.002x + 48.8$; (d) $R'(10) = \$50$ per unit, $C'(10) = \$1.22$ per unit, $P'(10) = \$48.78$

28. (a) $\dfrac{dM}{dt} = -0.003t^2 + 0.2t$; (b) 9; (c) 1.7 words/min　**29.** 12　**30.** $-2\left(\dfrac{1 + 3x}{1 - 3x}\right)^{1/3} + \left(\dfrac{1 - 3x}{1 + 3x}\right)^{2/3}$

CHAPTER 3

Margin Exercises, Section 3.1, pp. 140–157

1. (a) $[a, b], [c, d]$; (b) $(a, b), (c, d)$; (c) $[b, c], [d, e]$; (d) $(b, c), (d, e)$　**2.** (a) $[b, c], [d, e]$; (b) $[a, b], [c, d]$
3. (a) $(0, b]$; (b) $[a, 0)$,　**4.** R, T, V　**5.** (a) $x_1, x_3, x_5, x_6, x_8, x_{10}$; (b) x_4, x_7, x_9; (c) $x_1, x_3, x_4, x_5, x_6, x_7, x_8, x_9, x_{10}$
6. Answers will vary on the graph, but it must have at least one critical point.

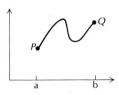

7. It is not possible.　**8.** (a) Maximum at c_2, minimum at b; (b) maximum at c_1, minimum at c_2.　**9.** (a) c_2; (b) c_1; (c) c_1;
(d) c_2　**10.** Maximum $= 4$ at $x = 2$; minimum $= 1$ at $x = 1$ and $x = -1$　**11.** Maximum $= 176$ at $x = 6$; minimum $= 97$
at $x = 5$　**12.** Minimum $= -4$ at $x = 2$; there is no maximum　**13.** Minimum $= -4$ at $x = 2$; maximum $= 0$ at $x = 0$
and $x = 4$　**14.** There are none.　**15.** (a) Maximum $= 8$ at $x = 2$; minimum $= -8$ at $x = -2$

16. Minimum $= \dfrac{10}{\sqrt{10}} + \sqrt{10}$ at $x = \dfrac{1}{\sqrt{10}}$

Exercise Set 3.1, pp. 158–159

1. (a) 41 mph; (b) 80 mph; (c) 13.5 mpg; (d) 16.5 mph; (e) about 22%　**3.** Max. $= 5\dfrac{1}{4}$ at $x = \dfrac{1}{2}$; min. $= 3$ at $x = 2$

5. Max. = 4 at $x = 2$; min. = 1 at $x = 1$ **7.** Max. = $\frac{59}{27}$ at $x = -\frac{1}{3}$; min. = 1 at $x = -1$ **9.** Max. = 1 at $x = 1$;

min. = -5 at $x = -1$ **11.** None   **13.** Max. = 1225 at $x = 35$ **15.** Min. = 200 at $x = 10$ **17.** Max. = $\frac{1}{3}$ at $x = \frac{1}{2}$

19. Max. = $\frac{289}{4}$ at $x = \frac{17}{2}$ **21.** Max. = $2\sqrt{3}$ at $x = -\sqrt{3}$; min. = $-2\sqrt{3}$ at $x = \sqrt{3}$ **23.** Max. = 5700 at $x = 2400$

25. Min. = $-55\frac{1}{3}$ at $x = 1$ **27.** Max. = 2000 at $x = 20$; min. = 0 at $x = 0$ and $x = 30$ **29.** Min. = 24 at $x = 6$

31. Min. = 108 at $x = 6$ **33.** Max. = 3 at $x = -1$; min. = $-\frac{3}{8}$ at $x = \frac{1}{2}$ **35.** Max. = 2 at $x = 8$; min. = 0 at $x = 0$

37. None   **39.** 22506; \$150,000   **41.** (a) 179 ft at 32°; (b) 0 ft at $-57.5°$ **43.** 61.25 mph   **45.** Max. = $3\sqrt{6}$ at
$x = 3$; min. = -2 at $x = -2$ **47.** Max. = 1 at $x = -1$ and $x = 1$; min. = 0 at $x = 0$ **49.** None

51. Max. = $-\frac{10}{3} + 2\sqrt{3}$ at $x = 2 - \sqrt{3}$; min. = $-\frac{10}{3} - 2\sqrt{3}$ at $x = 2 + \sqrt{3}$ **53.** Min. = -1 at $x = -1$ and $x = 1$
55. 7

Margin Exercises, Section 3.2, pp. 162–168

1. Rel. max. at $\left(-1, \frac{13}{6}\right)$; rel. min. at $\left(2, -\frac{7}{3}\right)$ **2.** Rel. max. at $(0, 2)$; rel. min. at $(-1, 0)$ and $(1, 0)$

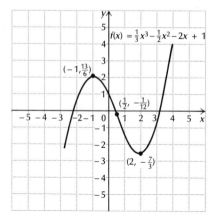

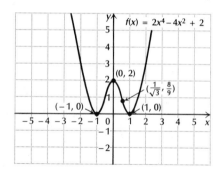

4. Rel. max. at $(-1, -2)$; rel. min. at $(1, 2)$

3. Rel. min. at $(0, 0)$

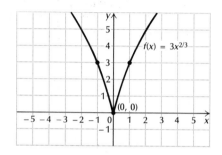

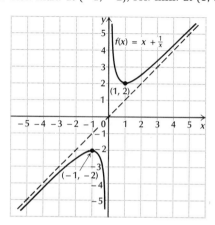

Exercise Set 3.2, p. 168

1. Rel. max. at $(0, 2)$

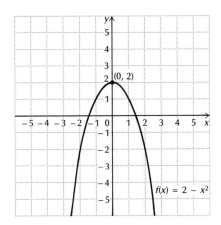

3. Rel. min. at $\left(-\dfrac{1}{2}, -\dfrac{5}{4}\right)$

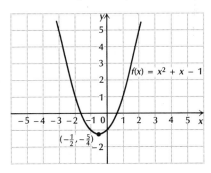

5. Rel. max. at $\left(-\dfrac{1}{2}, 1\right)$; rel. min. at $\left(\dfrac{1}{2}, -\dfrac{1}{3}\right)$

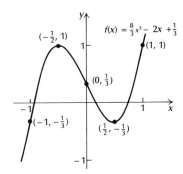

7.

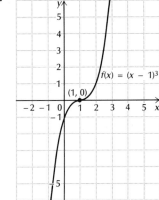

9. Rel. min. at $(-1, 0)$

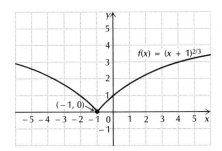

11. Rel. max. at $(0, 0)$; rel. min. at $(-\sqrt{3}, -9)$ and $(\sqrt{3}, -9)$

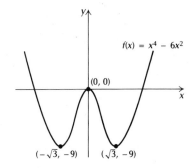

13. Rel. max. at $(-3, -6)$; rel. min. at $(3, 6)$

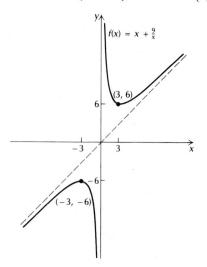

15. Rel. max. at $\left(-\dfrac{2}{3}, \dfrac{121}{27}\right)$; rel. min. at $(2, -5)$

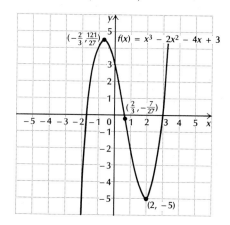

17. Rel. min. at $(-1, -1)$

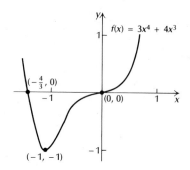

19.

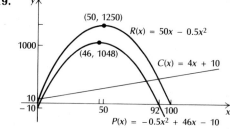

21.

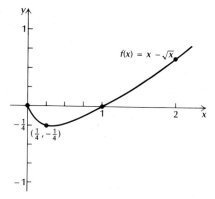

23.

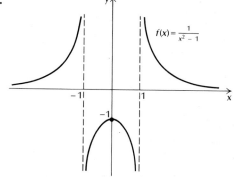

25.

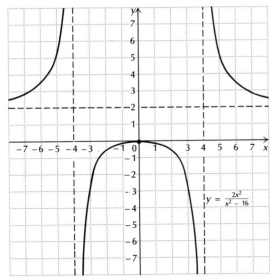

$$y = \frac{2x^2}{x^2 - 16}$$

Margin Exercises, Section 3.3, pp. 168–178

1. (a) y: 20, 16, 13.5, 12, 10, 8, 6.8, 0; A: 0, 64, 87.75, 96, 100, 96, 89.76, 0;

(b) 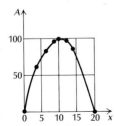   (c) yes; (d) max. = 100 ft² at x = 10 ft

2. 25 ft by 25 ft; 625 ft²   **3.** (a) R(x) = x(200 − x) = 200x − x²; (b) P(x) = −x² + 192x − 5000; (c) 96; (d) $4216;
(e) 200 − 96, or $104 per unit   **4.** (a) h: 4, 3.5, 3, 2.5, 2, 1.7, 1.5, 1, 0.6, 0.5, 0; V: 0, 3.5, 12, 22.5, 32, 35.972, 37.5, 36,

27.744, 24.5, 0; (b) 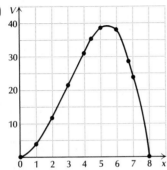   (c) about 38; between 5 and 6, maybe 5.2

5. Max. volume $= 74\frac{2}{27}$ in^3; dimensions $6\frac{2}{3}$ in. by $6\frac{2}{3}$ in. by $1\frac{2}{3}$ in.   **6.** Min. area $= 300$ cm^2; dimensions 10 cm by 10 cm by 5 cm   **7.** 40¢

Exercise Set 3.3, pp. 178–182

1. 25 and 25; max. prod. $= 625$ **3.** No; $Q = x(50 - x)$ has no minimum   **5.** 2 and -2; min. prod. $= -4$ **7.** $x = \frac{1}{2}$, $y = \sqrt{\frac{1}{2}}$; max. $= \frac{1}{4}$ **9.** $x = 10$, $y = 10$; min. $= 200$ **11.** $x = 2$, $y = \frac{32}{3}$; max. $= \frac{64}{3}$ **13.** 30 yd by 60 yd; max.

area $= 1800$ yd^2 **15.** 13.5 ft by 13.5 ft; 182.25 ft^2 **17.** 46 units; max. profit $= \$1048$ **19.** 70 units; max. profit $= \$19$ **21.** Approx. 1667 units; max. profit $\approx \$5500$ **23.** (a) $R(x) = 150x - 0.5x^2$; (b) $P(x) = -0.75x^2 + 150x - 4000$; (c) 100; (d) \$3500; (e) \$100   **25.** 20 in. by 20 in. by 5 in.; max. $= 2000$ in^3 **27.** 5 in. by 5 in. by 2.5 in.; min. $= 75$ in.^{2}

29. \$5.75, 72,500 (Will the stadium hold that many?)   **31.** 25   **33.** 14 in. by 14 in. by 28 in.   **35.** $\sqrt[3]{\frac{1}{10}}$ **37.** 4 ft by 4 ft by 20 ft   **39.** $x = y = \frac{24}{4 + \pi}$ **41.** 9%   **43.** Min. at $x = \frac{24\pi}{\pi + 4} \approx 10.55$, $24 - x = \frac{96}{\pi + 4} \approx 13.45$; there is no maximum if the string is to be cut. One would interpret the maximum to be at the endpoint, with the string uncut and used to form a circle.   **45.** S should be about 4.25 miles down shore from A **47.** (a) $C'(x) = 8 + \frac{3x^2}{100}$; (b) $A(x) = 8 + \frac{20}{x} + \frac{x^2}{100}$; (c) $A'(x) = \frac{x}{50} - \frac{20}{x^2}$; (d) min. $= 11$ at $x_0 = 10$, $C'(10) = 11$; (e) $A(10) = 11$, $C'(10) = 11$ **49.** Min. $= 6 - 4\sqrt{2}$ at $x = 2 - \sqrt{2}$ and $y = -1 + \sqrt{2}$

Margin Exercises, Section 3.4, pp. 182–186

1.

x	$\frac{2500}{x}$	$\frac{x}{2}$	$10 \cdot \frac{x}{2}$	$20 + 9x$	$(20 + 9x)\frac{2500}{x}$	$10 \cdot \frac{x}{2} + (20 + 9x)\frac{2500}{x}$
2500	1	1250	\$12,500	\$22,520	\$22,520	\$35,020
1250	2	625	\$ 6250	\$11,270	\$22,540	\$28,790
500	5	250	\$ 2500	\$ 4520	\$22,600	\$25,100
250	10	125	\$ 1250	\$ 2270	\$22,700	\$23,950
167	15	84	\$ 840	\$ 1523	\$22,845	\$23,685
125	20	63	\$ 630	\$ 1145	\$22,900	\$23,530
100	25	50	\$ 500	\$ 920	\$23,000	\$23,500
90	28	45	\$ 450	\$ 830	\$23,240	\$23,690
50	50	25	·\$ 250	\$ 470	\$23,500	\$23,750

2. 15 times per year at lot size 40   **3.** 19 times per year at lot size 31

Exercise Set 3.4, p. 187

1. Reorder 5 times per year; lot size = 20   **3.** Reorder 12 times per year; lot size = 30   **5.** Reorder about 13 times per year; lot size = 28   **7.** $x = \sqrt{\dfrac{2bQ}{a}}$

Margin Exercises, Section 3.5, pp. 187–193

1. (a)

(b)

2. (a)

(b) oscillating

3. 4.0, 4.08, 4.16, 4.24, 4.32, 4.40, 4.48, 4.57, 4.66, 4.75, 4.84; these were rounded each time to the nearest hundredth of a billion   **4.** $f(P) = 1.0825P$ **5.** $f(P) = 1.01P$ **6.** 3500; maximum sustainable harvest = 12,250

Exercise Set 3.5, pp. 193–194

1. 9500; max. sus. harvest = 90,250   **3.** 60,000; max. sus. harvest = 90,000   **5.** 400,000; max. sus. harvest = 400,000
7. There is none.

Margin Exercises, Section 3.6, pp. 194–198

1. $\Delta y = -0.59$ **2.** $\Delta y = 19$ **3.** 8.1875; 8.185 in Table 1   **4.** (a) $\Delta C = \$4.11$, $C'(5) = \$4.10$; (b) $\Delta C = \$6.01$, $C'(100) = \$6$

Exercise Set 3.6, pp. 198–199

1. $\Delta y = 0.0401$; $f'(x)\,\Delta x = 0.04$ **3.** 0.2816, 0.28   **5.** −0.556, −1   **7.** 6, 6   **9.** $\Delta C = \$2.01$, $C'(70) = \$2$
11. $\Delta R = \$2$, $R'(70) = \$2$ **13.** (a) $P(x) = -0.01x^2 + 1.4x - 30$; (b) $\Delta P = \$-0.01$, $P'(70) = 0$ **15.** 4.375   **17.** 10.1
19. 2.167   **21.** 2.512 cm³

Margin Exercises, Section 3.7, pp. 199–204

1. $\dfrac{dy}{dx} = \dfrac{2}{5y^4}$ **2.** (a) $\dfrac{dy}{dx} = \dfrac{-3x^2 - 2xy^4}{3y^2 + 4x^2y^3}$; (b) 0   **3.** $\dfrac{dp}{dx} = \dfrac{-\sqrt{p}}{50}$ **4.** $24\pi\dfrac{\text{ft}^2}{\text{sec}} \approx 75$ square feet per sec
5. (a) $\dfrac{dR}{dt} = \$0$ per day;   (b) $\dfrac{dC}{dt} = \$64$ per day; (c) $\dfrac{dP}{dt} = -\$64$ per day

Exercise Set 3.7, pp. 204–206

1. $\dfrac{1-y}{x+2}$; $-\dfrac{1}{9}$ **3.** $-\dfrac{x}{y}$; $-\dfrac{1}{\sqrt{3}}$ **5.** $\dfrac{6x^2 - 2xy}{x^2 - 3y^2}$; $-\dfrac{36}{23}$ **7.** $-\dfrac{y}{x}$ **9.** $\dfrac{x}{y}$ **11.** $\dfrac{3x^2}{5y^4}$ **13.** $\dfrac{-3x^2y^4 - 2xy^3}{3x^2y^2 + 4x^3y^3}$ **15.** $\dfrac{dp}{dx} = \dfrac{-2}{2p+1}$
17. $\dfrac{dp}{dx} = -\dfrac{p+4}{x+3}$ **19.** 0.1728π cm/day $\approx$ 0.54 cm/day   **21.** \$400 per day, \$80 per day, \$320 per day

23. \$16 per day, \$8 per day, \$8 per day   **25.** 65 mph   **27.** (a) $1000R \cdot \dfrac{dR}{dt}$; (b) −0.01125 mm/min²   **29.** $-\dfrac{3}{4}$
31. $-\dfrac{\sqrt{y}}{\sqrt{x}}$ **33.** $\dfrac{2}{3y^2(x+1)^2}$ **35.** $-\dfrac{9}{4}\sqrt[3]{y}\,\sqrt{x}$ **37.** $\dfrac{dy}{dx} = \dfrac{1+y}{2-x}$, $\dfrac{d^2y}{dx^2} = \dfrac{2+2y}{(2-x)^2}$ **39.** $\dfrac{dy}{dx} = \dfrac{x}{y}$, $\dfrac{d^2y}{dx^2} = \dfrac{y^2 - x^2}{y^3}$

Chapter 3 Test, pp. 206–207

1. $\dfrac{d^3y}{dx^3} = 24x$ **2.** Max. = 9 at $x = 3$ **3.** Max. = 2 at $x = -1$; min. = −1 at $x = -2$ **4.** Max. = 28.49 at $x = 4.3$

5. Max. = 7 at $x = -1$; min. = 3 at $x = 1$ **6.** None   **7.** Min. = $-\dfrac{13}{12}$ at $x = \dfrac{1}{6}$ **8.** Min. = 48 at $x = 4$ **9.** 4 and −4
10. $x = 5$, $y = -5$; min. = 50   **11.** Max. profit = \$24,980; 500 units   **12.** 40 in. by 40 in. by 10 in.; max. volume = 16,000 in³   **13.** 35 times at lot size 35   **14.** 49,500; max. sus. harvest = 2,450,250   **15.** $\Delta y = 1.01$, $f'(x)\,\Delta x = 1$

16. 10.2   **17.**

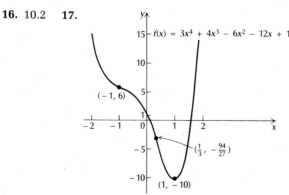

$f(x) = 3x^4 + 4x^3 - 6x^2 - 12x + 1$

18. $-\dfrac{x^2}{y^2}, -\dfrac{1}{4}$ **19.** -0.96 ft/sec   **20.** Max. $= \dfrac{1}{3} \cdot 2^{2/3}$ at $x = \sqrt[3]{2}$; min. $= 0$ at $x = 0$

21. (a) $A(x) = \dfrac{C(x)}{x} = 100 + \dfrac{100}{\sqrt{x}} + \dfrac{\sqrt{x}}{100}$; (b) min. $= 102$ at $x = 10{,}000$

CHAPTER 4

Margin Exercises, Section 4.1, pp. 210–219

1. 8, 8.574188, 8.815241, 8.821353, 8.824411, 8.824962; $2^\pi \approx 8.82$

2. (a)

x	0	$\frac{1}{2}$	1	2	-1	-2
3^x	1	1.7	3	9	$\frac{1}{3}$	$\frac{1}{9}$

(b)

3. (a)

x	0	$\frac{1}{2}$	1	2	-1	-2
$\left(\frac{1}{3}\right)^x$	1	0.6	$\frac{1}{3}$	$\frac{1}{9}$	3	9

(b)

4. (a) 0.8284, 0.7568, 0.7241, 0.7083, 0.7008; (b) 0.7   **5.** (a) 1.4641, 1.2642, 1.1776, 1.1372, 1.1177; (b) 1.1

6. (a)

x	-3	-2	-1	0	1	2	3
2^x	0.125	0.25	0.5	1	2	4	8
$(0.7)2^x$	0.09	0.18	0.35	0.7	1.4	2.8	5.6

(b) See Figure 1 in text.

7. (a)

x	-2	-1	0	1	2
3^x	0.11	0.33	1	3	9
$(1.1)3^x$	0.12	0.36	1.1	3.3	9.9

(b) See Figure 2 in text.

8. \$2, \$2.25, \$2.370370, \$2.441406, \$2.488320, \$2.704814, \$2.714567, \$2.718121   **9.** $6e^x$ **10.** $e^x(x^3 + 3x^2)$, or $x^2e^x(x + 3)$ **11.** $\dfrac{e^x(x - 2)}{x^3}$ **12.** $-4e^{-4x}$ **13.** $(3x^2 + 8)e^{x^3+8x}$ **14.** $\dfrac{xe^{\sqrt{x^2+5}}}{\sqrt{x^2 + 5}}$

15.

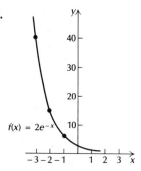

$f(x) = 2e^{-x}$

16. (a)

x	0	$\frac{1}{2}$	1	2	3	4
$f(x)$	0	0.39	0.63	0.86	0.95	0.98

(b)

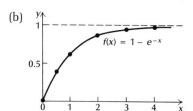

$f(x) = 1 - e^{-x}$

Exercise Set 4.1, p. 220

1.

$y = 4^x$

3.

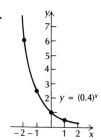

$y = (0.4)^x$

5.

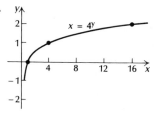

$x = 4^y$

7. $3e^{3x}$ **9.** $-10e^{-2x}$ **11.** e^{-x} **13.** $-7e^x$ **15.** e^{2x} **17.** $x^3 e^x(x + 4)$ **19.** $\dfrac{e^x(x - 4)}{x^5}$ **21.** $(-2x + 7)e^{-x^2+7x}$

23. $-xe^{-x^2/2}$ **25.** $\dfrac{e^{\sqrt{x-7}}}{2\sqrt{x - 7}}$ **27.** $\dfrac{e^x}{2\sqrt{e^x - 1}}$ **29.** $(1 - 2x)e^{-2x}$ $-e^{-x} + 3x^2$ **31.** e^{-x} **33.** ke^{-kx}

35.

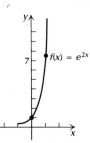

$f(x) = e^{2x}$

37.

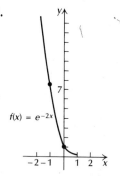

$f(x) = e^{-2x}$

39.

$f(x) = 1 - e^{-x}$

41. (a) $C'(t) = 50e^{-t}$; (b) \$50 million; (c) \$.916 million   **43.** $15(e^{3x} + 1)^4 e^{3x}$ **45.** $-e^{-t} - 3e^{3t}$ **47.** $\dfrac{e^x(x - 1)^2}{(x^2 + 2)^2}$

49. $\dfrac{e^{\sqrt{x}}}{2\sqrt{x}} + \dfrac{1}{2}\sqrt{e^x}$ **51.** $\dfrac{1}{2}e^{x/2}\left[\dfrac{x}{\sqrt{x - 1}}\right]$ **53.** $\dfrac{4}{(e^x + e^{-x})^2}$ **55.** 2, 2.25, 2.48832, 2.59374, 2.71692

57. Max. $= 4e^{-2} \approx 0.54$ at $x = 2$

59.

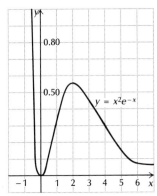

Margin Exercises, Section 4.2, pp. 221–236

1. (a) $b^T = P$; (b) $9^{1/2} = 3$; (c) $10^3 = 1000$; (d) $10^{-1} = 0.1$ **2.** (a) $\log_e T = k$; (b) $\log_{16} 2 = \frac{1}{4}$; (c) $\log_{10} 10{,}000 = 4$;

(d) $\log_{10} 0.001 = -3$

3.

4. (a) 1000; (b) 3; (c) 0.699; (d) ≈ 5.000 **5.** 0.602

6. 1   **7.** -0.398 **8.** 0.398   **9.** -0.699 **10.** $\frac{3}{2}$ **11.** 1.699   **12.** 1.204   **13.** (a) 4.4977; (b) -0.0419; (c) 1.8954;
(d) 0.8954; (e) -0.1046; (f) -1.1046; (g) -2.1046 **14.** 2.3025   **15.** -0.9163 **16.** 0.9163   **17.** 2.7724   **18.** 2.6094

19. $\frac{1}{2}$ **20.** -1.6094 **21.** 0.693147   **22.** 2.995732   **23.** 4.605170   **24.** -2.599375 **25.** 0.076961   **26.** -0.000100

27. 2.0956   **28.** 11.3059   **29.** -2.5096 **30.** 7.6009   **31.** -9.2103 **32.** $t \approx 4$ **33.** $t \approx 44$

34. (a)

x	0.5	1	2	3	4
ln x	-0.7	0	0.7	1.1	1.4

(b) same as text figure

35. $\frac{5}{x}$ **36.** $x^2(1 + 3 \ln x) + 4$ **37.** $\frac{1 - 2 \ln x}{x^3}$ **38.** $\frac{1}{x}$ **39.** $\frac{6x}{3x^2 + 4}$ **40.** $\frac{1}{x \ln 5x}$ **41.** $\frac{4x^5 + 2}{x(x^5 - 2)}$ **42.** (a) 500;

(b) $N'(a) = \frac{200}{a}$; (c) min. $= 500$ at $a = 1$ **43.** 87 days

Exercise Set 4.2, pp. 236–239

1. $2^3 = 8$ **3.** $8^{1/3} = 2$ **5.** $a^J = K$ **7.** $b^v = T$ **9.** $\log_e b = M$ **11.** $\log_{10} 100 = 2$ **13.** $\log_{10} 0.1 = -1$

15. $\log_M V = p$ **17.** 2.708   **19.** 0.51   **21.** -1.609 **23.** $\dfrac{3}{2}$ **25.** 2.609   **27.** 3.218   **29.** 2.9957   **31.** 0.2231

33. -1.3863 **35.** 2.3863   **37.** 4   **39.** 2.7726   **41.** 8.681690   **43.** -4.006334 **45.** 0.631272   **47.** -3.973898

49. 6.809039   **51.** -4.509860 **53.** $t \approx 5$ **55.** $t \approx 4$ **57.** $t \approx 2$ **59.** $t \approx 141$ **61.** $-\dfrac{6}{x}$ **63.** $x^3(1 + 4 \ln x) - x$

65. $\dfrac{1 - 4 \ln x}{x^5}$ **67.** $\dfrac{1}{x}$ **69.** $\dfrac{10x}{5x^2 - 7}$ **71.** $\dfrac{1}{x \ln 4x}$ **73.** $\dfrac{x^2 + 7}{x(x^2 - 7)}$ **75.** $e^x\left(\dfrac{1}{x} + \ln x\right)$ **77.** $\dfrac{e^x}{e^x + 1}$ **79.** $\dfrac{2 \ln x}{x}$

81. (a) 68%; (b) 36%; (c) 3.6%; (d) 5%; (e) $-\dfrac{20}{t + 1}$; (f) max. = 68% at $t = 0$ **83.** (a) 1000; (b) $N'(a) = \dfrac{200}{a}$, $N'(10) = 20$;

(c) min. = 1000 at $a = 1$ **85.** (a) 2.37 ft/sec; (b) 3.37 ft/sec; (c) $v'(p) = \dfrac{0.37}{p}$, $v'(p) = $ the acceleration of the walker

87. 58 days   **89.** (a) 18.1%, 69.9%; (b) $P'(t) = 0.2e^{-0.2t}$; (c) 11.5   **91.** (a) \$58.69, \$78.00; (b) \$63.80 $e^{-1.1t}$; (c) 2.7

93. $\dfrac{-4 (\ln x)^{-5}}{x}$ **95.** $\dfrac{15t^2}{t^3 + 1}$ **97.** $\dfrac{4 [\ln (x + 5)]^3}{x + 5}$ **99.** $\dfrac{5t^4 - 3t^2 + 6t}{(t^3 + 3)(t^2 - 1)}$,   **101.** $\dfrac{24x + 25}{8x^2 + 5x}$ **103.** $\dfrac{2(1 - \ln t^2)}{t^3}$ **105** $x^n \ln x$

107. $\dfrac{1}{\sqrt{1 + t^2}}$ **109.** $\dfrac{3}{x \ln x}$ **111.** 1   **113.** $\sqrt[e]{e} \approx 1.444667861$; $\sqrt[e]{e} > \sqrt[x]{x}$ for any $x > 0$ such that $x \neq e$ **115.** $-\dfrac{1}{2e}$

117. $t = \dfrac{\ln P - \ln P_0}{k}$ **119.** ∞ **121.** Let $a = \ln x$; then $e^a = x$, so $\log x = \log e^a = a \log e$;

then $a = \ln x = \dfrac{\log x}{\log e} \approx \dfrac{\log x}{0.4343} \approx 2.3026 \log x$.

Margin Exercises, Section 4.3, pp. 240–247

1. (a) $\dfrac{dy}{dx} = 20e^{4x}$; (b) $\dfrac{dy}{dx} = 4y$ **2.** (a) $N(t) = ce^{kt}$; (b) $f(t) = ce^{kt}$ **3.** Should be about $\dfrac{1}{8}$ in., or 0.125 in., (b) 0.032, 0.064,

0.128; (c) about 2 mi   **4.** (a) $P(t) = P_0e^{0.13t}$; (b) \$1138.83; (c) 5.3 yr   **5.** 2%, 34.7; 6.9%, 10; 14%, 5.0; 4.6%, 15; 1%, 69.3
6. (a) $P(t) = 225e^{0.008t}$; (b) 242 million; (c) 86.6 yr (2067).   **7.** (a) $E(t) = 800e^{kt}$; (b) $k = 0.08$; (c) $E(t) = 800e^{0.08t}$;
(d) 8818.5 billion kWh

Exercise Set 4.3, pp. 248–250

1. $Q(t) = Q_0e^{kt}$ **3.** (a) $P(t) = P_0e^{0.09t}$; (b) \$1094.17, \$1197.22; (c) 7.7 yr   **5.** 19.8 yr   **7.** 6.9%   **9.** (a) $P(t) = 209e^{0.01t}$;
(b) 312 million; (c) 69.3 yr   **11.** 0.20%   **13.** (a) $N(t) = 50e^{0.1t}$; (b) 369; (c) 6.9 yr   **15.** 6.9 yr (1986)   **17.** (a)
$k = 0.081374$, $P(t) = 25e^{0.081374t}$; (b) 207 thousand   **19.** (a) $k = 0.07$, $P(t) = \$90e^{0.07t}$; (b) \$2980 million; (c) 9.9 yr (1983)

21. (a) $k = 0.061248$, $P(t) = \$100e^{0.061248t}$; (b) \$340.40; (c) 11.3 yr (1978)   **23.** 15.03%   **25.** 9%   **27.** $T_3 = \dfrac{\ln 3}{k}$
29. Answers depend on particular data.   **31.** $\approx\$66,000,000,000,000$ **33.** $\approx 2.2\%$ **35.** \$16.64, \$27.71.
37. $\dfrac{\ln 2}{24} \approx 2.9\%$ per hr

Margin Exercises, Section 4.4, pp. 251–259

1. See Exercises 35 and 37 of Exercise Set 4.1.   **2.** (a) $N(t) = N_0e^{-0.14t}$; (b) 246.6 gr; (c) 5 days   **3.** 30 yr   **4.** 5.3%
5. 13,412 yr   **6.** (a) $a = 130°$; (b) $k = 0.01$; (c) 188°; (d) 256 min   **7.** Answers wll vary. "Theoretically" it is never possible for the temperature of the water to be the same as the room temperature.   **8.** 10:00 A.M.

Exercise Set 4.4, pp. 260–262

1. 7.2 days   **3.** 23% per min   **5.** 10.1 g   **7.** 19,188 yr   **9.** 7636 yr   **11.** (a) $A = A_0 e^{-kt}$; (b) 9 hr   **13.** (a) 0.8%;
(b) 79% W_0 **15.** (a) 11 watts; (b) 173 days; (c) 402 days; (d) 50 watts   **17.** (a) \$40,000; (b) \$5413   **19.** (a) 25% I_0, 6% I_0,
1.5% I_0; (b) 0.00008%   **21.** (a) $a = 25°$; (b) $k = 0.05$; (c) 84.2°; (d) 32 min   **23.** 7 P.M.   **25.** (a) $k = 0.010470$,
$P(t) = 503{,}000 e^{-0.010470t}$, where $t = $ years since 1960; (b) 367,414; (c) 1254 yr   **27.** \$2018.97
29. (a) $k = 0.03$, $P(t) = \$100 e^{-0.03t}$, where $t = $ years that precede 1967; (b) \$13.40

Margin Exercises, Section 4.5, pp. 262–266

1. $e^{6.9315}$ **2.** $e^{0.6931x}$ **3.** $(\ln 5)5^x$ **4.** $(\ln 4)4^x$ **5.** $(\ln 4.3)(4.3)^x$ **6.** $\dfrac{1}{\ln 2} \cdot \dfrac{1}{x}$ **7.** $-\dfrac{7}{x \ln 10}$ **8.** $x^5\left(\dfrac{1}{\ln 10} + 6 \log x\right)$

Exercise Set 4.5, pp. 266–268

1. $e^{6.4376}$ **3.** $e^{12.238}$ **5.** $e^{k \cdot \ln 4}$ **7.** $e^{kT \cdot \ln 8}$ **9.** $(\ln 6)6^x$ **11.** $(\ln 10)10^x$ **13.** $(6.2)^x[x \ln 6.2 + 1]$
15. $10^x x^2[x \ln 10 + 3]$ **17.** 7.85   **19.** (a) $I = 10^7 \cdot I_0$; (b) $I = 10^8 \cdot I_0$; (c) the intensity in (b) is 10 times that in part (a);
(d) $\dfrac{dI}{dR} = (I_0 \cdot \ln 10) \cdot 10^R$ **21.** $\dfrac{1}{\ln 4} \cdot \dfrac{1}{x}$ **23.** $\dfrac{2}{\ln 10} \cdot \dfrac{1}{x}$ **25.** $\dfrac{1}{\ln 10} \cdot \dfrac{1}{x}$ **27.** $x^2\left[\dfrac{1}{\ln 8} + 3 \log_8 x\right]$ **29.** $\dfrac{1}{\ln 10} \cdot \dfrac{1}{I}$
31. $\dfrac{m}{\ln 10} \cdot \dfrac{1}{x}$ **33.** $-250{,}000 (\ln 4)\left(\dfrac{1}{4}\right)^t$ **35.** $2(\ln 3)3^{2x}$ **37.** $(\ln x + 1)x^x$ **39.** $x^{e^x} e^x\left(\ln x + \dfrac{1}{x}\right)$ **41.** $\dfrac{1}{\ln a} \cdot \dfrac{f'(x)}{f(x)}$

Margin Exercises, Section 4.6, pp. 268–271

1. (a) $E(x) = \dfrac{300 - x}{x}$; (b) $E(100) = 2$, $E(200) = \dfrac{1}{2}$; (c) $x = 150$; (d) $R(x) = 300x - x^2$; (e) $x = 150$

Exercise Set 4.6, p. 272

1. (a) $E(x) = \dfrac{400 - x}{x}$; (b) $x = 200$ **3.** (a) $E(x) = \dfrac{50 - x}{x}$; (b) $x = 25$ **5.** (a) $E(x) = 1$, for all $x > 0$; (b) total

revenue $= R(x) = 400$, for all $x > 0$. It has 400 as a maximum for all $x > 0$.   **7.** (a) $E(x) = \dfrac{1000 - 2x}{x}$; (b) $x = \dfrac{1000}{3}$

9. (a) $E(x) = \dfrac{4}{x}$; (b) $x = 4$ **11.** (a) $E(x) = \dfrac{x + 3}{2x}$; (b) $x = 3$ **13.** (a) $E(x) = \dfrac{1}{n}$; (b) no, E is a constant $\dfrac{1}{n}$; (c) only when

$n = 1$ **15.** $E(x) = -\dfrac{1}{x} \cdot \dfrac{1}{L'(x)}$

Chapter 4 Test, pp. 272–273

1. e^x **2.** $\dfrac{1}{x}$ **3.** $-2xe^{-x^2}$ **4.** $\dfrac{1}{x}$ **5.** $e^x - 15x^2$ **6.** $3e^x\left(\dfrac{1}{x} + \ln x\right)$ **7.** $\dfrac{e^x - 3x^2}{e^x - x^3}$ **8.** $\dfrac{\dfrac{1}{x} - \ln x}{e^x}$, or $\dfrac{1 - x \ln x}{xe^x}$ **9.** 2.639
10. -1.2528 **11.** 2.9459   **12.** $M(t) = M_0 e^{kt}$ **13.** 17.3% per hr   **14.** 10 yr   **15.** (a) $F(t) = 3e^{0.12t}$; (b) 86 billion gal;
(c) 5.8 yr (1965)   **16.** (a) $A(t) = 3e^{-0.1t}$; (b) 1.1 cc; (c) 6.9 hr   **17.** 63 days   **18.** 0.000069% per yr   **19.** $(\ln 20)20^x$
20. $\dfrac{1}{\ln 20} \cdot \dfrac{1}{x}$ **21.** (a) $E(x) = \dfrac{5}{x}$; (b) $x = 5$ **22.** $(\ln x)^2$ **23.** Max. $= 256e^{-4} \approx 4.69$ at $x = 4$; min. $= 0$ at $x = 0$

CHAPTER 5

Margin Exercises, Section 5.1, pp. 276–282

1. $y = 7x$, $y = 7x - \dfrac{1}{2}$, $y = 7x + C$ (answers can vary)   **2.** $y = -2x$, $y = -2x + 27$, $y = -2x + C$ (answers can vary)

3. $\dfrac{x^2}{2} + C$ **4.** $\dfrac{x^4}{4} + C$ **5.** $e^x + C$ **6.** $\ln x + C$ **7.** $\dfrac{x^4}{4} + C$ **8.** $\dfrac{x^2}{2} + C$ **9.** $\ln x + C$ **10.** $\dfrac{7}{5}x^5 + x^2 + C$

11. $e^x - \dfrac{5}{7}x^{7/5} + C$ **12.** $5\ln x - 7x - \dfrac{1}{5}x^{-5} + C$ **13.**

14. $f(x) = \dfrac{x^3}{3} + \dfrac{23}{3}$

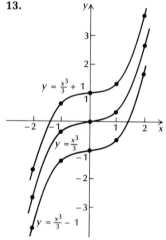

$y = \dfrac{x^3}{3} + 1$

$y = \dfrac{x^3}{3}$

$y = \dfrac{x^3}{3} - 1$

15. $g(x) = x^2 - 4x + 13$ **16.** $C(x) = \dfrac{1}{3}x^3 + \dfrac{5}{2}x^2 + 35$ **17.** $s(t) = t^4 + 13$ **18.** $s(t) = 2t^4 - 6t^2 + 7t + 8$

Exercise Set 5.1, pp. 283–284

1. $\dfrac{x^7}{7} + C$ **3.** $2x + C$ **5.** $\dfrac{4}{5}x^{5/4} + C$ **7.** $\dfrac{x^3}{3} + \dfrac{x^2}{2} - x + C$ **9.** $\dfrac{t^3}{3} - t^2 + 3t + C$ **11.** $5e^x + C$ **13.** $\dfrac{x^4}{4} - \dfrac{7}{15}x^{15/7} + C$

15. $1000\ln x + C$ **17.** $-x^{-1} + C$ **19.** $f(x) = \dfrac{x^2}{2} - 3x + 13$ **21.** $f(x) = \dfrac{x^3}{3} - 4x + 7$ **23.** $C(x) = \dfrac{x^4}{4} - x^2 + 100$

25. (a) $R(x) = \dfrac{x^3}{3} - 3x$; (b) If you sell no products you make no money.   **27.** $s(t) = t^3 + 4$ **29.** $v(t) = 2t^2 + 20$

31. $s(t) = -\dfrac{1}{3}t^3 + 3t^2 + 6t + 10$ **33.** $s(t) = -16t^2 + v_0t + s_0$ **35.** $\dfrac{1}{4}$mi   **37.** (a) $E(t) = 30t - 5t^2 + 32$; (b) $E(3) = 77\%$,

$E(5) = 57\%$ **39.** (a) $A(t) = 43.4t^{-1} - 3.7$; (b) 2.5 cm²   **41.** $f(t) = \dfrac{t^{\sqrt{3}+1}}{\sqrt{3}+1} + 8$ **43.** $\dfrac{x^6}{6} - \dfrac{2}{5}x^5 + \dfrac{1}{4}x^4 + C$

45. $\dfrac{2}{5}t^{5/2} + 4t^{3/2} + 18\sqrt{t} + C$ **47.** $\dfrac{t^4}{4} + t^3 + \dfrac{3}{2}t^2 + t + C$ **49.** $\dfrac{b}{a}e^{ax} + C$ **51.** $\dfrac{12}{7}x^{7/3} + C$ **53.** $\dfrac{t^3}{3} - t^2 + 4t + C$

Margin Exercises, Section 5.2, pp. 285–294

1. (a) $A(x) = 3x$; (b) $A(1) = 3$, $A(2) = 6$, $A(5) = 15$; (c)

$A(x) = 3x$

(d) $A(x)$ is an antiderivative of $f(x)$

2. (a) $C(x) = 50x$; (b) $A(x) = 50x$; (c) as x increases, the area over $[0, x]$ increases **3.** (a) $5000 + 4000 + 3000 + 2000$, or

$14,000; (b) 14,000, same **4.** (a) $A(x) = \frac{3}{2}x^2$; (b) $A(1) = \frac{3}{2}$; (c) $A(2) = 6$; (d) $A(3.5) = 18.375$; (e)

(f) $A(x)$ is an antiderivative of $f(x)$ **5.** (a) $C(x) = -0.05x^2 + 50x$; (b) $A(x) = -0.05x^2 + 50x$; (c) $12,000; less than in Margin Exercise 3 **6.** $5\frac{1}{3}$ **7.** $13\frac{1}{2}$ **8.** Yes; (b) $4100

Exercise Set 5.2, pp. 294–295

1. 8 **3.** 8 **5.** $41\frac{2}{3}$ **7.** $\frac{1}{4}$ **9.** $10\frac{2}{3}$ **11.** $e^3 - 1 \approx 19.086$ **13.** $\ln 3 \approx 1.0986$ **15.** 51 **17.** An antiderivative, velocity **19.** An antiderivative, energy used in time t **21.** An antiderivative, total revenue **23.** An antiderivative, amount of drug in blood **25.** (a) $s(t) = t^3 + t^2$; (b) 150 **27.** (a) $C(x) = 100x - 0.1x^2$; (b) $R(x) = 100x + 0.1x^2$; (c) $P(x) = 0.2x^2$; (d) $P(1000) = \$200,000$ **29.** $3\frac{1}{2}$ **31.** $359\frac{7}{15}$ **33.** $6\frac{3}{4}$

Margin Exercises, Section 5.3, pp. 296–303

1. $\int_a^b 2x\, dx = b^2 - a^2$ **2.** $\int_a^b e^x\, dx = e^b - e^a$ **3.** 8 **4.** $1 - \frac{1}{e^2}$ **5.** $\frac{2}{3}$ **6.** $e + e^3 - 3$ **7.** $5\frac{1}{3}$ **8.** $13\frac{1}{2}$ **9.** $\ln 7$
10. $-\frac{1}{3}b^{-3} + \frac{1}{3}$ **11.** 75 **12.** $17\frac{1}{3}$ **13.** (a) $595,991.60; (b) $k = 7.3$, so it is the 8th day; (c) $4,185,966.53

Exercise Set 5.3, pp. 304–305

1. $\frac{1}{6}$ **3.** $\frac{4}{15}$ **5.** $e^b - e^a$ **7.** $b^3 - a^3$ **9.** $\frac{e^2}{2} + \frac{1}{2}$ **11.** $\frac{2}{3}$ **13.** $\frac{5}{34}$ **15.** 4 **17.** $9\frac{5}{6}$ **19.** 12 **21.** $e^5 - \frac{1}{e}$ **23.** $7\frac{1}{3}$
25. $17\frac{1}{3}$ **27.** (a) $2948.26; (b) $2913.90; (c) $k = 6.9$, so it will be on the 7th day **29.** 90 **31.** $3600 **33.** 9
35. $\frac{307}{6}$ **37.** $\frac{15}{4}$ **39.** 8 **41.** 12

Margin Exercises, Section 5.4, pp. 306–312

1. 4 **2.** 4 **3.** (a) $-\frac{1}{6}$; (b) $\frac{5}{6}$; (c) $\frac{4}{6}$ **4.** (a) -4; (b) 4; (c) 0 **5.** (a) 0; (b) $A - A$, or 0 **6.** (a) Negative; (b) $A - 2A$, or $-A$ **7.** (a) Positive; (b) $3A - A$, or $2A$ **8.** (a) Negative; (b) $-3A$ **9.** $\frac{1}{6}$ **10.** (a) $P(k) = 5k^2 + 10k$; (b) $P(5) = 175$

Exercise Set 5.4, pp. 312–313

1. $\frac{1}{4}$ **3.** $\frac{9}{2}$ **5.** $\frac{125}{6}$ **7.** $\frac{9}{2}$ **9.** $\frac{1}{6}$ **11.** $\frac{125}{3}$ **13.** $\frac{32}{3}$ **15.** 3 **17.** (a) B; (b) 2 **19.** $\frac{(e-1)^2}{e}$ **21.** $\frac{2}{3}$ **23.** 16

Margin Exercises, Section 5.5, pp. 314–318

1. $dy = (12x + 1)\, dx$ 2. $du = dx$ 3. $e^{x^3} + C$ 4. $\ln(5 + x^2) + C$ 5. $-\dfrac{1}{3 + x^2} + C$ 6. $\dfrac{(\ln x)^2}{2} + C$ 7. $\dfrac{1}{3}e^{x^3} + C$

8. $\dfrac{1}{5}e^{5x} + C$ 9. $\dfrac{1}{0.02}e^{0.02x} + C$, or $50e^{0.02x} + C$ 10. $-e^{-x} + C$ 11. $\dfrac{1}{80}(x^4 + 5)^{20} + C$

12. $\displaystyle\int \dfrac{\ln x\, dx}{x} = \dfrac{(\ln x)^2}{2} + C$, so $\displaystyle\int_1^e \dfrac{\ln x\, dx}{x} = \left[\dfrac{(\ln x)^2}{2}\right]_1^e = \dfrac{(\ln e)^2}{2} - \dfrac{(\ln 1)^2}{2} = \dfrac{1}{2} - 0 = \dfrac{1}{2}$

Exercise Set 5.5, pp. 319–320

1. $\ln(7 + x^3) + C$ 3. $\dfrac{1}{4}e^{4x} + C$ 5. $2e^{x/2} + C$ 7. $\dfrac{1}{4}e^{x^4} + C$ 9. $-\dfrac{1}{3}e^{-t^3} + C$ 11. $\dfrac{(\ln 4x)^2}{2} + C$ 13. $\ln(1 + x) + C$

15. $-\ln(4 - x) + C$ 17. $\dfrac{1}{24}(t^3 - 1)^8 + C$ 19. $\dfrac{1}{8}(x^4 + x^3 + x^2)^8 + C$ 21. $\ln(4 + e^x) + C$ 23. $\dfrac{1}{4}(\ln x^2)^2 + C$, or

$(\ln x)^2 + C$ 25. $\ln(\ln x) + C$ 27. $\dfrac{2}{3a}(ax + b)^{3/2} + C$ 29. $\dfrac{b}{a}e^{ax} + C$ 31. $e - 1$ 33. $\dfrac{21}{4}$

35. $\ln 4 - \ln 2 = \ln\dfrac{4}{2} = \ln 2$ 37. $\ln 19$ 39. $1 - \dfrac{1}{e^5}$ 41. $1 - \dfrac{1}{e^{mb}}$ 43. $\dfrac{208}{3}$ 45. (a) $\dfrac{100,000}{0.025}(e^{2.2} - 1) \approx 32{,}100{,}054$;

(b) $\dfrac{100,000}{0.025}(e^{2.2} - e^2) \approx 6{,}543{,}830$ 47. $-\dfrac{5}{12}(1 - 4x^2)^{3/2} + C$ 49. $-\dfrac{1}{3}e^{-x^3} + C$ 51. $-e^{1/t} + C$ 53. $-\dfrac{1}{3}(\ln x)^{-3} + C$

55. $\dfrac{2}{9}(x^3 + 1)^{3/2} + C$ 57. $\dfrac{3}{4}(x^2 - 6x)^{2/3} + C$ 59. $\dfrac{1}{8}[\ln(t^4 + 8)]^2 + C$ 61. $x + \dfrac{9}{x + 3} + C$ 63. $t - \ln(t - 4) + C$

65. $-\ln(1 + e^{-x}) + C$ 67. $\dfrac{1}{n + 1}(\ln x)^{n+1} + C$

Margin Exercises, Section 5.6, pp. 321–324

1. $xe^{3x} - \dfrac{1}{3}e^{3x} + C$ 2. $\dfrac{x^2}{2}\ln x - \dfrac{x^2}{4} + C$; let $u = \ln x$, $dv = x\, dx$ 3. $\dfrac{2}{3}x(x + 3)^{3/2} - \dfrac{4}{15}(x + 3)^{5/2} + C$ 4. $2\ln 2 - \dfrac{3}{4}$

5. $\dfrac{1}{10}\ln\left(\dfrac{x - 5}{x + 5}\right) + C$ 6. $x^3e^x - 3x^2e^x + 6xe^x - 6e^x + C$

Exercise Set 5.6, pp. 325–326

1. $xe^{5x} - \dfrac{1}{5}e^{5x} + C$ 3. $\dfrac{1}{2}x^6 + C$ 5. $\dfrac{x}{2}e^{2x} - \dfrac{1}{4}e^{2x} + C$ 7. $-\dfrac{x}{2}e^{-2x} - \dfrac{1}{4}e^{-2x} + C$ 9. $\dfrac{x^3}{3}\ln x - \dfrac{x^3}{9} + C$

11. $\dfrac{x^2}{2}\ln x^2 - \dfrac{x^2}{2} + C$ 13. $(x + 3)\ln(x + 3) - x + C$. Let $u = \ln(x + 3)$, $dv = dx$, and choose $v = x + 3$ for an

antiderivative of v.   15. $\left(\dfrac{x^2}{2} + 2x\right)\ln x - \dfrac{x^2}{4} - 2x + C$ 17. $\left(\dfrac{x^2}{2} - x\right)\ln x - \dfrac{x^2}{4} + x + C$

19. $\dfrac{2}{3}x(x + 2)^{3/2} - \dfrac{4}{15}(x + 2)^{5/2} + C$ 21. $\dfrac{x^4}{4}\ln 2x - \dfrac{x^4}{16} + C$ 23. $x^2e^x - 2xe^x + 2e^x + C$

25. $\dfrac{1}{2}x^2e^{2x} + \dfrac{1}{4}e^{2x} - \dfrac{1}{2}xe^{2x} + C$ 27. $\dfrac{8}{3}\ln 2 - \dfrac{7}{9}$ 29. $9\ln 9 - 5\ln 5 - 4$ 31. 1 33. $\dfrac{1}{9}e^{-3x}(-3x - 1) + C$

35. $\dfrac{5^x}{\ln 5} + C$ 37. $\dfrac{1}{8}\ln\left(\dfrac{4 + x}{4 - x}\right) + C$ 39. $5 - x - 5\ln(5 - x) + C$ 41. $\dfrac{1}{5(5 - x)} + \dfrac{1}{25}\ln\left(\dfrac{x}{5 - x}\right) + C$

43. (a) $10[e^{-T}(-T - 1) + 1]$; (b) ≈ 9.085 45. $\dfrac{2}{9}x^{3/2}[3\ln x - 2] + C$ 47. $\dfrac{e^t}{t + 1} + C$ 49. $2\sqrt{x}\ln x - 4\sqrt{x} + C$

51. Let $u = x^n$ and $dv = e^x\, dx$. Then $du = nx^{n-1}\, dx$ and $v = e^x$. Then use integration by parts.

Margin Exercises, Section 5.7, pp. 326–333

1. $\sum_{i=1}^{6} i^2$ **2.** $\sum_{i=1}^{4} e^i$ **3.** $\sum_{i=1}^{38} P(x_i)\,\Delta x$ **4.** $4^1 + 4^2 + 4^3$, or 84 **5.** $e^1 + 2e^2 + 3e^3 + 4e^4 + 5e^5$
6. $t(x_1)\,\Delta x + t(x_2)\,\Delta x + \cdots + t(x_{20})\,\Delta x$ **7.** \$11,250 **8.** (a) 3.0667; (b) 2.4501; (c) $\int_1^7 \frac{1}{x}\,dx = \ln 7 \approx 1.9459$ **9.** 2
10. (a) $\frac{95}{3} \approx 31.7°$; (b) $-10°$; (c) 46.25° **11.** \$257

Exercise Set 5.7, pp. 333–335

1. (a) 1.4914; (b) 0.8571 **3.** 0 **5.** $e - 1$ **7.** $\frac{4}{3}$ **9.** 13 **11.** $\frac{1}{n + 1}$ **13.** (a) 100 after 10 hr; (b) $33\frac{1}{3}$ **15.** 225 million
17. (a) 3 cc; (b) $\approx$2.7 cc **19.** 47.5

Chapter 5 Test, pp. 336–337

1. $x + C$ **2.** $200x^5 + C$ **3.** $e^x + \ln x + \frac{8}{11}x^{11/8} + C$ **4.** $\frac{1}{6}$ **5.** $4 \ln 3$ **6.** An antiderivative, total number of words
typed in t minutes **7.** 12 **8.** $-\frac{1}{2}\left(\frac{1}{e^2} - 1\right)$ **9.** $\ln b - \ln a$ **10.** 0 **11.** Negative **12.** Positive **13.** $\ln (x + 8) + C$
14. $-2e^{-0.5x} + C$ **15.** $\frac{1}{40}(t^4 + 1)^{10} + C$ **16.** $\frac{x}{5}e^{5x} - \frac{e^{5x}}{25} + C$ **17.** $\frac{x^4}{4}\ln x^4 - \frac{x^4}{4} + C$ **18.** $\frac{2^x}{\ln 2} + C$
19. $\frac{1}{7}\ln\left(\frac{x}{7 - x}\right) + C$ **20.** 6 **21.** $\frac{1}{3}$ **22.** 95 **23.** \$49,000 **24.** 94 words **25.** $\frac{(\ln x)^4}{4} - \frac{4}{3}(\ln x)^3 + 5\ln x + C$
26. $(x + 3)\ln (x + 3) - (x + 5)\ln (x + 5) + C$

CHAPTER 6

Margin Exercises, Section 6.1, pp. 340–343

1. (2, \$16) **2.** \$18.67 **3.** \$7.33

Exercise Set 6.1, p. 343

1. (a) (6, \$5); (b) \$15; (c) \$9 **3.** (a) (1, \$9); (b) \$3.33; (c) \$1.67 **5.** (a) (3, \$9); (b) \$36; (c) \$18 **7.** (a) (5, \$0.61);
(b) \$86.36; (c) \$2.45

Margin Exercises, Section 6.2, pp. 343–350

1. \$1161.83 **2.** \$29,786.67 **3.** \$215.41 **4.** \$7577.50 **5.** 1,136,884 million barrels **6.** After 7 yrs (1987)

Exercise Set 6.2, pp. 350–351

1. \$131 **3.** \$5,610.67 **5.** \$340,754.12 **7.** \$949.94 **9.** \$259.37 **11.** 1,250,427,016 tons **13.** After 31 years (2011)
15. $\int_0^{365} Pe^{0.08 \cdot t/365}\, dt = \dfrac{365P}{0.08}(e^{0.08} - 1);$ \$379,997.25 **17.** \$33,535.73 **19.** \$125,289.70

Margin Exercises, Section 6.3, pp. 351–354

1. \$2231.30 **2.** \$24,261.23 **3.** \$15,563.96

Exercise Set 6.3, pp. 354–355

1. \$826.50 **3.** \$22,973.57 **5.** \$17,802.91 **7.** \$511,471.15 **9.** 19.994 lb **11.** \$74.23 **13.** \$68.83
15. $\int_0^{365} Pe^{-(0.12/365)t}\, dt = \dfrac{365P}{0.12}(1 - e^{-0.12});$ \$4815.30

Margin Exercises, Section 6.4, pp. 355–359

1. $\dfrac{2}{3}, \dfrac{9}{10}, \dfrac{99}{100}, \dfrac{199}{200}$ **2.** $\dfrac{1}{2}$ **3.** ∞ **4.** Convergent, 1 **5.** Divergent **6.** \$20,000

Exercise Set 6.4, pp. 360–361

1. $\dfrac{1}{3}$ **3.** Divergent **5.** 1 **7.** $\dfrac{1}{2}$ **9.** Divergent **11.** 5 **13.** Divergent **15.** Divergent **17.** Divergent **19.** 1
21. \$36,000 **23.** 33,333 lb **25.** Divergent **27.** 2 **29.** $\dfrac{1}{2}$ **31.** \$0.93 **33.** $\dfrac{1}{k^2}$; the total dose of the drug

Margin Exercises, Section 6.5, pp. 361–371

1. (a) $\dfrac{7}{20}$; (b) $\dfrac{6}{20}$, or $\dfrac{3}{10}$; (c) $\dfrac{4}{20}$, or $\dfrac{1}{5}$; (d) 0 **2.** (a) $\dfrac{1}{2}$; (b) $\dfrac{1}{8}$; (c) $\dfrac{1}{4}$; (d) $\dfrac{1}{8}$ **3.** [15, 25] **4.** [0, ∞) **5.** $\dfrac{2}{3}$
6. $\int_1^2 \dfrac{2}{3}x\, dx = \left[\dfrac{2}{3} \cdot \dfrac{1}{2}x^2\right]_1^2 = \dfrac{1}{3}(2^2 - 1) = 1$ **7.** (a) $\int_3^6 \dfrac{24}{t^3}\, dt = \left[24 - \left(-\dfrac{t^{-2}}{2}\right)\right]_3^6 = -12\left(\dfrac{1}{6^2} - \dfrac{1}{3^2}\right) = 1$; (b) $\dfrac{64}{75}$; (c) $\dfrac{5}{12}$
8. $\dfrac{3}{26}$ **9.** 4 **10.** $\dfrac{1}{4}$ **11.** $\dfrac{1}{4}$ **12.** 0.3297

Exercise Set 6.5, pp. 372–373

1. $\int_0^1 2x\, dx = [x^2]_0^1 = 1^2 - 0^2 = 1$ **3.** $\int_4^7 \dfrac{1}{3}\, dx = \left[\dfrac{1}{3}x\right]_4^7 = \dfrac{1}{3}(7 - 4) = 1$ **5.** $\int_1^3 \dfrac{3}{26}x^2\, dx = \left[\dfrac{3}{26} \cdot \dfrac{x^3}{3}\right]_1^3 = \dfrac{1}{26}(3^3 - 1^3) = 1$
7. $\int_1^e \dfrac{1}{x}\, dx = [\ln x]_1^e = \ln e - \ln 1 = 1 - 0 = 1$ **9.** $\int_{-1}^1 \dfrac{3}{2}x^2\, dx = \left[\dfrac{3}{2} \cdot \dfrac{1}{3}x^3\right]_{-1}^1 = \dfrac{1}{2}(1^3 - (-1)^3) = \dfrac{1}{2}(1 + 1) = 1$
11. $\int_0^\infty 3e^{-3x}\, dx = \lim_{b \to \infty} \int_0^b 3e^{-3x}\, dx = \lim_{b \to \infty}\left[\dfrac{3}{-3}e^{-3x}\right]_0^b = \lim_{b \to \infty}[-e^{-3x}]_0^b = \lim_{b \to \infty}[-e^{-3b} - (-e^{-3 \cdot 0})] = \lim_{b \to \infty}\left(1 - \dfrac{1}{e^{3b}}\right) = 1$

13. $k = \frac{1}{4}$ **15.** $k = \frac{3}{2}$ **17.** $k = \frac{1}{5}$ **19.** $k = \frac{1}{2}$ **21.** $k = \frac{1}{\ln 3}$ **23.** $k = \frac{1}{e^3 - 1}$ **25.** $\frac{8}{25}$ **27.** $\frac{1}{2}$ **29.** 0.3297
31. 0.99995 **33.** 0.9502 **35.** 0.3935 **37.** $b = \sqrt[4]{4}$, or $\sqrt{2}$

Margin Exercises, Section 6.6, pp. 374–382

1. $E(x) = \frac{2}{3}$, $E(x^2) = \frac{1}{2}$ **2.** $\mu = \frac{2}{3}$, $\sigma^2 = \frac{1}{18}$, $\sigma = \sqrt{\frac{1}{18}} = \frac{1}{3}\sqrt{\frac{1}{2}}$ **3.** (a) 0.4850; (b) 0.4608; (c) 0.9559; (d) 0.2720;
(e) 0.1102; (f) 0.0307 **4.** 0.1210

Exercise Set 6.6, pp. 383–384

1. $\mu = E(x) = \frac{7}{2}$, $E(x^2) = 13$, $\sigma^2 = \frac{3}{4}$, $\sigma = \frac{1}{2}\sqrt{3}$ **3.** $\mu = E(x) = 2$, $E(x^2) = \frac{9}{2}$, $\sigma^2 = \frac{1}{2}$, $\sigma = \sqrt{\frac{1}{2}}$ **5.** $\mu = E(x) = \frac{14}{9}$,
$E(x^2) = \frac{5}{2}$, $\sigma^2 = \frac{13}{162}$, $\sigma = \sqrt{\frac{13}{162}} = \frac{1}{9}\sqrt{\frac{13}{2}}$ **7.** $\mu = E(x) = -\frac{5}{4}$, $E(x^2) = \frac{11}{5}$, $\sigma^2 = \frac{51}{80}$, $\sigma = \sqrt{\frac{51}{80}} = \frac{1}{4}\sqrt{\frac{51}{5}}$
9. $\mu = E(x) = \frac{2}{\ln 3}$, $E(x^2) = \frac{4}{\ln 3}$, $\sigma^2 = \frac{4\ln 3 - 4}{(\ln 3)^2}$, $\sigma = \frac{2}{\ln 3}\sqrt{\ln 3 - 1}$ **11.** 0.4964 **13.** 0.3665 **15.** 0.6442
17. 0.0078 **19.** 0.1716 **21.** 0.0013 **23.** (a) 0.6826; (b) 68.26% **25.** 0.2898 **27.** 0.4514 **29.** (a) 0.2956; (b) 0.1731;
(c) 0.3015 **31.** 0.62% **33.** $\mu = E(x) = \frac{a+b}{2}$, $E(x^2) = \frac{b^3 - a^3}{3(b-a)}$, or $\frac{b^2 + ba + a^2}{3}$, $\sigma^2 = \frac{(b-a)^2}{12}$, $\sigma = \frac{b-a}{2\sqrt{3}}$
35. $\sqrt{2}$ **37.** $\frac{\ln 2}{k}$

Margin Exercises, Section 6.7, pp. 384–386

1. $\frac{\pi}{3}$ **2.** $\frac{\pi}{2}(e^2 - e^{-4})$

Exercise Set 6.7, p. 387

1. $\frac{9\pi}{2}$ **3.** $\frac{7\pi}{3}$ **5.** $\frac{\pi}{2}(e^{10} - e^{-4})$ **7.** $\frac{2\pi}{3}$ **9.** $\pi \ln 3$ **11.** 32π **13.** $\frac{32\pi}{5}$ **15.** 56π **17.** $\frac{32\pi}{3}$ **19.** $2\pi e^3$
21. (a) Divergent; (b) π

Chapter 6 Test, pp. 387–388

1. \$29,192 **2.** \$344.66 **3.** 13,941,765 thousand tons **4.** After 33 years (2013) **5.** \$2465.97 **6.** \$30,717.71
7. \$34,545.46 **8.** Convergent, $\frac{1}{4}$ **9.** Divergent **10.** $k = \frac{1}{4}$ **11.** 0.8647 **12.** $E(x) = \frac{3}{4}$ **13.** $E(x^2) = \frac{3}{5}$ **14.** $\mu = \frac{3}{4}$
15. $\sigma^2 = \frac{3}{80}$ **16.** $\sigma = \frac{1}{4}\sqrt{\frac{3}{5}}$ **17.** 0.4332 **18.** 0.4420 **19.** 0.9071 **20.** 0.4207 **21.** (3, \$16) **22.** \$45 **23.** \$22.50
24. $\pi \ln 5$ **25.** $\frac{5\pi}{2}$ **26.** $b = \sqrt[6]{6}$ **27.** Convergent. $-\frac{1}{4}$

CHAPTER 7

Margin Exercises, Section 7.1, pp. 390–401

1. (a) 128; the profit from selling 14 items of the first product and 12 of the second is \$128; (b) 48; the profit from selling none of the first product and 8 items of the second is \$48　　**2.** 7.2, 94.5, 0.60　　**3.** 2000　　**4.** (a) 352; (b) 15.4.
5. \$125,656.83　　**6.** (a) 4; (b) 4　　**7.** (a) $f(x, 4) = -x^2 - 15$; (b) $-2x$　　**8.** $\dfrac{\partial f}{\partial x} = -2x$　　**9.** (a) $\dfrac{\partial z}{\partial x} = 6xy + 15x^2$;
(b) $\dfrac{\partial z}{\partial y} = 3x^2$　　**10.** (a) $\dfrac{\partial t}{\partial x} = y + z + 2x$; (b) $\dfrac{\partial t}{\partial y} = x + 3y^2$; (c) $\dfrac{\partial t}{\partial z} = x$　　**11.** (a) $f_x = 9x^2y + 2y$; (b) $f_x(-4, 1) = 146$;
(c) $f_y = 3x^3 + 2x$; (d) $f_y(2, 6) = 28$　　**12.** $f_x = \dfrac{1}{x} + ye^x,\ f_y = \dfrac{1}{y} + e^x$　　**13.**

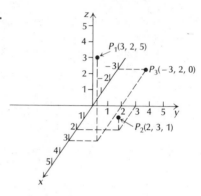

14. (a) 108,000 units; (b) $\dfrac{\partial p}{\partial x} = 600\left(\dfrac{y}{x}\right)^{1/4},\ \dfrac{\partial p}{\partial y} = 200\left(\dfrac{x}{y}\right)^{3/4}$; (c) 1000, 43.2

Exercise Set 7.1, pp. 402–403

1. $f(0, -2) = 0, f(2, 3) = -8, f(10, -5) = 200$　　**3.** $f(0, -2) = 1, f(-2, 1) = -13\dfrac{8}{9}, f(2, 1) = 23$
5. $f(e, 2) = \ln e + 2^3 = 1 + 8 = 9, f(e^2, 4) = 66, f(e^3, 5) = 128$　　**7.** $f(-1, 2, 3) = 6, f(2, -1, 3) = 12$　　**9.** 105, 95
11. 7.7%　　**13.** $\dfrac{\partial z}{\partial x} = 2 - 3y,\ \dfrac{\partial z}{\partial y} = -3x,\ \dfrac{\partial z}{\partial x}\bigg|_{(-2,-3)} = 11,\ \dfrac{\partial z}{\partial y}\bigg|_{(0,-5)} = 0$　　**15.** $\dfrac{\partial z}{\partial x} = 6x - 2y,\ \dfrac{\partial z}{\partial y} = -2x + 1,$
$\dfrac{\partial z}{\partial x}\bigg|_{(-2,-3)} = -6\ \dfrac{\partial z}{\partial y}\bigg|_{(0,-5)} = 1$　　**17.** $f_x = 2, f_y = -3, f_x(-2, 4) = 2, f_y(4, -3) = -3$　　**19.** $f_x = \dfrac{x}{\sqrt{x^2 + y^2}}, f_y = \dfrac{y}{\sqrt{x^2 + y^2}},$
$f_x(-2, 1) = \dfrac{-2}{\sqrt{5}}, f_y(-3, -2) = \dfrac{-2}{\sqrt{13}}$　　**21.** $f_x = 2e^{2x+3y}, f_y = 3e^{2x+3y}$　　**23.** $f_x = ye^{xy}, f_y = xe^{xy}$　　**25.** $f_x = \dfrac{y}{x + y},$
$f_y = \dfrac{y}{x + y} + \ln(x + y)$　　**27.** $f_x = 1 + \ln xy, f_y = \dfrac{x}{y}$　　**29.** $f_x = \dfrac{1}{y} + \dfrac{y}{x^2}, f_y = -\dfrac{x}{y^2} - \dfrac{1}{x}$　　**31.** $f_x = 12(2x + y - 5),$
$f_y = 6(2x + y - 5)$　　**33.** $\dfrac{\partial f}{\partial b} = 12m + 6b - 30,\ \dfrac{\partial f}{\partial m} = 28m + 12b - 64$　　**35.** $f_x = 3y - 2\lambda, f_y = 3x - \lambda,$
$f_\lambda = -(2x + y - 8)$　　**37.** $f_x = 2x - 10\lambda, f_y = 2y - 2\lambda, f_\lambda = -(10x + 2y - 4)$　　**39.** (a) 3,888,064 units;
(b) $\dfrac{\partial p}{\partial x} = 1117.8\left(\dfrac{y}{x}\right)^{0.379}, \dfrac{\partial p}{\partial y} = 682.2\left(\dfrac{x}{y}\right)^{0.621}$; (c) 965.8, 866.8　　**41.** (a) 0°; (b) −10°; (c) −22°; (d) −64°　　**43.** $f_x = \dfrac{-4xt^2}{(x^2 - t^2)^2},$
$f_t = \dfrac{4x^2t}{(x^2 - t^2)^2}$　　**45.** $f_x = \dfrac{1}{\sqrt{x}(1 + 2\sqrt{t})}, f_t = \dfrac{-1 - 2\sqrt{x}}{\sqrt{t}(1 + 2\sqrt{t})^2}$　　**47.** $f_x = 4x^{-1/3} - 2x^{-3/4}t^{1/2} + 6x^{-3/2}t^{3/2},$
$f_t = -4x^{1/4}t^{-1/2} - 18x^{-1/2}t^{1/2}$

Margin Exercises, Section 7.2, pp. 403–405

1. (a) $\dfrac{\partial z}{\partial y} = 6xy + 2x$; (b) $\dfrac{\partial}{\partial x}\left(\dfrac{\partial z}{\partial y}\right) = 6y + 2$; (c) $\dfrac{\partial}{\partial y}\left(\dfrac{\partial z}{\partial y}\right) = 6x$ **2.** (a) $f_y = 6xy + 2x$; (b) $f_{yx} = 6y + 2$; (c) $f_{yy} = 6x$

3. $\dfrac{\partial^2 f}{\partial x^2} = 2$, $\dfrac{\partial^2 f}{\partial y\,\partial x} = 6y + 2 + \dfrac{1}{y}$, $\dfrac{\partial^2 f}{\partial x\,\partial y} = 6y + 2 + \dfrac{1}{y}$, $\dfrac{\partial^2 f}{\partial y^2} = 6x - \dfrac{x}{y^2}$

Exercise Set 7.2, pp. 405–406

1. $\dfrac{\partial^2 f}{\partial x^2} = 6$, $\dfrac{\partial^2 f}{\partial y\,\partial x} = \dfrac{\partial^2 f}{\partial x\,\partial y} = -1$, $\dfrac{\partial^2 f}{\partial y^2} = 0$ **3.** $\dfrac{\partial^2 f}{\partial x^2} = 0$, $\dfrac{\partial^2 f}{\partial y\,\partial x} = \dfrac{\partial^2 f}{\partial x\,\partial y} = 3$, $\dfrac{\partial^2 f}{\partial y^2} = 0$ **5.** $\dfrac{\partial^2 f}{\partial x^2} = 20x^3y^4 + 6xy^2$,

$\dfrac{\partial^2 f}{\partial y\,\partial x} = \dfrac{\partial^2 f}{\partial x\,\partial y} = 20x^4y^3 + 6x^2y$, $\dfrac{\partial^2 f}{\partial y^2} = 12x^5y^2 + 2x^3$ **7.** $f_{xx} = 0, f_{yx} = 0, f_{xy} = 0, f_{yy} = 0$

9. $f_{xx} = 4y^2e^{2xy}, f_{yx} = f_{xy} = 4xye^{2xy} + 2e^{2xy}, f_{yy} = 4x^2e^{2xy}$ **11.** $f_{xx} = 0, f_{yx} = f_{xy} = 0, f_{yy} = e^y$

13. $f_{xx} = -\dfrac{y}{x^2}, f_{yx} = f_{xy} = \dfrac{1}{x}, f_{yy} = 0$ **15.** $f_{xx} = \dfrac{-6y}{x^4}, f_{yx} = f_{xy} = \dfrac{2(y^3 - x^3)}{x^3y^3}, f_{yy} = \dfrac{6x}{y^4}$ **17.** $\dfrac{\partial^2 f}{\partial x^2} = \dfrac{2y^2 - 2x^2}{(x^2 + y^2)^2}$,

$\dfrac{\partial^2 f}{\partial y^2} \ne \dfrac{2x^2 - 2y^2}{(x^2 + y^2)^2}$, so the sum is 0 **19.** (a) $-y$; (b) x; (c) $f_{yx}(0, 0) = 1, f_{xy}(0, 0) = -1$; so $f_{yx}(0, 0) \ne f_{xy}(0, 0)$

Margin Exercises, Section 7.3, pp. 406–413

1. Min. $= -9$ at $(-3, 3)$ **2.** Max. $= \dfrac{1}{108}$ at $\left(\dfrac{1}{6}, \dfrac{1}{6}\right)$ **3.** 5 thousand of $15 calculator; 7 thousand of $20 calculator

Exercise Set 7.3, pp. 414–415

1. Min. $= -\dfrac{1}{3}$ at $\left(-\dfrac{1}{3}, \dfrac{2}{3}\right)$ **3.** Max. $= \dfrac{4}{27}$ at $\left(\dfrac{2}{3}, \dfrac{2}{3}\right)$ **5.** Min. $= -1$ at $(1, 1)$ **7.** Min. $= -7$ at $(1, -2)$

9. Min. $= -5$ at $(-1, 2)$ **11.** None **13.** 6 (thousand) of the $17 radio and 5 (thousand) of the $21 radio **15.** Max. of
$P = 35$ (million dollars) when $a = 10$ (million dollars) and $p = \$3$ **17.** (a) $R(p_1,$
$p_2) = 78p_1 - 6p_1^2 - 6p_1p_2 + 66p_2 - 6p_2^2$; (b) $p_1 = 5$ ($50), $p_2 = 3$ ($30); (c) $q_1 = 78 - 6\cdot 5 - 3\cdot 3 = 39$ (hundreds), $q_2 = 33$ (hundreds); (d) $R = 50\cdot 3900 +$
$30\cdot 3300 = \$294{,}000$ **19.** None **21.** Min. $= \dfrac{1}{6}$ at $\left(\dfrac{211}{3}, \dfrac{3}{2}\right)$

Margin Exercises, Section 7.4, pp. 415–421

1. $dz = e^y dx + xe^y dy$ **2.** $\dfrac{1}{x}dx + \dfrac{1}{y}dy$ **3.** $9dx + 12dy$ **4.** -0.66 **5.** $dw = \left(-\dfrac{1}{x^2} - y\right)dx - xdy + 2dz$

6. $\dfrac{xdx + ydy + zdz}{(x^2 + y^2 + z^2)^{1/2}}$ **7.** $2(x_1 dx_1 + x_2 dx_2 + x_3 dx_3 + x_4 dx_4)$ **8.** $dw = -2dx + 3dy - 2dz$ **9.** 0.3

10. $\Delta z = 0.7947$ **11.** 0.78 **12.** $\Delta w = 0.2226, dw = 0.22$ **13.** $\ln([1.01][0.97]) \approx -0.02$ **14.** 0.04 **15.** 28.875 psi

Exercise Set 7.4, p. 422

1. $dz = 2xdx + 8ydy$ **3.** $e^{4y}dx + 4xe^{4y}dy$ **5.** $\left(4x^3 + \dfrac{1}{y}\right)dx + \left(3y^2 - \dfrac{x}{y^2}\right)dy$ **7.** $2xdx - 2ydy + 6zdz$
9. $yz^3[yzdx + 2xzdy + 4xydz]$ **11.** $dw = 2x_1dx_1 + x_3dx_2 + x_2dx_3 - 2x_4dx_4$ **13.** $\Delta z = 0.5805$, $dz = 0.58$
15. $\Delta z = 0.015553$, $dz = 0.0105$ **17.** $\Delta w = 0.0801$, $dw = 0.08$ **19.** $\Delta w = -0.062608$, $dw = -0.06$
21. 167.66   **23.** 4.09   **25.** 4.53025 sq ft   **27.** 39.06 sq ft

Margin Exercises, Section 7.5, pp. 423–430

1. (a) \$5.3 billion; (b) \$5.5 billion; (c) differ by \$0.2 billion   **2.** (a) $y = 53,600x + 124,500$; (b) \$499,700; \$767,700
3. Same answers   **4.** (a) $y = 148,241.87e^{0.211321x}$; (b) \$650,730; \$1,871,883

Exercise Set 7.5, pp. 430–432

1. (a) $y = 23.71x - 0.59$; (b) \$165.38 million, \$283.93 million   **3.** (a) $y = 1.068421x - 1.236842$; (b) 85.3
5. (a) $y = -0.005861x + 15.423439$; (b) 3:47.8; (c) According to the regression line the record should have been 3:49.2,
so Ovett also beat the regression line prediction.   **7.** (a) $y = 18.105703e^{0.402491x}$; (b) \$302.97 million, \$2,266.73 million

Margin Exercises, Section 7.6, pp. 432–439

1. (a) $A(x, y) = xy$, subject to $x + y = 50$; (b) max. $= 625$ at $(25, 25)$ **2.** Max. $= 125$ at $(2.5, 10)$ **3.** $r \approx 1.8$ in.,
$h \approx 3.6$ in.; surface area is about 61.04 in^2

Exercise Set 7.6, pp. 439–441

1. Max. $= 8$ at $(2, 4)$ **3.** Max. $= -16$ at $(2, 4)$ **5.** Min. $= 20$ at $(4, 2)$ **7.** Min. $= -96$ at $(8, -12)$ **9.** Min. $= \dfrac{3}{2}$
at
$\left(1, \dfrac{1}{2}, -\dfrac{1}{2}\right)$ **11.** 35 and 35   **13.** 3 and -3 **15.** $9\dfrac{3}{4}$ in., $9\dfrac{3}{4}$ in.; $95\dfrac{1}{16}$ in.^{2}; no   **17.** $r = \sqrt[3]{\dfrac{27}{2\pi}} \approx 1.6$ ft;
$h = 2 \cdot r \approx 3.2$ ft;
min. surface area ≈ 48.3 ft^2 **19.** Max. of $S = 800$ at $L = 20$, $M = 60$ **21.** (a) $C(x, y, z) = 7xy + 6yz + 6xz$;
(b) $x = 60$ ft, $y = 60$ ft, $z = 70$ ft; \$75,600   **23.** 10,000 on A, 100 on B   **25.** Min. $= -\dfrac{155}{128}$ at $\left(-\dfrac{7}{16}, -\dfrac{3}{4}\right)$
27. Max. $= \dfrac{1}{27}$ at $\left(\dfrac{1}{\sqrt{3}}, \dfrac{1}{\sqrt{3}}, \dfrac{1}{\sqrt{3}}\right)$ and $\left(-\dfrac{1}{\sqrt{3}}, -\dfrac{1}{\sqrt{3}}, -\dfrac{1}{\sqrt{3}}\right)$ **29.** Max. $= 2$ at $\left(\dfrac{1}{2}, \dfrac{1}{2}, \dfrac{1}{2}, \dfrac{1}{2}\right)$
31. Min. $= \dfrac{1}{30}$ at $\left(\dfrac{1}{30}, -\dfrac{1}{15}, \dfrac{1}{6}\right)$

Margin Exercises, Section 7.7, pp. 441–446

1. $h \approx 184$ ft, $k \approx 74$ ft; dimensions are 74 ft by 74 ft by 194 ft

Exercise Set 7.7, p. 447

1. $h = \sqrt[3]{10,240,000} \approx 217$ ft, $k \approx 54$ ft; dimensions are 54 ft by 54 ft by 225 ft **3.** $h = \sqrt[3]{\dfrac{Aca^2}{b^2}}$, $k = \sqrt[3]{\dfrac{Abc}{a}}$, dimensions are k by k by $h + c$

Margin Exercises, Section 7.8, pp. 447–457

1. $4y$ **2.** $\dfrac{1}{2}y^2$ **3.** $\dfrac{1}{3}x$ **4.** $\dfrac{3}{2}x + 3x^2$ **5.** $2rx^2 + 2y$ **6.** $\dfrac{1}{6}$ **7.** $\dfrac{25}{6}$ **8.** $\dfrac{1}{6}$; they are equal **9.** $\dfrac{25}{6}$; equal **10.** (a) $4x^2$
(b) $\dfrac{224}{3}$ **11.** 72 **12.** $\dfrac{3}{2}(e - 1)$ **13.** $\dfrac{15}{8}$ **14.** $\dfrac{5}{6}$

Exercise Set 7.8, pp. 457–458

1. $\dfrac{1}{3}y$ **3.** $\dfrac{8}{3}$ **5.** $\dfrac{4}{3}$ **7.** 1 **9.** 12 **11.** $\dfrac{1}{3}\ln 2$ **13.** 2 **15.** $8 \ln 2$ **17.** $f_{av} = 1$

Margin Exercises, Section 7.9, pp. 458–469

1. $1 - x^3$ **2.** $8x$ **3.** $\dfrac{3}{4}$ **4.** 4 **5.** (a) (b) $\dfrac{3}{4}$ **6.** (a) (b) 4 **7.** (a)

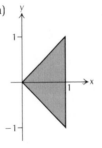

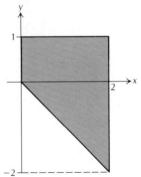

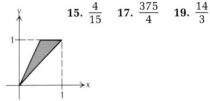

(b) 28 **8.** $\dfrac{3}{4}$
9. $\dfrac{1}{2}(e^2 - 1)$ **10.** $\dfrac{3}{2}$ **11.** $\dfrac{3}{8}$ **12.** $\dfrac{27}{32}$ **13.** $\dfrac{8}{3}$ **14.** $\dfrac{3}{4}$ **15.** 4

Exercise Set 7.9, pp. 470–472

1. 1 **3.** 0 **5.** 6 **7.** $\dfrac{3}{20}$ **9.** 4 **11.** 2; **13.** $\dfrac{1}{2}$; **15.** $\dfrac{4}{15}$ **17.** $\dfrac{375}{4}$ **19.** $\dfrac{14}{3}$

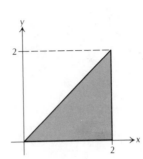

21. (a) $\int_1^{e^2}\left[\int_{\ln y}^2 dx\right]dy$ (b) $e^2 - 3$ **23.** 1   **25.** $\dfrac{5}{12}$ **27.** 39   **29.** $\dfrac{13}{240}$ **31.** $\dfrac{1}{8}$

Chapter 7 Test, p. 472–473

1. $\dfrac{\partial f}{\partial x} = e^x + 6x^2y$ **2.** $\dfrac{\partial f}{\partial y} = 2x^3 + 1$ **3.** $\dfrac{\partial^2 f}{\partial x^2} = e^x + 12xy$ **4.** $\dfrac{\partial^2 f}{\partial x\,\partial y} = 6x^2$ **5.** $\dfrac{\partial^2 f}{\partial y\,\partial x} = 6x^2$ **6.** $\dfrac{\partial^2 f}{\partial y^2} = 0$

7. Min. $= -\dfrac{7}{16}$ at $\left(\dfrac{3}{4},\dfrac{1}{2}\right)$ **8.** None   **9.** $dz = 3x^2\,dx + e^y\,dy$ **10.** $dz = 3y^2\,dx + 6xy\,dy$ **11.** $dV = 0.59\ \text{ft}^3$

12. (a) $y = \dfrac{9}{2}x + \dfrac{17}{3}$; (b) \$23.67   **13.** Max. $= -19$ at $(4, 5)$ **14.** 1   **15.** $\dfrac{4}{3}$ **16.** $\dfrac{3}{20}$ **17.** \$400,000 for labor, \$200,000 for capital   **18.** The cylindrical one

CHAPTER 8

Margin Exercises, Section 8.1, pp. 476–486

1. (a) I; (b) III; (c) IV; (d) III; (e) I; (f) IV; (g) I   **2.** (a) $\dfrac{3}{4}\pi$; (b) $\dfrac{7}{4}\pi$; (c) $-\dfrac{\pi}{2}$; (d) 4π; (e) $-\dfrac{5\pi}{4}$; (f) $-\dfrac{7\pi}{4}$; (g) $\dfrac{9\pi}{4}$; (h) $\dfrac{8\pi}{3}$

3. (a) 60°; (b) 135°; (c) 450°; (d) 1800°; (e) −210°; (f) 54,000°; (g) −48,600°; (h) 1125°   **4.** (a) $-\dfrac{\sqrt{3}}{2}$; (b) $\dfrac{1}{2}$; (c) $-\dfrac{\sqrt{3}}{2}$;

(d) $\dfrac{1}{2}$ **5.** (a) 1; (b) $\sqrt{3}$; (c) 0; (d) $\sqrt{2}$; (e) 2; (f) 1   **6.** (a) $\sqrt{3}$; (b) undefined; (c) $\dfrac{\sqrt{3}}{3}$; (d) 2; (e) undefined; (f) $\sqrt{2}$

7. $1 + \cot^2 t = \csc^2 t$ **8.** $\dfrac{\sqrt{6} - \sqrt{2}}{4}$ **9.** $\dfrac{\sqrt{2} + \sqrt{6}}{4}$ **10.** $2\sin u \cos u$

Exercise Set 8.1, pp. 486–487

1. I   **3.** III   **5.** $\dfrac{\pi}{6}$ **7.** $\dfrac{\pi}{3}$ **9.** $\dfrac{5\pi}{12}$ **11.** 270°   **13.** −45°   **15.** 1440°   **17.** $\left(\dfrac{180}{\pi}\right)^\circ \approx 57.3°$ **19.** $\dfrac{\sqrt{3}}{2}$ **21.** 0   **23.** $\dfrac{1}{2}$

25. Undefined   **27.** $\dfrac{\sqrt{3}}{3}$ **29.** −1   **31.** $\tan(u + v) = \dfrac{\sin(u + v)}{\cos(u + v)}$. Then use identities (5) and (3)   **33.** 0.5210

35. −0.8632   **37.** −0.3327   **39.** Undefined   **41.** 101.6°, 102.7°, 103.7°, 104.6°, 98.6°

Margin Exercises, Section 8.2, pp. 487–496

1. $\cot x = \dfrac{\cos x}{\sin x}$, so $\dfrac{d}{dx}\cot x = [\sin x(-\sin x) - \cos x(\cos x)] \div \sin^2 x = \dfrac{-(\sin^2 x + \cos^2 x)}{\sin^2 x} = -\dfrac{1}{\sin^2 x} = -\csc^2 x$

2. $\dfrac{3\sin^2 x}{\cos^4 x}$ **3.** $(4x^3 + 6x)\cos(x^4 + 3x^2)$ **4.** $-2xe^{x^2}\sin(e^{x^2})\sin x^3 + 3x^2\cos(e^{x^2})\cos x^3$ **5.** (a) Amplitude $= 4$,

period $= \pi$, phase shift $= \dfrac{\pi}{4}$; (b)   (c) $8\cos\left(2t - \dfrac{\pi}{2}\right)$

6. $\dfrac{dy}{dx} = 0.03594\pi \cos 1.198\pi x$ **7.** $\dfrac{dL}{dt} = 2\pi Af \cos\left(2\pi fT - \dfrac{d}{\omega}\right)$ **8.** (a) $L = (a^{2/3} + b^{2/3})^{3/2}$; (b) 27

Exercise Set 8.2, pp. 496–498

1. $\sec x = \dfrac{1}{\cos x}$, so $\dfrac{d}{dx}\sec x = \dfrac{\cos x \cdot 0 - (-\sin x)\cdot 1}{\cos^2 x} = \dfrac{\sin x}{\cos^2 x} = \dfrac{\sin x}{\cos x}\cdot\dfrac{1}{\cos x} = \tan x \sec x$ **3.** $x \cos x + \sin x$

5. $e^x(\cos x + \sin x)$ **7.** $\dfrac{x \cos x - \sin x}{x^2}$ **9.** $2 \sin x \cos x$ **11.** $\cos^2 x - \sin^2 x$ **13.** $\dfrac{1}{1 + \cos x}$ **15.** $\dfrac{2 \sin x}{\cos^3 x}$

17. $\dfrac{-\sin x}{2\sqrt{1 + \cos x}}$ **19.** $-x^2 \sin x$ **21.** $\cos x \cdot e^{\sin x}$ **23.** $-\sin x$ **25.** $(2x + 3x^2) \cos (x^2 + x^3)$

27. $(4x^3 - 5x^4) \sin (x^5 - x^4)$ **29.** $-\dfrac{1}{2}x^{-1/2} \cdot \sin \sqrt{x}$ **31.** $-\sin x \cdot \cos (\cos x)$ **33.** $\dfrac{-2 \sin 4x}{\sqrt{\cos 4x}}$

35. $\dfrac{2}{3}[\csc^2 (5 - 2x)^{1/3}](5 - 2x)^{-2/3}$ **37.** $-12(\tan 3x)(\sec 3x)^2$ **39.** $\cot x$ **41.** $\dfrac{3\pi}{8} \cos \dfrac{\pi}{8}t$ **43.** $-\dfrac{1000\pi}{9}\left[\sin \dfrac{\pi}{45}(t - 10)\right]$

45. (a) Amp. $= 5$, period $= \dfrac{\pi}{2}$, phase shift $= -\dfrac{\pi}{4}$; (b) $20 \cos (4t + \pi)$ **47.** $40{,}000(\cos t - \sin t)$ **49.** $8\sqrt{8}$ **51.** 16

53. $f_x = -ye^x$, $f_y = 2 \cos 2y - e^x$, $f_{xx} = -ye^x$, $f_{yx} = f_{xy} = -e^x$, $f_{yy} = -4 \sin 2y$ **55.** $f_x = -2 \sin (2x + 3y)$, $f_y = -3 \cdot \sin (2x + 3y)$, $f_{xx} = -4 \cos (2x + 3y)$, $f_{yx} = f_{xy} = -6 \cos (2x + 3y)$, $f_{yy} = -9 \cos (2x + 3y)$ **57.** $f_x = 5x^3y \cdot \sec^2 (5xy) + 3x^2 \cdot \tan (5xy)$, $f_y = 5x^4 \cdot \sec^2 (5xy)$, $f_{xx} = 50x^3y^2 \cdot \sec^2 (5xy) \cdot \tan (5xy) + 30x^2y \sec^2 (5xy) + 6x \tan (5xy)$, $f_{yx} = f_{xy} = 50x^4y \cdot \sec^2 (5xy) \cdot \tan (5xy) + 20x^3 \cdot \sec^2 (5xy)$, $f_{yy} = 50x^5 \cdot \sec^2 (5xy) \cdot \tan (5xy)$

59. $\dfrac{\cos y}{1 + x \sin y}$

Margin Exercises, Section 8.3, pp. 498–501

1. 2 **2.** $\dfrac{(\sin x)^4}{4} + C$ **3.** $\dfrac{1}{4} \sin 4x + C$ **4.** $\ln |x^2 + \cos x| + C$ **5.** $\sin x - x \cos x + C$
6. $-\ln |\csc x + \cot x| + C$

Exercise Set 8.3, pp. 501–502

1. $\dfrac{1}{2}$ **3.** $\dfrac{(\sin x)^5}{5} + C$ **5.** $\dfrac{(\cos x)^3}{3} + C$ **7.** $\sin (x + 3) + C$ **9.** $-\dfrac{1}{2} \cos 2x + C$ **11.** $\dfrac{1}{2} \sin x^2 + C$ **13.** $-\cos (e^x) + C$

15. $-\ln |\cos x| + C$, or $\ln |\sec x| + C$ **17.** $\dfrac{1}{4}x \sin 4x + \dfrac{1}{16} \cos 4x + C$ **19.** $3x \sin x + 3 \cos x + C$ **21.** $2x \sin x - (x^2 - 2) \cos x + C$ **23.** $\tan x - x + C$ **25.** \$84 thousand. **27.** $\dfrac{1}{2}x[\sin (\ln x) - \cos (\ln x)] + C$

29. $\dfrac{e^x}{2}(\cos x + \sin x) + C$ **31.** $\dfrac{1}{7} \tan 7x + C$ **33.** $\tan x + \sec x + C$ **35.** $-\csc u + C$
37. $\cos x - \ln |\csc x + \cot x| + C$ **39.** $x + 2 \ln |\sec x + \tan x| + \tan x + C$

Margin Exercises, Section 8.4, pp. 503–505

1. $\dfrac{\pi}{6}$ **2.** 0 **3.** $-\dfrac{\pi}{3}$ **4.** $\dfrac{\pi}{3}$ **5.** $\dfrac{\pi}{6}$ **6.** $\dfrac{2\pi}{3}$ **7.** $\dfrac{1}{5} \tan^{-1} (5x) + C$ **8.** $\dfrac{1}{2} \cos^{-1} (2x) + C$

A–40

ANSWERS

Exercise Set 8.4, pp. 505–506

1. $\frac{\pi}{4}$ **3.** $\frac{\pi}{2}$ **5.** $\frac{\pi}{4}$ **7.** $-\frac{\pi}{6}$ **9.** $\tan^{-1}(e^t) + C$ **11.** Use substitution. Let $u = \sin^{-1} x$, then $x = \sin u$ and $dx = \cos u\, du$.

Then the integral $\int \frac{1}{\sqrt{1-x^2}}\, dx = \int \frac{1}{\sqrt{1-\sin^2 u}}\cos u\, du = \int \frac{1}{\cos u}\cdot \cos u\, du = \int du = u + C = \sin^{-1} x + C.$

13. 1.412 radians   **15.** 0.049 radian   **17.** $x \tan^{-1} x - \frac{1}{2}\ln(1+x^2) + C$

Chapter 8 Test, pp. 506–507

1. $\frac{2\pi}{3}$ **2.** $150°$ **3.** $\frac{1}{2}$ **4.** -1 **5.** $-\sin t$ **6.** $(6x-5)\cos(3x^2-5x)$ **7.** $\frac{\sin x - x\cdot\cos x}{\sin^2 x}$ **8.** $2\tan t\cdot\sec^2 t$

9. $\frac{\cos x - \sin x}{2\sqrt{\sin x + \cos x}}$ **10.** $\frac{-2}{1 - 2\sin x \cos x}$ **11.** $S'(t) = \frac{5\pi}{2}\cos\frac{\pi}{8}t$ **12.** $1 - \frac{\sqrt{3}}{2}$ **13.** $\frac{1}{10}(\sin x)^{10} + C$

14. $\frac{1}{5}\sin 5t + C$ **15.** $\frac{1}{5}\sin 5x - x\cos 5x + C$ **16.** $\frac{1}{4}\tan^{-1} 4t + C$ **17.** $\ln|2x - \sin x| + C$

18. $\frac{dy}{dx} = -2(\sin x)e^{\cos x}$, $\frac{d^2y}{dx^2} = 2e^{\cos x}(\sin^2 x - \cos x)$ **19.** $f_x = -\frac{\sin x}{\cos y}$, $f_{xy} = \frac{-\sin x \sin y}{\cos^2 y}$

CHAPTER 9

Margin Exercises, Section 9.1, pp. 510–512

1. $y = x^3 + C$ **2.** (a) $y = x^3 + C$, (b) $y = x^3 - 7$, $y = x^3 + \frac{1}{2}$, $y = x^3$. Answers may vary.   **3.** $y = x^2 + 6$

4. $f(x) = \ln x - x^2 + \frac{2}{3}x^{3/2} + \frac{13}{3}$ **5.** $y' = 2e^x - 21e^{3x}$, $y'' = 2e^x - 63e^{3x}$.

Then

$y'' - 4y' + 3y = 0$	
$2e^x - 63e^{3x} - 4(2e^x - 21e^{3x}) + 3(2e^x - 7e^{3x})$	0
$2e^x - 63e^{3x} - 8e^x + 84e^{3x} + 6e^x - 21e^{3x}$	
	0

6. $\frac{dy}{dx} = 2xe^{2x} + e^{2x}$. Then

$\frac{dy}{dx} - 2y = e^{2x}$	
$(2xe^{2x} + e^{2x}) - 2(xe^{2x})$	e^{2x}
e^{2x}	

Exercise Set 9.1, p. 513

1. $y = x^4 - C$; $y = x^4 + 3$, $y = x^4$, $y = x^4 - 796$; answers may vary.   **3.** $y = \frac{1}{2}e^{2x} + \frac{1}{2}x^2 + C$; $y = \frac{1}{2}e^{2x} + \frac{1}{2}x^2 - 5$, $y = \frac{1}{2}e^{2x} + \frac{1}{2}x^2 + 7$; $y = \frac{1}{2}e^{2x} + \frac{1}{2}x^2$; answers may vary.   **5.** $y = 3\ln x - \frac{1}{3}x^3 + \frac{1}{6}x^6 + C$;

$y = 3\ln x - \frac{1}{3}x^3 + \frac{1}{6}x^6 - 15$, $y = 3\ln x - \frac{1}{3}x^3 + \frac{1}{6}x^6 - 7$, $y = 3\ln x - \frac{1}{3}x^3 + \frac{1}{6}x^6$; answers may vary.

7. $y = \frac{1}{3}x^3 + x^2 - 3x + 4$ **9.** $y = \frac{3}{5}x^{5/3} - \frac{1}{2}x^2 - \frac{61}{10}$ **11.** $y'' = \frac{1}{x}$. Then $y'' - \frac{1}{x} = 0$

$$\begin{array}{c|c} \frac{1}{x} - \frac{1}{x} & 0 \end{array}$$

13. $y' = 4e^x + 3xe^x$, $y'' = 7e^x + 3xe^x$. Then

$$\begin{array}{c|c} y'' - 2y' + y = 0 & \\ \hline (7e^x + 3xe^x) - 2(4e^x + 3xe^x) + (e^x + 3xe^x) & 0 \\ 7e^x + 3xe^x - 8e^x - 6xe^x + e^x + 3xe^x & \\ & 0 \end{array}$$

15. $C(x) = 2.6x - 0.01x^2 + 120$, $A(x) = 2.6 - 0.01x + \frac{120}{x}$ **17.** (a) $P(C) = \frac{400}{(C + 3)^{1/2}} - 40$; (b) \$97

Margin Exercises, Section 9.2, pp. 513–520

1. $y = C_1 e^{x^3}$, where $C_1 = e^C$ **2.** $y = \sqrt[3]{x^2 + C}$ **3.** $y = \sqrt{10x + C_1}$, $y = -\sqrt{10x + C_1}$, where $C_1 = 2C$
4. $y = -2 + C_1 e^{x^2/2}$, where $C_1 = e^C$ **5.** $p = Cx^{-1/3}$ **6.** $R = C_1 \cdot S^k$, where $C_1 = e^C$

Exercise 9.2, pp. 519–520

1. $y = C_1 e^{x^4}$, where $C_1 = e^C$ **3.** $y\sqrt[3]{\frac{5}{2}x^2 + C}$ **5.** $y = \sqrt{2x^2 + C_1}$, $y = -\sqrt{2x^2 + C_1}$, where $C_1 = 2C$ **7.** $y = \sqrt{6x + C_1}$,
$y = -\sqrt{6x + C_1}$, where $C_1 = 2C$ **9.** $y = -3 + C_1 e^{x^2/2}$, where $C_1 = e^C$ **11.** $y = \sqrt[3]{15x + C_1}$, where $C_1 = 3C$
13. $y = C_1 e^{3x}$, where $C_1 = e^C$ **15.** $P = C_1 e^{2t}$, where $C_1 = e^C$ **17.** (a) $P = C_1 e^{kt}$ where $C_1 = e^C$; (b) $P = P_0 e^{kt}$
19. (a) $R = k \cdot \ln(S + 1) + C$; (b) $R = k \cdot \ln(S + 1)$; (c) No units, no pleasure from them   **21.** $p(x) = 200 - x$
23. $p(x) = C_1 x^{-1/2}$

Margin Exercises, Section 9.3, pp. 520–528

1. (a)

t	-3	-2	-1	0	1	2	3
$P(t)$	0.05	0.12	0.27	0.5	0.73	0.88	0.95

(b)

2. (a) $P = \dfrac{\dfrac{P_0}{L - P_0} \cdot Le^{Lkt}}{1 + \dfrac{P_0}{L - P_0} e^{Lkt}} = \dfrac{P_0 Le^{Lkt}}{(L - P_0) + P_0 e^{Lkt}}$ (b) $\dfrac{P_0 Le^{Lkt}}{(L - P_0) + P_0 e^{Lkt}} \cdot \dfrac{e^{-Lkt}}{e^{-Lkt}} = \dfrac{P_0 L}{P_0 + e^{-Lkt}(L - P_0)}$ (add exponents)

3. 3312   **4.** (a) $P(t) = \dfrac{80{,}000}{20 + 3980e^{-2t}}$ (b)

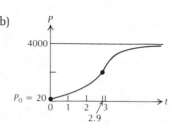

(c) $t \approx 1.95$ mo   **5.** $Q(y) = \dfrac{Q_0 M}{Q_0 + e^{-Mry}(M - Q_0)}$

Exercise Set 9.3, pp. 529–530

1. 900   **3.** (a) $P(t) = \dfrac{7000}{10 + 690e^{-0.56t}}$ (b)

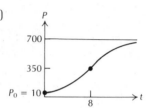

(c) $t \approx 9.197$ hr

5. (a) $N(t) = \dfrac{4800}{6 + 794e^{-800kt}}$, (b) $k = 0.0005$; (c) $t \approx 17.08$ min   **7.** $P(w) = \dfrac{0.04}{0.04 + 0.96e^{-kw}}$; (b) $k = 1.06$;

(c) $P(w) = \dfrac{0.04}{0.04 + 0.96e^{-1.06w}}$, (d) $w \approx 4.3$ (about 106 g)

Margin Exercises, Section 9.4, pp. 533–535

1. (a) Undefined, 5%, 2.5%, 1.2%, 0.7%, 0.3%. (b)

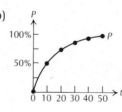   (c)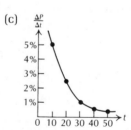

(d) No
(e) No
(f) 100%
(g) Decreasing

2. (a)

t	0	1	2	3	4	5
$P(t)$	0	1.9	2.6	2.85	2.95	2.98

(b)

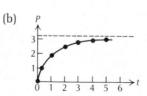

3. (a) $k = 0.05$; (b) $P(t) = 1 - e^{-0.05t}$; (c) $P(60) = 0.9502$, or 95.02%; (d) $t \approx 92$

Exercise Set 9.4, pp. 536–538

1.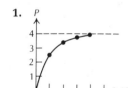

3.

5. (a) $P(t) = 0.5(1 - e^{-kt})$; (b) $k = 0.07$; (c) $P(t) = 0.5(1 - e^{-0.07t})$; (d) $P(30) = 0.439$, or 43.9%; (e) $t \approx 56$

7. (a) $P(t) = 1 - e^{-kt}$; (b) $k = 0.18$; (c) $P(t) = 1 - e^{-0.18t}$; (d) $P(10) = 0.8347$; (e) 13   **9.** $M(t) = \frac{p}{k}(1 - e^{-kt})$

11. $C(t) = (C_0 - C_T)e^{-kAt/V} + C_T$

Margin Exercises, Section 9.5, pp. 538–542

1. It checks.   **2.** $x\,dx + y\,dy = 0$ **3.** $P_y = 0 = Q_x$ **4.** $P_y = 6x - 2 = Q_x$ **5.** (a) $2y\,dx + x\,dy = 0$; (b) $2 = P_y \neq Q_x = 1$

6. $x^2 y = C$ **7.** $xy = C$ **8.** $(3x^2 - 2x + 1)y = C$ **9.** $y = \dfrac{x \pm \sqrt{x^2 - 4x^2(5x - 7)}}{2x^2}$ **10.** $x^2 y^3 = 8$ **11.** $y = \dfrac{-(x + 1)}{2x}$

Exercise Set 9.5, p. 542

1. Exact   **3.** Exact   **5.** Exact   **7.** Exact   **9.** $x^2 + 3x + 3y^2 = C$ **11.** $x \sin y = C$ **13.** $x^3 + 3xy + y^3 = 5$
15. $x \cos y + y \cos x = 1$ **17.** $x^2 - 2xy + 2y^2 = 1$

Margin Exercises, Section 9.6, pp. 543–548

1. $y = C_1 x$ **2.** It checks.   **3.** $y = C_1 e^{-x^2}$ **4.** $y = 2 + Cx^{-4}$, $x > 0$ **5.** $y = \dfrac{1}{5}e^{3x} + Ce^{-2x}$ **6.** $A(t) = 1200 - Ce^{-t/240}$
7. $A(t) = 720 - 320e^{-t/240}$; $A(300) \approx 628.32$ lbs

Exercise Set 9.6, p. 549

1. $y = Ce^{3x}$ **3.** $y = Cx^{-3}$ **5.** $m(x) = e^{-3x}$ **7.** $m(x) = e^{5/x}$ **9.** $y = Ce^{3x}$ **11.** $y = Ce^{x^2}$ **13.** $y = 7e^{-5x}$
15. $y = Ce^{x^2} - 1$ **17.** $y = \dfrac{2x^2 + x^4 + 4C}{4(1 + x^2)}$ **19.** $y = \dfrac{x^3}{4} + \dfrac{6}{x}$ **21.** (a) 201.16 lbs; (b) 249.99 lbs

Margin Exercises, Section 9.7, pp. 550–560

1. They all check.   **2.** $y = C_1 e^{4x} + C_2 e^{2x}$ **3.** $y = C_1 e^{(1/2)x} + C_2 e^{3x}$ **4.** $y = C_1 e^x + C_2 e^{-2x} + C_3$ **5.** $r^2 - 7r + 12 = 0$
6. $r^3 - 6r^2 + 11r - 6 = 0$ **7.** $y = C_1 e^{3x} + C_2 e^{4x}$ **8.** $y = C_1 e^{2x} + C_2 x e^{2x}$ **9.** $y = C_1 e^{3x} + C_2 x e^{3x}$

10. $y = C_1e^{-(1/2)x} + C_2xe^{-(1/2)x}$ **11.** $y = C_1e^{-x} + C_2xe^{-x} + C_3x^2e^{-x}$ **12.** $y = C_1 \sin 3x + C_2 \cos 3x$
13. $y = C_1e^{2x} \sin 5x + C_2e^{2x} \cos 5x$ **14.** $y = C_1e^{-x} \sin 2x + C_2e^{-x} \cos 2x$
15. $y = C_1 + C_2e^x \sin(\sqrt{7}x) + C_2e^x \cos(\sqrt{7}x)$ **16.** $y = C_1 \sin 2x + C_2 \cos 2x + C_3x \sin 2x + C_4 x \cos 2x$

17. It checks. **18.** $y = C_1e^{-(1/2)x} + C_2xe^{-(1/2)x} + 3$ **19.** $y = C_1 \sin 3x + C_2 \cos 3x + \frac{1}{9}x^2 - \frac{2}{9}x - \frac{2}{81}$

20. $y = C_1 + C_2e^{-x} + 7x$ **21.** $y = C_1 + C_2x + C_3e^{-2x} + \frac{1}{2}x^3 - \frac{7}{4}x^2$ **22.** $y = \frac{9}{2}e^x - \frac{17}{2}e^{-x} + 5$

Exercise Set 9.7, pp. 560–561

1. $y = C_1e^x + C_2e^{5x}$ **3.** $y = C_1e^{-x} + C_2e^{2x}$ **5.** $y = C_1e^{-x} + C_2e^{-2x}$ **7.** $y = C_1e^{(1/2)x} + C_2e^{2x}$ **9.** $y = C_1e^{3x} + C_2e^{-3x}$
11. $y = C_1e^{-5x} + C_2xe^{-5x}$ **13.** $y = C_1e^{-(3/2)x} + C_2xe^{-(3/2)x}$ **15.** $y = C_1e^{2x} + C_2xe^{2x} + C_3e^{-3x}$
17. $y = C_1e^{-x} + C_2xe^{-x} + C_3x^2e^{-x} + C_4$ **19.** $y = C_1e^x + C_2e^{4x} + C_3 + C_4x$ **21.** $y = C_1 \sin 6x + C_2 \cos 6x$
23. $y = C_1e^{-4x} \sin 5x + C_2e^{-4x} \cos 5x$ **25.** $y = C_1e^{-x} \sin 2x + C_2e^{-x} \cos 2x + C_3$
27. $y = C_1e^x + C_2e^{-x} \sin 4x + C_3e^{-x} \cos 4x$ **29.** $y = C_1 \sin x + C_2 \cos x + 7$ **31.** $y = C_1e^x + C_2xe^x + 3$

33. $y = C_1e^{-2x} + C_2xe^{-2x} - 3x + 5$ **35.** $y = C_1e^{-x} + C_2e^{-3x} + 2x^2 - \frac{16}{3}x + \frac{40}{9}$ **37.** $y = C_1 + C_2e^{3x} - \frac{4}{3}x$

39. $y = C_1 + C_2x + C_3e^{-x} - x^2$ **41.** $y = C_1 + C_2e^{-2x} \sin 4x + C_3e^{-2x} \cos 4x + x^2 - x$ **43.** $y = 1 - e^x$
45. $y = 2e^x - e^{-x}$ **47.** $y = 6e^x - 2e^{-x} - 3$ **49.** $y = \sin 3x + 2 \cos 3x - x + 1$
55. $y = C_1e^{7x} + C_2e^{9x} + C_3e^{2x} \sin 4x + C_4e^{2x} \cos 4x + C_5xe^{2x} \sin 4x + C_6xe^{2x} \cos 4x$

Margin Exercises, Section 9.8, pp. 562–572

1. $x = C_1e^{4t}, y = C_2e^t$ **2.** $x = C_1e^{4t} + C_2e^{-4t}, y = 2C_1e^{4t} - 2C_2e^{-4t}$
3. $x = C_1 \sin 2t + C_2 \cos 2t, y = -2C_2 \sin 2t + 2C_1 \cos 2t$ **4.** They check. **5.** $x = C_1e^{2t} + C_2e^{-2t}$,
$y = -\frac{2}{3}C_1e^{2t} - 2C_2e^{-2t}$ **6.** $x = e^{2t} + 4te^{2t}, y = 3e^{2t} - 4te^{2t}$ **7.** $x = 2e^t + e^{2t}, y = 4e^t + e^{2t}$ **8.** $x = C_1e^{2t} + C_2e^{5t} + 5$,
$y = -2C_1e^{2t} + C_2e^{5t} - 20$ **9.** $A(t) = 1200 + 400e^{-t}, B(t) = 1200 - 400e^{-t}$ **10.** $A(t) = 1100, B(t) = 1100$
11. $P(t) = \sin 8t + 4 \cos 8t + 20, Q(t) = 40 - 2 \cos 8t + 8 \sin 8t$

Exercise Set 9.8, pp. 572–573

1. $x = C_1e^{2t}, y = C_2e^{3t}$ **3.** $x = C_1e^t + C_2e^{2t}, y = C_1e^t + 2C_2e^{2t}$ **5.** $x = C_1e^{5t} + C_2e^t, y = 3C_1e^{5t} - C_2e^t$
7. $x = C_1e^{(1/2)t} + C_2e^{-t}, y = C_1e^{(1/2)t} - \frac{1}{2}C_2e^{-t}$ **9.** $x = C_1e^{3t} + C_2te^{3t}, y = (C_2 - C_1)e^{3t} - C_2te^{3t}$
11. $x = C_1 \sin 6t + C_2 \cos 6t, y = -3C_2 \sin 6t + 3C_1 \cos 6t$ **13.** $x = C_1e^t \sin t + C_2e^t \cos t$,
$y = \left(\frac{2}{5}C_1 + \frac{1}{5}C_2\right)e^t \sin t + \left(-\frac{1}{5}C_1 + \frac{2}{5}C_2\right)e^t \cos t$ **15.** $x = e^{3t} + 2e^{4t}, y = -e^{3t} - 4e^{4t}$
19. $x = \sin 2t + \cos 2t, y = -2 \sin 2t + 2 \cos 2t$ **21.** $x = C_1e^{3t} + C_2e^{4t} - 1, y = 2C_1e^{3t} + C_2e^{4t}$
23. $x = C_1 \sin t + C_2 \cos t + 4t, y = -C_2 \sin t + C_1 \cos t - 2t + 1$ **25.** $x = e^{3t} + e^{-3t} - 2, y = e^{3t} - 5e^{-3t} + 1$
27. $A' = -\frac{1}{5}A + \frac{1}{5}B, B' = \frac{1}{5}A - \frac{1}{5}B, A(0) = 200, B(0) = 100$ $A(t) = 150 + 50e^{-(2/5)t}, B(t) = 150 - 50e^{-(2/5)t}$
Limiting concentration in both is 0.3 lbs/gal **29.** $P(t) = 13 \sin 5t + 13 \cos 5t + 36, Q(t) = -6 \sin 5t + 4 \cos 5t + 28$
31. $x = C_1e^{2t} + C_2e^{-3t}, y = 2C_1e^{2t} - \frac{1}{2}C_2e^{-3t} - e^t$

Chapter 9 Test, pp. 573–574

1. $y = \frac{1}{3}x^3 + \frac{3}{2}x^2 - 5x + 7$ **2.** $y = C_1 e^{x^8}$, where $C_1 = e^C$ **3.** $y = \sqrt{18x + C_1}$, $y = -\sqrt{18x + C_1}$, where $C_1 = 2C$

4. $y = C_1 e^{6t}$, where $C_1 = e^C$.   **5.** 1360   **6.** (a) $P(t) = \dfrac{16{,}000}{20 + 780e^{-0.72t}}$; (b)   (c) $t = 5.798$

7.

8. (a) $S(t) = 20(1 - e^{-kt})$; (b) $k = 0.11$; (c) $S(t) = 20(1 - e^{-0.11t})$; (d) \$13.343 million; (e) $t = 12.6$ yr, or the 13th yr.

9. $x^3 + y^2 = 1$ **10.** $x^3 + 2x^2y - y^5 = -1$ **11.** $y = \dfrac{4}{x^3}$, $x > 0$ **12.** $y = e^{4x} + Ce^{3x}$ **13.** $y = C_1 e^{2x} + C_2 e^{3x}$

14. $y = C_1 e^{6x} + C_2 x e^{6x}$ **15.** $y = C_1 \sin 6x + C_2 \cos 6x + 2$ **16.** $x = C_1 e^{5t} + C_2 e^{-3t}$, $y = \frac{3}{5}C_1 e^{5t} - C_2 e^{-3t}$

17. $x = C_1 \sin 12t + C_2 \cos 12t$, $y = -2C_2 \sin 12t + 2C_1 \cos 12t$ **18.** $x = C_1 e^{-3t} + C_2 t e^{-3t}$, $y = (C_1 + \frac{1}{3}C_2)e^{-3t} + C_2 t e^{-3t}$

19. $x = e^{-5t} - e^{-4t}$, $y = e^{-5t} - 2e^{-4t}$

CHAPTER 10

Margin Exercises, Section 10.1, pp. 576–585

1. $2(x - 3) + 4 + \dfrac{1}{x - 3}$ **2.** $x - 2 + \dfrac{2}{x - 2} + \dfrac{4}{(x - 2)^2}$ **3.** $(x - 3)^2 + 2(x - 3) - 1$ **4.** (a) 24; (b) 8; (c) -5; (d) 1
5. $2(x - 1)^2 + 3(x - 1) + 4$ **6.** $(x - 3)^3 + 9(x - 3)^2 + 26(x - 3) + 24$ **7.** $(x - 4)^2 + 5(x - 4) + 5$
8. $(x - 2)^2 + (x - 2) - 1$ **9.** $(x - 1)^2 - (x - 1) + 2$ **10.** $3(x + 1)^3 + 2(x + 1)^2 - \frac{1}{2}(x + 1) + 1$
11. $-\frac{1}{2}x^5 + \frac{1}{2}x^4 + 7x^3 - \frac{1}{2}x^2 + 2x + 1$

Exercise Set 10.1, pp. 585–586

1. $x - 4 + \dfrac{3}{x}$ **3.** $2x - 12 + \dfrac{7}{x - 3}$ **5.** $(x - 2)^2 + 7 + \dfrac{4}{x - 2}$ **7.** $(x - 2)^2 + 2(x - 2) + 3$ **9.** $(x - 2)^2 - (x - 2) + 1$
11. $(x + 1)^2 - 7(x + 1) + 13$ **13.** $(x + 2)^2 - 7(x + 2) + 14$ **15.** $(x - 1)^3 + (x - 1)^2 + 3(x - 1) + 4$

17. $(x + 1)^3 - 5(x + 1)^2 + 11(x + 1) - 6$ **19.** $2(x - 3)^2 + 6(x - 3) + 3$
21. $3(x - 1)^4 + 12(x - 1)^3 + 11(x - 1)^2 - 2(x - 1) - 3$ **23.** $(x - 1)^5 - 10(x - 1)^3 - 20(x - 1)^2 - 10(x - 1)$
27. $x + 1 - \dfrac{4}{x - 1}$ **29.** $x - 1 + \dfrac{2}{x + 1}$ **31.** $(x - 1)^4 + 3$

Margin Exercises, Section 10.2, pp. 586–592

1. $2 - 6x + 9x^2$ **2.** (a) $(x - 4)^2 + 5(x - 4) + 5$; (b) They are equal.   **3.** $p_2(x) = x$, $p_3(x) = x - \dfrac{1}{6}x^3$ **4.** $1 + x^2$
5. (a) $1 + x + \dfrac{1}{2}x^2 + \dfrac{1}{6}x^3$; (b) 1.105167   **6.** 0.100333

Exercise Set 10.2, pp. 592–593

1. $1 - x + \dfrac{1}{2}x^2$ **3.** x^2 **5.** $-4x^2 + x - 1$ **7.** $1 + \dfrac{1}{2}x^2$ **9.** x^2 **11.** $1 + \dfrac{1}{2}x + \dfrac{1}{8}x^2 + \dfrac{1}{48}x^3$
13. $1 + 3(x - 1) + 3(x - 1)^2 + (x - 1)^3$ **15.** $-x - \dfrac{1}{2}x^2 - \dfrac{1}{3}x^3$ **17.** $p_4(x) = x - \dfrac{1}{6}x^3$; $p_5(x) = x - \dfrac{1}{6}x^3 + \dfrac{1}{120}x^5$
19. $p_4(x) = 1 - \dfrac{1}{2}x^2 + \dfrac{1}{24}x^4 = p_5(x)$

21. $p_0(x) = 0$; $p_1(x) = x = p_2(x)$

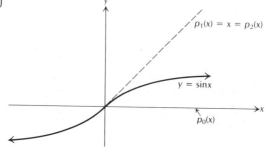

23. $p_0(x) = 1$, $p_1(x) = 1 - x$, $p_2(x) = 1 - x + \dfrac{1}{2}x^2$

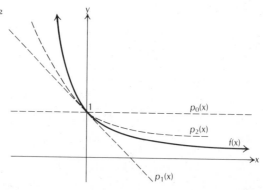

25. 0.1296338   **27.** 0.9512292   **29.** -0.0512917 **31.** (a) $P(1) \approx 0.16473$, $P(2) \approx 0.30232$, $P(3) \approx 0.41725$;
(b) $p_3(t) = 0.18t - 0.0162t^2 + 0.000972t^3$; (c) $p_3(1) = 0.164772$, $p_3(2) = 0.302976$, $p_3(3) = 0.420444$
33. (a) $p_2(t) = 5 + 0.1(t - 3) - 0.001(t - 3)^2$; (b) $P(3.5) \approx 5.04975$, $P(4) \approx 5.099$

Margin Exercises, Section 10.3, pp. 593–603

1. (a) $a_n = \frac{2}{3^n}$; (b) $s_1 = \frac{2}{3} \approx 0.667$, $s_2 = \frac{8}{9} \approx 0.889$, $s_3 = \frac{26}{27} \approx 0.963$, $s_4 = \frac{80}{81} \approx 0.988$, $s_5 = \frac{242}{243} \approx 0.996$, $s_6 = \frac{728}{729} \approx 0.999$

2. (a) $a_n = 2n - 1$; (b) $s_1 = 1$, $s_2 = 4$, $s_3 = 9$, $s_4 = 16$, $s_5 = 25$; (c) $s_n = n^2$; (d) No, it diverges.

3. (a) $a_n = \frac{2}{5^{n-1}} = 2\left(\frac{1}{5}\right)^{n-1}$; (b) $a_5 = \frac{2}{625}$, $a_6 = \frac{2}{3,125}$ **4.** $a = \frac{1}{5}$, $r = \frac{1}{3}$ **5.** $s_n = 3\left[1 - \left(\frac{1}{3}\right)^n\right]$ **6.** $\frac{5}{2}$ **7.** $\frac{3}{10}$

8. Diverges ($r = -2$) **9.** $\frac{5}{7}$ **10.** \$20,000 **11.** \$6,666.67 **12.** $\frac{4}{9}$ **13.** $\frac{17}{37}\left(= \frac{459}{999}\right)$ **14.** 2, 1.3333, 1.5556, 1.4815,

1.5062, and 1.4979 **15.** Converges **16.** $0.9498457 \approx s_3 \geq s \geq s_4 \approx 0.9459394$ **17.** (a) $0.8998629 \leq s \leq 0.9012346$;

(b) $\frac{9}{10}$

Exercise Set 10.3, pp. 603–606

1. $a_n = 5^{-n} = \left(\frac{1}{5}\right)^n$ **3.** $\frac{n}{n+1}$ **5.** $\left(\frac{2}{3}\right)^n$ **7.** $n(-1)^{n+1} = n(-1)^{n-1}$ **9.** $a = 1$, $r = \frac{1}{3}$, $s_{15} = \frac{3}{2}\left[1 - \left(\frac{1}{3}\right)^{15}\right]$

11. $a = 1$, $r = -\frac{1}{3}$, $s_{15} = \frac{3}{4}\left[1 + \left(\frac{1}{3}\right)^{15}\right]$ **13.** $a = \frac{1}{8}$, $r = -\frac{2}{3}$, $s_{15} = \frac{3}{40}\left[1 + \left(\frac{2}{3}\right)^{15}\right]$

15. $a = \frac{5}{3}$, $r = -\frac{2}{5}$, $s_{15} = \frac{25}{21}\left[1 + \left(\frac{2}{5}\right)^{15}\right]$ **17.** Converges; $s = \frac{8}{7}$ **19.** Converges; $s = \frac{9}{8}$ **21.** Converges; $s = \frac{3}{5}$

23. Diverges **25.** $\frac{7}{9}$ **27.** $2 + \frac{38}{99} = \frac{236}{99}$ **29.** $\frac{1}{7}$ **31.** Diverges **33.** Converges **35.** Converges **37.** Converges

39. Converges **41.** $0.7489712 \leq s \leq 0.7530864$ **43.** $0.9459394 \leq s \leq 0.9475394$ **45.** $0.0791736 \leq s \leq 0.0874381$

47. 32 billion dollars **49.** (a) 5 mg, $5 + 5(0.4) = 7$ mg, $5 + 5(0.4) + 5(0.4)^2 = 7.8$ mg (b) 8.3311488 mg

(c) $\frac{25}{3} \approx 8.333333$ mg **51.** $V \approx \$37.04$ **55.** $\frac{d}{p}$ mg **57.** $\frac{r}{i}$

Margin Exercises, Section 10.4, pp. 606–616

1. $1 - x + \frac{x^2}{2!} - \frac{x^3}{3!} + \cdots = \sum_{j=0}^{\infty} \frac{(-1)^j x^j}{j!}$ **2.** $\sum_{j=0}^{\infty} \frac{(-1)^j x^{2j}}{(2j)!}$ **3.** $4 + 2x - 5x^2 + 7x^3$

4. $\sum_{j=0}^{\infty} (-1)^j x^{2j}$ **5.** $\sum_{j=0}^{\infty} (-1)^j (j+1)x^j$, $|x| < 1$ **6.** $1 + 3x + 6x^2 + 10x^3 + \cdots$, $|x| < 1$

7. $1 + x^3 + x^6 + \cdots = \sum_{j=0}^{\infty} x^{3j}$, $|x| < 1$ **8.** $x^2 - \frac{x^6}{3!} + \frac{x^{10}}{5!} - \frac{x^{14}}{7!} + \cdots = \sum_{j=0}^{\infty} (-1)^j \frac{x^{4j+2}}{(2j+1)!}$, all x

9. $e^{-5x} = 1 - 5x + \frac{25}{2}x^2 - \frac{125}{3!}x^3 + \cdots$, all x **10.** $\ln(1 - x) = -x - \frac{x^2}{2} - \frac{x^3}{3} - \frac{x^4}{4} \cdots = \sum_{j=0}^{\infty} -\frac{x^{j+1}}{j+1}$,

11. 0.494 **12.** $x^2 \cos x = x^2 - \frac{1}{2!}x^4 + \frac{1}{4!}x^6 - \frac{1}{6!}x^8 + \cdots$ **13.** $x \tan^{-1} x = x^2 - \frac{x^4}{3} + \frac{x^6}{5} - \frac{x^8}{7} + \cdots$, $|x| \leq 1$

14. $f(x) = 1 + \frac{1}{2!}x + \frac{1}{3!}x^2 + \frac{1}{4!}x^3 + \cdots$

Exercise Set 10.4, pp. 616–617

1. $1 + 4x + 8x^2 + \frac{32}{3}x^3 + \frac{4^4}{4!}x^4 + \cdots$ **3.** $1 - (x - 1) + \frac{1}{2!}(x - 1)^2 - \frac{1}{3!}(x - 1)^3 + \cdots$

5. $-2 + 3(x + 1) - 3(x + 1)^2 + (x + 1)^3 + \cdots$ **7.** $1 + 2x + 4x^2 + 8x^3 + \cdots$ **9.** $e^{6x} = 1 + (6x) + \dfrac{1}{2!}(6x)^2 + \dfrac{1}{3!}(6x)^3 + \cdots$

11. $\dfrac{1}{1 - x^5} = 1 + x^5 + x^{10} + x^{15} + \cdots, |x| < 1$ **13.** $\ln(1 + 4x) = 4x - 8x^2 + \dfrac{64}{3}x^3 - 64x^4 + \cdots, |x| < \dfrac{1}{4}$

15. $\dfrac{1}{2 + x} = 1 - (x + 1) + (x + 1)^2 - (x + 1)^3 + \cdots, -2 < x < 0$

17. $\ln(2 + x) = (x + 1) - \dfrac{1}{2}(x + 1)^2 + \dfrac{1}{3}(x + 1)^3 - \dfrac{1}{4}(x + 1)^4 + \cdots, -2 < x < 0$ **19.** $\dfrac{5x^4}{1 - x^5} = 5x^4 + 5x^9 + 5x^{14} + \cdots,$

$|x| < 1$ **21.** $f(x) = -\dfrac{1}{2} + \dfrac{1}{4!}x^2 - \dfrac{1}{6!}x^4 + \cdots,$ all x **23.** $x^5 \sin(x^3) = x^8 - \dfrac{1}{3!}x^{14} + \dfrac{1}{5!}x^{20} - \dfrac{1}{7!}x^{26} + \cdots,$ all x

25. $\displaystyle\int_0^1 \cos(x^2)\,dx = 1 - \dfrac{1}{10} + \dfrac{1}{216} - \dfrac{1}{9360} + \cdots \approx 0.90463$ (error $< \dfrac{1}{9360} \le 0.0002$) **27.** 0.82143 (error $< \dfrac{1}{60} < 0.02$)

29. 0.0002167 (error $< 10^{-13}$) **31.** (a) $F(t) = t - \dfrac{1}{6}t^3 + \dfrac{1}{40}t^5 - \dfrac{1}{336}t^7 + \cdots$ (b) $P(0 \le t \le 0.1) \approx 0.0398278$
(c) It checks.

Margin Exercises, Section 10.5, pp. 617–626

1. $\dfrac{4}{11}$ **2.** 0 **3.** 0 **4.** $-12.356, -109.12,$ and -998.00 **5.** Does not exist **6.** -7 **7.** -2 **8.** $\dfrac{1}{2}$ **9.** 2 **10.** -2
11. 1 **12.** $-\dfrac{1}{2}$ **13.** e^2(L'Hôpital's Rule doesn't apply.) **14.** Does not exist. **15.** $-\dfrac{5}{3}$ **16.** Does not exist. **17.** 3

Exercise Set 10.5, pp. 626–627

1. 1 **3.** Does not exist. **5.** 0 **7.** Does not exist. **9.** 7 **11.** $\dfrac{1}{4}$ **13.** 0 **15.** 2 **17.** $\dfrac{1}{3}$ **19.** -1 **21.** $-\dfrac{1}{4}$ **23.** -1
25. 1 **27.** 0 **29.** Does not exist. **31.** $\dfrac{1}{2}$ **33.** $\dfrac{1}{3}$ **35.** Does not exist. **37.** Does not exist. **39.** ap **41.** $f(0)$

Chapter 10 Test, p. 628

1. $p_2(x) = -6(x + 2)^2 + 12(x + 2) - 8; p_3(x) = (x + 2)^3 - 6(x + 2)^2 + 12(x + 2) - 8$ **2.** $p_2(x) = 1; p_3(x) = 1 + x^3$
3. $p_2(x) = x - 1; p_3(x) = (x - 1) - \dfrac{1}{6}(x - 1)^3$ **4.** 0.039989 **5.** 1.072507 **6.** $\dfrac{9}{25}$ **7.** $\dfrac{7}{10}\left[1 - \left(\dfrac{2}{7}\right)^{13}\right]$ **8.** $\dfrac{2407}{990}$
9. Converges; $s \approx s_3 \approx 0.929167$ **10.** Diverges **11.** Diverges **12.** Converges;
$s = \dfrac{16}{17} \approx 0.941176$ **13.** $y = x^3 - \dfrac{x^5}{3!} + \dfrac{x^7}{5!} - \dfrac{x^9}{7} + \cdots,$ all x **14.** $y = 1 + (x - 1)^2 + \dfrac{(x - 1)^4}{2!} + \dfrac{(x - 1)^6}{3!} + \cdots,$ all x
15. $y = 1 + 5x + 25x^2 + 125x^3 + \cdots, |x| < \dfrac{1}{5}$ **16.** $\dfrac{1}{2}$ **17.** 2 **18.** 0 (not indeterminate) **19.** -1 **20.** 0 **21.** $\dfrac{1}{2}$
22. \$20 **23.** \$$\dfrac{D}{i}$ **24.** $\displaystyle\sum_{n=1}^{\infty} \dfrac{x^n}{n \cdot n!}$

CHAPTER 11

Margin Exercises, Section 11.1, pp. 630–634

1. 1 **2.** The iterates diverge. **3.** $\bar{x} = 0$ (all 3 cases) **4.** 2 **5.** $2 - \sqrt{3} \approx 0.267949$ **6.** $2 + \sqrt{3} \approx 3.732051$

Exercise Set 11.1, p. 634

1. $\bar{x} = \dfrac{3}{2} - \dfrac{1}{2}\sqrt{5} \approx 0.381966$ **3.** 1.497300   **5.** They diverge.   **7.** They diverge.   **9.** 0.567143   **11.** 0.395489

Margin Exercises, Section 11.2, pp. 635–641

1. $x_2 = -\dfrac{1}{3}$, $x_3 = -\dfrac{1}{15}$, $x_4 = \dfrac{-1}{255}$, $x_5 = \dfrac{-1}{65{,}535}$, $x_6 \approx x_7 \approx 0 = \bar{x}$ **2.** 0.450184   **3.** -2.103803 **4.** 2.154435

Exercise Set 11.2, p. 642

1. $\bar{x} = \dfrac{3}{2} - \dfrac{1}{2}\sqrt{5} \approx 0.381966$ **3.** 1.497300   **5.** -2.103803 **7.** 0.772914   **9.** $\bar{x}_1 \approx -0.152368$; $\bar{x}_2 \approx 1.109643$; $\bar{x}_3 \approx -2.957275$ **11.** -1.518662 **13.** 0.6345599   **15.** (b) $\sqrt{19} \approx 4.358899$ **17.** (b) $\sqrt[10]{20} \approx 1.349283$

Margin Exercises, Section 11.3, pp. 642–649

1. \$10,147.84   **2.** \$982.14   **3.** \$759.34   **4.** $i \approx 7.9156\%$ **5.** $i \approx 9.7011\%$

Exercise Set 11.3, pp. 649–650

1. \$974.36   **3.** \$29,959.92   **5.** \$10,062.66   **7.** \$751.31   **9.** \$148.64   **11.** \$422.41   **13.** \$1000   **15.** \$1690.05
17. \$1833.39   **19.** \$758.16   **21.** (a) $5800\,v^2 + 6000v - 10000 = 0$ (b) $i \approx 11.853\%$ **23.** $i \approx 5.3445\%$
25. The first one (12.044% against 11.775%)   **27.** (a) $2500v^2 - 4300v + 2000 = 0$ (b) No real roots.

Margin Exercises, Section 11.4, pp. 650–661

1. Approx. = 1.375; Exact = 1.5   **2.** Approx. = 1.625; Exact = 1.5   **3.** $T_4 = 1.5$; Exact = 1.5
4. Error = 0 since $f'' = 0$ **5.** $T_{10} = 0.2525$; $E \le \dfrac{1}{200} = 0.005$ **6.** $S_5 = 0.25$; $E = 0$

Exercise Set 11.4, p. 662

1. (a) 1.9 (b) 2.1   **3.** (a) 0.833732 (b) 0.733732   **5.** $T_{10} = 2$ **7.** $T_5 = 0.7837315$ **9.** $T_{10} \approx 0.170825$; $E \le 0.01667$
11. $T_{10} \approx 0.385878$; $|E| < 0.000833$ **13.** $S_5 = 2$ **15.** 0.7853982   **17.** 0.25; $E = 0$ **19.** 0.3862934; $E \le 0.0000533$

Margin Exercises, Section 11.5, pp. 663–668

1. $x_2 \approx 3.741657$; $x_3 \approx 3.564868$; $x_4 \approx 3.181632$; $x_5 \approx 3.147723$; $x_6 \approx 3.120675$; $x_7 \approx 3.014054$ **2.** $T_5 \approx 0.695635$;
$T_{10} \approx 0.6937714$; $I \approx 0.6931502$ **3.** $S_2 \approx 0.6932539$; $S_4 \approx 0.6931545$; $R_2 \approx 0.6931479$

Exercise Set 11.5, p. 669

1. $x_2 = \frac{1}{3}$, $x_3 = \frac{10}{27}$, $x_4 = \frac{3}{8}$ **3.** $x_2 \approx 1.341471$, $x_3 \approx 1.4738199$, $x_4 \approx 1.557581$ **5.** $x_2 \approx -1.817121$, $x_3 \approx -2.036927$, $x_4 \approx -2.117815$ **7.** $T_2 = \frac{3}{8} = 0.375$; $T_4 = \frac{11}{32} \approx 0.34375$; $T_8 = \frac{43}{128} \approx 0.3359375$; $S_2 = \frac{1}{3}$; $S_4 = \frac{1}{3}$; $R_2 = \frac{1}{3}$ **9.** $T_2 = 0.775$; $T_4 \approx 0.7827941$; $T_8 \approx 0.7847471$; $S_2 \approx 0.78539216$; $S_4 \approx 0.7853981$; $R_2 \approx 0.7853985$

Margin Exercises, Section 11.6, pp. 669–677

1. (a) $y = g(x) = e^{-x^2}$ (b) $y_0 = 1$, $y_1 = 1$, $y_2 = 0.92$, $y_3 = 0.7728$, $y_4 = 0.587328$, $y_5 \approx 0.399383$ (c) $y_0 = 1$, $y_1 = 1$, $y_2 = 0.98$, $y_3 = 0.9408$, $y_4 = 0.884352$, $y_5 \approx 0.813604$, $y_6 \approx 0.7322435$, $y_7 \approx 0.644374$, $y_8 \approx 0.554162$, $y_9 \approx 0.465496$, $y_{10} \approx 0.381707$ **2.** $y'' = y^2 \cos(xy) + xy^2 \sin(xy) \cos(xy) + y \sin^2(xy)$ **3.** (a) $y'' = y(4x^2 - 2)$ (b) $y_{n+1} = y_n[0.96 - (0.4)x_n + (0.08)x_n^2]$ (c) $y_0 = 1$, $y_1 = 0.96$, $y_2 = 0.847872$, $y_3 \approx 0.689150$, $y_4 \approx 0.5160358$, $y_5 \approx 0.356684$

4.

	Euler	Three-Term	Exact
$y_0 = y(0)$	1	1	1
$y_1 = y(0.2)$	1	0.96	0.960789
$y_2 = y(0.4)$	0.92	0.847872	0.852144
$y_3 = y(0.6)$	0.7728	0.689150	0.697676
$y_4 = y(0.8)$	0.587328	0.516036	0.527292
$y = y(1)$	0.399383	0.356684	0.367879

5. (b) $y(0.4) \approx 1.5425366$; $x(0.6) \approx 2.8347761$; $y(1) \approx 0.7540745$ **6.** $x_1 = 2.4$, $y_1 = -0.8$, $x_2 = 3.2$, $y_2 = -1.92$

Exercise Set 11.6, pp. 677–680

1. 1, 0.9, 0.8, 0.7, 0.6, 0.5, 0.4, 0.3, 0.2, 0.1, and 0   **3.** 0, 0, 0.125, 0.375, and 0.75   **5.** 1, 1, 1.08, 1.2528, 1.553472, and 2.050583   **7.** 1, 1, 1.04, 1.121584, 1.248670, and 1.427460   **9.** -1, -0.99, -0.96, -0.91, -0.84, -0.75, -0.64, -0.51, -0.36, -0.19, and 0   **11.** 1, 0.8, 0.648, 0.592, 0.68, and 0.96   **13.** 1, 0.78125, 0.6103516, 0.476837, and 0.372529   **15.** $y = g(x) = 1 - x$. Both methods yield the exact values.   **17.** $y = g(x) = e^{x^2}$. For Euler, see number 5 above. Three-term gives 1, 1.04, 1.168218, 1.4167056, 1.854184, and 2.616625.   **19.** $x = 2e^{2t} + e^{-2t}$, $y = 2e^{2t} - 3e^{-2t}$ are the exact solutions. Euler gives $x_0 = 3$, $y_0 = -1$, $x_1 = 3.4$, $y_1 = 1$, $x_2 = 4.28$, $y_2 = 2.84$, $x_3 = 5.704$, $y_3 = 4.84$, $x_4 = 7.8128$, $y_4 = 7.2944$, $x_5 = 10.83424$, and $y_5 = 10.5232$.   **21.** $x = e^{2t} - e^t$, $y = 2e^{2t} - e^t$ are the exact solutions. Euler gives $x_0 = 0$, $y_0 = 1$, $x_1 = 0.2$, $y_1 = 1.6$, $x_2 = 0.52$, $y_2 = 2.48$, $x_3 = 1.016$, $y_3 = 3.76$, $x_4 = 1.768$, $y_4 = 5.6096$, $x_5 = 2.88992$, and $y_5 = 8.26816$.   **23.** $x_0 = x_1 = \cdots = x_{10} = 0$. $y_0 = 1$, $y_1 = 0.9$, $y_2 = 0.81$, $y_3 = 0.729$, $y_4 = 0.6561$, $y_5 = 0.59049$, $y_6 = 0.531441$, $y_7 = 0.478297$, $y_8 = 0.430467$, $y_9 = 0.387420$, and $y_{10} = 0.348678$ **25.** \$903.11.   **27.** (a) 557.18; (b) 527.09   (c) 528   **29.** $x(5) \approx 1349$; $y(5) \approx 428$ **31.** (b) $F(10) = \sum_{i=0}^{9} f(i)$.   **33.** (a) $x(5) \approx 56.138$, $y(5) \approx 17.399$ (b) $x(10) \approx 72.204$, $y(10) \approx 18.782$

Chapter 11 Test, pp. 680–681

1. 2   **2.** -2 **3.** 0.8977725   **4.** 0.372559   **5.** The iterates diverge.   **6.** $\bar{x}_1 \approx 0.2033642$; $\bar{x}_2 \approx 1.469617$ **7.** 0.8977725   **8.** $\bar{x}_1 \approx 0.50049263$; $\bar{x}_2 \approx 1.1027159$ **9.** $x_E \approx 1.1366384$ **10.** \$17,043.98

11. (a) $6000v^4 = 3000 + 1000v$ (b) $i \approx 11.386791\%$ **12.** $i_1 \approx 14.47\%$; $i_2 \approx 14.99\%$ **13.** $T_2 \approx 0.9397933$;
$T_4 \approx 0.94451352$; $T_8 \approx 0.94569087$ **14.** $S_2 \approx 0.94608694$; $S_4 \approx 0.94608331$ **15.** $x_{10} \approx 1.1105271$ **16.** $x_{10} \approx 0.3725596$
17. $R_2 \approx 0.9460837$ **18.** $y(0) = 1$, $y(0.2) = 0.8$, $y(0.4) = 0.672$, $y(0.6) = 0.581683$, $y(0.8) = 0.514012$, and
$y(1) = 0.461170$ **19.** $y(0) = 1$, $y(0.2) = 0.84$, $y(0.4) = 0.722588$, $y(0.6) = 0.633253$, $y(0.8) = 0.563209$, and
$y(1) = 0.506914$ **20.** $y = (1 + x)^{-1}$ **21.** $x_0 = 1$, $y_0 = 2$, $x_1 = 1.2$, $y_1 = 1.8$, $x_2 = 1.2$, $y_2 = 1.56$, $x_3 = 1.104$, $y_3 = 1.32$,
$x_4 = 0.9696$, $y_4 = 1.0992$, $x_5 = 0.82752$, and $y_5 = 0.90528$.

CUMULATIVE REVIEW, pp. 683–686

1. $y = -4x - 27$ **2.** $x^2 + 2xh + h^2 - 5$ **3.** (a)

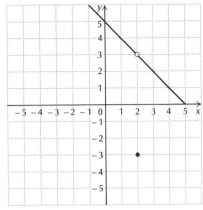

(b) 3; (c) −3; (d) no

4. −8 **5.** 3 **6.** Does not exist **7.** 4 **8.** $3x$ **9.** −9 **10.** $2x - 7$ **11.** $\frac{1}{4}x^{-3/4}$ **12.** $-6x^{-7}$

13. $(x - 3)5(x + 1)^4 + (x + 1)^5$, or $2(x + 1)^4(3x - 7)$ **14.** $\frac{5 - 2x^2}{x^6}$ **15.** $\frac{1 - x}{e^x}$ **16.** $\frac{2x}{x^2 + 5}$ **17.** 1 **18.** $3e^{3x} + 2x$

19. $e^{-x}[2x \cos (x^2 + 3) - \sin (x^2 + 3)]$ **20.** $-\tan x$ **21.** $2(\sec^2 x)(\tan x)$ **22.** $3xy^2 + \frac{y}{x}$ **23.** Min. = −7 at $x = 1$

24. None **25.** Max. $= 6\frac{2}{3}$ at $x = -1$; min. $= 4\frac{1}{3}$ at $x = -2$ **26.** 15 **27.** 30 times; lot size 15

28. $P = 199{,}500$; max. sus. harvest = 39,800,250 **29.** $\Delta y = 3.03$, $f'(x) \Delta x = 3$

30.

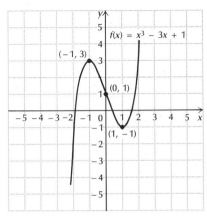

31. (a) $A(t) = 66{,}164e^{0.1t}$; (b) 179,852.4 million barrels **32.** $E(x) = \dfrac{(x+1)[80 - 30 \ln (x+1)]}{x}$ **33.** $\dfrac{1}{2}x^6 + C$

34. $3 - \dfrac{2}{e}$ **35.** $\dfrac{7}{9(7 - 3x)} + \dfrac{1}{9} \ln (7 - 3x) + C$ **36.** $\dfrac{1}{4}e^{x^4} + C$ **37.** $\left(\dfrac{x^2}{2} + 3x\right) \ln x - \dfrac{x^2}{4} - 3x + C$

38. $-\cos (x^2 + 3) + C$ **39.** $\sin (e^x) + C$ **40.** $\dfrac{232}{3}$ **41.** \$525.23 **42.** \$49,444.55 **43.** Convergent, $\dfrac{1}{4374}$

44. $\dfrac{3}{2} \ln 3$ **45.** 0.6826 **46.** (7, \$169); \$751.33 **47.** $-\dfrac{\pi}{2}\left(\dfrac{1}{e^{10}} - 1\right)$ **48.** $y = C_1 e^{x^2/2}$, where $C_1 = e^C$

49. (a) $P(t) = 1 - e^{-kt}$; (b) $k = 0.12$; (c) $P(t) = 1 - e^{-0.12t}$; (d) $P(20) = 0.9093$, or 90.93% **50.** $8xy^3 + 3$ **51.** $e^y + 24x^2y$

52. None **53.** Max. $= 4$ at (3, 3) **54.** $3(e^3 - 1)$ **55.** $\dfrac{\sqrt{2}}{2}$ **56.** $\dfrac{1}{2}$ **57.** 0 **58.** $\dfrac{1}{8} \ln |\sec 8x + \tan 8x| + C$

59. $\dfrac{1}{3} \tan^{-1} (3x) + C$ **60.** $[-\csc^2 (\sin (x^3))][\cos (x^3)][3x^2]$ **61.** $\dfrac{1}{\sqrt{1 - x^2}}$ **62.** $dw = 15x^4y^2 \, dx + 6x^5y \, dy$

63. $dw = 7e^x[\sin(yz) \, dx + z \cos(yz) \, dy + y \cos(yz) \, dz]$ **64.** 13.81 cubic feet. **65.** $\dfrac{1}{12}$ **66.** $\dfrac{2}{3}$ **67.** 0

68. $xy - 3x + y^3 = C$ **69.** $y = Ce^{-\sin x}$ **70.** $y = C_1 + C_2 \sin 2x + C_3 \cos 2x$ **71.** $y = C_1 e^{7x} + C_2 x e^{7x}$

72. $y = 5 + C_1 e^{-x} + C_2 e^{-3x}$ **73.** $x = C_1 e^t + C_2 e^{-t}$; $y = 5C_1 e^t + 3C_2 e^{-t}$ **74.** $5(x - 1)^3 + 2(x - 1)^2 - 3(x - 1) - 5$

75. $\sin(0.07) \approx (0.07) - \dfrac{(0.07)^3}{6} = 0.069942833$ **76.** $\dfrac{7}{40} = 0.175$ **77.** $\dfrac{679}{999}$ **78.** Converges

79. $x^5 - \dfrac{x^{13}}{3!} + \dfrac{x^{21}}{5!} - \dfrac{x^{29}}{7!} + \cdots = \displaystyle\sum_{k=0}^{\infty} \dfrac{(-1)^k x^{8k+5}}{(2k + 1)!}$ **80.** 0.32180075 **81.** Does not exist. **82.** -2 **83.** 4 **84.** 1

85. 0.3725595 **86.** 0.7320508 **87.** \$466.51 **88.** 13.7% **89.** $T_5 \approx 0.4059274$; Exact $= \ln(1.5) \approx 0.4054651$

90. $s_5 \approx 0.4054653$ **91.** (a) $s_2 \approx 0.1173604$; $s_4 \approx 0.1115501$; $I = \dfrac{1}{9} \approx 0.1111111$ (b) $R_2 \approx 0.1111627$

INDEX

CHAPTER 5

18. $\int k\,dx = kx + C$

19. $\int x^r\,dx = \dfrac{x^{r+1}}{r+1} + C, \qquad r \neq -1;$

$\int (r+1)x^r\,dx = x^{r+1} + C, \qquad r \neq -1$

20. $\int \dfrac{dx}{x} = \ln x + C, \qquad x > 0; \qquad \int \dfrac{dx}{x} = \ln|x| + C, \qquad x < 0$

21. $\int e^x\,dx = e^x + C$

22. $\int k\,f(x)\,dx = k \int f(x)\,dx$

23. $\int [f(x) \pm g(x)]\,dx = \int f(x)\,dx \pm \int g(x)\,dx$

24. $\int u\,dv = uv - \int v\,du$

Further integration formulas occur in Table 5 at the back of the book.